The Rules of Counting

- The Multiplication Rule: If one choice can be made in m ways, and if for each of these m choices a second choice can be made in n ways, the total number of ways of making both choices is $m \times n$.

- Choosing k Elements from an n-Element Set:

	repetition allowed	repetition not allowed
order matters	n^k	$n(n-1)\ldots(n-k+1)$
order does not matter	$\binom{n+k-1}{k}$	$\binom{n}{k}$

- Permutations of an n-Element Set: $n! = n(n-1)\ldots2.1.$

n:	0	1	2	3	4	5	6	7	8	9	10
$n!$:	1	1	2	6	24	120	720	5040	40320	362880	3628800

- Pascal's Triangle for Binomial Coefficients:

$$\binom{n}{0} = 1; \quad \binom{n}{k} = \binom{n}{n-k}; \quad \binom{n}{k} = \binom{n-1}{k} + \binom{n-1}{k-1}.$$

1
1 1
1 2 1
1 3 3 1
1 4 6 4 1
1 5 10 10 5 1
1 6 15 20 15 6 1
1 7 21 35 35 21 7 1
1 8 28 56 70 56 28 8 1
1 9 36 84 126 126 84 36 9 1
1 10 45 120 210 252 210 120 45 10 1
⋮

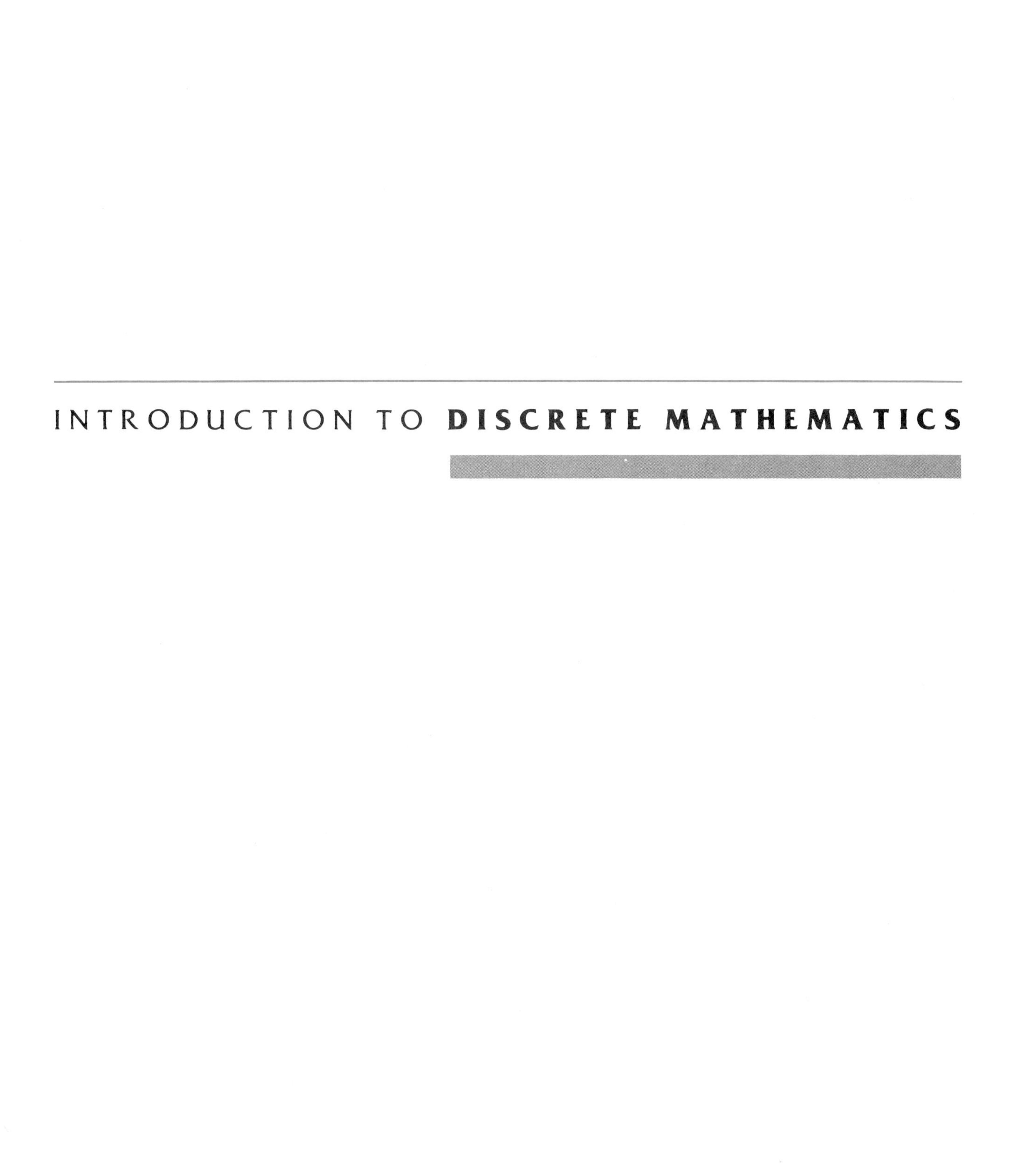

INTRODUCTION TO **DISCRETE MATHEMATICS**

INTRODUCTION TO DISCRETE MATHEMATICS

California Institute of Technology
ROBERT J. McELIECE

University of Illinois-Urbana
ROBERT B. ASH

University of Illinois-Urbana
CAROL ASH

RANDOM HOUSE New York

First Edition

9 8 7 6 5 4 3 2 1

Library of Congress Cataloging-in-Publication Data

McEliece, Robert J.
Introduction to discrete mathematics/Robert J. McEliece, Robert B. Ash, Carol Ash.—1st ed.
p. cm.
Includes index.
ISBN 0-394-35819-8
1. Mathematics—1961-2. Electronic data processing—Mathematics.
I. Ash, Robert B. II. Ash, Carol, 1935- III. Title.
QA39.2.M315 1989
510—dc19 88-29211
CIP

Manufactured in the United States of America

To Norma and Dick

PREFACE

...A line will take us hours maybe;
Yet if it does not seem a moment's thought,
Our stitching and unstitching has been naught.

W. B. Yeats, *Adam's Curse*

Discrete mathematics is the theoretical foundation for much of today's advanced technology. If you wish to understand modern computer hardware or software, communication systems, digital signal processing, information theory, neural networks, control systems, operations research, etc., you will need to learn at least a little, and possibly a lot, about discrete mathematics. At the same time, discrete mathematics is the key prerequisite for advanced work in many branches of mathematics and theoretical computer science. If you wish to pursue these topics, you will also have to study discrete mathematics.

But what exactly is discrete mathematics? According to the *Random House Dictionary* (our favorite reference book), discrete mathematics is mathematics that uses only arithmetic and algebra, and does not involve calculus. This is true, but it's a little like saying *Moby Dick* is a large book about whaling! We can't think of a short definition (our table of contents is as close as we can come), but we have been enjoying discrete mathematics for many years, through teaching, writing, research, and consulting for government and industry, and we think we know what discrete mathematics is. Anyway, we have written a large book giving our view of the subject! We have been guided in our choice of topics largely by our wish to give a solid introduction to all parts of the subject that are essential for applications in computer science and engineering, but we have also included many topics which don't have any applications (as far as we know), but which are too much fun to omit.

Pedagogy

Accessibility. The only prerequisite for this book is a good high school algebra course. This means that many of our readers won't have a lot of mathematical experience, and so we have written in an informal and even light-hearted style. But don't be misled by this; the book is serious, complete, and rigorous.

Algorithms. A characteristic feature of modern discrete mathematics is its emphasis on algorithms, and algorithms form a unifying theme for our book. Beginning with Chapter 3, we have included dozens of algorithms to illustrate the basic material in the text. Almost every algorithm is presented in three forms:

informally, by flowchart, and by pseudocode. We have devised a simple, structured pseudolanguage, called "pseudo C," especially for this book; it is described in detail in the Appendix and summarized on the back inside cover. In Section 4.4, we give an introduction to the analysis of algorithms, and we give examples of natural algorithms whose complexity is $O(n)$, $O(n^2)$, $O(n^3)$, and $O(2^n)$. Furthermore, the Instructor's Manual contains suggestions for several dozen programming projects which can be used to supplement the problems in the text.

Examples. In writing this book, we have been guided by a simple teaching philosophy: *An undergraduate mathematics textbook should teach the student to solve problems.* An ideal textbook would consist entirely of a sequence of carefully chosen worked examples, whose details illustrate all of the features and difficulties that can possibly be encountered in the general case. Of course, we have fallen short of this ideal, but our book contains hundreds of carefully chosen worked examples, which introduce and clarify every new concept.

Problems. The text contains approximately 1600 exercises, ranging from routine practice problems to challenging explorations. Solutions to the odd-numbered problems are available at the back of the book and solutions to all the problems are available in the Instructor's Manual. We tell our students that the best way to study is to try to do the homework problems before reading the text, and to read the text only when they get stuck. It's amazing how efficiently you can learn mathematics when you're motivated by the need to solve a specific problem!

Accuracy. Robert Crawford has carefully prepared the solutions to all of the problems, and we have ourselves re-checked every solution provided by him. The instructor can therefore be fully confident in the accuracy of the solutions, both in the back of the book and in the Instructor's Manual.

Figures and Tables. In discrete mathematics, more than in most other branches of mathematics, a picture is worth a thousand words, and recognizing this, we have included more than 600 figures and tables in the book. There are dozens of Venn diagrams, Hasse diagrams, tree diagrams, graphs (directed, undirected, bipartite), flowcharts, state diagrams, truth tables, state tables, etc., which are used to explain, motivate, and otherwise clarify the text and the solutions to the problems. Most of the figures are in two colors; whenever possible, we have used the second color to make important pedagogical differentiations, many of which would have been extremely difficult to make verbally.

Special Features

Chapter Overviews and Summaries. Each chapter begins with an overview, and ends with a summary. The overviews tells the reader exactly what's going to be covered, and why; the summaries give a short descriptive list of the chapter's key results, formulas, and concepts, to help students review the material.

Historical Material. Throughout the text, historical footnotes give brief biographical sketches for dozens of discrete mathematicians who have contributed to our subject. For the pre-twentieth century mathematicians, most of this material was gleaned from the wonderful *Dictionary of Scientific Biography*; for many of the living contributors, the material is drawn from personal correspondence, and has not previously appeared in print.

Endpapers. There are four endpapers which give handy summaries of (1) the notation used in discrete mathematics; (2) the basic rules of counting; (3) the laws of Boolean algebra; and (4) the definition of our structured pseudolanguage, "pseudo C".

Enrichment Material

Besides the core material in Chapters 1-7, this book contains a lot of enrichment material, i.e., material which is not often covered in introductory courses. Some of this material is included to illustrate or amplify the more standard parts of the text, some is included because of its importance for applications, and some is included just because we like it. However, each of these special topics has been chosen to fit thematically with the surrounding material, and is no more difficult to learn.

For example, scattered throughout Chapters 1-7, you will find the following enrichment material: Hamming codes, counting *onto* functions, the edge-vertex inequality for planar graphs, Hall's matching theorem, a decision algorithm for the propositional calculus, Turing machines, Viterbi's decoding algorithm, Hofstadter's *MIU*-language, lattice-path counting, and Kleene's theorem for regular languages.

In a certain sense, Chapters 8, 9, and 10 can be considered as entirely enrichment material, although in the case of Chapters 9 and 10 we would argue that the material is basic. In any case, here is a brief summary of these important chapters:

- Chapter 8. This chapter presents the whimsical Berlekamp-Conway-Guy theory of impartial (Nim-like) games. This material has never appeared in a textbook before, but it is not difficult and we think you will agree that it is a charming addition to discrete mathematics.
- Chapter 9. This chapter contains the basics of discrete probability theory, which is the subject that historically led to the development of the counting techniques in Chapter 2. This material is nowadays often omitted in courses in discrete mathematics, which we feel is a pity, given the beauty of the material and its importance to modern technology (digital communications, reliability theory, and so on).
- Chapter 10. This chapter discusses finite difference equations, which in discrete mathematics are analogous to differential equations in "continuous" mathematics. Like probability theory, this is a topic of great theoretical and practical importance (with applications to digital signal processing and error-correcting codes, for example) we feel is too often neglected in modern discrete mathematics courses.

Course Organization and Flexibility

The text is divided into 10 chapters, and subdivided into 49 sections, which together include enough material for a one-quarter, one-semester, two-quarter, or even a full-year course. The approximate logical dependence of the chapters is shown in the accompanying figure. As we mentioned above, the only prerequisite for the book is high school algebra; we have been careful not to assume the student knows more than this. In particular, calculus and analytic geometry *are not* needed!

In general, we recommend that any course taught from this book begin with the material in Chapters 1 through 5, which is a little more than can be comfortably covered in a quarter. (A class not emphasizing algorithms could skip Sections 4.3 and 4.4; and a class not emphasizing logical design could skip Section 5.5.) However, note that after Chapters 1 and 2 are covered, both Chapters 9 (Discrete Probability) and 10 (Finite Difference Equations) are accessible. This sequence makes perfect sense, and for many students, e.g., engineering students headed toward digital communications or digital signal processing, it is a good choice.

Once Chapters 1–5 are covered, the instructor has many options. We think the best choice at this point is Chapter 8 (Mathematical Games). It has absolutely no practical value, but it is nevertheless serious mathematics and also great fun! Chapter 8 can be covered in 4–5 one-hour lectures. The material in Chapters 6 and 7 leads to Kleene's Theorem about regular languages (Section 7.4), and, as the accompanying figure shows, a minimal sequence of sections leading to this interesting and important result is 6.2, 6.4, 7.1, 7.3, and 7.4. This sequence should take about 10 one-hour lectures, although if time permits, some or all the material on Turing machines (Section 6.5), Viterbi decoding (Section 6.6), and counting lattice paths (Section 7.2) should also be included. And, as already mentioned, the material in Chapters 9 and 10 can be taught any time after Chapters 1 and 2 are covered. Chapter 9 will take about 10 lectures, and Chapter 10 will take about 6, assuming the students already know a little about complex numbers. (Most high school algebra courses now include more than enough about complex numbers for Chapter 10.) If this is not the case, the supplementary material on complex numbers in Section 10.4 can be covered in one or two additional lectures.

Supplements

Instructor's Manual. The Instructor's Manual was prepared by Robert Crawford and provides solutions to all the problems. Enough detail is given to aid the instructor in judging the suitability of the problems for his/her class. In addition, the Instructor's Manual provides 4–5 programming project suggestions for each chapter. These projects provide a way for computer-literate students to get an in-depth understanding of the material, beyond what can be achieved by simply solving the problems.

Test Bank. The Test Bank was also prepared by Robert Crawford. It contains three chapter tests for each chapter in the book, and three final examinations. Complete solutions to these problems are provided.

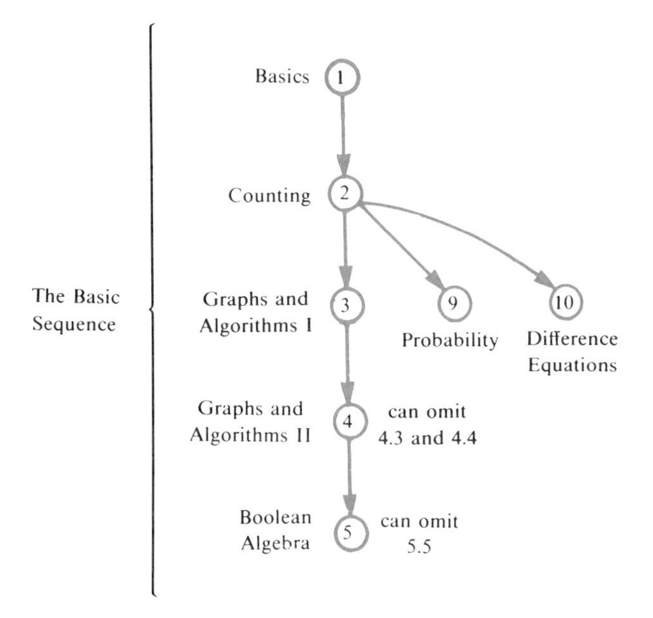

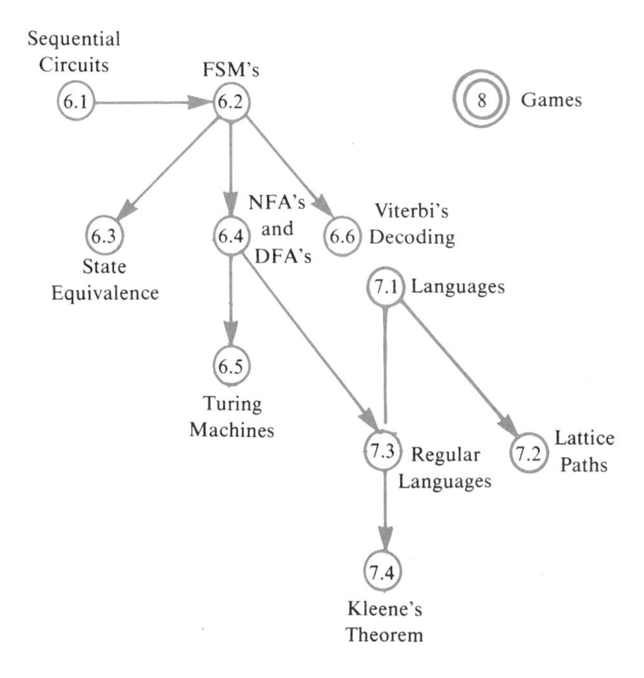

Chapter Dependence for *Discrete Mathematics*

Acknowledgments

It is a pleasure, after all this time, for us to thank some of the people who have helped to make this book possible: Yaser Abu-Mostafa, Oliver Collins, Rod Goodman, and Laif Swanson at Caltech, for many suggestions about what to include and what to omit; Robert Crawford at Western Kentucky University for his magnificent work on the Instructor's Manual and the Test Bank; Joanne Clark at Caltech for her patient and expert typing; Daniel Bargerstock at the University of Illinois for help in solving the problems; and Virginia Eaton at Western Kentucky University for help in preparing the solutions.

Our sincere thanks go to our reviewers for showing us how to improve our original manuscript in a thousand ways:

David M. Arnold	New Mexico State University
Richard H. Austing	University of Maryland
David M. Berman	University of New Orleans
Duncan R. Buell	Supercomputing Research Center (Lanham, MD) (formerly at Louisiana State University)
Douglas Campbell	Brigham Young University
Robert R. Crawford	Western Kentucky University
Tom Cheatham	Western Kentucky University
Thomas A. Dowling	Ohio State University
Donald K. Friesen	Texas A&M University
Richard M. Grassl	University of New Mexico
Evan G. Houston	University of North Carolina—Charlotte
Keith L. Phillips	New Mexico State University
Wayne B. Powell	University of Kansas
George B. Purdy	University of Cincinnati
K. Brooks Reid	Louisiana State University
Douglas R. Shier	William & Mary College
Wallace L. Terwilliger	Bowling Green State University

Our thanks and admiration go to the talented editorial and production people who somehow transformed our dog-eared manuscript into a really handsome book: Margaret Pinette, Michael Weinstein, Alexa Barnes at Random House, and Barbara Gracia and Carol Beal at Woodstock Publishers' Services.

And finally, we'd like to thank our editor, John Martindale at Random House, for his unflagging enthusiasm for this project from the beginning. Without his cheerful, expert, and firm guidance, we would no doubt still be working on a much inferior version of this book, instead of eagerly looking forward to our next one! Thank you, John!

CONTENTS

CHAPTER ONE

CHAPTER TWO

CHAPTER THREE

CHAPTER EIGHT

MATHEMATICAL GAMES 332

CHAPTER NINE

DISCRETE PROBABILITY THEORY 372

CHAPTER TEN

FINITE DIFFERENCE EQUATIONS 417

Appendix:

INTRODUCTION TO **DISCRETE MATHEMATICS**

CHAPTER ONE

PRELIMINARIES

In this first chapter we will introduce you to the basic concepts of discrete mathematics. We will learn about *sets, functions, relations, partial orderings, mathematical induction*, and many of the symbols which represent things discrete mathematicians like to do. This material is important but by no means difficult or advanced. Still, in the last section we will already be in a position to tell you something about one of the most important practical applications of discrete mathematics: *error-correcting codes*.

1.1 Basic Set Theory Terminology and Notation

The basic object in mathematics is the **set**. A set is just a collection of **elements**. These elements can be anything: numbers, points, words, etc. In this section we will introduce you to the standard terminology and notation that mathematicians use to describe sets and their properties. The contents of this section are essential for understanding the rest of the book.

The elements of a set are usually displayed inside a pair of matching **braces** ("{" and "}"). Thus the set containing the numbers $1, 3, 5$ and nothing else is denoted by $\{1, 3, 5\}$, and the set containing as elements the words *yes* and *no* is denoted by $\{yes, no\}$.

It is often convenient to represent a set by a special symbol, usually a capital letter. For example, we might write "Let A be the set consisting of the elements $1, 2, 3, 7$, i.e., $A = \{1, 2, 3, 7\}$, and let $B = \{2, 4, 6, 8\}$."

The symbol "$\in$" is used to denote set membership. If x belongs to the set B, we write $x \in B$ (which is usually read "x is in B"). For example, $4 \in \{2, 4, 6, 8\}$.

If x is not an element of B, however, we write $x \notin B$ ("x is not in B"). For example, $3 \notin \{2, 4, 6, 8\}$.

A **finite set** is a set consisting of a finite number of elements. An **infinite set** is a set which contains an infinite number of elements. For example, $X = \{-2, 0, 1\}$ is finite, but

$$N = \{1, 2, 3, 4, 5, 6, \ldots\}$$

(the set of all positive integers) is infinite. In this expression we have used three dots ("$\cdots$"), sometimes called *ellipsis marks*, to indicate "and so on," i.e., not all of the set's elements have been listed. In mathematical writing it is often necessary to talk about sets with a large, indefinite, or even infinite number of elements, and these three dots come in handy in such cases.

If X is a finite set, the symbol $|X|$ denotes the number of elements in X. For example:

$$|\{1, 2\}| = 2, \qquad |\{0, \pi\}| = 2, \qquad |\{-1, -2, -3, -8\}| = 4.$$

If the set X is infinite, we sometimes write $|X| = \infty$ ("∞" is the mathematical symbol for "infinity").

Sometimes a set is described in terms of its *properties*. The usual notation for a set defined this way is

$$A = \{x: x \text{ has property } P\},$$

which is read "A equals the set of x's such that x has property P." For example, if

$$A = \{x: x \text{ is an even integer between 1 and 8}\},$$

then in fact,

$$A = \{2, 4, 6, 8\}, \qquad |A| = 4.$$

Similarly, if

$$X = \{p: p \text{ is a positive prime number}\},$$

then

$$X = \{2, 3, 5, 7, 11, 13, 17, \ldots\}, \qquad |X| = \infty.$$

It is strange, but true, that one of the most important sets is the **empty set**, which is the set with *no* elements! It is denoted by the symbol "$\varnothing$":

$$\varnothing = \{\ \}, \qquad |\varnothing| = 0.$$

We will have more to say about this interesting set later on.

There are many operations that can be performed on sets, operations which combine and change the sets in various ways. The three most important operations are called *union*, *intersection*, and *complement*. These three set operations are closely related to the operations which can be performed on English phrases with the connecting words *or*, *and*, and *not*.

If A and B are sets, we can combine them to form a new set, called the **union** of A and B, which is denoted by the symbol $A \cup B$, pronounced "A union

B." The set $A \cup B$ is defined to be the set of elements which belong to *A or B or both.**

Similarly, the **intersection** of *A* and *B*, which is denoted by the symbol $A \cap B$, pronounced "*A* intersect *B*," is defined to be the set of elements belonging to both *A and B*.

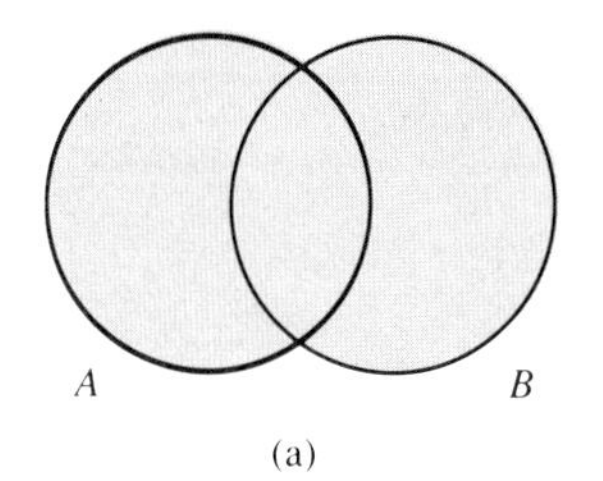

(a)

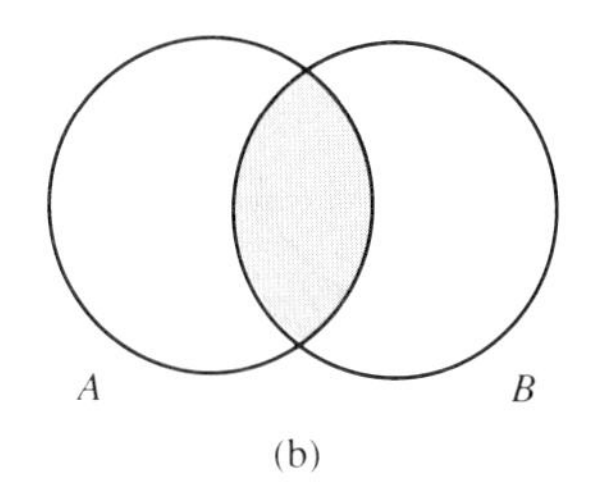

(b)

Figure 1.1 Venn diagram of (a) $A \cup B$ and (b) $A \cap B$.

EXAMPLE 1.1 Let $A = \{1, 3, 5\}$, $B = \{3, 5, 7, 8\}$, $C = \{1, 5\}$, and $D = \{2, 3, 4\}$. Find $A \cup B$, $|A \cup B|$, $A \cap B$, $|A \cap B|$, $C \cap D$, and $|C \cap D|$.

SOLUTION

$$A \cup B = \{1, 3, 5, 7, 8\}; \qquad |A \cup B| = 5.$$

$$A \cap B = \{3, 5\}; \qquad |A \cap B| = 2.$$

$$C \cap D = \varnothing; \qquad |C \cap D| = 0.$$

(Notice that if two sets have nothing in common, their intersection is the empty set $\varnothing$.)

A convenient representation of the sets $A \cup B$ and $A \cap B$ is by means of a picture called a *Venn† diagram.* Look at Figure 1.1. If you think of the sets *A* and *B* as the insides of the corresponding circles, then the shaded area in Figure 1.1(a) represents $A \cup B$, and the shaded area in Figure 1.1(b) represents $A \cap B$.

It is possible to take the union or intersection of more than two sets at a time. For example, in Figure 1.2(a) we see a Venn diagram representing the union of the three sets *A*, *B*, and *C*. More generally, the union of the *n* sets $A_1, A_2, \ldots, A_n$ is defined to be the set of elements belonging to *at least one* of the *n* sets; this set is denoted by the symbol

$$A_1 \cup A_2 \cup \cdots \cup A_n.$$

Similarly, in Figure 1.2(b) we see the Venn diagram for the intersection of the three sets *A*, *B*, and *C*. The intersection of *n* sets $A_1, A_2, \ldots, A_n$ is defined to be the set of elements belonging to *all* of the *n* sets; it is denoted by the symbol

$$A_1 \cap A_2 \cap \cdots \cap A_n.$$

Unions and intersections can even be defined for an infinite number of sets—see Problems 29–31 at the end of this section.

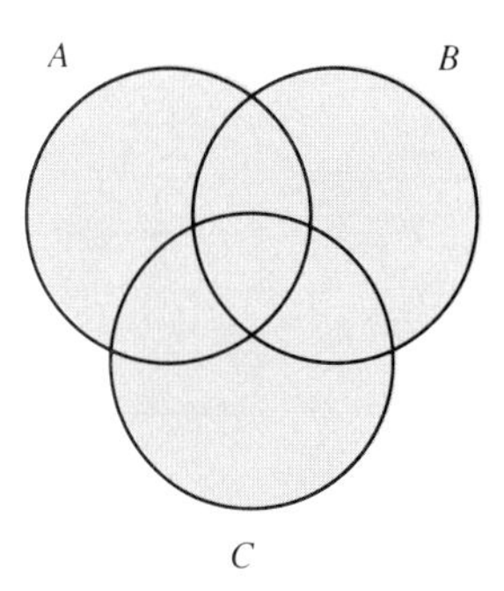

(a)

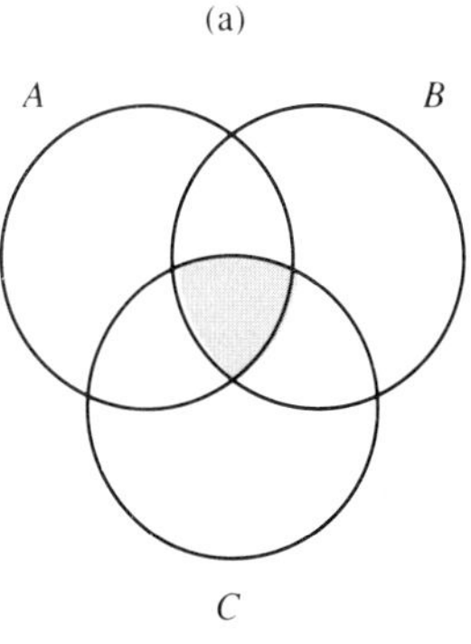

(b)

Figure 1.2 (a) Union of *A*, *B*, and *C*. (b) Intersection of *A*, *B*, and *C*.

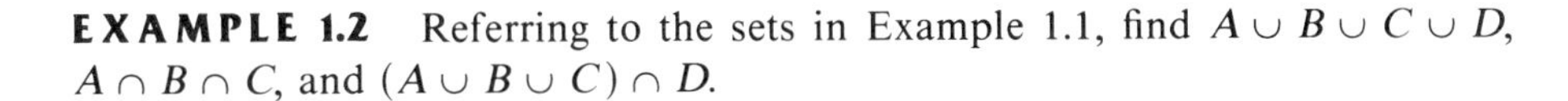

EXAMPLE 1.2 Referring to the sets in Example 1.1, find $A \cup B \cup C \cup D$, $A \cap B \cap C$, and $(A \cup B \cup C) \cap D$.

* In most mathematical writing, and certainly in this book, the word *or* is taken to have its *inclusive* connotation—thus the phrase "*A* or *B*" will always mean "*A* or *B or both*."

† John Venn (1834–1923), English mathematician. His book *Symbolic Logic* (1881) introduced the logical diagrams we shall meet often in this chapter.

SOLUTION

$$A \cup B \cup C \cup D = \{1, 2, 3, 4, 5, 7, 8\}$$
$$A \cap B \cap C = \{5\}$$
$$(A \cup B \cup C) \cap D = \{1, 3, 5, 7, 8\} \cap \{2, 3, 4\} = \{3\}.$$

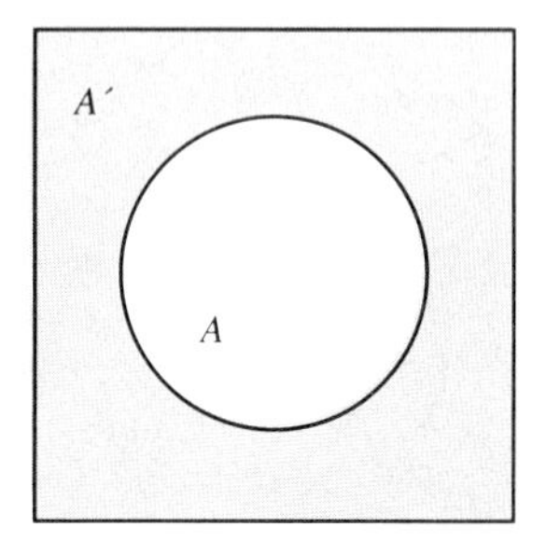

Figure 1.3 Complement of A.

Next we come to the operation of **complementation**. Informally, the complement of a set A, denoted by A', is the set of elements *not in* A.* For example, consider $A = \{1, 2, 3\}$. What is A'? If A' is supposed to be the set of all elements not in A, it must be a very large set indeed; it contains all negative numbers, all positive numbers bigger than three, every word in every dictionary, every point in the plane, every animal on Noah's ark, all of Shakespeare's works, . . . ! Of course this is silly. The point is that complementation makes sense only if we define all of our sets relative to a given bigger set S, which is called the **universe of discourse**, or just the **universe**. Then with respect to this universe, A' is the set of elements in S but not in A. Figure 1.3 illustrates the idea.

EXAMPLE 1.3 Let $A = \{1, 3, 5\}$ and $B = \{2, 4, 7, 8\}$. Find A' and B' with respect to the universe $\{1, 2, 3, 4, 5, 6, 7, 8\}$.

SOLUTION

$$A' = \{2, 4, 6, 7, 8\}; \qquad |A'| = 5.$$
$$B' = \{1, 3, 5, 6\}; \qquad |B'| = 4.$$

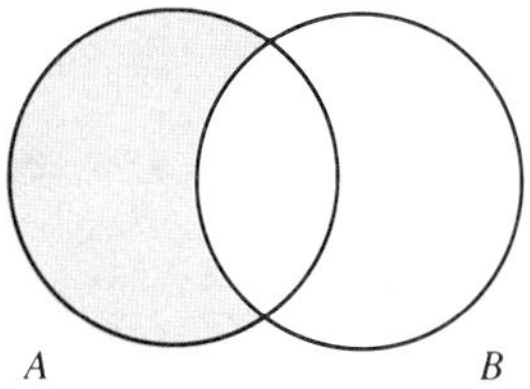

A B

Figure 1.4 The difference $A - B$.

The notation for complementation hides the fact that it is really a *two-set operation*. In fact, complementation is a special case of a more general two-set operation, the **difference operation**. The **difference** $A - B$ of the two sets A and B is defined to be the set of elements which are in A but not in B (see Figure 1.4). For example, if $A = \{1, 2, 3\}$ and $B = \{3, 4, 5\}$, then $A - B = \{1, 2\}$, and $B - A = \{4, 5\}$. In these terms the complement of a set A with respect to the universe S is just the set $S - A$. If A and B both belong to the same universe S, then (again, see Figure 1.4) we have

$$A - B = A \cap B',$$

and so the operation "−" can always be expressed in terms of intersection and complementation.

EXAMPLE 1.4 Let $A = \{1, 2, 3, 4\}$, $B = \{2, 3, 4, 5\}$, and $C = \{3, 4, 5, 6\}$. Find $A - B$, $B - C$, $A - (B - C)$, and $(A - B) - C$.

* In some texts the notation $\bar{A}$ or A^c is used to denote the complement of A.

SOLUTION

$$A - B = \{1\}; \qquad B - C = \{2\};$$
$$A - (B - C) = A - \{2\} = \{1, 3, 4\};$$
$$(A - B) - C = \{1\} - C = \{1\}.$$

Given two or more sets, we may wish to compare them, to see how they are related. The simplest possible relationship is **equality**. Two sets A and B are said to be **equal**, written $A = B$, if A and B contain exactly the same elements. If A and B are not equal, we write $A \neq B$. Hence

$$\{1, 2, 3\} = \{3, 1, 2\}; \qquad \{1, 2, 3\} \neq \{2, 3, 4\}.$$

Another important relation is the **subset** relation. We say that A is a *subset* of B, and write $A \subseteq B$, if every element of A is also an element of B. If, on the other hand, A is not a subset of B, we write instead $A \not\subseteq B$. For example,

$$\{1, 2, 3\} \subseteq \{1, 2, 3, 4\}; \qquad \{1, 2, 3\} \subseteq \{1, 2, 3\};$$
$$\{1, 2, 3\} \not\subseteq \{2, 3, 4\}.$$

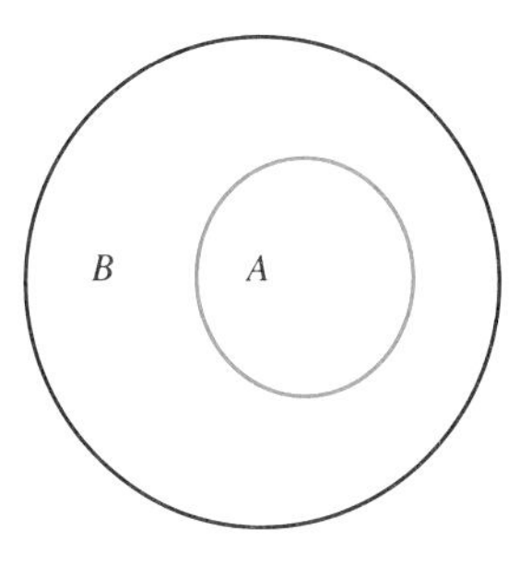

Figure 1.5 A is a subset of B.

Figure 1.5 is a Venn diagram showing A as a subset of B.

Incidentally, notice that if $A \subseteq B$ and $B \subseteq A$, then A and B must be equal, and vice-versa. In other words,

$$A = B \qquad \text{if and only if} \qquad A \subseteq B \text{ and } B \subseteq A.$$

This rather obvious fact often provides a useful way of proving that two sets are equal, as we shall see in Section 1.2.

It seems at first confusing, but it is nevertheless true, that the empty set should be considered to be a subset of *all sets.* For example,

$$\varnothing \subseteq \{1, 2, 3\}; \qquad \varnothing \subseteq \{a, b\}; \qquad \varnothing \subseteq \varnothing.$$

The first of these peculiar-looking statements says, literally, that "every element of $\varnothing$ is an element of $\{1, 2, 3\}$." Mathematicians regard statements of this type as being *true by default,* since $\varnothing$ has no elements! We will discuss this idea in greater detail in Chapter 5.

If $A \subseteq B$, but $A \neq B$, then B must contain at least one element which is not in A, in which case we say that A is a **proper subset** of B, and write $A \subset B$. Hence

$$\{1, 2, 3\} \subset \{1, 2, 3, 4\}; \qquad \varnothing \subset \{1\}.$$

If $A \not\subseteq B$, and $B \not\subseteq A$, we say that A and B are **incomparable**. The two sets A and B in Figure 1.1 are incomparable. Finally, two sets A and B are said to be **disjoint** if they have nothing in common, i.e., if $A \cap B = \varnothing$. Thus $\{1, 3, 5\}$ and $\{2, 4, 6\}$ are disjoint, but $\{1, 3, 5\}$ and $\{3, 8\}$ are not. A collection of sets $A_1, A_2, A_3, \ldots$ is said to be **pairwise disjoint** if each pair of them is disjoint. Thus

$$\{1, 2\}, \{3, 4\}, \{5, 6\} \quad \text{are pairwise disjoint.}$$
$$\{1, 2\}, \{3, 4\}, \{1, 6\} \quad \text{are not pairwise disjoint.}$$

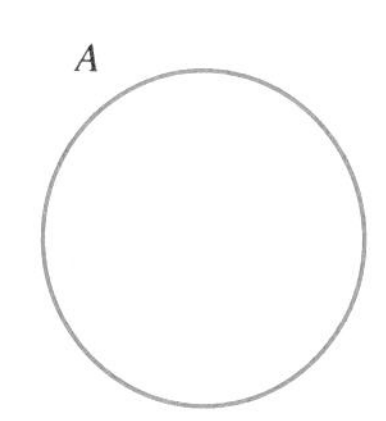

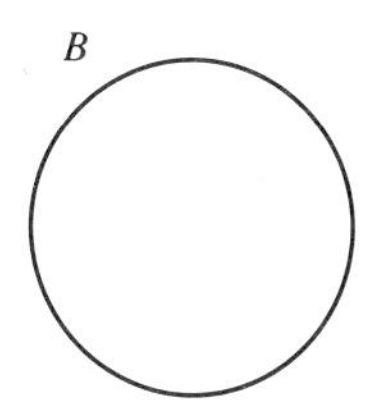

Figure 1.6 A and B are disjoint.

Figure 1.6 is a Venn diagram showing two disjoint sets.

One of the most important skills in discrete mathematics is the ability to count things, and throughout this book we will constantly ask the question "How many?" Here is an example of this question.

EXAMPLE 1.5 Let $A = \{1, 2, 3, 4\}$ and $B = \{1, 3\}$.

(a) How many subsets does B have?
(b) How many *proper* subsets does B have?
(c) How many subsets of A are *incomparable* with B?
(d) How many subsets of A are *disjoint* from B?

SOLUTION

(a) The answer is *four.* The subsets of B are $\emptyset, \{1\}, \{3\}, \{1, 3\}$.
(b) The answer is *three.* The proper subsets of B are $\emptyset, \{1\}, \{3\}$.
(c) The answer is *nine.* The subsets of A not comparable with B are

$$\{2\}, \{4\}, \{1, 2\}, \{1, 4\}, \{2, 3\}, \{2, 4\}, \{3, 4\}, \{2, 3, 4\}, \{1, 2, 4\}.$$

(d) The answer is *four.* The subsets of A disjoint from B are $\emptyset, \{2\}, \{4\}, \{2, 4\}$.

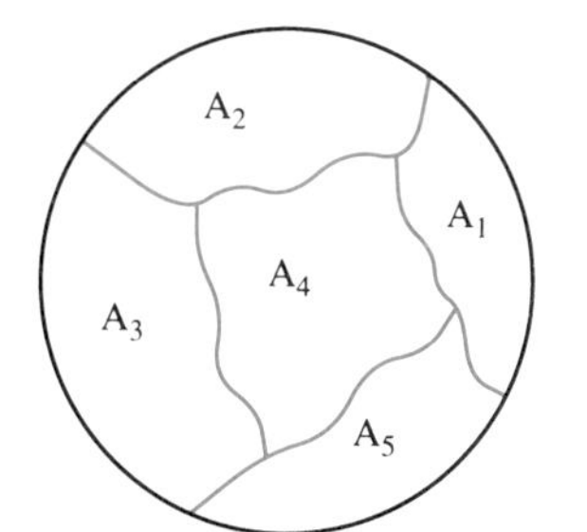

Figure 1.7 A partition of the set A.

If a set A is written as the union of n nonempty pairwise disjoint sets, viz.,

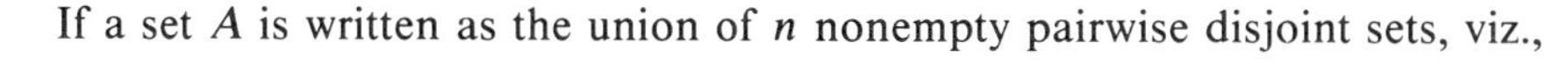

$$A = A_1 \cup A_2 \cup \cdots \cup A_n \qquad \text{(the } A_i\text{'s are pairwise disjoint and not empty)},$$

A is said to be **partitioned** into n subsets. Figure 1.7 shows a Venn diagram of a set partition.

Here is another typical counting problem.

EXAMPLE 1.6 Let $X = \{a, b, c\}$. How many ways are there to partition X?

SOLUTION The answer is *five*:

$$\begin{aligned} X &= \{a\} \cup \{b\} \cup \{c\} \\ &= \{a\} \cup \{b, c\} \\ &= \{a, b\} \cup \{c\} \\ &= \{a, c\} \cup \{b\} \\ &= \{a, b, c\}. \end{aligned}$$

(Note that, for example, $\{a\} \cup \{b, c\}$ and $\{b, c\} \cup \{a\}$ count as identical partitions. The order in which the subsets of the partition appear does not matter.)

Problems for Section 1.1

1. Refer to Example 1.1 and find $A \cup C$, $A \cup D$, $A \cap C$, and $A \cap D$.

2. In Example 1.1, find $B \cup C$, $B \cup D$, $B \cap C$, and $B \cap D$.

3. In Example 1.1, find $(A \cup B) \cap (B \cup C \cup D)$.

4. In Example 1.1, find $(A \cap B \cap D) \cup C$.

5. Referring to Example 1.3, find the complements of the following sets with respect to the universe $S = \{0, 1, 2, 3, 4, 5, 6, 7, 8, 9\}$:
(a) A; (b) B; (c) S.

6. As in Problem 5, find the complements of the following sets:
(a) $A \cup B$; (b) $A \cap B$; (c) $\varnothing$.

7. Let $A = \{1, 2, 3, 4, 5, 6\}$, $B = \{2, 4, 6, 8, 10\}$, and $C = \{3, 4, 5, 7, 9, 11\}$. Find the following:
(a) $A - B$; (c) $(B - A) - C$.
(b) $B - (A - C)$;

8. In Problem 7, find the following:
(a) $B - A$; (c) $(B - A) \cup C$.
(b) $B - (A \cup C)$;

9. Let $A = \{1, 2, 3, 4, 5\}$ and $B = \{1, 3, 5\}$.
(a) List all subsets of A that are disjoint from B.
(b) List at least three subsets of A that are incomparable with B.

10. Let A and B be defined as in Problem 9.
(a) How many proper subsets of B are there?
(b) How many subsets of B are disjoint from A?

11. Let $A = \{1, 2\}$. List all ways of partitioning A.

12. Let $A = \{1, 2, 3, 4\}$. How many partitions of A are there?

13. Let $A = \{1, 2, 3, 5\}$, $B = \{1, 3, 4, 6\}$, and $C = \{1, 2, 4, 7\}$. Calculate $|A \cup B \cup C|$, $|A \cap (B \cup C)|$, and $|A \cap B' \cap C|$. (Take as the universe the set $S = \{1, 2, 3, 4, 5, 6, 7\}$.)

14. Let A be the set of even integers between 0 and 10, and let B be the set of all integers between 7 and 13. Find $A \cup B$ and $|A \cap B|$.

15. If A and B are disjoint sets, and C is any other set, show with a Venn diagram that $A \cap C$ and $B \cap C$ are disjoint.

16. Consider the two-element subsets of $\{1, 2, 3, 4, 5\}$. For example, $\{1, 2\}$ and $\{3, 5\}$ are two such subsets. How many such subsets are there?

17. In Problem 16, what is the largest number of two-element subsets having the property that *no two* are disjoint? In other words, we wish to find distinct two-element subsets $A_1, \ldots, A_N$ such that $A_i \cap A_j \neq \varnothing$ whenever $i \neq j$. What is the largest possible value of N? For example, $N \geq 3$ since we can use $\{1, 2\}$, $\{2, 4\}$, and $\{1, 4\}$. Can you do better?

18. Find three sets A, B, and C such that $A \cap B \cap C = \varnothing$, but none of the three sets $A \cap B$, $A \cap C$, or $B \cap C$ is empty.

19. In some ways, the *union* of sets is similar to the *addition* of numbers, the *intersection* of sets is similar to the *multiplication* of numbers, the set $\varnothing$ is like the number 0, and the universal set S is like the number 1. For each of the following six set theory statements, tell whether or not the corresponding statement about numbers is correct:
(a) $A \cup \varnothing = A$; (d) $A \cap S = A$;
(b) $A \cap \varnothing = \varnothing$; (e) $A \cup A = A$;
(c) $A \cup S = S$; (f) $A \cap A = A$.

20. Let $X = \{1, 2, 3\}$. What is the largest collection of pairwise disjoint subsets $A_1, A_2, \ldots$ of X that you can find for the following conditions?
(a) None of the sets is allowed to be the empty set.
(b) One of the A_i can be empty.
(c) Two of the A_i can be empty.

21. Sets can themselves be members of other sets. For example, if $A = \{a, b, c\}$ and $B = \{A, 7\}$, then the set A belongs to the set B. Note that the only members of B are A and 7, and a, b, and c do *not* belong to B. Now let $A = \{1, 4, 8\}$ and $B = \{A, 4, \{3\}\}$; find $A \cup B$ and $A \cap B$.

22. In Problem 21, identify an element c and a set C such that $c \in C$, $C \in B$, but $c \notin B$.

23. Let $A = \{1, \{1, 2\}\}$, so that A is a set with two elements, viz., 1 and $\{1, 2\}$. If $B = \{A, 1, 2\}$, what is $|A \cap B|$?

24. Refer to Problem 23.
(a) Is A a subset of B? (b) What is $|A \cup B|$?

25. We saw in the text that the empty set $\varnothing$ is a subset of every set. Can you think of a set that has $\varnothing$ as *an element*, i.e., a set A such that $\varnothing \in A$?

26. Let $A = \varnothing$, $B = \{\varnothing\}$, and $C = \{\varnothing, \{\varnothing\}\}$.
(a) Which of these three sets has $\varnothing$ as a subset?
(b) Which of these three sets has $\varnothing$ as an element?

27. The **power set** of a set X, denoted by $P(X)$, is the set of all subsets of X. For example, if $X = \{0, 1\}$, then $P(X) = \{\varnothing, \{0\}, \{1\}, \{0, 1\}\}$.
(a) What is $P(X)$ if $X = \{0, 1, 2\}$?
(b) What is $P(\varnothing)$?

28. The notation $[a, b]$ is used for the set of real numbers between a and b, with a and b *included*, i.e.,

$$[a, b] = \{x\colon a \leq x \leq b\}.$$

Similarly, (a, b) denotes the set of real numbers between a and b, with a and b *excluded*, i.e.,

$$(a, b) = \{x: a < x < b\}.$$

We also write

$$[a, b) = \{x: a \leq x < b\}$$

and

$$(a, b] = \{x: a < x \leq b\}.$$

Now let $A = [0, 2]$ and $B = (1, 3)$. Find $A \cup B$ and $A \cap B$.

Problems 29–33 use the notation of Problem 28.

29. Let $A_n = [0, 1 - 1/n]$, $n = 1, 2, \ldots$. (Thus $A_1 = [0, 0] = \{0\}$, the set consisting of 0 alone; $A_2 = [0, \frac{1}{2}]$; $A_3 = [0, \frac{2}{3}]$; and so on.) Find the union of these sets, i.e., the set of elements belonging to *at least one* of the sets A_n.

30. Let $A_n = (0, 1 + 1/n)$, $n = 1, 2, \ldots$. Find the union of this collection of sets.

31. With the A_n defined as in Problem 30, find the intersection of the collection, i.e., the set of elements belonging to *all* of the sets A_n.

32. Suppose that A and B are **open intervals**, i.e., sets of the form (a, b). Is $A \cap B$ an open interval? How about $A \cup B$?

33. Suppose that A and B are **closed intervals**, i.e., sets of the form $[a, b]$. Is $A \cap B$ a closed interval? How about $A \cup B$?

34. We have seen that every set is a subset of itself. Can you find a finite set that is an *element* of itself?

1.2 Venn Diagrams, Truth Tables, and Proof

There are many important interrelationships among the set operations we introduced in the previous section. For example, there are the **associative laws**:

(1.1) $(A \cup B) \cup C = A \cup (B \cup C)$ (associative law for union);

(1.2) $(A \cap B) \cap C = A \cap (B \cap C)$ (associative law for intersection).

Simply *stating* such a rule does not make it true, however! For example, here is the "associative law for set *difference*":

$$(A - B) - C \stackrel{?}{=} A - (B - C),$$

which, as we saw in Example 1.4, isn't true. In this section we'll present techniques which can be used to see whether a given identity is true. We leave the verification of the associative laws as problems, and we will illustrate these techniques by establishing the important **distributive law**, which holds for any three sets, A, B, and C:

(1.3) $$A \cap (B \cup C) = (A \cap B) \cup (A \cup C).$$

For example if $A = \{1, 2\}$, $B = \{2, 3\}$, and $C = \{3, 4\}$, then $A \cap (B \cup C) = \{1, 2\} \cap \{2, 3, 4\} = \{2\}$, and $(A \cap B) \cup (A \cap C) = \{2\} \cup \varnothing = \{2\}$. This verifies Equation (1.3) in one particular case. But we need to *prove* that (1.3) always holds, for *any* three sets. How can we do this? There are at least three possible methods of proof, and we will briefly discuss them all in this section.

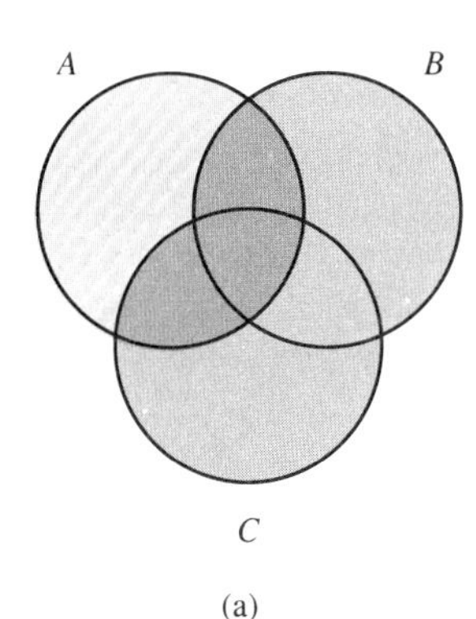

(a)

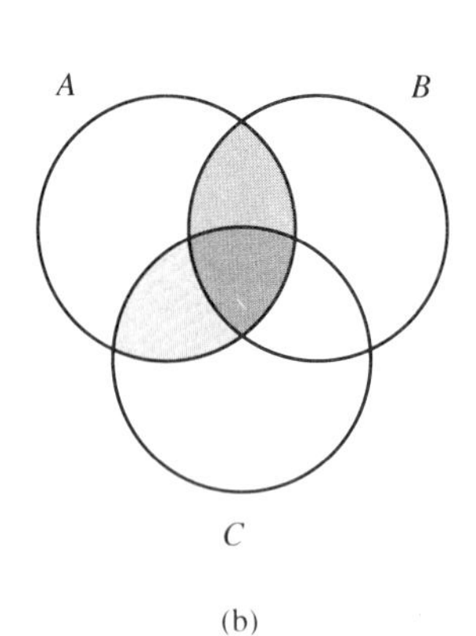

(b)

Figure 1.8 Verifying the distributive law with a Venn diagram.

Method 1 (Venn Diagrams)

If three or fewer sets are involved in the given identity, we can draw a Venn diagram for each side and verify that the sets in question are equal. In Figure 1.8(a), for example, we have computed, by Venn diagram, the set $A \cap (B \cup C)$ (A is shown in color, $B \cup C$ in grey, and $A \cap (B \cup C)$ is in color and grey). It is seen to consist of three of the seven compartments in the standard ABC Venn diagram. Similarly, in Figure 1.8(b) we see that $(A \cap B) \cup (A \cap C)$ consists of the same three compartments. This is one proof of the distributive law.

If intersection and union are interchanged in the distributive law, another valid identity is obtained:

$$(1.4) \qquad A \cup (B \cap C) = (A \cup B) \cap (A \cup C).$$

Identity (1.4) is also referred to as the distributive law. Equation (1.3) says that *intersection distributes over union*; (1.4) says that *union distributes over intersection.* These facts can also be verified by using Venn diagrams (see Figure 1.9).

Incidentally, the process of going from (1.3) to (1.4) is a special case of the general **duality principle** to be considered in Chapter 5. It says that if you have a valid identity involving sets, and replace $\cup$ by $\cap$, $\cap$ by $\cup$, the universal set S by the empty set $\varnothing$, and $\varnothing$ by S, the result will also be a valid identity.

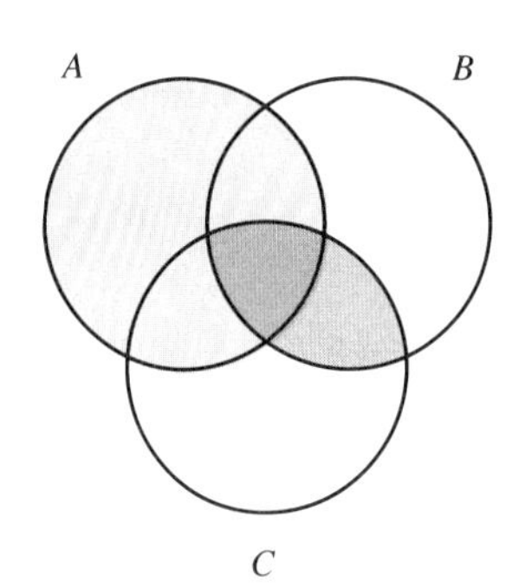

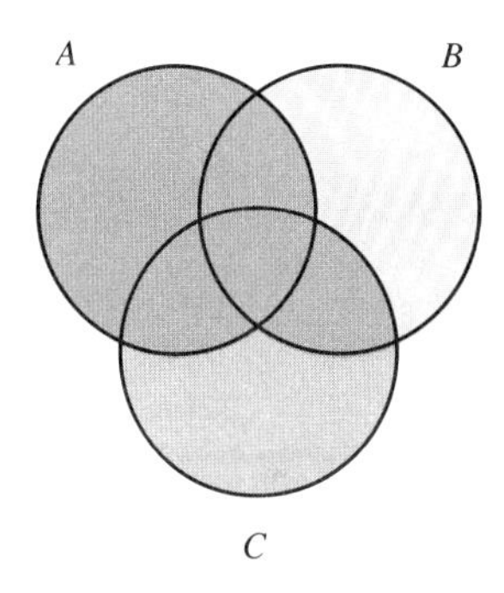

Figure 1.9 $A \cup (B \cap C) = (A \cup B) \cap (A \cup C)$.

EXAMPLE 1.7 Use Venn diagrams to show that $(A - B) \cap C = (A \cap C) \cap B'$.

SOLUTION In Figure 1.10(a) we have shown $A - B$ in grey and C in color. The set $(A - B) \cap C$ is then shown in color and grey.

On the other hand, in Figure 1.10(b) we have colored the part of $A \cap C$ not in B. This represents $(A \cap C) \cap B'$. We see that the colored and grey region in Figure 1.10(a) is the same as the colored region in Figure 1.10(b), and this proves the given equality.

We now come to the second method that can be used to prove set-theory identities.

Method 2 (Truth Tables)

Method 2 is really just a generalization of the Venn diagram method that can be used if more than three sets are involved. (Venn diagrams, for more than three sets, become quite unwieldy; see Problem 34.) We again use the distributive law (1.3) to illustrate the general method. The idea now is that we classify each point x in the universe S according to its membership in the three sets A, B, and C. That is, for each point x each of three statements "$x \in A$," "$x \in B$," and "$x \in C$" is either *true* (T) or *false* (F).

Each point will then fall into one of the eight classifications shown in the first three columns of Table 1.1. For example, the classification TFF represents those points x satisfying $x \in A$, $x \notin B$, and $x \notin C$, i.e., the set $A \cap B' \cap C'$. These eight classifications correspond to the eight compartments in the standard ABC Venn diagram (see Figure 1.2): the seven *interior* compartments and the *exterior* compartment, which corresponds to $A' \cap B' \cap C'$.

Now to describe a compound set, any $A \cap B$, we give its eight **truth values**, one for each of the eight basic classifications. A point x belongs to $A \cap B$ if and only if $x \in A$ and $x \in B$, and so any classification with TT in the first two columns describes points in $A \cap B$. There are just two such classifications, TTT and TTF,

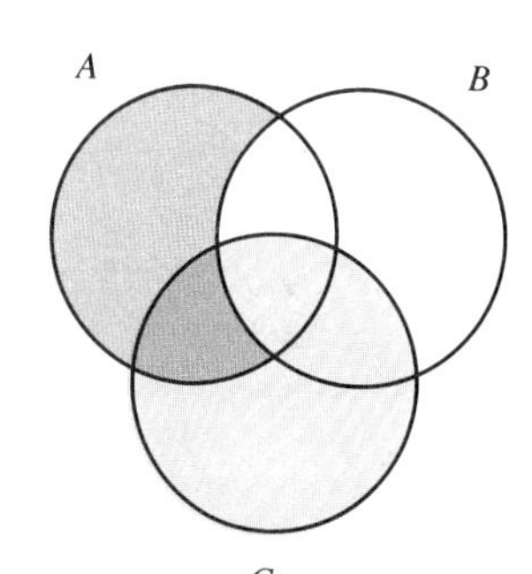

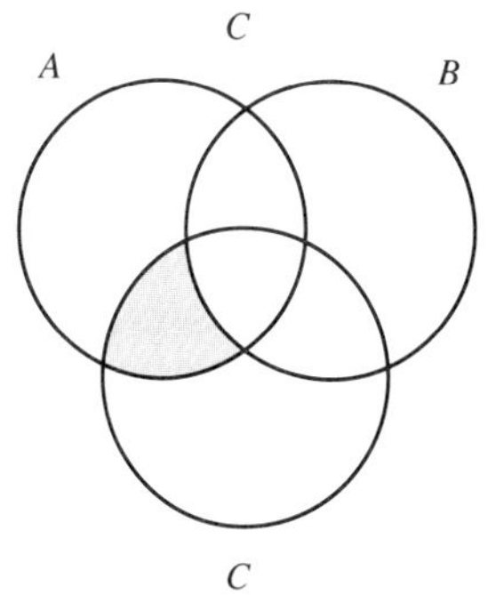

Figure 1.10 Solution to Example 1.7.

TABLE 1.1 A truth table proof of the distributive law

A	B	C	$A \cap B$	$A \cap C$	$B \cup C$	$A \cap (B \cup C)$	$(A \cap B) \cup (A \cap C)$
T	T	T	T	T	T	T	T
T	T	F	T	F	T	T	T
T	F	T	F	T	T	T	T
T	F	F	F	F	F	F	F
F	T	T	F	F	T	F	F
F	T	F	F	F	T	F	F
F	F	T	F	F	T	F	F
F	F	F	F	F	F	F	F

and so the column labelled "$A \cap B$," which contains T's opposite TTT and TTF, and all F's otherwise, represents the set $A \cap B$. Similarly, Table 1.1 gives truth table representation for the sets $A \cap C$ and $B \cup C$.

To represent a more complicated set, e.g., $A \cap (B \cup C)$ or $(A \cap B) \cup (A \cap C)$, we proceed as follows: If $x \in A \cap (B \cup C)$, then $x \in A$ *and* $x \in (B \cup C)$. Thus, we put a T in the column labelled $A \cap (B \cup C)$ if and only if there is a T in column A *and* in column $(B \cup C)$. Comparing these columns, we get the representation of $A \cap (B \cup C)$ shown in Table 1.1. Similarly, $x \in (A \cap B) \cup (A \cap C)$ if and only if $x \in (A \cap B)$ *or* $x \in (A \cap C)$. Thus, we put a T in the column $(A \cap B) \cup (A \cap C)$ if and only if there is a corresponding T in column $(A \cap B)$ *or* in column $(A \cap C)$. This procedure leads to the representation of $(A \cap B) \cup (A \cap C)$ given in Table 1.1.

After all this work, we see that the last two columns in Table 1.1 are identical. In other words, $x \in A \cap (B \cup C)$ if and only if $x \in (A \cap B) \cup (A \cap C)$, which again proves the distributive law (1.3).

EXAMPLE 1.8 Use the truth table method to prove the **DeMorgan laws.***

(1.5) $$(A \cup B)' = A' \cap B';$$

(1.6) $$(A \cap B)' = A' \cup B'.$$

SOLUTION The proof of (1.5) via a truth table is given below. The proof of (1.6) is very similar, and is considered in the problems at the end of this section.

A	B	$A \cup B$	$(A \cup B)'$	A'	B'	$A' \cap B'$
T	T	T	F	F	F	F
T	F	T	F	F	T	F
F	T	T	F	T	F	F
F	F	F	T	T	T	T

* Augustus DeMorgan (1806–1871), English mathematician. He was the author of many textbooks, including *Formal Logic*, in which he established "DeMorgan's law" and many other rules of set theory.

The truth table method is a very powerful "mechanical" procedure which can *always* be used to prove an identity involving a finite number of sets. Truth tables will prove to be an important tool in our study of the propositional calculus in Chapter 5. However, if n sets are involved, the number of entries in each column of the truth table is 2^n (another counting problem!), a number which is unwieldy if n is more than 4, unmanageable if $n \geq 8$, and unthinkable if $n \geq 16$. The next method can give shorter proofs, but requires more creativity on the part of the student.

Method 3 (Direct Use of the Definitions)

We can always fall back on the basic definitions! Rather than give a formal description of this method, we'll just illustrate it with one example. This technique will be greatly expanded upon in Chapter 5.

EXAMPLE 1.9 Use Method 3 to prove DeMorgan's law (1.5) again.

SOLUTION As we observed in Section 1.1, proving that the two sides of (1.5) are equal is equivalent to showing two things:

(1.7) $$(A \cup B)' \subseteq A' \cap B';$$

(1.8) $$A' \cap B' \subseteq (A \cup B)'.$$

To prove (1.7), suppose that $x \in (A \cup B)'$, i.e., it is *not* the case that x belongs to A or B. In other words, x belongs to neither A nor B, i.e.,

$$x \notin A \quad \text{and} \quad x \notin B.$$

This is the same as saying

$$x \in A' \quad \text{and} \quad x \in B'.$$

And this, in turn, is the same as saying

$$x \in A' \cap B'.$$

Therefore every element of $(A \cup B)'$ is also an element of $A' \cap B'$, so that (1.7) holds. To complete the proof, we must show the opposite inequality (1.8). In this case the previous argument is reversible, for if $x \in A' \cap B'$, then

$$x \in A' \quad \text{and} \quad x \in B',$$

so x belongs to *neither* A nor B. Thus,

$$x \notin A \cup B,$$

which is the same as saying

$$x \in (A \cup B)'.$$

This shows that (1.8) is true and, combined with (1.7), completes the proof.

Problems for Section 1.2

In Problems 1–7, use Venn diagrams to establish the given identities.

1. The associative law for union:

$$A \cup (B \cup C) = (A \cup B) \cup C.$$

2. The associative law for intersection:

$$A \cap (B \cap C) = (A \cap B) \cap C.$$

3. $A = (A')'$.

4. Identity (1.6): $(A \cap B)' = A' \cup B'$.

5. $A \cap (B - C) = (A \cap B) - (A \cap C)$.

6. $A - (B \cup C) = (A - B) - C$.

7. $((A \cap B) \cup (A' \cap C))' = (A \cap B') \cup (A' \cap C')$.

In Problems 8–14, use truth tables to establish the given identities.

8. The identity of Problem 3.

9. The identity of Problem 4.

10. $A \cup (A' \cap B) = A \cup B$.

11. $(A \cup B) \cap (A \cup C) = A \cup (B \cap C)$.

12. The identity of Problem 5.

13. The identity of Problem 6.

14. The identity of Problem 7.

In Problems 15–21, use basic definitions to establish the given identities.

15. The identity of Problem 1.

16. The identity of Problem 2.

17. The identity of Problem 3.

18. $(A \cup B \cup C)' = A' \cap B' \cap C'$.

19. $(A \cap B \cap C)' = A' \cup B' \cup C'$.

20. $(A \cap B) \cup (A' \cap C) \cup (B \cap C) = (A \cap B) \cup (A' \cap C)$.

21. The identity of Problem 5.

22. Prove the following statement in three ways: "If A and B are disjoint, then $A \cap X$ and $B \cap X$ are also disjoint."

23. In each case, decide if possible from the given information about x, whether or not $x \in A$.
(a) $x \in (A \cup B \cup C)$.
(b) $x \notin (A \cup B \cup C)$.
(c) $x \in (A \cap B \cap C)$.
(d) $x \notin (A \cap B \cap C)$.

24. If $A \subseteq B$, show that $A \cup B = B$ and illustrate with a Venn diagram.

25. If $A \subseteq B$, show that $A \cap B = A$ and illustrate with a Venn diagram.

26. See if you can find nonempty sets A, B, and C for which the "associative law for difference," $A - (B - C) = (A - B) - C$, *is* true.

27. Is it true $(A - B) \cup C = (A \cup C) - B$?

28. The **symmetric difference** between sets A and B, denoted $A \oplus B$, is defined by

$$A \oplus B = (A - B) \cup (B - A).$$

(a) Draw a Venn diagram for $A \oplus B$.
(b) Show that $x \in A \oplus B$ if and only if x belongs to *exactly one* of the two sets A and B.
(c) Show that $A \oplus (B \oplus C) = (A \oplus B) \oplus C$, and describe in words what it means for x to be an element of the set $A \oplus B \oplus C$.
(d) Use a Venn diagram or truth table to simplify the expression $A \oplus B \oplus (A \cap B)$.

29. If $A_n \subseteq A_{n-1} \subseteq \cdots \subseteq A_1$, show that

$$A_1 \cap A_2 \cap \cdots \cap A_n = A_n;$$
$$A_1 \cup A_2 \cup \cdots \cup A_n = A_1.$$

30. The identities in Problems 18 and 19 give the DeMorgan laws for *three* sets. Can you see a pattern? State and prove the appropriate statements for arbitrary sets $A_1, \ldots, A_n$.

31. In a group of 100 people, 65 are men, 30 are registered voters, and 10 are male registered voters. Draw a Venn diagram with a universe of 100 and two sets labelled M and R, and fill in numbers until you find the number of non-registered women.

32. Repeat Problem 31, but this time use two sets labelled M and N (=nonregistered).

33. Among the 100 people in Problem 31, it is further known that 40 are tall, 20 are tall men, 15 are tall and registered, and 6 are tall, registered men. How many are short, nonregistered women?

34. See if you can draw a four-set Venn diagram. It should have 15 interior compartments, representing the 15 sets $A \cap B \cap C \cap D, A \cap B \cap C \cap D', \ldots, A' \cap B' \cap C' \cap D'$.

1.3 Functions and Relations

As a student of mathematics, you have certainly already worked with functions. Familiar examples include polynomials, trigonometric functions, exponentials, and logarithms. Informally, we usually think of a function f as a rule that associates with each "input" x an "output" $f(x)$. For example, in Figure 1.11 we illustrate one function which **maps** the set $A = \{a_1, a_2, a_3, a_4\}$ to the set $B = \{b_1, b_2, b_3, b_4, b_5\}$.

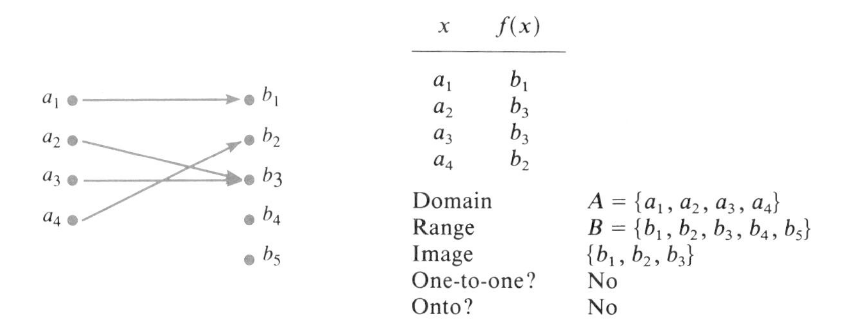

x	$f(x)$
a_1	b_1
a_2	b_3
a_3	b_3
a_4	b_2

Domain	$A = \{a_1, a_2, a_3, a_4\}$
Range	$B = \{b_1, b_2, b_3, b_4, b_5\}$
Image	$\{b_1, b_2, b_3\}$
One-to-one?	No
Onto?	No

Figure 1.11 A function $f: A \to B$.

We will use the notation $f: A \to B$ to mean that the function f maps the set A to the set B.

DEFINITIONS A **function** $f: A \to B$ is a rule that assigns to each element x in A a unique element in B, denoted $f(x)$. The set A is called the **domain** of f, and the set B is called the **range** of f. The **image** of f is the set $\{f(x): x \in A\}$. The image is sometimes denoted by the symbol $f(A)$.

The symbol "f" has no special significance. It is the first letter in the word "function," and mathematicians seem to prefer it to other letters. But any other letter or symbol would do as well, e.g., $g(x)$, $h(x)$, $F(x)$, $\phi(x)$, $\exp(x)$, etc. The symbol "$f(x)$" is pronounced "f of x," "$h(a)$" is pronounced "h of a," etc.

The function f in Figure 1.11 has two different inputs which lead to the same output: $f(a_2) = f(a_3) = b_3$. On the other hand, for the function in Figure 1.12 no output is duplicated; in other words, different inputs yield different outputs. Functions of this type are said to be **one-to-one**.* Mathematically, the one-to-one condition can be expressed as follows:

(1.9) If $x \neq y$, then $f(x) \neq f(y)$.

Equivalently:

(1.10) If $f(x) = f(y)$, then $x = y$.

* Or, in some texts, **injective**.

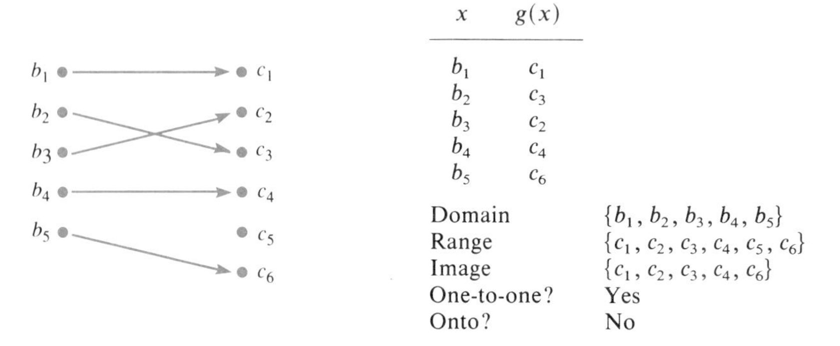

Figure 1.12 A one-to-one function.

Statements (1.9) and (1.10) are equivalent because of the important **contrapositive law** of logic, which says that if A and B are any two statements, then the statement

"If A, then B"

is logically equivalent to the statement

"If not B, then not A."

We will have much more to say about the contrapositive law, and many other laws of logic, in Chapter 5.

In the functions f and g of Figures 1.11 and 1.12 the image isn't the whole range, i.e., not every point in the range is covered by the function. If every point *is* covered, the function is called **onto**.* It is always possible to make a function onto by cutting down its range. For example, the function f of Figure 1.11, with range $\{b_1, b_2, b_3\}$, is onto; and the function g of Figure 1.12 would have been onto if its range had been $\{c_1, c_2, c_3, c_4, c_6\}$.

EXAMPLE 1.10 Let $A = \{0, 1, 2\}$ and $B = \{a, b\}$. Find all functions $f: A \to B$ and $g: B \to A$. For each such function, determine whether it is one-to-one, onto, both, or neither.

SOLUTION There are *eight* functions mapping $A \to B$; in tabular form they are listed below:

x	f_1	f_2	f_3	f_4	f_5	f_6	f_7	f_8
0	a	a	a	a	b	b	b	b
1	a	a	b	b	a	a	b	b
2	a	b	a	b	a	b	a	b
One-to-one?	N	N	N	N	N	N	N	N
Onto?	N	Y	Y	Y	Y	Y	Y	N

For example, f_3 is defined as $f_3(0) = a$; $f_3(1) = b$; $f_3(2) = a$. The function f_3 is *not* one-to-one, since $f_3(0) = f_3(2) = a$. The function f_3 *is* onto, however, because

* Or, in some texts, **surjective**.

its image is $\{a, b\} = B$. Notice that none of these eight functions is one-to-one. This is so because in each case there are only two elements of B (a and b) available for the three values of f [$f(0)$, $f(1)$, and $f(2)$], so that some element of B must be repeated. By the same reasoning, we can say that if $f: A \to B$ and $|A| > |B|$, then f cannot be one-to-one. Incidentally, this simple use of common sense is sometimes called the **pigeonhole principle**, and stated as follows.

> If m pigeons are placed in n pigeonholes, and $m > n$, then some pigeonhole must contain more than one pigeon.

Some other applications of the pigeonhole principle are given in Problems 38–43.

Similarly, there are *nine* functions $g: B \to A$:

x	g_1	g_2	g_3	g_4	g_5	g_6	g_7	g_8	g_9
a	0	0	0	1	1	1	2	2	2
b	0	1	2	0	1	2	0	1	2
One-to-one?	N	Y	Y	Y	N	Y	Y	Y	N
Onto?	N	N	N	N	N	N	N	N	N

We see from Example 1.10 that there can be functions that are one-to-one but not onto, and vice versa. However, if the sets A and B are finite and have the same number of elements, this is impossible. You can visualize this result by drawing a typical mapping diagram of a function on a finite set (Figure 1.13). If f is one-to-one, then no two arrows can point to the same element of B. Thus, if $A = \{a_1, \ldots, a_n\}$ and $B = \{b_1, \ldots, b_n\}$, the list of elements $f(a_1), \ldots, f(a_n)$ must be a rearrangement of $b_1, \ldots, b_n$. (In Figure 1.13, b_1, b_2, b_3, b_4 is rearranged by f to b_2, b_4, b_1, b_3.) It follows that all elements of B are listed, so that f is onto. (*Warning*: This result is false if the sets are infinite. See Problem 26.) If $f: A \to B$ is one-to-one and onto, we say that A and B are in **one-to-one correspondence**.

On the other hand, if f is not one-to-one, then (as in Figure 1.11) two arrows (at least) point to the same element of B. There is no way for the remaining $n - 2$ arrows to cover the remaining $n - 1$ points of B, so f is not onto. Hence by the contrapositive law, if f is onto, then f is one-to-one.

In summary, if $f: A \to B$ and $|A| = |B|$, then f is one-to-one if and only if f is onto.

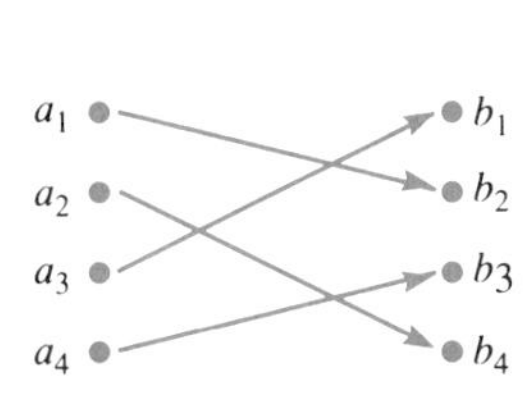

x	$f(x)$
a_1	b_2
a_2	b_4
a_3	b_1
a_4	b_3

Domain	$\{a_1, a_2, a_3, a_4\}$
Range	$\{b_1, b_2, b_3, b_4\}$
Image	$\{b_1, b_2, b_3, b_4\}$
One-to-one?	Yes
Onto?	Yes

Figure 1.13 A function whose domain and range have the same number of elements is one-to-one if and only if it is onto.

EXAMPLE 1.11 Let $A = \{1, 2, 3\}$. Find all one-to-one (and therefore onto) functions $f: A \to A$.

SOLUTION There are exactly *six* such functions:

x	f_1	f_2	f_3	f_4	f_5	f_6
1	1	1	2	2	3	3
2	2	3	1	3	1	2
3	3	2	3	1	2	1

(*Note:* It is important to be able to count the number of functions of various kinds from a given set A to a given set B, and in fact, much of Chapter 2 is devoted to a study of such counting problems. However, without peeking at the results in Chapter 2, you should be able to solve problems like Examples 1.10 and 1.11 if $|A|$ and $|B|$ aren't too large; see Problems 18–25.)

Two functions f and g can often be combined to produce a third function, by an operation called **composition**. For example, the functions $f(x) = x^2$ and $g(x) = \sin(x)$ can be combined to give the function $h(x) = \sin(x^2)$. The function $h(x)$ is computed by first applying the "squaring function" f and then the "sine" function g. In general, if

$$f: A \to B \qquad \text{and} \qquad g: B \to C,$$

we define the **composition** of f and g to be the function $h: A \to C$, by

$$h(x) = g(f(x)).$$

The composition of f and g is denoted by the symbol $g \circ f$, which means "first f, then g." Continuing the illustration above, we have

$$g \circ f(x) = \sin(x^2) \qquad \text{but} \qquad f \circ g(x) = (\sin(x))^2.$$

Thus $g \circ f$ and $f \circ g$ are not the same, as a rule.

EXAMPLE 1.12 Let f be as defined in Figure 1.11 and g as defined in Figure 1.12. Find $h = g \circ f$ and $h' = f \circ g$.

SOLUTION The function $h = g \circ f$ is given in Figure 1.14. For example, $h(a_3)$ is computed as follows:

$$h(a_3) = g(f(a_3)) = g(b_3) = c_2.$$

However, the function $h' = f \circ g$ *doesn't make sense*. For example, let's try to compute $h'(b_2)$:

$$\begin{aligned} h'(b_2) &= f(g(b_2)) \\ &= f(c_3) \qquad \text{(from Figure 1.12)} \\ &= ? \end{aligned}$$

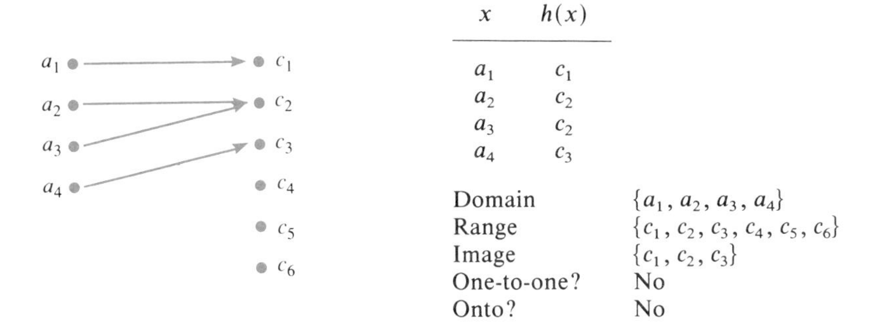

x	$h(x)$
a_1	c_1
a_2	c_2
a_3	c_2
a_4	c_3

Domain	$\{a_1, a_2, a_3, a_4\}$
Range	$\{c_1, c_2, c_3, c_4, c_5, c_6\}$
Image	$\{c_1, c_2, c_3\}$
One-to-one?	No
Onto?	No

Figure 1.14 The composition $h = g \circ f$ of the functions in Figures 1.11 and 1.12.

This is illegal, because the domain of f is $\{a_1, a_2, a_3, a_4\}$—$f(c_3)$ isn't defined! Here is the moral: If

$$f : A \to B \qquad \text{and} \qquad g : C \to D,$$

then the composition $f \circ g$ makes sense only if the image $g(C)$ of g is a subset of the domain A of f.

If the function $f : A \to B$ is one-to-one and onto, then we can make a new function $g : B \to A$ simply by reversing the arrows in the diagram! This function is called the **inverse** of f, and is usually denoted by the symbol f^{-1}. If the function g of Figure 1.12 is made onto by cutting down its range to $\{c_1, c_2, c_3, c_4, c_6\}$, there is an inverse function g^{-1}, which is shown in Figure 1.15. Notice that the domain of f^{-1} is the image of f, and that the image of f^{-1} is the domain of f. The composition of the two functions f and f^{-1} is a function which maps each element to itself:

$$(f^{-1} \circ f)(a) = f^{-1}(f(a)) = a, \qquad \text{for all } a \in A.$$

A function $i : A \to A$ with the property that $i(a) = a$ for all $a \in A$ is called the **identity function** on A, and so $f \circ f^{-1}$ is the identity function on A. Similarly, the composition $f \circ f^{-1}$ is the identity function on B, i.e.,

$$f(f^{-1}(b)) = b \qquad \text{for all } b \in B.$$

As further examples, if $A = \{x : x \geq 0\}$ and $f(x) = x^2$, the inverse function is $f^{-1}(x) = \sqrt{x}$; if $A = \{x : -\pi/2 \leq x \leq \pi/2\}$ and $g(x) = \sin(x)$, the inverse function is $g^{-1}(x) = \arcsin(x)$.

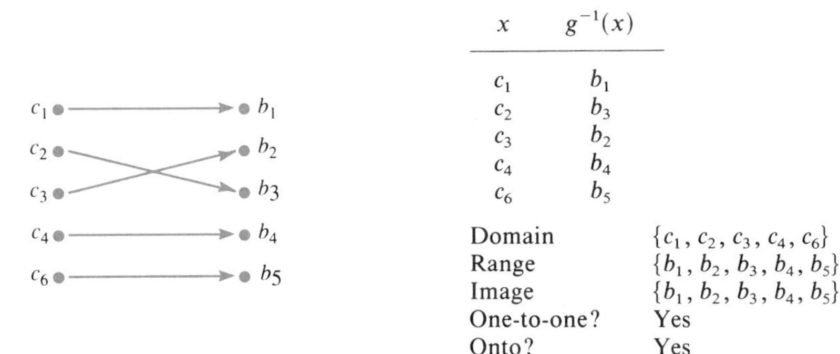

x	$g^{-1}(x)$
c_1	b_1
c_2	b_3
c_3	b_2
c_4	b_4
c_6	b_5

Domain	$\{c_1, c_2, c_3, c_4, c_6\}$
Range	$\{b_1, b_2, b_3, b_4, b_5\}$
Image	$\{b_1, b_2, b_3, b_4, b_5\}$
One-to-one?	Yes
Onto?	Yes

Figure 1.15 The inverse of the function g (with cut-down range) in Figure 1.12.

EXAMPLE 1.13 For each of the one-to-one functions from B to A in Example 1.10, cut the range down appropriately and find the corresponding inverse function.

SOLUTION In Example 1.10 we counted six one-to-one functions $g: B \to A$, and they were labelled g_2, g_3, g_4, g_6, g_7, and g_8. Here we display their inverses in a little table:

y	g_2^{-1}	g_3^{-1}	g_4^{-1}	g_6^{-1}	g_7^{-1}	g_8^{-1}
0	a	a	b	·	b	·
1	b	·	a	a	·	b
2	·	b	·	b	a	a

For instance, this table tells us that the inverse of g_7, which is called g_7^{-1}, has domain $\{0, 2\}$ and range $\{a, b\}$, with $g_7^{-1}(0) = b$ and $g_7^{-1}(2) = a$. ∎

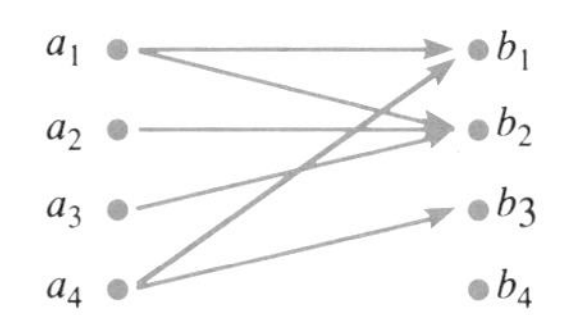

Figure 1.16 A relation between the sets $A = \{a_1, a_2, a_3, a_4\}$ and $B = \{b_1, b_2, b_3, b_4\}$.

In the definition of a function we require that each input have a unique output. Thus, in the arrow diagrams of Figures 1.11–1.15 we never allowed *two* arrows to originate at the same point in the domain. You may have wondered why. The answer is that we can allow it, but the resulting object is no longer called a function. Instead, it is called a **relation** (see Figure 1.16).

A relation may be described just by listing the "head" and the "tail" of each arrow in its diagram. For example, the relation of Figure 1.16 corresponds to the set

(1.11) $$R = \{(a_1, b_1), (a_1, b_2), (a_2, b_2), (a_3, b_2), (a_4, b_1), (a_4, b_3)\}.$$

In (1.11) the elements of the set R are called **ordered pairs**. The set of *all* ordered pairs (a, b), where $a \in A$ and $b \in B$, is called the **Cartesian product*** of the two sets, and it is denoted $A \times B$.

EXAMPLE 1.14 Let $A = \{0, 1\}$ and $B = \{x, g, z\}$. Find $A \times B$.

SOLUTION

$$A \times B = \{(0, x), (0, y), (0, z), (1, x), (1, y), (1, z)\}.$$ ∎

Formally, then, a relation between the sets A and B is just a subset of $A \times B$. For example, the set R in (1.11), which represents the relation of Figure 1.16, is just one of the possible subsets of $\{a_1, a_2, a_3, a_4\} \times \{b_1, b_2, b_3, b_4\}$. There are 65,535 other possible relations between A and B, as you will be able to verify after studying Chapter 2. A subset of $A \times A$ is called simply a **relation on** A.

* Sometimes called the **combinatorial**, or **direct product**.

Any relation on a set A can be represented by a picture called a **directed graph** or **digraph**. For example, suppose $A = \{1, 2, 3, 4\}$ and

$$R = \{(1, 2), (1, 3), (2, 1), (2, 4), (3, 3), (3, 4), (4, 1)\}.$$

We draw points (called **vertices**, one for each element of A) and arrows (called **directed edges**, one for each member of R) so that if $(a, b) \in R$, then vertex a is connected to vertex b by a directed edge from a to b; see Figure 1.17.

Conversely, every digraph determines a relation, so that we may recover R from the picture in Figure 1.17. For example, since there is a directed edge from vertex 4 to vertex 1, we have $(4, 1) \in R$. We will have more to say about graphs and digraphs in Chapters 3 and 4.

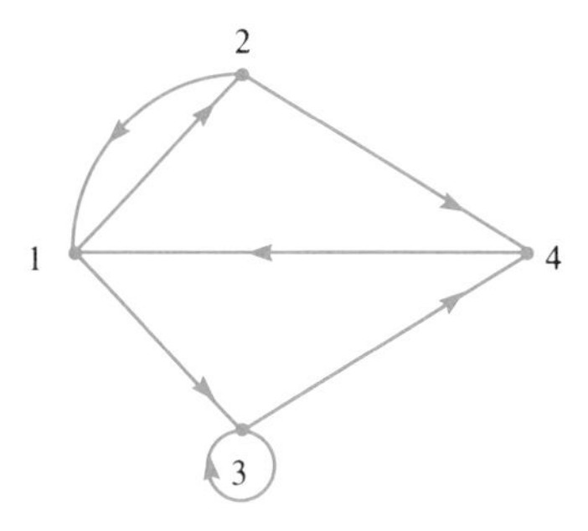

Figure 1.17 Digraph of a relation.

Problems for Section 1.3

1. Let $A = \{1, 2\}$ and $B = \{a, b\}$. Find all functions $f: A \to B$, and for each such function, determine whether it is one-to-one, onto, both, or neither.

2. Repeat Problem 1 for $A = \{1, 2\}$ and $B = \{a, b, c\}$.

3. With $A = \{1, 2\}$ and $B = \{a, b, c\}$ as in Problem 2, find all functions $g: B \to A$, and for each such function, determine whether it is one-to-one, onto, both, or neither.

4. Let $A = \{1, 2, 3, 4\}$. Find all one-to-one (and therefore onto) functions $f: A \to A$.

5. If $f(x) = x^3$ and $g(x) = \cos x$, find the compositions $h = g \circ f$ and $h' = f \circ g$.

6. Functions f and g are given by the following tables:

x	$f(x)$
a_1	b_1
a_2	b_2
a_3	b_2

x	$g(x)$
b_1	c_3
b_2	c_1
b_3	c_2
b_4	c_2

Find the composition $g \circ f$.

7. State the contrapositives of each of the following statements:
(a) If $x = y$, then $f(x) = f(y)$.
(b) If $x \neq y$, then $f(x) = f(y)$.
(c) If $x = y$, then $f(x) \neq f(y)$.
(d) If you are smart, you will do your homework.
(e) If $1 + 1 = 3$, then elephants have wings.

8. In Problem 6, can you find the composition $f \circ g$? If not, why not?

9. For each of the one-to-one onto functions in Example 1.11, find the corresponding inverse function.

10. For each of the one-to-one functions in Problem 2, cut the range down appropriately and find the corresponding inverse function.

11. Let $A = \{1, 2, 3\}$ and $B = \{a, b\}$. Find $A \times B$.

12. If $A = \{1, 2, 3\}$, find $A \times A$.

13. Let $A = \{1, 2, 3\}$ and let R be the relation $\{(1, 1), (1, 2), (1, 3), (2, 1), (2, 3), (3, 2), (3, 3)\}$. Draw the digraph of R.

14. Let $A = \{1, 2, 3, 4\}$ and let the relation R be defined by $(a, b) \in R$ if and only if $a \leq b$. List the elements of R, and draw the associated digraph.

15. Repeat Problem 14 if the relation R is defined by $(a, b) \in R$ if and only if $a = b$.

16. We have seen how to construct the inverse of a one-to-one, onto function $f: A \to B$. A function is a special kind of relation, and it is natural to think about the notion of *inverse of an arbitrary relation.* Show that the same procedure, i.e., reversing all arrows in the diagram, can be used to define the inverse of *any* relation between A and B. Illustrate your result by finding the inverse of the function in Figure 1.11. (Note that the inverse of a non-one-to-one function will not be a function, but it will be a relation.)

17. Continuing Problem 16, find the inverses of the relations given in Figures 1.16 and 1.17.

In Problems 18–25, let $A = \{1, 2, 3\}$ and $B = \{1, 2, 3, 4\}$. In each case, compute the number of things involved. (See also Problem 2.6.24.)

18. Functions $f: A \to B$.

19. Functions $f: B \to A$.

20. One-to-one functions $f: A \to B$.

21. One-to-one functions $f: B \to A$.

22. Onto functions $f: A \to B$.

23. Onto functions $f: B \to A$.

24. Elements in $A \times B$.

25. Relations between A and B.

26. Let $A = \{1, 2, 3, \ldots\}$, the set of positive integers. Consider the function $f: A \to A$, defined by $f(x) = 2x$.
(a) Show that f is one-to-one but not onto.
(b) Find a function $g: A \to A$ which is onto but not one-to-one.

27. Consider the function f: $\{1, 2, 3, 4\} \to \{1, 2, 3, 4\}$ given by $f(1) = 2$, $f(2) = 4$, $f(3) = 1$, $f(4) = 3$ (cf. Figure 1.13). Compute the following in tabular form:
(a) $f \circ f$.
(b) $f \circ f \circ f$.
(c) $f \circ f \circ f \circ f$.
(d) What about higher "powers" of f?

28. Let $f(x) = 1 - x^2$ and $g(x) = \sqrt{x}$. Show that $f \circ g$ only makes sense if $x \geq 0$. For which values of x does $g \circ f$ make sense?

29. If f and g are one-to-one, show that $g \circ f$ is one-to-one.

30. If f and g are onto, show that $g \circ f$ is onto.

In Problems 31–34, if the answer to the question is yes, explain why. If the answer is no, give a counterexample. For example, in Problem 31, if you decide that f need not be one-to-one, you must produce *explicit* functions f and g such that $g \circ f$ is one-to-one, but f is not one-to-one.

31. If $g \circ f$ is one-to-one, must f be one-to-one?

32. If $g \circ f$ is one-to-one, must g be one-to-one?

33. If $g \circ f$ is onto, must f be onto?

34. If $g \circ f$ is onto, must g be onto?

35. In Example 1.10 we saw that there were no *onto* functions from $B \to A$. Can you state a general rule that will allow you to predict when there are no onto functions from one set to another?

36. If $f: A \to B$ and $g: B \to C$ are one-to-one functions (so that $g \circ f$ is one-to-one by Problem 29), show that

$$(g \circ f)^{-1} = f^{-1} \circ g^{-1}.$$

Use this rule to compute the inverse of the functions $h(x) = \sin(x^2)$ and $k(x) = [\sin(x)]^2$. (The domains of h and k must be restricted so that the functions are one-to-one.)

37. Let $f: A \to A$ and let f_n be the composition of f with itself n times (see Problem 27). Suppose that for some n, f_n is the identity function [$i(x) = x$ for all x]. Show that f is one-to-one, and find its inverse.

Problems 38–43 illustrate typical applications of the pigeonhole principle.

38. Multiple-Choice:

If m pigeons are placed in n pigeonholes, and $m < n$,

(a) Some pigeonhole must contain more than one pigeon.
(b) Some pigeonhole must be empty.
(c) Two pigeonholes must contain the same number of pigeons.
(d) Every pigeonhole must be occupied.

39. Multiple-Choice:

If $g: B \to A$, and $|B| < |A|$, then . . . .

(a) Function g must be one-to-one.
(b) Function g must be onto.
(c) Function g cannot be one-to-one.
(d) Function g cannot be onto.

40. Suppose that five points are chosen in a square whose sides have length 2. Show that there must be at least two points P and Q such that the distance from P to Q is less than or equal to $\sqrt{2}$. (*Suggestion*: Divide the square into four equal subsquares, and take these subsquares as pigeonholes.)

41. Let $A = \{1, 2, \ldots, n\}$, where n is even. Suppose we make the following claim: If k numbers are selected from A, at least two of them have sum $n + 1$. Verify directly that the claim is true when $n = 6$ and $k \geq 4$.

42. Continuing Problem 41, show that the claim holds in general with $k \geq n/2 + 1$. (*Suggestion:* Take as pigeonholes the $n/2$ two-element sets $\{1, n\}, \{2, n-1\}, \ldots, \{n/2, n/2 + 1\}$.)

43. Continuing Problem 42, if n is *odd*, what is the appropriate statement about k for the claim to hold? Justify your answer.

1.4 Partial Orderings and Equivalence Relations

In the previous section we introduced the idea of a **relation**; formally, a relation R on a set A is just a subset of the Cartesian product $A \times A$. This definition is admittedly a bit abstract and possibly not very enlightening, but as a student, you have already worked with many special kinds of relations, without knowing it. For example, consider the symbol "$\le$," which means *less than or equal to.* We usually think of "$\le$" as a condition that is satisfied by certain pairs of integers, but not others. For example, $3 \le 5$, $-9 \le 0$, $-2 \le -2$, $6 \not\le 2$, $-1 \not\le -2$, etc. But "$\le$" can also be thought of as a relation R on the set $Z = \{0, \pm 1, \pm 2, \ldots\}$ of all integers. The relation R consists of all pairs (m, n) of integers such that $m \le n$:

$$R = \{(0,0), (0,1), (0,2), (1,2), (-1,-1), (2,2), \ldots\}.$$

Because of this important example, and similar examples, mathematicians usually write $a\,R\,b$ instead of $(a, b) \in R$. The particular relation "$\le$" has three important properties which are not shared by all relations (see Problem 34): For every x, y, and z, we have, for $R =$ "$\le$,"

$x\,R\,x$	(R is **reflexive**)
if $x\,R\,y$ and $y\,R\,x$, then $x = y$	(R is **antisymmetric**)
if $x\,R\,y$ and $y\,R\,z$, then $x\,R\,z$	(R is **transitive**).

Any relation that is reflexive, antisymmetric, and transitive is called a **partial ordering**, and may be denoted by the symbol "$\le$," even though it may not be the usual inequality relation for numbers.

A convenient way to represent a partial ordering on a finite set A of moderate size is via a **Hasse* diagram**. Each element of A is represented by a point in the diagram, and we connect x to y (with x below y) with a line if and only if $x < y$ but there is nothing between x and y, i.e., there is no $z \in A$ such that $x < z < y$. The symbol $x < y$ means that $x \le y$ but $x \ne y$. If a, $b \in A$, then $a < b$ if and only if there is a rising path in the Hasse diagram joining a to b. This will become a bit clearer after the following examples.

EXAMPLE 1.15 Let $A = \{a, b, c, d\}$, and consider the relation

$$R = \{(a,a), (a,b), (a,c), (a,d), (b,b), (b,d), (c,c), (c,d), (d,d)\}.$$

Show that R is a partial ordering, and draw its Hasse diagram.

SOLUTION We need to check the three conditions:

1. The relation R is **reflexive**, because (a, a), (b, b), (c, c), and (d, d) are all in R.
2. The relation R is **antisymmetric**, because there is no pair (x, y) in R for which (y, x) is also in R—except for the four pairs (a, a), (b, b), (c, c), (d, d).
3. The relation R is **transitive** because we have $a\,R\,b$ and $b\,R\,d$ and also $a\,R\,d$. Similarly, $a\,R\,c$ and $c\,R\,d$ and also $a\,R\,d$. (Technically, there are some other "degenerate" cases to consider, e.g., $a\,R\,a$, $a\,R\,d$, and $a\,R\,d$; but these cases present no difficulties.)

* Helmut Hasse (1898–1979), German mathematician. In his 1926 textbook *Höhere Algebra* he introduced these diagrams as an aid to the study of the solution of polynomial equations.

Thus R is indeed a partial ordering. Since R is reflexive, the Hasse diagram only needs to include the pairs (a, b), (a, c), (c, d), and (b, d) and is shown in Figure 1.18. The Hasse diagram makes the structure of R slightly clearer!

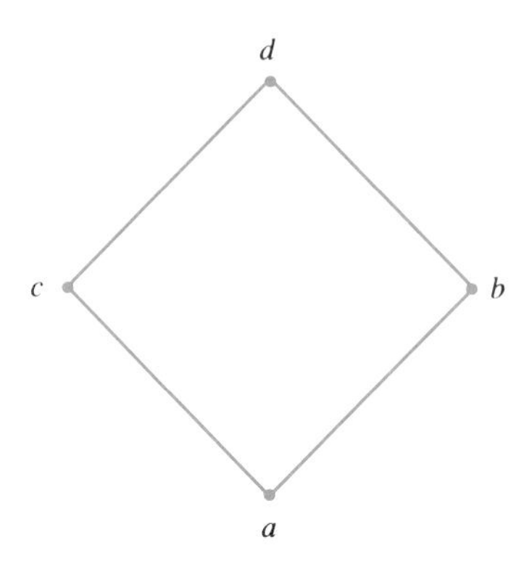

Figure 1.18 A Hasse diagram for the partial ordering of Example 1.15.

EXAMPLE 1.16 If a and b are positive integers, $a|b$ means that a is a **divisor** of b, i.e., $b = ac$ for some integer c. Show that "$|$" is a partial ordering of the set of positive integers. Also, draw the Hasse diagram for "$|$" when it is restricted to the set A of divisors of 24, viz., $A = \{1, 2, 3, 4, 6, 8, 12, 24\}$.

SOLUTION By definition, the $a|b$ means that the number b/a is an integer. We have to verify reflexivity, antisymmetry, and transitivity.

1. **Reflexivity**: For all positive integers n, $n|n$. (Because $n/n = 1$.)
2. **Antisymmetry**: If $n|m$ and $m|n$, then m/n and n/m are both integers. Since $n/m = (m/n)^{-1}$, the integer n/m has the property that its reciprocal is also an integer. The only such positive integer is 1, and so $n/m = 1$, i.e., $n = m$.
3. **Transitivity**: If $n|m$ and $m|p$, then $p/n = (p/m) \times (m/n)$ is an integer, since it is the product of two other integers.

It follows that "$|$" is a partial ordering. The Hasse diagram for A, which takes a little work to find, is shown in Figure 1.19. For example, we connect 4 to 12 because 4 divides 12 and there is no divisor z of 24 such that 4 divides z and z divides 12.

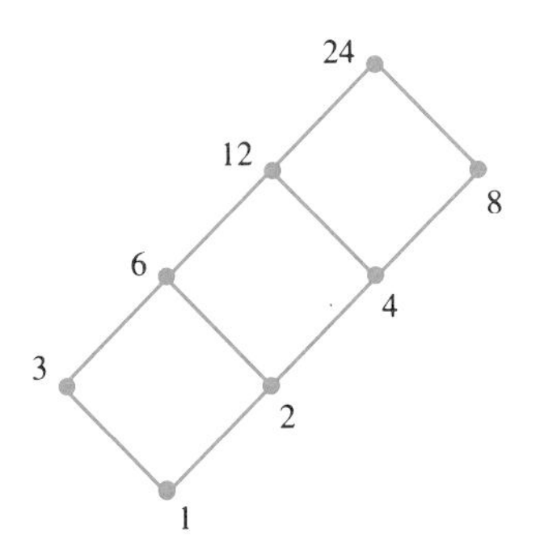

Figure 1.19 Hasse diagram of divisors of 24.

An important example of a partial ordering is the partial ordering on the subsets of a set determined by the relation "$\subseteq$". (Recall from Section 1.1 that $A \subseteq B$ means that A is a subset of B.) If S is an arbitrary set, and $P(S)$ is the collection of subsets of S, then the relation "$\subseteq$" is in fact a partial ordering of $P(S)$ (Problem 8). The Hasse diagram for the case $S = \{a, b, c, d\}$ is shown in Figure 1.20. Hasse diagrams can be complicated and beautiful!

On the other hand, some partial orderings have Hasse diagrams of a very simple type; see Figure 1.21. This kind of an ordering is called a **total ordering**, because any two elements x and y are **comparable**: either $x \leq y$ or $y \leq x$. In Figure 1.18, on the other hand, c and b are **incomparable**; i.e., neither $c \leq b$ nor $b \leq c$ is true. In Figure 1.19 elements 3 and 8 are incomparable; and in Figure 1.20 the subsets $\{a, b\}$ and $\{a, c\}$ are incomparable. But in Figure 1.21 every pair *is* comparable, and the ordering is total. As another example, the ordering "$\leq$" on the integers is a total ordering.

A set on which a partial ordering is defined is called a **partially ordered set** or **poset**; if the ordering is total, we have a **totally ordered set**. You should study the following example carefully.

EXAMPLE 1.17 Let $A = \{1, 2, 3\}$. How many different partial orderings are there on A? How many are total orderings?

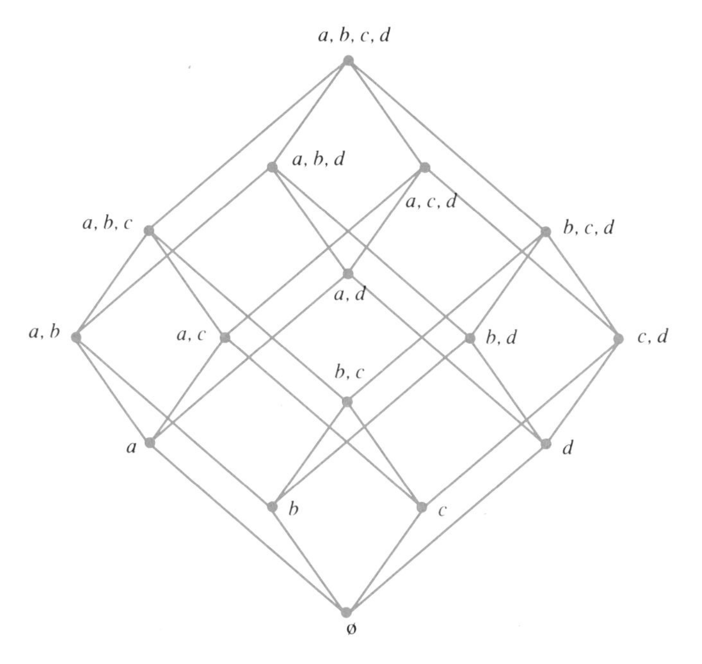

Figure 1.20 Inclusion relation on subsets of a set with four members.

SOLUTION There are 19 different partial orderings, of which six are total orderings. This can be seen by examining the possible Hasse diagrams with three points, of which there are five types (Figure 1.22).

How do we get these five types? We reason as follows: If no element is comparable with any other element, we get type (a); if however, there is one pair of comparable elements, say x and y, with the third element z not comparable to anything else, we get type (b); the remaining cases [(c), (d), (e)] occur when there are two comparable elements, say $x < y$, and the remaining element z is comparable to one or both of x and y:

- *Case (c)*: z is comparable to both x and y (e.g. $z < x < y$, as illustrated in Figure 1.22(c))
- *Case (d)*: $z < y$ but z not comparable to x
- *Case (e)*: $z > x$ but z not comparable to y.

Figure 1.21 The Hasse diagram for a total ordering on $A = \{a, b, c, d, e, f\}$.

Each of these five types can occur several different ways, depending on which elements from $A = \{1, 2, 3\}$ are called x, y, and z. For type (a) there is only *one*

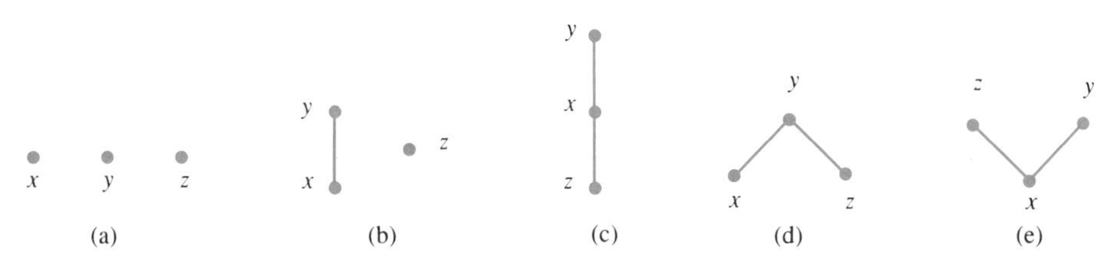

Figure 1.22 The five possible Hasse diagrams on a set with three elements.

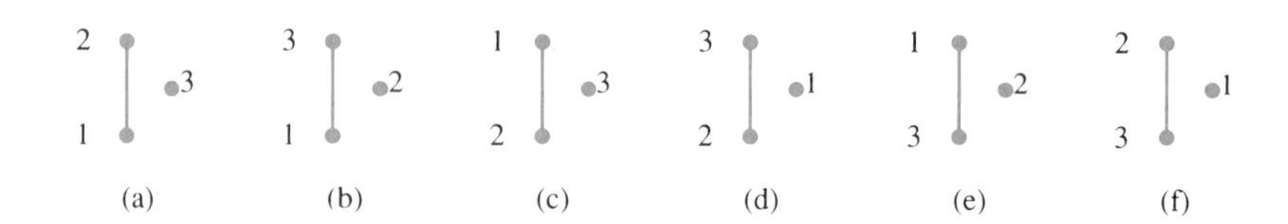

Figure 1.23 The six labellings of Figure 1.22(a).

possibility, and the only partial order with Hasse diagram of this type is

$$R = \{(1, 1), (2, 2), (3, 3)\}.$$

However, for type (b) there are six possibilities (see Figure 1.23).

For example, the Hasse diagram in Figure 1.23(a) represents the partial order

$$R = \{(1, 1), (2, 2), (3, 3), (1, 2)\},$$

and that in Figure 1.23(b) represents

$$R = \{(1, 1), (2, 2), (3, 3), (1, 3)\},$$

and so on. Similarly, there are *six* partial orderings on A leading to the Hasse diagram in Figure 1.23(c), three leading to Figure 1.23(d), and three leading to Figure 1.23(e). The total is $1 + 6 + 6 + 3 + 3 = 19$, and all of them are shown in Figure 1.24. Finally, only Figure 1.22(c) represents a total ordering, and there are just *six* total orderings.

If the first and third conditions in the definition of a partial ordering are retained, but the second is changed, we obtain another important class of relations,

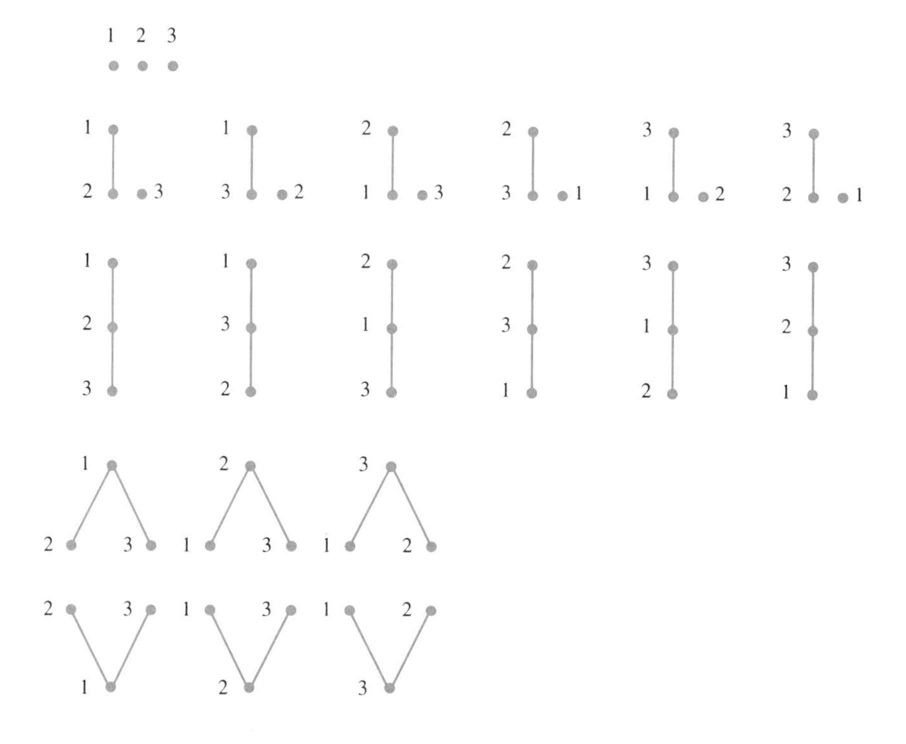

Figure 1.24 The 19 partial orderings of the set $\{1, 2, 3\}$.

the **equivalence relations**. An equivalence relation is often denoted by "$\sim$." It is defined to be a relation on A that satisfies, for all x, y, and $z \in A$,

$$x \sim x \qquad (\sim \text{ is reflexive})$$

$$\text{if } x \sim y, \text{ then } y \sim x \qquad (\sim \text{ is symmetric})$$

$$\text{if } x \sim y \text{ and } y \sim z, \text{ then } x \sim z \qquad (\sim \text{ is transitive}).$$

If $x \sim y$, we say that x is **equivalent** to y or that x and y are **equivalent**. The **equivalence class** of x is the set $C(x)$ consisting of all elements equivalent to x. Reflexivity guarantees that the sets $C(x)$ are never empty, since always $x \in C(x)$.

EXAMPLE 1.18 Let $A = \{\text{all words in the English language}\}$. Let us say that two words are "equivalent" if they *begin with the same letter*, or in symbols,

$$x \sim y \text{ if and only if } x \text{ and } y \text{ begin with the same letter.}$$

Show that this is an equivalence relation. Find the equivalence class of the word *quiet*.

SOLUTION We have three things to check:

1. **Reflexivity**: Yes, because any word starts with the same letter as itself!
2. **Symmetry**: Yes, because if x starts with the same letter as y, then y starts with the same letter as x!
3. **Transitivity**: Yes, because if x starts with the same letter as y, and y starts with the same letter as z, then x starts with the same letter as z.

Thus we do have an equivalence relation. The equivalence class of *quiet* is the set of words beginning with q. There are, in all, 26 equivalence classes, one for each letter in the alphabet.

EXAMPLE 1.19 Let A = the set of all integers, and let us say that $x \sim y$ if and only if $x - y$ is a multiple of 3. Show that this is an equivalence relation. What is the equivalence class of 0? How many equivalence classes are there?

SOLUTION The verification of reflexivity and symmetry is easy, but transitivity is a little tricky.

1. **Reflexivity**: Yes, because $x - x = 0$ is a multiple of 3.
2. **Symmetry**: Yes, because if $x - y$ is a multiple of 3, say $x - y = 3 \cdot n$, then $y - x = 3 \cdot (-n)$ is also a multiple of 3.
3. **Transitivity**: Yes, if $x - y = 3a$ and $y - z = 3b$, then $x - z = (x - y) + (y - z) = 3a + 3b = 3(a + b)$, which is a multiple of 3.

The equivalence class of 0 contains all integers x such that $x - 0$ is a multiple of 3, i.e.,

$$C(0) = \{0, \pm 3, \pm 6, \pm 9, \pm 12, \ldots\}.$$

The number 1 isn't in $C(0)$. What is its equivalence class? The class $C(1)$ consists of all x such that $x - 1$ is a multiple of 3, i.e., $x = 1 + 3a$ for some a. Hence

$$C(1) = \{\ldots, -11, -8, -5, -2, 1, 4, 7, 10, 13, \ldots\}.$$

And finally (check this result),

$$C(2) = \{\ldots, -10, -7, -4, -1, 2, 5, 8, 11, \ldots\}.$$

Are there any more equivalence classes? No, because every integer is already accounted for, in one of $C(0)$, $C(1)$, or $C(2)$. For example, $C(119) = C(2)$, because $119 \in C(2)$, i.e., $2 \sim 119$. Thus there are *three* equivalence classes. ■

Note: Example 1.19 is a special case of a very important and famous equivalence relation, invented by Carl Gauss* in 1801. If m is a positive integer, we say that x and y are **congruent modulo** m, and write

$$x \equiv y \pmod{m},$$

if $x - y$ is a multiple of m. Example 1.19 is the case $m = 3$. For every m, "congruence modulo m" is an equivalence relation, and every integer is congruent (mod m) to one and only one of the numbers $0, 1, 2, \ldots, m - 1$. Indeed, if n is an integer, and if r is the remainder when n is divided by m, then

$$n \equiv r \pmod{m}.$$

This is true because if q is the quotient when n is divided by m, we have $n = q \cdot m + r$, which means that

$$n - r = qm,$$

i.e., that $n - r$ is a multiple of m.

For example, let $m = 12$ and $n = 1000$. Dividing 1000 by 12, we find $1000 = 83 \cdot 12 + 4$, and so $1000 \equiv 4 \pmod{12}$. This process of dividing n by m and keeping the remainder is sometimes called "reducing n mod m." For example, we could also write

$$4 = 1000 \bmod 12.$$

Our next example of an equivalence relation may surprise you!

EXAMPLE 1.20 Let X be a set, and let us call two elements x and y equivalent if and only if x and y are *equal*, i.e.,

$$x \sim y \quad \text{if and only if} \quad x = y.$$

Show that this is an equivalence relation, and find the equivalence classes.

* Carl Friedrich Gauss (1777–1855), German mathematician and physicist. He began as an amazing child prodigy (see Example 1.23 in the next section) and matured into one of the greatest and most productive geniuses the human race has produced. The notion of congruence first appeared in Gauss's celebrated work *Disquisitiones Arithmeticae*, published in 1801, when Gauss was 24. The actual idea had occurred to him several years earlier, when he was still in his teens.

SOLUTION Showing that equality is an equivalence relation isn't hard, since the reflexive, symmetric, and transitive laws for equality are simply

$$x = x \quad \text{(reflexive);}$$
$$\text{if } x = y, \text{ then } y = x \quad \text{(symmetric);}$$
$$\text{if } x = y, \text{ and } y = z, \text{ then } x = z \quad \text{(transitive);}$$

which are the basic axioms about equality.

What about the equivalence classes? The equivalence class $C(x)$ consists of those things which are *equal to* x! Therefore $C(x)$ contains only one element, viz., x itself. The equivalence classes are therefore simply the sets containing exactly one element from X. For example, if $X = \{1, 2, 3, 4\}$, the equivalence classes are

$$C(1) = \{1\}, \qquad C(2) = \{2\}, \qquad C(3) = \{3\}, \qquad C(4) = \{4\}.$$

These one-element subsets are called **singletons**, incidentally. ■

In Example 1.20 we observed that *equality* is an equivalence relation. This is not a profound observation, but it is an important one, since any equivalence relation behaves in many ways like the equality relation. Perhaps the most important similarity is that if in the well-known bromide

Things *equal* to the same thing are *equal* to each other.

we change *equal* to *equivalent*, i.e.,

(1.12) Things *equivalent* to the same thing are *equivalent* to each other,

the statement remains true. To see why this is so, let x and x' be two things equivalent to the same thing, viz.

$$x \sim y, \qquad \text{and} \qquad x' \sim y.$$

Then since "$\sim$" is symmetric, we also have $y \sim x'$. But now since $x \sim y$ and $y \sim x'$, the transitive property of "$\sim$" implies that $x \sim x'$, i.e., that x and x' are equivalent to each other.

An important consequence of this discussion is that any two equivalence classes must either be *disjoint* or *equal* (see Figure 1.25). To see this, we reason as follows. Let $C(x)$ and $C(y)$ be two equivalence classes, defined by

$$C(x) = \{x' \colon x' \sim x\} \qquad \text{and} \qquad C(y) = \{y' \colon y' \sim y\}.$$

If $C(x)$ and $C(y)$ are disjoint, as shown in Figure 1.25(a), well and good. However, if they are *not* disjoint, i.e., if they have any elements in common, as shown in Figure 1.25(c), then we can prove that they are equal, as shown in Figure 1.25(b). We can do this by showing that $C(x) \subseteq C(y)$, and $C(y) \subseteq C(x)$.

To show that $C(x) \subseteq C(y)$, we need to show that any element in $C(x)$ is necessarily in $C(y)$. Therefore let $x' \in C(x)$. Then, by definition,

(1.13) $$x' \sim x.$$

Now, let z be an element which is common to both $C(x)$ and $C(y)$. Then,

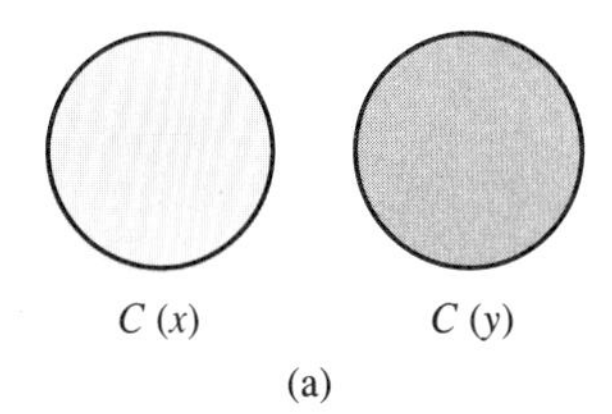

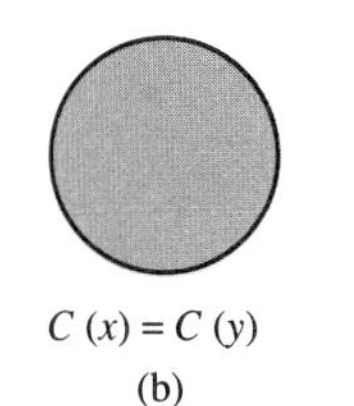

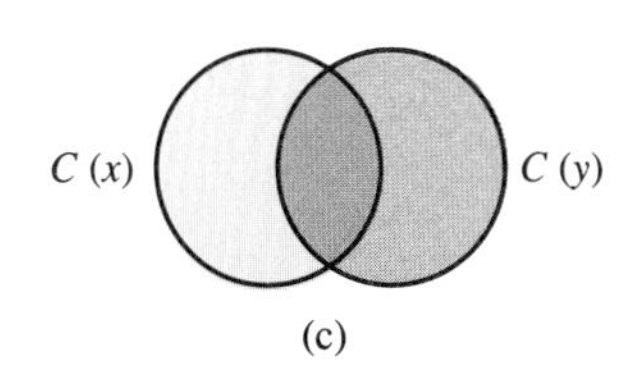

Figure 1.25 Possible relationships between the equivalence classes $C(x)$ and $C(y)$. (a) $C(x)$ and $C(y)$ are *disjoint*; (b) $C(x)$ and $C(y)$ are *equal*; (c) This can't happen.

since $z \in C(x)$, we have

(1.14) $$z \sim x.$$

Therefore x' and z, being equivalent to the same thing (in this case x), must be equivalent to each other:

(1.15) $$x' \sim z.$$

But since $z \in C(y)$, we also have

(1.16) $$z \sim y.$$

It follows from (1.15) and (1.16) and the transitivity law that

(1.17) $$x' \sim y,$$

i.e., that $x' \in C(y)$. This proves that $C(x) \subseteq C(y)$. The proof that $C(y) \subseteq C(x)$ is similar and is left as Problem 20.

The equivalence relations in Examples 1.18, 1.19, and 1.20 are quite different, but they do share one important feature in common. In each case the equivalence classes form a **partition** of the underlying set. In Example 1.18 the set of all English words is partitioned by the equivalence relation into 26 pairwise disjoint subsets; in Example 1.19 the set of all integers is partitioned into the three pairwise disjoint subsets $C(0)$, $C(1)$, and $C(2)$; and in Example 1.20 the set X is partitioned into the singletons. In fact, this feature is shared by *all* equivalence relations: *The equivalence classes always form a partition of the underlying set.*

To see why, let R be an equivalence relation on the set A, and let the distinct equivalence classes be $A_1, A_2, \ldots, A_n$. Then every element of A belongs to one of these equivalence classes, and as we have seen, the equivalence classes are pairwise disjoint. Therefore (using the definition we learned in Section 1.1), $A_1, A_2, \ldots, A_n$ do indeed form a partition of A.

Thus if we are given an *equivalence relation* on a set A, we can always produce a *partition.* The next example shows that conversely, if we are given a *partition* of A, we can always produce an *equivalence relation.* It follows that there is no real difference between partitions and equivalence relations; they are just two ways of looking at the same thing!

EXAMPLE 1.21 Let A be any set, and let $A_1, A_2, \ldots, A_n$ be nonempty pairwise disjoint subsets of A, such that

$$A = A_1 \cup A_2 \cup \cdots \cup A_n,$$

in other words, a partition of A. Let us now say that $x \sim y$ if and only if x and y belong to the same set of the partition. Show that this is an equivalence relation. Find the number of equivalence classes.

SOLUTION Again we need to check three things:

1. **Reflexivity**: x and x both belong to the same subset.
2. **Symmetry**: If x and y belong to the same subset, then y and x belong to the same subset(!).

3. **Transitivity**: If x and $y \in A_i$ and y and $z \in A_j$, then $A_i = A_j$, since y can only belong to one of the sets in the partition. Hence x and z are both in A_i.

Thus "$\sim$" is an equivalence relation; the equivalence classes are the original sets $A_1, A_2, \ldots, A_n$, and the number of equivalence classes is n, the number of sets in the partition. As a specific case of this construction, let $A = \{1, 2, 3, 4, 5, 6, 7\}$, $A_1 = \{1, 4, 7\}$, $A_2 = \{2, 3, 6\}$, and $A_3 = \{5\}$. Then 1 and 4 are equivalent, $2 \sim 6$, but $4 \nsim 5$, etc. Also, $C(7) = \{1, 4, 7\} = A_1$, $C(2) = \{2, 3, 6\} = A_2$, etc.

Our final example shows the other side of the coin we admired in Example 1.20.

EXAMPLE 1.22 Continuing Example 1.21, suppose the sets A_i are all singletons, i.e., contain only one element. Describe the resulting equivalence relation.

SOLUTION We have $x \sim y$ if and only if x and y belong to the same set of the partition. But each set contains only one element, so x and y must be *equal*. In other words, the resulting equivalence relation is the *equality* relation!

Problems for Section 1.4

1. (a) Let $A = \{1, 2, 3, 4\}$ and consider the relation

$$R = \{(1, 1), (1, 2), (1, 3), (2, 2), (3, 2), (3, 3), (4, 2), (4, 3), (4, 4)\}.$$

Show that R is a partial ordering, and draw its Hasse diagram.
(b) Repeat part (a) with

$$R = \{(1, 1), (2, 1), (2, 2), (3, 1), (3, 3), (3, 4), (4, 4)\}.$$

2. Consider $A = \{1, 2, 3, 4, 5\}$ and $R = \{(1, 1), (1, 2), (1, 3), (1, 4), (1, 5), (2, 2), (2, 4), (2, 5), (3, 3), (3, 4), (3, 5), (4, 4), (4, 5), (5, 5)\}$. Verify that R is a partial ordering, and draw the Hasse diagram.

3. Consider the partial ordering on the set $A = \{1, 2, 3, 4, 5, 6, 10, 12, 15, 20, 30, 60\}$ of integers that are divisors of 60, where $a \leq b$ if a is a divisor of b. Draw the Hasse diagram.

4. Repeat Problem 3 for divisors of 84. Can you see why the diagrams in Problems 3 and 4 are similar?

5. The accompanying diagram shows the original Hasse diagram, as given by Hasse himself in his textbook. Write, in set-theoretic notation, the corresponding relation.

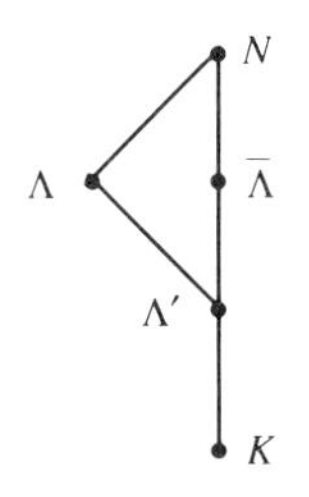

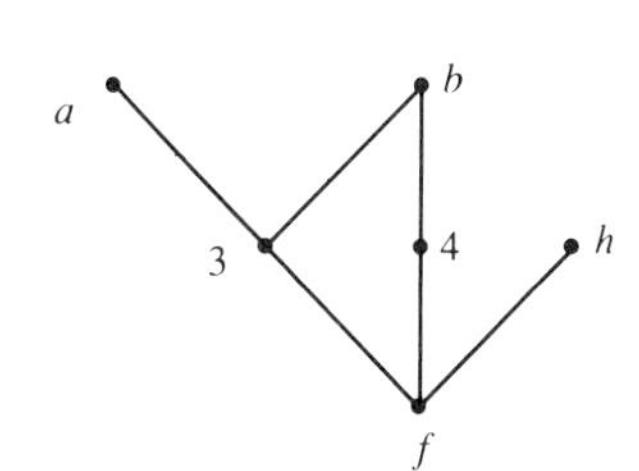

6. For each Hasse diagram in the accompanying figure, list all pairs (x, y) such that $x < y$.

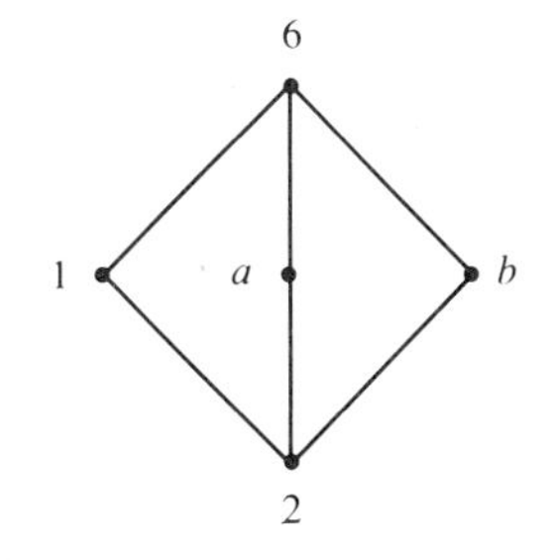

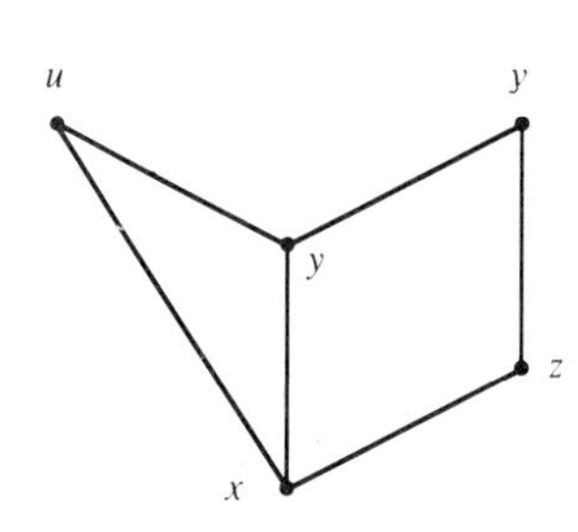

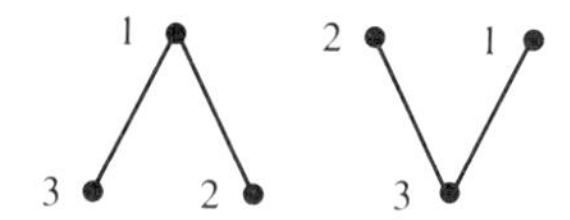

7. Why is the accompanying figure *not* a Hasse diagram?

8. Show that the relation "$\subseteq$" is a partial ordering of the subsets of a given set.

9. An alert student once carefully examined Figure 1.24 and complained that many partial orderings were missing. For example, he claimed that the following two partial orderings should have been included:

Did the alert student have a point?

10. Below we give Hasse diagrams which might occur for partial orderings on a four-element set, e.g., $\{1, 2, 3, 4\}$:

For each of these Hasse diagrams, find out how many partial orderings on $\{1, 2, 3, 4\}$ will give to rise the diagram. (*Hint:* Read Example 1.17 carefully.)

11. How many partial orderings on the set $\{$*yes, no, maybe*$\}$ are there?

12. Which (if any) of the following relations on $A = \{1, 2, 3\}$ is a partial ordering?
(a) $R = \varnothing$.
(b) $R = \{(1, 1), (2, 2), (3, 3)\}$.
(c) $R = \{(1, 1), (2, 2), (3, 3), (1, 2), (2, 1)\}$.
(d) $R = \{(1, 2), (2, 3), (1, 3)\}$.
(e) $R = \{(1, 1), (2, 2), (3, 3), (1, 2), (2, 3)\}$.
(f) $R = A \times A$.

13. Let $A = \{\text{all words in the English language}\}$, as in Example 1.18. If $x, y \in A$, define $x \sim y$ if and only if x and y have the same number of letters. Show that this is an equivalence relation, and find the equivalence class of the word *student*.

14. With A as defined in Problem 13, define $x \, R \, y$ if and only if x and y have at least one letter in common. Is R an equivalence relation? Explain.

15. Referring to Example 1.19, consider **congruence mod 7**, i.e., congruence mod m with $m = 7$. Verify that this is an equivalence relation. How many equivalence classes are there? Describe them.

16. Fill in the details of the argument that for any positive integer m, congruence mod m is an equivalence relation.

17. Let $A = \{1, 2, 3, 4, 5, 6, 7, 8\}$, and consider the partition defined by the sets

$$A_1 = \{1, 5\},\ A_2 = \{2, 6\},\ A_3 = \{3, 7\},\ A_4 = \{4, 8\}.$$

As we know (Example 1.21), the corresponding equivalence relation is given by $x \sim y$ if and only if x and y belong to the same set A_i. Can you give a more illuminating description, in other words, can you replace "x and y belong to the same set A_i" by something that gives more insight?

18. Decide whether each of the following statements is true or false:
(a) $71 \equiv 20 \pmod 3$
(b) $451 \equiv -101 \pmod 3$
(c) $10{,}000 \equiv 5 \pmod{12}$
(d) $211 \equiv 210 \pmod 1$
(e) $2^7 \equiv 2 \pmod 7$.

19. Calculate the following remainders:
(a) 71 mod 3
(b) 451 mod 2
(c) 10,000 mod 12
(d) 211 mod 1
(e) 2^7 mod 7.

20. In the course of our proof that any two equivalence classes must either be equal or disjoint, we showed that if $z \in C(x) \cap C(y)$, then $C(x) \subseteq C(y)$. Prove the reverse inequality, i.e., show that if $z \in C(x) \cap C(y)$, then $C(y) \subseteq C(x)$.

21. Let R be an equivalence relation on the set A, and let A_i be one of the equivalence classes. Can A_i be the empty set?

22. Let F be the collection of all functions $f: \{1, 2, 3, 4\} \to \{1, 2, 3, 4\}$. If f and g belong to F, define $f \sim g$ if and only if $f(3) = g(3)$. Show that this is an equivalence relation, and describe the corresponding partition.

23. Is the relation "$\geq$" on numbers a partial ordering? What about "$>$"?

24. Let F be the set of all real-valued functions on the set A. If $f, g \in F$, write $f \leq g$ if $f(x) \leq g(x)$ for all $x \in A$. Show that "$\leq$" is a partial ordering. Is it total?

25. Below is a list of relations *among people.* For each relation, tell whether it is (i) reflexive, (ii) symmetric, (iii) antisymmetric, and (iv) transitive. Also, identify the relations, if any, that are partial orderings, total orderings, or equivalence relations.
(a) "Is taller than."
(b) "Is the same sex as."
(c) "Lives within 30 miles of."
(d) "Recognizes."
(e) "Lives in the same house."

26. Repeat Problem 25 for the following relations.
(a) "Is a direct descendent of."
(b) "Is an ancestor of."
(c) "Is married to."
(d) "Is friends with" (assume everyone is friendly with self!).
(e) "Is secretly in love with."

27. Consider a partial ordering and its Hasse diagram. Determine whether or not the following statements are true or false:
(a) If x is not directly connected to y in the Hasse diagram, then we do not have $x < y$.
(b) If $x < y$, then x is directly connected to y in the Hasse diagram.

28. Consider a partial ordering or a set A. We say that x is a **maximal element** in A if there is no y such that $x < y$. Similarly, x is a **minimal element** in A if there is no y such that $y < x$. We say that x is a **greatest element** in A if $y \leq x$ for every y in A. Similarly, x is a **least element** in A if $x \leq y$ for every y in A.
(a) Show that every greatest element is maximal. (Similarly, every least element is minimal.)
(b) Give an example to show that a maximal element need not be greatest. (Similarly, a minimal element need not be least.)

29. Continuing Problem 28, describe how you would find maximal, minimal, greatest, and least elements (if any) by looking at a Hasse diagram.

30. (a) Can a partial ordering have more than one maximal element? Explain. (See Problem 28.)
(b) Can a partial ordering have more than one greatest element? Explain.

31. Consider the following two very special relations on $A: \varnothing$ and $A \times A$ (both are subsets of $A \times A$ and so qualify as relations.) Which of the following properties are enjoyed by these relations

	$\varnothing$	$A \times A$
Reflexive	?	?
Symmetric	?	?
Antisymmetric	?	?
Transitive	?	?

32. Let $A = \{a, b, c\}$. How many different *equivalence* relations on A are there?

33. What can be said about a relation R that is *both* a partial ordering *and* an equivalence relation? Give an example of such a relation on the set $\{1, 2, 3\}$.

34. Consider the set $\{1, 2, 3\}$. A relation R either has or does not have each of the three properties *reflexivity, antisymmetry,* and *transitivity.* Give an example of a relation for each of the eight combinations of "has/does not have." (You are supposed to find eight different examples, from a relation having *all* to one having *none* of the properties.)

35. Given a set A, if $\mathscr{A}$ and $\mathscr{B}$ are partitions of A, define $\mathscr{A} \leq \mathscr{B}$ if and only if every set in $\mathscr{A}$ is a subset of some set in $\mathscr{B}$ (e.g., $\{\{1, 2\}, \{3\}, \{4\}\} \leq \{\{1, 2, 4\}, \{3\}\}$). Show that this definition gives a partial ordering of the set of partitions of A. Draw the Hasse diagram of this partial order for $A = \{1, 2, 3\}$.

36. Let R be a partial ordering on the set A. The **inverse relation** R^{-1} is defined as follows (see Problem 1.3.16):

$$x R^{-1} y \qquad \text{if and only if} \qquad y R x.$$

If R is denoted by $\leq$, then the inverse is denoted by $\geq$; thus $x \geq y$ if and only if $y \leq x$. For example, the inverse of "is a descendant of" is "is an ancestor of."
(a) Find the inverse of "is taller than" (for people).
(b) Find the inverse of "divides" (for integers).
(c) Describe how to construct the Hasse diagram of the inverse relation, and illustrate for the diagram of Figure 1.19.

1.5 Mathematical Induction

Have you ever noticed that if you add up the first few odd numbers, the result is always a perfect square? For example,

$$1 + 3 = 4$$
$$1 + 3 + 5 = 9$$
$$1 + 3 + 5 + 7 = 16$$
$$1 + 3 + 5 + 7 + 9 = 25 \ldots .$$

Do you see the pattern? The number on the right side of each equation is the square of the number of terms on the left side. It's natural to assume that this pattern will continue forever, so we might predict, for example, that the sum of the first one thousand odd numbers equals one million:

$$\underbrace{1 + 3 + 5 + \cdots + 1999}_{1000 \text{ odd numbers}} \stackrel{?}{=} \underbrace{1{,}000{,}000}_{(1000)^2}.$$

This prediction is correct, as you can verify by writing a short computer program. But it's important to realize that merely predicting that this pretty pattern continues isn't the same as rigorously *proving* that it does. In this section we will describe an important tool of discrete mathematics, called **mathematical induction**, or sometimes just **induction**, which can sometimes be used to prove that an empirically observed pattern like the one above goes on forever.

Here's how it works. Let $S(n)$ be a conjectured statement about the integer n, for example, "The sum of the first n odd positive integers equals n^2." Suppose we want to prove that $S(n)$ is true for all positive integers n. The *principle of mathematical induction* says that we can succeed if we can perform the following two steps:

I1. Prove that $S(1)$ is true.
I2. Prove that if $S(n)$ is true, then $S(n + 1)$ is also true.

(I1 is sometimes referred to as the **basis**, and I2 as the **induction step**.)

In our example the statement $S(n)$ is

$$1 + 3 + \cdots + (2n - 1) = n^2.$$

To carry out step I1, we must verify that $S(1)$ is true. But $S(1)$ is just the statement $1 = 1$, so there is no problem. To carry out I2, we assume that the statement $S(n)$ (which is called the **induction hypothesis**) is true, and try to prove $S(n + 1)$, which in this case is

$$1 + 3 + \cdots + (2n - 1) + (2n + 1) = (n + 1)^2.$$

Notice, however, that the sum $1 + 3 + \cdots + (2n - 1)$ equals n^2, by *assumption*, so to prove $S(n + 1)$, we need only prove that

$$n^2 + (2n + 1) = (n + 1)^2.$$

But $n^2 + 2n + 1 = (n + 1)^2$ is true by elementary algebra, and thus we can accomplish step I2 as well. This proves that $S(n)$ is *always* true, and so we can say with absolute confidence, for example, that

$$1 + 3 + \cdots + 1{,}999{,}999 = 1{,}000{,}000{,}000{,}000.$$

Before we proceed, we should indicate why this powerful proof technique works. It's like knocking over a row of dominos. If $S(1)$ is known to be true, then by the induction step I2, $S(2)$ must also be true. Since now $S(2)$ is true, the induction step shows that $S(3)$ is true. Once $S(3)$ is true, the induction step proves $S(4)$; and so on. The analogy with dominos is quite strong; the induction step I2 lines the dominos up, and the basis step I1 starts them falling. Indeed, once we set them up with I2, we can start them falling anywhere, not necessarily with $S(1)$. We will return to this point in Example 1.27.

We continue with several examples.

EXAMPLE 1.23 There is a legend that the great German mathematician Carl Gauss (see footnote in Section 1.4), at the age of eight, was asked by his teacher to add up the first one hundred positive integers. This assignment was probably busy work intended to keep the class quiet, but little Carl astonished the teacher by instantly giving the right answer, 5050. Gauss may have reasoned as follows:

$$\begin{array}{rcl} \text{sum} &=& 1 + 2 + \cdots + 100 \\ \text{sum} &=& 100 + 99 + \cdots + 1 \\ \hline 2 \times \text{sum} &=& 101 + 101 + \cdots + 101. \end{array}$$

Therefore sum $= (101 \times 100)/2 = 5050$. Based on this reasoning, we might guess that more generally

$$1 + 2 + \cdots + n = \frac{n(n+1)}{2}.$$

Prove that this is true for all values of n, by induction.

SOLUTION We denote by $S(n)$ the statement "$1 + 2 + \cdots + n = n(n+1)/2$." To prove $S(n)$ by induction, we first verify $S(1)$: $1 = 1 \cdot 2/2$. So far, so good. To complete the proof, we must assume that $S(n)$ is true and show that this assumption implies $S(n+1)$, i.e., $1 + 2 + \cdots + (n+1) = (n+1)(n+2)/2$. Here are the needed steps:

$$\begin{aligned} 1 + 2 + \cdots + (n+1) &= (1 + 2 + \cdots + n) + (n+1) \\ &= \frac{n(n+1)}{2} + (n+1) && \text{(induction hypothesis)} \\ &= \frac{n^2 + 3n + 2}{2} \\ &= \frac{(n+1)(n+2)}{2} && \text{(algebra)}. \end{aligned}$$

This completes the proof by induction.

EXAMPLE 1.24 Let X be a set containing n elements. Show that X has 2^n subsets.

SOLUTION The given statement, which we will call $S(n)$, is true for $n = 1$, since a set with only *one* element, say $X = \{a\}$, has only *two* subsets, viz., $\varnothing$ and $\{a\}$. Now suppose $S(n)$ is true, and let X be a set containing $n+1$ elements, say $X = \{a_1, a_2, \ldots, a_n, a_{n+1}\}$. The subsets of X can be classified into two categories:

- *Category I*: Subsets not containing a_{n+1}.
- *Category II*: Subsets containing a_{n+1}.

For example, let $n = 2$ and $X = \{a_1, a_2, a_3\}$. Then the subsets of X are categorized as follows:

- *Category I*: $\varnothing, \{a_1\}, \{a_2\}, \{a_1, a_2\}$.
- *Category II*: $\{a_3\}, \{a_1, a_3\}, \{a_2, a_3\}, \{a_1, a_2, a_3\}$.

We see that every subset in Category I is in fact a subset of $Y = \{a_1, a_2, \ldots, a_n\}$. But Y has 2^n subsets, by the induction hypothesis. On the other hand, every subset in Category II contains the element a_{n+1}, and so is of the form $\{a_{n+1}\} \cup Z$, where Z is a subset of Y. Thus for $n = 2$, we have the alternative representation for the sets in Category II:

- *Category II*: $\{a_3\} \cup \varnothing, \{a_3\} \cup \{a_1\}, \{a_3\} \cup \{a_2\}, \{a_3\} \cup \{a_1, a_2\}$.

Therefore there are exactly as many subsets in Category II as there are subsets of Y—and this number is again 2^n, by the induction hypothesis. To summarize:

Category I:	2^n subsets
Category II:	2^n subsets
Total:	$2^n + 2^n = 2^{n+1}$ subsets.

Thus the set X with $|X| = n + 1$ has 2^{n+1} subsets; i.e., $S(n+1)$ is true, and this completes the proof. ■

EXAMPLE 1.25 Prove by induction that $5^n - 4n - 1$ is exactly divisible by 16 for $n = 1, 2, 3, \ldots$.

SOLUTION When $n = 1$, we have $5^n - 4n - 1 = 5 - 4 - 1 = 0$, which is divisible by 16. Thus $S(1)$ is true. If $S(n)$ is true, we want to prove $S(n+1)$, i.e., that $5^{n+1} - 4(n+1) - 1$ is divisible by 16. By slightly clever algebra we have

$$5^{n+1} - 4(n+1) - 1 = 5(5^n) - 4n - 5$$
$$= 5(5^n - 4n - 1) + 20n + 5 - 4n - 5.$$

The purpose of the unusual factoring, which forced us to add $20n + 5$ as a balance, is to make the expression $5^n - 4n - 1$ appear, since the induction hypothesis is about this expression. Collecting terms, we get

$$5^{n+1} - 4(n+1) - 1 = 5(5^n - 4n - 1) + 16n.$$

Since $5^n - 4n - 1$ is divisible by 16 by the induction hypothesis, and $16n$ is clearly divisible by 16, it follows that $5^{n+1} - 4(n+1) - 1$ is divisible by 16, as desired. ■

EXAMPLE 1.26 Show that $3^n > 2n$ for all $n = 1, 2, \ldots$.

SOLUTION When $n = 1$, our statement is $3 > 2$, which is certainly true. Assume that the result holds for n, i.e., $3^n > 2n$. We must show that $3^{n+1} > 2(n+1)$. But

$$3^{n+1} = 3(3^n) > 3(2n) \qquad \text{(by the induction hypothesis)},$$

so we will be finished if $3(2n) \geq 2(n+1)$. But this amounts to $6n \geq 2n + 2$, i.e., $n \geq \frac{1}{2}$, which *does* hold for all positive integers! ■

The next example shows how to use induction to prove statements $S(n)$ which are *false* for the first few values of n.

EXAMPLE 1.27 Prove that $2^n \geq n + 10$, for $n \geq 4$.

SOLUTION The interesting thing here is that the statement $S(n)$: "$2^n \geq n + 10$" is false for $n = 1$, 2, and 3. Nevertheless, we can perform step I2, i.e., show that $S(n)$ implies $S(n+1)$:

$$\begin{aligned} 2^{n+1} &= 2 \cdot 2^n \\ &\geq 2 \cdot (n+10) \qquad \text{(induction hypothesis)} \\ &= 2n + 20 \\ &\geq (n+1) + 10 \qquad \text{(algebra)}. \end{aligned}$$

This "sets the dominos up." We can't verify $S(1)$, $S(2)$, or $S(3)$, since these statements are false. However, $S(4)$ is true: $2^4 \geq 4 + 10$. This starts the dominos falling from $S(4)$ onward, and proves that $S(n)$ is true for all $n \geq 4$. ■

It is sometimes convenient to replace the induction hypothesis $S(n)$ by the stronger assumption that *all* the statements $S(1), S(2), \ldots, S(n)$ are true. The resulting principle is called **strong induction**, and is characterized by the following two steps:

I1. Prove that $S(1)$ is true.
I2′. Prove that if $S(1), \ldots, S(n)$ are all true, then $S(n+1)$ is true.

If I1 and I2′ can be carried out, then $S(n)$ is guaranteed to be true for all positive integers.

The intuitive argument we gave to justify the principle of mathematical induction works just as well for strong induction. The reason why strong induction is sometimes used is that by strengthening the inductive step, in other words, assuming $S(1), \ldots, S(n)$ rather than just $S(n)$, proofs can be simplified. The next example will illustrate the idea.

EXAMPLE 1.28 The numbers in the sequence 1, 1, 2, 3, 5, 8, 13, 21, 34, ..., in which each new term is the sum of the previous two, are called the **Fibonacci*** **numbers**. (We'll have more to say about these famous numbers in Chapter 10.) If we denote the nth Fibonacci number by F_n, we have $F_1 = 1$, $F_2 = 1$, and for $n \geq 2$,

$$F_{n+1} = F_n + F_{n-1}.$$

This is called a **recursive definition**, in which each element of the sequence is defined in terms of previous numbers in the sequence. Use strong induction to prove the inequality

$$F_n \leq \left(\frac{1+\sqrt{5}}{2}\right)^{n-1}, \qquad n = 1, 2, 3, \ldots.$$

SOLUTION Let $S(n)$ be the statement that the inequality is true for the integer n. Then (step I1) $S(1)$ is certainly true: $1 \leq 1$. In this case it is also useful to verify $S(2)$ directly:

$$1 \leq \frac{1+\sqrt{5}}{2},$$

which holds because $\sqrt{5}$ is greater than 1.

Now assume that $S(1), \ldots, S(n)$ are true ($n \geq 2$), and consider $S(n+1)$. Let $\phi = (1+\sqrt{5})/2$ and observe that ϕ is one of the solutions of the quadratic equation $x^2 - x - 1 = 0$, so we have

$$\phi^2 = \phi + 1.$$

Now consider the following sequence of steps:

$$F_n \leq \phi^{n-1} \qquad \text{[by assumption, } S(n) \text{ is true]}$$

$$F_{n-1} \leq \phi^{n-2}$$

[by assumption, $S(n-1)$ is true; in this step we need the strong induction hypothesis I2′; also note that the statement $F_{n-1} \leq \phi^{n-2}$ makes sense *only* for $n \geq 2$, which is why we verified $S(2)$ directly]

$$\begin{aligned} F_{n+1} = F_n + F_{n-1} &\leq \phi^{n-1} + \phi^{n-2} && \text{(by addition)} \\ &= \phi^{n-2} \times (\phi + 1) && \text{(by algebra)} \\ &= \phi^{n-2} \times \phi^2 && \text{(by } \phi^2 = \phi + 1\text{)} \\ &= \phi^n && \text{(by gosh!).} \end{aligned}$$

* Also known as Leonardo of Pisa, Fibonacci (1180–1250) was possibly the greatest European mathematician before the Renaissance. He made many original contributions to mathematics, but today is remembered only for this famous sequence of numbers.

The argument in Example 1.28 is quite typical of proofs by induction in that the statement $S(n)$ is quite mysterious. Mathematical induction gives no clue as to how to *discover* plausible statements $S(n)$; it is only a tool that can be used to try to prove such statements once they are discovered. In fact, mathematics has been defined as the art of discovering patterns!

Problems for Section 1.5

All proofs are to be done by induction.

1. Prove that

$$1^2 + 2^2 + 3^2 + \cdots + n^2 = \frac{n^3}{3} + \frac{n^2}{2} + \frac{n}{6} = \frac{n(n+1)(2n+1)}{6}.$$

2. Show that the sum of the first n even integers is $n^2 + n$.

3. Using some of the ideas from Example 1.24, show that a set containing n elements has exactly $n(n-1)/2$ two-element subsets.

4. Let A be a set containing n elements. Experiment with $n = 1, 2, 3$, guess a general formula for the number of elements in $A \times A$, and prove your guess.

5. Show that $n^3 - 4n + 6$ is divisible by 3 for all positive integers n.

6. Show that $11^n - 4^n$ is divisible by 7 for $n = 1, 2, 3, \ldots$.

7. Show that $2^n > n$ for $n \geq 1$.

8. Show that $n^2 > n + 1$ for $n \geq 2$.

9. Notice that every third Fibonacci number is even:

$$1, 1, \underline{2}, 3, 5, \underline{8}, 13, 21, \underline{34}, \ldots.$$

Show that this pattern continues indefinitely.

10. In Example 1.28 we used the solution $(1 + \sqrt{5})/2$ to the quadratic equation $x^2 - x - 1 = 0$. Suppose we used the other solution, $(1 - \sqrt{5})/2$, and attempted to show that

$$F_n \leq \left(\frac{1-\sqrt{5}}{2}\right)^{n-1}, \qquad n = 1, 2, \ldots.$$

What would go wrong?

11. Notice that

$$1 = 1$$
$$1 - 4 = -(1 + 2)$$
$$1 - 4 + 9 = 1 + 2 + 3$$
$$1 - 4 + 9 - 16 = -(1 + 2 + 3 + 4).$$

Guess the pattern, and prove it by induction.

In Problems 12–14, establish the indicated summation formula by induction.

12. $1^3 + 2^3 + 3^3 + \cdots + n^3 = \dfrac{n^4}{4} + \dfrac{n^3}{2} + \dfrac{n^2}{4} = \dfrac{n^2(n+1)^2}{4}.$

13. $1^4 + 2^4 + 3^4 + \cdots + n^4 = \dfrac{n^5}{5} + \dfrac{n^4}{2} + \dfrac{n^3}{3} + kn,$

for a certain constant k. Find k.

14. $1^2 + 3^2 + 5^2 + \cdots + (2n-1)^2 = \dfrac{4n^3 - n}{3}.$

15. Consider the DeMorgan law

$$(A_1 \cup A_2 \cup \cdots \cup A_n)' = A_1' \cap A_2' \cap \cdots \cap A_n'$$

for arbitrary sets $A_1, \ldots, A_n$. This law can be proved directly from basic definitions of sets; in particular, see Problem 1.2.30. Give a proof of the identity using mathematical induction.

16. Prove the other DeMorgan law

$$(A_1 \cap \cdots \cap A_n)' = A_1' \cup A_2' \cup \cdots \cup A_n'.$$

17. Show that $2^{2n} - 1$ is divisible by 3 for all positive integers n.

18. Show that $x^n - 1$ is divisible by $x - 1$ for all positive integers $n \geq 2$.

19. Consider an attempt to prove that $n^2 + 5n + 1$ is even for $n = 1, 2, 3, \ldots$.

(a) Show that the statement is false.

(b) Show that part I2 of the principle of mathematical induction can be done anyway, but part I1 fails.

20. Suppose a post office sells only 2¢ and 3¢ stamps. Show that any postage of 2¢ or over can be paid for using only these stamps. (*Suggestion*: For the inductive step, consider two possibilities. *Case 1*: n cents of postage can be paid for by using only 2¢ stamps. *Case 2*: n cents of postage requires at least one 3¢ stamp.)

21. If the post office of Problem 19 sells only 5¢ and 9¢ stamps, show that any postage of 35¢ or over can be paid for using only these stamps.

22. Guess a general rule that simplifies the product

$$\left(1-\frac{1}{4}\right)\left(1-\frac{1}{9}\right)\left(1-\frac{1}{16}\right)\cdots\left(1-\frac{1}{n^2}\right),$$

and prove it by induction.

23. Prove by induction that if $n \geq 10$, then $2^n \geq n^3$. [Here the statement $S(n)$ is *false* for $n = 1, 2, \ldots, 9$, and so a slight modification of the proof by induction is required.]

24. There must be a flaw in the following proof. What is it?
Theorem: Everybody is rich.
Proof: Consider the statement $S(n)$: "Given any collection of n persons. If at least one of them is rich, then they're all rich." We prove $S(n)$ by induction. The statement $S(1)$ is obviously true. We illustrate that $S(n)$ implies $S(n+1)$ by taking, for example, $n = 3$. Thus let $\{P_1, P_2, P_3, P_4\}$ be any four people, at least one of whom is rich. Suppose, for example, that P_1 is rich. Then $\{P_1, P_2, P_3\}$ is a set of $n = 3$ persons, and one is rich, so by $S(3)$, P_2 and P_3 are also rich. Similarly, by considering the set $\{P_1, P_2, P_4\}$, we can conclude that P_4 is rich, and so P_1, P_2, P_3, P_4 are *all* rich. Thus $S(4)$ is true. Hence $S(n)$ implies $S(n+1)$, and so by induction $S(n)$ is true for all n. Since there certainly exists at least one rich person, it follows that all persons are rich.

25. Consider the statement $S(n)$: $n^2 - n + 41$ is prime (i.e., not divisible by any positive integer except itself and 1) for all $n = 0, 1, 2, \ldots$.
(a) Verify that $S(0)$, $S(1)$, $S(2)$, and $S(3)$ are true.
(b) Why must an attempt to prove $S(n)$ by induction fail?

26. Define $n!$ ("n factorial") as the product of all positive integers from 1 to n. Thus,

$$1! = 1,\ 2! = 1(2) = 2;\ 3! = 1(2)(3) = 6,\ 4! = 1(2)(3)(4) = 24,$$

etc. For convenience, define $0!$ to be 1. Guess a general rule for the sum

$$1(1!) + 2(2!) + \cdots + n(n!)$$

and prove it by induction. (*Hint*: The sum can be expressed as a difference of two factorials.)

27. Imagine an infinite number of hotel rooms labelled $1, 2, \ldots$. Assume that at least one room is occupied, and let A be the set of numbers corresponding to occupied rooms. Show that A has a smallest element, i.e., there is a smallest "occupied number." [*Suggestion*: Take $S(n)$ to be the statement that n does *not* belong to A. If A does not have a smallest element, use strong induction to prove that $S(n)$ holds for all n. Conclude that A is empty, which is a contradiction.]

28. The result of Problem 27 may be translated into slightly more abstract language. Let A be any nonempty subset of positive integers. Show that A has a smallest element. (This statement is known as the *well-ordering principle*.)

29. Here is the problem, originally posed by Fibonacci in 1202, which leads to the Fibonacci numbers: "How many pairs of rabbits can be produced from a single pair in a year's time?" (Assume that all pairs of rabbits born in the month of June, for example, become fertile in July, start reproducing in August, and produce exactly one new pair of offspring each month from then on. Also, rabbits live forever.)

30. Answer Fibonacci's problem (Problem 29), only now assume that each new pair becomes fertile at the age of *two* months. (Once fertile, each pair produces a new pair every month.) Give a recursive formula for the number of pairs of rabbits after n months.

31. Experiment with the bound $F_n \leq [(1+\sqrt{5})/2]^{n-1}$ for the nth Fibonacci number. How accurate is it? Can you improve it?

1.6 Some Useful Mathematical Notation

In the previous section we needed in several places to indicate that an indefinite number of things were being added together. To do this we used three dots (ellipsis marks) to represent the omitted material. For example, we denoted the sum of the first n odd integers by

(1.18) $$1 + 3 + 5 + \cdots + (2n - 1).$$

There is another notation that mathematicians use in situations like this; it is called the $\sum$ notation. For example, in $\sum$ notation (1.18) can be written as

(1.19) $$\sum_{i=1}^{n} (2i - 1).$$

This formula is usually read as follows: "the sum from $i = 1$ to n of $2i - 1$."

In (1.19) "$\sum$" is the capital Greek letter *sigma*, the first letter in the Greek word for sum. To understand the rest of the formula (1.19), consider the function $i \to (2i - 1)$, which maps the integer i to the integer $(2i - 1)$:

i	1	2	3	4	$\cdots$	n
$2i - 1$	1	3	5	7	$\cdots$	$2n - 1$

The formula in (1.19) tells us to *add together* the values assumed by the function $(2i - 1)$ as the integer i runs from 1 to n. The letter i is called the **index**, or **dummy variable**. It can be replaced by any other convenient symbol. For example, the following sums all mean exactly the same thing as (1.19):

(1.20) $$\sum_{j=1}^{n} (2j - 1)$$

(1.21) $$\sum_{x=1}^{n} (2x - 1)$$

(1.22) $$\sum_{1 \le k \le n} (2k - 1).$$

In (1.22) the limits of summation appear entirely below the $\sum$. This is a fairly common alternative to (1.19). We might also add that the use of "x" in (1.21), while correct, is a bit unorthodox. Mathematicians have traditionally used the letters i, j, k, l, m, and n for dummy variables in summations.*

EXAMPLE 1.29 What is the value of

(1.23) $$\sum_{k=1}^{100} k^2 ?$$

SOLUTION This expression represents the sum of the squares of the first 100 integers, i.e., $1^2 + 2^2 + 3^2 + \cdots + 100^2$. According to Problem 1.5.1, the value of the sum is $(100)(101)(201)/6 = 338{,}350$.

It is not an exaggeration to say that the following example illustrates one of the most important facts in all of mathematics.

EXAMPLE 1.30 Evaluate the sum

(1.24) $$S = \sum_{k=0}^{n-1} x^k.$$

SOLUTION By our rules, $S = 1 + x + x^2 + \cdots + x^{n-1}$, which is the sum of a *geometric series*. You may recall from your algebra background that

* This is why the letters I, J, K, L, M, and N were selected to represent integers in the programming language FORTRAN!

if $x \neq 1$, then

$$S = \frac{1 - x^n}{1 - x};$$

whereas if $x = 1$, the sum is $1 + 1 + \cdots + 1 = n$.

To derive the formula for S, write

$$S = 1 + x + x^2 + \cdots + x^{n-1}$$
$$xS = \quad x + x^2 + \cdots + x^{n-1} + x^n$$

and subtract the second equation from the first to obtain

$$(1 - x)S = 1 - x^n.$$

The next example shows how the $\sum$ notation can be used to express a simple fact in a complicated way!

EXAMPLE 1.31 Evaluate the following sum: $S = \sum_{k=3}^{6} 1$.

SOLUTION This expression is at first confusing, because the thing inside the summation is "1," which doesn't depend on the index k. But that's okay; there are four terms in the sum, corresponding to the four index values $k = 3, 4, 5, 6$, and each of these terms is equal to 1:

$$S = \underset{(k=3)}{1} + \underset{(k=4)}{1} + \underset{(k=5)}{1} + \underset{(k=6)}{1} \quad as = 4.$$

Examples 1.29, 1.30, and 1.31 have all been special cases of the following general rule. If $\{\ldots, a_{-1}, a_0, a_1, a_2, \ldots\}$ is a set of numbers, then the symbol

$$\sum_{i=r}^{t} a_i$$

represents the sum of all those numbers whose subscripts i satisfy the inequality $r \leq i \leq t$. For example,

$$\sum_{i=2}^{4} a_i = a_2 + a_3 + a_4, \qquad \sum_{k=-1}^{0} b_k = b_{-1} + b_0,$$

$$\sum_{j=-1}^{1} f(j) = f(-1) + f(0) + f(1), \qquad \sum_{l=6}^{6} b(l) = b(6).$$

But here's a tricky one: What is

(1.25)
$$\sum_{i=4}^{2} a_i = ?$$

There are *no* subscripts i that satisfy $4 \leq i \leq 2$. The sum in (1.25) is said to be *empty*, and its value is *defined* to be zero. Thus $\sum_{i=4}^{2} a_i = 0$.

In all of the sums we have considered so far, the index variable assumed *consecutive* integer values. Sometimes we need to add the values assumed by a function at *irregular* intervals, and for this a new notation is needed. For example, suppose $\{a_1, a_2, a_3, a_4, a_5, a_6, a_7, a_8\}$ is a set of eight numbers. If we wanted to add *all* of them, we could write $S = \sum_{k=1}^{8} a_k$. But suppose we only wanted to add a_2, a_3, a_5, and a_7? Of course we could write simply $S = a_2 + a_3 + a_5 + a_7$, but alternatively, we could define the **index set** $I = \{2, 3, 5, 7\}$, and write

(1.26) $$S = \sum_{k \in I} a_k.$$

EXAMPLE 1.32 Let O_n denote the set of the first n odd integers, i.e., $O_n = \{1, 3, 5, \ldots, 2n - 1\}$. Evaluate

(1.27) $$S_n = \sum_{j \in O_n} j.$$

SOLUTION Equation (1.27) is just a fancy (unnecessarily fancy, probably) way of writing

$$S_n = 1 + 3 + 5 + \cdots + (2n - 1).$$

Thus $S_n = n^2$, as we saw at the beginning of the previous section. ■

If the index set I is the empty set $\varnothing$, then (1.26) doesn't really make sense. However, by convention such a sum is *defined* to be 0:

(1.28) $$\sum_{k \in \varnothing} a_k = 0.$$

For example, with this convention, the sum S_n in Example 1.32 with $n = 0$ is 0. This convention is consistent with our previous discussion of (1.25), since (1.25) could be rewritten as $\sum_{i \in I} a_i$, with $I = \{i: 4 \le i \le 2\} = \varnothing$.

There is also a compact mathematical notation for *products*, called the $\prod$ notation. For example, the product

$$\frac{1}{2} \cdot \frac{2}{3} \cdot \frac{3}{4} \cdots \frac{n-1}{n}$$

can also be denoted by

(1.29) $$\prod_{i=2}^{n} \left(\frac{i-1}{i} \right).$$

In (1.29) "$\prod$" is the capital Greek letter *pi*, which is the Greek alphabet version of "*P*." It is treated in exactly the same way as $\sum$, except that an *empty product* is defined to be 1 rather than 0.

Similarly, the union and intersection of the n sets $A_1, A_2, \ldots, A_n$, which until now we have been denoting by

$$A_1 \cup A_2 \cup \cdots \cup A_n \quad \text{and} \quad A_1 \cap A_2 \cap \cdots \cap A_n,$$

can also be denoted by

$$\bigcup_{i=1}^{n} A_i \quad \text{and} \quad \bigcap_{i=1}^{n} A_i.$$

Before closing this section, we want to discuss two more symbols that mathematicians like to use, namely, "$\exists$" and "$\forall$." These symbols are mathematical shorthand for the phrases "there exists" and "for all," respectively. Thus, for example, the statement

$$x + 0 = x, \qquad \text{for all values of } x$$

can also be written as

$$\forall x\colon x + 0 = x.$$

Similarly, the statement

$$\text{there exists a solution } x \text{ to the equation } x^2 = 4$$

can be written as

$$\exists x\colon x^2 = 4.$$

These symbols can appear more than once in a statement. For example,

$$\forall x \exists y\colon x + y = 0$$

says that for any value of x, there exists a value of y such that $x + y = 0$ (the value of y is $-x$). Similarly,

$$\exists x \forall y\colon x + y = y$$

means that there is a value of x such that for any value of y, $x + y = y$ (the value of x is 0).

All working mathematicians use the symbols $\forall$ and $\exists$ occasionally in handwritten work. However, these symbols rarely appear in print (with the exception mentioned below), because they are quite cryptic and make mathematics unnecessarily hard to read, as the examples above amply demonstrate. Beginning mathematicians tend to overuse them, and we strongly recommend writing out *there exists* or *for all* in words except in rare circumstances (for example, to impress your nonmathematical friends!).

Technically, the symbols $\exists$ and $\forall$ are called **quantifiers**. The symbol $\exists$ is the **existential quantifier**, and $\forall$ is the **universal quantifier**. The mathematical sentence following the quantifier is called a **predicate**. There is a branch of discrete mathematics called **predicate calculus** (no relation to ordinary calculus) which uses quantifiers and predicates to study the logical foundations of mathematics. We will not study predicate calculus in this book, but in Chapter 5 we will study the **propositional calculus**, which is a sort of poor man's version of predicate calculus, not using quantifiers.

Problems for Section 1.6

In Problems 1–14, evaluate the sums.

1. $\sum_{i=1}^{2} (2i+1)$.

2. $\sum_{n=-1}^{+1} n^2$.

3. $\sum_{k=-2}^{2} 4^k$.

4. $\sum_{n=1}^{m} n$.

5. $\sum_{k=0}^{n-1} 3^k$.

6. $\sum_{k=0}^{n-1} (-2)^k$.

7. $\sum_{l=0}^{9} 9x^l$.

8. $\sum_{i=0}^{10} (2x)^i$.

9. $\sum_{k=3}^{9} 2$.

10. $\sum_{n=1}^{m} m$.

11. $\sum_{3 \le k \le 9} k$.

12. $\sum_{k \in E_n} (k/2)$, where E_n is the set of the first n *even* numbers, starting with 2.

13. $\sum_{j=1}^{n} j^3$ (see Problem 1.5.12).

14. $\sum_{k=1}^{n} (-1)^{k-1}k^2$ (see Problem 1.5.11).

In Problems 15–20, evaluate the products.

15. $\prod_{i=1}^{5} i^2$.

16. $\prod_{i=2}^{4} (-1)$.

17. $\prod_{k=1}^{3} (1+k)$.

18. $\prod_{k=1}^{3} (1+x)$.

19. $\prod_{j=1}^{n} \left(\frac{j}{j+1}\right)$.

20. $\prod_{i=0}^{-1} (i+1)$.

21. Let A and B be two subsets of the integers. Under what (if any) circumstances will the following statement be true?

$$\sum_{i \in A} a_i + \sum_{j \in B} a_j = \sum_{k \in A \cup B} a_k.$$

[A complete answer to this question will lead you to a better appreciation of the convention (1.28) about empty sums.]

22. Refer to Problem 21. Let A and B be two subsets of the integers. Under what circumstances will the following be true?

$$\left(\prod_{i \in A} a_i\right)\left(\prod_{j \in B} a_j\right) = \prod_{k \in A \cup B} a_k.$$

Use this result to explain the convention, mentioned in the text, that an empty product is 1.

23. Evaluate the following product (in terms of n):

$$\prod_{k=2}^{n} \left(1 - \frac{1}{k^2}\right).$$

24. Use induction to prove that

$$\sum_{i=1}^{n} (x_i + y_i) = \sum_{i=1}^{n} x_i + \sum_{i=1}^{n} y_i.$$

25. State and prove a result analogous to Problem 24 for products.

26. The expression $\sum_{i=1}^{3} \sum_{j=1}^{4} a_{ij}$ is called a **double sum**. It is evaluated by proceeding "inside out." In other words, for each i we compute $b_i = \sum_{j=1}^{4} a_{ij}$, and the double sum is given by $\sum_{i=1}^{3} b_i$. Evaluate $\sum_{i=1}^{3} \sum_{j=1}^{4} (i+j)$.

27. Continuing Problem 26, verify that the same answer is obtained by simply adding the 12 numbers $i+j$, $1 \le i \le 3$, $1 \le j \le 4$.

28. *Evaluate* $\sum_{n=1}^{4} \prod_{i=1}^{n} i$. (Proceed "inside out" as in Problem 26. For each n, let $b_n = \prod_{i=1}^{n} i$; then compute $\sum_{n=1}^{4} b_n$.)

29. Let $f(i, j)$ be any function which maps pairs of integers into real numbers. Write out all terms in the expansion of

$$\begin{aligned}&[f(1,1)+f(1,2)+f(1,3)+f(1,4)]\\ &\quad \times [f(2,1)+f(2,2)+f(2,3)+f(2,4)]\\ &\quad \times [f(3,1)+f(3,2)+f(3,3)+f(3,4)].\end{aligned}$$

30. Use the insight gained from Problem 29 to prove the following result, called the **generalized distributive law**: Let I and J be sets of integers, and let F be the set of *functions* mapping I into J. If $f(i, j)$ is any function which maps pairs of integers into real numbers, show that

$$\prod_{i \in I} \sum_{j \in J} f(i,j) = \sum_{g \in F} \prod_{i \in I} f(i, g(i)).$$

31. If $A_1, A_2, \ldots, A_N$ are N sets, evaluate

$$\bigcap_{n=1}^{N} \bigcup_{i=1}^{n} A_i \quad \text{and} \quad \bigcup_{n=1}^{N} \bigcap_{i=1}^{n} A_i.$$

32. Assuming that x, y, and z represent integers (positive, negative, or zero), determine whether each of the following statements is true or false:

(a) $\forall x \exists y$: $x + y = 10$

(b) $\forall z \forall x \exists y$: $x + y = z$.

33. Repeat Problem 32 for the following statements:

(a) $\exists x \forall y$: $x + y = 10$

(b) $\forall y \exists x$: $x + y = 10$.

1.7 An Application: Hamming Codes

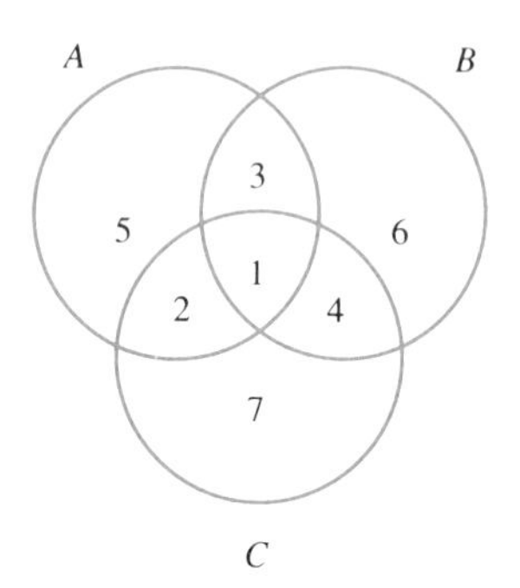

Figure 1.26

One of the most important applications of discrete mathematics is to the theory of **error-correcting codes**. Much of this theory depends on somewhat advanced mathematics, but we can already describe one of the most important examples of an error-correcting code, the **Hamming*** **code**.

Consider again the Venn diagram associated with three sets *A*, *B*, and *C*. There are seven compartments in this diagram, which are numbered 1 through 7 in Figure 1.26.

Suppose we are given four **bits** (binary digits) of information, say 1110, and we wish to transmit these bits reliably from one point to another. (These bits could be part of a computer program, or data from an experiment, for example.) But also suppose that some of these bits may be *garbled* in transmission, i.e., a transmitted 0 might be received as a 1, or vice versa. It is possible to gain some protection against such transmission errors, using the Venn diagram, as follows.

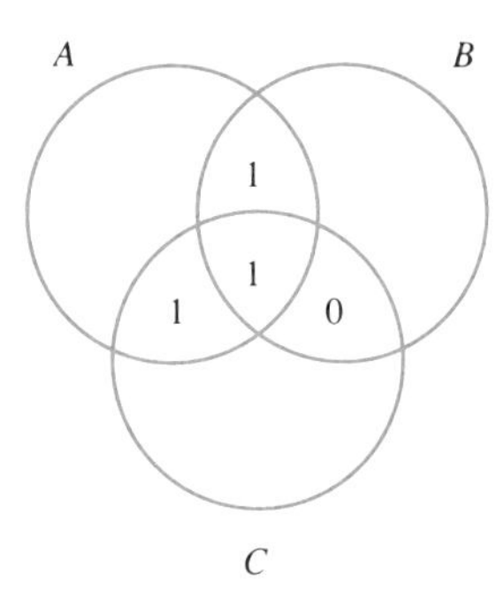

Figure 1.27

We prepare the four data bits for transmission by first placing them in the first four compartments of the Venn diagram, as shown in Figure 1.27. We then fill the remaining three compartments with what are called **parity check**, or just **parity**, bits. The rule for generating the parity check bits is that the total number of 1s in each of the three sets *A*, *B*, and *C* must be *even*. For example, the parity bit in compartment 5 must be 1, since in Figure 1.27 the set *A* already contains three 1s. Similarly, the parity bits in compartments 6 and 7 should be 0.

When this process is completed (Figure 1.28), each of the seven compartments in the Venn diagram will contain one bit: Four bits are information bits, and three are parity bits.

Now we transmit all seven bits, to the desired destination, via the noisy channel. This packet of seven bits, in this case 1110100 (we list the bits in the order given in Figure 1.26), is called a **codeword**. The procedure which generates the seven-bit data word is called the **encoding algorithm**.

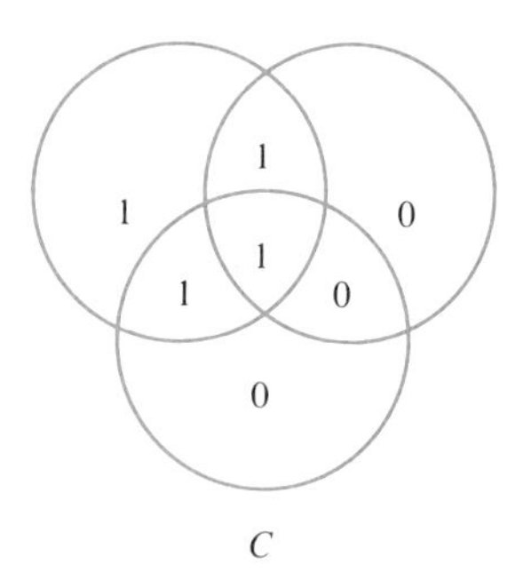

Figure 1.28

EXAMPLE 1.33 Encode the four bits 1111.

SOLUTION 1 If we place the four given bits into the four compartments of the Venn diagram as shown in Figure 1.29, we see that there are three 1s in each of the sets *A*, *B*, and *C*. To make the total number of 1s in each of these sets even, we must put a 1 in each of the remaining three compartments. Therefore the resulting codeword is 1111111. ■

SOLUTION 2 It gets a bit tiresome to keep showing the Venn diagram every time; here's an alternative approach that would be simpler, e.g., for a computer to learn. Make a list of the seven bits of the codeword:

$$\underset{1}{\_}\ \underset{2}{\_}\ \underset{3}{\_}\ \underset{4}{\_}\ \underset{5}{\_}\ \underset{6}{\_}\ \underset{7}{\_}.$$

* Richard Hamming (1915–), American mathematician and computer scientist. After helping with the development of the atomic bomb during World War II, he joined the research staff at Bell Laboratories, where, inspired by the work of Claude Shannon, he invented Hamming codes (1948).

Then, place the four given information bits in the first four places:

$$\underset{1}{\underline{1}}\quad \underset{2}{\underline{1}}\quad \underset{3}{\underline{1}}\quad \underset{4}{\underline{1}}\quad \underset{5}{\underline{\ }}\quad \underset{6}{\underline{\ }}\quad \underset{7}{\underline{\ }}.$$

Notice that set A occupies compartments 1, 2, 3, and 5:

$$\begin{array}{ccccccc} A & A & A & & A & & \\ \downarrow & \downarrow & \downarrow & & \downarrow & & \\ \underline{1} & \underline{1} & \underline{1} & \underline{1} & \underline{?} & \underline{\ } & \underline{\ } \\ 1 & 2 & 3 & 4 & 5 & 6 & 7. \end{array}$$

We see that so far A has *three* 1s; to make the total even, we must put a fourth 1 in compartment 5. Similarly, the set B occupies compartments 1, 3, 4, and 6:

$$\begin{array}{ccccccc} B & & B & B & & B & \\ \downarrow & & \downarrow & \downarrow & & \downarrow & \\ \underline{1} & \underline{1} & \underline{1} & \underline{1} & \underline{1} & \underline{?} & \underline{\ } \\ 1 & 2 & 3 & 4 & 5 & 6 & 7, \end{array}$$

so 1 goes in compartment 6. Finally, C occupies compartments 1, 2, 4, and 7:

$$\begin{array}{ccccccc} C & C & & C & & & C \\ \downarrow & \downarrow & & \downarrow & & & \downarrow \\ \underline{1} & \underline{1} & \underline{1} & \underline{1} & \underline{1} & \underline{1} & \underline{?} \\ 1 & 2 & 3 & 4 & 5 & 6 & 7, \end{array}$$

and so 1 goes in compartment 7, too. The final codeword, then, is

$$\underset{1}{\underline{1}}\quad \underset{2}{\underline{1}}\quad \underset{3}{\underline{1}}\quad \underset{4}{\underline{1}}\quad \underset{5}{\underline{1}}\quad \underset{6}{\underline{1}}\quad \underset{7}{\underline{1}}$$

just as we saw in Solution 1. ■

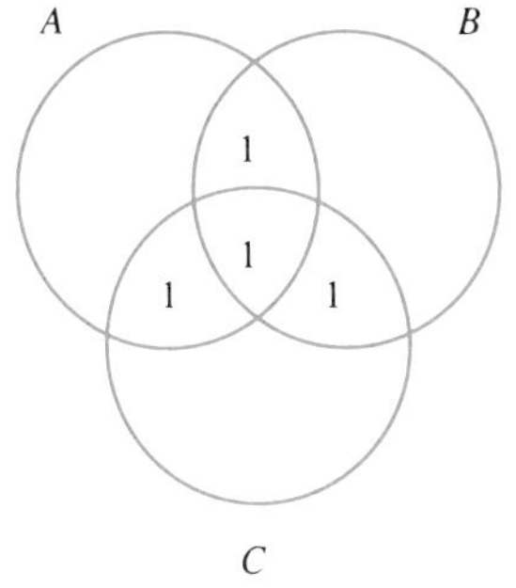

Figure 1.29

Now suppose that one of the bits in the codeword 1110100 is garbled in transmission; for example, suppose 1010100 is received. The receiver can detect and correct this error, using the following **decoding algorithm**.

The decoder's first step is to place the seven received bits back into the seven compartments, as shown in Figure 1.30. The decoder then *rechecks the three parity bits*. In this case, the set A is in error since it contains three 1s; set B is okay, but set C is also in error. The conclusion is that an error occurs in $A \cap B' \cap C$, i.e., in compartment 2. So the decoder changes the bit in compartment 2, and decides that the transmitted bits were actually 1110. The decoder has corrected the error.

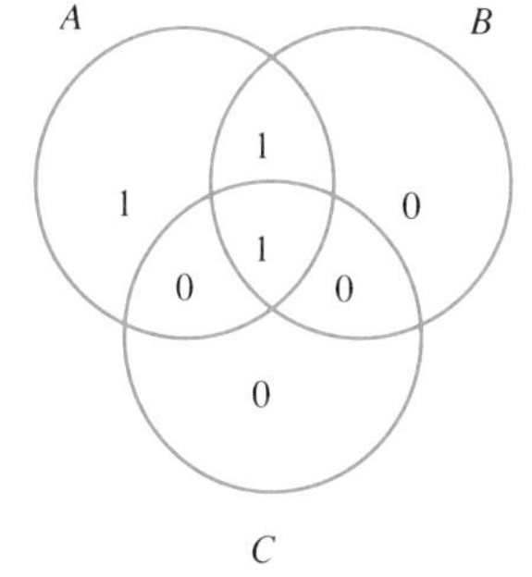

Figure 1.30

This procedure will locate any *single* error, even if it occurs in one of the parity locations. However, if two or more errors occur, it will never succeed (see Problems 5–9). For this reason the procedure just described is called a **single-error-correcting code**. Since each codeword contains seven bits, including four data bits, it is more formally called the (7, 4) **Hamming code**.

EXAMPLE 1.34 Decode the received word 1110000.

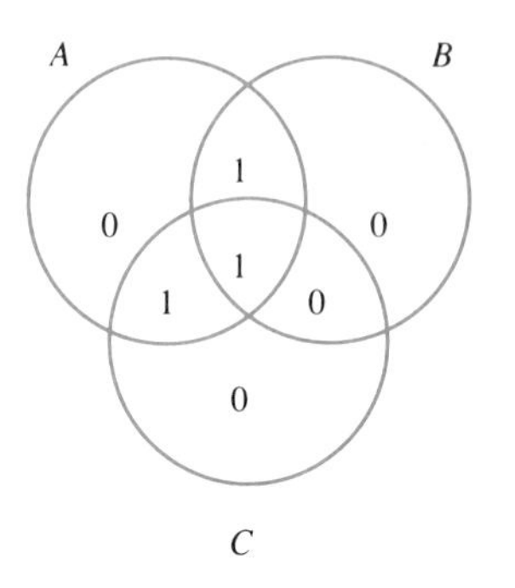

Figure 1.31

SOLUTION 1 If we carefully place the seven received bits into the seven compartments of the Venn diagram as shown in Figure 1.31, we see that the parity of the number of 1s in the three sets A, B, and C, is as follows:

$$A : odd, \qquad B : even; \qquad C : even.$$

We conclude that an error occurred in the set A, but not B or C, i.e., in compartment $A \cap B' \cap C'$, which is numbered 5 in Figure 1.26. Hence the 0 in compartment 5 should be changed to a 1, and the decoder's output is 1110100. Note that the four information bits actually sent were 1110, and there was an error in a parity check bit, namely, digit 5.

SOLUTION 2 As in Solution 2 to Example 1.33, we can avoid the Venn diagrams by using instead a numbered list of the received bits:

$$\begin{array}{ccccccc} A & A & A & & A & & \\ \downarrow & \downarrow & \downarrow & & \downarrow & & \\ \underline{1} & \underline{1} & \underline{1} & \underline{0} & \underline{0} & \underline{0} & \underline{0} \\ 1 & 2 & 3 & 4 & 5 & 6 & 7 \end{array}.$$

We see that the parity in set A is *odd*; but

$$\begin{array}{ccccccc} B & & B & B & & B & \\ \downarrow & & \downarrow & \downarrow & & \downarrow & \\ \underline{1} & \underline{1} & \underline{1} & \underline{0} & \underline{0} & \underline{0} & \underline{0} \\ 1 & 2 & 3 & 4 & 5 & 6 & 7 \end{array}$$

and

$$\begin{array}{ccccccc} C & C & & C & & & C \\ \downarrow & \downarrow & & \downarrow & & & \downarrow \\ \underline{1} & \underline{1} & \underline{1} & \underline{0} & \underline{0} & \underline{0} & \underline{0} \\ 1 & 2 & 3 & 4 & 5 & 6 & 7 \end{array}$$

show that the parity in sets B and C is *even*. Again we see that only in set A do we need to change a bit (compartment 5)—and so the decoded word is

$$\begin{array}{ccccccc} \underline{1} & \underline{1} & \underline{1} & \underline{0} & \underline{1} & \underline{0} & \underline{0} \\ 1 & 2 & 3 & 4 & 5 & 6 & 7 \end{array}$$

as before.

There are many other Hamming codes. For example, there is one which is based (in principle) on a Venn diagram with *four* sets A, B, C, and D. In this code each codeword has 15 bits, of which 11 are data and 4 are parity checks. It is called a $(15, 11)$ Hamming code (see Problems 28–30). This procedure, the encoding and decoding, can be further generalized to produce the $(31, 26)$, $(63, 57)$, $(127, 120)$, $(255, 247)$, ... Hamming codes. Although it's difficult to work with these codes by hand, they can easily be implemented on a computer. And they are! Practically every large computer manufactured since 1965 has had each word in its memory protected against errors by a Hamming code (see also Problems 12–17).

Problems for Section 1.7

Problems 1–11 refer to the (7, 4) Hamming code, with the ordering suggested by Figure 1.26.

1. Encode the four-bit data words 0000 and 0101.

2. Encode the four-bit data words 1010 and 0011.

3. Decode each of these (possibly garbled) codewords: 0110101, 1110110.

4. Decode the following received words: 1011000, 1010100.

Problems 5–9 consider the problem of double errors in transmission.

5. Consider the codeword 1110100 used in the analysis at the beginning of the section. Suppose that *two* errors are made in transmission, in locations 3 and 4, so that the received codeword is 1101100. Decode this word, and show that the decoding algorithm will actually insert another error. In which location will the error occur?

6. Look at Figure 1.26, and visualize the process of changing the bit in compartments 3 and 4. An error in compartment 3 causes a change in the parity of sets A and B, while compartment 4 will affect sets B and C. Show that if bits 3 and 4 are both changed in transmission, the decoder will declare an error in $A \cap B' \cap C$, i.e., compartment 2, and thus another error will be inserted.

7. As in Problem 6, analyze what the decoder will do when digits 1 and 6 are garbled.

8. Repeat Problem 7 for digits 5 and 7.

9. In general, when two errors occur, will the decoder ever correct one of them, or will it *always* make an extra error?

10. Suppose that the codeword 0000000 is sent, and errors are made in positions 5, 6, and 7. Decode the received word, and show that the decoder has introduced still another error.

11. Suppose that the codeword 0000000 is sent, and errors are made in positions 1, 2, and 6. What will the decoder do?

There is no known simple way to modify the Hamming code so that it can correct *two* errors. However, the disaster described in Problems 5–9 can easily be avoided by making a simple modification. If we label the *exterior* compartment in Figure 1.26, i.e., the one corresponding to $A' \cap B' \cap C'$, with the number 8, and use this compartment to define an eighth bit in each codeword, using the rule that the total number of 1s in the whole codeword must be even, the result is called the **extended** (8, 4) Hamming code. Problems 12–17 will explore this code.

12. Encode the four-bit data words 0100 and 1101.

13. Consider the four-bit data word 1100, with corresponding encoded word 11000101. Suppose that a transmission error occurs in position 2, so that 10000101 is received. *If a single error is known to have occurred,* verify that the decoding algorithm described in the text, *when applied to the first seven digits* of the received word, will correct the error.

14. Continuing Problem 13, notice that the number of 1s in the received word 10000101 is *odd,* and a parity check of the entire codeword will reveal this fact. Verify that the overall parity check will distinguish between the two possibilities (a) single error in transmission, and (b) double error (or no errors).

15. Use the results of Problems 13 and 14 to show that the (8, 4) extended Hamming code can correct any pattern of zero or one errors; and if two errors occur, it can always detect this fact. Describe an appropriate decoding algorithm.

16. Using the algorithm developed in Problem 15, decode the words 0110000 and 10001001.

17. Repeat Problem 16 for the words 10011110 and 11101110.

On some communication channels a transmitted bit can be *erased,* i.e., received in such poor shape that it doesn't look like a 0 or a 1. An erasure is not as serious as an error, and the Hamming code can correct *two* erasures. Problems 18–21 will explore this idea for the (7, 4) Hamming code.

18. Suppose that the codeword 1101001 is transmitted, and bits 2 and 6 are erased, so the received word is 1?010?1. Show that in this case, the erased digits are immediately recoverable using the decoding procedure given in the text.

19. Now suppose again that 1101001 is sent, but digits 1 and 3 are erased, so that the received word is ?1?1001. In this case, the transmitted word is not immediately recoverable, but there are only four possibilities:

0101001, 0111001, 1101001, 1111001.

Apply our decoding procedure to each of these candidates, and verify that a correct decision is reached.

20. Look at the four possibilities given in Problem 19, and notice that they correspond to, respectively, a single error in digit 1, double errors in digits 1 and 3, correct transmission, and a single error in digit 3. In view of what we know about the correction of single errors and what happens when there is a double error (Problem 9), explain why a correct decision was reached in Problem 19.

21. Show that with an appropriate modification of the decoding algorithm in the text, the (7, 4) Hamming code will correct one or two erasures.

22. Again, add an extra parity bit to obtain the (8, 4) extended Hamming code, as in Problems 12–17, and consider the four-bit data word 1100, with corresponding encoded word 11000101. Suppose that an error is made in digit 2, and digit 5 is erased, so that 1000?101 is received. Assume that you know that there is at most a single error in a digit other than the fifth. Then there are four possibilities for the transmitted word:

1. Digit 5 = 0, no error in another digit.
2. Digit 5 = 0, single error in another digit.
3. Digit 5 = 1, no error in another digit.
4. Digit 5 = 1, single error in another digit.

Show that the decoding algorithm developed in Problems 12–17 will rule out possibilities 1, 3, and 4, and will correctly identify the error in digit 2.

23. Continuing Problem 22, devise a decoding algorithm so that (8, 4) extended Hamming code will correct one error and/or one erasure.

24. Use the decoding algorithm of Problem 23 to decode the following received sequences:

?0101000, 1001110?, 001?1101.

25. Suppose we had defined the (7, 4) Hamming code by requiring that the parity check bits be chosen so that the parity in each set is *odd* instead of even. (The codeword in Figure 1.28 would then be 1110011, for example.) Would this "odd parity" code be capable of correcting one error?

26. Try to draw a four-set Venn diagram. It should have 15 compartments (and should convince you that a different approach to Hamming codes with more than four information bits might be profitable).

27. Consider Solution 2 to Example 1.33, and represent the fact that A occupies compartments 1, 2, 3, and 5 by writing a sequence of seven 0s and 1s, with 1s in positions 1, 2, 3, and 5: $A \rightarrow 1110100$. Similarly, write $B \rightarrow 1011010$ and $C \rightarrow 1101001$. Now arrange the sequences in the form of a matrix with three rows and seven columns:

$$\begin{array}{cccccccc} A & \rightarrow & 1 & 1 & 1 & 0 & 1 & 0 & 0 \\ B & \rightarrow & 1 & 0 & 1 & 1 & 0 & 1 & 0 \\ C & \rightarrow & 1 & 1 & 0 & 1 & 0 & 0 & 1 \end{array}$$

Notice that the columns of this matrix contain all possible binary sequences of length three, except the all-zero sequence. In Example 1.34 we found that parity check A indicated an odd number of 1s, while B and C indicated an even number, which predicted an error in compartment $A \cap B' \cap C'$. Verify that this error pattern corresponds to the column of the above matrix with entries 100, i.e., column 5.

Problems 28–30 consider the (15, 11) Hamming code.

28. As in Problem 27, write down a matrix whose columns yield all possible binary sequences of length four, except the all-zero sequence. One example is

$$\begin{array}{ccccccccccccccccc} & & 1 & 2 & 3 & 4 & 5 & 6 & 7 & 8 & 9 & 10 & 11 & 12 & 13 & 14 & 15 \\ A & \rightarrow & 1 & 1 & 1 & 1 & 0 & 1 & 1 & 0 & 1 & 0 & 0 & 1 & 0 & 0 & 0 \\ B & \rightarrow & 1 & 1 & 1 & 0 & 1 & 1 & 0 & 1 & 0 & 1 & 0 & 0 & 1 & 0 & 0 \\ C & \rightarrow & 1 & 1 & 0 & 1 & 1 & 0 & 1 & 1 & 0 & 0 & 1 & 0 & 0 & 1 & 0 \\ D & \rightarrow & 1 & 0 & 1 & 1 & 1 & 0 & 0 & 0 & 1 & 1 & 1 & 0 & 0 & 0 & 1 \end{array}$$

In the (15, 11) Hamming code, the first 11 bits are information bits and the check bits 12, 13, 14, 15 are found exactly as in Example 1.33. Digit 12 (the A bit) is determined by the requirement that the total number of 1s in positions 1, 2, 3, 4, 6, 7, 9, and 12 must be even. Similarly, digit 13 (the B bit) scans positions 1, 2, 3, 5, 6, 8, 10, and 13; digit 14 (the C bit) scans positions 1, 2, 4, 5, 7, 8, 11, and 14; and finally, digit 15 (the D bit) scans positions 1, 3, 4, 5, 9, 10, 11, and 15.

Using the algorithm outlined above, show that the 11-bit data word 00001011001 is encoded as 000010110011000.

29. Suppose that the 15-bit encoded word of Problem 28 is transmitted, and a single error occurs in digit 7, so that the received word is 000010010011000. Apply the decoding algorithm of Example 1.34; in other words, recheck the four parity bits. Verify that checks A and C indicate an odd number of 1s and that checks B and D indicate an even number of 1s. This predicts an error in compartment $A \cap B' \cap C \cap D'$, which corresponds to the column of the matrix in Problem 28 with entries 1010, i.e., column 7.

30. After doing Problems 28 and 29, can you see why the given decoding algorithm corrects all single errors in the (15, 11) Hamming code? (We are stretching our modest equipment to the limit here. A systematic study of error-correcting codes requires some background in abstract algebra. One purpose of this text is to bring you to the point where you can begin to study this area.)

Summary

The following concepts and ideas have been introduced:

- *Sets and operations on sets*: union, intersection, complement, difference, subset, disjoint and pairwise disjoint sets, partitions.
- *Proof techniques for verifying set identities*: Venn diagrams, truth tables, direct reasoning.
- *Functions and relations and their properties*: domain, range, and image; one-to-one functions, onto functions, pigeonhole principle, and composition; representation of a relation by a digraph.
- *Partial orderings*: relations that are reflexive, antisymmetric, and transitive; representation by Hasse diagrams; total orderings.
- *Equivalence relations*: relations that are reflexive, symmetric, and transitive; important examples (congruence modulo m, equality); equivalence classes and partitions.
- *Mathematical induction*: basis and induction step; strong induction.
- *Useful notation*: sums and products; union and intersection of n sets; quantifiers.

Finally, we discussed how Venn diagrams can be used to describe error-correcting codes.

CHAPTER TWO

COMBINATORICS: THE THEORY OF COUNTING

Problems which involve counting things are very common in discrete mathematics. Such problems range from very simple (how many even numbers are in the set $\{1, 2, 3, 4, 5, 6, 7, 8, 9, 10\}$?) to horribly difficult (how many partial orderings are there on a set with 10 elements?). The formal study of counting problems is called *combinatorics*, or *combinatorial analysis*. There are many advanced and highbrow combinatorial techniques which mathematicians use to solve counting problems, but most of them are based, ultimately, on a few basic rules, which we will discuss in this chapter.

2.1 The Multiplication Rule

The grandfather of all counting rules is the simple multiplication rule, which we state below.

THE MULTIPLICATION RULE

If one choice can be made in m ways, and if for each of these m choices a second choice can be made in n ways, then the total number of ways of making *both* choices is $m \times n$.

EXAMPLE 2.1 If a child can draw *two* kinds of faces and *three* kinds of hats, how many cartoons can she produce?

SOLUTION The answer is *six* $(2 \times 3 = 6)$; see Figure 2.1.

The multiplication rule can in principle always be illustrated by what is called a *tree diagram.* Figure 2.1 is a tree diagram illustrating the solution to Example 2.1. The two main branches of the tree represent the two possibilities for the face; each main branch has three smaller branches attached to it, representing the three possibilities for the hat. The multiplication rule can also be extended to cover situations in which there are more than two choices allowed. If t choices are made, with n_1 possibilities for the first choice, n_2 possibilities for the second choice, etc., then the overall number of possibilities is $n_1 \times n_2 \times \cdots \times n_t$.

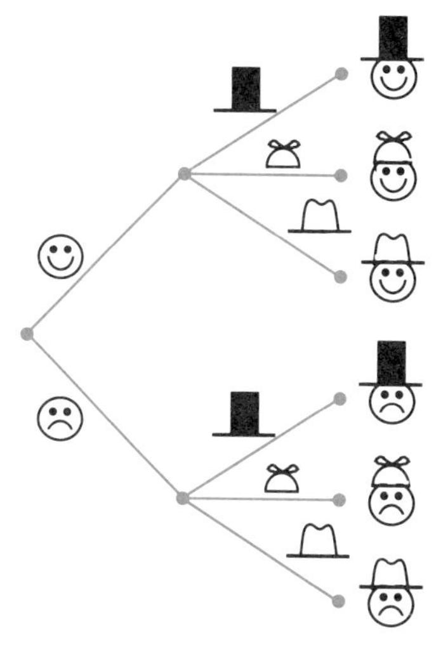

Figure 2.1 A tree diagram for Example 2.1.

EXAMPLE 2.2 How many four-letter "words" can be made from the two letters A and B?

SOLUTION There are *two* possibilities for the first letter, *two* for the second, *two* for the third, and *two* for the fourth. There are therefore $2 \times 2 \times 2 \times 2 = 16$ such words. Figure 2.2 shows a tree diagram for this problem.

The multiplication rule can be also stated in the language of set theory. Suppose the possibilities for the first choice all lie in the set S, and the possibilities for the second choice all lie in the set T. In the childish Example 2.1, for example,

S = { ☺, ☹ }

T = { , , }

We can represent the choice of the element s from S and the element t from T as the ordered pair (s, t). The total number of possible ways to make both choices is then the same as the number of such ordered pairs. In Section 1.3 we learned that the set of all ordered pairs is called the **Cartesian product** of the sets S and T, denoted $S \times T$. Since there are $|S|$ choices for S, and $|T|$ choices for t, we have the following reformulation of the multiplication rule:

$$|S \times T| = |S| \cdot |T|.$$

(The multiplication rule we take as "obviously true," but it is subject to proof; see Problem 30.)

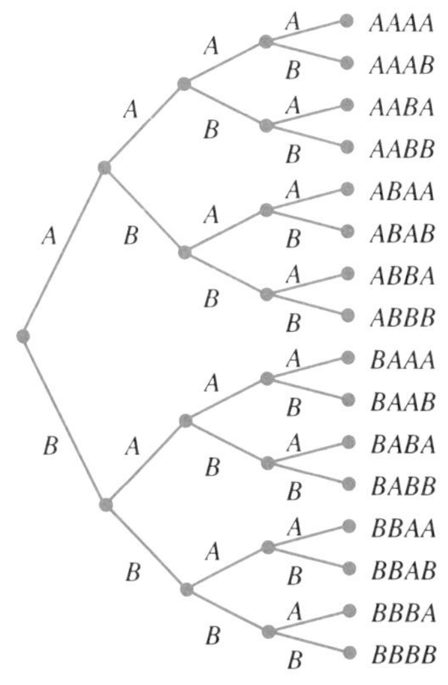

Figure 2.2 A tree diagram for Example 2.2.

EXAMPLE 2.3 Reformulate the problem in Example 2.2 in set-theoretic terms.

SOLUTION Let $S = \{A, B\}$. Example 2.2 asks for all 4-*tuples* (s_1, s_2, s_3, s_4), with each s_i in S. If E, F, G, and H are arbitrary sets, the collection of all 4-tuples (e, f, g, h) with e in E, f in F, g in G, h in H, is called the Cartesian product of the sets E, F, G, and H, and is denoted by $E \times F \times G \times H$. Thus in Example 2.2, we are asked to calculate $|S \times S \times S \times S|$.

In the first three examples, the set of possibilities for the *second* choice did not depend on the *first* choice. However, this is not always the case. The next example shows how the multiplication rule can be used to handle more general problems.

EXAMPLE 2.4 A living room contains an armchair, a desk chair, and a reclining chair. A photographer wants to photograph the room with a child seated in each chair. If there are seven children available, how many possible photographs are there?

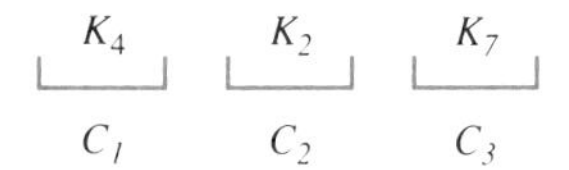

Figure 2.3 Slot-filling visualization of Example 2.4.

SOLUTION To solve this problem, and many similar problems, it is helpful to think in terms of *filling slots* (see Figure 2.3).

We denote the three chairs to be filled by C_1, C_2, and C_3, and the seven children by $K_1, K_2, \ldots, K_7$. We think of the chairs as *slots*, and the children as *markers* to be put in the slots. Slot C_1 can be filled with any one of the seven markers $K_1, K_2, \ldots, K_7$; slot C_2 can then be filled with any of the six remaining markers. Finally, slot C_3 can be filled in five ways, and so, by the multiplication rule, the total number of photographs is $7 \times 6 \times 5 = 210$.

We conclude this brief section by noting an important difference between Example 2.2 and Example 2.4. In Example 2.2, the letters (A and B) could be *repeated*; in effect, we have an inexhaustible supply of each letter. However, in Example 2.4 the children $K_1, K_2, \ldots, K_7$ *cannot* be repeated; a child assigned to a particular chair cannot also sit in another chair! In combinatorial problems, it is important to know whether repetition *is*, or *is not*, allowed. If repetition were not allowed in Example 2.2, no words could have been formed; and if repetition were allowed in Example 2.4 (by using multiple-exposure techniques, perhaps), there would have been $7 \times 7 \times 7 = 343$ possible photographs.

Another possible variation on Example 2.4 would be to consider two photographs to be different only if the photos involved a different *set* of children. Thus, the photograph $K_4K_2K_7$ of Figure 2.3 would be considered the same as the photograph $K_2K_4K_7$, or $K_2K_7K_4$, etc. In this case the total number of different photographs would not be 210, but a much smaller number (35, it turns out). In Example 2.4 as given, *order matters*; but in the revised problem, order *doesn't matter*.

In the next three sections we will discuss the techniques needed for problems in which order does and does not matter, with repetition allowed or not allowed.

Problems for Section 2.1

1. Suppose the child cartoonist can make two kinds of face, three kinds of hat, and four types of body. How many possible cartoons are there?

2. The library has four golf books, three tennis books, and six track books. In how many ways can you bring home three books, one for each sport?

3. How many possible four-letter English "words" are there? (All sequences, e.g., *AAAA*, *XYRQ*, *FACE*, *NJDA*, etc., are legal.)

4. How many four-digit numbers are there with no digit lower than a 3?

5. Calculate the Cartesian product of the "childish" sets S and T in the text. Are the elements of $S \times T$ the same as the cartoons in Figure 2.1?

6. Reformulate the problem in Example 2.4 in set-theoretic terms.

7. An agency has 10 available foster families, $F_1, \ldots, F_{10}$, and 6 children, $C_1, \ldots, C_6$, to place. If no family can get more than one child, in how many ways can the children be placed?

8. Suppose you have 26 Scrabble chips, one for each of the letters $A, B, C, \ldots, Z$. How many four-letter words can you make? (Now *AABC* is illegal, since you don't have two *A* chips.)

9. In computer terminology, a **byte** is a pattern of eight bits (i.e., 0s and 1s). For example, 01011110 and 00001001 are possible bytes. How many different bytes are there?

10. Suppose you take a test with five questions. Each question can be answered true (T) or false (F), or can be omitted (?). A student's answer sheet will then be a sequence of five T's, F's, and ?'s. For example, some possible answers are as follows:

T T F ? F
T F ? ? F
F T ? F ?

etc. How many possible answer sets are there?

11. (a) How many four-letter words begin with Z?
(b) How many four-letter words begin and end with Z?
(c) How many four-letter words begin with Z and then have no more Z's in them?

12. Suppose 10 people share a house. In how many ways can a cook be appointed for Monday, Tuesday, and Wednesday under the following conditions?
(a) A person can't get stuck with the job for more than one day.
(b) A person can be assigned for more than one day.

13. In Problem 7, find the number of placements if a family is allowed more than one child. (Here it is legal to place all the children in one family, or all in different families, or anything in between.)

14. Four students are going to enroll in college and there are 11 schools available to them. For example, one possibility is that John, Mary, and Bill go to school S_5 while Betty goes to S_{11}.
(a) How many possibilities are there?
(b) How many possibilities would there be if no two students can attend the same school?

15. There are five roads between A and B and four roads between B and C. (Assume that you can't travel from A to C without going through B.)
(a) In how many ways can you drive from A to B to C?
(b) In how many ways can you drive from A to B to C and then back to A again, making a round trip?
(c) How many ways are there in part (b) if no road can be used more than once?

16. Consider all six-digit numbers which do not repeat a digit and do not begin with 0. For example, 876592 and 103249 are acceptable, but 045678 is not. How many such numbers are there?

17. In Problem 16, how many of the numbers are divisible by 5? (*Suggestion*: To be divisible by 5, the number must end in 0 or 5, but the number of possibilities for the first digit depends on whether the last digit is 0 or 5. So count those that end in 0 and those that end in 5 separately, and then add.)

18. How many six-digit numbers (no leading zeros allowed) have no repeated digits and are even?

19. Five people P_1, P_2, P_3, P_4, and P_5 make plane reservations but do not necessarily show up. For example, one possibility is that P_3 and P_5 show but P_1, P_2, and P_4 are no-shows. Another is that P_1 and P_2 show, but P_3, P_4, and P_5 do not. How many possibilities are there?

20. In Problem 19, suppose the airline doesn't care about the *names* of the people who show up, but just *how many* do. For example, the two possibilities mentioned in Problem 19 would not be counted separately because both amount to two shows and three no-shows. From this point of view, how many possibilities are there?

21. In how many ways can the letters A, B, C, d, e, f, g be lined up if all the capitals must come first?

22. In a certain programming language, a name may consist of a letter (26 letters are available) or a letter followed by up to 7 symbols, which may be either letters or digits (10 digits are available). How many distinct names are there?

23. In a certain state every license plate has three letters followed by three numbers (e.g., XYZ123 or RVF174). There are 10 million cars in the state. Are there enough license plates to go around?

24. Reconsider Example 2.4, when one of the children is Fat Albert, who needs two adjacent chairs to support him. Now how many photographs are possible? What if all seven children are fat?

25. Suppose a certain family has n members, and at Christmas each family member gives each other member a present. How many presents will be under the tree?

26. In Problem 25, suppose that in addition, Santa Claus also gives each family member a present; now how many presents will there be?

27. Consider the question "How many ways can k objects be selected from a set of n objects?" This question is ambiguous, since we have not specified whether order *does* or *does not* matter and whether repetitions *are* or *are not* allowed. How many different unambiguous forms of this question are there?

28. A number like 43034, which reads the same forward as backward, is called a **palindrome**. There are, for example, only 9 two-digit palindromes, viz., 11, 22, 33, . . . , 99. (Zero is not allowed as a first digit.)
(a) How many three-digit palindromes are there?
(b) How many four-digit palindromes are there?

29. In Problem 28, give a formula for the number of n-digit palindromes. (*Hint:* Separate the cases of even and odd n.)

30. Prove by induction that the multiplication rule is correct. (*Hint:* Let S be a given set, and consider the statement A_n: $|S \times \{1, 2, \ldots, n\}| = |S| \cdot n$.)

31. Verify the statement made at the end of this section that if Example 2.4 were modified so that *order doesn't count*, then there are only 35 possibilities. (This problem will be easier after you study Section 2.3.)

32. If all possible words of five or fewer letters are arranged alphabetically, i.e., *A, AA, AAA, AAAA, AAAAA, AAAAB, . . . , ZZZZZ*, how many words appear between *DONT* and *PANIC*?

2.2 Ordered Samples and Permutations

Sometimes the order in which things occur matters, and sometimes it doesn't. For example, the word *STOP* is different from the word *POTS*, even though both words are formed from the letters in the set $\{O, P, S, T\}$. On the other hand, the sum of the numbers $1 + 2 + 3$ is the same as the sum $2 + 1 + 3$, even though the order of the three numbers has been changed. In this section we will learn how to solve certain combinational problems *when order matters*, and in the next section, we will see how things change *when order doesn't matter.* Along the way, we will see that it makes a difference whether or not *repetition is allowed.*

Before beginning our formal studies, let's consider a simple example, which we encourage you to study carefully.

EXAMPLE 2.5 Consider the set $\{A, B, C, D\}$. In how many ways can we select two of these letters?

SOLUTION There are *four* possible answers to this question, depending on whether order matters, and whether repetitions are allowed.

(a) If order matters and repetition is allowed, there are 16 possible selections:

$$\begin{array}{cccc} AA & BA & CA & DA \\ AB & BB & CB & DB \\ AC & BC & CC & DC \\ AD & BD & CD & DD \end{array}$$

(b) If order matters but repetition is not allowed, there are 12 possibilities:

$$\begin{array}{cccc} AB & BA & CA & DA \\ AC & BC & CB & DB \\ AD & BD & CD & DC \end{array}$$

(c) If order doesn't matter but repetitions are allowed, there are 10 possibilities:

$$\begin{array}{cccc} AA & BB & CC & DD \\ AB & BC & CD & \\ AC & BD & & \\ AD & & & \end{array}$$

(d) Finally, if order doesn't matter, and repetitions aren't allowed, there are only 6 possibilities:

$$\begin{array}{ccc} AB & BC & CD \\ AC & BD & \\ AD & & \end{array}$$

■

Example 2.5 illustrates four of the most important types of combinatorial problems. In this chapter, we will learn how to solve all four types *in general*, i.e., when the problem is to choose k objects from a set of n objects. In this section we'll generalize cases (a) and (b); cases (c) and (d) will be covered in Sections 2.3 and 2.4.

Ordered Samples with Repetition Allowed

Suppose we are given a set with n elements, say $A = \{1, 2, \ldots, n\}$. An *ordered sample of size k, with repetition allowed,* is just a sequence $x_1 x_2 \cdots x_k$ of k elements from A, with no restriction on the number of times a particular element may occur. There are n possibilities for x_1, viz., any element in A; n possibilities for $x_2, \ldots$; and n possibilities for x_n. Hence by the multiplication rule the total number of possibilities is

$$\overbrace{n \times n \times \cdots \times n}^{k \text{ factors}} = n^k.$$

This is the simplest of the four cases, but still important:

ORDERED SAMPLES WITH REPETITION

The number of ordered samples of size k, repetition allowed, from a set of n elements, is n^k.

In Example 2.5, $n = 4$, $k = 2$, and indeed we counted $4^2 = 16$ possibilities in case (a).

Problems involving ordered samples with repetition are very common (e.g., Example 2.6 and many problems in Section 2.1), but the terminology used to describe these problems varies considerably. The following example illustrates this.

EXAMPLE 2.6 Let X be a set containing k elements, and let Y be a set containing n elements. How many functions are there from X to Y?

SOLUTION Let $X = \{x_1, x_2, \ldots, x_k\}$; let us construct such a function f. There are n possibilities for $f(x_1)$, viz., the n elements in Y; n possibilities for $f(x_2), \ldots$; and n possibilities for $f(x_n)$. Hence by the multiplication rule there are $n \times n \times \cdots \times n = n^k$ such functions. Alternatively, the function can be viewed as an ordered sample of size k, repetition allowed, from Y, viz., $f(x_1), f(x_2), \ldots, f(x_k)$, and now the rule for ordered samples with repetition gives n^k as the answer. By either argument we see that the number of functions mapping X to Y is $|Y|^{|X|}$. As an illustration, take $X = \{1, 2\}$ and $Y = \{A, B, C, D\}$. The $4^2 = 16$ functions mapping X to Y are in one-to-one correspondence with the 16 pairs in part (a) of the solution to Example 2.5. ■

We move on to the next case, which is a bit harder.

Ordered Samples Without Repetition

Here again we are given a set with n elements, $A = \{1, 2, \ldots, n\}$, and asked to select k of the elements, say $x_1, x_2, \ldots, x_k$, but this time no element is allowed to appear more than once. How many possibilities are there? Looking at solution (b) to Example 2.5 for guidance, we see that there are still n possibilities for x_1 (anything in A is legal) but only $n - 1$ possibilities for x_2 (since x_1 can't be used again), $n - 2$ possibilities for x_3 (since x_1 and x_2 are now eliminated), etc. Thus by the multiplication rule, the total number of possibilities is

$$\overbrace{n \times (n-1) \times (n-2) \times \cdots \times (n-k+1)}^{k \text{ factors}} = \prod_{i=0}^{k-1} (n-i).$$

Just as for ordered samples with repetition, the answer is the product of k numbers, but now the numbers decrease by 1 each time. With $n = 4$ and $k = 2$, this number is $4 \times 3 = 12$, in agreement with solution (b) to Example 2.5.

ORDERED SAMPLES WITHOUT REPETITION

If repetition is not allowed, the number of ordered samples of size k from a set with n elements, is

$$n(n-1)\cdots(n-k+1).$$

EXAMPLE 2.7 Let $X = \{A, B, C\}$. Count the number of ordered samples of size four from X, where repetition (a) is allowed, (b) is not allowed.

SOLUTION Here $n = 3$, $k = 4$. In case (a) the rule for ordered samples with repetition tells us that there are $3^4 = 81$ possibilities. In case (b) the rule for ordered samples without repetition tells us that there are $3 \times 2 \times 1 \times 0 = 0$ possibilities! This is because there is no way to form a four-letter word from the three letters A, B, C without using some letter more than once. ■

EXAMPLE 2.8 Refer to Example 2.6. Let X be a set containing k elements, and let Y be a set containing n elements. How many one-to-one functions from X to Y are there?

SOLUTION Let $X = \{x_1, x_2, \ldots, x_k\}$, and let f be a one-to-one function from X to Y, with $f(x_i) = y_i$. Then f is completely specified by the list $(y_1, y_2, \ldots, y_k)$, which is an ordered sample of size k from Y, with repetition not allowed, since f is one-to-one. Hence, by the rule for ordered samples without repetition, there are $n(n-1)\cdots(n-k+1)$ one-to-functions. ■

There is an important special case of the rule for ordered samples without repetition that deserves special mention, and that is the case when n and k are *equal.* When $n = k$, we are counting the number of ordered samples using *all* the elements in the set. For example, when $n = k = 3$ and the set is $\{1, 2, 3\}$, there are $3 \times 2 \times 1 = 6$ possibilities: 123, 132, 213, 231, 312, 321. These are just the possible rearrangements of the set, or, in the usual mathematical jargon, the **permutations** of the set. There are then $n \times (n-1) \times (n-2) \times \cdots \times 2 \times 1$ permutations of a set with n elements. This product is usually denoted by the symbol $n!$ (pronounced "*n factorial*"), and so we have the following definition.

PERMUTATIONS

The number of permutations of a set with n elements is $n!$.

It turns out that we can use this new "factorial" notation to express the rule for ordered samples without repetition in a slightly different way. For example, with $n = 4$ and $k = 2$, the number of ordered samples without repetition is, as we have seen, $4 \times 3 = 12$. However, we can also write this product as

$$4 \times 3 = \frac{4 \times 3 \times 2 \times 1}{2 \times 1} = \frac{4!}{2!}.$$

In general, we will have

$$\text{number of ordered samples without repetition} = n(n-1)\cdots(n-k+1)$$
$$= \frac{n(n-1)\cdots 2 \cdot 1}{(n-k)(n-k-1)\cdots 2 \cdot 1}$$
$$= \frac{n!}{(n-k)!}.$$

We conclude this section with two more examples.

EXAMPLE 2.9 A coin is tossed five times and the results recorded in order (for example, HTTHH is regarded as distinct from THHTH, even though there are three heads and two tails in each case). How many possible outcomes are there?

SOLUTION We are counting ordered samples of size five (with repetition allowed) using the alphabet H, T. By the rule for ordered samples with repetition, the number of outcomes is $2^5 = 32$. In general, if a coin is tossed n times, there are 2^n possible outcomes.

EXAMPLE 2.10 Given 10 people $P_1, \ldots, P_{10}$.

(a) In how many ways can the people be lined up in a row?
(b) How many lineups are there if P_2, P_6, and P_9 want to stand together (in any order)?
(c) How many lineups are there in which P_2, P_6, and P_9 do *not* stand together?

SOLUTION

(a) We are asked to compute the number of permutations of the set $\{P_1, P_2, \ldots, P_{10}\}$. By the permutation rule, the answer is $10! = 3628800$.
(b) This problem cannot be solved by a direct application of any of our rules. We proceed indirectly as follows: Replace the three friends P_2, P_6, and P_9 by the symbol G (for "group"). A typical legal arrangement of the 10 is

$$P_1P_3P_5P_4\overbrace{P_2P_9P_6}^{G}P_{10}P_8P_7$$

or

$$\overbrace{P_9 P_6 P_2}^{G} P_1 P_3 P_4 P_5 P_{10} P_7 P_8,$$

corresponding to a permutation of the set $\{P_1, P_3, P_4, P_5, P_7, P_8, P_{10}, G\}$, which contains eight objects. There are $8! = 40320$ such permutations. However, within each such permutation there are $3! = 6$ ways of rearranging the group G. Hence, by the multiplication rule there are

$$8!3! = 40320 \times 6 = 241920$$

possibilities.

There is often more than one way to solve a given counting problem. Here is another approach to (b). First note that the group G can occupy any one of eight positions in the overall lineup, namely, positions (1 2 3), (2 3 4), (3 4 5), (4 5 6), (5 6 7), (6 7 8), (7 8 9), or (8 9 10). Once the position for G is selected there are 3! ways to permute the friends P_2, P_6, P_9 with G and 7! ways to permute the other seven. Hence, by the multiplication rule there are $8 \cdot 7! \cdot 3! = 241920$ possibilities. (There might be many ways to solve a problem, but there is only supposed to be one answer!)

(c) Finally, to solve this part, we subtract: There are 10! total arrangements and 8!3! arrangements in which the friends are together. Hence there are $10! - 8!3! = 3386880$ possibilities. ■

Part (c) of Example 2.10 illustrates another basic counting rule. The number of possibilities in which a certain event *does not occur* equals the total number of possibilities *minus* the number of possibilities in which the event *does* occur. In set-theoretic notation this amounts to saying that if A is a subset of the universal set S, then

$$|A'| = |S| - |A|.$$

Problems for Section 2.2

1. As in Example 2.5, list all ways of selecting two letters from the set $\{A, B, C\}$, depending on whether order matters and whether repetitions are allowed.

2. Repeat Problem 1 for a selection of three letters from $\{A, B, C\}$.

3. If X is a set with six elements and Y a set with five elements, how many functions are there from X to Y?

4. If $X = \{1, 2\}$ and $Y = \{A, B, C\}$, verify that the functions from X to Y are in one-to-one correspondence with the ordered samples with replacement found in Problem 1.

5. Three cards are drawn from an ordinary deck and the results recorded so that the order in which they were drawn counts.

(a) If the drawing is done with replacement, how many ordered samples are there?

(b) If the drawing is done without replacement, how many ordered samples are there?

6. In Problem 5, in each case how many samples contain king, queen, and jack in that order?

7. If X is a set with six elements and Y a set with five elements, how many one-to-one functions are there from X to Y?

8. If X is a set with six elements, how many one-to-one functions are there from X to X?

9. If a coin is tossed four times and the results recorded in order, there are $2^4 = 16$ possible outcomes (see Example 2.9). How many outcomes have exactly one head?

10. In Problem 9, how many outcomes have *at least* one head?

11. Given four bands, seven floats, and three equestrian units.
(a) Suppose they are going to parade down the street with the bands first, then the floats, and then the equestrian units. How many such parades are possible?
(b) Suppose they parade with each type still sticking together, but with no requirement about which type goes first, second, or third. Now how many parades are possible?
(c) Suppose they forget it and just parade. How many parades are there?

12. Find the number of ways of lining up the letters A, B, C, D, E, F if B has to be directly between A and C, i.e., the lineup must include ABC or CBA.

13. A single die is tossed n times. If the results are recorded in order, how many possible outcomes are there?

14. Given 10 adults and 6 children, how many photographs can be made consisting of 2 adults with 1 child between them?

15. You have 15 friends. Find the number of ways you can
(a) choose 4 to be president, vice president, secretary, and treasurer of your club.
(b) lend them four books titled A, B, C, and D so that no person gets more than one book.
(c) lend them four books A, B, C, and D so that a person can get more than one book.

16. Consider 17 people $A_1, \ldots, A_{17}$ and 3 errands E_1, E_2, E_3.
(a) In how many ways can the people do the errands if a person can get stuck with 2 or even all 3 of the errands?
(b) In how many ways can the people do the errands if a person can't do more than 1 errand?

17. Find the number of permutations of the 26 letters of the alphabet that contain the string *TOPEKA*.

18. Given 26 Scrabble chips, one for each letter. Find the number of six-letter words that
(a) begin with a vowel.
(b) contain no vowel.

19. In Problem 18, find the number of six-letter words that contain exactly one vowel.

20. A club has five members M_1, M_2, M_3, M_4, M_5. They are going to form a finance committee but haven't decided how large it should be. They may pick everyone on it, or just M_3, or M_1 and M_2, etc. How many possibilities are there?

21. Suppose that we have distinguishable boxes $B_1, \ldots, B_n$, and we toss k balls into them. Assume that the balls are distinguishable, and are labelled $b_1, \ldots, b_k$. If, for example, ball b_1 goes into box B_3 and ball b_2 into box B_5, this is not the same as b_1 into B_5 and b_2 into B_3. The assumption of *distinguishable balls* means that if we write down the sequence of boxes into which $b_1, \ldots, b_k$ are tossed, *order counts.* If each box is allowed to contain any number of balls, show that the number of possible ways of distributing the balls is n^k (cf. the rule for ordered samples with repetition).

22. In Problem 21, if no box can contain more than one ball, show that the number of possible ways of distributing the balls is $n(n-1)\cdots(n-k+1)$ (cf. the rule for ordered samples without repetition).

23. Find the number of ways of lining up 10 people $P_1, \ldots, P_{10}$ if no two people from the set $\{P_2, P_6, P_9\}$ can be seated next to each other. (*Suggestion*: Line up the 7 other people, and then consider the spaces between them and at the ends.)

24. How many ways are there of permuting the set $\{A, B, C, D, E\}$ so that A comes before B?

25. How many permutations of $\{A, B, C, D, E, F\}$ have the B between the A and the C? (Thus *DFAEBC* and *CBAEDF* are allowable, but *ADECFB* is not.)

26. Let S be a set of n elements. Use the rule for ordered samples with repetition to count the number of subsets of S. (*Hint:* Let $x_1, x_2, \ldots, x_n$ be the elements of the set S. To identify an unknown subset X, you could ask the following ordered sequence of questions: "Is $x_1 \in X$?"; "Is $x_2 \in X$?"; . . . ; "Is $x_n \in X$?" Note that the sequence of yes/no answers completely determines the subset X.)

27. X and Y are certain sets. Suppose you know that there are exactly 103 functions from X to Y. What can you conclude about $|X|$ and $|Y|$?

28. X and Y are certain sets. Suppose you know that there are exactly 120 one-to-one functions from X to Y. What can you say about $|X|$ and $|Y|$?

29. How do you think that the rule for ordered samples with repetition and the rule for ordered samples without repetition should be modified if n and/or k is *zero*?

30. Let X be a set containing k elements, and Y be a set containing n elements. How many *one-to-one and onto* functions from X to Y are there?

31. The poem "Combinations" by Mary Ann Hoberman begins with this verse:

> A flea flew by a bee. The bee
> To flee the flea flew by a fly.
> The fly flew high to flee the bee
> Who flew to flee the flea who flew
> To flee the fly who now flew by.

There are several other verses, obtained by interchanging the roles of the flea, the bee, and the fly in all possible ways.
(a) How many verses does "Combinations" have?
(b) Is the poem aptly named? Can you think of a better title?

2.3 Unordered Samples Without Repetition; Binomial Coefficients

In the previous section, we learned how to count ordered samples with and without repetition. In this section, we will learn how to count *unordered* samples *without* repetition. In simplest terms, this is the problem of counting the number of ways of choosing k elements from a set of n objects, when repetition isn't allowed and order doesn't matter.*

We have already solved one instance of this problem in Example 2.5(d); when $n = 4$ and $k = 2$, there are *six* possibilities. Notice that six is *also* the number of two-element subsets of a four-element set, say $\{A, B, C, D\}$:

$$\{A, B\}, \{A, C\}, \{A, D\}, \{B, C\}, \{B, D\}, \{C, D\}.$$

This is not a coincidence, since when we specify a subset, order *doesn't* matter and repetition *isn't* allowed! Hence, the problem of counting the unordered samples of size k, repetition not allowed, from an n-element set is really just the problem of counting the k-element subsets of an n-element set, thinly disguised.

EXAMPLE 2.11 How many three-element subsets are there of the five-element set $\{1, 2, 3, 4, 5\}$?

SOLUTION The answer can be found by careful enumeration, even though we don't yet know the general solution to our problem. It is 10, and here are the sets:

$$\{1, 2, 3\}, \{1, 2, 4\}, \{1, 2, 5\}, \{1, 3, 4\}, \{1, 3, 5\}, \{1, 4, 5\},$$
$$\{2, 3, 4\}, \{2, 3, 5\}, \{2, 4, 5\}, \{3, 4, 5\}.$$

While "careful enumeration" is a feasible approach to a problem like Example 2.11, if we wanted to know the number of 11-element subsets of a set with 55 elements, for example, another approach would clearly be helpful! And, in fact, there is a very useful general formula for the number of k-element subsets, which does not require us to actually list all of the possibilities. To derive this formula, we will exploit the close relationship between ordered and unordered samples without repetition.

Let us again consider samples of size two from the set $\{A, B, C, D\}$. By the results of Section 2.2, there are $4 \times 3 = 12$ ordered samples without repetition. These samples were already listed in the solution to Example 2.5(b), but here we list them again in a different arrangement:

$$\begin{array}{cc} AB & BC \\ BA & CB \\ AC & BD \\ CA & DB \\ AD & CD \\ DA & DC \end{array}$$

* This is often called the problem of counting the number of **combinations** of k objects out of n.

These 12 possibilities form six pairs, with the members of each pair representing the possible *rearrangements* of one of the two-element subsets of $\{A, B, C, D\}$. The pair (AB, BA), for example, corresponds to the subset $\{A, B\}$; (CD, DC) corresponds to $\{C, D\}$, etc. Thus in the list of the *ordered* samples, each *unordered* sample appears twice.

Similarly, with $n = 4$ and $k = 3$, we can divide the $4 \times 3 \times 2 = 24$ ordered samples of size three into four groups of six each, with each group representing the number of rearrangements of one particular three-element subset of $\{A, B, C, D\}$. See Table 2.1. In Table 2.1 there are six rearrangements of each three-element subset, because of the rule for permutations: $3! = 6$.

After studying these two simple examples, we can see what happens in general. The total number of *ordered samples without repetition* is

$$n(n-1)\cdots(n-k+1).$$

This set of samples can be viewed as all possible *permutations* of all possible *unordered samples without repetition.* Each unordered sample has, by the permutation rule, $k!$ permutations.

Thus we can view the process of choosing an ordered sample without repetition as a two-stage process. In the first state, an *unordered* sample without repetition, i.e., the underlying subset, is chosen. In the second stage a particular *permutation* of the subset is chosen. By the multiplication rule, then, we have

$$\text{ordered samples without repetition} = \text{unordered samples without repetition} \times \text{permutations}.$$

But we already know how to count ordered samples without repetition and permutations. Therefore

$$n(n-1)\cdots(n-k+1) = \text{unordered samples without repetition} \times k!.$$

Thus we have derived the following general formula.

UNORDERED SAMPLES WITHOUT REPETITION
(THE SUBSET RULE)

If repetition is not allowed, the number of unordered samples of size k, from a set with n elements, is

$$\frac{n(n-1)\cdots(n-k+1)}{k!}.$$

With $n = 4$, $k = 2$, application of this rule gives $(4 \times 3)/(2 \times 1) = 6$; and with $n = 4$, $k = 3$ it gives $(4 \times 3 \times 2)/(3 \times 2 \times 1) = 4$. These numbers agree, as they should, with our earlier long-winded calculations.

There is a special notation used for the formula in the rule for unordered samples without repetition. It is the symbol $\binom{n}{k}$:

$$\binom{n}{k} = \frac{n(n-1)\cdots(n-k+1)}{k!}. \tag{2.1}$$

TABLE 2.1 24 ordered samples and 4 unordered samples: $24 = 4 \times 3!$

$\{A, B, C\}$	$\{A, B, D\}$	$\{A, C, D\}$	$\{B, C, D\}$
ABC	ABD	ACD	BCD
ACB	ADB	ADC	BDC
BAC	BAD	CAD	CBD
BCA	BDA	CDA	CDB
CAB	DAB	DAC	DBC
CBA	DBA	DCA	DCB

This symbol is pronounced "n choose k," and is called a **binomial coefficient**. (We'll soon give an explanation of this terminology.) A useful alternative formula for $\binom{n}{k}$ can be obtained by multiplying the numerator and denominator of the fraction in Equation (2.1) by $(n - k)!$. For example, consider the case $n = 5$, $k = 3$. Then by Equation (2.1), we have

$$\binom{5}{3} = \frac{5 \cdot 4 \cdot 3}{3!} = \frac{5 \cdot 4 \cdot 3 \cdot 2 \cdot 1}{3!2 \cdot 1} = \frac{5!}{3!2!}.$$

In general, the formula is

(2.2)
$$\binom{n}{k} = \frac{n!}{k!(n-k)!}.$$

Binomial coefficients occur constantly in discrete mathematics. This is not surprising since, as we have seen, the binomial coefficient $\binom{n}{k}$ counts the number of k-element subsets of an n-element set—a very fundamental quantity! Binomial coefficients have many interesting and useful properties, and in the remainder of this section we will give the most important ones.

EXAMPLE 2.12 Let X be a set with $|X| = 8$. How many three-element subsets does X have? How many five-element subsets?

SOLUTION By the subset rule, there are $\binom{8}{3} = (8 \times 7 \times 6)/(1 \times 2 \times 3) = 56$ three-element subsets, and $\binom{8}{5} = (8 \times 7 \times 6 \times 5 \times 4)/(5 \times 4 \times 3 \times 2 \times 1) = 56$ five-element subsets.

Is the fact that both answers in Example 2.12 are 56 a coincidence? No, because by Equation (2.2),

$$\binom{8}{3} = \frac{8!}{3!5!} \quad \text{and} \quad \binom{8}{5} = \frac{8!}{5!3!}.$$

The general rule is this:

(2.3)
$$\binom{n}{k} = \binom{n}{n-k}.$$

Another way to see that (2.3) is true is to notice that (with respect to the n-element universe) the *complement of a* k-element subset is an $(n-k)$-element subset, and vice versa. Thus the k-element subsets are in one-to-one correspondence with the $(n-k)$-element subsets, and so there must be an equal number of each.

EXAMPLE 2.13 How many five-element subsets are there of the set $\{0, 1, 2, 3\}$?

SOLUTION The answer, obviously, is zero! And the Formula (2.1) even covers this peculiar case, since

$$\binom{4}{5} = \frac{4 \cdot 3 \cdot 2 \cdot 1 \cdot 0}{5 \cdot 4 \cdot 3 \cdot 2 \cdot 1} = 0.$$

Generalizing from Example 2.13, we have the rule

(2.4) $$\binom{n}{k} = 0 \qquad \text{if} \qquad k > n.$$

EXAMPLE 2.14 How many zero-element subsets does a set have?

SOLUTION The only set with no elements is the empty set $\varnothing$, which is a subset of every set, and so the answer is *one*:

(2.5) $$\binom{n}{0} = 1 \qquad \text{for all } n.$$

The rule (2.5) cannot be deduced directly from Formula (2.1) or (2.2). However, by (2.1) we see that $\binom{n}{n} = 1$; in order for (2.3) to hold for $k = n$, we must have $\binom{n}{0} = \binom{n}{n} = 1$. Thus (2.5) is the correct definition of a binomial coefficient when $k = 0$. [In fact, even the case $k = 0$ *is* covered by (2.2) if we define $0!$ to be $1(!)$.]

$$\begin{array}{c}
1 \\
1 \quad 1 \\
1 \quad 2 \quad 1 \\
1 \quad 3 \quad 3 \quad 1 \\
1 \quad 4 \quad 6 \quad 4 \quad 1 \\
1 \quad 5 \quad 10 \quad 10 \quad 5 \quad 1 \\
1 \quad 6 \quad 15 \quad 20 \quad 15 \quad 6 \quad 1 \\
1 \quad 7 \quad 21 \quad 35 \quad 35 \quad 21 \quad 7 \quad 1
\end{array}$$

Figure 2.4 Pascal's triangle (up to $n = 7$).

A basic property of binomial coefficients is illustrated by **Pascal's triangle,*** a simple triangular array of binomial coefficients displayed as follows:

$$\begin{array}{c}
\binom{0}{0} \\
\binom{1}{0} \quad \binom{1}{1} \\
\binom{2}{0} \quad \binom{2}{1} \quad \binom{2}{2} \\
\binom{3}{0} \quad \binom{3}{1} \quad \binom{3}{2} \quad \binom{3}{3} \\
\binom{4}{0} \quad \binom{4}{1} \quad \binom{4}{2} \quad \binom{4}{3} \quad \binom{4}{4} \\
\binom{5}{0} \quad \binom{5}{1} \quad \binom{5}{2} \quad \binom{5}{3} \quad \binom{5}{4} \quad \binom{5}{5} \\
\binom{6}{0} \quad \binom{6}{1} \quad \binom{6}{2} \quad \binom{6}{3} \quad \binom{6}{4} \quad \binom{6}{5} \quad \binom{6}{6} \\
\binom{7}{0} \quad \binom{7}{1} \quad \binom{7}{2} \quad \binom{7}{3} \quad \binom{7}{4} \quad \binom{7}{5} \quad \binom{7}{6} \quad \binom{7}{7} \\
\vdots
\end{array}$$

The rows are numbered $0, 1, 2, \ldots,$ and the entries in each row are also numbered $0, 1, 2, \ldots.$ Thus entry number 2 in row number 6 of Pascal's triangle is $\binom{6}{2} = (6 \cdot 5)/(1 \cdot 2) = 15$. The actual numerical values of rows 0 through 7 of Pascal's triangle are displayed in Figure 2.4.

Pascal's triangle has a very interesting and useful property: Each number in the triangle is the sum of the two numbers above it, i.e., the number just above it and to the *right*, and the number just above it and to the *left*. Thus for example with $n = 5$ and $k = 3$, we have $\binom{5}{3} = \binom{4}{3} + \binom{4}{2}$. Looking at Pascal's triangle in Figure 2.4, we see that this is in fact so: $10 = 4 + 6$. This important property, called the *rule of Pascal's triangle*, can be written symbolically as follows:

(2.6)
$$\binom{n}{k} = \binom{n-1}{k} + \binom{n-1}{k-1}.$$

Using this rule, and the fact that Pascal's triangle is bordered with 1s [because of Equation (2.5)], you can compute any desired binomial coefficient using only *addition*, even though Formula (2.1) requires multiplication and division!

Why is Equation (2.6) true? It is possible to prove (2.6) by direct use of the definition (2.1) of the binomial coefficients (Problem 3.1). Such a proof is not difficult, but neither is it very enlightening. It is however also possible to give a *set-theoretic* proof, using the fact that $\binom{n}{k}$ is the number of k-element subsets of the set $\{1, 2, \ldots, n\}$. This proof, as you will see, gives a perfect explanation of why Equation (2.6) is true.

* Blaise Pascal (1623–1662), French mathematician and physicist. He made many important scientific contributions, including a serious investigation of the "arithmetical triangle" that bears his name. But today he is best remembered as the inventor of the first working mechanical calculating device (a computer in 1645!). In his honor, a popular programming language is called **Pascal**.

To illustrate this proof, let's consider an example, say $n = 5$ and $k = 3$. The three-element subsets of $\{1, 2, 3, 4, 5\}$ can be divided into two types, those subsets which *don't* contain the element 5, and those that *do*:

Subsets not containing 5:

$$\{1, 2, 3\}, \{1, 2, 4\}, \{1, 3, 4\}, \{2, 3, 4\}.$$

Subsets containing 5:

$$\{1, 2, 5\}, \{1, 3, 5\}, \{1, 4, 5\}, \{2, 3, 5\}, \{2, 4, 5\}, \{3, 4, 5\}.$$

Those that don't contain 5 are three-element subsets of $\{1, 2, 3, 4\}$, and there are $\binom{4}{3}$ of them. Those that do contain 5 must consist of 5 plus a *two*-element subset of $\{1, 2, 3, 4\}$, and there are $\binom{4}{2}$ of them. In total, then, the number of three-element subsets of $\{1, 2, 3, 4, 5\}$ equals the number of three-element subsets of $\{1, 2, 3, 4\}$ plus the number of two-element subsets of $\{1, 2, 3, 4\}$. Thus $\binom{5}{3} = \binom{4}{3} + \binom{4}{2}$, as Equation (2.6) says.

In general, the proof of the rule of Pascal's triangle is just the same. Each k-element subset of $\{1, 2, \ldots, n\}$ either contains the element n or it doesn't. The sets that don't contain n are k-element subsets of $\{1, 2, \ldots, n-1\}$, and there are $\binom{n-1}{k}$ of them. The sets which do contain n must consist of n plus a $(k-1)$-element subset of $\{1, 2, \ldots, n-1\}$, and there are $\binom{n-1}{k-1}$ of them. Thus $\binom{n}{k} = \binom{n-1}{k} + \binom{n-1}{k-1}$, as advertised.

We promised to explain why the numbers $\binom{n}{k}$ are called binomial coefficients. The explanation is that they also occur in the *binomial theorem*, as we shall now demonstrate.

A **binomial** is a polynomial with only two terms. For example, $x + y$ is a binomial. The **binomial theorem** is a formula for expressing powers of $x + y$ in terms of x and y. For example,

$$(x + y)^2 = x^2 + 2xy + y^2$$
$$(x + y)^3 = x^3 + 3x^2y + 3xy^2 + y^3$$
$$(x + y)^4 = x^4 + 4x^3y + 6x^2y^2 + 4xy^3 + y^4,$$

etc.

Notice that the coefficients in the expansions of $(x + y)^2$, $(x + y)^3$, and $(x + y)^4$, viz.,

$$\begin{array}{ccccccccc} & & 1 & & 2 & & 1 & & \\ & 1 & & 3 & & 3 & & 1 & \\ 1 & & 4 & & 6 & & 4 & & 1 \end{array}$$

are identical to the *binomial coefficients* in rows 2, 3, and 4 of Pascal's triangle. This is not an accident! The *binomial theorem* states that this pattern continues forever; in other words, for all $n \geq 2$,

$$(x + y)^n = \binom{n}{0}x^n + \binom{n}{1}x^{n-1}y + \cdots + \binom{n}{n-1}xy^{n-1} + \binom{n}{n}y^n.$$

Since $\binom{n}{k} = \binom{n}{n-k}$, the above equation can also be written as

$$(x + y)^n = \binom{n}{0} y^n + \binom{n}{1} xy^{n-1} + \cdots + \binom{n}{n-1} x^{n-1}y + \binom{n}{n} x^n,$$

or using the summation notation of Section 1.6,

$$(x + y)^n = \sum_{k=0}^{n} \binom{n}{k} x^k y^{n-k}.$$

It is possible to give a *combinatorial* proof of the binomial theorem, using the fact that $\binom{n}{k}$ is the number of k-element subsets of an n-element set. The proof can best be appreciated by first considering a special case, say $n = 5$.

Thus we consider the expansion of $(x + y)^5$, which is the product of five identical binomials:

$$(x + y)^5 = \overbrace{(x + y)}^{1}\overbrace{(x + y)}^{2}\overbrace{(x + y)}^{3}\overbrace{(x + y)}^{4}\overbrace{(x + y)}^{5}.$$

If we multiplied out these five binomials by hand, we would obtain a large number of terms. For example, we could multiply the x from the first binomial, the y from the second and third, and the x from the fourth and fifth binomials, thereby obtaining the term $xyyxx$, or x^3y^2. In fact, the multiplication rule tells us that we would obtain 32 terms in all, since there are two choices (x or y) from each of the five binomials, and $2 \times 2 \times 2 \times 2 \times 2 = 32$.

However, many of these 32 terms turn out to be the same, because in multiplication *order doesn't matter.* For example, $xyyxx$ and $xxyxy$ both simplify to x^3y^2. How many of the 32 terms do simplify to x^3y^2? Well, a string of five symbols (involving x's and y's) will simplify to x^3y^2 if and only if it contains exactly three x's and two y's. The three x's could be chosen from any three of the five binomials, and so if we numbered the binomials 1, 2, 3, 4, and 5, we would get x^3y^2 if and only if the x's occurred in one of the three-element subsets of $\{1, 2, 3, 4, 5\}$. But we know that there are exactly $\binom{5}{3} = 10$ such subsets, and therefore exactly 10 terms which simplify to x^3y^2. It follows that the coefficient of x^3y^2 is 10.

An exactly similar argument shows that in general, of the 2^n terms involved in the expansion of $(x + y)^n$, exactly $\binom{n}{k}$ simplify to $x^k y^{n-k}$, which proves the binomial theorem.

We conclude this section with two more examples illustrating the usefulness of binomial coefficients in counting problems.

EXAMPLE 2.15 Given a deck of 52 cards, a "poker hand" is a subset of 5 of the cards. There are $\binom{52}{5}$ poker hands, by the subset rule and (2.1). How many of them contain the queen of spades?

SOLUTION We must select the queen of spades as one of our cards, and choose 4 more from the remaining 51. The answer is $1 \cdot \binom{51}{4} = 249900$. ■

EXAMPLE 2.16 Given a group of 6 Samurai, 7 Lords, and 8 Ninjas.

(a) How many teams of 5 members can be formed?
(b) How many with exactly 3 Samurai?
(c) How many 10-member teams with 3 Samurai, 2 Lords, and 5 Ninjas?

SOLUTION

(a) There are $6 + 7 + 8 = 21$ characters, and so $\binom{21}{5} = 20349$ teams.
(b) If there must be exactly 3 Samuari, there are $\binom{6}{3}$ ways to choose them, and $\binom{15}{2}$ ways to choose the Lords and Ninjas. Thus by the multiplication rule, the number is

$$\binom{6}{3} \cdot \binom{15}{2} = 20 \cdot 105 = 2100.$$

(c) For a 10-member team, there are $\binom{6}{3}$ ways of choosing the Samurai, $\binom{7}{2}$ ways of choosing the Lords, and $\binom{8}{5}$ ways of choosing the Ninjas. The total number is, by the multiplication rule, then,

$$\binom{6}{3} \cdot \binom{7}{2} \cdot \binom{8}{5} = 20 \cdot 21 \cdot 56 = 23520.$$

A final remark: In the last displayed equation notice the pattern of the numbers on the top and bottom of the binomial coefficients:

$6 + 7 + 8 = 21 =$ number of characters *available*
$3 + 2 + 5 = 10 =$ number of characters *selected.*

This pattern suggests a general formula (see Problem 27), which will be encountered again in Section 2.5 when we study *multinomial coefficients.*

Problems for Section 2.3

1. List all two-element subsets of $\{1, 2, 3, 4, 5\}$, and verify that the subset rule gives the correct number of subsets.

2. List all three-element subsets of $\{1, 2, 3, 4\}$, and verify that Formula (2.1) [or (2.2)] gives the correct number of subsets.

3. How many 11-element subsets does a 55-element set contain?

4. How many four-element subsets does the set $\{A, B, C, D\}$ have?

5. Verify that $\binom{n}{n} = 1$ and $\binom{n}{1} = n$ for all n.

6. Verify that $\binom{n}{2} = n(n-1)/2$.

7. Compute the binomial coefficients $\binom{52}{5}$ and $\binom{51}{4}$ that occurred in Example 2.15.

8. Compute the next four rows of Pascal's triangle, i.e., up to $n = 11$.

9. Find the number of poker hands containing
(a) the ace of spades and the ace of diamonds;
(b) the ace of spaces and the ace of diamonds but no other aces.

10. Find the number of poker hands containing
(a) all hearts;
(b) all hearts, including the three of hearts.

11. In Example 2.16, how many five-member teams are there consisting of one Samurai, two Lords, and two Ninjas?

12. An urn contains 6 white, 10 red, and 8 green balls. How many (unordered) selections can be made containing exactly 3 white, 4 red, and 5 green balls?

13. How many 10-letter words can be formed using 4 A's and 6 B's?

14. How many 12-letter words can be formed from 3 A's, 4 B's, and 5 C's?

15. Consider a set of 10 people $\{P_1, \ldots, P_{10}\}$.
(a) How many 6-member teams can be formed?
(b) How many do not contain both $\{P_2, P_4\}$?

16. Consider an urn containing 34 colored balls, 10 yellow, 15 green, and 9 blue.
(a) How many (unordered) selections of 4 balls can be made?
(b) How many selections are there containing 1 yellow ball and no more than 2 green balls?

17. Find the number of poker hands containing
(a) a flush (all cards of the same suit).
(b) a full house (three of one kind plus two of another kind, e.g., three jacks and two sevens).
(c) a straight (five cards in numerical sequence, regardless of suit; ace can be high but not low).

18. Find the number of poker hands containing
(a) three of a kind [exactly three cards of the same face value, e.g., three queens, but not a full house (see Problem 17(b)].
(b) two pairs (two of one kind plus two of another kind, but not a full house).
(c) one pair.

19. A group of n people are having dinner together, and a toast is proposed. If everyone must "clink glasses" with everybody else, how many "clinks" will there be?

20. In how many ways can 20 books be distributed among 4 children so that
(a) each child gets 5 books?
(b) the two oldest get 7 books and the two youngest get 3 books each?

21. Given three men M_1, M_2, M_3, seven women $W_1, \ldots, W_7$, and eight children $C_1, \ldots, C_8$, find the number of committees of size five that can be formed containing
(a) no men.
(b) M_3 and no other men.
(c) M_3 and exactly two women.
(d) M_3, W_1, and W_7.
(e) exactly one man.

22. In Problem 21, find the number of committees of size five that can be formed containing
(a) at least one man; (b) at least two men.

23. The president invites 25 senators to the White House. In how many ways can this be done so that 25 different states are represented?

24. Given four couples C_1, C_2, C_3, C_4 (four men and four women), how many committees of size three can be formed not containing a pair of spouses?

25. Five people are to be chosen from a group of 20 to take a trip.
(a) In how many ways can this be done?
(b) In how many ways can this be done if A and B refuse to travel together?
(c) In how many ways can this be done if A and B refuse to travel without each other?

26. Suppose seven people visit a certain city in which there is only one hotel with four double rooms (labelled A, B, C, D). How many ways are there of assigning people to rooms? (Each room can be empty, singly occupied, or doubly occupied.)

27. Generalize the result of Example 2.16 by solving the following problem: Suppose there are n_1 Samurai, n_2 Lords, and n_3 Ninjas. In how many ways can a team consisting of k_1 Samurai, k_2 Lords, and k_3 Ninjas be selected?

28. In Problem 2.2.26 we counted the number of subsets of a set S with n elements. Here is another method.
(a) Show that the number of subsets can be expressed as $\sum_{k=0}^{n} \binom{n}{k}$.
(b) Use the binomial theorem to show that $\sum_{k=0}^{n} \binom{n}{k} = 2^n$.

29. Let A and B be sets, with $B \subseteq A$. Suppose that $|A| = 75$, and $|B| = 5$. How many 10-element subsets of A are there which are *disjoint* from B?

30. Consider property (2.5) of binomial coefficients. Is it true for $n = 0$? That is, does the empty set have *one* subset with no elements? Explain.

31. Using Formula (2.2) for binomial coefficients, verify algebraically that the rule of Pascal's triangle (2.6) is valid.

32. The accompanying diagram is a street map of the town of O'prime–R'prime. You are supposed to find your way from point A to point B by following the streets. You are allowed to move either eastward or southward, and can change direction at any intersection. How many possible paths are there from A to B?

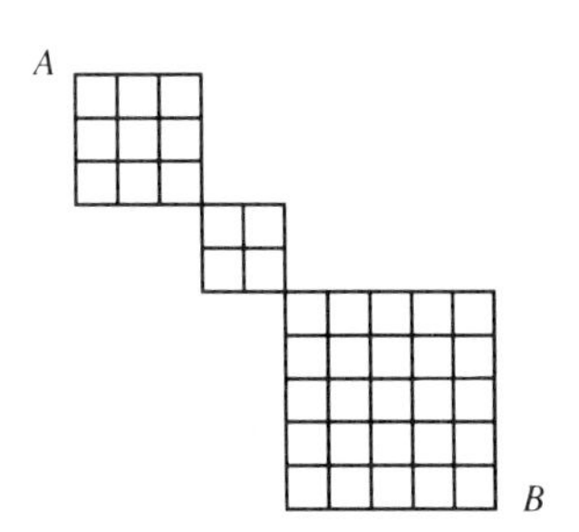

33. Refer to Problems 2.2.21 and 2.2.22. Suppose that we have n distinguishable boxes $B_1, \ldots, B_n$ and we toss k indistinguishable balls into them. Thus the result "first ball into box 3, second ball into box 8, third ball into box 5" is regarded as the same as "first ball into box 8, second ball into box 5, third ball into box 3." In other words, if we write down the sequence of boxes into which the balls are tossed, *order does not count.* If no box can contain more than one ball, show that the number of possible ways of distributing the balls is $\binom{n}{k}$.

34. In the text we sketched a combinatorial proof of the binomial theorem. In this problem, we ask you to give a proof by *induction*; that is, prove that

$$(x + y)^n = \sum_{k=0}^{n} \binom{n}{k} x^k y^{n-k} \qquad \text{for all } n \geq 1,$$

by induction. [*Hint:* $(x + y)^{n+1} = (x + y)^n(x + y)$. Now use the induction hypothesis and the rule of Pascal's triangle.]

2.4 Unordered Samples with Repetition

In this section we will consider the last and most difficult of the four problems about samples: How many unordered samples of size k can be formed from a set of n elements, if repetition is allowed? In Example 2.5(c) we found that with $n = 4$ and $k = 2$, the answer is 10. The answer in general is given by the following rule.

UNORDERED SAMPLES WITH REPETITION

If repetition is allowed, the number of unordered samples of size k from an n-element set is

$$\frac{n(n+1)\cdots(n+k-1)}{k!} = \binom{n+k-1}{k}.$$

This rule is very similar to the rule for unordered samples *without* repetition (Section 2.3). In both cases we find a product of k numbers in the numerator and $k!$ in the denominator. The difference, however, is that when repetition *isn't* allowed, the factors in the numerator start at n and *decrease* by one each time, whereas when repetition is permitted, the factors start at n and *increase.*

EXAMPLE 2.17 Calculate the number of unordered samples with $n = 6$ and $k = 3$ when (a) repetition isn't allowed and (b) when repetition is allowed.

SOLUTION

(a) According to the rule for unordered samples *without* repetition, the answer is

$$\frac{6 \cdot 5 \cdot 4}{3 \cdot 2 \cdot 1} = 20.$$

(b) According to the rule for unordered samples *with* repetition, the answer is

$$\frac{6 \cdot 7 \cdot 8}{3 \cdot 2 \cdot 1} = 56.$$

Of course it's easy to plug numbers into the given formula for unordered samples with repetition. But why is the formula true? To see why, let's consider the case $n = 4$, $k = 2$ again, from a new point of view. The problem is to choose an unordered sample of size two from the set $\{A, B, C, D\}$, with repeats allowed. Since order doesn't matter, all we need to know is how many A's, B's, C's, and D's are present, in order to specify the sample.

Now look at Figure 2.5, where we have listed the ten possibilities, in three different ways. We have listed the possibilities $AA, AB, \ldots, DD$, as before. We have also imagined a box which has been partitioned into four compartments, labelled A, B, C, and D. Into these four compartments we have placed two balls, corresponding to the particular sample being considered. For example, BD corresponds to one ball in compartment B and one ball in compartment D. Finally, we have given an abstract and simplified representation of the balls and partitions, in which a "0" represents a ball, and a "1" represents a partition wall. For example, the picture for CC reads, from left to right (ignoring the *exterior* wall of the box, which is the same for all the pictures), "partition, partition, ball, ball, partition," i.e., 11001. In this way we have set up a *one-to-one correspondence* between the samples and the sequences of 0s and 1s of length five with exactly two 0s; every sample leads to such a sequence, and every sequence leads to a sample.

Sample	A	B	C	D	Code
AA	••				00111
AB	•	•			01011
AC	•		•		01101
AD	•			•	01110
BB		••			10011
BC		•	•		10101
BD		•		•	10110
CC			••		11001
CD			•	•	11010
DD				••	11100

Figure 2.5 Counting unordered samples with repetition allowed; $0 =$ ball and $1 =$ partition wall.

The point of making this correspondence is that we *already* know how to count the number of length-five sequences of 0s and 1s with exactly two 0s. There are $\binom{5}{2}$ ways of selecting the locations of the 0s (the unordered sample without repetition rule), and $\binom{5}{2} = 10$.

The same argument will work in general. If we are counting unordered samples of size k, there will be k balls (0s) and $n - 1$ internal partition walls (1s). Each sample will be encoded into a sequence of $(n + k - 1)$ 0s and 1s, with k 0s and $(n - 1)$ 1s. The total number of such sequences is, by the rule for unordered samples without repetition, $\binom{n+k-1}{k}$, as advertised.

EXAMPLE 2.18 In each of the following three cases, find the number of unordered samples of size four (repetition allowed) from the set $\{A, B, C, D, E\}$.

(a) No further restrictions.
(b) All samples must contain *at least* two A's.
(c) All samples must contain *exactly* two A's.

SOLUTION

(a) We use the rule for unordered samples with repetition with $n = 5$ and $k = 4$. The answer is therefore

$$\binom{8}{4} = \frac{5 \cdot 6 \cdot 7 \cdot 8}{4 \cdot 3 \cdot 2 \cdot 1} = 70.$$

(b) Some possible samples containing at least two A's are $AACE$, $AABD$, $AABB$, $AAAE$, $AAAA$, etc. Thus we are asked to count samples of the form $AAXY$, where XY is an unordered sample of size two (repetition allowed) from the five-element set $\{A, B, C, D, E\}$. By our rule for unordered samples with repetition, the answer is

$$\frac{5 \cdot 6}{2 \cdot 1} = 15.$$

(c) As in part (b), these samples are of the form $AAXY$; but now neither X nor Y can be the letter A. Thus, XY is an unordered sample of size two (repetition allowed) from the four-element set $\{B, C, D, E\}$. Therefore our rule gives the answer

$$\frac{4 \cdot 5}{2 \cdot 1} = 10.$$

The next example illustrates a typical application of the rule for counting unordered samples with repetition allowed.

EXAMPLE 2.19

(a) Consider the equation $x_1 + x_2 + x_3 = 5$, where x_1, x_2, and x_3 must all be nonnegative integers. How many distinct solutions are there? (*Note*: The solutions $2 + 2 + 1 = 5$ and $1 + 2 + 2 = 5$, for example, are considered to be distinct.)

(b) More generally, how many distinct solutions are there to the equation $x_1 + x_2 + \cdots + x_n = k$, where each x_i must be a nonnegative integer?

SOLUTION

(a) The answer is 21; the following table lists all the solutions:

x_1	5	4	4	3	3	3	2	2	2	2	1	1	1	1	1	0	0	0	0	0	0
x_2	0	1	0	2	1	0	3	2	1	0	4	3	2	1	0	5	4	3	2	1	0
x_3	0	0	1	0	1	2	0	1	2	3	0	1	2	3	4	0	1	2	3	4	5

However, it is possible to see that 21 is the answer *without* having to list all the solutions, using the ideas of this section, by encoding each solution to the equation $x_1 + x_2 + x_3 = 5$ into a sequence of five x_1's, x_2's, and x_3's. For example, the solution $x_1 = 1$, $x_2 = 3$, $x_3 = 1$ is encoded into the sequence $x_1x_2x_2x_2x_3$, and the solution $x_1 = 0$, $x_2 = 1$, $x_3 = 4$ is encoded into $x_2x_3x_3x_3x_3$. Each encoded sequence is an unordered sample of size five (repetition allowed) from the three-element set $\{x_1, x_2, x_3\}$. Our rule for counting unordered samples with repetition tells us that there are

$$\frac{3 \cdot 4 \cdot 5 \cdot 6 \cdot 7}{5 \cdot 4 \cdot 3 \cdot 2 \cdot 1} = 21$$

such sequences, which is also the number of solutions to $x_1 + x_2 + x_3 = 5$.

(b) Exactly the same argument used in part (a) shows that the number of distinct solutions to the equation $x_1 + x_2 + \cdots + x_n = k$ equals the number of unordered samples of size k (repetition allowed) from the n-element set $\{x_1, x_2, \ldots, x_n\}$, which is $\binom{n+k-1}{k}$.

EXAMPLE 2.20 The Yreka Bakery sells only four kinds of cookies: chocolate chip, oatmeal, peanut butter, and raisin.

(a) How many different combinations of eight cookies can we buy?
(b) How many different combinations of eight cookies can we buy if we must have at least one chocolate chip and at least two oatmeal cookies?

SOLUTION

(a) Our purchase is an unordered sample of size eight with repetition allowed, from the four-element set {chocolate chip, oatmeal, peanut butter, raisin}. Therefore the answer is $\binom{11}{8} = 165$.
(b) Here our purchase is a list of the form $(C, O, O, *, *, *, *, *)$, where the five $*$'s are an unordered sample of size five from the four-element set $\{C, O, P, R\}$. Therefore the answer is $\binom{8}{5} = 56$.

Problems for Section 2.4

1. For $n = 5$, $k = 4$, and then for $n = 7$, $k = 8$, compute the number of samples of size k from a set of n objects, under the following conditions.
(a) order counts, repetition allowed.
(b) order counts, repetition not allowed.
(c) order doesn't count, repetition not allowed.
(d) order doesn't count, repetition allowed.

2. Repeat Problem 1 for $n = 9$, $k = 5$, and for $n = 5$, $k = 9$.

3. Consider the set $\{A, B, C, D, E, F\}$. Count the number of samples of size six (unordered, repetition allowed), with the following restrictions.
(a) no further restrictions.
(b) B occurs at least 3 times.
(c) exactly three different letters are used.

4. Repeat Problem 3 with the following restrictions:
(a) every letter occurs once.
(b) one letter occurs twice and four letters each occur once.
(c) two letters occur twice and two letters each occur once.

5. In Example 2.19, complete the given table by writing out the unordered sample with repetition corresponding to each solution.

6. In Example 2.19, find the number of solutions satisfying
(a) $x_1 \leq x_2 \leq x_3$; (b) $x_1 < x_2 < x_3$.

7. A concert promoter has 1000 unreserved grandstand seats to distribute free among alumni, students, faculty, public. For example, one possibility is to give half to the students and half to the faculty. How many possibilities are there?

8. A donut store offers 15 different kinds of donuts. In how many ways can we buy 7?

9. An executive has \$1000 in \$100 bills to distribute as bonuses among five assistants. Find how many distributions there are under the following conditions.
(a) There are no restrictions.
(b) Each assistant must get at least \$100.
(c) Each assistant must get at least \$100 and assistant A_1 must get at least \$300.

10. In how many ways can seven baseballs and six bats be distributed among four children so that each child receives at least one ball?

11. A child has a repertoire of four piano pieces P_1, P_2, P_3, P_4. Her parents expect her to practice 10 pieces each day. For example, one practice session might consist of P_3 played 7 times and P_1 played 3 times. How many possible practice sessions are there?

12. In Problem 11, how many possible sessions are there under the following conditions?

(a) She must play each piece at least twice.
(b) She must play P_3 exactly 4 times.

13. Consider buying six company cars from a dealer who carries five models M_1, M_2, M_3, M_4, M_5. For example, one possibility is to buy three M_1's, one M_2, and two M_5's.
(a) How many possible ways are there to buy the six cars?
(b) How many possibilities include exactly two models?
(c) How many possibilities include as much variety as possible?

14. Find the number of solutions of the equation $x + y + z + w = 18$ where x, y, z, and w must be nonnegative integers.

15. In Problem 14, find the number of solutions satisfying the restriction that $x \geq 1$ and $z \geq 3$.

16. A message is to consist of the 5 symbols U, V, W, X, and Y (in any order) plus 20 blank spaces (denoted by b) distributed between the symbols so that there are at least two blanks between successive symbols. For example, one possibility is

W bbb X bbbbbbb-
U bb Y bbbbbbbb V.

Find the number of possible messages.

17. An experiment with five outcomes, E_1, E_2, E_3, E_4, E_5, is repeated 19 times. A **histogram** records how many times each outcome turns up. One possibility is shown in the following diagram:

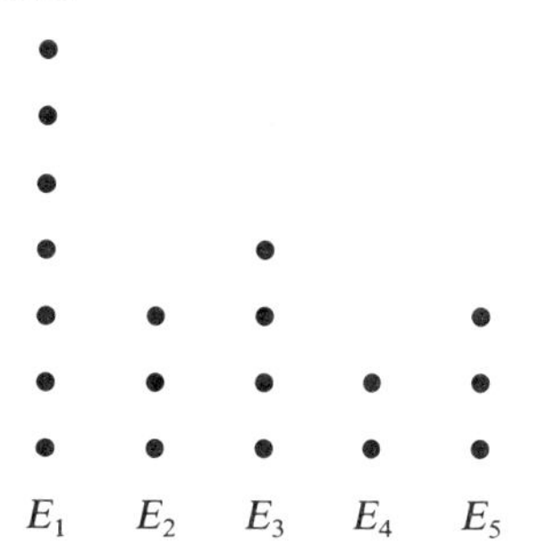

How many possible histograms are there?

18. Continuing Problem 17, find the number of histograms if an experiment with n possible outcomes is repeated r times.

19. A **multiset** is just like a set, except that it may contain a given element several times. Thus $\{a, a, b, c\}$ is a multiset. Find a formula for the number of k-element multisets that can be formed from the elements of the set $\{1, 2, 3, \ldots, n\}$.

20. Refer to Problem 2.3.33. If k indistinguishable balls are tossed into n distinguishable boxes, and each box can contain any number of balls, show that the number of possible ways of distributing the balls is $\binom{n+k-1}{k}$.

21. Find the number of ways that 15 balls can be tossed into 8 distinguishable boxes under the following conditions:
(a) The balls are indistinguishable;
(b) The balls are all different colors.

22. Find the number of ways that nine identical balls can be tossed into five distinguishable boxes under the following conditions:
(a) At least one box is to be left empty;
(b) The third box gets an even number of balls (regard 0 as an even number).

23. Suppose we are to distribute one quarter, two dimes, three nickels, and five pennies to four children. Find the number of possible ways this can be done if
(a) there is no restriction on the distribution
(b) Jeanne must get 15¢.

24. In how many ways can eight chocolate donuts and seven jelly donuts be distributed among three people if each person wants at least one donut of each kind?

25. As in Problem 3, find the number of samples of size six (unordered, repetition allowed) from the set $\{A, B, C, D, E, F\}$, with the following restrictions:
(a) Exactly four different letters are used;
(b) At most four different letters are used.

26. Repeat Problem 25 with the following restrictions:
(a) At least two different letters are used;
(b) No letter occurs more than twice;
(c) Some letter occurs at least four times.

27. A restaurant serves 10 different varieties of soup. Find the number of ways we can place 6 orders under the following conditions:
(a) All six soups are different;
(b) Exactly four soups are cream of mushroom;
(c) At least four soups are cream of mushroom.

28. In Problem 27, find the number of ways we can place the six orders so that no variety can be ordered as many as six times. (*Suggestion*: Consider the number of cases in which this condition does *not* occur.)

29. Consider the inequality (1) $x_1 + x_2 + x_3 < 14$, where the x_i are nonnegative integers; also consider the related equation (2) $x_1 + x_2 + x_3 + x_4 = 14$, where x_1, x_2, x_3 are again nonnegative integers but x_4 is a *positive* integer. Show that (1) and (2) have the same number of solutions. [Here is the basic idea: Look at a typical solution of (1), say $x_1 = 2$, $x_2 = 5$, $x_3 = 4$. Since $x_1 + x_2 + x_3 = 11$, we can get a solution of (2) by setting $x_4 = 3$; x_4 "takes up the slack" between 11 and 14, and sometimes it is referred to as a **slack variable**.]

30. Continuing Problem 29, find the number of solutions to $x_1 + x_2 + x_3 < 14$, where the x_i are nonnegative integers. (By Problem 29, we are counting solutions of $x_1 + x_2 + x_3 + x_4 = 14$ with a restriction that $x_4 \geq 1$.)

31. Continuing Problem 30, find the number of solutions to $x_1 + x_2 + x_3 \leq 14$, where the x_i are nonnegative integers.

2.5 Permutations Involving Indistinguishable Objects; Multinomial Coefficients

Consider the English word *facetiously*.* The number of ways of permuting the letters in this 11-letter word is $11! = 39916800$, as we know from the permutation rule (Section 2.2). For example, the permutation that interchanges the third and fourth letters and does nothing else produces *faectiously*.

Now consider the word *mississippi*. In how many ways can we permute these 11 letters? The answer is *not* $11!$ because the word *mississippi* contains repeated letters. For example, the permutation that interchanges the third and fourth letters, and does nothing else, would not produce a new arrangement. On the other hand, the permutation that interchanges the first two letters produces *imssissippi*, which is different. In this section we will learn how to count the number of different permutations of a list of objects in which, as in *mississippi*, some of the objects are repeated, or *indistinguishable*.

In *mississippi*, there are 4 i's, 4 s's, 2 p's, and 1 m. To count the number of different arrangements, visualize 11 slots:

$$\frac{\ }{1}\;\frac{\ }{2}\;\frac{\ }{3}\;\frac{\ }{4}\;\frac{\ }{5}\;\frac{\ }{6}\;\frac{\ }{7}\;\frac{\ }{8}\;\frac{\ }{9}\;\frac{\ }{10}\;\frac{\ }{11}.$$

We need to place the 11 letters in *mississippi* in these 11 slots. We can begin by placing the 4 i's. The locations of the i's form a 4-element subset of the 11-element set $\{1, 2, 3, \ldots, 11\}$, and so by the subset rule (Section 2.3), there are $\binom{11}{4} = 330$ ways of placing the i's. If, for example, we choose positions $\{2, 5, 9, 10\}$ for the i's, our slots will look like this:

$$\frac{\ }{1}\;\frac{i}{2}\;\frac{\ }{3}\;\frac{\ }{4}\;\frac{i}{5}\;\frac{\ }{6}\;\frac{\ }{7}\;\frac{\ }{8}\;\frac{i}{9}\;\frac{i}{10}\;\frac{\ }{11}.$$

We next place the 4 s's. There are 7 remaining slots and so $\binom{7}{4} = 35$ ways of placing the s's. If we select slots 1, 3, 6, and 11 for the s's, we get

$$\frac{s}{1}\;\frac{i}{2}\;\frac{s}{3}\;\frac{\ }{4}\;\frac{i}{5}\;\frac{s}{6}\;\frac{\ }{7}\;\frac{\ }{8}\;\frac{i}{9}\;\frac{i}{10}\;\frac{s}{11}.$$

There are 3 slots left, and so $\binom{3}{2} = 3$ ways of placing the p's. If we choose slots 7 and 8, we get

$$\frac{s}{1}\;\frac{i}{2}\;\frac{s}{3}\;\frac{\ }{4}\;\frac{i}{5}\;\frac{s}{6}\;\frac{p}{7}\;\frac{p}{8}\;\frac{i}{9}\;\frac{i}{10}\;\frac{s}{11}.$$

Now, there is only 1 slot left, and so $\binom{1}{1} = 1$ way of placing the m:

$$\frac{s}{1}\;\frac{i}{2}\;\frac{s}{3}\;\frac{m}{4}\;\frac{i}{5}\;\frac{s}{6}\;\frac{p}{7}\;\frac{p}{8}\;\frac{i}{9}\;\frac{i}{10}\;\frac{s}{11}.$$

By the multiplication rule, then, there are in all

$$\binom{11}{4}\cdot\binom{7}{4}\cdot\binom{3}{2}\cdot\binom{1}{1} = 330\cdot 35\cdot 3\cdot 1 = 34{,}650$$

different arrangements of the letters in the word *mississippi*.

* A remarkable word. It contains all five vowels, *a*, *e*, *i*, *o*, *u*, and *y*, in order! If *y* is not considered a vowel, then *facetious* will do.

However, our decision to place the i's, then the s's, then the p's, then the m, was completely arbitrary. If we had placed the letters in the order m, i, s, p, we would, using the same argument as above, conclude that there were instead

$$\binom{11}{1} \cdot \binom{10}{4} \cdot \binom{6}{4} \cdot \binom{2}{2}$$

arrangements. But luckily this number is

$$11 \cdot 210 \cdot 15 \cdot 1 = 34{,}650,$$

as before. This is not really luck at all, but rather a particular instance of a general rule. Note that

$$\binom{11}{4}\binom{7}{4}\binom{3}{2}\binom{1}{1} = \frac{11!}{4!7!} \cdot \frac{7!}{4!3!} \cdot \frac{3!}{2!1!} \cdot \frac{1!}{1!} = \frac{11!}{4!4!2!1!}$$

and

$$\binom{11}{1}\binom{10}{4}\binom{6}{4}\binom{2}{2} = \frac{11!}{1!10!} \cdot \frac{10!}{4!6!} \cdot \frac{6!}{4!2!} \cdot \frac{2!}{2!} = \frac{11!}{1!4!4!2!}$$

The general result, sometimes called the rule of *generalized permutations*, follows.

GENERALIZED PERMUTATIONS

The number of distinguishable arrangements of n objects, in which there are k_1 objects of type 1, k_2 objects of type $2, \ldots, k_m$ objects of type m (we require $k_1 + \cdots + k_m = n$), is given by

$$\frac{n!}{k_1!k_2! \cdots k_m!}$$

The number $n!/k_1!k_2! \cdots k_m!$ appearing in the generalized permutations rule is often called a **multinominal coefficient**, and is denoted by the symbol

$$\binom{n}{k_1, k_2, \ldots, k_m},$$

which is pronounced "n choose $k_1, k_2, \ldots, k_m$." For example, the number of arrangements of the letters in *mississippi* is given by

$$\binom{11}{4, 4, 2, 1}.$$

The order of the numbers 4, 4, 2, 1 doesn't matter, so we could equally well write

$$\binom{11}{1, 4, 4, 2}, \qquad \binom{11}{4, 2, 1, 4},$$

etc. (For an explanation of the term *multinomial*, see Problem 26.)

TABLE 2.2 The correspondence between subsets and binary sequences

Subset		A	B	C	D
$\{A, B\}$	$\leftrightarrow$	1	1	0	0
$\{A, C\}$	$\leftrightarrow$	1	0	1	0
$\{A, D\}$	$\leftrightarrow$	1	0	0	1
$\{B, C\}$	$\leftrightarrow$	0	1	1	0
$\{B, D\}$	$\leftrightarrow$	0	1	0	1
$\{C, D\}$	$\leftrightarrow$	0	0	1	1

It turns out that the rule for generalized permutations can be used to derive two of our previous rules, the permutation rule and the subset rule. To get the permutation rule, just take $k_1 = k_2 = \cdots = k_m = 1$, which means that there is only one element of each kind among the n elements to be analyzed, as in the word *facetiously*. Here the number of distinguishable arrangements is the same as the number of permutations, and indeed the rule for generalized permutations gives $n!/1!1! \cdots 1! = n!$ as the number of possibilities, in agreement with the permutation rule.

On the other hand, to get the rule for unordered samples without repetition from the generalized permutation rule, notice that the *binomial coefficients* introduced in Section 2.3 are also *multinomial coefficients* with $m = 2$, i.e., when there are only two distinguishable types of objects. For example, we can once again count the number of two-element subsets of the four-element set $\{A, B, C, D\}$ as follows. Let us write a sequence of length four, containing exactly two 0s and two 1s, with 0 denoting "is not an element of" and 1 denoting "is an element of," corresponding to each such subset (see Table 2.2). We see that there are exactly as many two-element subsets of $\{A, B, C, D\}$ as there are distinguishable arrangements of the symbols in the sequence 1100. According to the generalized permutation rule with $n = 4$, $k_1 = 2$, $k_2 = 2$, the number of distinguishable arrangements of 1100 is

$$\frac{4!}{2!2!} = 6$$

This number can be written either as

$$\binom{4}{2, 2} \qquad \text{(multinomial coefficient notation)}$$

or as

$$\binom{4}{2} \qquad \text{(binomial coefficient notation).}$$

The binomial coefficient notation, however, is preferable.

EXAMPLE 2.21 A coin is tossed 20 times.

(a) How many possible outcomes (i.e., sequences of heads and tails) are there?
(b) How many possible outcomes with exactly 13 heads and 7 tails?
(c) How many with exactly 5 heads in the first 10 tosses?

SOLUTION

(a) Since there are two possibilities for the first toss (H or T), two for the second, . . . , and two for the 20th, by the multiplication rule, the answer is $2 \cdot 2 \cdot \cdots \cdot 2 = 2^{20} = 1048576$.
(b) The number of possibilities in this case is equal to the number of sequences of length 20 with 13 H's and 7 T's. We can count this number in two ways, depending on our point of view. On one hand, the location of the H's is a subset of size 13 from a 20-element set, and so by the subset rule, the required number is $\binom{20}{13} = 77520$. On the other hand, the required number is also the number of distinguishable rearrangements of the sequence HHHHHHHHHHHHHTTTTTTT, which is, by the generalized permutation rule, $20!/13!7! = 77520$.
(c) Using reasoning similar to that in part (b), we find (using either the subset rule or the generalized permutation rule) that the number of ways of having exactly 5 H's in the first 10 tosses is $\binom{10}{5} = 252$. After the 10th toss there are two possibilities for the 11th toss, two for the 12th, . . . , and two for the 20th. Thus by the multiplication rule, the total number of possibilities is $252 \times 2^{10} = 258048$. ∎

EXAMPLE 2.22 (Labelled groupings.) Suppose 26 people are sent to six numbered rooms so that

4 people go to room R_1	2 people go to room R_4
4 people go to room R_2	6 people go to room R_5
4 people go to room R_3	6 people go to room R_6

If the people are named Art, Betty, Charles, . . . , Zenobia, then one possibility is given by

$$\frac{A}{3}\ \frac{B}{6}\ \frac{C}{2}\ \frac{D}{1}\ \frac{E}{5}\ \frac{F}{2}\ \frac{G}{5}\ \frac{H}{6}\ \frac{I}{1}\ \frac{J}{6}\ \frac{K}{2}\ \frac{L}{3}\ \frac{M}{3}\ \frac{N}{6}\ \frac{O}{4}\ \frac{P}{1}\ \frac{Q}{1}\ \frac{R}{5}\ \frac{S}{2}\ \frac{T}{4}\ \frac{U}{6}\ \frac{V}{5}\ \frac{W}{5}\ \frac{X}{6}\ \frac{Y}{5}\ \frac{Z}{3},$$

which corresponds to the following arrangement:

Room R_1: $\{D, I, P, Q\}$	Room R_4: $\{O, T\}$
Room R_2: $\{C, F, K, S\}$	Room R_5: $\{E, G, R, V, W, Y\}$
Room R_3: $\{A, L, M, Z\}$	Room R_6: $\{B, H, J, N, U, X\}$

How many possible arrangements are there?

SOLUTION We need to assign a room to each person so that the specified conditions are satisfied. This amounts to writing down a sequence of length 26 containing exactly 4 1s, 4 2s, 4 3s, 2 4s, 6 5s, and 6 6s. By the generalized permutation rule, the total number of possibilities is

$$\frac{26!}{4!4!4!2!6!6!} \cong 2.81 \times 10^{16}.$$

EXAMPLE 2.23 (Unlabelled groupings.) Suppose the 26 people of the previous example are instead to be divided up into 6 *teams*, 3 teams of size 4, 1 team of size 2, and 2 teams of size 6. Now how many possibilities are there?

SOLUTION At first glance this problem looks like Example 2.22. We just put the people in room R_1 on one team, the people in room R_2 on another team, etc., and apparently the answer should again be $26!/4!4!4!2!6!6!$. But this is wrong. The problem is that different *room* assignments might lead to the same *team* assignments. For example, consider the following room assignment:

Room R_1: $\{A, L, M, Z\}$ Room R_4: $\{O, T\}$

Room R_2: $\{D, I, P, Q\}$ Room R_5: $\{B, H, J, N, U, X\}$

Room R_3: $\{C, F, K, S\}$ Room R_6: $\{E, G, R, V, W, Y\}$

This is a different room assignment than the one we saw in Example 2.22, yet it leads to the same set of teams. On reflection, we can see that this is because it doesn't matter which of the rooms R_1, R_2, and R_3 we assign to the 4-member teams, and which of the rooms R_5 and R_6 we assign to the 6-member teams. In Example 2.22, order matters, but in this example, order doesn't matter. In Example 2.22 we were asked to count *labelled groups*—the room numbers labelled the groups; whereas in Example 2.23 we are asked to count *unlabelled groups.*

To solve the given problem, we can reason as follows: Let's suppose we are given a team assignment and asked to make a room assignment. How many room assignments are possible? The 4-member teams must be put in rooms R_1, R_2, and R_3; and by the permutation rule, this can be done in $3! = 6$ ways. The 2-member team must go in room R_4, so there's only one possibility for that. Finally, the 6-member teams must be put in rooms R_5 and R_6; again by the permutation rule, there are $2! = 2$ ways to do this. Thus by the multiplication rule, there are $6 \times 1 \times 2 = 12$ ways to make a room assignment once a team assignment is given. Therefore, if N denotes the number of *team* assignments, the total number of *room* assignments is (by the multiplication rule again!) $N \times 12$. But we already know how many room assignments there are, from Example 2.22. In summary,

$$N \cdot 3! \cdot 2! = \frac{26!}{4!4!4!2!6!6!},$$

$$N = \frac{26!}{4!4!4!2!6!6!3!2!} \cong 2.34 \times 10^{15}.$$

[Notice the pattern in the denominator of N: There are 3 teams of one size (4), 2 teams of another size (6), and 1 team of a third size (2); and a "correction factor" of $3!2!1!$. If no two groups are the same size, this correction factor will be $1!1!\cdots 1! = 1$, and the number of labelled and unlabelled groupings will be the same. See Problem 23.]

To solve Example 2.23, we needed to use the multiplication rule, the permutation rule, and the generalized permutation rule. We can solve some pretty hard problems now!

Problems for Section 2.5

1. For each of the following words, calculate the number of distinguishable arrangements of the letters:
(a) *appletree* (b) *banana.*

2. Repeat Problem 1 for the following words:
(a) *candy* (b) *daddy* (c) *eek.*

3. Refer to Table 2.2, and write down the correspondence between *all* subsets of $\{A, B, C, D\}$ and binary sequences of length four.

4. Exhibit the correspondence between all subsets of $\{A, B, C\}$ and binary sequences of length three.

5. In how many distinguishable ways can the letters A, A, A, B, B, C, C, C, C, C, C, D, D, D, D, D, D, D, E, F be arranged?

6. In how many ways can the complete works of Shakespeare in 10 volumes and 6 identical copies of *War and Peace* be arranged on a shelf?

7. A coin is tossed 50 times.
(a) How many possible outcomes are there?
(b) How many of these have exactly 20 heads and 30 tails?

8. Refer to Problem 7.
(a) How many outcomes have 30 heads and 20 tails?
(b) How many begin with a sequence of 7 heads?

9. Suppose 20 people are divided into 6 (numbered) committees so that 3 people each serve on committees C_1 and C_2, 4 people each on committees C_3 and C_4, 2 people on committee C_5, and 4 people on committee C_6. How many possible arrangements are there?

10. If the 20 people in Problem 9 are divided into 4 committees C_1, C_2, C_3, C_4, each consisting of 5 people, how many arrangements are there?

11. If the people in Problem 9 are divided into *teams* of size 3, 3, 4, 4, 2, 4, find the number of possibilities.

12. If the people in Problem 9 are divided into 4 teams of 5 members each, find the number of possibilities.

13. For each of the words in Problem 1, count the number of permutations of the letters that leave the word unchanged. (For example, the permutation $\binom{123456789}{132486759}$ doesn't change *appletree.*)

14. Repeat Problem 13 for the words in Problem 2.

15. (a) Find the number of permutations of the letters in *sociological.*
(b) Find the number of permutations that have a and g adjacent.
(c) Find the number of permutations that have all the vowels adjacent.

16. Refer to Problem 15.
(a) Find the number of permutations with the vowels adjacent and in alphabetical order.
(b) Find the number of permutations with the vowels in alphabetical order, but not necessarily adjacent as in (a).

17. Consider 4-letter words containing two pairs, e.g., *ADAD, QQSS, DAAD.*
(a) What is wrong with the following attempt to count them? Pick two letters of a kind, e.g., pick two Z's; this can be done in 26 ways. Pick two more of another kind, e.g., 2 D's; this can be done in 25 ways. Now we have four letters, two of one kind and two of another kind, e.g., *ZZDD.* The letters can be permuted in $4!/2!2!$ ways. The answer is $26 \times 25 \times (4!/2!2!)$.
(b) Count the words correctly.

18. Consider permutations of four checks, four crosses, and three circles. One possibility is

$$\surd\surd\surd\times\bigcirc\times\times\surd\bigcirc\times\bigcirc.$$

(a) How many permutations are there?

(b) How many of the permutations in (a) have a cross in the middle?

19. A single die is rolled 30 times, and the results are recorded in order. Let n_i denote the number of times the number i appeared in the 30 trials. Suppose it is known that $n_1 = 7$, $n_2 = n_3 = 4$, $n_4 = 8$, $n_5 = 2$, $n_6 = 5$. How many possibilities are there for the sequence of outcomes?

20. (a) In how many ways can eight policemen be divided into four teams of two and sent to patrol areas A, B, C, D?

(b) In how many ways can eight policemen be divided into four teams of two?

21. Suppose 21 people show up at the chess club to play. There will be 10 matches to take place at numbered tables $1, 2, \ldots, 10$, and one bye. In how many ways can the matches and bye be arranged?

22. In Problem 21, suppose that the tables are not numbered, so that, for example, A vs. B at table 1 and C vs. D at table 2 is regarded as the same as C vs. D at table 1 and A vs. B at table 2. Now how many arrangements are there?

23. Suppose 26 people are divided into groups of size 8, 7, 6, and 5.

(a) If these groups are assigned to rooms R_1, R_2, R_3, R_4, respectively, how many arrangements are possible?

(b) How many arrangements are possible if it doesn't matter which group goes into which room?

24. A library is moving 90 books to a new location. There are 10 large cartons available, each of which can hold 7 books, and there are 3 small cartons, each of which holds 6 books. Thus, the 90 books must be divided into 13 groups for the cartons, plus 2 leftovers to be carried by hand. In how many ways can the books be divided?

25. Suppose 15 people are going to divide up so as to play a basketball game, with 5 left over to watch. In how many ways can the two teams and the watchers be chosen?

26. Recall from Section 2.3 that an expression of the form $x_1 + x_2$ is a *binomial*; a **multinomial** is an expression of the form $x_1 + x_2 + \cdots + x_n$ with $n \geq 3$. Just as binomial coefficients appear in the expansion of powers of a binomial, multinomial coefficients appear when a power of a multinomial is expanded. We'll give a sequence of exercises to indicate how this works. First, write out the expansion of $(x_1 + x_2 + x_3)^4$ and show that

$$\begin{aligned}(x_1 + x_2 + x_3)^4 = {} & x_1^4 + x_2^4 + x_3^4 + 4x_1^3x_2 + 4x_1^3x_3 + 4x_1x_2^3 \\ & + 4x_2^3x_3 + 4x_1x_3^3 + 4x_2x_3^3 \\ & + 6x_1^2x_2^2 + 6x_1^2x_3^2 + 6x_2^2x_3^2 + 12x_1^2x_2x_3 \\ & + 12x_1x_2^2x_3 + 12x_1x_2x_3^2.\end{aligned}$$

27. Suppose that in Problem 26 we want to find the coefficient of $x_1^2x_2x_3$ without writing out the expansion. Notice that

$$\begin{aligned}(x_1 + x_2 + x_3)^4 = {} & (x_1 + x_2 + x_3) \\ & \times (x_1 + x_2 + x_3) \\ & \times (x_1 + x_2 + x_3) \\ & \times (x_1 + x_2 + x_3).\end{aligned}$$

Refer to the derivation of the binomial theorem in Section 2.3, and show that the number of terms of the form $x_1^2x_2x_3$ is the number of distinguishable arrangements of $x_1x_1x_2x_3$, namely, $4!/2!1!1! = 12$.

28. Find the coefficient of $x_1^4x_2^4x_3^2x_4$ in the expansion of $(x_1 + x_2 + x_3 + x_4)^{11}$.

29. In Problem 26, notice that there are 15 terms in the expansion of $(x_1 + x_2 + x_3)^4$. By setting up a correspondence between terms in the expansion and unordered samples of size 4 with repetition, using the symbols x_1, x_2, x_3, obtain the number 15 by using the results of Section 2.4.

30. Find the number of terms in the expansion of $(x_1 + \cdots + x_n)^k$.

31. State and prove a "multinomial theorem" for the expansion of $(x_1 + \cdots + x_n)^k$. Your result should reduce to the binomial theorem when $n = 2$.

32. Let $X = \{1, 2, 3, 4, 5\}$, and consider partitions of X into one one-element set and two two-element sets [so-called "partitions of type (1, 2, 2)"]. There are 15 such partitions:

$$\begin{gathered}\{\{1\},\{2,3\},\{4,5\}\}, \{\{1\},\{2,4\},\{3,5\}\}, \{\{1\},\{2,5\},\{3,4\}\}, \\ \{\{2\},\{1,3\},\{4,5\}\}, \{\{2\},\{1,4\},\{3,5\}\}, \{\{2\},\{1,5\},\{3,4\}\}, \\ \{\{3\},\{1,2\},\{4,5\}\}, \{\{3\},\{1,4\},\{2,5\}\}, \{\{3\},\{1,5\},\{2,4\}\}, \\ \{\{4\},\{1,2\},\{3,5\}\}, \{\{4\},\{1,3\},\{2,5\}\}, \{\{4\},\{1,5\},\{2,3\}\}, \\ \{\{5\},\{1,2\},\{3,4\}\}, \{\{5\},\{1,3\},\{2,4\}\}, \{\{5\},\{1,4\},\{2,3\}\}.\end{gathered}$$

Obtain the number of partitions of type (1, 2, 2) by identifying each partition with an unlabelled grouping of five people into two groups of size two and one group of size one.

33. Let $X = \{1, 2, \ldots, n\}$ and let $k_1, k_2, \ldots, k_m$ be positive integers such that $k_1 + k_2 + \cdots + k_m = n$. A **partition of** X **of type** $(k_1, k_2, \ldots, k_m)$ is a decomposition of X into pairwise disjoint subsets of sizes $k_1, k_2, \ldots, k_m$. Find a general formula for the number of partitions of type $(k_1, k_2, \ldots, k_m)$.

2.6 The Principle of Inclusion and Exclusion

We begin immediately with an example, which illustrates the kind of counting problem we will study in this section.

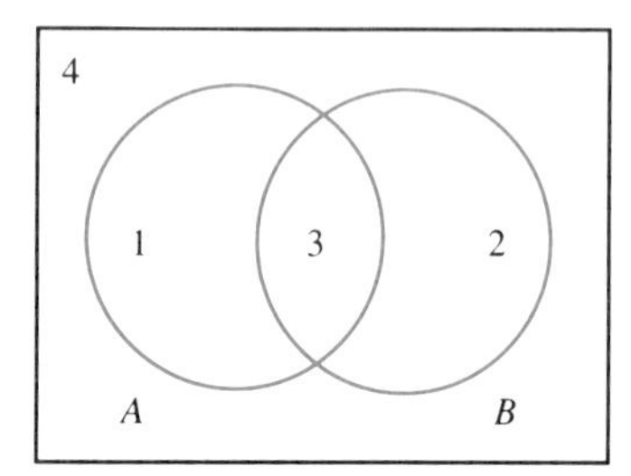

Figure 2.6 A Venn diagram for Example 2.24.

EXAMPLE 2.24 There are 100 people in a certain room. In this group, 60 are men, 30 are young, and 10 are young men. How many are old women?

SOLUTION You don't really need advanced mathematics to solve this problem, but let's proceed in a systematic way that will allow us to solve any problem of this same general type, with no extra effort.

Let's denote the set of all the people in the room by S. The set S is the universe for this problem. Denote the set of *men* by A, and the set of *young people* by B. We can draw a simple Venn diagram to represent the situation; see Figure 2.6.

In Figure 2.6, the two circles divide the universe S into four disjoint regions, labelled 1, 2, 3, and 4:

$$\begin{aligned} \text{region } 1 &= A \cap B' && \text{(old men)} \\ \text{region } 2 &= A' \cap B && \text{(young women)} \\ \text{region } 3 &= A \cap B && \text{(young men)} \\ \text{region } 4 &= A' \cap B' && \text{(old women).} \end{aligned}$$

In the given problem, we're asked to count the people in region 4. The best way to do this is to first count the people in the *complementary* region, which consists of regions 1, 2, and 3. Together, these three regions form $A \cup B$, and to count the people in $A \cup B$, it is tempting to say

$$|A \cup B| \stackrel{?}{=} |A| + |B|,$$

and conclude that $|A \cup B| = 60 + 30 = 90$. But this is wrong, because when we add $|A|$ (regions 1 and 3) to $|B|$ (regions 2 and 3), we count region 3 ($A \cap B$) *twice*. To correct things we must *subtract* the number of things in region 3, which is $|A \cap B|$. So the correct formula for $|A \cup B|$ is

(2.7)
$$|A \cup B| = |A| + |B| - |A \cap B|.$$

Now we can solve Example 2.24. By Equation (2.7), the number of people who are *not* old women is $60 + 30 - 10 = 80$. Since the total number of people is 100, there must be exactly 20 old women.

The specific numbers in Example 2.24 aren't of any importance; but the rule given in Equation (2.7) for computing $|A \cup B|$ is one of the most important tools of combinatorics. Its derivation can be summarized as follows: The term $|A|$ *includes* regions 1 and 3 of Figure 2.6; the term $|B|$ *includes* regions 2 and 3; and the term $-|A \cap B|$ *excludes* region 3. The net result is that each of regions 1, 2, and 3 is counted just *once*, which is exactly what is needed to compute $|A \cup B|$.

For this reason, Equation (2.7) is called the **principle of inclusion and exclusion,** or *PIE* for short.

PIE: PRINCIPLE OF INCLUSION AND EXCLUSION (FIRST FORM)

$$|A \cup B| = |A| + |B| - |A \cap B|.$$

We also saw in Example 2.24 that it is a simple matter to compute $|A' \cap B'|$, once $|A \cup B|$ is known:

$$|A' \cap B'| = |S| - |A \cup B|.$$

Combining this fact with PIE, we arrive at another popular formulation of the principle of inclusion and exclusion.

PIE′: PRINCIPLE OF INCLUSION AND EXCLUSION (SECOND FORM)

$$|A' \cap B'| = |S| - |A| - |B| + |A \cap B|.$$

One useful way to interpret the PIE′ Formula is as follows: Suppose the objects in the universe S can have either one (or both) of two *properties*, say *property 1* and *property 2*. If A denotes the set of objects with property 1, and B denotes the set of objects with property 2, PIE′ says that

the number of elements with *neither* property
= (the total number of objects)
− (the number of elements with *property 1*)
− (the number of elements with *property 2*)
+ (the number of elements with *both* properties).

Returning to Example 2.24, let property 1 = "maleness" and property 2 = "youth." PIE′ leads to a one-line solution:

$$\text{the number of old women} = 100 - 60 - 30 + 10 = 20.$$

We now consider a slightly harder problem of the same general kind.

EXAMPLE 2.25 Among the 100 people described in Example 2.24, it is known further that 40 are Republicans, 20 are Republican men, 15 are young Republicans, and 5 are young Republican men. Assuming each person is either a Republican or a Democrat, how many are old Democratic women?

SOLUTION 1 TO EXAMPLE 2.25

This is another job for the principle of inclusion and exclusion, only this time there are *three* properties present (maleness, youth, Republicanism), so our previous formulas won't work. What is required is a formula for $|A \cup B \cup C|$ in terms of $|A|$, $|B|$, $|C|$, $|A \cap B|$, etc. To derive such a formula, consider a three-set Venn diagram (see Figure 2.7).

We begin by estimating $|A \cup B \cup C|$ as follows:

$$|A \cup B \cup C| \stackrel{?}{=} |A| + |B| + |C|.$$

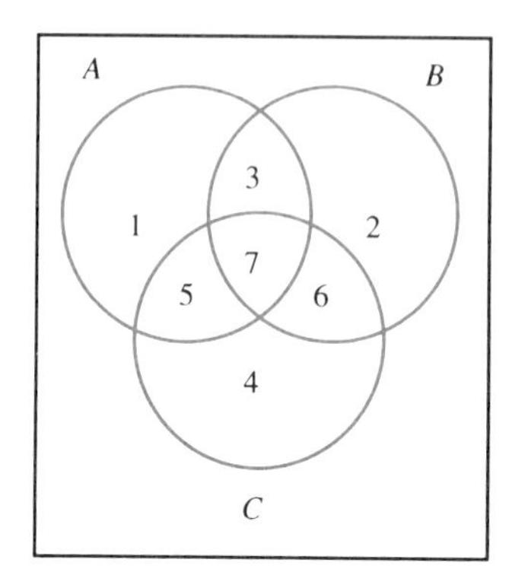

Figure 2.7 A Venn diagram for Example 2.25.

But as before, this is wrong. In fact, the formula $|A| + |B| + |C|$ *includes* each of the terms $|A \cap B|$, $|A \cap C|$, and $|B \cap C|$ twice; we can correct this by *excluding* these terms:

$$|A \cup B \cup C| \stackrel{?}{=} |A| + |B| + |C| - |A \cap B| - |A \cap C| - |B \cap C|.$$

But this is still not correct! The problem is that now region 7 in Figure 2.7, viz., $A \cap B \cap C$, has been included three times and *excluded* three times, so it's not counted at all! The remedy is to *include* it again, as follows:

$$|A \cup B \cup C| = |A| + |B| + |C| - |A \cap B| - |A \cap C| - |B \cap C| + |A \cap B \cap C|. \quad \textbf{(2.8)}$$

And this formula does correctly calculate $|A \cup B \cup C|$. We can now solve the given problem.

Let A = the men, B = the young people, and C = the Republicans. Then applying (2.8) we get for the number of people who are men *or* young *or* Republican $60 + 30 + 40 - 10 - 20 - 15 + 5 = 90$. Since there are 100 people in all, we find that the number of old Democratic women is 10. ■

The formula we found above (Equation 2.8) is also called the principle of inclusion and exclusion; but to distinguish it from PIE, we will call it PIE_3, to indicate that three sets are involved. Corresponding to PIE_3, we also have PIE'_3:

$$|A' \cap B' \cap C'| = |S| - |A| - |B| - |C| + |A \cap B| + |A \cap C| + |B \cap C| - |A \cap B \cap C|. \quad \textbf{(2.9)}$$

You will not be surprised to learn that there is a generalization of PIE and PIE′ that works for an *arbitrary* number of sets. Here's how it goes: If $A_1, A_2, \ldots, A_n$ are n subsets of a universal set S, then the generalized principle of inclusion and exclusion is a formula that lets us calculate $|A_1 \cup A_2 \cup \cdots \cup A_n|$ in terms of the numbers $|A_1|, |A_2|, \ldots, |A_n|, |A_1 \cap A_2|, |A_1 \cap A_2|, \ldots, |A_1 \cap A_2 \cdots \cap A_n|$. Here it is.

PIE_n: THE GENERALIZED PRINCIPLE OF INCLUSION AND EXCLUSION (FIRST FORM)

$$\begin{aligned}|A_1 \cup A_2 \cup \cdots \cup A_n| = {} & |A_1| + |A_2| + \cdots + |A_n| \\ & - |A_1 \cap A_2| - |A_1 \cap A_3| - \cdots - |A_{n-1} \cap A_n| \\ & + |A_1 \cap A_2 \cap A_3| + \cdots + |A_{n-2} \cap A_{n-1} \cap A_n| \\ & \vdots \\ & + (-1)^{n-1}|A_1 \cap A_2 \cap \cdots \cap A_n|.\end{aligned}$$

The formula for PIE_n really isn't as horrible as it looks. It just says that in order to compute $|A_1 \cup A_2 \cup \cdots \cup A_n|$ you need to compute the number of elements in all possible intersections of collections of the sets A_i, and add and subtract these numbers. If an *odd* number of sets are involved, you add; if an *even* number of subsets are involved, you subtract. Of course, it isn't exactly obvious why this should be so, and we will shortly give a proof. But before we do, let's state the alternative form of PIE_n, viz., PIE'_n.

PIE'_n: THE GENERALIZED PRINCIPLE OF INCLUSION AND EXCLUSION (SECOND FORM)

$$\begin{aligned}|A'_1 \cap A'_2 \cap \cdots \cap A'_n| = |S| &- |A_1| - |A_2| - \cdots - |A_n| \\ &+ |A_1 \cap A_2| + |A_1 \cap A_3| + \cdots + |A_{n-1} \cap A_n| \\ &- |A_1 \cap A_2 \cap A_3| + \cdots + |A_{n-2} \cap A_{n-1} \cap A_n| \\ &\vdots \\ &+ (-1)^n |A_1 \cap A_2 \cap \cdots \cap A_n|.\end{aligned}$$

As with PIE_n, the formula for PIE'_n isn't as bad as it looks. In PIE'_n, the number of elements in $A'_1 \cap A'_2 \cap \cdots \cap A'_n$ is obtained by starting with the number of elements in the universe, viz. $|S|$, and then adding and subtracting the number of elements in all possible intersections of combinations of the sets $A_1, A_2, \ldots, A_n$. In PIE'_n, however, the signs are reversed: If the number of sets in the intersection is *odd*, the term is *subtracted*, and if the number of sets is *even*, the term is *added*.

For example, PIE'_3 is the formula (compare with Equation 2.9)

$$\begin{aligned}|A'_1 \cap A'_2 \cap A'_3| = |S| &- |A_1| - |A_2| - |A_3| \\ &+ |A_1 \cap A_2| + |A_1 \cap A_3| + |A_2 \cap A_3| - |A_1 \cap A_2 \cap A_3|.\end{aligned} \tag{2.10}$$

As we have already seen, PIE_n and PIE'_n are very closely related. In fact, by DeMorgan's law we have

$$A'_1 \cap A'_2 \cap \cdots \cap A'_n = (A_1 \cup \cdots \cup A_n)' = S - (A_1 \cup \cdots \cup A_n),$$

and so by the simple result $|B'| = |S| - |B|$, PIE'_n follows from PIE_n. Notice that PIE'_3 is exactly what we needed to solve Example 2.25, as the following solution indicates.

SOLUTION 2 FOR EXAMPLE 2.25

Let property 1 = "maleness," property 2 = "youth," property 3 = "Republicanism." Then the number of old women Democrats is, by PIE'_3, i.e., Equation (2.10),

$$100 - 60 - 30 - 40 + 10 + 20 + 15 - 5 = 10.$$

The time has now come for us to consider the proof of PIE_n. We already know that PIE_2 is true; it is Equation (2.7). It is possible to prove it for *all* $n \geq 3$, using the method of mathematical induction of Section 1.5.

The strategy is to *assume* that PIE_n has been proved for some value of n and to use this assumption to prove PIE_{n+1}. Then the principle of mathematical induction guarantees that PIE_n is true for all $n \geq 2$. So we assume that PIE_n is true, and consider the $n + 1$ sets $A_1, A_2, \ldots, A_{n+1}$. The idea of the proof is to apply PIE_n to the n sets

$$A_1, A_2, \ldots, A_{n-1}, (A_n \cup A_{n+1})$$

and then "expand" the terms involving $(A_n \cup A_{n+1})$, using the distributive law and PIE_2.

Rather than actually going through the general induction step, viz. $\mathrm{PIE}_n \to \mathrm{PIE}_{n+1}$, we will just do the case $\mathrm{PIE}_2 \to \mathrm{PIE}_3$, which contains all the important ideas of the general proof, without suffering from the notational problems of the general case.

Thus to prove PIE_3, i.e. Equation (2.8), we begin by observing that $A \cup B \cup C = A \cup (B \cup C)$ and applying PIE_2, i.e. Equation (2.7) to the two sets A and $B \cup C$. The result is as follows:

(2.11) $$|A \cup B \cup C| = |A \cup (B \cup C)| = |A| + |B \cup C| - |A \cap (B \cup C)|.$$

The next step is to expand the term $|B \cup C|$ that appears in (2.11). Applying PIE_2, we get

(2.12) $$|B \cup C| = |B| + |C| - |B \cap C|.$$

Next, we expand the term $|A \cap (B \cup C)|$ that appears in (2.11). By the distributive law (refer to Section 1.2 if you need to review the distributive law), we have

(2.13) $$A \cap (B \cup C) = (A \cap B) \cup (A \cap C).$$

Now expand the term $(A \cap B) \cup (A \cap C)$ which appears in (2.13) by applying PIE_2 to the two sets $(A \cap B)$ and $(A \cap C)$. The result is

(2.14) $$|A \cap (B \cup C)| = |A \cap B| + |A \cap C| - |A \cap B \cap C|.$$

[We have used the fact that $(A \cap B) \cap (A \cap C) = A \cap B \cap C$.] Finally, return to (2.11) and replace the term $|B \cup C|$ using Equation (2.12), and the term $|A \cap (B \cup C)|$ using Equation (2.14). After a little rearranging, we obtain in this way

$$|A \cup B \cup C| = |A| + |B| + |C| - |A \cap B| - |A \cap C| - |B \cap C| + |A \cap B \cap C|,$$

which is the same as Equation (2.8), i.e., PIE_3.

The next example illustrates one possible use of PIE, and shows again that there are usually several ways to solve a given problem.

EXAMPLE 2.26 Suppose 100 people, including the 4-member Smith family, buy lottery tickets (one apiece). Three winning tickets will be drawn (without replacement) from a fishbowl, so there are $\binom{100}{3}$ possible outcomes. How many outcomes will make the Smith family happy?

SOLUTION 1 If A_i is the set of outcomes for which family member i wins, then $A_1 \cup A_2 \cup A_3 \cup A_4$ is the set of outcomes that make the Smith family happy (at least one of the Smiths wins). To find $|A_1 \cup A_2 \cup A_3 \cup A_4|$, we can use PIE_4, which in principle requires computing the 15 numbers $|A_1|, |A_2|, \ldots, |A_1 \cap A_2 \cap A_3 \cap A_4|$. But it's not really as hard as all that, since many of these numbers turn out to be the same. For example, to compute $|A_1|$, $|A_2|$, $|A_3|$, or $|A_4|$, note that if family member i has a winning ticket, there are $\binom{99}{2}$ ways to choose the other 2 tickets. Hence

$$|A_i| = \binom{99}{2} = 4851 \qquad \text{for} \qquad i = 1, 2, 3, 4.$$

Similarly, to compute any one of the $\binom{4}{2} = 6$ terms of the form $|A_i \cap A_j|$, note that if both i and j are winners, there are $\binom{98}{1} = 98$ ways to choose the other ticket. Hence

$$|A_i \cap A_j| = \binom{98}{1} = 98 \qquad \text{for all} \qquad i \neq j.$$

To compute the $\binom{4}{3} = 4$ terms of the form $|A_i \cap A_j \cap A_k|$, i.e., the number of ways all 3 winners are Smiths, simply note that since there are only 3 winning tickets, for a given set of 3 Smiths, there is only one way for this to happen. Hence,

$$|A_i \cap A_j \cap A_k| = 1 \qquad \text{for all distinct } i, j, k.$$

Finally, since there is no way for all 4 Smiths to win if there are only 3 winners, we have

$$|A_1 \cap A_2 \cap A_3 \cap A_4| = 0.$$

Putting all these facts together using PIE_4, we get

$$\begin{aligned} |A_1 \cup A_2 \cup A_3 \cup A_4| &= 4 \cdot \binom{99}{2} - \binom{4}{2} \cdot \binom{98}{1} + \binom{4}{3} \cdot 1 \\ &= 19404 - 588 + 4 = 18820. \end{aligned}$$

■

SOLUTION 2 In this solution we will use PIE_3. Let B_i represent the set of outcomes in which *exactly* i Smiths win. Then $B_1 \cup B_2 \cup B_3$ is the set of outcomes in which at least 1 Smith wins. But here the application of PIE_3 is especially easy, because the 3 sets B_1, B_2, and B_3 are pairwise disjoint! For example, $B_1 \cap B_2$ represents the set of outcomes in which exactly 1 Smith wins and exactly 2 Smiths win. Plainly, this is the empty set! The only terms in the PIE_3 formula for $|B_1 \cup B_2 \cup B_3|$ which are not empty, in fact, are $|B_1|$, $|B_2|$, and $|B_3|$. Hence PIE_3 in this case simplifies to

$$|B_1 \cup B_2 \cup B_3| = |B_1| + |B_2| + |B_3|.$$

It remains to calculate these three numbers. To calculate $|B_1|$, for example, note that for exactly 1 Smith to win, we must choose 1 of the Smiths and 2 non-Smiths. This can be done in $\binom{4}{1} \cdot \binom{96}{2}$ ways (use the subset rule twice, and the multiplication rule.) Thus

$$|B_1| = \binom{4}{1} \cdot \binom{96}{2} = 18240.$$

Similarly,

$$|B_2| = \binom{4}{2} \cdot \binom{96}{1} = 576.$$

Finally,

$$|B_3| = \binom{4}{3} \cdot \binom{96}{0} = 4.$$

Therefore, by PIE_3, the number of outcomes in which the Smiths win is 18240 + 576 + 4 = 18820, as before. ■

SOLUTION 3 The number of outcomes N in which at least 1 Smith wins is the total number of outcomes minus the number in which no Smiths win, i.e., all 3 winners are non-Smiths. Thus,

$$N = \binom{100}{3} - \binom{96}{3} = 161700 - 142880 = 18820.$$

Pretty easy if you do it this way. ■

The last example in this section, which is also the last example in Chapter 2, gives a much more serious application of our rules of counting.

EXAMPLE 2.27 Let X and Y be sets with $|X| = n$ and $|Y| = m$. Specifically, suppose that $X = \{1, 2, \ldots, n\}$ and $Y = \{1, 2, \ldots, m\}$. There are, by Example 2.6, m^n functions mapping X to Y. How many of these functions are *onto*?

SOLUTION This is a difficult question. But if we proceed in small steps, we can make steady progress toward a complete solution.

In hopeful anticipation of final success, let us denote by **onto**(n, m) the total number of onto functions from an n-element set to an m-element set. First, recall from Section 1.3 that unless $n \geq m$, there are no onto functions at all; the image of X won't be big enough to completely cover Y. Thus our first step towards a solution is the observation that

$$\textbf{onto}(n, m) = 0 \qquad \text{if } n < m.$$

Next, recall that if $|X| = |Y| = n$, then a function is onto if and only if it is one-to-one (see Figure 1.13). Such a function amounts to nothing more than a permutation of the set Y, and so by the permutation rule, there are $n!$ such functions. Thus we have another partial result:

$$\textbf{onto}(n, n) = n! \qquad \text{for all } n \geq 1.$$

Here's one more easy case: $m = 1$. In this case there is only one possible function from X to Y, viz. $f(x) = 1$ for all $x \in X$, and it is onto. Using our **onto**(n, m) notation this simple fact becomes

$$\textbf{onto}(n, 2) = 2^n - 2 \qquad \text{for all } n \geq 1.$$

Next, let's examine the case $m = 2$, i.e., the case where $Y = \{1, 2\}$. There are a total of 2^n functions mapping $\{1, 2, \ldots, n\}$ to $\{1, 2\}$, and almost all of them are onto. In fact, the only ones that are *not* onto are the functions which are either always equal to 1 or always equal to 2, i.e., the two functions $f_1(x) = 1$, for all $x \in X$, and $f_2(x) = 2$, for all $x \in X$. So here is another fact about **onto**(n, m):

$$\textbf{onto}(n, 2) = 2^n - 2 \qquad \text{for all } n \geq 1.$$

The next special case to consider is $m = 3$, i.e., $Y = \{1, 2, 3\}$. There are a total of 3^n functions mapping X to Y in this case, and again most of them are onto, although now it is a bit harder to count them. It turns out that we can use PIE'_3 to advantage here, as follows.

Let us denote the set of functions which do not take on the value 1 by A_1, those that do not take on the value 2 by A_2, and those that do not take on the value 3 by A_3. Then the set $A'_1 \cap A'_2 \cap A'_3$ is the set of functions that *do* take on the values 1, 2, and 3, i.e., the onto functions. Thus PIE'_3, which is stated explicitly in Equation (2.10), is just what we need. We have

$$|S| = 3^n,$$

since for this problem the universe S is the set of all functions mapping X to Y. Also,

$$|A_1| = |A_2| = |A_3| = 2^n,$$

since, for example, A_1 is the set of all functions mapping X to the two-element set $\{2, 3\}$. Also,

$$|A_1 \cap A_2| = |A_1 \cap A_3| = |A_2 \cap A_3| = 1,$$

since, for example, $A_2 \cap A_3$ is the set of all functions mapping X to the one-element set $\{1\}$. Finally,

$$|A_1 \cap A_2 \cap A_3| = 0,$$

since there are no functions at all in $A_1 \cap A_2 \cap A_3$. Thus from PIE'_3, i.e., Equation (2.10), we have

$$\textbf{onto}(n, 3) = 3^n - 3 \cdot 2^n + 3 \qquad \text{for all } n \geq 1.$$

After considering all these special cases, we are now ready to tackle the general case of **onto**(n, m). If $Y = \{1, 2, \ldots, m\}$, let us denote by A_i the set of functions mapping X to Y which do not take on the value i. Then, generalizing the observation we made when we solved the $m = 3$ case, we see that

$$\textbf{onto}(n, m) = |A'_1 \cap A'_2 \cap \cdots \cap A'_m|,$$

and we can therefore use PIE'_m to find **onto**(n, m). The term $|S|$ is equal to m^n, the total number of functions from X to Y. The m terms $|A_1|, |A_2|, \ldots, |A_m|$ are all equal to $(m - 1)^n$, since each of them counts the number of functions from X to an $(m - 1)$-element set, i.e., $Y - \{i\}$. There are $\binom{m}{2}$ terms of the form $|A_i \cap A_j|$, and each has the value $(m - 2)^n$, since each of them counts the number of functions from X to an $(m - 2)$-element set, i.e., $Y - \{i, j\}$. Continuing this reasoning, we arrive at the following formula for **onto**(n, m);

$$\textbf{onto}(n, m) = m^n - m(m - 1)^n + \binom{m}{2}(m - 2)^n - \binom{m}{3}(m - 3)^n + \cdots.$$

Alternatively, if we use the summation notation introduced in Section 1.6, we have the impressive formula

$$\textbf{onto}(n, m) = \sum_{k=0}^{m-1} (-1)^k \binom{m}{k} (m-k)^n. \tag{2.15}$$

[Incidentally, if **onto**(n, m) is divided by $m!$, the resulting integer is called a **Stirling number of the second kind** and is denoted by $\{ {n \atop m} \}$. These numbers occur in the problem of counting partitions of an n-element set into m pairwise disjoint subsets. Some properties of these numbers are considered in Problems 28–31.]

Problems for Section 2.6

1. In Example 2.24, how many people are old *or* women?

2. In a meeting of 50 scientists and poets, 35 are scientists, 30 have short hair, and 25 are scientists with short hair. How many long-haired poets are there?

3. Out of 100 cups of coffee, 25 are too hot, 35 are too cold, 45 are too bitter, 15 are too hot and too bitter, and 5 are too cold and too bitter. How many are just right?

4. Continuing Problem 3, suppose we have 3 sets H, C, and B, all subsets of a universe S. Assume that

$$|S| = 100; \quad |H| = 25, \quad |C| = 35, \quad |B| = 45;$$
$$|H \cap B| = 15, \quad |C \cap B| = 5.$$

From this data, is it possible to find $|H' \cap C' \cap B'|$? Explain.

5. Generalize Example 2.26 by considering a lottery with n tickets, k of which are winning, and an m-member Smith family. Find the number of outcomes that will make the Smith family happy, using the method of Solution 2 in the text.

6. Solve Problem 5, using the method of solution 3 to Example 2.26 in the text.

7. Compute **onto**$(6, 5)$.

8. (a) Verify that Formula (2.15) gives the correct result **onto**$(n, n) = n!$ for $n = 2, 3, 4$.
(b) Use Formula (2.15) to give an explicit formula for **onto**$(n, 4)$.

9. Find the number of poker hands (5 cards from a 52-card deck) containing the following:
(a) three aces or two kings [i.e., *exactly* three aces or *exactly* two kings (or both)];
(b) three aces or three kings.

10. How many poker hands contain two aces or two kings or one queen?

11. Find the number of poker hands containing *all spades*, or *the ace of spades*, or *the ace of hearts*.

12. Find the number of bridge hands (13 cards from an ordinary 52-card deck) containing the AKQJ10 of at least one suit.

13. Find the number of permutations of the 26 letters of the alphabet that do not contain either of the strings *cat* or *dog*. In other words, count permutations not containing *cat* and not containing *dog*. (Hint: think of the opposite event.)

14. Form committees of size five from a group of six men, seven women, and five children. How many committees contain the following?
(a) Exactly one woman.
(b) At least one woman.
(c) At most one woman.

15. A committee of size three is picked from a group with six children, seven teenagers, and eight adults. Find the number of committees containing only one age group.

16. In how many ways can a committee of five senators be formed so that it contains at least one senator each from Hawaii, Massachusetts, and Pennsylvania? (*Hint*: This problem can be done by making a direct list of all possibilities, e.g., two from Hawaii, one from Massachusetts, one from Pennsylvania, one from another state, etc. Alternatively, find the total number of committees and subtract the number in which at least one of the states of Hawaii, Massachusetts, and Pennslyvania is missing.)

17. Derive a formula for $|A - B|$ in terms of $|A|$, $|B|$, and $|A \cap B|$. Does the formula simplify if $B \subseteq A$? If $A \subseteq B$? If A and B are *disjoint*?

18. Consider the formula for PIE_n.
(a) How many of the $2^n - 1$ terms involve *exactly* k of the subsets $A_1, A_2, \ldots, A_n$?

(b) How many "+" signs and how many "−" signs are there?
(c) What is the sign of the term $|A_1 \cap A_2 \cap \cdots \cap A_n|$?

19. Verify Equation (2.9) directly by showing that each of the regions 1–7 in Figure 2.7 is counted the same number of times positively as it is negatively in the given formula for $|A' \cap B' \cap C'|$, while the exterior region is counted just once.

20. Consider the following inequality:

$$|A \cup B \cup C| \le |A \cup B| + |A \cup C|.$$

Is this always true? If so, give a proof. If not so, give a counterexample, i.e., three sets A, B, and C for which the inequality is violated.

21. Answer the same question as in Problem 20 for the inequality

$$|A \cup B \cup C| \le |A \cup B| + |A \cup C| - |A \cap B \cap C|.$$

22. (The Union Bound)
(a) Show that if $A_1, A_2, \ldots, A_n$ are arbitrary sets,

$$|A_1 \cup \cdots \cup A_n| \le |A_1| + |A_2| + \cdots + |A_n|.$$

(b) Show that if the sets $A_1, \ldots, A_n$ are pairwise disjoint, then *equality* holds in part (a).

23. (Lewis Carroll.) In a very hotly contested battle, among 100 combatants, 80 lost an arm, 85 lost a leg, 70 lost an eye, 75 lost an ear. How many (at least) lost all four? (Use Problem 22.)

24. Return to Section 1.3 and do Problems 1.3.18–1.3.25.

25. Suppose X and Y are sets such that there are exactly 36 onto functions from X to Y. What can you conclude about $|X|$ and $|Y|$?

26. Consider the set of $n!$ one-to-one and onto functions from $X \to X$, where $|X| = n$. If f is one of these functions, a *fixed point* of f is an element $x \in X$ such that $f(x) = x$. Denote by $\mathbf{nofix}(n)$ the number of such functions with no fixed points. For example, $\mathbf{nofix}(3) = 2$, as you can verify by referring to Example 1.11, where the functions f_4 and f_5 are the two without fixed points.
(a) Calculate $\mathbf{nofix}(n)$ for $n = 1, 2, 3$, and 4.
(b) Find a general formula for $\mathbf{nofix}(n)$.

27. A group of n people check their hats at a restaurant. During the meal, the hats are scrambled, and as a result, the people (call them $1, 2, \ldots, n$) do not receive hats $1, 2, \ldots, n$ but instead get back a *permutation* of $1, 2, \ldots, n$. Of the $n!$ possible permutations of $1, 2, \ldots, n$, find the number in which *every* integer has been moved to a different position; in other words, find the number of permutations in which *no* customer gets his own hat. (Permutations of this type are called *derangements* of $A = \{1, 2, \ldots, n\}$.)

28. Consider a partition of the set $\{1, 2, 3, 4, 5, 6, 7\}$ into three pairwise disjoint sets, for example, $\{1, 2, 3\}$, $\{4, 5\}$, $\{6, 7\}$. Observe that we can associate with this partition an onto function f from A to $B = \{1, 2, 3\}$ by taking

$$f(1) = f(2) = f(3) = 1,$$
$$f(4) = f(5) = 2, \qquad f(6) = f(7) = 3.$$

This is not the only possibility; we can take $f(1) = f(2) = f(3) = 2$, $f(4) = f(5) = 1$, $f(6) = f(7) = 3$. There are $3! = 6$ such functions, corresponding to the six permutations of a set with three members. In general, show that each partition of $\{1, 2, \ldots, n\}$ into m disjoint sets gives rise to $m!$ onto functions from $\{1, 2, \ldots, n\}$ to $\{1, 2, \ldots, m\}$.

29. Let $S(n, m)$ be the number of partitions of $\{1, 2, \ldots, n\}$ into m pairwise disjoint subsets. Show that

$$S(n, m) = \frac{1}{m!}\,\mathbf{onto}(n, m),$$

so that $S(n, m) = \left\{ {n \atop m} \right\}$, i.e., the $S(n, m)$ are the Stirling numbers of the second kind mentioned at the end of the chapter. (Use Problem 28.)

30. This problem and the next examine some properties of the Stirling numbers of the second kind.
(a) Arrange the numbers $\left\{ {n \atop m} \right\}$ for $n = 0, 1, 2, 3, 4$ into a triangle like Pascal's triangle for binomial coefficients.
(b) Show that the Stirling numbers satisfy the following "rule of Stirling's triangle":

$$\left\{ {n \atop m} \right\} = m \cdot \left\{ {n-1 \atop m} \right\} + \left\{ {n-1 \atop m-1} \right\}.$$

(c) Use the result of part (b) to extend Stirling's triangle to $n = 5, 6, 7$.

31. Show that $\left\{ {n \atop m} \right\}$ can be interpreted as the number of ways of distributing n distinguishable balls into m indistinguishable boxes, with each box containing at least one ball.

Summary

In this chapter we have studied eight of the basic rules of counting, which, for convenience, we repeat here:

- *The multiplication rule*: If one choice can be made in m ways, and if for each of these m choices a second choice can be made in n ways, then the total number of ways of making *both* choices is $m \times n$.
- *Ordered samples with repetition*: If repetition is allowed, the number of ordered samples of size k, from an n-element set, is n^k.
- *Ordered samples without repetition*: If repetition is not allowed, the number of ordered samples of size k, from an n-element set, is $n(n-1)\cdots(n-k+1)$.
- *The permutation rule*: The number of permutations of an n-element set is $n!$.
- *Unordered samples without repetition (the subset rule)*: If repetition is not allowed, the number of unordered samples of size k, from an n-element set, is

$$\frac{n(n-1)\cdots(n-k+1)}{k!} = \binom{n}{k},$$

  which is the same as the number of k-element subsets of an n-element set.
- *Unordered samples with repetition*: If repetition is allowed, the number of unordered samples of size k, from an n-element set, is

$$\frac{n(n+1)\cdots(n+k-1)}{k!} = \binom{n+k-1}{k}.$$

- *Generalized permutations*: The number of distinguishable arrangements of n objects, in which there are k_1 objects of type 1, k_2 objects of type $2, \ldots, k_m$ objects of type m, is

$$\frac{n!}{k_1!k_2!\cdots k_m!}.$$

- *The principle of inclusion and exclusion (PIE)*:

$$\begin{aligned}|A_1 \cup A_2 \cup \cdots \cup A_n| = {} & |A_1| + |A_2| + \cdots + |A_n| \\ & - |A_1 \cap A_2| - |A_1 \cap A_3| - \cdots - |A_{n-1} \cap A_n| \\ & \vdots \\ & + (-1)^{n-1}|A_1 \cap A_2 \cap \cdots \cap A_n|.\end{aligned}$$

CHAPTER THREE

GRAPHS AND ALGORITHMS I

In Chapters 1 and 2, we introduced you to the theory of sets. Another fundamental object in discrete mathematics is the *graph*. In this chapter and the next, we will introduce you to the theory of graphs and some of its many applications. We will find that graph theory contains many interesting and important problems which are easy to state, but difficult to solve. In many cases, the solution to a graph problem can be given in the form of an *algorithm*, which is an explicit set of instructions, suitable for computer programming, for solving the problem. Thus our study of graph theory will naturally lead us to the study of a number of important algorithms, including the grandfather of all graph algorithms, Euler's algorithm for finding "Euler cycles."

3.1 Leonhard Euler and the Seven Bridges of Königsburg

In the old Prussian city of Königsburg (now Kaliningrad in Soviet Lithuania), seven bridges crossed the River Pregel as shown in Figure 3.1. There was a traditional puzzle about these bridges: Is it possible to walk through the city, cross each bridge exactly once, and return to the starting point? No one in Königsburg ever succeeded in finding such a walk, but people didn't quite give up trying until 1736, when the great Swiss mathematician Leonhard Euler* *proved* once and for all that such a walk is impossible. Euler reasoned as follows: Suppose such a walk *were* possible, and suppose that a pedestrian starts out on the walk. If an observer were located on island A, this observer would see the pedestrian visit the island a certain number of times. How many times? *Once*? No, because if the pedestrian only visited the island once, he would have only used *two* of the *five* bridges on the island (one arriving and one leaving). *Twice*? Again no,

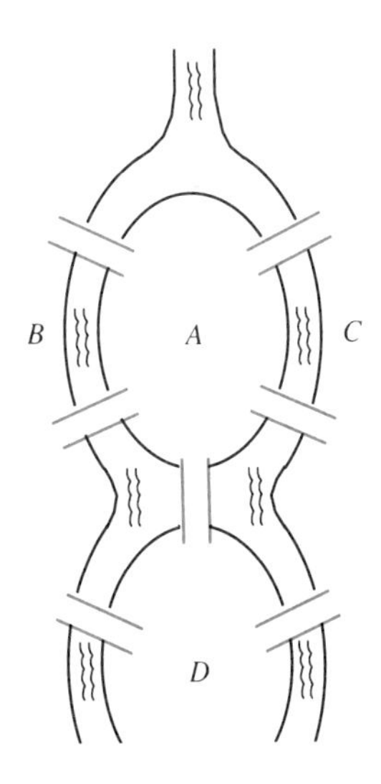

Figure 3.1 The seven bridges of Königsburg.

* Leonhard Euler (1707–1783), Swiss mathematician and physicist. A truly prodigious intellect, he made major and lasting contributions to number theory, the calculus of variations, differential equations, geometry, and many other branches of mathematics. His collected works run to more than 70 volumes; and much of this work was done after 1771, when Euler became completely blind!

since *two* visits would only use *four* bridges. *Three* times? Still no: Three visits would now need *six* bridges, and only five are available. Plainly then, no such walk is possible! That's how easily Euler solved the ancient problem.

Of course the Königsburg bridge problem is itself of no special importance. But the *ideas* Euler introduced to solve the problem gave birth to an important branch of mathematics called **graph theory**, which is the subject of this chapter. So let us examine Euler's proof in a little more detail.

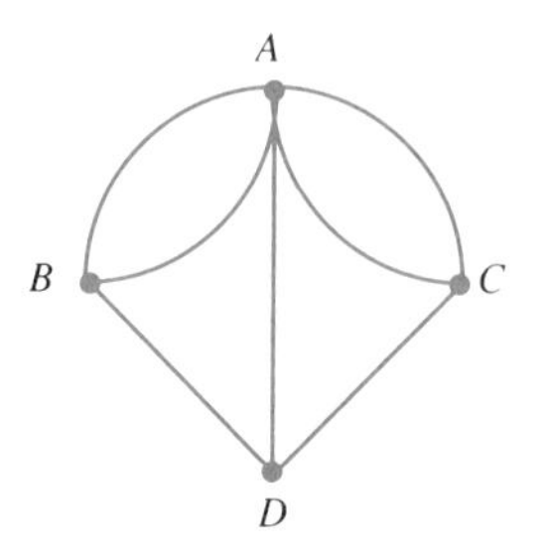

Figure 3.2 "Essence of Königsburg"—a *graph*.

Euler realized that the Königsburg bridge problem has nothing to do with bridges at all, but is really a question about the abstract diagram shown in Figure 3.2. Figure 3.2 forces us to focus our attention on the essentials of the problem—there are *four* locations linked by *seven* bridges, as shown. Euler's proof boils down simply to the observation that a walk through the city must necessarily use an *even* number of bridges on island A; but island A has an odd number of available bridges.

Ever since Euler introduced them, pictures like Figure 3.2 have turned out to be extremely useful in discrete mathematics; they are called **graphs**.* In general, a graph is just a set of points called **vertices** connected by lines called **edges**. In Figure 3.3 we have drawn six more graphs.

Sometimes the edges in a graph are **directed**, in the sense that they "point" from one vertex to another. A graph whose edges are all directed is called a **directed graph**. In Figure 3.4 we show a simple directed graph. Directed graphs have many uses. For example, if the bridges of Königsburg were all "one-way" bridges, a directed graph would be appropriate. In Chapter 8 we will see that

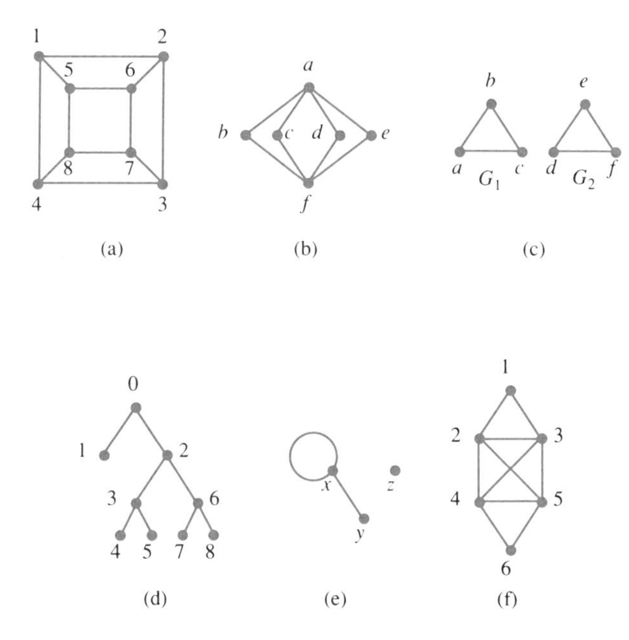

Figure 3.3 Six more graphs.

* There is, unfortunately, another mathematical object called a graph, viz., a wiggly line that represents a function of a real variable. These two kinds of graphs are unrelated and should really have been given different names. Luckily, however, they usually arise in such totally different contexts that no confusion is likely to arise.

directed graphs can be used to represent the positions and legal moves in certain types of games, and in Chapters 6 and 7 we will see that directed graphs are used to describe finite state machines.

As this chapter develops, we will see many more graphs, directed and undirected, and we will see that they are useful for all sorts of things; but for now, let's just get used to this new idea by introducing some of the jargon used in graph theory, and then restate the Königsburg bridge problem, using this jargon. For simplicity, we will only worry about *un*directed graphs in the rest of this section.

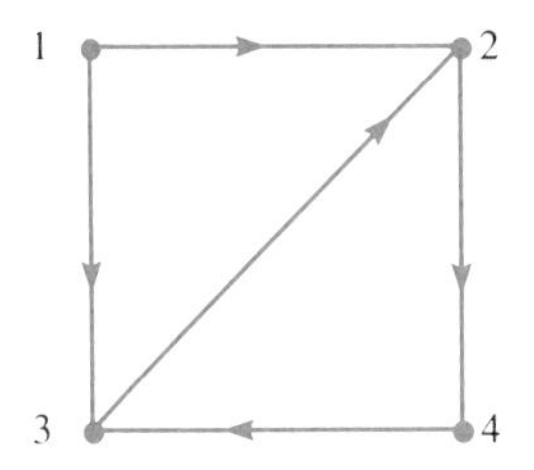

Figure 3.4 A simple directed graph.

A **path** in a graph is a sequence of vertices $v_1, v_2, \ldots, v_n$, such that $\overline{v_1v_2}$, $\overline{v_2v_3}, \ldots, \overline{v_{n-1}v_n}$ are edges in the graph. A vertex can appear more than once in a path, but an edge can't. For example, the vertices in the graph in Figure 3.3(a) are labelled with the integers 1 through 8, and the vertex sequence 567341 represents a path consisting of the five edges $\overline{56}$, $\overline{67}$, $\overline{73}$, $\overline{34}$, and $\overline{41}$. (Note that this notation isn't appropriate for a graph like the one in Figure 3.2, which has **multiple edges**, i.e., two or more edges connecting the same pair of vertices. For such a graph, a notation like $\overline{AB}^{(1)}$, $\overline{AB}^{(2)}$, etc., would be needed to describe the edges connecting vertices A and B. Luckily, graphs with multiple edges are relatively rare, and we won't be needing this clumsy notation.)

A graph is said to be **connected** if every pair of its vertices is connected by a path. The graphs in Figures 3.3(a), 3.3(b), 3.3(d), and 3.3(f) are connected, but those in Figures 3.3(c) and 3.3(e) aren't. The graph in Figure 3.3(c), for example, isn't connected because there is no path from vertex a to vertex e. Indeed, the graph in Figure 3.3(c) consists of two **connected components**, G_1 and G_2.

A **cycle** is a path that begins and ends at the same place. In the graph of Figure 3.3(b), the path *abfea* is a cycle with four edges, or a **cycle of length 4**, for short.

A connected graph with no cycles is called a **tree**. The graph in Figure 3.3(d) is a tree; the other graphs in Figure 3.3 aren't trees. Trees are especially important in graph theory, and we'll look at them in detail in the next three sections.

The number of edges connected to a given vertex is called the **degree** of that vertex. For example, in Figure 3.2 the vertex A, corresponding to island A in Figure 3.1, has degree 5; in Figure 3.3(a), all vertices have degree 3; in Figure 3.3(d), vertex 0 has degree 2; and in Figure 3.3(e), vertex z has degree 0—it is called an **isolated vertex**. (Isolated vertices are rarely encountered in practice, and for simplicity we will usually assume there aren't any.) In Figure 3.3(e), the peculiar edge which goes from x back to x is called a **loop**; loops, when they are present, contribute 2 to a vertex's degree, and thus the vertex x in Figure 3.3(e) has degree 3.

Finally, an **Euler cycle** is a cycle which uses every edge of the graph exactly once. Thus the Königsburg bridge problem can be restated as follows: Does the graph in Figure 3.2 have an Euler cycle? Of course we already know the answer is no; but *some* graphs do have Euler cycles. For example, in the graph of Figure 3.3(b), *abfcadfea* is an Euler cycle. Euler cycles are, of course, named in honor of Leonhard Euler, and in his study of the Königsburg problem Euler found a simple and foolproof way to tell whether or not any given graph has one. Here is his famous theorem, the theorem that gave birth to graph theory.

EULER's THEOREM FOR EULER CYCLES If the graph G doesn't have any isolated vertices, then it has an Euler cycle if and only if the following two conditions are both satisfied:

(a) G is *connected.*
(b) The degree of each vertex of G is *even.*

We will now prove Euler's theorem. Our proof will be in two halves:

1. If G has an Euler cycle, then (a) and (b) are satisfied.
2. If (a) and (b) are satisfied, then G has an Euler cycle.

We begin by proving the first half. Thus suppose G has an Euler cycle. If x and y are two distinct vertices, then both x and y appear in certain (possibly different) edges of G, since G has no isolated vertices. Since the Euler cycle passes through *all* the edges of G, the Euler cycle connects x and y. Therefore G is connected; i.e., (a) is satisfied.

Continuing to assume that G has an Euler cycle, we need next to show that (b) is true, i.e., that every vertex of G has even degree. So let x be any vertex of G, and imagine that an ant, starting at x, crawls along the edges of G in the order prescribed by the Euler cycle. Since after completing the Euler cycle the ant will again be at x, then in the course of its travels, the ant must have *departed* from x exactly as many times as it *returned* to x. Hence the edges which have x as one endpoint can be classified as "departure edges" or "return edges," and there will be an equal number of each. It follows that the total number of edges at x, i.e., the degree of x, will be twice the number of departure edges and so must be even. This completes the proof of the first half of Euler's theorem. As illustrations, note that none of the graphs in Figure 3.2, 3.3(a), 3.3(c), and 3.3(d) have Euler cycles, for the following reasons:

- Figure 3.2 has vertices of odd degree.
- Figure 3.3(a) has vertices of odd degree.
- Figure 3.3(c) is not connected.
- Figure 3.3(d) has vertices of odd degree.

Now we need to prove the second half of Euler's theorem, i.e., that if (a) and (b) are satisfied, then G has an Euler cycle. Our proof will be *constructive* and will show exactly how to find an Euler cycle. In fact, in Example 3.1, which follows this proof, we will see that this procedure is so simple that even a computer can learn it! Systematic and foolproof procedures that can be taught to a computer are usually called **algorithms.**

Here then is an algorithm for finding an Euler cycle. We start at any vertex, say vertex a, and begin traveling along edges at random, never repeating an edge, until we get stuck, say at vertex b. Why are we stuck? Because all the edges from b have already been chosen, that's why. If $a \neq b$, as shown in Figure 3.5, we will have *arrived* at b one more time than we have *departed.* This means that we have used an *odd* number of edges at b. But we know that the number of edges at b is even, and so we can't be stuck! Plainly, then, if we ever do get stuck, we will be back where we started, and the edges so far chosen will form a *cycle,* as shown in Figure 3.6.

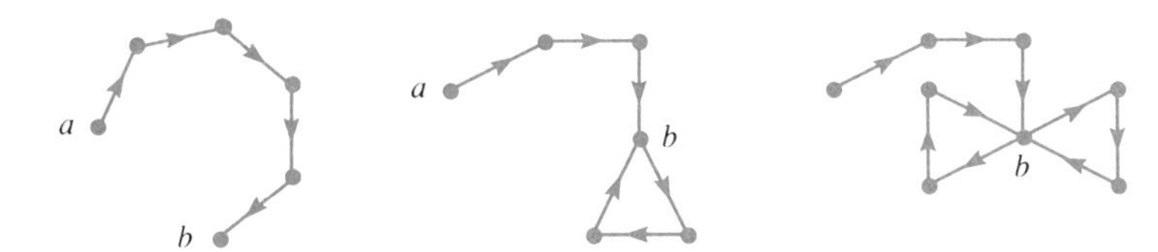

Figure 3.5 Stuck at b, with $b \neq a$ (which can't happen, since all vertices have even degree).

If some vertex, say vertex c, in this cycle has an attached edge that has not been used yet, then we can get a longer cycle in G by starting at c, traveling around the previously formed cycle back to c, and then continuing out along the new edge. In Figure 3.6, for example, if we started at a and got stuck after completing the cycle *abcdea* with five edges, we could replace this five-edge path with the path *cdeabcx* of length 6. We call this procedure a **breakout**.

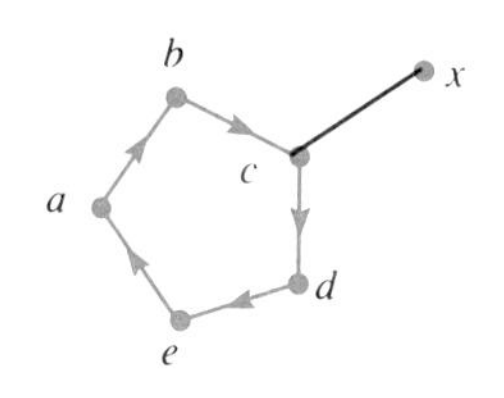

Figure 3.6 Stuck at a, with breakout possible at c.

We continue this algorithm, forming cycles and breaking out, as long as possible. We will finally arrive at a long cycle from which a breakout isn't possible, i.e., a cycle such that every edge from every vertex on the cycle has been used. Does this mean that we're stuck again? No, it means that we've found an Euler cycle! To see why this is so, i.e., the cycle formed contains every edge, we reason as follows: Let E be any edge of the graph, and suppose its endpoints are the vertices x and y. Then, since G is connected, there must be a path from a (the vertex where we started) to x, say $aa_1a_2 \cdots x$. The edge $\overline{aa_1}$ must be somewhere on the cycle; otherwise we could break out at a. Therefore the vertex a_1 is on the cycle, which in turn implies that the edge $\overline{a_1a_2}$ is on the cycle, since a breakout at a_1 isn't possible. Similarly, $a_2, a_3, \ldots, x$ must all be on the cycle. Finally, since a breakout at x isn't possible, the edge E connecting x and y must also be on the cycle. Thus, when a breakout is no longer possible, *every* edge must be included in the cycle, i.e., it is an Euler cycle.

This completes our constructive proof of the second half of Euler's theorem. The next example will illustrate the use of the algorithm we have just described.

EXAMPLE 3.1 Find an Euler cycle in the graph of Figure 3.3(f).

SOLUTION First note that the given graph is indeed connected and has all vertices of even degree; so Euler's theorem guarantees that there is an Euler cycle. Maybe you can look at Figure 3.3(f) and find an Euler cycle in your head; but let's proceed systematically, in order to better appreciate the proof of Euler's theorem. In fact, let's proceed *so* systematically that a computer could be programmed to carry out the steps.

We begin by representing the graph by what is called its **adjacency matrix*** (see Table 3.1). The adjacency matrix uses 1s to indicate the presence of edges and 0s to indicate the absence of edges. Thus, the (3, 4) entry is 1 because there is an edge connecting vertex 3 to vertex 4. Similarly, the (1, 5) entry is 0 because there is no edge connecting 1 and 5.

TABLE 3.1 Adjacency matrix for the graph in Figure 3f.

	1	2	3	4	5	6
1	0	1	1	0	0	0
2	1	0	1	1	1	0
3	1	1	0	1	1	0
4	0	1	1	0	1	1
5	0	1	1	1	0	1
6	0	0	0	1	1	0

* In actual computer implementation, another representation of the graph, called an **adjacency list**, is sometimes used instead. We refer you to the excellent book *Data Structures and Algorithms* by Aho, Hopcroft, and Ullman (Addison-Wesley, 1983) for more about adjacency lists.

In order to find an Euler cycle, we begin at an arbitrary vertex, say vertex 1, and choose any edge from 1, say $\overline{12}$.

Having chosen this edge, we change the (1, 2) and the (2, 1) entries in the adjacency matrix from 1 to 0, to indicate that this edge is no longer available. The result is a short path (12), and a new adjacency matrix:

Path so far: 12.
New adjacency matrix:

$$\begin{array}{c} \\ 1 \\ 2 \\ 3 \\ 4 \\ 5 \\ 6 \end{array} \begin{array}{c} \begin{array}{cccccc} 1 & 2 & 3 & 4 & 5 & 6 \end{array} \\ \begin{pmatrix} 0 & 0 & 1 & 0 & 0 & 0 \\ 0 & 0 & 1 & 1 & 1 & 0 \\ 1 & 1 & 0 & 1 & 1 & 0 \\ 0 & 1 & 1 & 0 & 1 & 1 \\ 0 & 1 & 1 & 1 & 0 & 1 \\ 0 & 0 & 0 & 1 & 1 & 0 \end{pmatrix} \end{array}$$

Next, from the vertex 2 we choose the first available edge $\overline{23}$ and delete the 1s corresponding to (2, 3) and (3, 2). Then, from vertex 3 we take the first available edge $\overline{31}$ and change the matrix. The situation is now as follows:

Path so far: 1231 (cycle).
New adjacency matrix:

$$\begin{array}{c} \\ 1 \\ 2 \\ 3 \\ 4 \\ 5 \\ 6 \end{array} \begin{array}{c} \begin{array}{cccccc} 1 & 2 & 3 & 4 & 5 & 6 \end{array} \\ \begin{pmatrix} 0 & 0 & 0 & 0 & 0 & 0 \\ 0 & 0 & 0 & 1 & 1 & 0 \\ 0 & 0 & 0 & 1 & 1 & 0 \\ 0 & 1 & 1 & 0 & 1 & 1 \\ 0 & 1 & 1 & 1 & 0 & 1 \\ 0 & 0 & 0 & 1 & 1 & 0 \end{pmatrix} \end{array}$$

And now we're stuck! We have arrived back at vertex 1, but there is no edge available to leave from (the first row of the adjacency matrix is all zeros). So now we *break out* by examining the *other* vertices in the cycle 1231 for an escape route. In this case we can escape via vertex 2 on the edge $\overline{24}$. So we now think of 2 as the beginning of the cycle, which becomes 2312:

Path so far: 2312.
New adjacency matrix: Same.

And continue as before, always visiting the lowest-numbered available vertex, to vertices 4, 3, 5, and 2, at which point the situation is as follows:

Path so far: 23124352 (cycle).
Adjacency matrix:

$$\begin{array}{c} \\ 1 \\ 2 \\ 3 \\ 4 \\ 5 \\ 6 \end{array} \begin{array}{c} \begin{array}{cccccc} 1 & 2 & 3 & 4 & 5 & 6 \end{array} \\ \begin{pmatrix} 0 & 0 & 0 & 0 & 0 & 0 \\ 0 & 0 & 0 & 0 & 0 & 0 \\ 0 & 0 & 0 & 0 & 0 & 0 \\ 0 & 0 & 0 & 0 & 1 & 1 \\ 0 & 0 & 0 & 1 & 0 & 1 \\ 0 & 0 & 0 & 1 & 1 & 0 \end{pmatrix} \end{array}$$

Here again we've formed a cycle and must break out. There is no escape from 3, 1, or 2; but the edge $\overline{45}$ is still available, so we choose 4 as the beginning of the cycle:

Path so far: 43523124.
New adjacency matrix: Same.

Now we continue without difficulty to 5, 6, and 4, at which point we have formed a third cycle:

Path so far: 43523124564 (cycle).
Adjacency matrix:

$$\begin{array}{c} \\ 1 \\ 2 \\ 3 \\ 4 \\ 5 \\ 6 \end{array} \begin{array}{c} \begin{array}{cccccc} 1 & 2 & 3 & 4 & 5 & 6 \end{array} \\ \begin{pmatrix} 0 & 0 & 0 & 0 & 0 & 0 \\ 0 & 0 & 0 & 0 & 0 & 0 \\ 0 & 0 & 0 & 0 & 0 & 0 \\ 0 & 0 & 0 & 0 & 0 & 0 \\ 0 & 0 & 0 & 0 & 0 & 0 \\ 0 & 0 & 0 & 0 & 0 & 0 \end{pmatrix} \end{array}$$

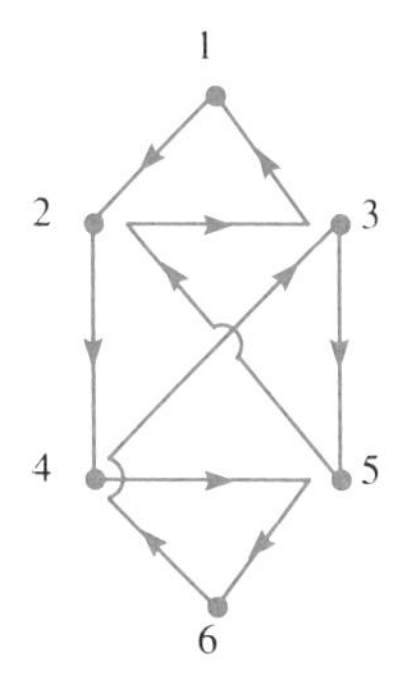

Figure 3.7 An Euler cycle for the graph of Figure 3.3(f).

There are no edges left (the adjacency matrix has all zeros) and so this last cycle is the desired Euler cycle (see Figure 3.7).

We'll conclude this section by mentioning some easy extensions of Euler's theorem.

Multiple Edges

Sometimes the graph under consideration will have *multiple edges*, i.e., more than one edge connecting a given pair of vertices. The Königsburg graph of Figures 3.1 and 3.2 has this feature, for example. In fact, Euler's Theorem covers this case, but the adjacency matrix approach used in Example 3.1 needs to be modified slightly. If multiple edges are present, the entries in the adjacency matrix should reflect the multiplicities, as illustrated in Figure 3.8. For example, the (3, 4) entry in the adjacency matrix in Figure 3.8 is 3 because there are 3 edges connecting vertex 3 and vertex 4.

EXAMPLE 3.2 Find a Euler cycle in the graph of Figure 3.8.

$$\begin{array}{c} \\ 1 \\ 2 \\ 3 \\ 4 \end{array} \begin{array}{c} \begin{array}{cccc} 1 & 2 & 3 & 4 \end{array} \\ \begin{pmatrix} 0 & 2 & 1 & 1 \\ 2 & 0 & 0 & 2 \\ 1 & 0 & 0 & 3 \\ 1 & 2 & 3 & 0 \end{pmatrix} \end{array}$$

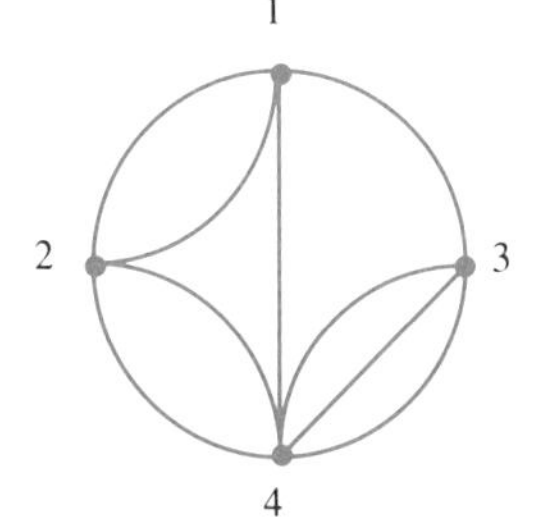

Figure 3.8 A graph with multiple edges and its adjacency matrix.

SOLUTION We proceed as in Example 3.1, starting with vertex 1 and at every stage choosing the lowest-numbered vertex that can be used to extend the current path, and breaking out as necessary. When we use an edge, we must remember to decrease the entry in the adjacency matrix by 1. Omitting many of the details, we can summarize the work by showing the situation just prior to each breakout:

First cycle: 121341.
Adjacency matrix:

$$\begin{array}{c} \\ 1 \\ 2 \\ 3 \\ 4 \end{array} \begin{array}{c} \begin{array}{cccc} 1 & 2 & 3 & 4 \end{array} \\ \begin{pmatrix} 0 & 0 & 0 & 0 \\ 0 & 0 & 0 & 2 \\ 0 & 0 & 0 & 2 \\ 0 & 2 & 2 & 0 \end{pmatrix} \end{array}$$

Breaking out at vertex 2, we continue:

Second cycle: 21341242.
Adjacency matrix:

$$\begin{array}{c} \\ 1 \\ 2 \\ 3 \\ 4 \end{array} \begin{array}{c} \begin{array}{cccc} 1 & 2 & 3 & 4 \end{array} \\ \begin{pmatrix} 0 & 0 & 0 & 0 \\ 0 & 0 & 0 & 0 \\ 0 & 0 & 0 & 2 \\ 0 & 0 & 2 & 0 \end{pmatrix} \end{array}$$

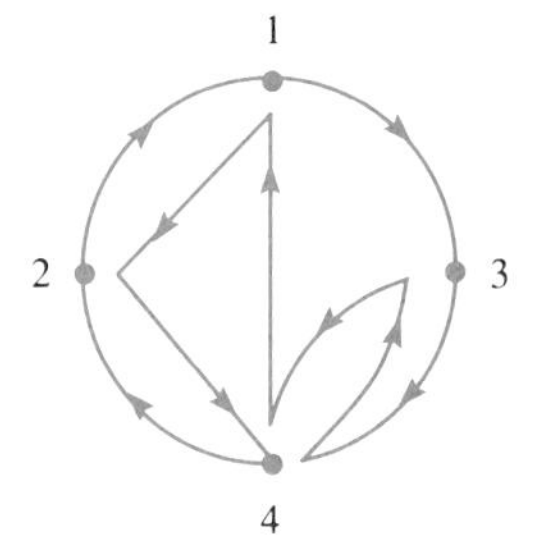

Figure 3.9 An Euler cycle for the graph of Figure 3.8.

Breaking out at vertex 3, we continue:

Third cycle: 3412421343.
Adjacency matrix:

$$\begin{array}{c} \\ 1 \\ 2 \\ 3 \\ 4 \end{array} \begin{array}{c} \begin{array}{cccc} 1 & 2 & 3 & 4 \end{array} \\ \begin{pmatrix} 0 & 0 & 0 & 0 \\ 0 & 0 & 0 & 0 \\ 0 & 0 & 0 & 0 \\ 0 & 0 & 0 & 0 \end{pmatrix} \end{array}$$

The third cycle uses all edges and so is an Euler cycle; see Figure 3.9. ■

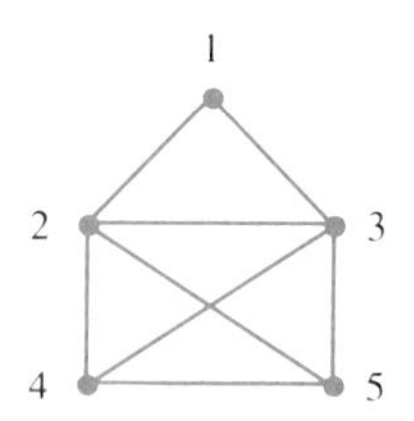

Figure 3.10 A graph with an Euler path but no Euler cycle.

Euler Paths

Sometimes we will want to know if the graph G has merely an **Euler path,** which is a path through G that covers all the edges, but which has *different starting and stopping points.* For example, the graph in Figure 3.10 has an Euler path (435231245) but no Euler cycle, because vertices 4 and 5 have odd degree.

The conditions for the existence of an Euler path are given in the following theorem.

EULER'S THEOREM FOR EULER PATHS If G doesn't have isolated vertices, then it has an Euler path if and only if the following two conditions are satisfied:

(a) G is connected.
(b) *Exactly two* vertices have odd degree.

PROOF The proof that a graph with an Euler path must have properties (a) and (b) is almost the same as the proof of the first part of the Euler cycle theorem. The only difference is that we must notice that the "begin" and "end" vertices in the path are different: The ant traveling the Euler path *leaves* the start one more time than he arrives there; and he *arrives* at the finish one more time than he leaves. So both these vertices have odd degree.

On the other hand, to prove that a graph with properties (a) and (b) must have an Euler path, we use a trick: Add an "imaginary" edge connecting the two vertices of odd degree. The resulting graph has *all* vertices of even degree—now find an *Euler cycle* in the new graph, starting with the imaginary edge and using the algorithm described above. Then delete the imaginary edge from the Euler cycle; and Bob's your uncle, the result is an Euler path in the original graph!

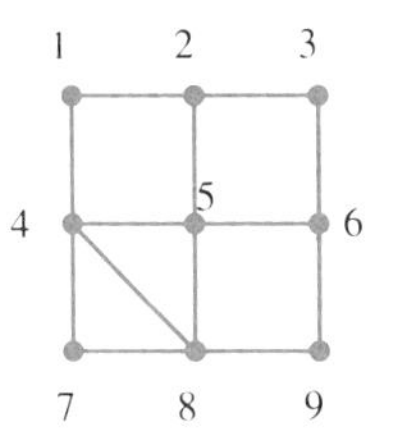

Figure 3.11 A graph with an Euler path.

EXAMPLE 3.3 Find an Euler path in the graph of Figure 3.11.

SOLUTION Note that the graph is connected, and all vertices have even degree except for 2 and 6; so Euler's theorem for Euler paths guarantees the existence of an Euler path. To find it, we add an imaginary edge from vertex 2 to vertex 6 (see Figure 3.12).

Using the Euler cycle algorithm, we find after some work that 625412365847896 is an Euler cycle in the graph of Figure 3.12. Deleting the imaginary edge 26 from this cycle, we get the path 25412365847896, which is the desired Euler path.

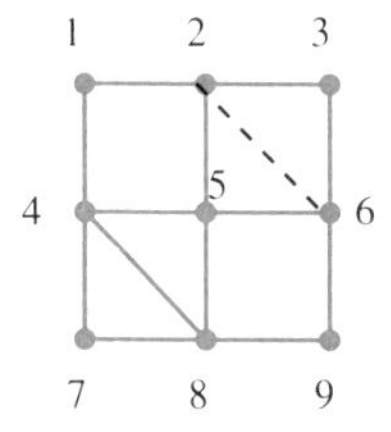

Figure 3.12 A graph with an Euler cycle.

Problems for Section 3.1

1. List all cycles in the accompanying graph.

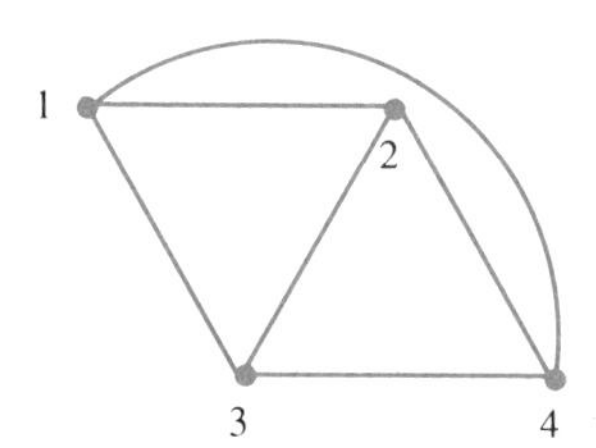

2. Draw a graph with four vertices such that if $d(v)$ is the degree of vertex v, then $d(v_1) = 3$, $d(v_2) = d(v_3) = 2$, and $d(v_4) = 1$.

3. Construct a disconnected graph with five vertices and three connected components.

4. In the graph of Problem 1, all vertices have degree 3. Draw a graph in which all vertices have degree 2.

5. Explain how to construct a graph in which all vertices have degree n, where n is an arbitrary positive integer.

6. Consider the tree of Figure 3.3(d). If an edge is added between vertices 1 and 6, verify that the resulting graph contains exactly one cycle.

7. In general, if an edge connecting a pair of vertices is added to a tree, explain why the resulting graph contains exactly one cycle. (This problem now will be much easier after we study trees in detail in Section 3.2.)

In Problems 8–12 we give the adjacency matrix for a graph (blank entries indicate 0s). In each case you are to *draw* the graph; decide whether or not it has an Euler cycle; if it has an Euler cycle, find it; if it doesn't, decide whether or not it has an Euler path; if it has an Euler path, find it.

8.

	a	b	c	d	e	f	g	h	i	j
a		1				1				
b	1		1			1		1		
c		1		1				1	1	
d			1		1				1	1
e				1		1				
f	1	1			1		1			
g						1		1		
h		1	1				1		1	
i			1	1				1		1
j				1					1	

9.

	A	B	C	D	E	F	G
A		1	1	1	1		
B	1			1			
C	1			1		1	1
D	1	1	1		1	1	1
E	1			1		1	1
F			1	1	1		1
G			1	1	1	1	

10.

	1	2	3	4	5	6	7
1		1				1	1
2	1		1				1
3		1		1			1
4			1		1		1
5				1		1	1
6	1				1		1
7	1	1	1	1	1	1	

11.

	1	2	3	4	5	6	7
1		1	1		1	1	
2	1		1			1	
3	1	1		1		1	
4			1		2		1
5	1			2		1	
6	1	1	1		1		
7				1			

12.

	0	1	2	3	4	5	6	7
0	2	1			1			
1	1		1	1	1			
2		1			1	2		
3		1				1	1	1
4	1	1	1				1	
5			2	1			1	
6				1	1	1		1
7				1			1	2

13. We learned in the text that the Königsburg bridge graph has no Euler cycle. Does it have an Euler path? If so, find it.

14. How many extra bridges would it be necessary to build in Königsburg so that an Euler cycle would exist?

15. What is the longest cycle you can find in the Königsburg graph? The longest path?

16. Is there a cycle through Königsburg that uses each bridge exactly twice?

17. In the statements of Euler's theorems, the phrase "with no isolated vertices" appears. Explain how to modify these theorems so that one can tell whether a graph *with* isolated vertices has an Euler cycle or path.

18. Does the graph in Figure 3.3(e) have an Euler cycle or path?

19. Show that every tree with at least two vertices has at least one vertex of odd degree.

20. In his original paper on graphs Euler described an imaginary town with islands and bridges as shown in the accompanying figure. Is there an Euler cycle or path in this town?

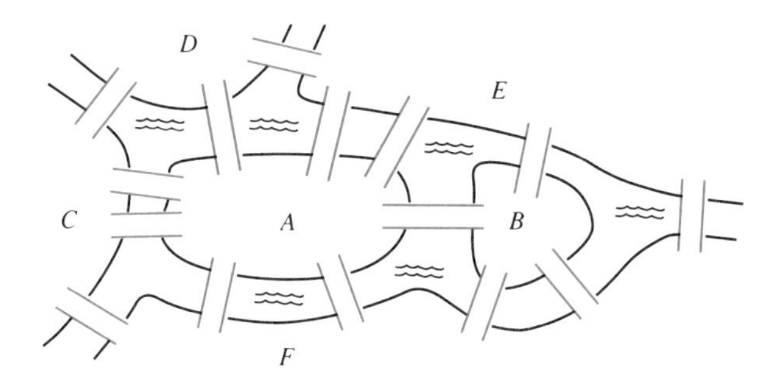

21. Describe the adjacency matrices for the graphs of Figure 3.2 and Figures 3.3(a)–3.3(e).

22. Do Example 3.1 again, but this time start at vertex 2 instead of vertex 1. Do you get the same result as before? (For definiteness, always proceed to the lowest-numbered available vertex; and when stuck, break out at the lowest-numbered possible vertex.)

23. Let G be an arbitrary graph. Let S denote the sum of the degrees of all vertices of G, and let N denote the number of vertices of odd degree. For example, in Figure 3.2, $S = 14$ and $N = 4$; in Figure 3.3(a), $S = 24$ and $N = 8$; in Figure 3.3(b), $S = 16$ and $N = 0$; in Figure 3.6, $S = 12$ and $N = 2$.
(a) Show that S and N are always both even numbers.
(b) Show that S is twice the number of edges.

24. How many different Euler cycles does the graph of Figure 3.3(b) have? (*Hint*: Start at vertex a.)

25. The accompanying figure shows a generalization of the graph in Figure 3.3(b). How many different Euler cycles does it have?

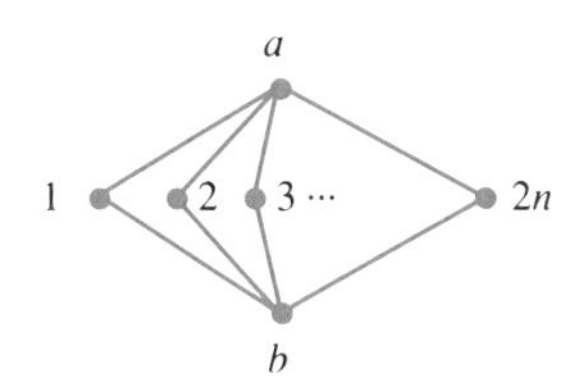

26. In this problem we ask you to prove Euler's theorem for *directed* graphs. In a directed graph a path must respect the "one-wayness" of each edge, so that for example in Figure 3.4, 13243 is a legal path but 13423 is not. In this case Euler's theorem for Euler cycles must be modified, and condition (b) becomes

(b) Each vertex is "balanced."

A "balanced" vertex is one such that the number of "in" directed edges equals the number of "out" directed edges; see the accompanying figures:

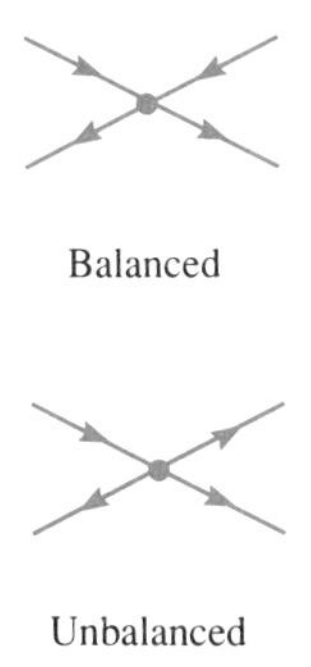

Balanced

Unbalanced

Prove Euler's theorem for directed graphs. *Hint*: The algorithm for finding an Euler cycle is almost the same as before; the only modification is in the breakout procedure, if the "breakout edge" happens to be directed *in toward* the cycle; see the accompanying figure:

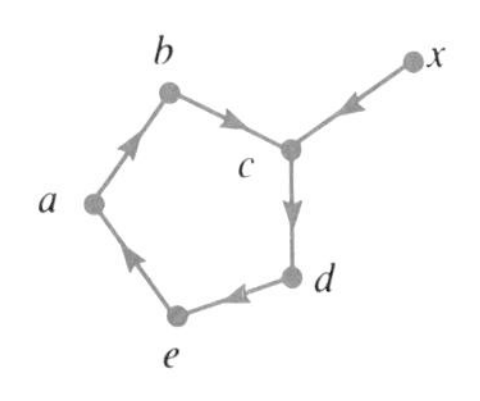

Also note that the adjacency matrix can reflect the directed edges; e.g., if $a \to b$ is a one-way edge, then the (a, b) entry is 1, but the (b, a) entry is 0. See the accompanying figure:

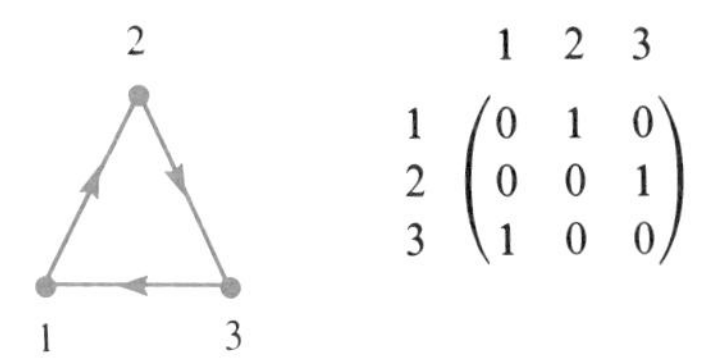

27. Continuing Problem 26, use the algorithm suggested there to find an Euler cycle for the graph in the accompanying figure.

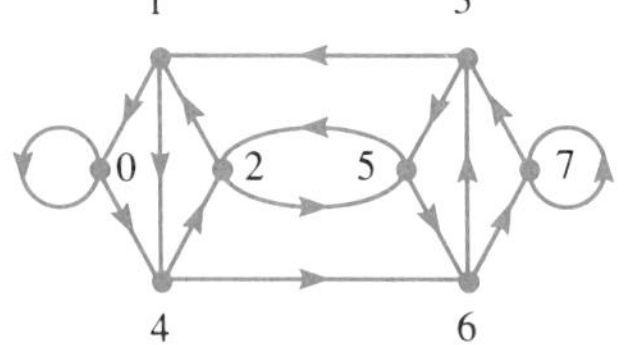

28. Consider a graph G with n vertices such that each vertex is directly connected by an edge to every other vertex. Under what conditions will G have an Euler cycle? Explain.

In Problems 29–32, let $V = \{1, 2, \ldots, n\}$ be a set of n vertices. Determine the number of different graphs with V as the set of vertices under the given conditions.

29. There are no loops or multiple edges.

30. There are no loops and at most two edges between each pair of vertices.

31. A single loop is allowed at each vertex, but no multiple edges.

32. Each edge must join an *even*-numbered vertex to an *odd*-numbered vertex.

3.2 Trees and Spanning Trees

In Section 3.1 we introduced the notion of a graph, and presented a simple algorithm for finding Euler paths and cycles in a given graph. It turns out that modern graph theory is filled with problems whose solution is best presented algorithmically, and in this chapter and the next we will discuss in detail several such problems, each of which has important practical applications. In each case we will present an efficient algorithm for the solution of the problem and give several examples of the application of the algorithm.

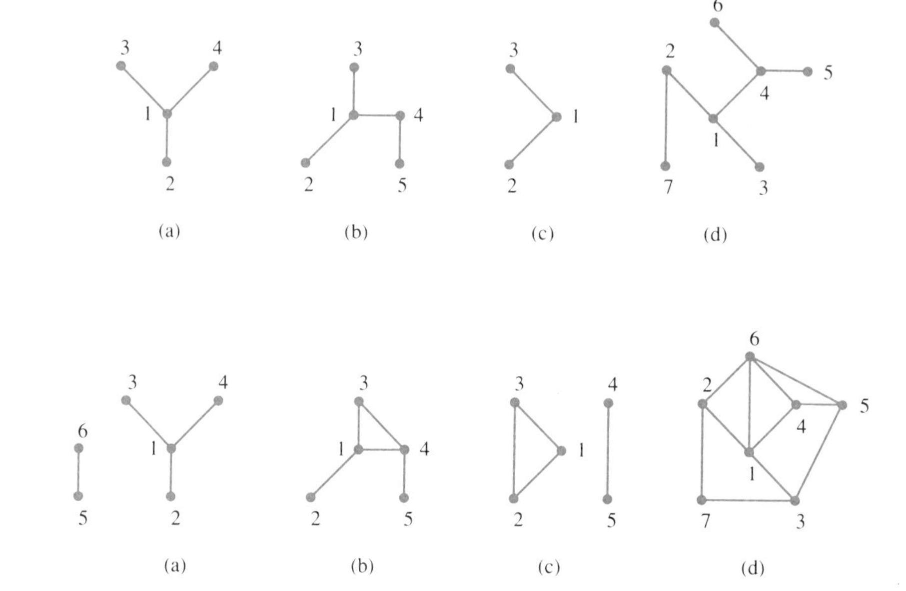

Figure 3.13 Four trees: (a) 4 vertices and 3 edges; (b) 5 vertices and 4 edges; (c) 3 vertices and 2 edges; (d) 7 vertices and 6 edges.

Figure 3.14 Four nontrees: (a) Not connected; (b) contains a cycle; (c) not connected and contains a cycle; (d) contains many cycles.

We mentioned briefly in Section 3.1 that a **tree** is a connected graph with no cycles. In this section, we'll learn more about trees. In Figure 3.3(d), we saw one example of a tree. In Figure 3.13, we show four others.

Not every graph is a tree, of course. Any graph which is *not connected*, or *contains a cycle*, or both, cannot be a tree. In Figure 3.14 we show four non-trees.

Notice that while the graph of Figure 3.14(d) isn't a tree, it does *contain* the tree of Figure 3.13(d) as a kind of skeleton. A skeleton like this is called a **spanning tree** of the original graph. The formal definition follows.

DEFINITION A **spanning tree** for a connected graph is a tree whose vertex set is the *same* as the vertex set of the given graph, and whose edge set is a *subset* of the edge set of the given graph.

Any connected graph will have a spanning tree, as we shall see.

EXAMPLE 3.4 Find another spanning tree for the graph of Figure 3.14(d).

SOLUTION Since the spanning tree in Figure 3.13(d) doesn't include the edge $\overline{26}$, let's try to grow a spanning tree, *starting* with the edge $\overline{26}$ (see Figure 3.15a). If we succeed in finding a spanning tree which includes the edge $\overline{26}$, it will definitely be different from the one in Figure 3.13(d).

Our strategy will be to add vertices to the tree, one at a time, until we arrive at a spanning tree. To implement this strategy, at every stage we will look for a vertex which isn't yet in the tree, but which is connected to some vertex *already* in the tree by an edge; and then we will grow the tree by adding that edge. We

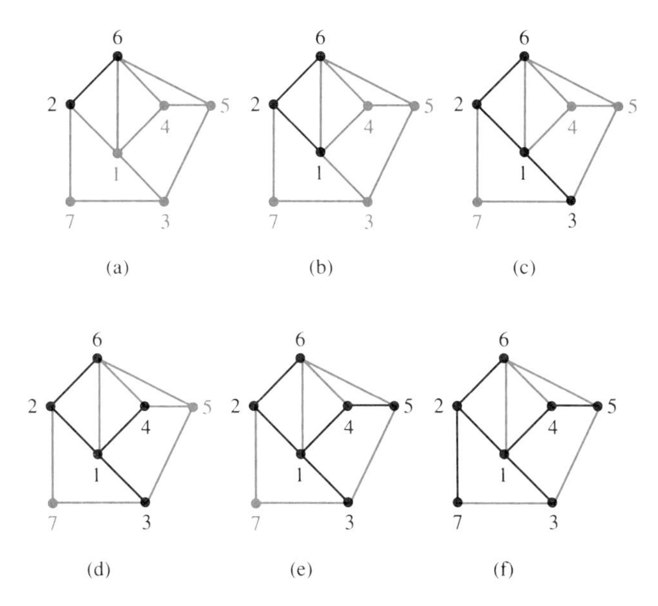

Figure 3.15 Growing a spanning tree for the graph of Figure 3.14(d).

begin, with the black one-edge tree of Figure 3.15(a). Looking for new vertices connected to this tree by an edge, we see that vertices 1, 4, 5, and 7 all qualify. For example, vertex 1 is connected to the growing tree in Figure 3.15(a) by the edge $\overline{12}$; if we add this edge to the tree, the two-edge tree of Figure 3.15(b) results. Continuing, we see that now *all* of the remaining vertices (3, 4, 5, 7) are connected to the tree of Figure 3.15(b) by an edge. So now we can add any one of the remaining vertices; but let's add vertex 3 via edge $\overline{13}$, obtaining the tree of Figure 3.15(c). We continue, adding successively vertices 4, 5, and 7, and the spanning tree of Figure 3.15(f) results. ■

The algorithm we used in Example 3.4 can be summarized as follows:

Spanning tree algorithm

1. Pick any vertex, and put it in the tree.
2. If all vertices are in the tree, stop.
3. Otherwise, find a vertex which isn't in the tree yet, but which is connected by an edge to a vertex already in the tree, and add this vertex and edge to the tree. Return to step 2.

When applied to an arbitrary connected graph, this algorithm will find a spanning tree. To see why this is so, we need to resolve two possible problems. First, we have no guarantee that it will always be possible to execute step 3, i.e., to find an edge with one vertex in, and one vertex out, of the growing tree. And second, when the algorithm ends, we don't know for sure that the edges selected will actually form a spanning tree. Fortunately, however, both of these concerns are groundless, as we will now see.

First, if at any stage of the algorithm it proves difficult to execute step 3, here is one procedure that will always work (see Figure 3.16): Denote the growing tree by T. Let v be a vertex in T, and w a vertex *not in* T. There may not be an

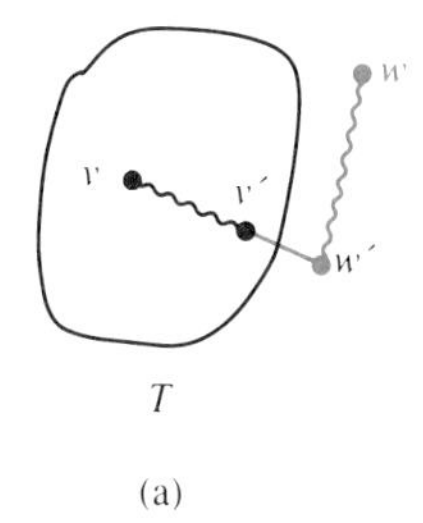

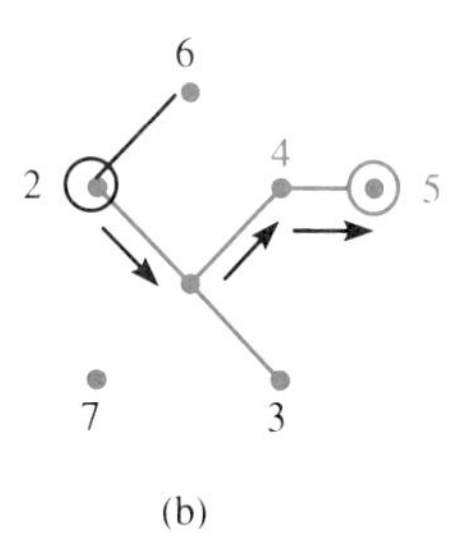

Figure 3.16 How to execute step 3 of the spanning tree algorithm: (a) In general. (b) In a specific case.

edge connecting v and w, but since the given graph is connected, there will at least be a *path* connecting v and w. Since the path *begins* inside T and *ends* outside T, at least one of its edges must be of the desired kind (one vertex in T, the other not in T). In Figure 3.16(a) this edge is $\overline{v'w'}$. Therefore, the vertex w' will qualify for addition to the growing tree at step 3.

If we apply this procedure to the growing tree in Figure 3.15(c) with vertex 2 playing the role of v and vertex 5 playing the role of w, we might find the path 2145 connecting 2 and 5. In this path, the edge $\overline{14}$ has the desired property (see Figure 3.16b); adding this edge to the tree of Figure 3.15(c), we arrive (as before) at the tree of Figure 3.15(d).

This argument shows that what we have been calling the "growing tree" will actually continue to grow until it contains all the vertices of the original graph. Now we need to show that it is in fact a *tree*! A good way to do this is to use mathematical induction, which was described in Section 1.5.

The idea is to let T_k denote the subgraph which has been produced by the algorithm after k executions of step 3, and to take as our induction statement

$$S_k\text{: "}T_k \text{ is a tree."}$$

First, note that S_1 is true, since T_1 consists of just one edge, the first edge selected to go in the spanning tree. Now we need to show that if S_k is true, then S_{k+1} is also true. To see this, consider Figure 3.17, which shows T_k growing into T_{k+1} via the addition of the new vertex v_{k+1}. For T_{k+1} to be a tree, it must (a) be *connected*, and (b) *contain no cycles.*

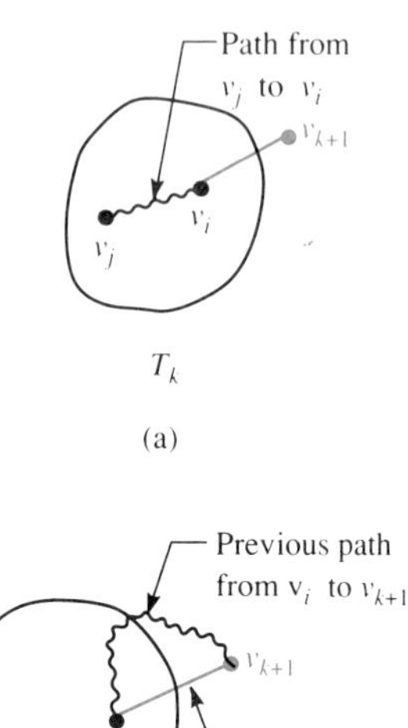

Figure 3.17 Proving that T_{k+1} is a tree, assuming that T_k is.

To prove that T_{k+1} is connected, note that the vertices in T_k are connected to each other, since T_k is a tree, by the induction assumption. When T_{k+1} is formed by the addition of v_{k+1}, the new vertex v_{k+1} is joined by an edge to one of the vertices in the tree T_k (vertex v_i in Figure 3.17a). Since v_i is connected to all of the other vertices in T_k, it follows that v_{k+1} is also connected to all of the vertices in T_k. Therefore T_{k+1} is connected.

To prove that T_{k+1} contains no cycles, suppose on the contrary that T_{k+1} *does* contain a cycle. Since T_k is by our induction assumption a tree, it doesn't have any cycles; and so if T_{k+1} does have a cycle, it must contain the new edge $\overline{v_i v_{k+1}}$ (see Figure 3.17b). But this would mean that v_i and v_{k+1} must have already been connected in T_k—which is plainly impossible, since v_{k+1} wasn't even in T_k. Therefore T_{k+1} contains no cycles.

This completes our induction proof that $T_1, T_2, \ldots,$ are all trees. The tree T_1 contains two vertices and one edge; T_2 contains three vertices and two edges; etc. Therefore, if the original graph contains n vertices, the algorithm will stop when T_{n-1}, containing all n vertices and $n - 1$ edges, is formed.

We apologize for belaboring the simple spanning tree algorithm, but we did wish to make the point that simply *stating* an algorithm does not make the algorithm work! In Problems 29–31 at the end of the section we give several other possible spanning tree algorithms; some of them work, and some don't. You're suppose to find out which is which.

EXAMPLE 3.5 Use the spanning tree algorithm to find a spanning tree for the graph of Figure 3.3(f).

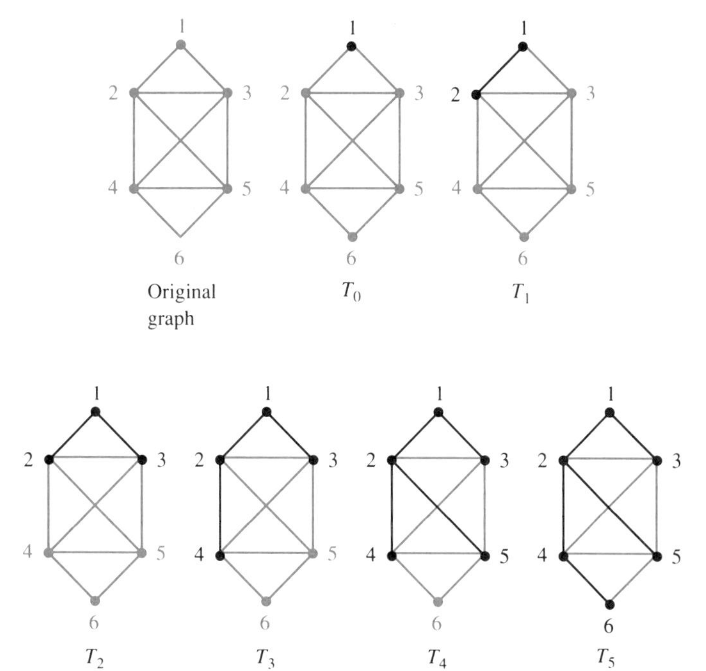

Figure 3.18 Solution to Example 3.5.

SOLUTION We begin by selecting some vertex, say vertex 1, and putting it into the spanning tree. This tree, which contains one vertex and no edges, is denoted by T_0 (in black) in Figure 3.18. There are now two vertices, vertex 2 and vertex 3, which aren't in the tree but are connected to the tree by an edge. (Vertex 2 by the edge $\overline{12}$, and vertex 3 by the edge $\overline{13}$.) If we add vertex 2 to the tree via the edge $\overline{12}$, we obtain the tree T_1 in Figure 3.18. Now we can add either vertex 3, vertex 4, or vertex 5; if we add vertex 3 via the edge $\overline{13}$, the tree T_2 results. We continue, adding vertices 4, 5, and 6, as shown in Figure 3.18, until we arrive at the tree T_5. This tree contains all of the vertices in the original graph, and is a spanning tree. ■

In the next section, we'll consider a practical problem about spanning trees, but just for fun we'll conclude this section by proving the famous "daisy chain theorem" about trees, which gives six different but equivalent definitions of a tree.

DAISY CHAIN THEOREM Let T be a graph with n vertices. The following six statements about T are equivalent, i.e., either T has *all six* of the properties, or *none* of them.

(a) T is a tree.
(b) T has no cycles and $n - 1$ edges.
(c) T is connected and has $n - 1$ edges.
(d) T is connected; but removing any edge disconnects it.
(e) There exists exactly one path between any two vertices.
(f) T has no cycles; but connecting two vertices with an edge will create a cycle.

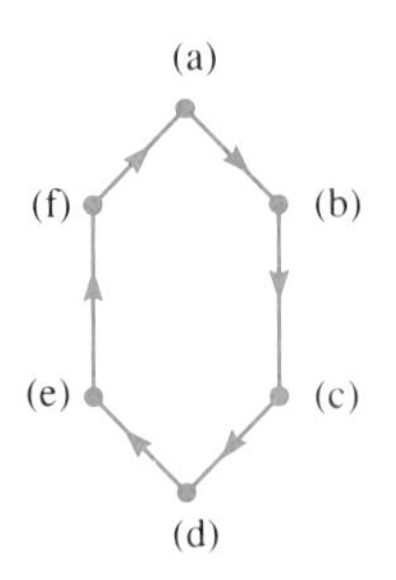

Figure 3.19 A directed graph that explains how the proof of the daisy chain theorem is organized.

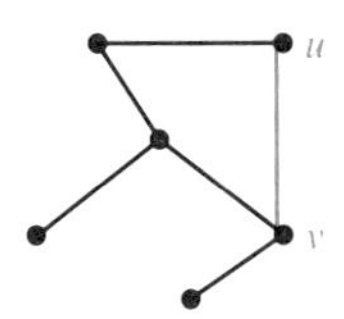

Figure 3.20. Adding an edge to a tree always produces a cycle. (Since u and v are already connected, adding a direct edge $\overline{uv}$ will create a cycle.)

PROOF The proof is in six parts. We will first show that if T has property (a), then it must also have property (b). This we call the (a) → (b) part of the proof. Then we will show that if T has property (b), then it must have property (c); this is the (b) → (c) part of the proof. Then, in succession, we will prove (c) → (d), (d) → (e), (e) → (f), and finally, (f) → (a), thereby completing the logical "daisy chain," illustrated in Figure 3.19, which proves that all six properties are equivalent. Here we go:

1. (a) → (b): Here we suppose that T is a tree, i.e., that it is a *connected graph with no cycles.* We need to prove that *T has no cycles and $n - 1$ edges.* We already know that T has no cycles—that's part of the definition of a tree—so all we need to show is that T must have $n - 1$ edges. To do this, let's apply the spanning tree algorithm to the graph T. Since T is connected, we know that the algorithm will produce a spanning tree for T with $n - 1$ edges. Furthermore, there can be no *other* edges in T, since adding an edge to a tree will always produce a cycle (see Figure 3.20), and T, being a tree, has no cycles.

2. (b) → (c): Here we assume that *T has no cycles and $n - 1$ edges*, and must prove that T is *connected and has $n - 1$ edges.* Since T has $n - 1$ edges by assumption, we just need to prove that T is connected.

The graph T, whether it is connected or not, will consist of one or more *connected components*, as shown in Figure 3.21. Furthermore, these connected components will all be *trees*, since T doesn't have any cycles. (Incidentally, a graph which consists of several disconnected trees is sometimes called a **forest**!) Now according to the (a) → (b) part of this proof, every tree has exactly *one more vertex than edge.* Thus in Figure 3.21, T_1 has one more vertex than edge, T_2 has one more vertex than edge, etc. Therefore the number of vertices exceeds the number of edges by the number of connected components. But we assumed that T has just one more vertex than edge (n vertices, $n - 1$ edges), and so T has just one connected component, i.e., T is connected.

3. (c) → (d): We assume now that T is connected and has $n - 1$ edges, and must prove that removing any edge from T disconnects T. So let us remove one edge from T, and call the resulting graph T'. The graph T' has n vertices and $n - 2$ edges. If T' is connected, then the spanning tree algorithm will produce a spanning tree for T' with n vertices and $n - 1$ edges; but this is impossible, since T' has only $n - 2$ edges. Therefore T' cannot be connected.

4. (d) → (e): Now we assume that T has property (d), and must prove that it has property (e), i.e., that there exists exactly one path between any two vertices.

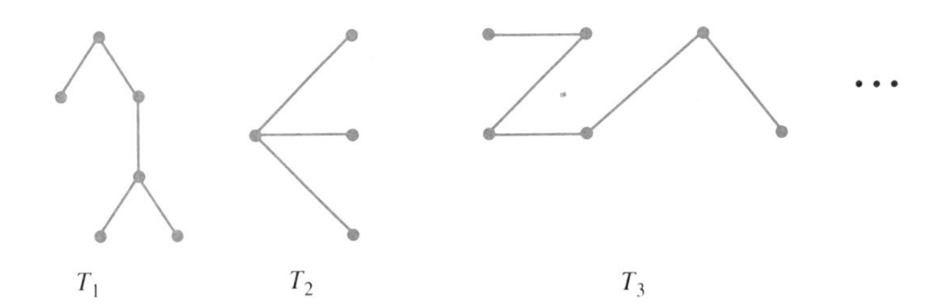

Figure 3.21 Connected components of a graph with no cycles. (A forest!)

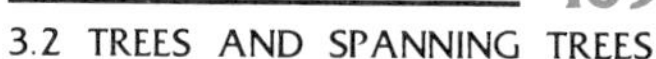

Thus let x and y be two vertices in T. Since by property (d), T is connected, there must be at least one path from x to y. Let us call one such path the "first path," as shown in Figure 3.22. Suppose that there is also a second path from x to y, also as shown in Figure 3.22. Any edge from the second path that doesn't belong to the first path can be removed from T without disconnecting it; but property (d) says that removing any edge from T *will* disconnect it. This is a contradiction, and so the second path doesn't exist: There is only one path from x to y.

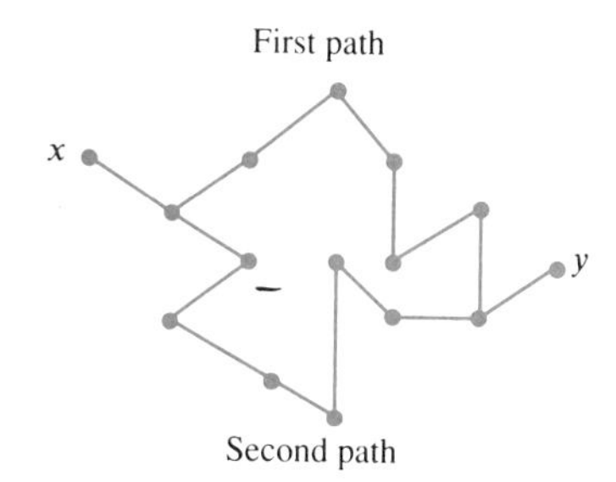

Figure 3.22 Proof of (d) → (e).

5. (e) → (f): We assume that T has property (e), and must prove that it has property (f), i.e., that T has no cycles, but adding an edge produces a cycle. First we will show that T has no cycles. This is easy: If T contains a cycle, and if u and v are vertices on the cycle, then there must be two paths between u and v (see Figure 3.23a). But by property (e), there cannot be two such paths; and so there cannot be a cycle.

Next we will show that adding an edge to T produces a cycle. Let x and y be two vertices in T which are not joined by an edge (see Figure 3.23b). By property (e), there is a path between x and y, so adding the edge $\overline{xy}$ will form a cycle.

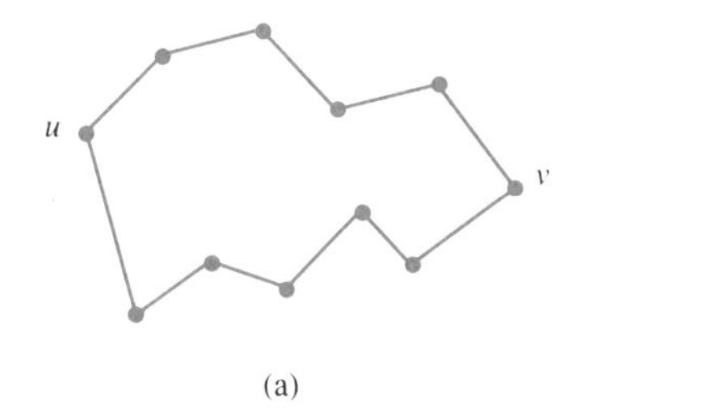

(a)

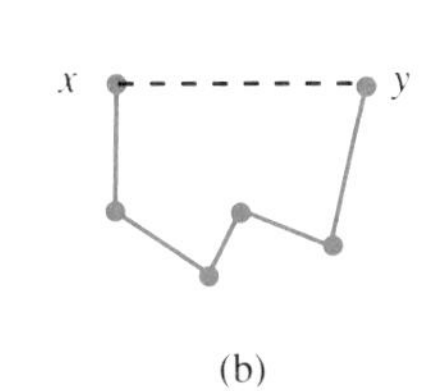

(b)

Figure 3.23 Proof of (e) → (f).

6. (f) → (a): Assuming property (f), we must show that T is connected, i.e., there is a path between any pair of vertices. If u and v are vertices of T, then adding the edge $\overline{uv}$ will create a cycle, by property (f) (see Figure 3.24). The edges of the cycle other than $\overline{uv}$ will form a path from u to v.

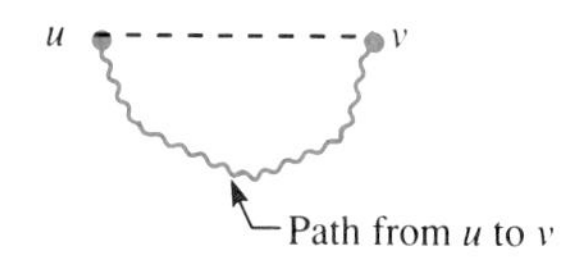

Figure 3.24 Adding the edge $\overline{uv}$ creates a cycle.

This completes the proof of the daisy chain theorem. For some practice using it, see Problems 16–23. ∎

Problems for Section 3.2

In Problems 1–3, find a spanning tree for each of the graphs.

1.

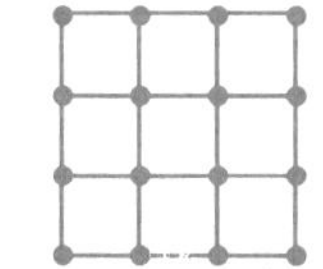

2.

3.

4. The following matrix is the adjacency matrix of a certain graph. Sketch the graph, and find a spanning tree.

	A	B	C	D	E
A	0	1	0	0	1
B	1	0	1	0	0
C	0	1	0	1	0
D	0	0	1	0	1
E	1	0	0	1	0

5. Repeat Problem 4, but use the following adjacency matrix.

	0	1	2	3	4	5	6	7	8	9
0	0	1	0	0	1	1	0	0	0	0
1	1	0	1	0	0	0	1	0	0	0
2	0	1	0	1	0	0	0	1	0	0
3	0	0	1	0	1	0	0	0	1	0
4	1	0	0	1	0	0	0	0	0	1
5	1	0	0	0	0	0	0	1	1	0
6	0	1	0	0	0	0	0	0	1	1
7	0	0	1	0	0	1	0	0	0	1
8	0	0	0	1	0	1	1	0	0	0
9	0	0	0	0	1	0	1	1	0	0

6. Consider the following set of four vertices:

Two possible spanning trees are shown below:

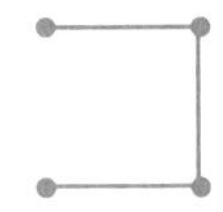

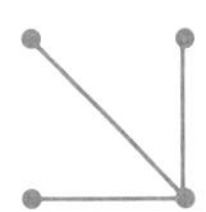

Draw all possible spanning trees.

7. Draw all possible spanning trees for a graph with five vertices.

8. Let C_n denote the *n-cycle* graph. The graphs C_3, C_4, C_5 are shown in the accompanying figure. Find a formula for the number of spanning trees in C_n, for $n = 2, 3, 4, \ldots$.

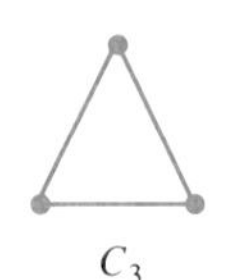

C_3

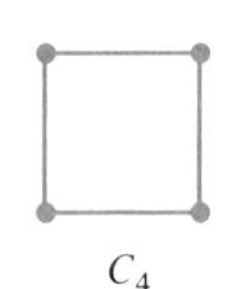

C_4

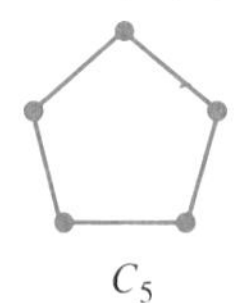

C_5

9. See how many cycles you can find in the graph of Figure 3.14(d).

10. How many spanning trees does the graph in the accompanying figure have?

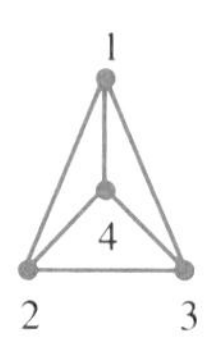

11. Redo Example 3.5, this time beginning with vertex 4, and always choosing the lowest-numbered available vertex in step 3. What spanning tree results?

In Problems 12–15, is there a tree with eight vertices satisfying the given conditions?

12. Each vertex has degree 1.

13. The vertices have degrees 2, 2, 2, 2, 2, 2, 1, 1.

14. All vertices have degree 2.

15. The vertices have degrees 7, 1, 1, 1, 1, 1, 1, 1.

In Problems 16 and 17, establish the result using the daisy chain theorem.

16. (a) A connected graph with one more vertex than edge is a tree.
(b) Removing an edge from a tree disconnects it.
(c) Adding an edge to a tree creates exactly one cycle.

17. (a) Every tree has one more vertex than edge.
(b) A spanning tree for a graph with n vertices has those n vertices and $n - 1$ edges.

In Problems 18–23, if the statement is true, prove it. If false, produce a counterexample.

18. If a graph G has one more vertex than edge, then it is a tree.

19. If G is connected, then it has no cycles.

20. If G is connected "edge minimally," i.e., removing an edge from G disconnects it, then G has no cycles.

21. If G has no cycles, 25 edges, and 26 vertices, then G is connected.

22. If G has 32 edges and 28 vertices, then G is not a tree.

23. If G is connected and has 10 edges and 10 vertices, then G contains at least one cycle.

24. Suppose G is connected, with 13 vertices and 17 edges. How many vertices are there in a spanning tree for G?

25. We know from the daisy chain theorem that (a) and (e) are equivalent, but the proof was indirect in that we proved (e) implies (f) implies (a) implies (b) implies (c) implies (d) implies (e). Try proving (a) and (e) equivalent directly, just for practice.

26. Suppose T_1 and T_2 are both spanning trees for G.
(a) Do T_1 and T_2 have the same vertices? The same edges?
(b) Do T_1 and T_2 have the same number of vertices? The same number of edges?

27. Show that a connected graph G with n vertices and at least one cycle must have at least n edges. (*Suggestion*: Consider a spanning tree for G.)

28. Suppose G has the same number of vertices as edges. Show that G must contain at least one cycle under the following conditions.
(a) G is connected.
(b) G is not connected.

In Problems 29–31, an algorithm for finding a spanning tree in a given connected graph is specified. In each case, determine whether or not the algorithm works, and justify your conclusion.

29. Begin with the given graph. If there is a cycle, remove one edge from it. If the resulting graph still has a cycle, remove one edge from it. Continue until no cycles remain.

30. Start with any edge and continue to add new edges one at a time until the resulting graph contains $n - 1$ edges.

31. Start with any edge and continue to add new edges one at a time, subject to the restriction that the addition of an edge is not allowed to create a cycle. Continue until no edge can be added without creating a cycle.

32. In the text we said that the spanning tree algorithm would only succeed in finding a spanning tree if the original graph was connected. Suggest a way to modify the algorithm so that
(a) If the graph is connected, it will produce a spanning tree.
(b) If the graph is not connected, this fact will be reported.

33. How many spanning trees does the graph of Figure 3.14(d) have?

3.3 Minimal Spanning Trees; Prim's Algorithm

In the previous section we discussed spanning trees in some detail. In this section, we'll see one reason why spanning trees are important.

Consider Figure 3.25; it is a connected graph with 7 vertices and 11 edges. It is identical to the graph in Figure 3.14(d), *except* that in Figure 3.25 each edge has a *number* attached to it. For definiteness, we could think of the 7 vertices as 7 *cities* located in a remote wilderness area, and the numbers on the edges as the *costs* (in millions of dollars) of building direct communications links between given pairs of cities. The fact that there is no edge between cities 1 and 5, for example, might be interpreted to mean that the cost of building a direct link between them is prohibitive, i.e., effectively infinite. It is of practical importance to know how to connect all these cities together with a communications network as cheaply as possible.

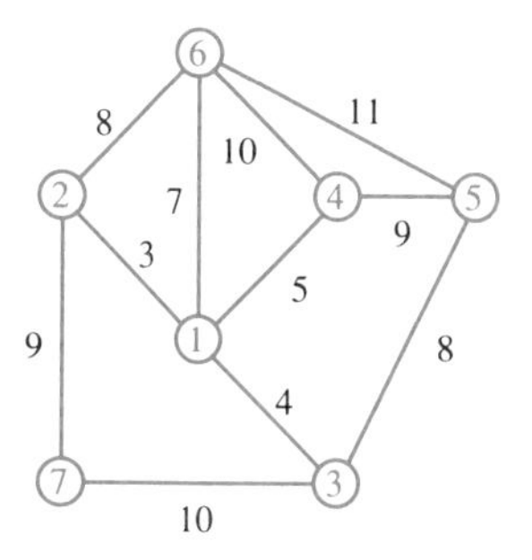

Figure 3.25 A connected graph with "edge-costs."

To do this, we should build a communications network consisting of a *subset* of the edges in Figure 3.25 which forms a *spanning tree.* Why a spanning tree? Because if the graph we build *isn't connected*, the cities won't all be able to communicate with each other; and if it *has a cycle*, we can remove an edge (thereby saving money), and the graph will still be connected. Connected + no cycles = tree. This much is easy!

Unfortunately, however, as we have seen, a given graph can have many spanning trees. In Figure 3.26, for example, we show three spanning trees for the graph of Figure 3.25 together with the "costs" of these trees, where the cost of a tree is the sum of the costs of its edges. Of the three spanning trees in Figure 3.26, the cheapest has cost 37. But there are many other spanning trees—is there a cheaper one?

This question is a special case of what is called the problem of the **minimal-cost spanning tree** (MST). Given a connected graph G in which each edge is labelled

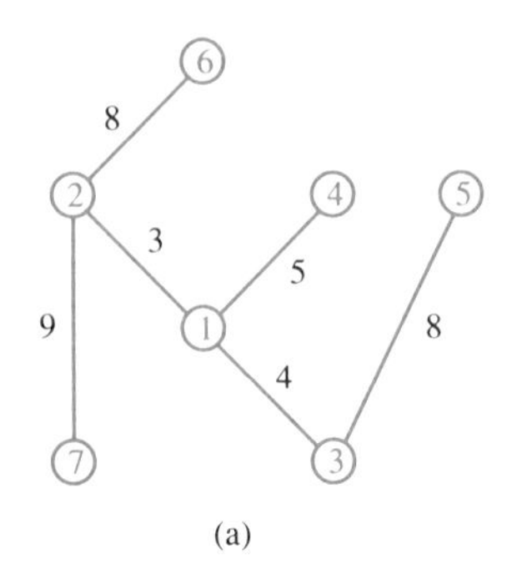

(a)

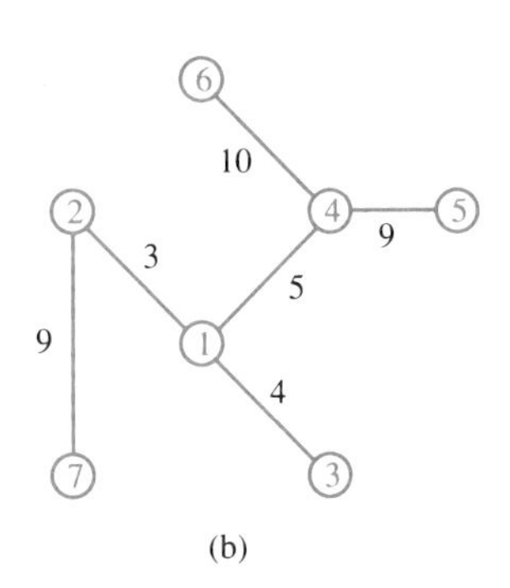

(b)

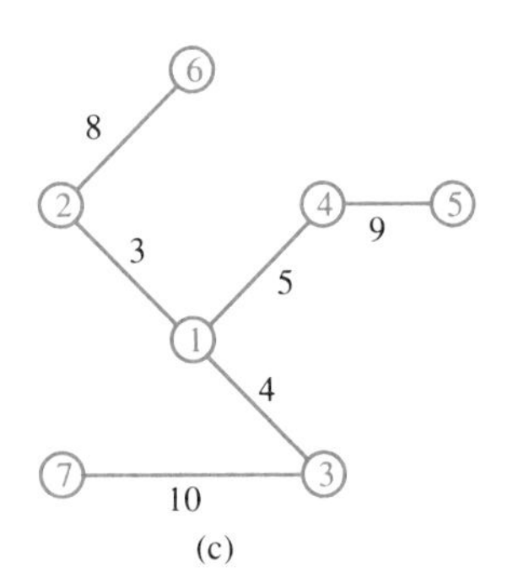

(c)

Figure 3.26 Some spanning trees for the graph of Figure 3.25. (a) cost = 37; (b) cost = 40; (c) cost = 39.

with a number, an MST is a spanning tree for which the sum of the edge labels is as small as possible. (In applications, the edge labels may signify *lengths*, *weights*, or something else other than costs.)

Surprisingly, there is a very simple algorithm for finding an MST—all we need to do is make a slight modification of the spanning tree algorithm of Section 3.2. In the spanning tree algorithm, any vertex not in the tree but connected to it by an edge can be added. To find a *minimal-cost* spanning tree, however, we must be more selective—we must always add a new vertex for which *the cost of the new edge is as small as possible.*

With this simple modification, the spanning tree algorithm becomes **Prim's algorithm*** for finding an MST. In Figure 3.27 we show Prim's algorithm in both a "flowchart" and "pseudocode" format, two useful ways for displaying algorithms. The flowchart format is more or less self-explanatory, with the various parts of the algorithm being represented by one of three types of boxes:

- *Oval-shaped* These are *terminal nodes*, where the algorithm either begins or ends.
- *Diamond-shaped* These are *decision nodes*, where a question is asked, and the answer determines which branch is taken.
- *Rectangular* These are *process nodes*, where some action takes place.

Pseudocode is a bit harder to explain; if you are not already familiar with it, you should skim the Appendix on pseudocode at the end of the book.†

Prim's algorithm is an example of a **greedy algorithm**, a term applied to any algorithm that always does the expedient thing without worrying about the long-term consequences. It is by no means obvious that by being greedy and at every stage picking the cheapest available edge, we will always end up with an MST—proof is required! However, Prim's algorithm does in fact always work. But before we give the proof, let's do an example.

EXAMPLE 3.6 Use Prim's algorithm to find an MST for the graph of Figure 3.25.

SOLUTION The solution can be seen in the seven graphs in Figure 3.28, in which we follow the growth of the MST, assuming that we begin with vertex 7. As you can see, the seven vertices are added to the tree in the order 7, 2, 1, 3, 4, 6, 5. For example, in Figure 3.28(a) the tree consists of a single vertex (vertex 7), and no edges. At this point, vertices 2 and 3 could be attached to the tree, via the edges $\overline{27}$ and $\overline{37}$. However, of these two edges, the one of *least cost* is $\overline{27}$ (cost 9). If we add the edge $\overline{27}$ to the tree of Figure 3.28(a), the tree of Figure

* Robert C. Prim (1921–), American mathematician. He discovered "Prim's algorithm" at Bell Laboratories in 1956, when his help was sought by an AT&T group trying to mechanize private-line tolling procedures.

† In Figure 3.27 the symbol ← represents the *assignment* operation. The statement $X \leftarrow 2$ means "assign the variable X the value 2." It is equivalent, for example, to the following programming statements:

BASIC: LET X = 2	C: X = 2
Pascal: X := 2	FORTH: 2 X !

3.28(b) results. Similarly, in Figure 3.28(d) the growing tree has four vertices (1, 2, 3, and 7) and three edges, and any one of the three remaining vertices 4, 5, or 6 could be attached to the tree, via the edges $\overline{41}$, $\overline{53}$, $\overline{61}$, or $\overline{62}$. Of these four edges, the one of least cost is $\overline{41}$ (cost 5), so we add it to the tree, and Figure 3.28(e) results. When finally the last vertex, vertex 5, is added to the tree, we arrive at the required minimal spanning tree of Figure 3.28(g). Its total cost is 36, which is, indeed, lower than the cost of any of the spanning trees in Figure 3.26. ■

We will conclude this section by giving a formal proof that Prim's MST algorithm does unfailingly produce a minimal spanning tree.

Thus, let G be a graph with n vertices, with costs attached to each edge, and let $e_1, e_2, \ldots, e_{n-1}$ be the sequence of $n - 1$ edges produced by Prim's algorithm. We want to show that the graph consisting of the $n - 1$ edges $\{e_1, \ldots, e_{n-1}\}$ is indeed an MST for G. In order to do this, for each $k = 0, 1, \ldots, n - 1$, we denote by G_k the graph consisting of the first k edges selected by Prim's algorithm. (The

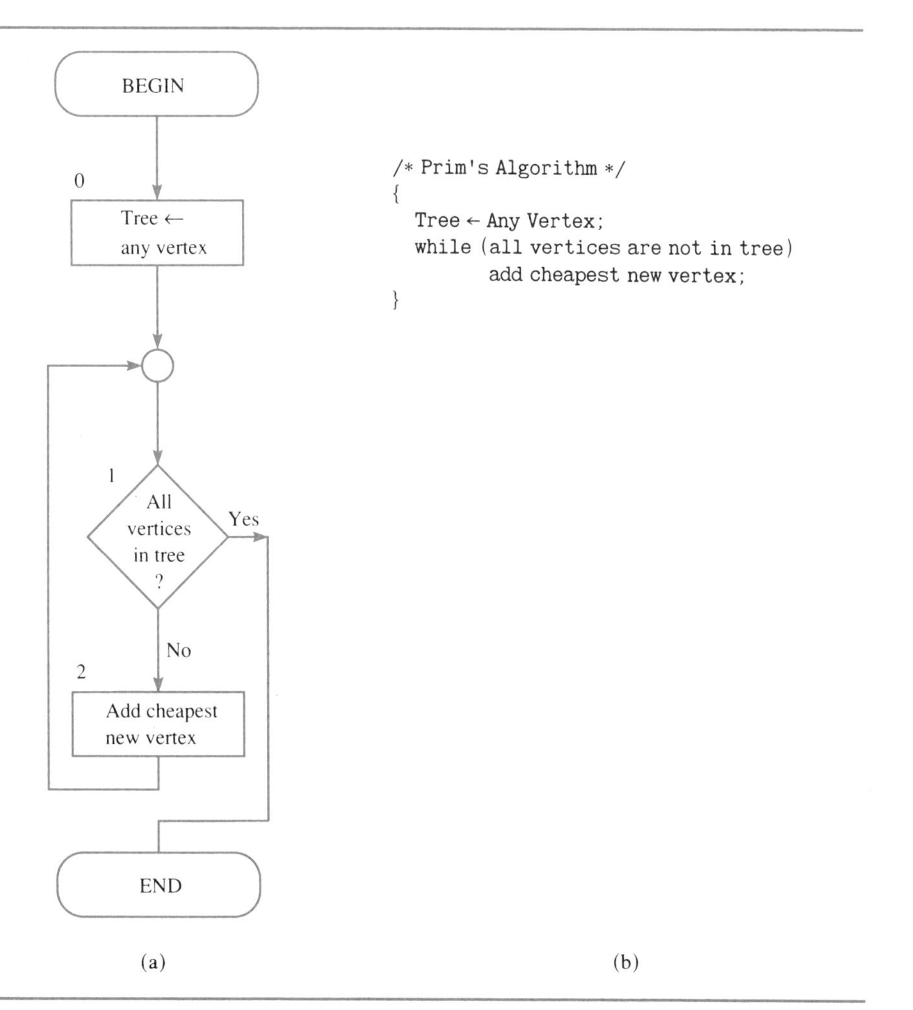

Figure 3.27 Prim's algorithm described in (a) structured flowchart form and (b) pseudocode.

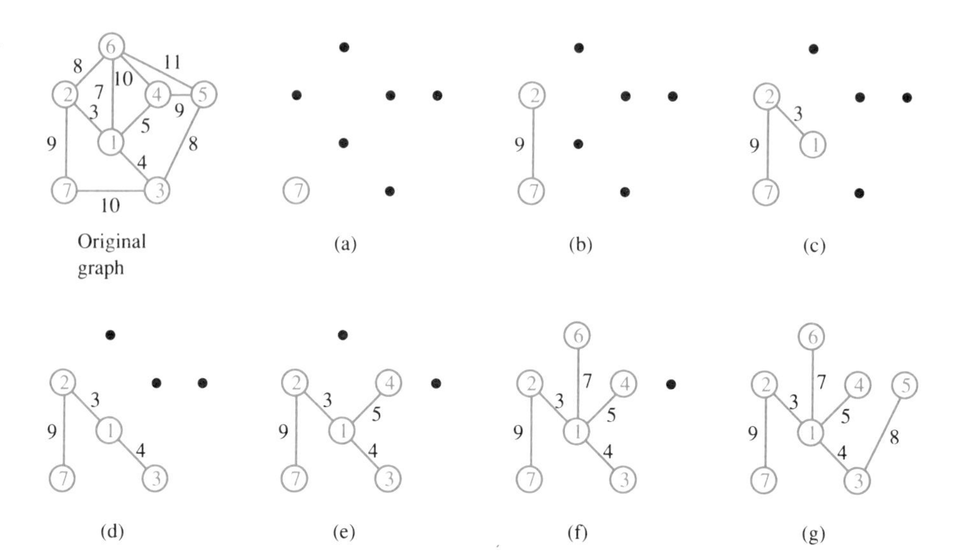

Figure 3.28 Solution to Example 3.6.

graph G_0 consists of the initial vertex v_0 and nothing else.) In Figures 3.28(a)–3.28(g), for example, we see the graphs $G_0, G_1, \ldots, G_6$ slowly growing into an MST for the seven-vertex graph G. The following assertion says that this happens in general:

> *Assertion* A_k: The edges in G_k form a subset of the edges in *some* MST.

Notice that assertion A_{n-1} is what we want to prove. The idea is to prove that the assertions $A_0, A_1, A_2, \ldots, A_{n-1}$ are all true, using the technique of mathematical induction. To do this, we proceed as follows: First, we show that A_0 is true; then we show that *if* A_k is true, *then* A_{k+1} is also true.

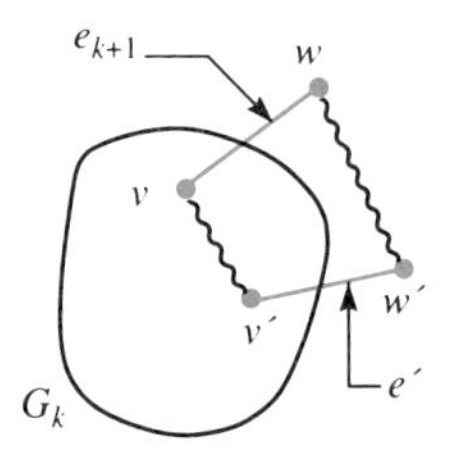

Figure 3.29 Induction step in the proof of the greedy tree algorithm.

The assertion A_0 is certainly true, because G_0 contains no edges at all!

Now let's assume that A_k is true, and show that A_{k+1} must also be true. Let T be an MST containing the subgraph G_k consisting of edges $e_1, e_2, \ldots, e_k$ (there must be such an MST, since we assume A_k is true). Because of the way the algorithm works (see Problem 15), one endpoint of e_{k+1}, call it v, lies in G_k and the other, call it w, lies outside of G_k. If the MST T also contains e_{k+1}, we have of course succeeded in finding an MST containing $\{e_1, e_2, \ldots, e_{k+1}\}$—$T$ itself! If T *doesn't* contain e_{k+1}, too bad. Nevertheless (see Figure 3.29), since T is connected, it must contain *some* path connecting v and w. This path *starts* inside G_k (at v) and *ends* outside G_k (at w), so it must contain an edge with one vertex in G_k and one not in G_k. In Figure 3.29 this edge is called e'; its vertices are v' and w'. Since e_{k+1} was chosen greedily, i.e., to have the smallest possible cost among *all* edges which have one vertex in G_k and one not in G_k, we know that cost $(e_{k+1}) \leq$ cost (e').

Now if we remove e' from T and add e_{k+1} to T, the resulting graph T' will still have $n - 1$ edges and still be connected. (Any two vertices are connected in T by a path; if this path uses the deleted edge e', replace each occurrence of e' with the longer path $v' \to v \to w \to w'$ in T'.) Thus, by property (c) of the daisy chain theorem in Section 3.2, T' is a tree. But the total cost of T' is less than or equal to the cost of T, since $\text{cost}(e_{k+1}) \le \text{cost}(e')$. Thus since T has smallest possible cost, so does T', and so T' is an MST containing $\{e_1, e_2, \ldots, e_{k+1}\}$. This completes the proof by induction that assertion A_k is true for all $k = 0, 1, \ldots, n-1$, and as we observed before, the assertion A_{n-1} is that Prim's algorithm produces an MST.

Problems for Section 3.3

Apply Prim's algorithm to each of the graphs in Problems 1–6.

1.

2.

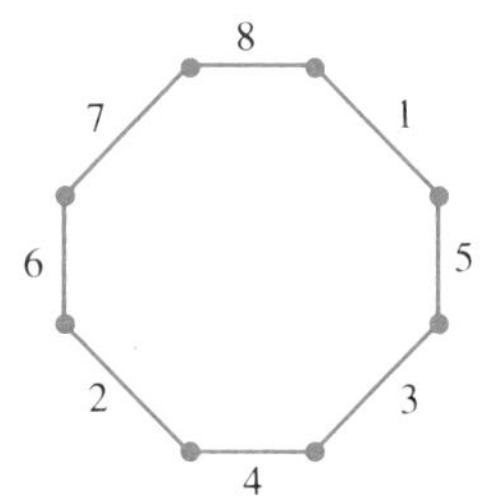

3.

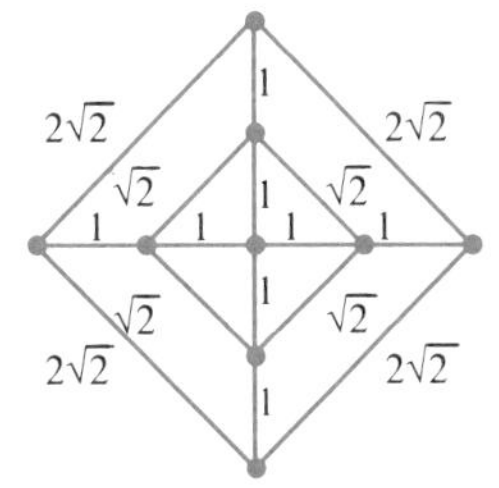

4.

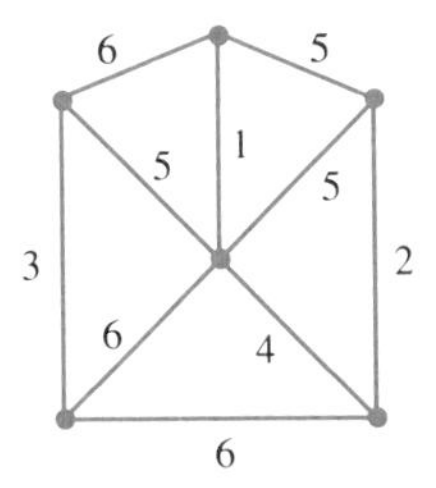

5. Graph described by the following cost matrix; for example, the entry 4 in row C, column D means that the cost of the edge CD is 4.

	A	B	C	D	E
A	0	2	∞	∞	3
B	2	0	1	∞	∞
C	∞	1	0	4	∞
D	∞	∞	4	0	5
E	3	∞	∞	5	0

6. Graph described by the following cost matrix.

	0	1	2	3	4	5	6	7	8	9
0	0	2	∞	∞	3	1	∞	∞	∞	∞
1	2	0	1	∞	∞	∞	2	∞	∞	∞
2	∞	1	0	4	∞	∞	∞	3	∞	∞
3	∞	∞	4	0	5	∞	∞	∞	4	∞
4	3	∞	∞	5	0	∞	∞	∞	∞	5
5	1	∞	∞	∞	∞	0	∞	2	3	∞
6	∞	2	∞	∞	∞	∞	0	∞	4	5
7	∞	∞	3	∞	∞	2	∞	0	∞	1
8	∞	∞	∞	4	∞	3	4	∞	0	∞
9	∞	∞	∞	∞	5	∞	5	1	∞	0

7. Find a graph with more than one minimal spanning tree.

8. In the accompanying figure there are 12 dots. Connect these dots into the structure of a tree with 11 edges, whose total length is as small as possible (You will need a ruler!)

9. Consider a graph with n vertices numbered $0, 1, \ldots, n-1$. Suppose the cost of the edge $\overline{uv}$ is $|u - v|$. The accompanying figures show the graphs for $n = 3$ and $n = 4$. Describe an MST for this graph, for general n.

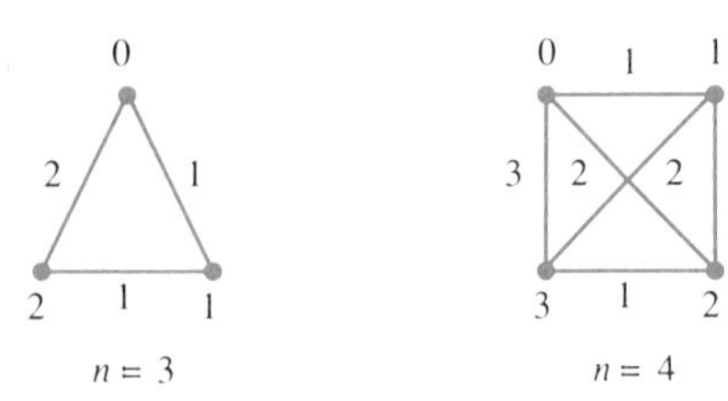

10. Repeat Problem 9, except use $c(u, v) = u + v$.

11. In Prim's algorithm it may be necessary to set certain costs equal to ∞. Suggest a practical way to implement "∞" in a computer program.

12. A **maximal spanning tree** for a graph with edge costs is a spanning tree whose total cost is as large as possible. Show how to modify Prim's algorithm to find a maximal spanning tree.

13. Continuing Problem 12, find a maximal spanning tree for the graph of Problem 1.

14. Consider again the graph for Problem 1. Suppose you are asked to find a spanning tree which includes the edge of cost 23 whose total cost is as small as possible. How well can you do?

15. In the proof of validity of Prim's algorithm, we stated that one endpoint of e_{k+1} lies inside G_k and the other endpoint lies outside G_k. Explain in detail why this is true.

3.4 Binary Trees and Tree Searching

In the previous two sections we have studied trees and some of their general properties. In this section we will study a very important special class of trees called **binary trees**, which arise in many applications, especially in computer algorithms.

In Figure 3.30, we see four examples of binary trees. In every case there is one vertex at the top; this vertex is called the **root** of the tree. (In graph theory, unlike nature, trees always have the root at the top!) From the root there may be one or two descending edges, one going left and one going right. The vertex connected to the root on the left is called the root's **leftchild**, and the one on the

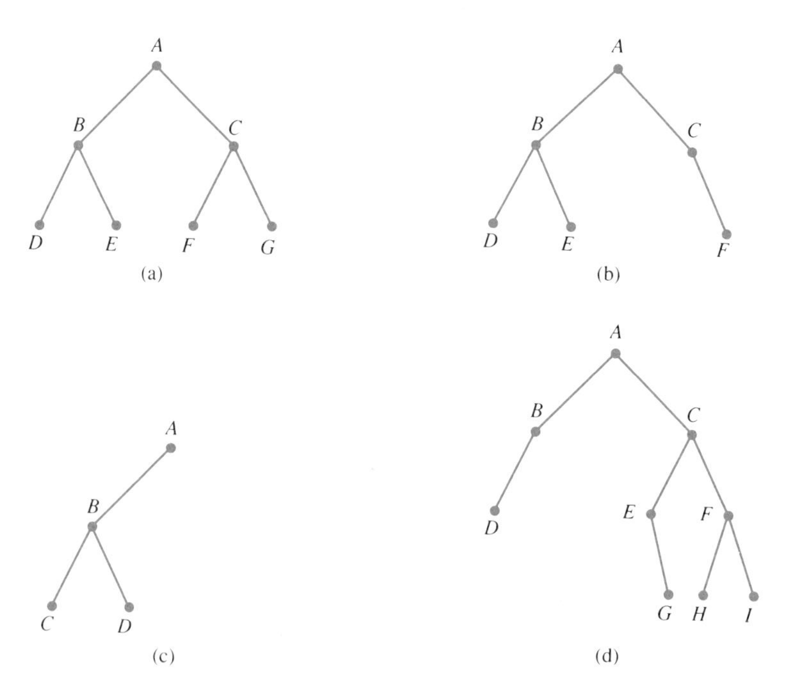

Figure 3.30 Four binary trees.

TABLE 3.2 Tabular representation of the four binary trees in Figure 3.30; in each case the root is designated with an arrow

Vertex	LCHILD	RCHILD
→A	B	C
B	D	E
C	F	G
D	—	—
E	—	—
F	—	—
G	—	—

(a)

Vertex	LCHILD	RCHILD
→A	B	C
B	D	E
C	—	F
D	—	—
E	—	—
F	—	—

(b)

Vertex	LCHILD	RCHILD
→A	B	—
B	C	D
C	—	—
D	—	—

(c)

Vertex	LCHILD	RCHILD
→A	B	C
B	D	—
C	E	F
D	—	—
E	—	G
F	H	I
G	—	—
H	—	—
I	—	—

(d)

right is called the **rightchild**. Each of these children may have left- and rightchildren, and so on, until at the bottom, all the vertices are childless. For example, in the binary tree of Figure 3.30(b) the root is A; A's leftchild is B, and its rightchild is C. B's leftchild is D, and its rightchild is E. C has no leftchild, but its rightchild is F. The vertices D, E, and F have no children at all.

A binary tree can be represented in a computer by a simple table, listing each of the vertices, together with the corresponding leftchild and rightchild. In Table 3.2, we give this representation for each of the binary trees in Figure 3.30.

Binary trees can also be defined **recursively**, in the following rather strange way (see Figure 3.31).

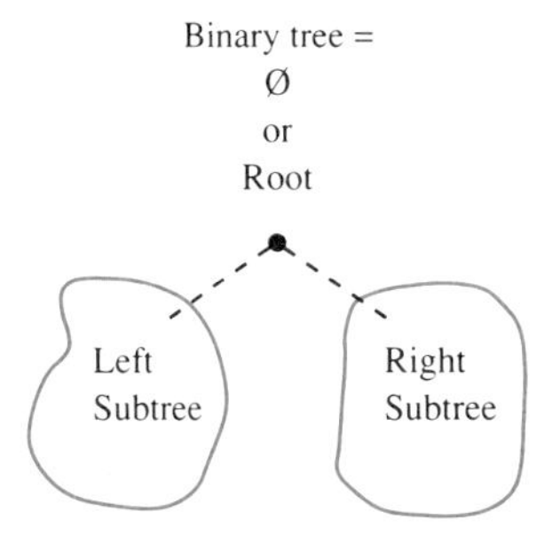

Figure 3.31 The recursive definition of a binary tree.

DEFINITION A binary tree is either empty or consists of a root and two disjoint binary trees, called the **left subtree** and the **right subtree**.

At first this definition may seem confusing and circular, since a binary tree is defined in terms of other binary trees. But it's really okay, since each subtree has fewer vertices than the original tree, and each sub-subtree involves fewer still, etc., until eventually the sub-. . . -subtrees are all empty. The decomposition is

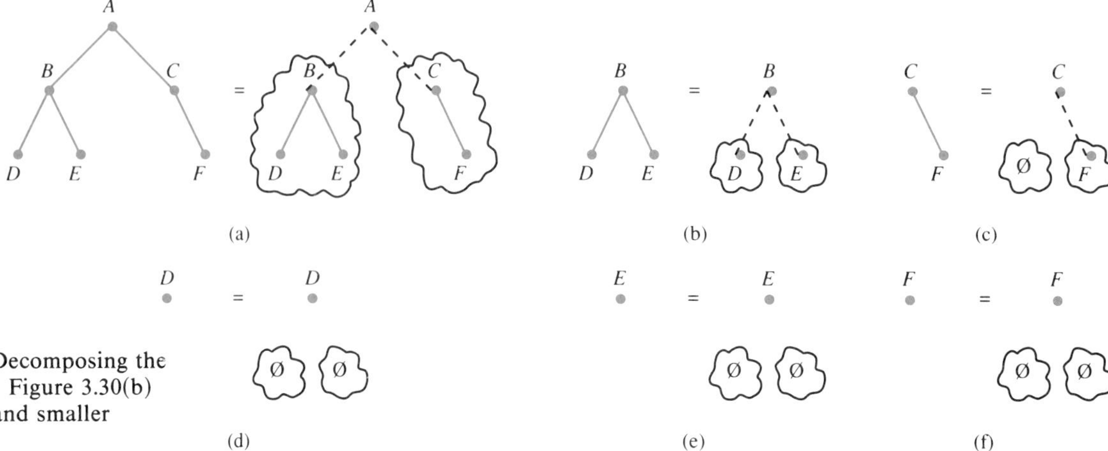

Figure 3.32 Decomposing the binary tree in Figure 3.30(b) into smaller and smaller subtrees.

illustrated in Figure 3.32, where we have dissected the binary tree in Figure 3.30(b) into smaller and smaller subtrees, using $\varnothing$ to denote the empty tree.

It is often necessary to "search" a binary tree, i.e., to travel through the tree, visiting the vertices, one at a time. Since the vertices in a tree are not arranged linearly, it is not clear in which order the vertices should be visited. For example, consider the tree in Figure 3.30(a). Assuming that we start at the root A, in which order should we visit the vertices? One natural answer is, of course, the alphabetical order A, B, C, D, E, F, G, but upon reflection this ordering may not seem so natural after all, since it requires several "jumps" between vertices that aren't connected by an edge, e.g., $B \to C$, $D \to E$, etc.

It turns out that there are several simple algorithms for tree searching, which allow us to visit each of the vertices in a binary tree without jumping all over the tree. We will study several of these algorithms in this section.

The most useful tree-searching algorithm is called **preorder search**.* Preorder search can be described in several different ways, but the simplest description is the following:

Preorder search

1. Visit the root.
2. Visit the left subtree, using preorder.
3. Visit the right subtree, using preorder.

This definition of preorder is recursive, since preorder is defined in terms of preorder. But as with the recursive definition of a binary tree, it isn't really circular. For example, let's search the binary tree in Figure 3.30(b) using preorder search. According to the definition, we must first visit the root A, then the left subtree, and then the right subtree. Thus the vertices of the tree will be visited in the order

$$A, [A\text{'s left subtree}], [A\text{'s right subtree}],$$

where [A's left subtree] denotes the preordering of the vertices in the left subtree, and [A's right subtree] denotes the preordering of the vertices in the right subtree.

* Preorder search is also called **depth-first search**, although this term is more often applied to algorithms that search more general graphs.

This reduces the problem to that of searching the subtrees of the original tree. Since both of these trees are smaller than the original tree, we have made progress.

The left subtree of A is shown in Figure 3.32(b). To visit this subtree, we first visit its root, which is B, then its left subtree, and then its right subtree. The left subtree, however, consists of the single vertex D, and the right subtree consists of the single vertex E, and so

$$[A\text{'s left subtree}] = B, D, E.$$

To finish up, we need to visit the right subtree of the original tree, which is shown in Figure 3.32(c). We first visit its root, which is C, then its left subtree, and then its right subtree. Its left subtree is empty, however, and its right subtree consists of the single vertex F, and so

$$[A\text{'s right subtree}] = C, F.$$

Putting this all together, we find that the preordering of the vertices in the tree of Figure 3.30(b) is

$$A, B, D, E, C, F.$$

Preorder search is also called **backtracking**, since another way to describe it is to say that at every stage you go left, unless this would cause you to visit an empty tree or to visit a vertex for the second time, at which stage you go right, unless this would also cause you to visit an empty tree or duplicate a visit, at which point you back up to the previous vertex. The algorithm halts when it reaches the root and can't move. To summarize, the backtracking version of preorder search is to start at the root, and at each successive vertex.

1. Go left, if possible; if not, go to step 2.
2. Go right, if possible; if not, go to step 3.
3. Back up, if possible; if not, go to step 4.
4. Halt.

Using the tree of Figure 3.30(b) as an example again, and starting at A, backtracking would proceed as follows (with brackets indicating a backtrack to a vertex previously visited):

$$A, B, D, [B], E, [B], [A], C, F, [C], [A], \text{Halt.}$$

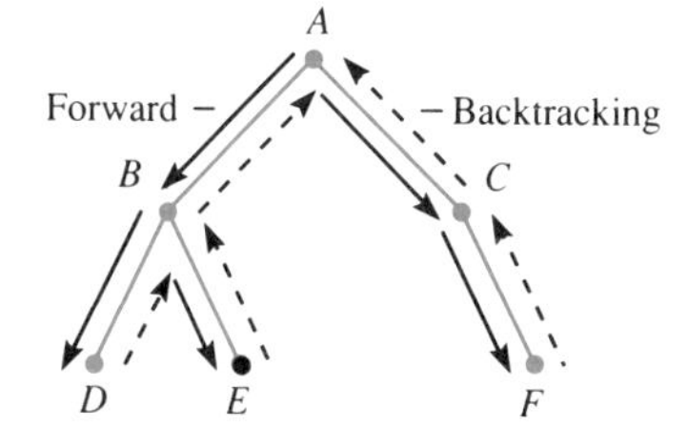

Figure 3.33 Backtracking (preorder) search of the tree in Figure 3.30(d).

In Figure 3.33, we show the path of the backtracking version of preorder search, as applied to the tree of Figure 3.30(b), with the forward moves shown with solid lines, and backward moves with dotted lines. If you imagine that the edges of the tree are walls, then preorder search simply amounts to starting at the root, putting your left hand on the wall, and following the walls until you reach the root again.

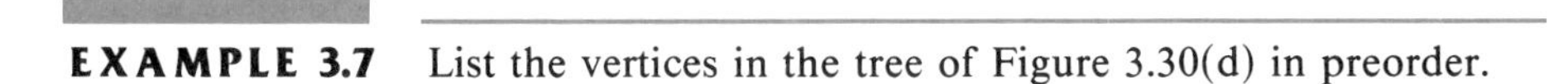

EXAMPLE 3.7 List the vertices in the tree of Figure 3.30(d) in preorder.

SOLUTION Let's do this in two ways, the recursive way and the backtrack way. If we proceed recursively, we start by writing

$$A, [A\text{'s left subtree}], [A\text{'s right subtree}].$$

The left subtree contains only the root B and its leftchild D, and so

$$[A\text{'s left subtree}] = B, D.$$

The right subtree is more complicated. Its root is C, so we can write

$$[A\text{'s right subtree}] = C, [C\text{'s left subtree}], [C\text{'s right subtree}].$$

Now, C's left subtree has root E and rightchild G, so that

$$[C\text{'s left subtree}] = E, G.$$

Similarly, C's right subtree has root F, leftchild H, and rightchild I; so that

$$[C\text{'s right subtree}] = F, H, I.$$

It follows that

$$[A\text{'s right subtree}] = C, E, G, F, H, I.$$

Thus we have found that the preordering of the vertices of the original tree is

$$A, B, D, C, E, G, F, H, I.$$

Now let's do the same thing using backtracking. We find that the backtrack search proceeds as follows:

$$A, B, D, [B], [A], C, E, G, [E], [C], F, H, [F], I, [F], [C], [A], \text{Halt.}$$

Besides preorder search, there are several other tree-searching algorithms which are occasionally useful. The most common are called **postorder** and **inorder**. They are defined recursively as follows:

Postorder search

1. Visit the left subtree, using postorder.
2. Visit the right subtree, using postorder.
3. Visit the root.

Inorder search

1. Visit the left subtree, using inorder.
2. Visit the root.
3. Visit the right subtree, using inorder.

EXAMPLE 3.8 Search the binary tree in Figure 3.30(d) in (a) postorder and (b) inorder.

SOLUTION

(a) According to the recursive definition, the postordering of the vertices will be

$$[A\text{'s left subtree}], [A\text{'s right subtree}], A,$$

where [A's left subtree] denotes the postordering of A's left subtree and [A's right subtree] denotes the postordering of A's right subtree. Similarly, since the root of A's left subtree is B, we have

$$[A\text{'s left subtree}] = [B\text{'s left subtree}], [B\text{'s right subtree}], B.$$

But since B's left subtree contains only the root D, and B's right subtree is empty, we have

$$[A\text{'s left subtree}] = D, B.$$

Next, since A's right subtree has root C, we have

$$[A\text{'s right subtree}] = [C\text{'s left subtree}], [C\text{'s right subtree}], C.$$

But since C's left subtree has root E, empty left subtree, and right subtree consisting only of the root G, we have

$$[C\text{'s left subtree}] = G, E.$$

Similarly, C's right subtree has root F, left subtree consisting of the root H, and right subtree consisting of the root I, so we have

$$[C\text{'s right subtree}] = H, I, F.$$

Hence,

$$[A\text{'s right subtree}] = G, E, H, I, F, C.$$

Finally, we see that the postordering of the vertices in the tree of Figure 3.30(d) is

$$D, B, G, E, H, I, F, C, A.$$

(b) According to the definition, the inordering of the vertices in the tree of Figure 3.30(d) is

$$[A\text{'s left subtree}], A, [A\text{'s right subtree}],$$

where [A's left subtree] denotes the inordering of A's left subtree, and [A's right subtree] denotes the inordering of A's right subtree. Since the root of A's left subtree is B, we have

$$[A\text{'s left subtree}] = [B\text{'s left subtree}], B, [B\text{'s right subtree}].$$

But B's left subtree contains only the root D, and B's right subtree is empty, so that

$$[A\text{'s left subtree}] = D, B.$$

Similarly,

$$[A\text{'s right subtree}] = [C\text{'s left subtree}], C, [C\text{'s right subtree}].$$

But C's left subtree has root E, empty left subtree, and right subtree consisting only of the root G, so that

$$[C\text{'s left subtree}] = E, G.$$

Similarly, C's right subtree has root F, left subtree consisting only of the root H, and right subtree consisting of the root I, so that

$$[C\text{'s right subtree}] = H, F, I.$$

Therefore

$$[A\text{'s right subtree}] = E, G, C, H, F, I,$$

and so finally we see that the inordering of the vertices of the tree of Figure 3.30(d) is

$$D, B, A, E, G, C, H, F, I.$$

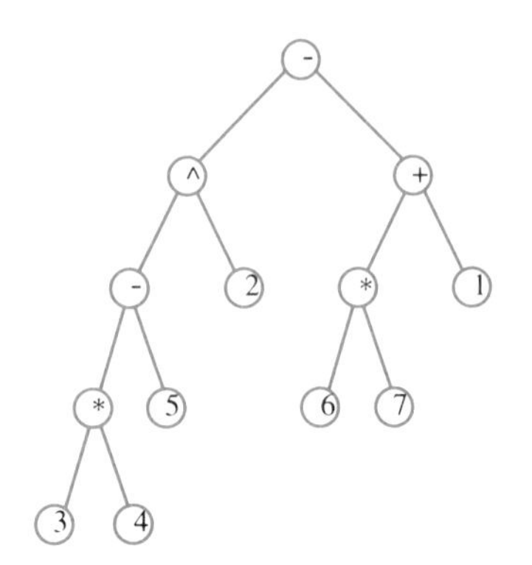

Figure 3.34 A binary tree representing the expression (((3 * 4) − 5) ^ 2) − ((6 * 7) + 1). The postordering of these vertices is 3, 4, *, 5, −, 2, ^, 6, 7, *, 1, +, −.

We said that preorder search is the most important tree-searching algorithm, as indeed it is. However, we should mention that postorder is also important, largely because of its relationship to **reverse Polish notation**, or RPN, which is used to manipulate algebraic expressions in many pocket calculators, and in the FORTH programming language.

For example, consider the algebraic expression

$$(((3 * 4) - 5)\hat{}2) - ((6 * 7) + 1),$$

in which "*" denotes multiplication and "^" denotes exponentiation. If you think for a moment about how such an expression is evaluated, you may agree that a binary tree like the one shown in Figure 3.34 is helpful. On a pocket calculator using RPN logic, in order to evaluate this expression, you would enter the following data, in order:

$$3, 4, *, 5, -, 2, \hat{}, 6, 7, *, 1, +, -,$$

after which the value 6 would appear on the screen. How does this work? The calculator stores the numbers you enter on a **stack**, which is a very simple data structure that allows the calculator to remember its place. Each time you enter a number, that number is pushed onto the stack. Each time you enter an operator like +, −, *, ÷, or ^, that operation is applied to the top two elements on the stack.* In Figure 3.35 we see how this works on the expression given above. The stack contents are shown below the line, and the symbols entered are shown above. Thus after 3 and 4 are entered, the stack contains both numbers, with the older of the two numbers on the bottom of the stack. But when the symbol * is entered, the 3 and 4 are multiplied, and the result 12 is placed on the stack, replacing the older numbers. This process continues, as shown, until the final result, in this case 6, is reached.

The important thing to notice here is that the order in which one enters the data in RPN logic corresponds exactly to the way in which postorder lists the vertices of the binary tree representing the given algebraic expression!

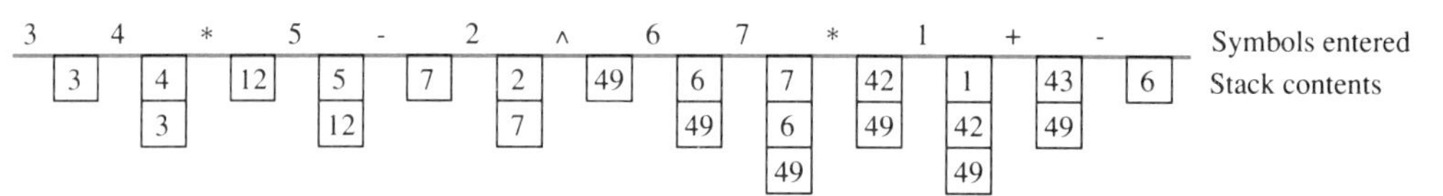

Figure 3.35 How the expression "3 4 * 5 − 2 ^ 6 7 * 1 + −" is evaluated in RPN logic.

* On most RPN calculators the exponentiation key is labelled "y^x", since "x" denotes the number most recently placed on the stack, and "y" denotes the previously entered number.

EXAMPLE 3.9 Consider the algebraic expression

$$(((7 - 5) * 2) \div 2) + (6 * (9 \div 3)).$$

(a) Draw the corresponding binary tree.
(b) Give the corresponding RPN expression.

SOLUTION

(a) The required binary tree is given in Figure 3.36.
(b) If we list the vertices of the tree in Figure 3.36 in postorder, we get the required RPN expression:

$$7, 5, -, 2, *, 2, \div, 6, 9, 3, \div, *, +.$$

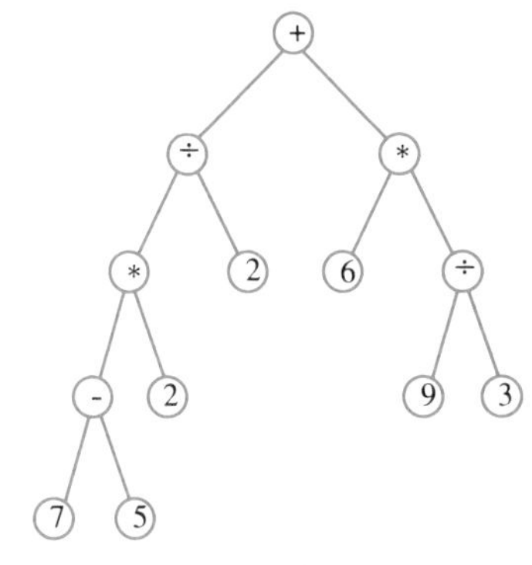

Figure 3.36 Solution to Example 3.9(a).

Problems for Section 3.4

The accompanying figure gives eight binary trees that will be referred to in some of the problems for this section.

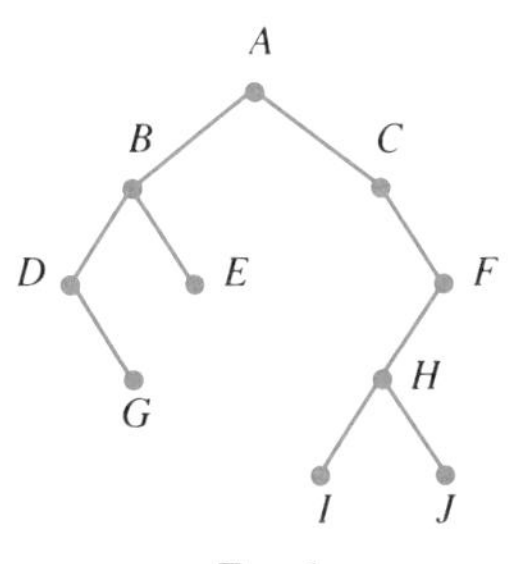

Tree 1

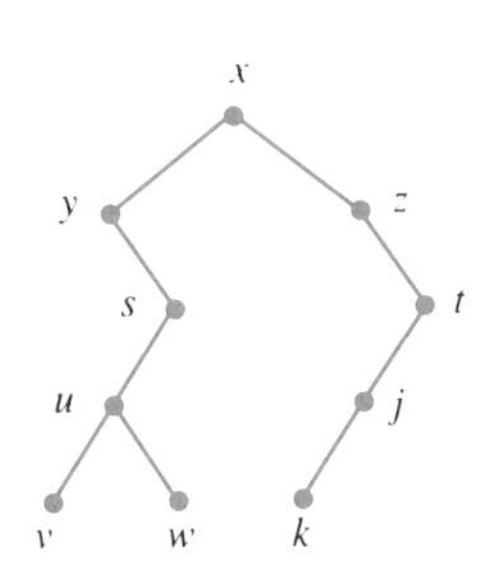

Tree 2

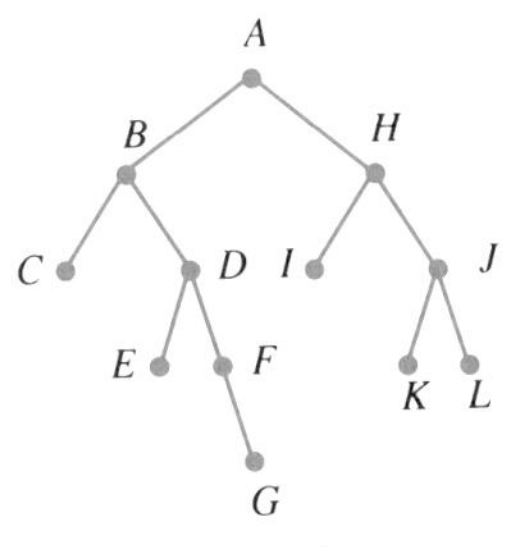

Tree 5

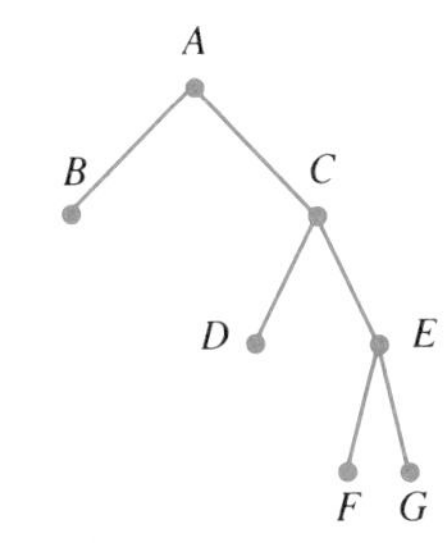

Tree 6

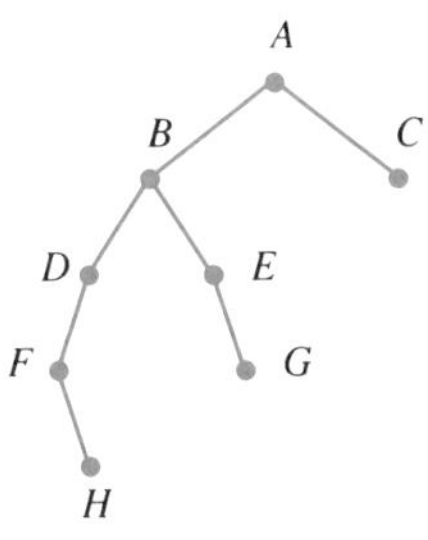

Tree 7

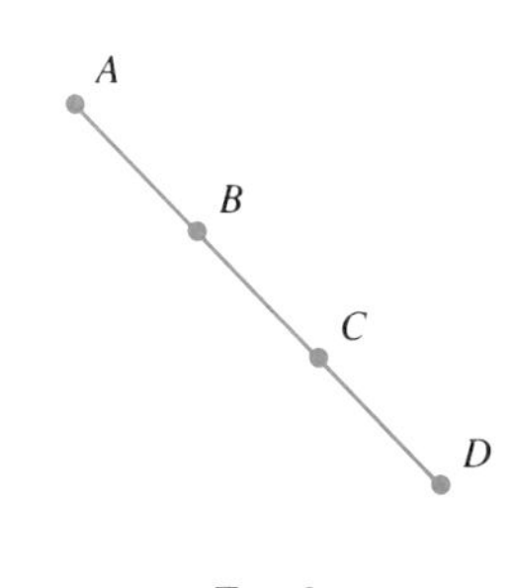

Tree 8

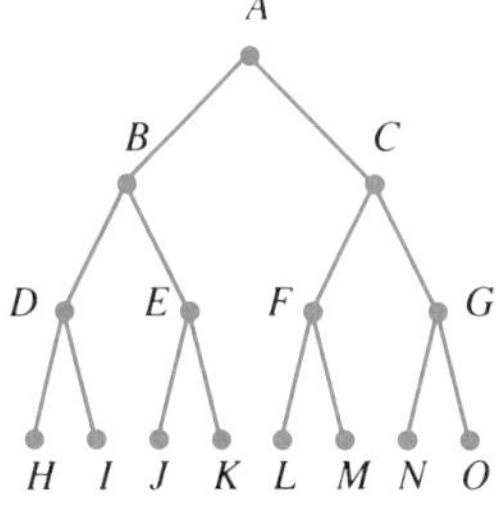

Tree 3

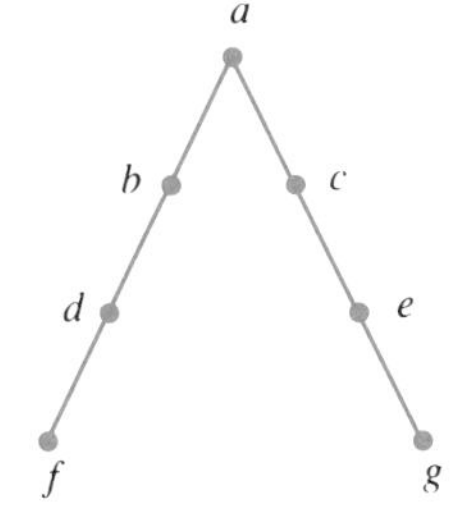

Tree 4

In Problems 1–8, give the tabular representation (cf. Table 3.2) of each of the indicated trees.

1. Tree 1.
2. Tree 2.
3. Tree 3.
4. Tree 4.
5. Tree 5.
6. Tree 6.
7. Tree 7.
8. Tree 8.

In Problems 9–12 you are given the tabular representation of a certain binary tree. In each case, draw the tree. (Remember that the root is indicated with an arrow.)

9.

Vertex	LCHILD	RCHILD
→A	B	C
B	D	E
C	—	F
D	—	—
E	—	—
F	—	—

10.

Vertex	LCHILD	RCHILD
→A	—	B
B	—	C
C	—	D
D	—	E
E	—	—

11.

Vertex	LCHILD	RCHILD
→1	4	5
2	—	—
3	—	—
4	2	3
5	—	—

12.

Vertex	LCHILD	RCHILD
A	—	C
B	—	—
C	—	—
D	B	—
→E	A	D

In Figure 3.32 we dissected the binary tree of Figure 3.30(b) into smaller and smaller binary trees. In Problems 13–15 you are asked to do the same for the other three binary trees in Figure 3.30.

13. Figure 3.30(a). **15.** Figure 3.30(d).

14. Figure 3.30(c).

In Problems 16–27, list the vertices in the given tree, using preorder search.

16. Tree 1. **19.** Tree 4. **22.** Tree 7.

17. Tree 2. **20.** Tree 5. **23.** Tree 8.

18. Tree 3. **21.** Tree 6.

24. The tree given in tabular form in Problem 9.

25. The tree given in tabular form in Problem 10.

26. The tree given in tabular form in Problem 11.

27. The tree given in tabular form in Problem 12.

In Problems 28–39, list the vertices in the given tree, using postorder search.

28. Tree 1. **31.** Tree 4. **34.** Tree 7.

29. Tree 2. **32.** Tree 5. **35.** Tree 8.

30. Tree 3. **33.** Tree 6.

36. The tree given in tabular form in Problem 9.

37. The tree given in tabular form in Problem 10.

38. The tree given in tabular form in Problem 11.

39. The tree given in tabular form in Problem 12.

In Problems 40–51, list the vertices in the given tree, using inorder search.

40. Tree 1. **43.** Tree 4. **46.** Tree 7.

41. Tree 2. **44.** Tree 5. **47.** Tree 8.

42. Tree 3. **45.** Tree 6.

48. The tree given in tabular form in Problem 9.

49. The tree given in tabular form in Problem 10.

50. The tree given in tabular form in Problem 11.

51. The tree given in tabular form in Problem 12.

52. Let us define "X-order search" for a binary tree, as follows:

X-order search

1. Visit the root.
2. Visit the right subtree, using X-order.
3. Visit the left subtree, using X-order.

List the vertices in the binary tree of Figure 3.30(b), using X-order.

53. Suppose the vertices of a certain binary tree are listed in preorder as A, B, D, F, E, C, and in postorder as F, D, E, B, C, A. Draw the tree.

54. Suppose a certain binary tree's vertices are listed in preorder as A, B, D, E, C, F, G, and in inorder as D, B, E, A, F, G, C. Draw the tree.

In Problems 55–60, represent the given algebraic expression as a binary tree, and then write the expression in RPN.

55. $((A - B) * C) + (D \hat{} E)$.

56. $(A \div (B * (C - D))) + E$.

57. $((A - D) * C)\hat{}((A + B) - D)$.

58. $A * ((B + (C \div (D \hat{} 2))) - E)$.

59. $(((N \hat{} N) \hat{} N) \hat{} N) * ((M * N) - (N * M))$.

60. $((A + B) * (C + D)) \div (((A - B) * C) + D)$.

In Problems 61–64, evaluate the given RPN expression.

61. $5\ 3 + 12\ 4 \div * 6\ 2 * +$.

62. $4\ 2\,\hat{}\,4 + 10 \div 16 +$.

63. $3\ 6 * 5\ 2\,\hat{}\, + 2\ 3 + -$.

64. $2\ 2\,\hat{}\,2\,\hat{}\,2\,\hat{}\,2\,\hat{}\,2\,\hat{}$.

3.5 Planar Graphs and Euler's Theorem

In Section 3.1 we saw how Euler's study of the Königsburg bridge problem led him to a theorem about graphs. Euler was also interested in the theory of polyhedra, and this too led him to an important theorem about graphs, which we will study in this section.

A polyhedron is a solid body which is bounded by polygons. In Figure 3.37 we see three simple polyhedra: a tetrahedron, a cube, and an octahedron. Every polyhedron has a certain number of vertices, edges, and faces. For example, the tetrahedron has 4 vertices, 6 edges, and 4 faces; the cube has 8 vertices, 12 edges, and 6 faces; and the octahedron has 6 vertices, 12 edges, and 8 faces. Euler studied many different polyhedra, and discovered a remarkable fact: The number of vertices minus the number of edges plus the number of faces is always equal to 2. This result is called Euler's theorem.

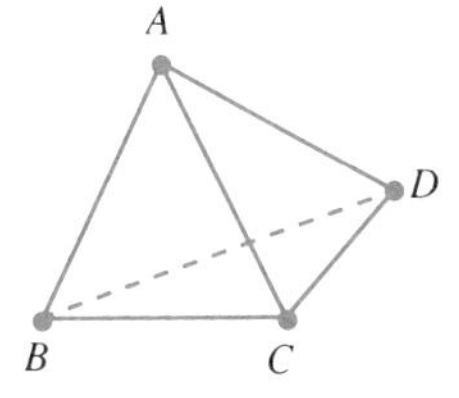

$V = 4, E = 6, F = 4$

(a)

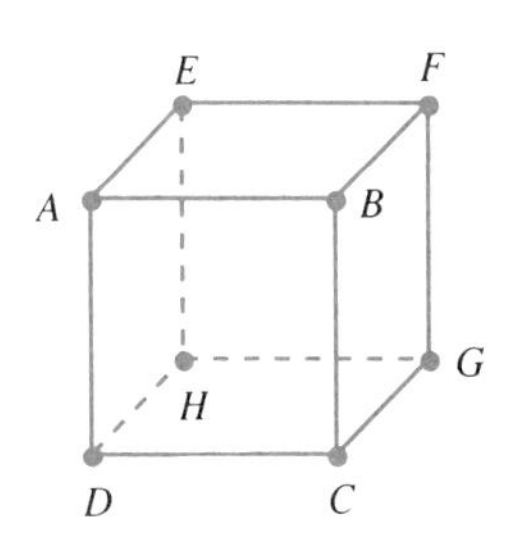

$V = 8, E = 12, F = 6$

(b)

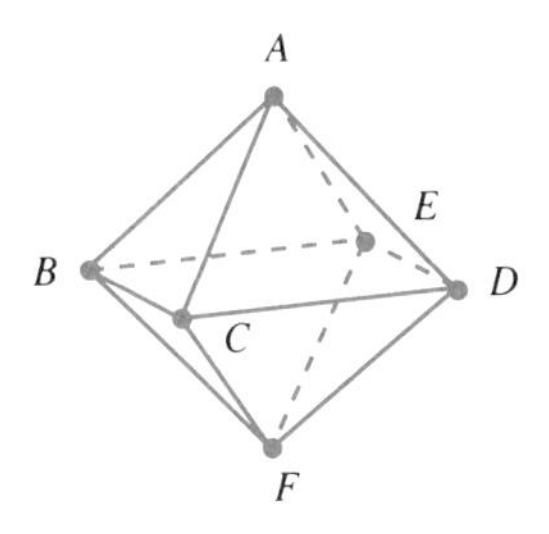

$V = 6, E = 12, F = 8$

(c)

Figure 3.37 Three simple polyhedra. (a) A tetrahedron. (b) A cube. (c) An octahedron. In each case $V - E + F = 2$.

EULER'S THEOREM FOR POLYHEDRA In any polyhedron, if V denotes the number of vertices, E the number of edges, and F the number of faces, then

$$V - E + F = 2.$$

Our goals in this section are to prove Euler's theorem and to investigate some of its consequences. Our first step in proving Euler's theorem is to restate it as a theorem about graphs. It's clear that the vertices and edges of the polyhedra in Figure 3.37 form connected graphs, but the fact that the polyhedra are three-dimensional makes the representations in Figure 3.37 somewhat awkward. In order to get a more convenient graphical representation of a polyhedron, imagine that the polyhedron is hollow and its faces are made of rubber. Cut a small hole in one of the faces, reach inside the hole, and pull the polyhedron apart until it lies flat, with the edges of the pulled-apart face forming the boundary of the new figure. The result is a connected graph, called a *planar representation* of the polyhedron. Except for the pulled-apart face, each face of the original polyhedron remains intact (though possibly distorted) in a planar representation. Even the pulled-apart face can be identified in a planar representation as the unbounded "outside" of the figure.

For example, consider the tetrahedron in Figure 3.37(a). If we pull apart the bottom face BCD, we get the planar representation shown in Figure 3.38(a). The faces ABC, ABD, and ACD are intact, but BCD has become the perimeter of the new figure. Similarly, if we pull apart the front face $ABCD$ of the cube in Figure 3.37(b), we get Figure 3.38(b), and if we pull apart the face ACD of the octahedron in Figure 3.37(c), we get Figure 3.38(c). Each of the graphs in Figure 3.38 has the same number of vertices, edges, and faces as its counterpart in Figure 3.37, if we interpret the outside of each planar representation as one of its faces.

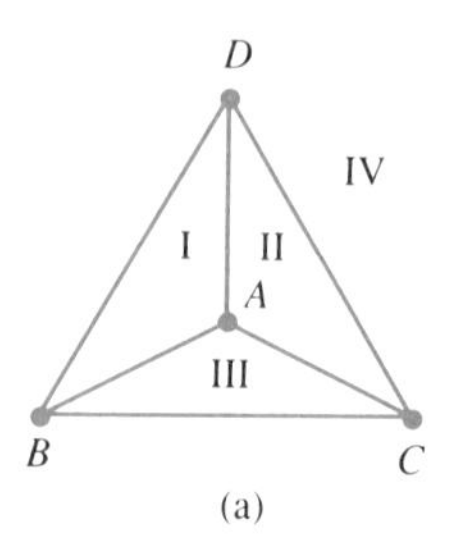

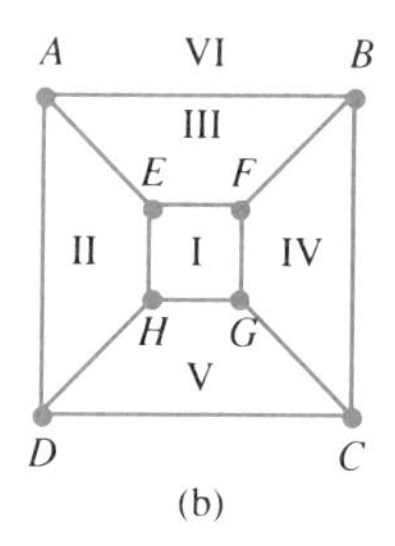

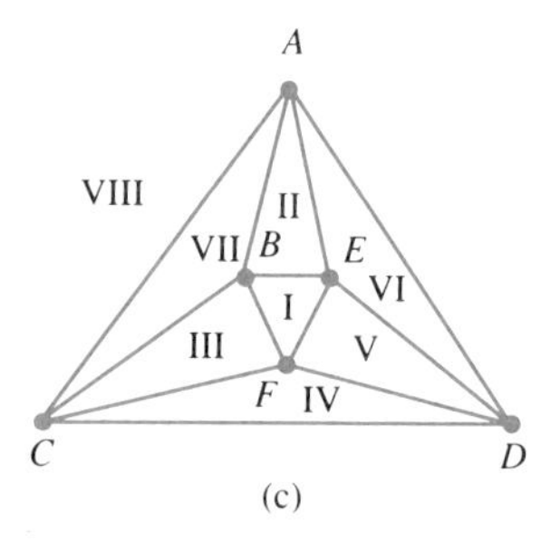

Figure 3.38 Planar representations of the polyhedra in Figure 3.37. In each case the faces are indicated with Roman numerals.

The graphs in Figure 3.38 are examples of **planar graphs**. A planar graph is a graph that can be drawn on the plane so that the edges don't cross each other. In Figure 3.39 we see some more examples of connected planar graphs, and in Figure 3.40 we see some graphs which do not (as they stand) qualify as planar, since they have edges that cross. However, the definition of graph planarity is a little subtler than you might think. For example, the graph in Figure 3.40(a), which is repeated in Figure 3.41(a), is apparently nonplanar, since edge AC crosses edge BD. However, if we redraw the graph by moving the edge AC outside the square, we get the graph in Figure 3.41(b), which obviously *is* planar. Since the graph in Figure 3.41(a) can be drawn so that the edges don't cross, it is a planar graph, despite its nonplanar appearance. Notice also that this same graph can be redrawn in yet another way, as shown in Figure 3.41(c). The three graphs in Figure 3.41, which are really the same graph in three different disguises, are said to be **isomorphic**.*

Any planar graph partitions the plane into a number of disjoint regions called the **faces** of the graph. (There is always one unbounded face outside the graph.) It turns out that Euler's theorem ($V - E + F = 2$) applies to any planar graph, not just those which are planar representations of polyhedra. For example, here is a table of V, E, and F for the six planar graphs in Figure 3.39:

Figure	V	E	F
3.39(a)	3	3	2
3.39(b)	6	12	8
3.39(c)	6	5	1
3.39(d)	20	30	12
3.39(e)	7	10	5
3.39(f)	7	7	2

As you can see, $V - E + F = 2$ in every case, even in the peculiar case of Figure 3.39(c), which is a tree, and plainly not related to a polyhedron! Now we can state and prove Euler's theorem, as it applies to connected planar graphs.

EULER'S THEOREM FOR PLANAR GRAPHS If the number of vertices, edges, and faces of a planar connected graph are denoted by V, E, and F, respectively, then invariably

$$V - E + F = 2.$$

PROOF The proof of this theorem is a perfect illustration of the power of mathematical induction. We will use induction on F, the number of faces. We begin with the case $F = 1$. (Figure 3.39c is an example of a connected planar graph with $F = 1$.) If there is only one face, then the graph cannot have any cycles, since a cycle has an inside and an outside, which produces at least two faces (see Figure 3.39f, for example). A connected graph with no cycles is, by definition, a tree. Thus if $F = 1$, the graph is a tree, and so we know by the daisy

* From the Greek roots *isos*, "equal," and *morphe*, "form."

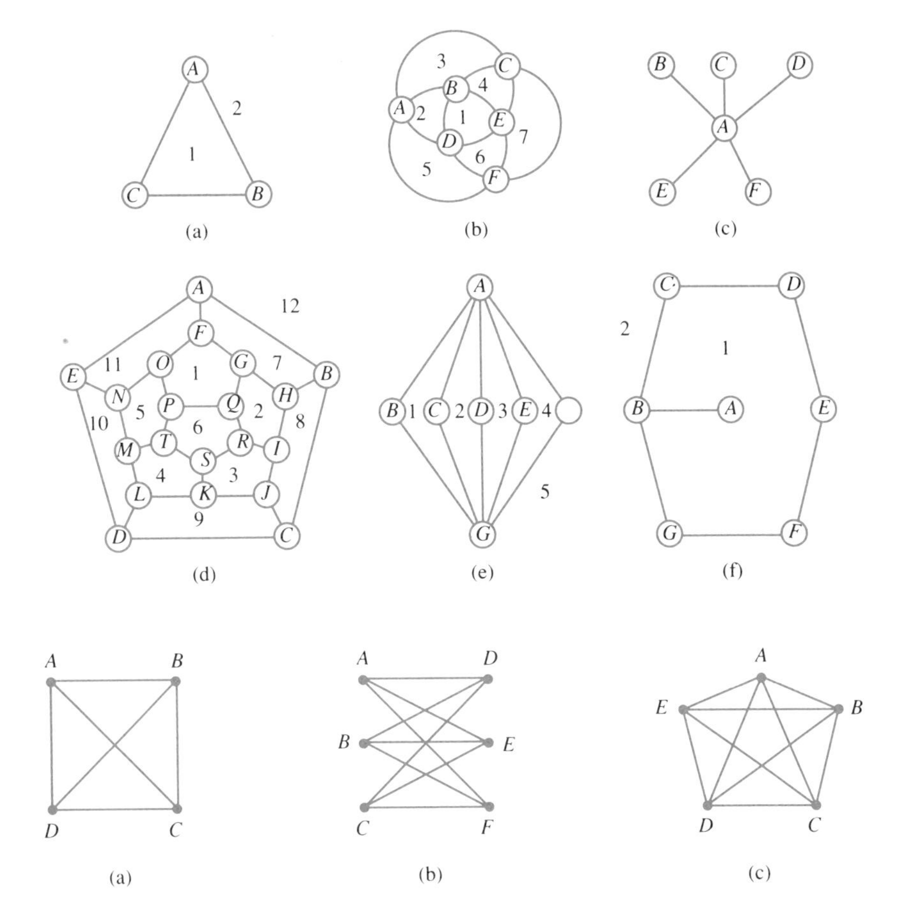

Figure 3.39 Six connected planar graphs, with their faces numbered.

Figure 3.40 Three graphs with crossing edges: (b) and (c) are nonplanar; but, despite appearances, (a) is planar.

chain theorem of Section 3.2 that the number of edges is one less than the number of vertices. Thus a planar connected graph with $F = 1$ also has $E = V - 1$, and so

$$V - E + F = V - (V - 1) + 1 = 2.$$

This proves Euler's theorem for $F = 1$. To complete the induction proof, we assume that $F \geq 2$, and that Euler's theorem has been proved for all planar connected graphs with $F - 1$ faces. Our goal is now to prove it for planar connected graphs with F faces.

Thus, let G be a planar connected graph with V vertices, E edges, and F faces. Since G has more than one face, we know that it must have a cycle. Let e be any edge of G which is on a cycle (see Figure 3.42). The edge e must lie on the boundary between two faces of the graph. If we remove e from the graph, these two faces will merge into one, and the resulting graph will be a connected planar graph with V vertices, $E - 1$ edges, and $F - 1$ faces. Since our induction hypothesis says that Euler's theorem is true for all planar graphs with $F - 1$ faces, then

$$V - (E - 1) + (F - 1) = 2.$$

But this is the same as

$$V - E + F = 2,$$

which is what we set out to prove. This completes the inductive proof of Euler's theorem. ∎

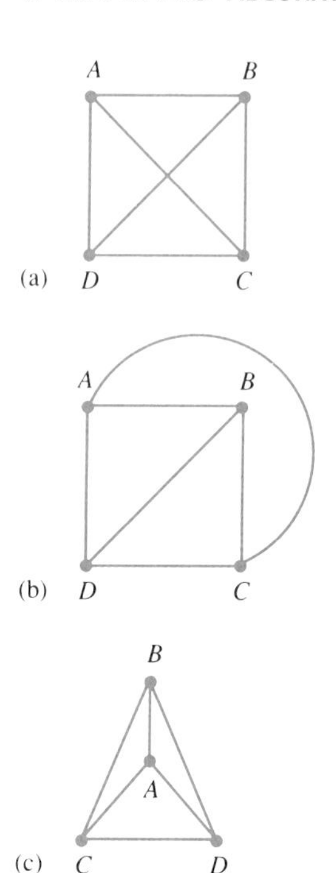

Figure 3.41 Three isomorphic graphs: (b) and (c) are planar representations of (a).

We have seen that sometimes a graph with crossing edges can be redrawn without crossing edges. This raises a question: Are all graphs planar? The next example shows that the answer is no.

EXAMPLE 3.10 Show that the graph in Figure 3.40(b) is nonplanar.*

SOLUTION We begin by noting that the given graph contains a cycle, containing all six vertices, viz., $A \to D \to B \to E \to C \to F \to A$. If the graph could be drawn with no edges crossing, then these six edges would form a hexagon, as shown in Figure 3.43(a). The only edges from the original graph that are missing from Figure 3.43(a) are AE, BF, and CD. If we put AE on the inside of the hexagon, we are forced to put BF on the outside, as shown in Figure 3.43(b). But then how can we draw CD? If we try to draw it on the inside, it will cross AE, and if we put it on the outside, it will cross BF. We conclude that the graph in Figure 3.40(b) is indeed nonplanar. ∎

The argument in Example 3.10 is a bit tricky. However, there is a useful theorem that sometimes makes it easier to see that a graph is nonplanar. Before stating the theorem, we need to introduce a new concept, the **girth** of a graph.

The girth of a graph is the length of the shortest cycle in the graph. If the graph has no cycles, we define its girth to be the total number of edges in the graph. For example, the following table gives the girth of some graphs in this section:

Figure	Girth	
3.38(a)	3	
3.38(b)	4	
3.38(c)	3	
3.39(c)	5	(This graph is a tree!)
3.39(d)	5	
3.40(b)	4	
3.40(c)	3	

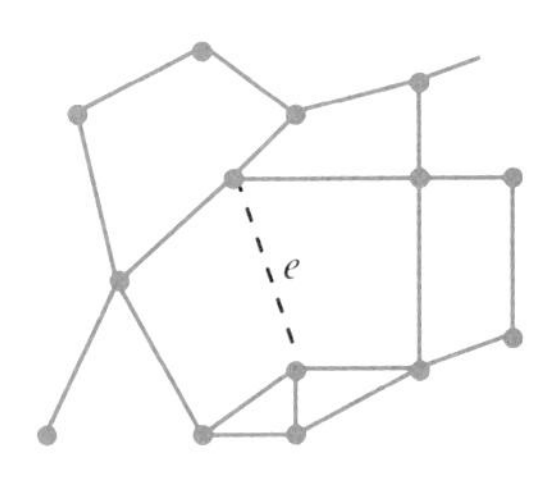

Figure 3.42 When the edge e is removed, E decreases by one and F decreases by one.

The important thing about the girth g is that if the graph is planar, then the boundary of every face has at least g edges. This is because the boundary of each face consists of a cycle and possibly some dangling edges. In Figure 3.44(a), for example, the graph has girth 3, and the boundary of each face has at least three edges. The faces numbered 1, 3, and 4 each have three-edge boundaries;

* This problem is a variation of a famous children's puzzle, in which vertices A, B, and C are houses, and vertices D, E, and F are utility companies—usually water, gas, and electricity. The puzzle is to draw a line from each house to each utility company without any lines crossing.

face 2 has a four-edge boundary (AC, CE, EA, and the dangler AF); and the exterior face 5 has a four-edge boundary.

The following theorem makes it easy to show that certain graphs are nonplanar.

THE EDGE-VERTEX INEQUALITY In any connected planar graph whose girth g is at least 3,

$$E \leq \frac{g}{g-2}(V-2).$$

PROOF We will illustrate the proof with the connected planar graph of Figure 3.44(a), which has $V = 6$, $E = 9$, $F = 5$, and $g = 3$. In Figure 3.44(b) we see the "edge-face incidence matrix" for this graph. In this matrix the rows are labelled by the edges, and the columns are labelled by the faces, of the graph. Whenever a given edge forms part of the boundary of a given face, a 1 appears in the corresponding position in the matrix. Thus, for example, since edge CE borders on face 4, there is a 1 in row CE and column 4. Similarly, since edge BD borders on face 5, there is a 1 in row BD and column 5.

The matrix of Figure 3.44(b) contains exactly 17 1s. But now let us *estimate* the number of 1s in two different ways, counting by rows and by columns.

Notice that each *row* of the matrix in Figure 3.44(b) has either one or two 1s, because each edge of the matrix borders only one or two faces. (In fact, only the "dangling" edge AF borders a single face.) Since there are nine rows, the matrix can contain at most $2 \times 9 = 18$ 1s.

On the other hand, notice that each *column* of the matrix in Figure 3.44(b) has at least three 1s. This is because each face of the graph has at least three edges on its boundary, and this in turn is because the graph has girth 3. Since there are five columns, the matrix must contain at least $3 \times 5 = 15$ 1s.

Of course, it is not surprising that the actual number of 1s in the matrix respects these upper and lower bounds: 17 is less than or equal to 18 and

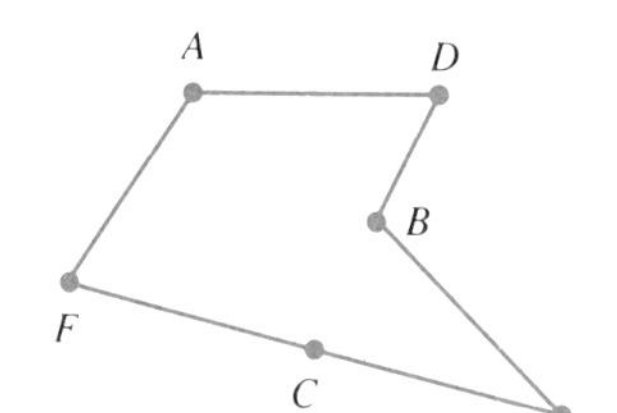

(a)

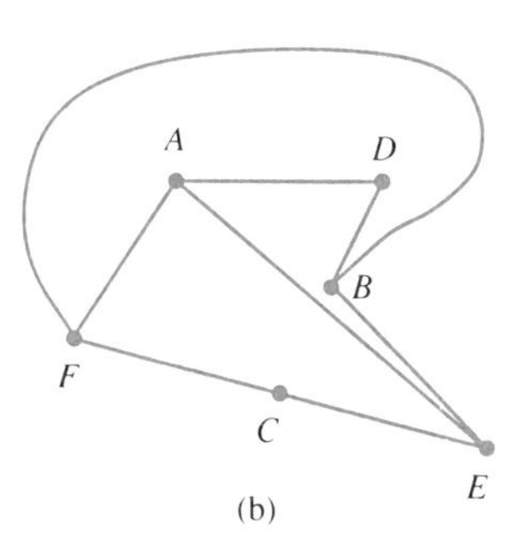

(b)

Figure 3.43 Solution to Example 3.10. (a) The vertices A, B, C, D, E, and F form a hexagon. (b) If AE is inside, and BF is outside, where does CD go?

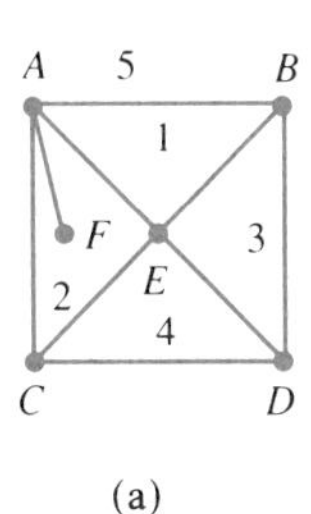

(a)

	1	2	3	4	5	
AB	1				1	2
AC		1			1	2
AE	1	1				2
AF		1				1
BD			1		1	2
BE	1		1			2
CD				1	1	2
CE		1		1		2
DE			1	1		2
	3	4	3	3	4	

(b)

Figure 3.44 (a) A planar graph; (b) its edge-face incidence matrix.

greater than or equal to 15. The important thing is that the same argument can be applied to any planar graph. If we form the edge-face incidence matrix, then each of the E rows will contain at most two 1s, and each of the F columns will contain at least g 1s. Therefore, the total number of 1s in the matrix will on the one hand be less than or equal to $2E$, and on the other hand be greater than or equal to gF. Thus we have proved that

$$gF \leq 2E$$

for any planar graph. If the graph is also connected, then by Euler's theorem, $F = E - V + 2$, and if we substitute this formula into the inequality $gF \leq 2E$, we get $g(E - V + 2) \leq 2E$, which, rearranged, becomes $(g-2)E \leq g(V-2)$. If $g \geq 3$, we can divide both sides of the inequality by the positive number $g - 2$ to obtain

$$E \leq \frac{g}{g-2}(V-2),$$

which is the edge-vertex (EV) inequality. ■

The following two examples show how the edge-vertex inequality can be used to show that certain graphs are nonplanar.

EXAMPLE 3.11 Use the edge-vertex inequality to prove again that the graph of Figure 3.40(b) is nonplanar.

SOLUTION This graph has $V = 6$ and $E = 9$. In order to use the EV inequality, we also need to know g, i.e., the length of the shortest cycle in the graph. Plainly there are no cycles of length one or two, since the graph has no loops or multiple edges. There are also no cycles of length 3, since every edge in the graph connects a vertex on the left (A, B, or C) to a vertex on the right (D, E, or F), so that any path of length 3 would start on the left and end on the right (or vice versa) and could not be a cycle. Since there are many cycles of length 4 (e.g., $ADBEA$), the graph has $g = 4$.

Now we are in a position to use the EV inequality. If the graph in Figure 3.40(b) were planar, then the EV inequality would apply. Since $V = 6$, $E = 9$, and $g = 4$, the EV inequality says that

$$9 \leq \tfrac{4}{2}(6-2), \qquad \text{or} \qquad 9 \leq 8,$$

which is false. Therefore the graph isn't planar. ■

EXAMPLE 3.12 Show that the graph of Figure 3.40(c) is not planar.

SOLUTION Here we have $V = 5$ and $E = 10$. The girth is 3, since there are no cycles of length 1 or 2 but many of length 3 (e.g., $ABCA$). If the graph were planar, then by the EV inequality,

$$10 \leq \tfrac{3}{1}(5-2), \qquad \text{or} \qquad 10 \leq 9.$$

which is false. Therefore the graph isn't planar. ■

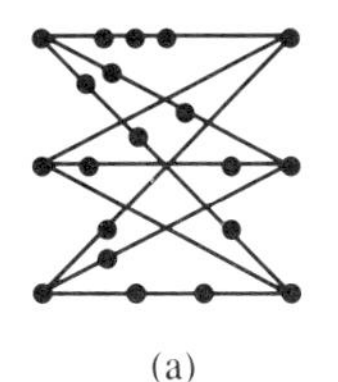

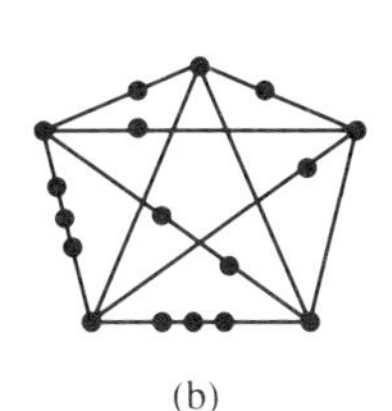

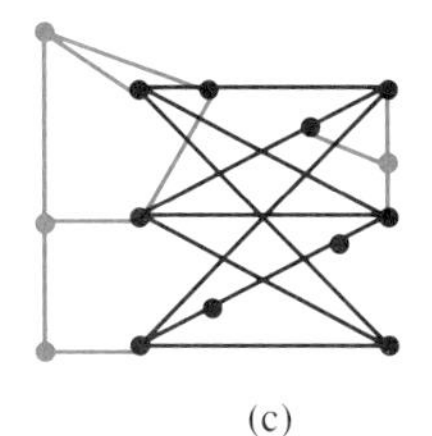

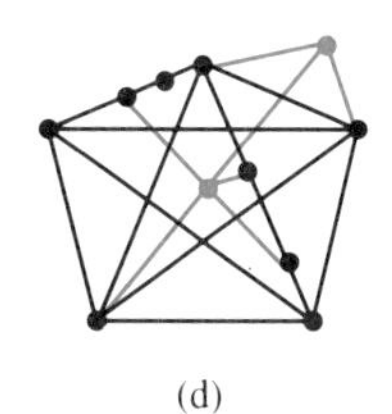

Figure 3.45 Some nonplanar graphs: (a) and (c) contain subgraphs of type $K_{3,3}$; (b) and (d) contain subgraphs of type K_5.

We now have two examples of nonplanar graphs, viz. those in Figure 3.40(b) and 3.40(c). These two graphs are special cases of important classes of graphs, the *complete bipartite graphs* and the *complete graphs.*

The **complete graph** K_n has n vertices, with every vertex connected to every other vertex. The nonplanar graph of Figure 3.40(c) is the complete graph K_5; K_3 appears in Figure 3.39(a), and K_4 appears in Figure 3.42. It turns out that K_n is planar for $n = 1, 2, 3$, and 4 but nonplanar for $n \geq 5$ (see Problem 19).

The **complete bipartite graph** $K_{m,n}$ is a graph with $m + n$ vertices, m vertices "on the left" and n vertices "on the right," with every vertex on the left connected to every vertex on the right. The nonplanar graph of Figure 3.40(b) is the complete bipartite graph $K_{3,3}$. The graph in Figure 3.39(e) is $K_{2,5}$. It turns out that $K_{m,n}$ is planar if and only if $m \leq 2$ or $n \leq 2$ (see Problems 20–22).

From the graphs K_5 and $K_{3,3}$ we can produce many more nonplanar graphs, merely by placing extra vertices on the edges of these two graphs, as shown in Figures 3.45(a) and 3.45(b), and then adding extra vertices and edges, as shown in Figures 3.45(c) and 3.45(d). Graphs like those in Figure 3.45 are said to contain K_5 or $K_{3,3}$ as **subgraphs**. Such graphs cannot be planar; for if there were a planar representation for such a graph, we could remove the extra vertices and edges and obtain a planar representation of K_5 or $K_{3,3}$, which we know does not exist.

Remarkably, it turns out that graphs with K_5 or $K_{3,3}$ as subgraphs are the *only* nonplanar graphs! This is because of *Kuratowski's theorem*, which says that every nonplanar graph must have a subgraph of type K_5 or $K_{3,3}$. (The proof of Kuratowski's theorem is quite difficult, and we won't give it.) In principle, this theorem allows one to test whether or not a given graph is planar. In practice, however, Kuratowski's theorem is not of much use, because it can be extremely difficult to identify these subgraphs, as the next example shows. (However, there are several known practical algorithms for testing for planarity, which can be found in more advanced texts.)

EXAMPLE 3.13 Show that the graph of Figure 3.46(a)—which is called the Petersen graph—is nonplanar, by finding a subgraph of type $K_{3,3}$.

SOLUTION You're not going to believe this, but the Petersen graph is isomorphic to the graph shown in Figure 3.46(b). (You can check that the graph in Figure 3.46(a) has exactly the same pairs of vertices connected by edges as the graph in Figure 3.46(b). If we remove the two edges *gi* and *ch*, we get a graph of type $K_{3,3}$. (For other approaches to proving the nonplanarity of the Petersen graph, see Problems 25 and 26.)

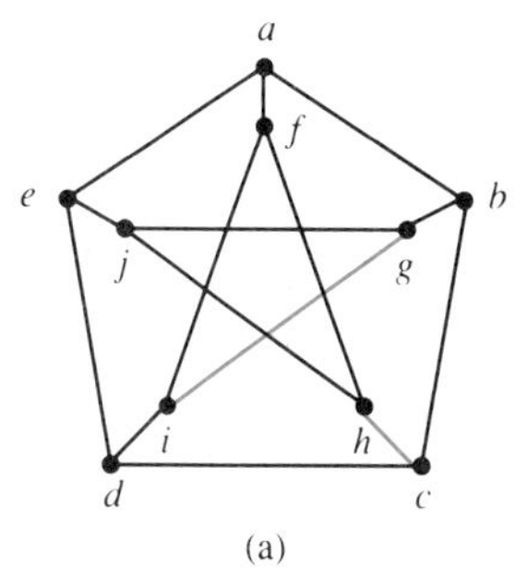

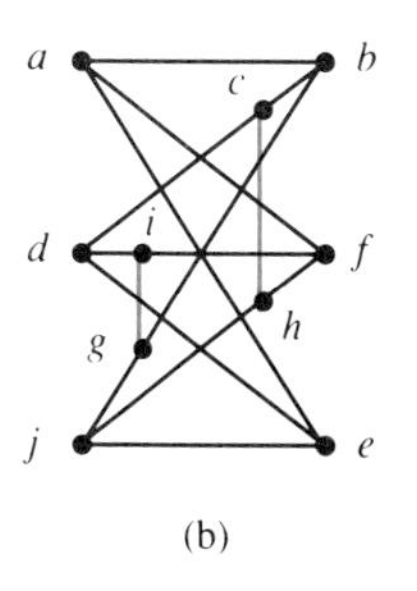

Figure 3.46 (a) The Petersen graph (b) The Petersen graph!

Problems for Section 3.5

1. For each of the polyhedra shown in the accompanying figure, draw a planar representation, with the given face forming the exterior face. Calculate V, E, and F in each case.

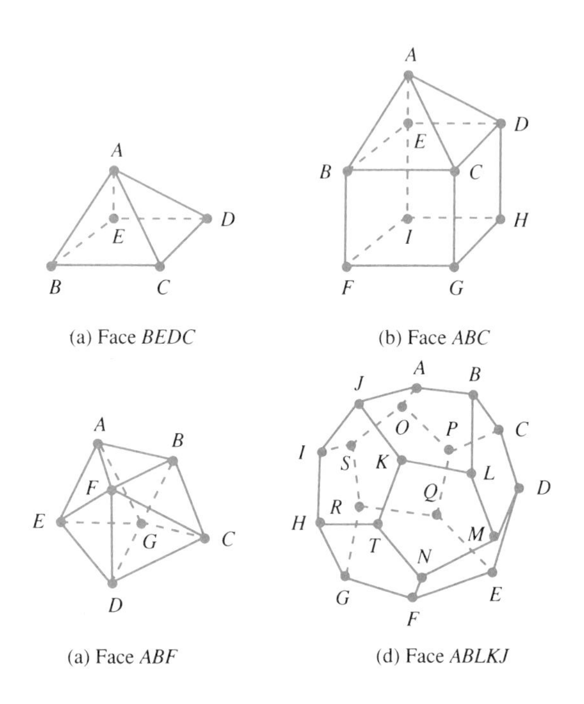

(a) Face *BEDC* (b) Face *ABC*

(a) Face *ABF* (d) Face *ABLKJ*

2. Show that the accompanying graphs are planar by redrawing the graphs so that there are no crossing edges.

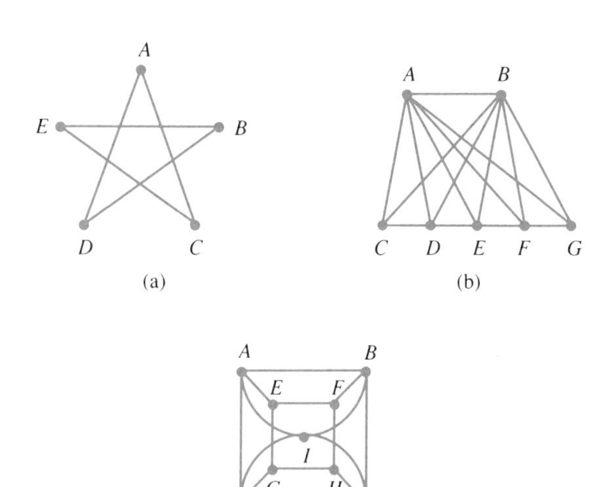

(a) (b)

(c)

In Problems 3–6 you will need to recall that the *degree* of a vertex in a graph is the number of edges which have that vertex as one of its endpoints. Thus for example in Figure 3.39(e) vertices A and B have degree 5, and the other five vertices have degree 2. (Problem 3.1.23(b) will also be helpful.)

3. Suppose that a planar connected graph has $V = 9$, with the degrees of the vertices being 2, 2, 2, 3, 3, 3, 4, 4, and 5, respectively. Find E and F.

4. Suppose a planar connected graph has $E = 16$, with each vertex having degree 4. Find F.

5. Let G be a connected graph with no loops or multiple edges.

(a) If G is planar, show that there must be some vertex with degree less than or equal to 5.

(b) If all vertices have degree greater than or equal to 6, show that G is nonplanar.

6. True or false?

(a) If G is a planar graph, and if any number of loops, multiple edges, degree 2 vertices, or danglers are added to G, then G remains planar.

(b) If G is a nonplanar graph, then removing any number of loops, multiple edges, degree 2 vertices, or danglers cannot make G planar.

7. Show that if the edge AD is removed from the graph of Figure 3.40(b), the resulting graph is planar.

8. Show that if any one of the edges is removed from the graph of Figure 3.40(c), the resulting graph is planar.

9. Compute the girth of the graphs in Figures 3.39(a), 3.39(b), 3.39(c), 3.39(d), 3.39(e), 3.39(f), and 3.40(a).

10. Why does the edge-vertex inequality fail if $g = 1$ or $g = 2$?

11. Find a planar graph with two vertices and 100,000 edges. Does this example contradict the edge-vertex inequality?

12. In the text, we considered only *connected* planar graphs. Prove that if a planar graph has m connected components, $V - E + F = m + 1$.

13. Redraw the graph in the accompanying figure so that the exterior face is bounded by the cycle $ABDCA$.

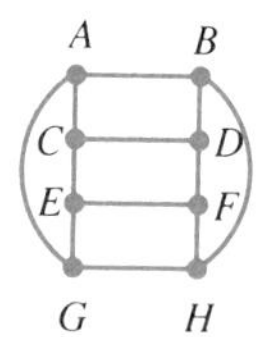

14. Give the edge-face incidence matrix for the following planar graphs.

(a) Figure 3.39(a). (d) Figure 3.39(d).
(b) Figure 3.39(b). (e) Figure 3.39(e).
(c) Figure 3.39(c). (f) Figure 3.39(f).

15. Prove the following theorem:

The Edge-Face Inequality. In any planar graph with at least two edges and no loops or multiple edges,

$$3F \leq 2E.$$

16. Prove the following theorem:

The Modified Edge-Vertex Inequality. In any connected planar graph with at least two edges and no loops or multiple edges,

$$E \leq 3V - 6.$$

17. Find V, E, and g for the complete graph K_n.

18. Find V, E, and g for the complete bipartite graph $K_{m,n}$.

19. Show that K_n is planar for $n = 1$, 2, 3, and 4, but for no larger values of n.

20. Show that the complete bipartite graph $K_{1,n}$ is planar for all $n \geq 1$.

21. Show that the complete bipartite graph $K_{2,n}$ is planar for all $n \geq 1$.

22. Show that the complete bipartite graph $K_{m,n}$ is nonplanar if $m \geq 3$ and $n \geq 3$.

23. The **complete tripartite graph** $K_{l,m,n}$ is defined as follows: There are three sets of vertices, X, Y, and Z, with $|X| = l$, $|Y| = m$, and $|Z| = n$. Two vertices are connected by an edge if and only if they lie in different sets. The graph $K_{1,2,3}$ is illustrated in the accompanying figure.

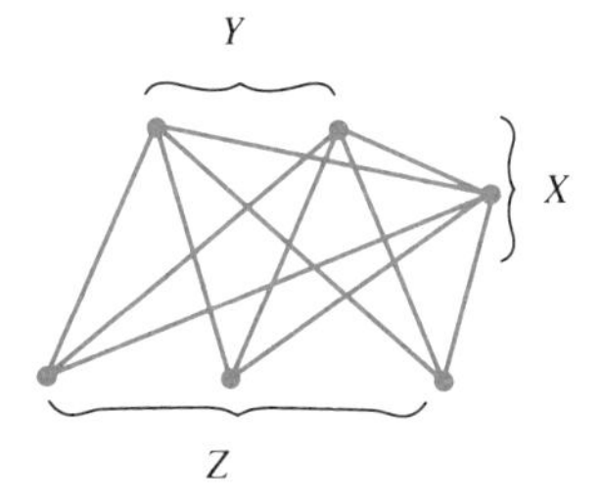

(a) How many edges does $K_{l,m,n}$ have?
(b) What is the girth of $K_{l,m,n}$?
(c) For what values of l, m, and n is $K_{l,m,n}$ planar?

24. In each case, given the adjacency matrix, determine whether or not the given graph is planar. (*Hint*: You may need to use Kuratowski's theorem.)

(a)

	A	B	C	D	E	F	G	H	I	J
A	0	1	0	0	1	0	0	0	0	0
B	1	0	1	0	0	0	0	0	1	0
C	0	1	0	1	0	0	0	1	0	0
D	0	0	1	0	0	1	0	0	0	0
E	1	0	0	0	0	1	1	0	0	0
F	0	0	0	1	1	0	0	0	0	1
G	0	0	0	0	1	0	0	1	0	0
H	0	0	1	0	0	0	1	0	1	0
I	0	1	0	0	0	0	0	1	0	1
J	0	0	0	0	0	1	0	0	1	0

(b)

	A	B	C	D	E	F
A	0	0	1	1	1	0
B	0	0	0	1	1	1
C	1	0	0	0	1	1
D	1	1	0	0	0	1
E	1	1	1	0	0	0
F	0	1	1	1	0	0

(c)

	A	B	C	D	E	F	G	H
A	0	1	0	1	1	0	0	0
B	1	0	1	0	0	1	0	0
C	0	1	0	1	0	0	0	1
D	1	0	1	0	0	0	1	0
E	1	0	0	0	0	1	0	1
F	0	1	0	0	1	0	1	0
G	0	0	0	1	0	1	0	1
H	0	0	1	0	1	0	1	0

(d)

	A	B	C	D	E	F
A	0	1	1	1	1	1
B	1	0	1	1	1	0
C	1	1	0	1	0	1
D	1	1	1	0	1	1
E	1	1	0	1	0	1
F	1	0	1	1	1	0

(e)

	A	B	C	D	E	F	G	H
A	0	1	0	1	0	1	1	0
B	1	0	1	0	1	0	0	1
C	0	1	0	1	0	1	1	0
D	1	0	1	0	1	0	0	1
E	0	1	0	1	0	1	1	0
F	1	0	1	0	1	0	0	1
G	1	0	1	0	1	0	0	1
H	0	1	0	1	0	1	1	0

25. The Petersen graph of Figure 3.46(a) bears a strong family resemblance to K_5. Can you show that it is nonplanar by finding a subgraph of type K_5?

26. (a) What is the girth of the Petersen graph?
(b) Use the edge-vertex inequality to show that the Petersen graph is nonplanar.

27. If one edge is removed from the Petersen graph, does the graph become planar?

28. Show that all cycles in $K_{m,n}$ are of even length.

29. How many length 4 cycles does $K_{3,3}$ have? Generalize to $K_{m,n}$.

30. How many length 3 cycles does K_5 have? Generalize to K_n.

31. The accompanying figure shows a three-dimensional solid with a *hole* in it. Calculate $V - E + F$ (don't forget the three interior faces), and comment.

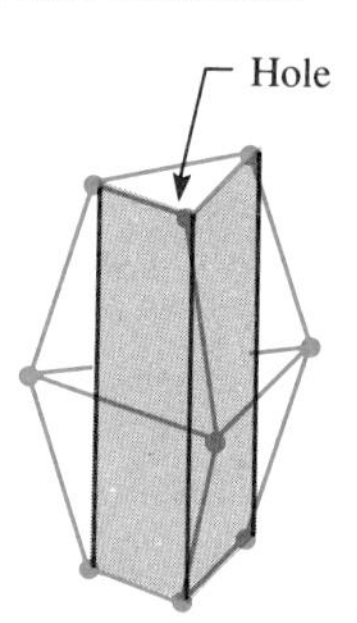

Summary

This chapter introduces various problems that can be formulated and solved using *graphs.*

- *Terminology*: Vertices, edges, directed graphs, undirected graphs, paths, multiple edges, loops, isolated vertices, degree of a vertex, connected graphs, cycles, trees, adjacency matrix of a graph.
- *Euler cycles and paths*: Conditions under which they exist (Euler's theorems) and how to find them (breakout algorithm).
- *Spanning trees* and an algorithm for finding them.
- *Equivalent definitions of a tree* (daisy chain theorem).
- *Minimal spanning trees*: Prim's algorithm, an example of a "greedy" algorithm.
- *Binary trees*: Recursive definitions; searching via preorder (depth-first search, or backtracking), postorder, and inorder; manipulation of algebraic expressions; reverse Polish notation.
- *Planar graphs*: $V - E + F = 2$ (Euler's theorem for planar graphs); basic nonplanar graphs (K_n, $n \geq 5$, and $K_{m,n}$, $m \geq 3$ and $n \geq 3$); conditions under which a graph will be nonplanar (Edge-vertex inequality, Kuratowski's theorem).

CHAPTER FOUR

GRAPHS AND ALGORITHMS II

In this chapter we will continue our study of graphs and algorithms. In Section 4.1 we will learn how to find the shortest route from one vertex in a graph to another, using *Dijkstra's algorithm*, which is closely related to Prim's algorithm, from Section 3.3. In Section 4.2 we will study *Floyd's algorithm*, which solves the similar problem of how to find the shortest paths between all pairs of vertices in a graph. In the same section we will study an earlier algorithm due to Warshall which solves a different problem (the "reachability problem"), yet is almost identical to Floyd's algorithm. In Section 4.3 we will study the famous "matching problem" and give its solution algorithmically. Finally, in Section 4.4 we will learn how to estimate the *running times* for many kinds of algorithms.

4.1 The Shortest-Path Problem; Dijkstra's Algorithm

In our discussion of Prim's algorithm in Section 3.3, we considered graphs with edge labels. In this section we will also consider graphs with edge labels, but whereas in Section 3.3 we thought of the labels as *costs*, here we will usually think of them as *lengths*.

For example, suppose that we were looking at a road map showing all the roads connecting a certain set of cities, and that we wished to find the shortest route between two of them. In such a case, we could boil the salient facts on the map down to a graph, with a vertex representing each city, and a labelled edge for each road, the label representing the length of the road. A graph like the one in Figure 4.1 might result. Alternatively, the edge labels might represent the *time* required to travel the roads, or perhaps the *cost* of air transportation between the cities. However, for definiteness, from now on we'll just call the edge labels *lengths*, and assume that the length of each edge in a graph is a positive number.

In a labelled graph, the length of a *path* is defined to be the sum of the lengths of its edges. For example, the length of the path *AFDEGH* from *A* to

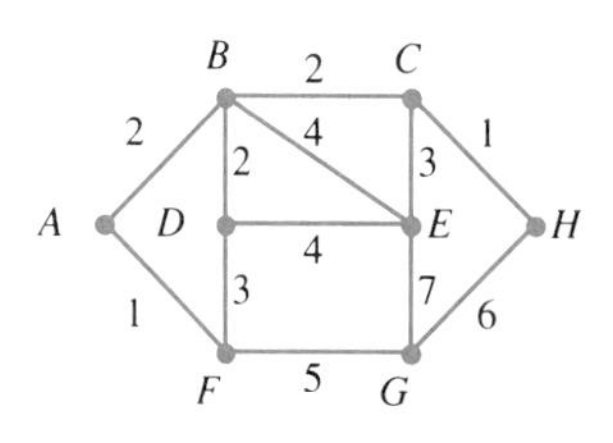

Figure 4.1 A labelled graph.

H in Figure 4.1 is $1 + 3 + 4 + 7 + 6 = 21$. If we are given a labelled graph, and if P and Q are two vertices, it is natural to want to find the *shortest* path between P and Q, and "shortest-path" problems arise constantly in discrete mathematics. (Note that the path $AFDEGH$ is *not* the shortest path from A to H in Figure 4.1; the path $AFGH$, for example, is shorter.) In this section, we will describe an algorithm for finding shortest paths which is very similar to Prim's algorithm for finding minimal spanning trees. It is called **Dijkstra's algorithm**, after its discoverer.*

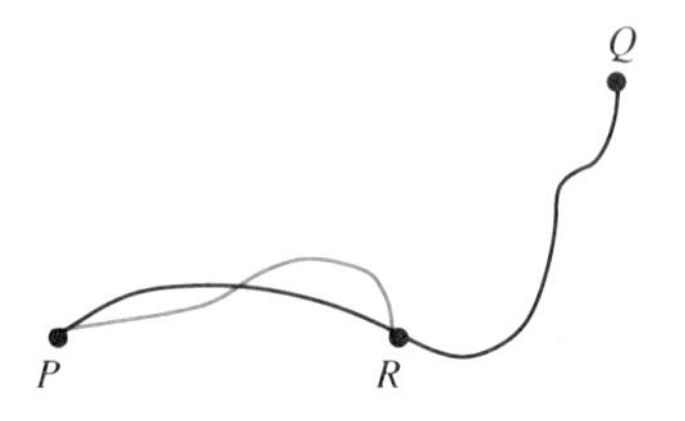

Figure 4.2 Illustrating the principle of optimality.

Dijkstra's algorithm takes a labelled graph and a pair of vertices P and Q, and finds the shortest path between them (or one of the shortest paths, if there is more than one). The strategy is to start at P and systematically build up a list of the shortest paths to all vertices which lie between P and Q, in order of increasing distance from P, until Q itself is reached. That this is a reasonable approach can be seen by studying Figure 4.2.

Suppose the black path in Figure 4.2 represents the shortest path from P to Q, and that R is some intermediate vertex along the path. Then the portion of the black line that goes from P to R *must be the shortest path from P to R.* This is because if there were a shorter path from P to R, say the colored path in Figure 4.2, then by taking the colored path from P to R and then the black path from R to Q, we would get a shorter path from P to Q. But this isn't possible, since the black path from P to Q is already as short as possible! Similarly, the black path from R to Q must be the shortest path from R to Q. This simple but important principle has been called the **principle of optimality**, and is the basis not only of Dijkstra's algorithm but many other algorithms in discrete mathematics as well. (For example, Viterbi's algorithm for decoding convolutional codes, to be discussed in Chapter 6, is based, ultimately, on the principle of optimality.)

Given the principle of optimality, it is clear that the strategy of Dijkstra's algorithm is sound: Since the shortest path from P to Q contains the shortest paths to all the intermediate vertices, if we can succeed in finding all the intermediate shortest paths, we will surely eventually find the shortest path to Q. All (!) that remains is to describe the details of Dijkstra's algorithm.

Our first example will give us an informal idea of how Dijkstra's algorithm works.

EXAMPLE 4.1 Use the principle of optimality to find the shortest path from A to H in the graph of Figure 4.1.

SOLUTION We begin with what might be called a "spectral" version of the given graph, in which all of the edges of the given graph are invisible (Figure 4.3a). We start with vertex A, and locate the vertex closest to it. This is vertex F (distance = 1), and so in Figure 4.3(b) we make the edge AF visible. Notice that

* Edsger Dijkstra (1930–), Dutch mathematician and physicist. One of the gurus of modern theoretical computer science (which Dijkstra himself prefers to call comput*ing* science), he invented structured programming, the language ALGOL, and has made many other fundamental contributions. It is said that he devised "Dijkstra's algorithm" simply as an example to illustrate one of his lectures in 1959!

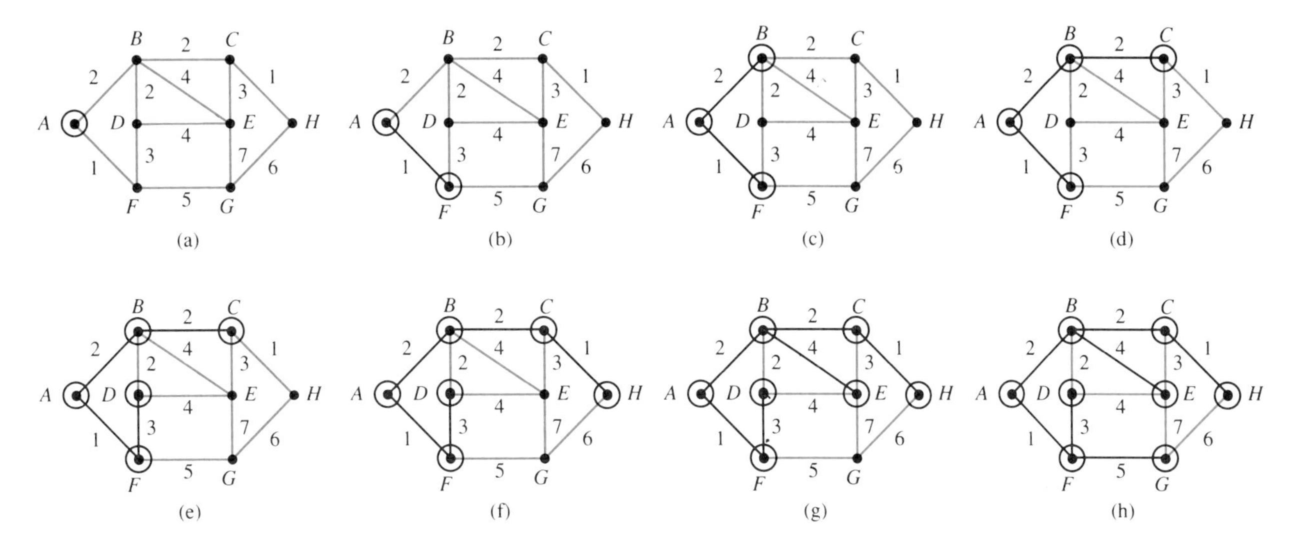

Figure 4.3 Solution to Example 4.1.

the path AF is the shortest possible path from A to F, since any other path from A to F would have to start out along the edge AB, which has length 2. Think of the edge AF as the first edge in a growing tree.

The next step is to look at all vertices which can be connected to the growing tree. These are the vertices B (via the edge AB), D (via the edge FD), and G (via the edge FG). We propose to add one of these three vertices to the growing tree, by choosing the one nearest to A. Here is the situation:

Proposed vertex	Path from A to proposed vertex	Length of path
B	AB	2
D	AFD	4
G	AFG	6

Since B is closest, we add B to the growing tree by making the edge AB visible, as shown in Figure 4.3(c). Notice that AB is definitely the shortest possible path in the original graph from A to B, since any other path would have to begin "AFD" or "AFG," and both of these partial paths have length greater than 2.

We continue in this way, at every stage adding one new vertex to the growing tree by choosing the vertex that, when connected to the tree, is closest to A. By part (e) of the daisy chain theorem (Section 3.2), we know that there will be *exactly one* path from A to each of the other vertices in the tree. The really important point, though, is that these paths will always be the *shortest possible* paths from A to these vertices in the original graph.

For example, to the tree of Figure 4.3(c) it is possible to add any one of the vertices C, D, E, or G. To determine which one should be added, we reason as follows:

Proposed vertex	Shortest path from *A* to proposed vertex	Length of path
C	*ABC*	4
D	*ABD* or *AFD*	4
E	*ABE*	6
G	*AFG*	6

Here the shortest path length is 4, and we have a choice: Either we can add *C* via the edge *BC*, or we can add *D* via either of the edges *BD* or *FD*. Let's choose *C*; the tree of Figure 4.3(d) results.

At the next stage, the candidate vertices are *D*, *E*, *G*, and *H*:

Proposed vertex	Shortest path from *A* to proposed vertex	Length of path
D	*ABD* or *AFD*	4
E	*ABE*	6
G	*AFG*	6
H	*ABCH*	5

(Notice that *E* can also be connected to the growing tree via the path *ABCE*, but since this path is longer than *ABE*, it is not included in the table.) Here the shortest path length is 4, and so we must add *D* to the tree, with one of the edges *BD* or *FD*. If we choose *FD*, the tree in Figure 4.3(e) results.

At the next stage we can add any of the three remaining vertices *E*, *G*, or *H*:

Proposed vertex	Shortest path from *A* to proposed vertex	Length of path
E	*ABE*	6
G	*AFG*	6
H	*ABCH*	5

Since the shortest of these paths is 5, we must add *H* to the tree via the edge *CH*, and the tree of Figure 4.3(f) results. Remembering that we were asked to find the shortest path from *A* to *H*, we see that we can stop! The path *ABCH* of length 5 is the shortest path from *A* to *H* in the graph of Figure 4.1. Indeed, the paths in the tree of Figure 4.3(f) from *A* to each of the vertices *B*, *C*, *D*, *F*, and *H* are the shortest paths from *A* to those vertices in the original graph.

But it seems a shame to stop when there are two vertices (*E* and *G*) not yet in the tree, so let's continue. With respect to the tree of Figure 4.3(f), the situation is now as follows:

Proposed vertex	Shortest path from *A* to proposed vertex	Length of path
E	*ABE*	6
G	*AFG*	6

Both of these vertices are the same distance from *A*, so it doesn't matter which we choose. If we choose *E*, however, the tree of Figure 4.3(g) results. Finally, only *G* remains to be added to the tree, via the edge *FG*, and if we add this edge we obtain a spanning tree for the original graph, as shown in Figure 4.3(h). We shall call this tree the "Dijkstra spanning tree." It shows all of the shortest paths from A to each of the other vertices in the original graph. ■

In Example 4.1 we have seen informally how Dijkstra's algorithm works. It is just an algorithm for finding a special kind of spanning tree for the original graph. The idea is to start with a given vertex v_0, and at every stage to add the vertex which is closest to v_0 when connected to the tree. Viewed in this way, like Prim's algorithm, Dijkstra's algorithm is a "greedy" algorithm in the sense that it always does what seems best without worrying about the long-term consequences. And once again, being greedy pays off! The spanning tree that results always contains one of the shortest paths from v_0 to each of the other vertices.

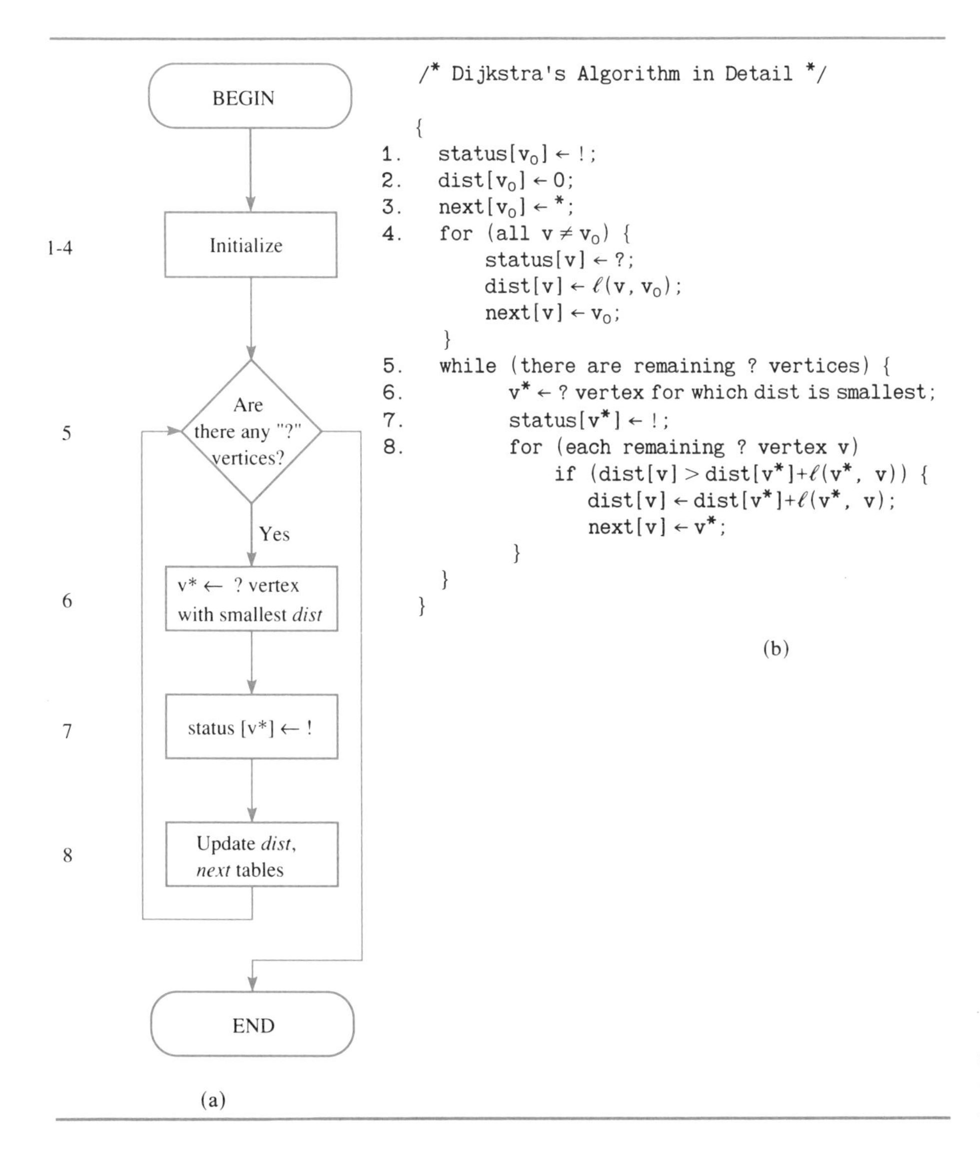

Figure 4.4 (a) Flowchart for Dijkstra's algorithm. (b) Pseudocode for Dijkstra's algorithm.

Our next task is to give a formal description of Dijkstra's algorithm, a description suitable for computer programming. We have done this in flowchart and pseudocode form in Figure 4.4.

The description of Dijkstra's algorithm in Figure 4.4 is probably not very enlightening at first reading; we will get to a practical example shortly. But before doing so, we should explain briefly what's going on. The algorithm is given the **length matrix** of the graph, which contains information about all the graph's edge lengths. In the description in Figure 4.4, the length of the edge from v_1 to v_2 is denoted by $\ell(v_1, v_2)$. Starting with an **initial vertex**, called v_0, the algorithm's job is to find a shortest path from v_0 to each of the other vertices in the graph. During the execution of the algorithm, these shortest paths will be found, one at a time. The vertex whose shortest path to v_0 has *most recently* been found will be denoted by v^*. Furthermore, at all times, each vertex v will have three things assigned to it: a **status**, denoted status[v]; a **distance**, denoted dist[v]; and a **next vertex**, denoted next[v]. The significance of these things is roughly as follows:

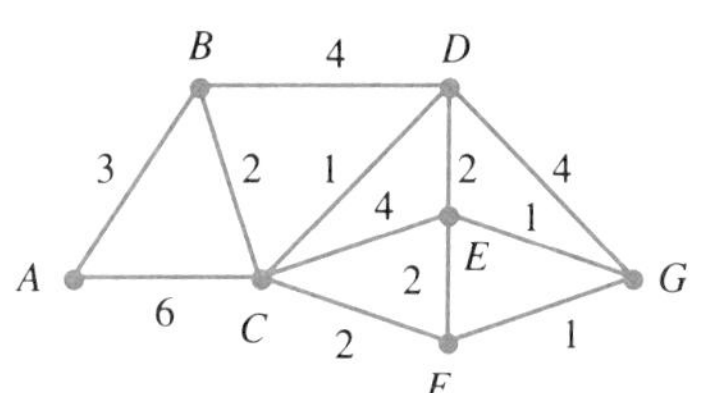

Figure 4.5 A labelled graph for Example 4.2.

- status[v] will be either "!," meaning that the shortest path from v to v_0 has definitely been found; or "?," meaning that it hasn't.
- dist[v] will be a number, representing the length of the shortest path from v to v_0 found so far. (In the early stages, this value might be ∞, meaning that *no path* has yet been found.)
- next[v] will be the first vertex on the way to v_0 along the shortest path found so far from v to v_0. (If dist[v] $= \infty$, no path has yet been found, and next[v] will still be equal to v_0, the value it was initially assigned in step 4. Also, next[v_0] is assigned the special symbol "*," indicating v_0's special role.)

When the algorithm eventually stops (after all the vertices have been given ! status), then for each vertex v, dist[v] will be the length of the shortest path from v to v_0, and the edges in the shortest path can be found by following the directions given by the nexts. The following example should make things clearer!

TABLE 4.1 Length matrix for the graph of Figure 4.5

	A	B	C	D	E	F	G
A	0	3	6	∞	∞	∞	∞
B	3	0	2	4	∞	∞	∞
C	6	2	0	1	4	2	∞
D	∞	4	1	0	2	∞	4
E	∞	∞	4	2	0	2	1
F	∞	∞	2	∞	2	0	1
G	∞	∞	∞	4	1	1	0

EXAMPLE 4.2 Use Dijkstra's algorithm, as detailed in Figure 4.4, to find the shortest paths from A to each of the other six vertices in the graph of Figure 4.5. (The distances between pairs of vertices are tabulated in the length matrix given in Table 4.1.)

SOLUTION We begin by executing the initialization steps 1–4, with A playing the role of v_0. After this initialization, the situation is summarized by the following table:

Vertex	A	B	C	D	E	F	G
status	!	?	?	?	?	?	?
dist	0	3	6	∞	∞	∞	∞
next	*	A	A	A	A	A	A

Pictorially, this situation is represented in Figure 4.6a, in which the dists are indicated inside the circles representing the vertices. A vertex whose status is "!" is shown in black, and one whose status is "?" is shown in color. The nexts are indicated with directed edges. (When the dist value is "∞," we don't bother indicating the nexts.) We then execute the control statement at step 5 (are there any remaining ? vertices?) and continue on, since there are many remaining ? vertices. At step 6, we see that among all ? vertices, *B* has the smallest dist

Figure 4.6 The progress of Dijkstra's algorithm on the graph of Figure 4.5, shown just before each execution of step 5.

(a)

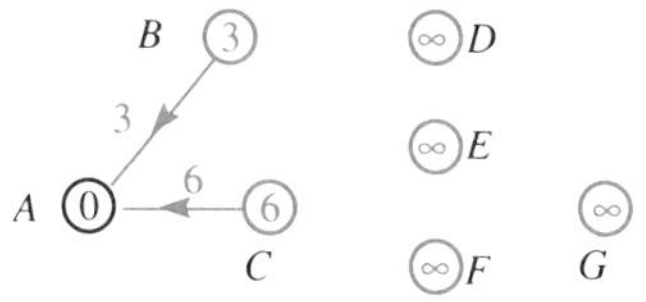

Vertex	*A*	*B*	*C*	*D*	*E*	*F*	*G*
status	!	?	?	?	?	?	?
dist	0	3	6	∞	∞	∞	∞
next	*	*A*	*A*	*A*	*A*	*A*	*A*

(b)

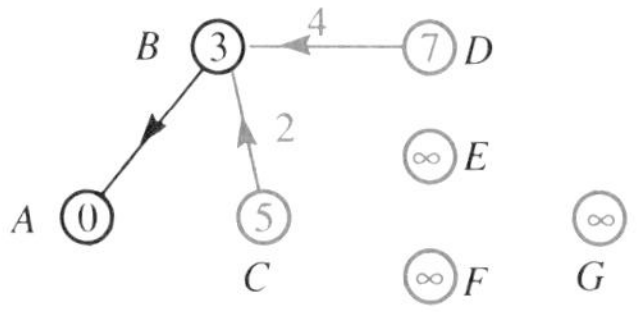

Vertex	*A*	*B*	*C*	*D*	*E*	*F*	*G*
status	!	!	?	?	?	?	?
dist	0	3	5	7	∞	∞	∞
next	*	*A*	*B*	*B*	*A*	*A*	*A*

(c)

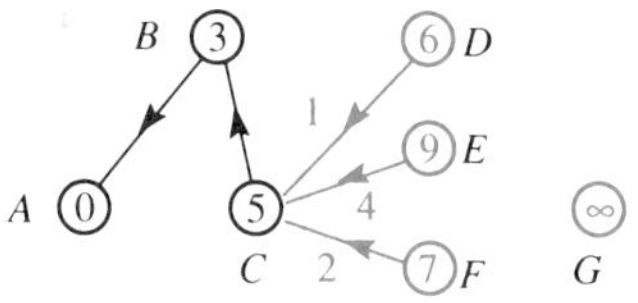

Vertex	*A*	*B*	*C*	*D*	*E*	*F*	*G*
status	!	!	!	?	?	?	?
dist	0	3	5	6	9	7	∞
next	*	*A*	*B*	*C*	*C*	*C*	*A*

(d)

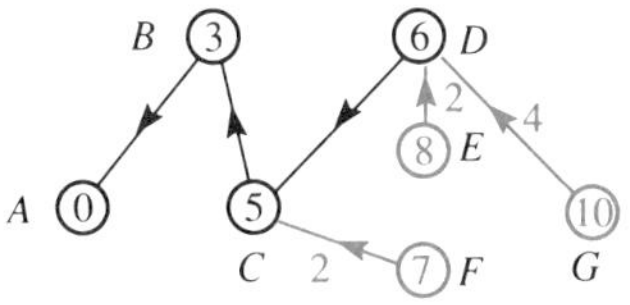

Vertex	*A*	*B*	*C*	*D*	*E*	*F*	*G*
status	!	!	!	!	?	?	?
dist	0	3	5	6	8	7	10
next	*	*A*	*B*	*C*	*D*	*C*	*D*

(e)

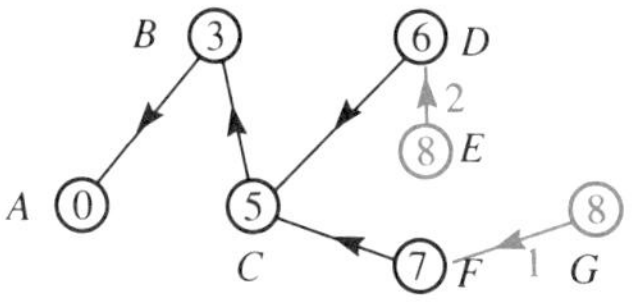

Vertex	*A*	*B*	*C*	*D*	*E*	*F*	*G*
status	!	!	!	!	?	!	?
dist	0	3	5	6	8	7	8
next	*	*A*	*B*	*C*	*D*	*C*	*F*

(f)

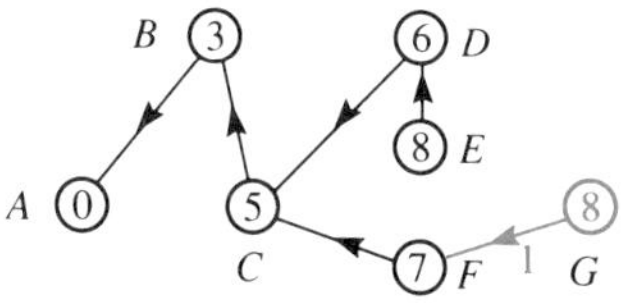

Vertex	*A*	*B*	*C*	*D*	*E*	*F*	*G*
status	!	!	!	!	!	!	?
dist	0	3	5	6	8	7	8
next	*	*A*	*B*	*C*	*D*	*C*	*F*

(g)

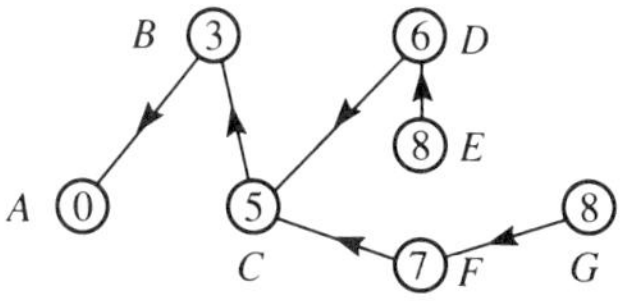

Vertex	*A*	*B*	*C*	*D*	*E*	*F*	*G*
status	!	!	!	!	!	!	!
dist	0	3	5	6	8	7	8
next	*	*A*	*B*	*C*	*D*	*C*	*F*

value, and so at step 7 we change its status to !. The situation is now as follows:

Vertex	A	B (v^* ↓)	C	D	E	F	G
status	!	!	?	?	?	?	?
dist	0	3	6	∞	∞	∞	∞
next	*	A	A	A	A	A	A

Next, we execute the "update" statement at step 8, which involves looking at each of the five remaining ? vertices, viz., C, D, E, F, and G. We must compare, for each of these vertices v, the value of dist$[v]$ to dist$[B] + \ell(B, v)$. Here is a table showing the results:

v	dist$[v]$	dist$[B] + \ell(B, v)$	Change?
C	6	$3 + 2 = 5$	Yes
D	∞	$3 + 4 = 7$	Yes
E	∞	$3 + \infty = \infty$	No
F	∞	$3 + \infty = \infty$	No
G	∞	$3 + \infty = \infty$	No

Thus after the execution of the *for* statement at step 8, the situation is as follows:

Vertex	A	B	C	D	E	F	G
status	!	!	?	?	?	?	?
dist	0	3	5	7	∞	∞	∞
next	*	A	B	B	A	A	A

The same situation is shown in Figure 4.6(b). Notice that in Figure 4.6(b) we have changed the color edge pointing from B to A to a black edge to indicate that it definitely (!) points the shortest way from B to A.

We now have to execute steps 6, 7, and 8 again, since there are still five ? vertices. (In computer lingo, these three steps are said to form a "loop," because the algorithm repeatedly executes them until there are no remaining ? vertices.) At step 6, C is identified as the ? vertex with the smallest value of dist, and at step 7, its status is changed to !. At step 8, we must compare dist$[v]$ to dist$[C] + \ell(C, v)$ for $v = D, E, F$, and G:

v	dist$[v]$	dist$[C] + \ell(C, v)$	Change?
D	7	$5 + 1 = 6$	Yes
E	∞	$5 + 4 = 9$	Yes
F	∞	$5 + 2 = 7$	Yes
G	∞	$5 + \infty = \infty$	No

So when we return again to step 5, the situation is as shown in Figure 4.6(c).

We continue, executing the loop (steps 6, 7, 8) three more times, until we arrive at the situation in Figure 4.6(f). Once more, we don't stop, since there is one remaining ? vertex, viz., G. At step 7, G's status is changed to !, and at step 8 there's nothing to do, since there are no remaining ? vertices. (This step produces Figure 4.6g.) Finally, we execute step 5 for one last time, and stop, since now all vertices have ! `status`.

The algorithm has terminated, and all the shortest paths from A have been found. For example, the shortest path from A to G has length 8, since `dist`[G] = 8, and that path can be obtained by repeatedly consulting the `next` values:

$$G \xrightarrow[1]{\text{next}} F \xrightarrow[2]{\text{next}} C \xrightarrow[2]{\text{next}} B \xrightarrow[3]{\text{next}} A$$

Notice also that the edges described by the `next` table form a spanning tree for the original graph; we saw this phenomenon in Example 4.1, too. ■

This completes our discussion of *how* Dijkstra's algorithm works—if you want to find the shortest paths, just follow the directions given in Figure 4.4.

The reason *why* Dijkstra's algorithm works is that just prior to each execution of the loop control statement (step 5), the following two assertions are true:

D1. For each vertex marked !, `dist`[v] is the length of the shortest path from v to v_0; and `next`[v] is the first vertex along this path.
D2. Furthermore, for each vertex marked ?, `dist`[v] is the length of the shortest path from v to v_0 that only uses intermediate vertices marked !, and `next`[v] is the first vertex along this path.

For example, consider Figure 4.6(d), which shows the situation just before the fourth execution of the loop. There we see that `dist`[D] = 6 and `status`[D] = !. It follows from assertion D1 that the length of the shortest path from D to A is 6; and following the `next`s, we find that this path is *DCBA*. On the other hand, in Figure 4.6(d) we see that `dist`[G] = 10, but `status`[G] = ?. According to assertion D2 this means that the shortest path from G to A that uses only ! vertices, i.e., vertices in the set $\{A, B, C, D\}$, has length 10. Indeed, by following the `next`s from G back to A, we find the path *GDCBA*, which has length 10. (This path is *not* the shortest path from G to A, since *GEDCBA* has length 9. However, since E has ? status, Dijkstra's algorithm cannot use it yet.)

In Figure 4.6(g) *all* vertices have ! status, and so by assertion D1, the table of `dist`s gives the lengths of the shortest paths from A to each of the other vertices, and the `next`s gives the actual paths.

It is possible to use mathematical induction to show that assertions D1 and D2 are true prior to each execution of the loop, thereby proving that Dijkstra's algorithm is correct. But the proof is a bit tricky, and we omit it.

However, we can gain further understanding of Dijkstra's algorithm by observing that it is very similar to Prim's MST algorithm, which we studied in Section 3.3. Indeed, in Figure 4.7 we give a detailed pseudocode version of Prim's algorithm which is nearly identical to Dijkstra's algorithm as described in Figure 4.4(b). In Figure 4.7, $C(u, v)$ denotes the *cost* of the edge from u to v, and the edge labels are interpreted as costs, rather than lengths. The only other difference (apart from the names of some of the variables) is in the "update" step 8. For this version of Prim's algorithm, the following two assertions are true just prior to each execution of the loop control statement (step 5):

P1. If `status`[v] = !, v is already in the growing tree. `cost`[v] is the cost of the edge by which v was added to the tree, and next[v] is the other endpoint of this edge.

```
/* Prim's Algorithm in Detail */

  {
1.  status[v0] ← !;
2.  cost[v0] ← 0;
3.  next[v0] ← *;
4.  for (all v ≠ v0) {
        status[v] ← ?;
        cost[v] ← C(v, v0);
        next[v] ← v0;
    }
5.  while (there are remaining ? vertices) {
6.        v* ← ? vertex of least cost;
7.        status[v*] ← !;
8.        for (each remaining ? vertex v)
              if(cost[v] > C(v*,v)) {
                 cost[v] ← C(v*,v);
                 next[v] ← v*;
              }
    }
  }
```

Figure 4.7 One way to implement Prim's algorithm (compare with Figure 4.4b).

P2. If status[v] = ?, v is not yet in the growing tree. cost[v] is the cost of the cheapest edge by which v can be added to the growing tree, and next[v] is the other endpoint of this edge.

The next example should clarify these points.

EXAMPLE 4.3 Use Prim's algorithm, as described in Figure 4.7, to find a minimal spanning tree for the graph of Figure 4.5, starting with the vertex A. [Here we interpret the matrix of Table 4.1 as the cost matrix for the graph; in Figure 4.7, $C(u, v)$ denotes the cost of the edge from u to v.]

SOLUTION The work is summarized in Figure 4.8. We begin by executing the initialization steps 1–4, at which point the situation is as shown in Figure 4.8(a). (The pictorial notation in Figure 4.8 is similar to that in Figure 4.6. The vertices with ! status are shown in black, and those with ? in color. The dist values are indicated inside the circles representing the vertices, and the nexts are indicated with directed edges.) Now vertex B has the smallest cost among all ? vertices, and so its status is changed to ! at step 7. We now must perform the update step 8. Here is the situation for all remaining ? vertices:

v	cost[v]	$C(v, B)$	Change?
C	6	2	Yes
D	∞	4	Yes
E	∞	∞	No
F	∞	∞	No
G	∞	∞	No

We see from this table that vertices C and D can now be added more cheaply to the growing tree (via vertex B) than previously, and so we change cost[C]

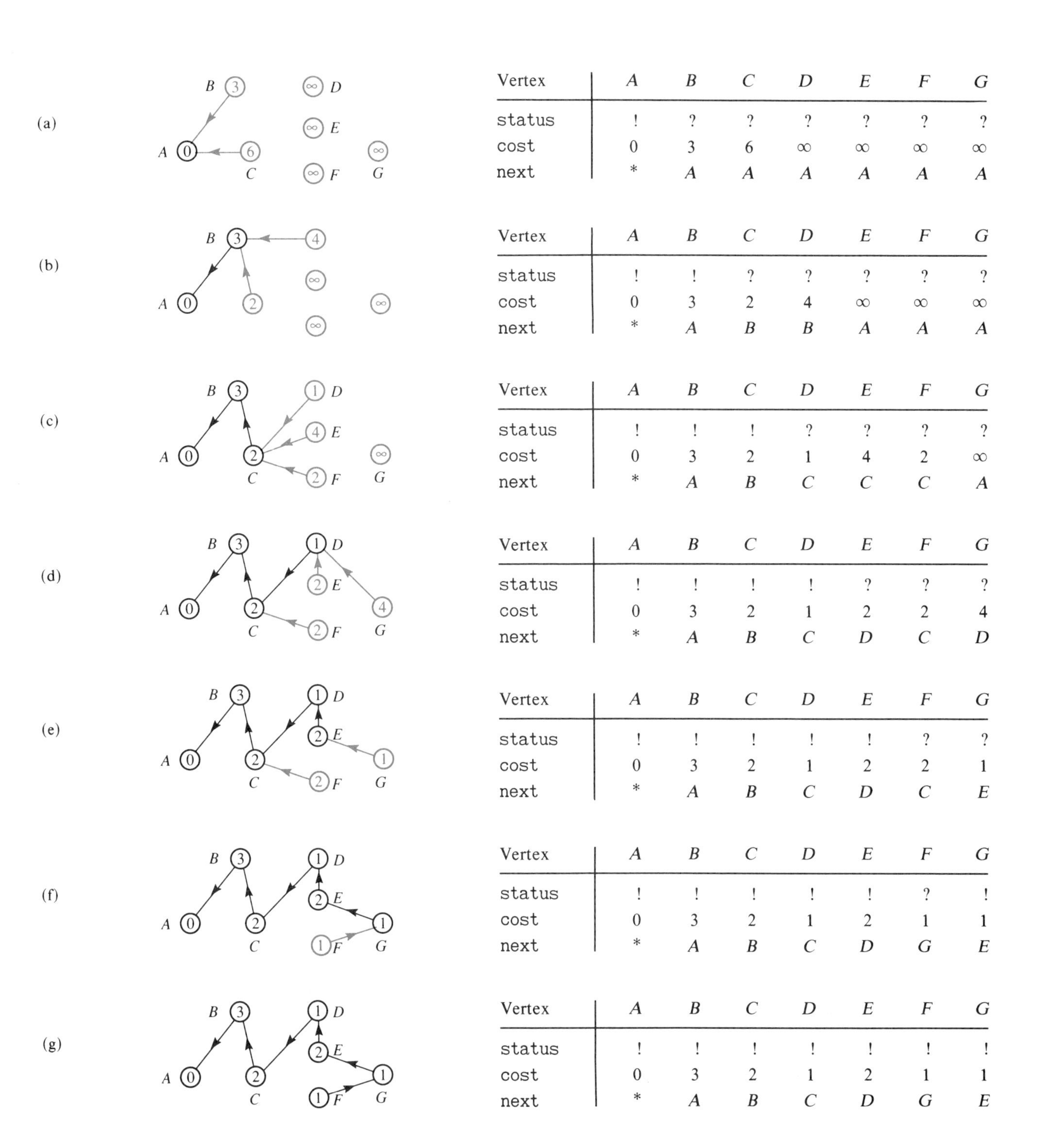

(a)

Vertex	A	B	C	D	E	F	G
status	!	?	?	?	?	?	?
cost	0	3	6	∞	∞	∞	∞
next	*	A	A	A	A	A	A

(b)

Vertex	A	B	C	D	E	F	G
status	!	!	?	?	?	?	?
cost	0	3	2	4	∞	∞	∞
next	*	A	B	B	A	A	A

(c)

Vertex	A	B	C	D	E	F	G
status	!	!	!	?	?	?	?
cost	0	3	2	1	4	2	∞
next	*	A	B	C	C	C	A

(d)

Vertex	A	B	C	D	E	F	G
status	!	!	!	!	?	?	?
cost	0	3	2	1	2	2	4
next	*	A	B	C	D	C	D

(e)

Vertex	A	B	C	D	E	F	G
status	!	!	!	!	!	?	?
cost	0	3	2	1	2	2	1
next	*	A	B	C	D	C	E

(f)

Vertex	A	B	C	D	E	F	G
status	!	!	!	!	!	?	!
cost	0	3	2	1	2	1	1
next	*	A	B	C	D	G	E

(g)

Vertex	A	B	C	D	E	F	G
status	!	!	!	!	!	!	!
cost	0	3	2	1	2	1	1
next	*	A	B	C	D	G	E

Figure 4.8 Solution to Example 4.3.

and cost[D] to 2 and 4, respectively, and next[C] and next[D] to B. This brings us to the situation in Figure 4.8(b).

We continue, as shown in Figures 4.8(c)–4.8(g), until we reach the MST shown in Figure 4.8(g). Notice that the MST (Figure 4.8g) and the tree of shortest paths from A (Figure 4.6g) are different. Prim's algorithm and Dijkstra's algorithm are similar in operation, but they solve quite different problems.

Problems for Section 4.1

In Example 4.1 we used an informal version of Dijkstra's algorithm to find the shortest paths from vertex A to each of the other vertices in the graph of Figure 4.1. In Problems 1–7 you are asked to do the same for one of the other vertices in the graph. In each case exhibit the resulting "Dijkstra spanning tree."

1. Vertex B.

2. Vertex C.

3. Vertex D.

4. Vertex E.

5. Vertex F.

6. Vertex G.

7. Vertex H.

8. Consider the graph in the accompanying figure.

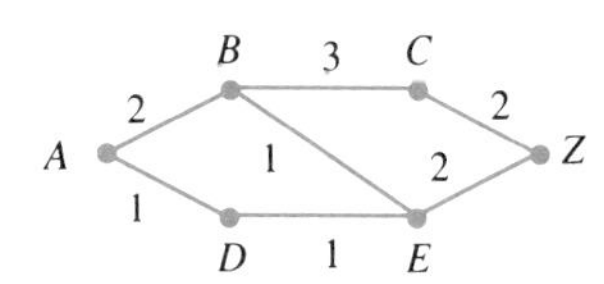

(a) Use Dijkstra's algorithm (but without formally keeping track of `status`, `dist`, and `next`) to find shortest paths from B to the other vertices. List all the shortest paths and their lengths, and draw the tree of shortest paths.
(b) For comparison, use Prim's algorithm starting with B to find a minimal spanning tree.
(c) Repeat part (a) but find shortest paths from A.

9. Use Dijkstra's algorithm to find the shortest path from A to B in the graph in the accompanying figure.

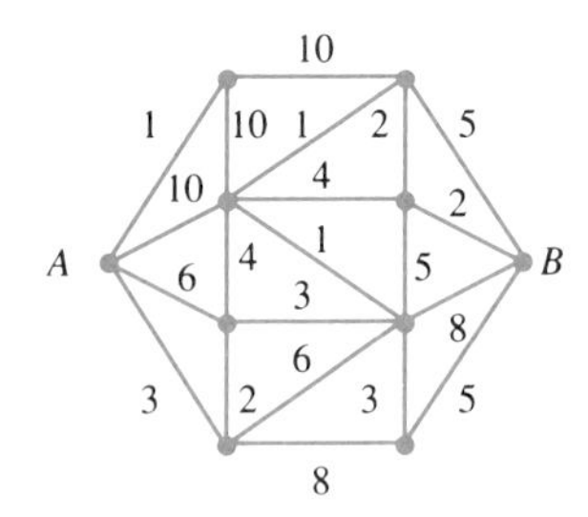

10. Relabel the edges in the graph of Figure 4.1 so that all the labels are different, and there are *two* shortest paths from A to H.

11. Consider the graph in the accompanying figure.

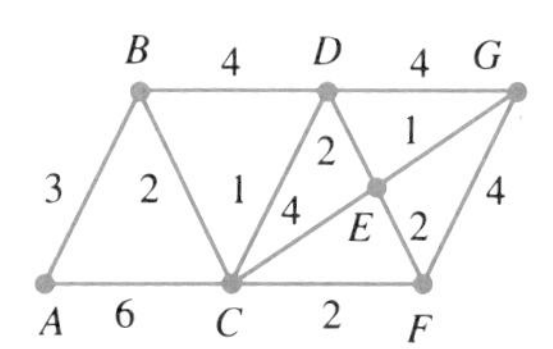

(a) Use Dijkstra's algorithm (without formally keeping track of `status`, `dist`, and `next`) to find a shortest path from B to E. (Even though the graph is small enough to find shortest paths by inspection, use the algorithm for practice.)
(b) Use Dijkstra's algorithm to find a tree of shortest paths from C to other vertices.

In Example 4.2 we used the formal version of Dijkstra's algorithm to find the shortest paths from vertex A to each of the other vertices in the graph of Figure 4.5. In Problems 12–17 you are asked to do the same for one of the other vertices in the graph.

12. Vertex B.

13. Vertex C.

14. Vertex D.

15. Vertex E.

16. Vertex F.

17. Vertex G.

18. For the graph in Problem 8, find shortest paths from A to the other vertices, this time keeping track of `status`, `dist`, and `next` at every step.

19. Without actually drawing the graph, find the shortest path from vertex 1 to vertex 4 in the graph whose length matrix is shown below.

	1	2	3	4	5
1	0	3.5	1.7	2.0	3.3
2	3.5	0	6.9	4.6	4.3
3	1.7	6.9	0	∞	∞
4	2.0	4.6	∞	0	∞
5	3.3	4.3	∞	∞	0

20. Given the following length matrix, keep track of `status`, `dist`, and `next` to find
(a) A shortest path from C to E;
(b) Shortest paths from B to all other vertices.

	A	B	C	D	E	F	G
A	0	3	6	∞	∞	∞	∞
B	3	0	2	4	∞	∞	∞
C	6	2	0	1	4	2	∞
D	∞	4	1	0	2	∞	4
E	∞	∞	4	2	0	2	1
F	∞	∞	2	∞	2	0	4
G	∞	∞	∞	4	1	4	0

21. Given the following length matrix, find a shortest path from A to D and its length. Keep track of `status`, `dist`, and `next`.

	A	B	C	D	E	F	G	Z
A	0	2	∞	∞	∞	1	∞	∞
B	2	0	2	2	4	∞	∞	∞
C	∞	2	0	∞	3	∞	∞	1
D	∞	2	∞	0	4	3	∞	∞
E	∞	4	3	4	0	∞	7	∞
F	1	∞	∞	3	∞	0	5	∞
G	∞	∞	∞	∞	7	5	0	6
Z	∞	∞	1	∞	∞	∞	6	0

22. Explain how Dijkstra's algorithm can be modified to find shortest paths in *directed* graphs, and use this modified algorithm to find the shortest paths from P to every other vertex in the graph in the accompanying figure.

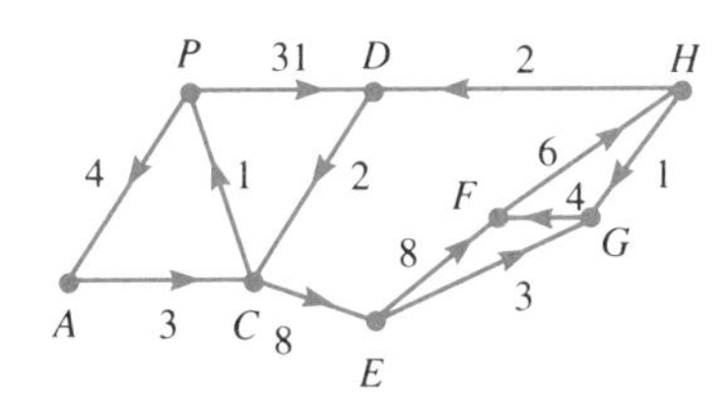

23. Given the following length matrix (this time for a directed graph), find a shortest path from P to D and its length. Keep track of `status`, `dist`, and `next`.

	A	B	C	D	E	F	G	P
A	0	∞	∞	∞	∞	∞	∞	∞
B	2	0	∞	∞	∞	∞	∞	∞
C	∞	1	0	∞	∞	∞	4	∞
D	∞	∞	3	0	∞	∞	8	∞
E	∞	∞	∞	∞	0	3	∞	∞
F	2	∞	∞	8	∞	0	∞	∞
G	∞	6	∞	∞	∞	∞	0	∞
P	31	∞	∞	∞	4	∞	∞	0

24. If, during the execution of Dijkstra's algorithm, we find that `next`$[C] = F$, what can be said about the relationship between `dist`$[C]$ and `dist`$[F]$?

25. What will happen in Dijkstra's algorithm if there is *no* path at all from v_0 to some other vertex v? For example, what happens when the algorithm is applied to the graph in the accompanying figure?

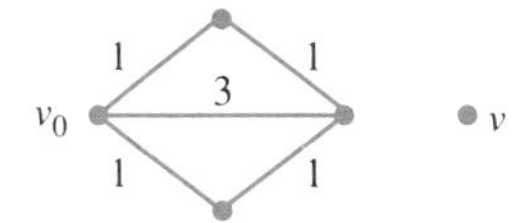

26. Explain how Dijkstra's algorithm can be modified to test whether or not a given undirected graph is *connected.* (Recall from Section 3.1 that a graph is said to be connected if, for any two vertices x and y, there is a path connecting x to y.) Use this modification to test for connectedness the graph whose length matrix is shown here.

	1	2	3	4	5	6	7	8
1	0	∞	4	2	∞	∞	1	∞
2	∞	0	∞	∞	3	3	∞	∞
3	4	∞	0	2	∞	∞	3	4
4	2	∞	2	0	∞	∞	∞	1
5	∞	3	∞	∞	0	5	∞	∞
6	∞	3	∞	∞	5	0	∞	∞
7	1	∞	3	∞	∞	∞	0	2
8	∞	∞	4	1	∞	∞	2	0

27. Referring to Figure 4.4 (or 4.7) if a graph G has n vertices, how many times will each of the steps 1, 2, and 3 be executed?

28. In Example 4.2 we saw that the vertices were marked ! in the order A, B, C, D, F, E, G, and that `dist`$[A] = 0$, `dist`$[B] = 3$, `dist`$[C] = 5$, `dist`$[D] = 6$, `dist`$[F] = 7$, `dist`$[E] = 8$, `dist`$[G] = 8$. Show that *in general,* if the vertices are marked ! in the order $v_0, v_1, \ldots, v_{n-1}$, then (when the algorithm stops) we have `dist`$[v_0] \le$ `dist`$[v_1] \le \cdots \le$ `dist`$[v_{n-1}]$.

29. In Dijkstra's algorithm, show that just before each execution of step 5, the following is true:
(a) For each vertex $v \neq v_0$, `dist[next`$[v]]$ < `dist`$[v]$.
(b) The path $v \to$ `next`$[v] \to$ `next[next`$[v]] \to \cdots$ eventually reaches v_0.

30. Suppose you wanted only to know the *length* of the shortest path from v_0 to the other vertices v. How could Dijkstra's algorithm be simplified to do this? Apply this simplified algorithm to the graph of Figure 4.1, using vertex G in the role of v_0.

31. Suppose Dijkstra's algorithm were modified so that steps 7 and 8 were interchanged. Would the algorithm still work? If not, why not? If so, what advantages and/or disadvantages would there be in doing this?

32. Suppose Dijkstra's algorithm were modified by deleting step 7. Would the algorithm still work? Why or why not?

33. Let $S = \{x_1, x_2, \ldots, x_n\}$ be a set of numbers. Consider the problem of *finding the smallest number* in the set S. Here is an algorithm for doing it:

```
     /* Find Minimum */
     {
1.     min ← ∞;
2.     for (i=1 to n)
           if (xi < min)
              min ← xi;
     }
```

(a) Show that this algorithm is correct. (*Hint*: Consider assertion M: Just before the ith execution of step 2, min is equal to the smallest number in the set $\{x_1, x_2, \ldots, x_{i-1}\}$.)

(b) How could you modify the algorithm so that when it stopped, it produced not only the *value* of the minimum, but also the index i for which x_i = smallest element in $\{x_1, x_2, \ldots, x_n\}$?

34. Write the procedure we gave in Section 3.1 for finding Euler cycles in the form of an algorithm like the one in Figure 4.4.

35. Consider the graph of Figure 3.25. If Prim's algorithm is implemented as in Figure 4.7, with v_0 = vertex 7, the first table of variables is as follows:

Vertex	1	↓ 2	3	4	5	6	7
status	?	?	?	?	?	?	!
next	7	7	7	7	7	7	*
cost	∞	9	10	∞	∞	∞	0

The arrow indicates that vertex 2 will be added to the growing tree. Carry out the details by finding the table of variables after each execution of the update step, and verify that the result agrees with that obtained in Example 3.6.

4.2 Two "All-Pairs" Algorithms; Floyd's Algorithm and Warshall's Algorithm

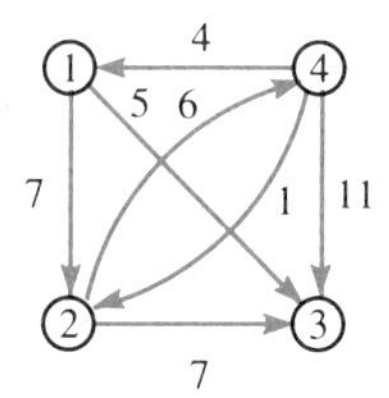

Figure 4.9 Graph for the Floyd example.

In the previous section we saw that Dijkstra's algorithm can be used to find the shortest path between a given pair of vertices in a given graph. In fact, we saw that what Dijkstra's algorithm really does is find the shortest path *from a given initial vertex to all the other vertices.* Often, however, one wishes to find the shortest distances between *all pairs* of vertices in a graph. Of course, one way to do this is to apply Dijkstra's algorithm, using each of the graph's vertices as the initial vertex, one at a time. But there is a more direct way to solve the "all-pairs" shortest-distance problem, called **Floyd's* algorithm**, which we shall now describe.

We begin by changing the rules a little. In the previous section we were dealing with labelled *undirected* graphs. In this section, however, we will consider instead labelled *directed* graphs, such as the one shown in Figure 4.9. If you like, you can think of the labels in Figure 4.9 as representing the cost of air transportation from one city to another. (Everyone knows that the cost of a one-way flight from Los Angeles to New York isn't necessarily the same as the cost from New York to Los Angeles, so sometimes we need directed graphs!)

Floyd's algorithm begins with the length matrix for the given graph. When it is finished, this matrix has been changed into a shortest-distance matrix which

* Robert W. Floyd (1936–), American computer scientist. Floyd was chosen chairman of the Stanford Computer Science Department in 1976, and he won the 1978 Turing Award (the top prize in computer science, named for Alan Turing, whom we shall meet in Chapter 6), despite the fact that his highest degree is a B.S. in physics.

$$\begin{array}{c} \\ 1 \\ 2 \\ 3 \\ 4 \end{array}\begin{array}{c} \begin{array}{cccc} 1 & 2 & 3 & 4 \end{array} \\ \begin{pmatrix} 0 & 7 & 5 & \infty \\ \infty & 0 & 7 & 6 \\ \infty & \infty & 0 & \infty \\ 4 & 1 & 11 & 0 \end{pmatrix} \\ \text{(a)} \end{array} \rightarrow \boxed{\text{Floyd's Algorithm}} \rightarrow \begin{array}{c} \\ 1 \\ 2 \\ 3 \\ 4 \end{array}\begin{array}{c} \begin{array}{cccc} 1 & 2 & 3 & 4 \end{array} \\ \begin{pmatrix} 0 & 7 & 5 & 13 \\ 10 & 0 & 7 & 6 \\ \infty & \infty & 0 & \infty \\ 4 & 1 & \infty & 0 \end{pmatrix} \\ \text{(b)} \end{array}$$

$$\begin{array}{c} \\ 1 \\ 2 \\ 3 \\ 4 \end{array}\begin{array}{c} \begin{array}{cccc} 1 & 2 & 3 & 4 \end{array} \\ \begin{pmatrix} 0 & 1 & 1 & 0 \\ 0 & 0 & 1 & 1 \\ 0 & 0 & 0 & 0 \\ 1 & 1 & 1 & 0 \end{pmatrix} \\ \text{(c)} \end{array} \rightarrow \boxed{\text{Warshall's Algorithm}} \rightarrow \begin{array}{c} \\ 1 \\ 2 \\ 3 \\ 4 \end{array}\begin{array}{c} \begin{array}{cccc} 1 & 2 & 3 & 4 \end{array} \\ \begin{pmatrix} 1 & 1 & 1 & 1 \\ 1 & 1 & 1 & 1 \\ 0 & 0 & 0 & 0 \\ 1 & 1 & 1 & 1 \end{pmatrix} \\ \text{(d)} \end{array}$$

Figure 4.10 (a) Length matrix for Figure 4.9. (b) All-pairs distance matrix for Figure 4.9. (c) Adjacency matrix for Figure 4.9. (d) All-pairs reachability matrix for Figure 4.9.

contains as entries the lengths of the shortest paths between each pair of vertices. For example, the length matrix for the graph of Figure 4.9 is shown in Figure 4.10(a), and the final all-pairs shortest-distance matrix is shown in Figure 4.10(b). (We draw your attention to two things about these matrices. First, the entries on the diagonal are all zero, because we assume that the distance between a vertex and itself is zero. Second, some of the entries in these matrices are "infinity." In the case of the length matrix an infinite entry means that there is no *edge* connecting the given pair of vertices; in the case of the shortest-distance matrix there is no *path* connecting them.)

Suppose that the graph has n vertices, numbered $1, 2, \ldots, n$. Floyd's algorithm updates the matrix n times, and then stops. Each updating of the matrix is called an **iteration** of the algorithm. The idea is that after the ith iteration, the entries in the matrix are the lengths of the shortest paths between pairs of vertices, where *the paths, apart from their endpoints, are restricted to pass only through vertices numbered 1 through i.* We shall call these paths the *shortest i-paths.* Thus after *no* interations, no intermediate vertices are allowed, and so the matrix simply gives the lengths of the direct edges between the vertices; and after n iterations, *all* intermediate vertices are allowed, and so the matrix gives the lengths of all the shortest paths.

Let's denote the matrix whose entries are the lengths of the shortest i-paths by A_i. Thus with respect to the graph of Figure 4.9, the matrix in Figure 4.10(a) is A_0, and the matrix of Figure 4.10(b) is A_4. Let us further denote the length of the shortest i-path between vertex x and vertex y by $A_i[x, y]$. The heart of Floyd's algorithm is the rule for computing the matrix A_i from the matrix A_{i-1}. In other words, for each of the n^2 pairs of vertices $[x, y]$, we need to compute $A_i[x, y]$, using the entries of the matrix A_{i-1}.

The idea behind the iteration rule for Floyd's algorithm is quite simple. The shortest i-path between x and y must use vertex i either *not at all,* or *exactly once.* If it uses it not at all, the length of the shortest i-path is the same as the shortest $(i-1)$-path. Otherwise, the shortest i-path consists of two parts: the shortest $(i-1)$-path from x to i, followed by the shortest $(i-1)$-path from i to y. These two possibilities are shown in Figure 4.11. (The "Warshall" parts of Figures 4.10 and 4.11 will be explained later, when we get to Warshall's algorithm.)

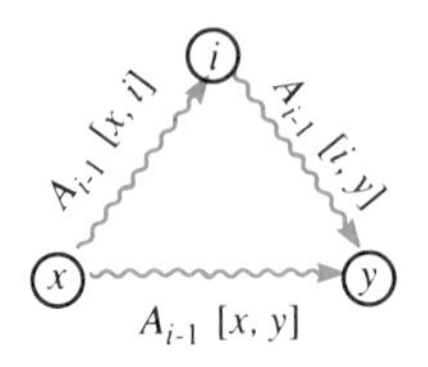

Figure 4.11 The two possibilities for the shortest i-path.

$$A_i[x, y] = \min\{A_{i-1}[x, y], A_{i-1}[x, i] + A_{i-1}[i, y]\} \quad \text{(Floyd)}$$

$$A_i[x, y] = A_{i-1}[x, y] + A_{i-1}[x, i]A_{i-1}[i, y] \quad \text{(Warshall)}$$

The direct path in Figure 4.11 represents the shortest $(i-1)$-path between x and y. Its length is $A_{i-1}[x, y]$. The indirect path has two parts. The first part is the shortest $(i-1)$-path from x to i; its length is $A_{i-1}[x, i]$. The second part is the shortest $(i-1)$-path from i to y. Its length is $A_{i-1}[i, y]$. Therefore the total

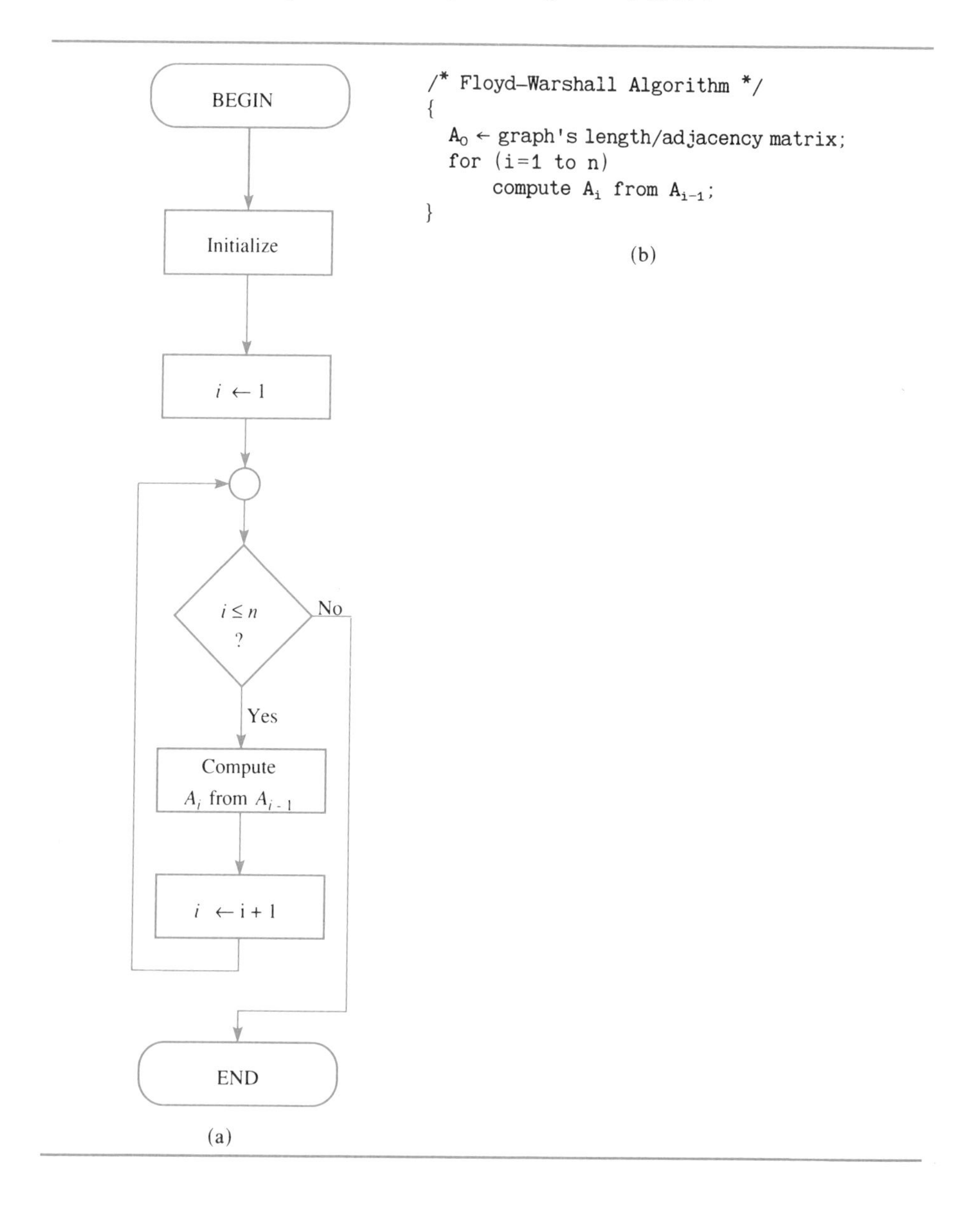

Figure 4.12 (a) Flowchart for Floyd's and Warshall's algorithms. (b) Floyd's and Warshall's algorithms in pseudocode. (See Figure 4.11 for details of how to compute A_i from A_{i-1}.)

length of the indirect path is $A_{i-1}[x, i] + A_{i-1}[i, y]$. As mentioned, one of these two paths must be the shortest i-path from x to y. Therefore, the rule for the ith iteration of Floyd's algorithm is to replace $A_{i-1}[x, y]$ with the smaller of these two values. In symbols, we have

$$A_i[x, y] = \min(A_{i-1}[x, y], A_{i-1}[x, i] + A_{i-1}[i, y]) \tag{4.1}$$

for each pair $[x, y]$. [In (4.1), if one of the numbers turns out to be ∞, we use the rule that $\infty + (\text{anything}) = \infty$.] Floyd's algorithm (flowchart and pseudocode) appears in Figure 4.12. The following example will give us some numerical experience with the algorithm.

EXAMPLE 4.4 Use Floyd's algorithm to compute the shortest distances between all pairs of vertices in the graph of Figure 4.9.

SOLUTION We begin with the length matrix of the graph, which appears as the matrix A_0 in Table 4.2. The first step is to compute the matrix A_1, which contains the lengths of the shortest 1-paths. There are 16 entries in this matrix, and each of them must be computed, using A_0 and the rule (4.1). For example, to compute $A_1[1, 2]$, i.e., the length of the shortest 1-path from vertex 1 to vertex 2, we reason as follows:

$$\text{direct path length} = A_0[1, 2] = 7;$$

$$\text{indirect path length} = A_0[1, 1] + A_0[1, 2] = 0 + 7 = 7.$$

Therefore the direct path is just as short as the indirect path, and so the [1, 2] entry doesn't change: $A_1[1, 2] = 7$. (This could have been predicted in advance—the computation amounts to comparing the paths 12 and 112.)

TABLE 4.2 Matrices A_0, A_1, A_2, A_3, and A_4 for Example 4.4.

A_0

	1	2	3	4
1	0	7	5	∞
2	∞	0	7	6
3	∞	∞	0	∞
4	4	1	11	0

A_1

	1	2	3	4
1	0	7	5	∞
2	∞	0	7	6
3	∞	∞	0	∞
4	4	1	9	0

A_2

	1	2	3	4
1	0	7	5	13
2	∞	0	7	6
3	∞	∞	0	∞
4	4	1	8	0

A_3

	1	2	3	4
1	0	7	5	13
2	∞	0	7	6
3	∞	∞	0	∞
4	4	1	8	0

A_4

	1	2	3	4
1	0	7	5	13
2	10	0	7	6
3	∞	∞	0	∞
4	4	1	8	0

We won't work through all of the entries in the matrix A_1 but content ourselves with two more entries, $A_1[3, 2]$ and $A_1[4, 3]$. To compute $A_1[3, 2]$, we make the following comparison:

$$\text{direct path length} = A_0[3, 2] = \infty;$$

$$\text{indirect path length} = A_0[3, 1] + A_0[1, 2] = \infty + 7 = \infty.$$

Thus no improvement is possible, and the [3, 2] entry remains ∞ in the matrix A_1. To compute $A_1[4, 3]$, we reason as follows:

$$\text{direct path length} = A_0[4, 3] = 11;$$

$$\text{indirect path length} = A_0[4, 1] + A_0[1, 3] = 4 + 5 = 9.$$

Aha! The indirect path is shorter, and so in the matrix A_1 the [4, 3] entry is reduced from 11 to 9. (And indeed if you look at Figure 4.9, you can see that the indirect path 413 is shorter than the direct path 43.)

After all 16 entries in A_0 have been updated, the matrix A_1 shown in Table 4.2 results. At the next iteration, A_1 is used to produce A_2. For example, we compute $A_2[1, 3]$, as follows:

$$\text{direct path length} = A_1[1, 3] = 5;$$

$$\text{indirect path length} = A_1[1, 2] + A_1[2, 3] = 7 + 7 = 14.$$

Hence by (4.1), $A_2[1, 3] = 5$. Similarly, $A_2[1, 4]$ is computed as follows:

$$\text{direct path length} = A_1[1, 4] = \infty;$$

$$\text{indirect path length} = A_1[1, 2] + A_1[2, 4] = 7 + 6 = 13.$$

Thus $A_2[1, 4] = 13$. Continuing in this way, we can compute the entire A_2 matrix as shown in Table 4.2. At the next iteration the matrix A_3 results (A_3 is identical to A_2 because vertex 3 can never be involved in a shortest path, since there is no way out of vertex 3!). Finally, after the fourth and last iteration the matrix A_4 results, and it gives the lengths of all 16 shortest paths in the graph of Figure 4.9. ■

If we look at the final matrix in Example 4.4 (the matrix in Figure 4.10b and the matrix A_4 in Table 4.2), we see that several entries are still equal to infinity. For example, $A_4[3, 4] = \infty$. This means that there is no path of any length from vertex 3 to vertex 4, and indeed every infinite entry corresponds to an ordered pair of vertices such that there is no path between them. On the other hand, every finite entry in the matrix corresponds to a pair of vertices connected by a path.

Sometimes all we want to do is determine, for each pair of vertices in a given directed graph, whether or not there is a path connecting them. As we have just seen, one way to do this is to run Floyd's algorithm on the given graph and only pay attention to whether or not the entries in the final matrix are infinite. However, this same information can be obtained by using a simplified version of Floyd's algorithm, called **Warshall's*** **algorithm**. We shall now describe Warshall's algorithm.

Warshall's algorithm doesn't need to know the lengths of the edges in the given directed graph; it only needs to know which edges exist and which do not. This information is most conveniently displayed in the *adjacency matrix* for the graph, in which a 1 indicates the existence of an edge and a 0 indicates nonexistence. In Figure 4.10(c) we show the adjacency matrix for the directed graph of Figure 4.9. This adjacency matrix is similar to the adjacency matrices we discussed in Section 3.1, but differs in one important respect. The adjacency matrices discussed in Section 3.1 corresponded to *undirected* graphs, so that the matrices were *symmetric*. Thus for example in Table 3.1, since there is a 1 in the $(1, 3)$ position (reflecting the existence of an edge from vertex 1 to vertex 3), we can tell without looking that there must be a 1 in the $(3, 1)$ position. However, in Figure 4.10(c) there is a 1 in the $(1, 3)$ position but a 0 in the $(3, 1)$ position, since the graph in Figure 4.9 has an edge from vertex 1 to vertex 3 but not from vertex 3 to vertex 1.

Warshall's algorithm is described by exactly the same flowchart and pseudocode as for Floyd's algorithm (see Figure 4.12). It begins with the adjacency matrix for the given graph, which is called A_0, and then updates the matrix n times, producing matrices called $A_1, A_2, \ldots, A_n$, and then stops. Whereas in Floyd's algorithm the matrix A_i contains the lengths of the shortest i-paths between all pairs of vertices, in Warshall's algorithm the matrix A_i merely contains information about the *existence* of i-paths. A 1 entry in the matrix A_i will correspond to the existence of an i-path, and a 0 entry will correspond to nonexistence. Thus when the algorithm stops, the final matrix, the matrix A_n, contains the desired connectivity information: A 1 entry indicates a pair of vertices which are connected, and a 0 entry indicates a pair which are not. This matrix we shall call the **reachability matrix** for the graph. In some texts, this matrix is called the **transitive closure** of the original adjacency matrix (see Problems 18–25).

The only difference between Warshall's algorithm and Floyd's algorithm is in the update rule. As shown in Figure 4.11, in Warshall's algorithm the rule for computing A_i from A_{i-1} is

(4.2) $$A_i[x, y] = A_{i-1}[x, y] + A_{i-1}[x, i] \cdot A_{i-1}[i, y].$$

In formula (4.2) all the arithmetic is *Boolean*, i.e., + represents the Boolean OR operation, and · represents the Boolean AND operation. We will discuss Boolean algebra in detail in Chapter 5, but for now we'll just give the definitions needed to interpret (4.2).

If x and y have possible values 0 and 1, then their *Boolean sum* and *product* are defined by

$$x + y = 1 \quad \text{if and only if} \quad x = 1 \quad \text{OR} \quad y = 1;$$

$$x \cdot y = 1 \quad \text{if and only if} \quad x = 1 \quad \text{AND} \quad y = 1.$$

* Stephen Warshall (1935–), American computer scientist and industrialist. As the founder and president of Massachusetts Computer Associates, Inc., he was Robert Floyd's boss when Floyd discovered his algorithm. As for Warshall's own algorithm, he produced it overnight on a bet with a colleague, thereby winning a bottle of rum!

Thus,

$$0 + 0 = 0, \qquad 0 + 1 = 1 + 0 = 1, \qquad 1 + 1 = 1$$

(so 1 plus anything = 1), and

$$0 \cdot 0 = 0 \cdot 1 = 1 \cdot 0 = 0, \qquad 1 \cdot 1 = 1$$

(so 0 times anything = 0).

The formula (4.2) states that there is an i-path from x to y if either of the following conditions is satisfied:

1. There is an $(i-1)$-path from x to y.
2. There is an $(i-1)$-path from x to i AND there is an $(i-1)$-path from i to y.

(This rule is simply a stripped-down version of the update rule for Floyd's algorithm as given in (4.1). Indeed, looking at (4.1), we can see that $A_i[x, y]$ will be finite if and only if either $A_{i-1}[x, y]$ is already finite, or if both $A_{i-1}[x, i]$ and $A_{i-1}[i, y]$ are both finite.) The following example will give us practice with Warshall's algorithm.

EXAMPLE 4.5 Use Warshall's algorithm to calculate the reachability matrix for the graph in Figure 4.9.

SOLUTION We begin with the adjacency matrix of the graph, which appears as the matrix in Figure 4.10(c) and as matrix A_0 in Table 4.3. The first step is to compute the matrix A_1. To do so, we use the updating rule (4.2), but before doing so we notice that any 1 entry in A_0 must remain a 1 in A_1, since in Boolean algebra 1 + (anything) = 1. Since there are only nine 0 entries in A_0,

TABLE 4.3 Matrices A_0, A_1, A_2, A_3, and A_4 for Example 4.5.

$$A_0 = \begin{array}{c} \\ 1 \\ 2 \\ 3 \\ 4 \end{array} \begin{array}{c} \begin{array}{cccc} 1 & 2 & 3 & 4 \end{array} \\ \begin{pmatrix} 0 & 1 & 1 & 0 \\ 0 & 0 & 1 & 1 \\ 0 & 0 & 0 & 0 \\ 1 & 1 & 1 & 0 \end{pmatrix} \end{array} \qquad A_1 = \begin{array}{c} \\ 1 \\ 2 \\ 3 \\ 4 \end{array} \begin{array}{c} \begin{array}{cccc} 1 & 2 & 3 & 4 \end{array} \\ \begin{pmatrix} 0 & 1 & 1 & 0 \\ 0 & 0 & 1 & 1 \\ 0 & 0 & 0 & 0 \\ 1 & 1 & 1 & 0 \end{pmatrix} \end{array} \qquad A_2 = \begin{array}{c} \\ 1 \\ 2 \\ 3 \\ 4 \end{array} \begin{array}{c} \begin{array}{cccc} 1 & 2 & 3 & 4 \end{array} \\ \begin{pmatrix} 0 & 1 & 1 & 1 \\ 0 & 0 & 1 & 1 \\ 0 & 0 & 0 & 0 \\ 1 & 1 & 1 & 1 \end{pmatrix} \end{array}$$

$$A_3 = \begin{array}{c} \\ 1 \\ 2 \\ 3 \\ 4 \end{array} \begin{array}{c} \begin{array}{cccc} 1 & 2 & 3 & 4 \end{array} \\ \begin{pmatrix} 0 & 1 & 1 & 1 \\ 0 & 0 & 1 & 1 \\ 0 & 0 & 0 & 0 \\ 1 & 1 & 1 & 1 \end{pmatrix} \end{array} \qquad A_4 = \begin{array}{c} \\ 1 \\ 2 \\ 3 \\ 4 \end{array} \begin{array}{c} \begin{array}{cccc} 1 & 2 & 3 & 4 \end{array} \\ \begin{pmatrix} 1 & 1 & 1 & 1 \\ 1 & 1 & 1 & 1 \\ 0 & 0 & 0 & 0 \\ 1 & 1 & 1 & 1 \end{pmatrix} \end{array}$$

there are only nine entries in A_0 that need to be updated. Thus for example to determine $A_1[1, 4]$, we make the following Boolean computation:

$$A_1[1, 4] = A_0[1, 4] + A_0[1, 1]A_0[1, 4] = 0 + 0 \cdot 0 = 0 + 0 = 0.$$

After eight similar calculations, the matrix A_1 shown in Table 4.3 results. (Notice that the off-diagonal entries in the matrix A_1 in Table 4.3 can also be obtained from the matrix A_1 in Table 4.2 by changing all the finite entries to 1 and all the infinite entries to 0; but the amount of work needed to go from A_0 to A_1 in Table 4.2 via Floyd's algorithm is a bit more than the amount needed to go from A_0 to A_1 in Table 4.3 via Warshall's algorithm.)

Next, A_2 must be calculated from A_1; but again we need only update the 0 entries. For example,

$$A_2[1, 4] = A_1[1, 4] + A_1[1, 2]A_1[2, 4] = 0 + 1 \cdot 1 = 0 + 1 = 1.$$

Similarly,

$$A_2[4, 4] = A_1[4, 4] + A_1[4, 2]A_1[2, 4] = 0 + 1 \cdot 1 = 0 + 1 = 1.$$

After seven similar calculations, the matrix A_2 shown in Table 4.3 results. This matrix has only seven 0 entries, and so to compute A_3, we need to do only seven computations. For example,

$$A_3[2, 1] = A_2[2, 1] + A_2[2, 3]A_2[3, 1] = 0 + 1 \cdot 0 = 0 + 0 = 0.$$

Once A_3 is calculated, we use the update rule (4.2) to calculate A_4 and stop. This matrix is the reachability matrix for the graph of Figure 4.9. The entire process of using Warshall's algorithm on this graph is summarized in Table 4.3.* ■

We'll conclude this section with a comment that can be used to simplify the update calculations in Warshall's algorithm. Suppose we are computing the xth row of the matrix A_i, i.e., the values $A_i[x, y]$, for $y = 1, 2, \ldots, n$. The update rule (4.2) is really not as complicated as it looks if we recall that $A_{i-1}[x, i]$ can assume only the two values 0 and 1. If $A_{i-1}[x, i] = 0$, then the update rule is

$$A_i[x, y] = A_{i-1}[x, y],$$

i.e., there is no change at all in the xth row in the step from i to $i + 1$. On the other hand, if $A_{i-1}[x, i] = 1$, the update rule is

$$A_i[x, y] = A_{i-1}[x, y] + A_{i-1}[i, y],$$

which says that the xth row of A_i equals the Boolean sum of the xth row of A_{i-1} and the ith row of A_{i-1}. The next example will illustrate the use of this trick.

* Note that according to the algorithm, vertex 3 is not reachable from itself! This is because, as can be seen in Figure 4.9, there is no path from vertex 3 back to itself. In some applications, it is better to regard every vertex as reachable from itself, and to change any 0s that appear on the main diagonal of the reachability matrix to 1s.

EXAMPLE 4.6 Use the trick just described to compute the reachability matrix for the graph whose adjacency matrix is

$$A_0 = \begin{pmatrix} 1 & 1 & 0 & 0 \\ 0 & 1 & 1 & 0 \\ 0 & 0 & 0 & 1 \\ 0 & 0 & 0 & 1 \end{pmatrix}.$$

SOLUTION To compute A_1, we need to focus our attention on the first row of A_0 $(1 \quad 1 \quad 0 \quad 0)$ and the first column of A_0 $(1 \quad 0 \quad 0 \quad 0)$. According to the trick described above, each row of A_1 is equal to the corresponding row of A_0, with $(1 \quad 1 \quad 0 \quad 0)$ added to those rows corresponding to the 1s in the first column. Thus

$$\begin{aligned} \text{1st row} &= (1 \quad 1 \quad 0 \quad 0) + (1 \quad 1 \quad 0 \quad 0) = (1 \quad 1 \quad 0 \quad 0) \\ \text{2nd row} &= (0 \quad 1 \quad 1 \quad 0) \qquad\qquad\qquad\qquad = (0 \quad 1 \quad 1 \quad 0) \\ \text{3rd row} &= (0 \quad 0 \quad 0 \quad 1) \qquad\qquad\qquad\qquad = (0 \quad 0 \quad 0 \quad 1) \\ \text{4th row} &= \underbrace{(0 \quad 0 \quad 0 \quad 1)}_{A_0} \qquad\qquad\qquad\qquad = \underbrace{(0 \quad 0 \quad 0 \quad 1)}_{A_1} \end{aligned}$$

To compute A_2, we look at the second row of A_1 $(0 \quad 1 \quad 1 \quad 0)$ and the second column of A_1 $(1 \quad 1 \quad 0 \quad 0)$. Each row of A_2 is equal to the corresponding row of A_1, with $(0 \quad 1 \quad 1 \quad 0)$ added to those rows corresponding to the 1s in the second column. Thus

$$\begin{aligned} \text{1st row} &= (1 \quad 1 \quad 0 \quad 0) + (0 \quad 1 \quad 1 \quad 0) = (1 \quad 1 \quad 1 \quad 0) \\ \text{2nd row} &= (0 \quad 1 \quad 1 \quad 0) + (0 \quad 1 \quad 1 \quad 0) = (0 \quad 1 \quad 1 \quad 0) \\ \text{3rd row} &= (0 \quad 0 \quad 0 \quad 1) \qquad\qquad\qquad\qquad = (0 \quad 0 \quad 0 \quad 1) \\ \text{4th row} &= \underbrace{(0 \quad 0 \quad 0 \quad 1)}_{A_1} \qquad\qquad\qquad\qquad = \underbrace{(0 \quad 0 \quad 0 \quad 1)}_{A_2} \end{aligned}$$

To compute A_3, we look at the third row of A_2 $(0 \quad 0 \quad 0 \quad 1)$ and the third column of A_2 $(1 \quad 1 \quad 0 \quad 0)$. Each row of A_3 is equal to the corresponding row of A_2, with $(0 \quad 0 \quad 0 \quad 1)$ added to those rows corresponding to the 1s in the third column. Thus

$$\begin{aligned} \text{1st row} &= (1 \quad 1 \quad 1 \quad 0) + (0 \quad 0 \quad 0 \quad 1) = (1 \quad 1 \quad 1 \quad 1) \\ \text{2nd row} &= (0 \quad 1 \quad 1 \quad 0) + (0 \quad 0 \quad 0 \quad 1) = (0 \quad 1 \quad 1 \quad 1) \\ \text{3rd row} &= (0 \quad 0 \quad 0 \quad 1) \qquad\qquad\qquad\qquad = (0 \quad 0 \quad 0 \quad 1) \\ \text{4th row} &= \underbrace{(0 \quad 0 \quad 0 \quad 1)}_{A_2} \qquad\qquad\qquad\qquad = \underbrace{(0 \quad 0 \quad 0 \quad 1)}_{A_3} \end{aligned}$$

Finally, to compute A_4, we use the fourth row of A_3 $(0 \quad 0 \quad 0 \quad 1)$ and the fourth column of A_3 $(1 \quad 1 \quad 1 \quad 1)$, to obtain

$$\begin{aligned} \text{1st row} &= (1 \quad 1 \quad 1 \quad 1) + (0 \quad 0 \quad 0 \quad 1) = (1 \quad 1 \quad 1 \quad 1) \\ \text{2nd row} &= (0 \quad 1 \quad 1 \quad 1) + (0 \quad 0 \quad 0 \quad 1) = (0 \quad 1 \quad 1 \quad 1) \\ \text{3rd row} &= (0 \quad 0 \quad 0 \quad 1) + (0 \quad 0 \quad 0 \quad 1) = (0 \quad 0 \quad 0 \quad 1) \\ \text{4th row} &= \underbrace{(0 \quad 0 \quad 0 \quad 1)}_{A_3} + (0 \quad 0 \quad 0 \quad 1) = \underbrace{(0 \quad 0 \quad 0 \quad 1)}_{A_4} \end{aligned}$$

Therefore the reachability matrix required is

$$A_4 = \begin{pmatrix} 1 & 1 & 1 & 1 \\ 0 & 1 & 1 & 1 \\ 0 & 0 & 0 & 1 \\ 0 & 0 & 0 & 1 \end{pmatrix}.$$

Problems for Section 4.2

In Problems 1–6, write the length matrix of the directed graph, and use Floyd's algorithm to find the matrix of shortest distances between all pairs of vertices.

1.

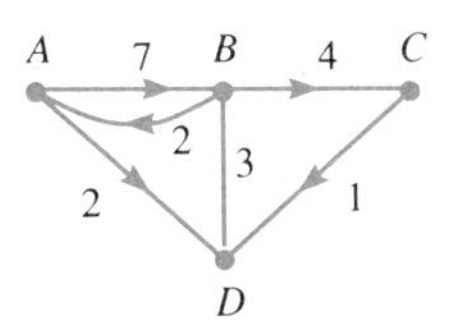

2.

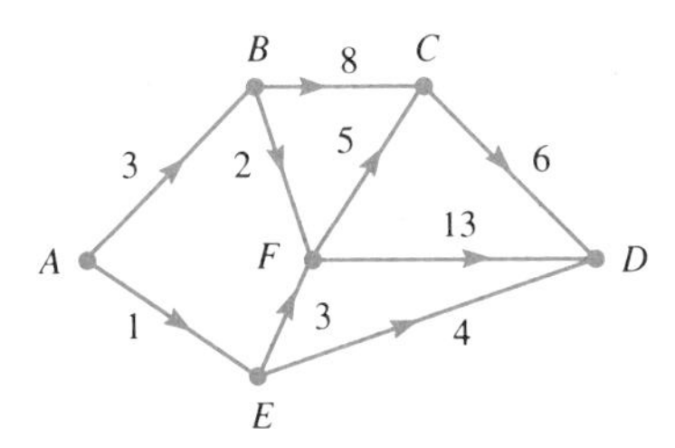

3.

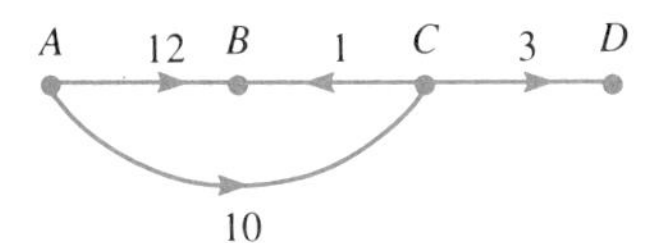

4.

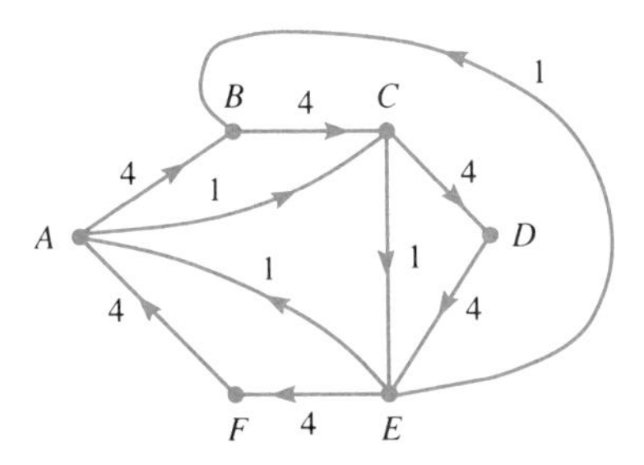

5. The graph of Problem 4.1.22.

6. The graph of Problem 4.1.23.

In Problems 7–12, write the adjacency matrix of the directed graph, and use Warshall's algorithm to find the reachability matrix of the graph.

7. The graph of Problem 1.

8. The graph of Problem 2.

9. The graph of Problem 3.

10. The graph of Problem 4.

11. The graph of Problem 5.

12. The graph of Problem 6.

13. Suppose that in step i of Floyd's algorithm (i.e., the computation of A_i from A_{i-1}) the $[x, y]$ entry does not change, i.e., $A_i[x, y] = A_{i-1}[x, y]$. What conclusion can be drawn?

14. Suppose that in step i of Warshall's algorithm the $[x, y]$ entry does not change, i.e., $A_i[x, y] = A_{i-1}[x, y]$. What conclusion can be drawn?

15. Suppose a graph contains an isolated vertex v_k without even a loop. What can you say about the matrix of shortest distances between pairs of vertices?

16. In Problem 15, what can you say about the reachability matrix?

17. At the end of the section, we discussed a way to speed up Warshall's algorithm. Show that there is a similar way to speed up Floyd's algorithm by checking to see whether or not the (x, i) entry is infinite.

Problems 18–25 explore the idea of "transitive closure."

18. Let R be the relation corresponding to the directed graph G, i.e., if a and b are vertices of G, then $a\,R\,b$ if there is an edge from a to b. If K is the reachability matrix of G, we can define a relation R^* associated with K by $a\,R^*\,b$ if and only if $K_{ab} = 1$, i.e., there is a path in G from a to b. Show that R^* is transitive.

19. Show that $R \subseteq R^*$; i.e., if $a\,R\,b$, then $a\,R^*\,b$.

20. Now let T be *any* transitive relation such that $R \subseteq T$. If there is a path $a_1 a_2 \ldots a_n$ passing through vertices $a_1, a_2, \ldots, a_n$ of G, show that $a_1\,T\,a_n$.

21. Show that $R^* \subseteq T$. Conclude that R^* is the *smallest* transitive relation containing R; R^* is called the *transitive closure* of R (and sometimes K is referred to as the transitive closure of the adjacency matrix of G).

22. If $T_1, T_2, \ldots, T_m$ are transitive relations, show that the intersection of the T_i is also transitive.

23. Let T^* be the intersection of all transitive relations T_i containing R (i.e., $R \subseteq T_i$). Use Problem 21 to show that T^* is also a transitive relation containing R.

24. Show that T^* is the smallest transitive relation containing R.

25. Conclude that $T^* = R^*$.

4.3 The Matching Problem and the Hungarian Algorithm

We begin with an example to illustrate the problem we'll be dealing with in this section. In solving the specific example, we will also develop an algorithm for solving the general problem.

EXAMPLE 4.7 At a certain party there are eight girls and eight boys. It is the girls' turn to choose partners, and when the music begins each girl will want to dance with a boy she likes. Here is a list of which boys are liked by which girls:

Girl	Boys she likes
Alice	Art, Eric, Frank, Gary, Hal
Betty	Bill, Carl, Hal
Cindy	Carl, Eric, Frank, Gary
Doris	Don, Hal
Edna	Bill, Carl, Don
Flo	Bill, Don
Gloria	Art, Bill, Carl, Don
Hazel	Bill, Carl, Don.

Question: Is it possible for each girl to dance with a boy she likes? (Maybe you should try this problem yourself, before reading further!)

SOLUTION This *particular* problem is of no real importance, of course. But it does illustrate a general *class* of problems, called **matching problems**, that *are* important. Here, for example, are some other situations that lead to matching problems:

- A football coach has 33 positions to fill (11 offense, 11 defense, 11 for special teams) from among 50 player candidates. Each candidate is qualified for only some (possibly none) of the positions. Will the coach be able to staff his team?
- A NASA administrator is in charge of tracking n satellites in earth orbit. She has m tracking antennas under her control, but because of weather, location, etc., each antenna can only "see" a subset of the satellites. How many of the satellites can be tracked simultaneously?
- A movie reviewer has been assigned by her editor to review a set of k movies during the next week. Each movie is only shown at certain hours; will she be able to see them all?

Although superficially quite different, these four problems have much in common. In each problem there are two sets involved, let us say A and B:

Problem	A	B
1	Girls	Boys
2	Positions	Players
3	Satellites	Antennas
4	Movies	Time slots

Furthermore, each element in A is "compatible" with certain elements in B. For example:

Problem	Element of A	Compatible subset of B
1	Edna	{Bill, Carl, Don}
2	Free safety	{Smith, Rashad, Yates, Henderson}
3	LANDSAT-5	{Antennas 66, 107, 111}
4	*Ghostbusters*	{Monday 2–4, Thursday 5–7, Saturday noon–2}

Finally, in each problem we are asked to *match* each element in A with a compatible element in B, so that no element in B is used more than once or, failing that, to match as many elements in A as possible.

These problems can all be stated in the language of graph theory. We construct a graph, with one vertex corresponding to each element of A and one to each element of B. The vertices a and b are joined by an edge if a and b are compatible. This kind of a graph is called a **bipartite graph.** Several examples are shown in Figure 4.13.

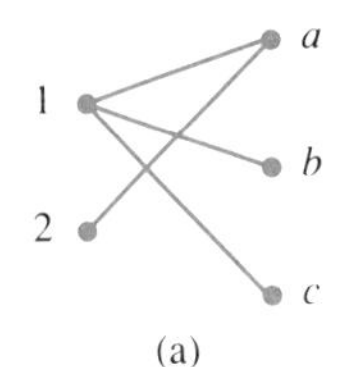

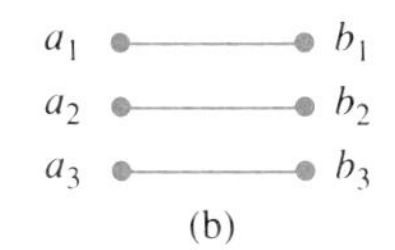

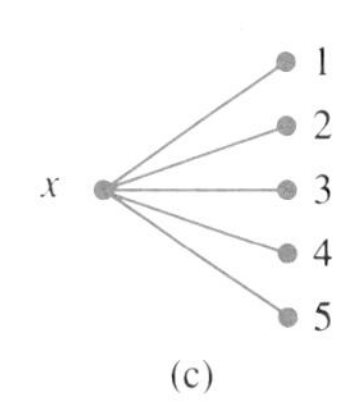

Figure 4.13 Three simple bipartite graphs.

A **matching** in a bipartite graph is a set of edges with no vertices in common—the endpoints are said to be **matched**. For example, in Figure 4.14 we show a matching for each of the graphs in Figure 4.13.

A **maximal matching** in a bipartite graph is a matching that contains as many edges as possible. The matchings in Figures 4.14(a) and 4.14(c) are maximal; no more edges can be added. But the matching in Figure 4.14(b) isn't maximal, because the edge $\overline{a_2b_2}$ can be added. A maximal matching represents the largest possible number of "compatible couples" in the graph. In the four examples cited above, we have:

Problem	Maximal matching: The largest possible number of
1	Happy girls
2	Positions filled
3	Satellites tracked
4	Movies seen

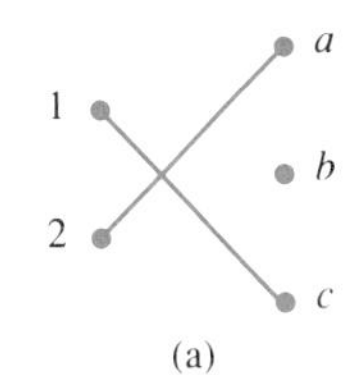

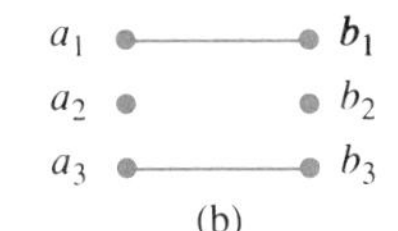

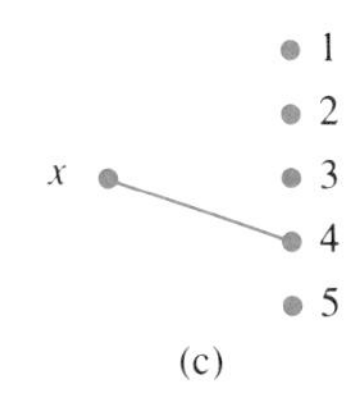

Figure 4.14 Examples of matchings for each of the graphs in Figure 4.13.

In this section we will describe an efficient algorithm for finding a maximal matching in a bipartite graph. This algorithm is called by several different names, but most often it is called the **Hungarian algorithm**, because its origins can be traced to the work of two Hungarian mathematicians (König and Egeváry).

Before we give the details of the Hungarian algorithm, let's return to our original example, the eight girls and boys at the dance. A corresponding bipartite graph and adjacency matrix are shown in Figure 4.15. Let's now attempt to find a maximal matching in this graph. (Maybe you have already solved this problem—or think you have—and the following discussion may seem slow and plodding.

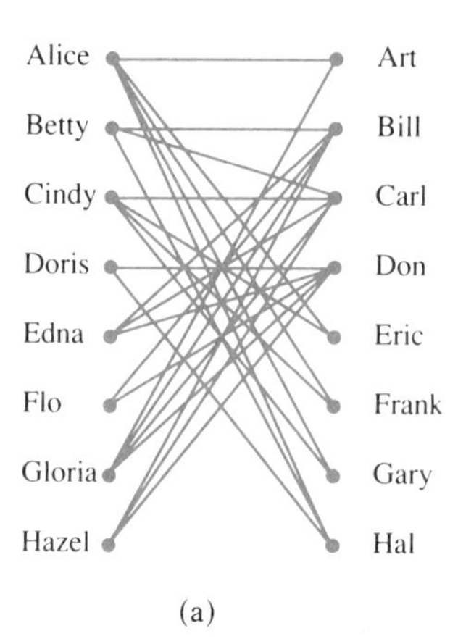

	a	b	c	d	e	f	g	h
A	1	0	0	0	1	1	1	1
B	0	1	1	0	0	0	0	1
C	0	0	1	0	1	1	1	0
D	0	0	0	1	0	0	0	1
E	0	1	1	1	0	0	0	0
F	0	1	0	1	0	0	0	0
G	1	1	1	1	0	0	0	0
H	0	1	1	1	0	0	0	0

(b)

Figure 4.15 (a) A larger bipartite graph. (b) The corresponding adjacency matrix.

Please bear with us; remember that our goal is not so much to solve this particular problem as it is to devise an algorithm which can be used to solve *any* matching problem.)

We have seen that being greedy pays off for the minimal spanning tree problem and the shortest-path problem, so let's be greedy in this problem and match each girl, in turn, with the *first* boy (alphabetically) that she likes, and who is not already matched. The result is the matching of Figure 4.16.

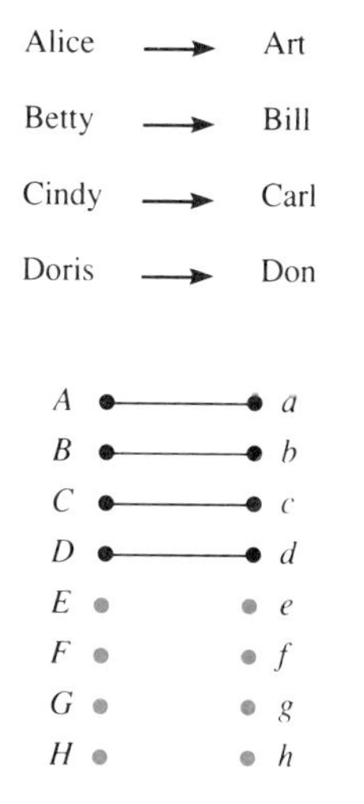

Figure 4.16 A "greedy" matching of the graph in Figure 4.13.

Now the first four girls are happy, but the last four aren't, because all the boys liked by the last four girls are already spoken for! However, improvements are possible. For example, we could match Alice with Eric instead of Art, and then Art could dance with Gloria; this would make five girls happy. A more systematic approach, however, would be to start with the *first unmatched girl*, Edna, and reason as follows:

> *Edna* likes *Bill*; but Bill is already matched with *Betty*; but Betty also likes *Hal*, who isn't matched yet!

Graphically, this situation is shown in Figure 4.17(b).

To improve the matching, we can now reverse the steps:

> So let's have *Hal* be *Betty*'s partner; then *Bill* can be *Edna*'s partner; and now five girls are happy!

In graphical terms, we have *added* the two edges *Eb* and *Bh* to the matching, while *removing* the edge *Bb*. The new matching is shown in Figure 4.17(c).

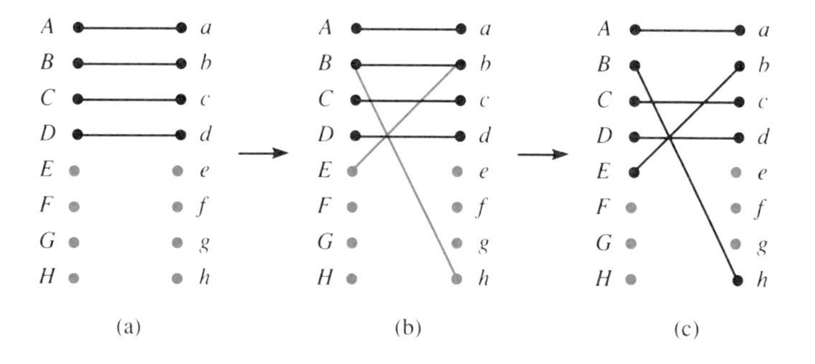

Figure 4.17 Improving the matching in Figure 4.16: (a) Original matching. (b) An alternating path. (c) The new matching.

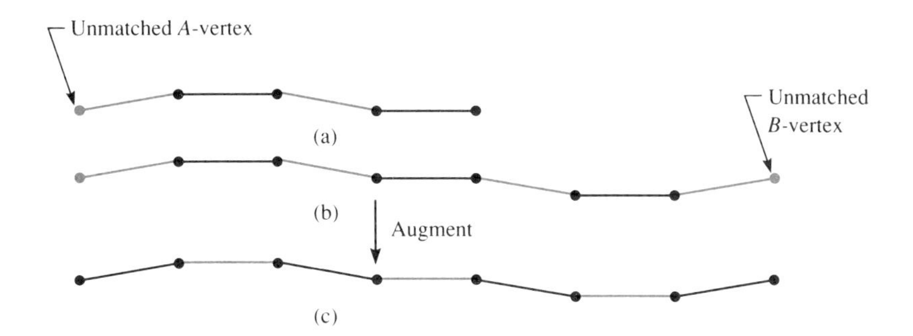

Figure 4.18 (a) An alternating path. (b) An augmenting alternating path. (c) After augmentation, there is one more edge in the matching than before.

This simple idea for enlarging, or *augmenting*, a given matching is the basis for the Hungarian algorithm. The idea is to build up a set of **alternating paths,** in hopes of finding one that can be used to augment the matching. By an alternating path we mean a path that begins with an unmatched A-vertex (i.e., a vertex in the set A which is not yet in the matching) and consists of edges which are alternately *not in* and *in* the matching (see Figure 4.18). An alternating path is said to be **augmenting** if it terminates in an unmatched B-vertex. This name reflects the fact that if an augmenting alternating path is found (as as we shall see, it won't always be possible to find such a path), then the matching can be augmented simply by *reversing* the matching membership of the edges in the path, i.e., if an edge is in the matching it is removed, and if it is not in, it is inserted. See Figures 4.18(b) and 4.18(c).

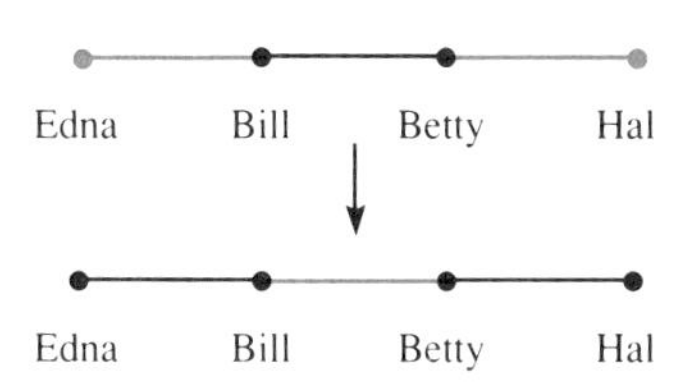

Figure 4.19 Another view of Figures 4.17(b) and 4.17(c).

For example, in Figure 4.17(b) we found a simple augmenting alternating path of length 3, and augmented the matching as shown in Figure 4.19.

Unsatisfied with making only five of the girls happy, let's continue, and look for a possible augmenting path for the matching in Figure 4.17(c). We begin with vertex F, which is the first unmatched A-vertex. Vertex F is compatible with both b and d; unfortunately, however, both of these B-vertices are already matched, to E and D, respectively. We can represent this situation with the *alternating tree* in Figure 4.20(a). Undiscouraged, we now attempt to grow the tree from vertex E. Vertex E is compatible with b, c, and d; b and d are *already* in the tree, but c isn't, so we can add the edge Ec to the tree. Unfortunately, the path from F to c isn't augmenting, since c is matched to C. Similarly, if we grow the tree from vertex D, we add vertices h and B, but we can't yet augment the matching. The situation is shown in Figure 4.20(b). Next, we extend the tree from C, and find that we can add any one of the vertices e, f, or g; and our patience is rewarded: *All* of these vertices are unmatched! (See Figure 4.20c). For example, the path from F to e is an augmenting path, and in Figure 4.20(d) we see this path after the edges have been reversed. The new, six-edge, matching is shown in Figure 4.21.

Continuing from the matching of Figure 4.21, we grow another tree. This time we omit the details and simply show the "augmenting tree" and the new matching in Figure 4.22.

Now all the girls are happy, except for Hazel. Can we find a partner for her? To find out, we grow an alternating tree starting at H. The result is shown in Figure 4.23.

The tree of Figure 4.23 is very curious: on one hand, it contains no augmenting path; and on the other hand, it *cannot grow further*, because all of the B-vertices

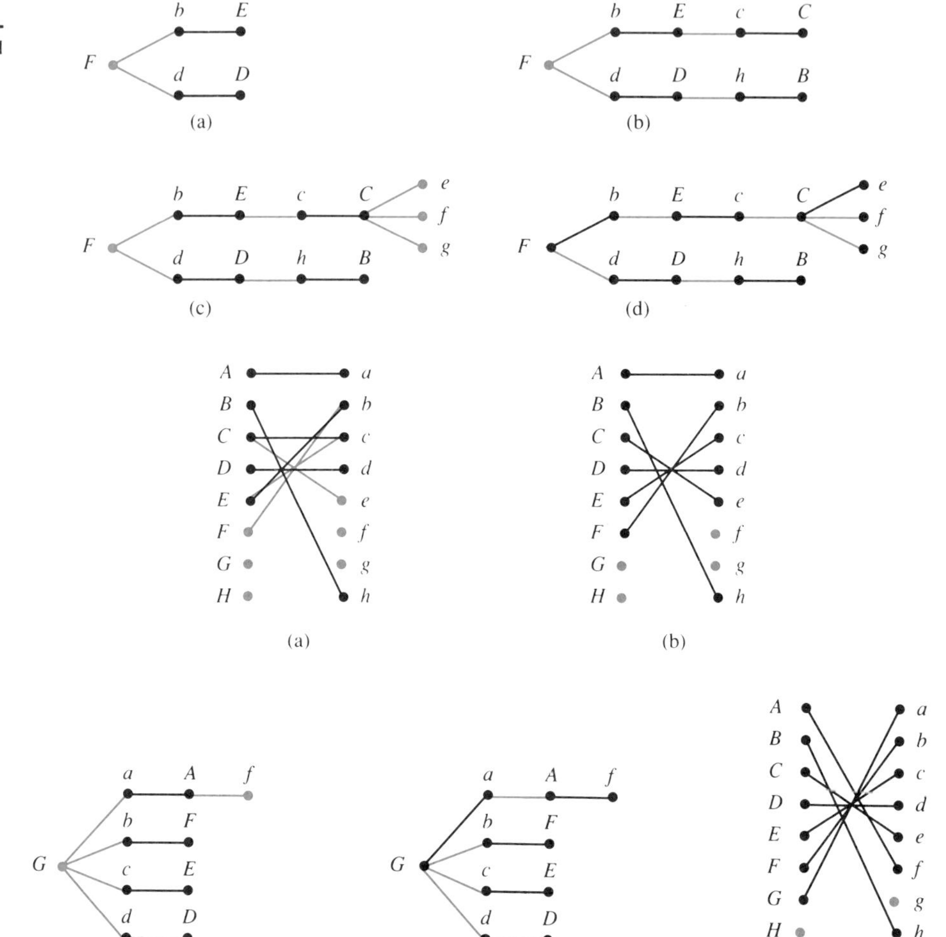

Figure 4.20 The growth of an alternating tree. In (c) the path from F to e is seen to be augmenting. In (d) the augmenting has been completed.

Figure 4.21 A bigger matching for the graph of Figure 4.15: (a) Before and (b) after the augmenting of Figure 4.20 has been completed.

Figure 4.22 The augmenting tree for the matching of Figure 4.21(b): (a) Before and (b) after the augmentation is complete. (c) The resulting matching.

(boys) compatible with its A-vertices (girls) are already on the tree. For example, we can't grow the tree from B since B is only compatible with b, c, and h, which are already present. Similarly, all vertices compatible with H, F, E, and D are also on the tree. We have unfortunately failed to find a partner for Hazel, and we are stuck.

Ironically though, we are now in a position to answer the original question: It is definitely *not* possible for all eight girls to dance with a compatible boy. Why? The tree of Figure 4.23 tells us why. The *five* girls on the tree, viz. {Betty, Doris, Edna, Flo, Hazel}, among them only like *four* boys: {Bill, Carl, Don, Hal}. Plainly, one of these five girls will have to sit out the dance. ■

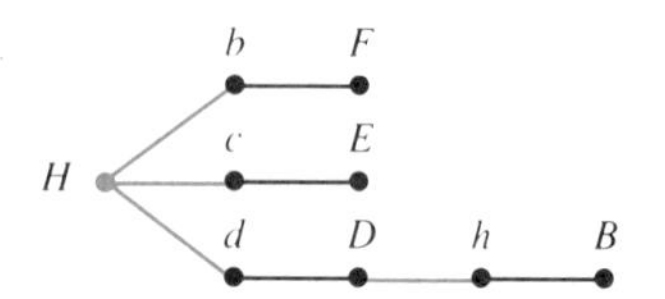

Figure 4.23 A "Hungarian tree." Vertex H is called a "Hungarian acorn."

In the language of the Hungarian algorithm, a tree like the one in Figure 4.23, i.e., one that contains no augmenting path and cannot be grown further, is called a **Hungarian tree**. When the algorithm encounters a Hungarian tree, it knows that there is no augmenting path starting at the initial vertex of the tree. In what follows, we shall call such a vertex, i.e., one which cannot be included in the matching via an augmenting path, a **Hungarian acorn**.

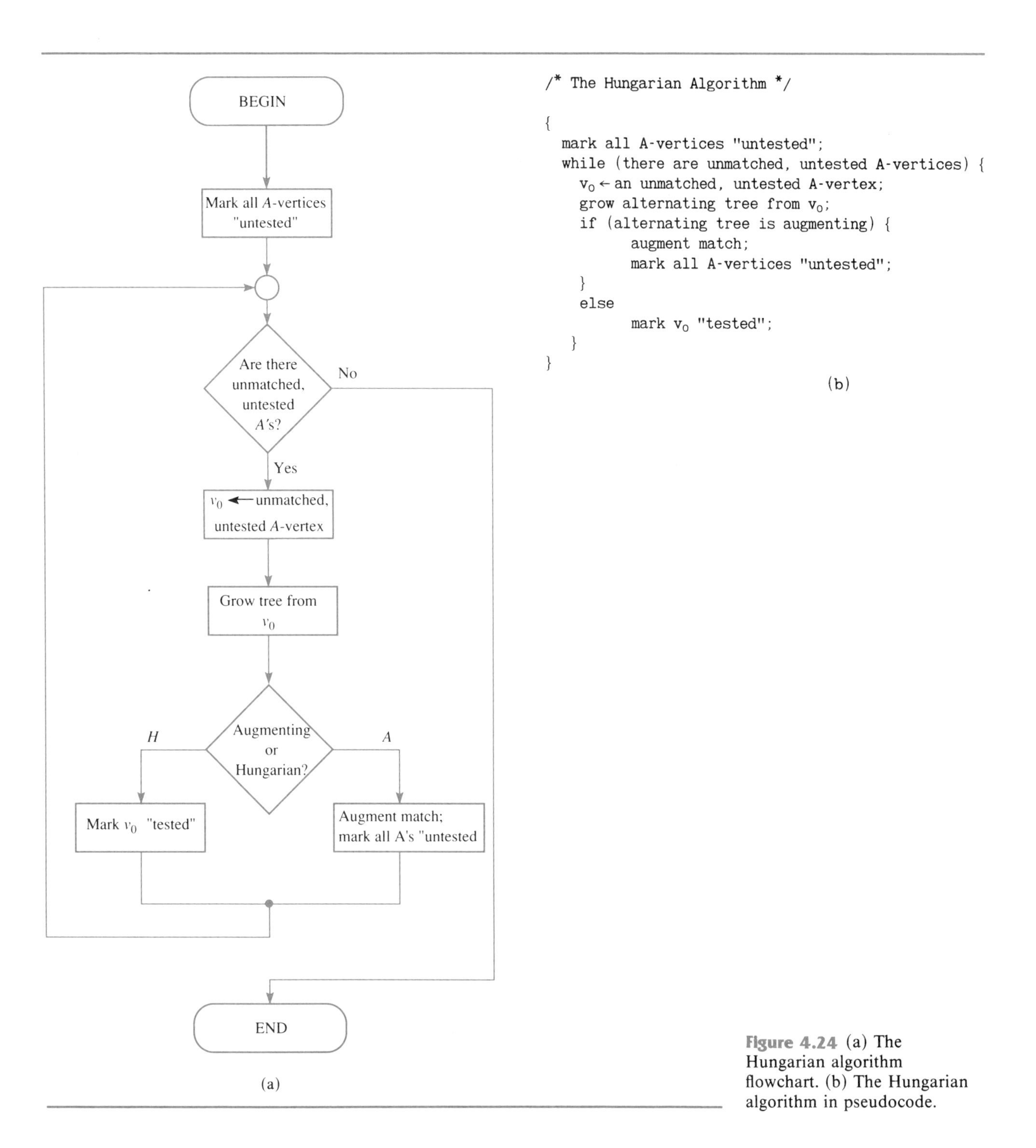

Figure 4.24 (a) The Hungarian algorithm flowchart. (b) The Hungarian algorithm in pseudocode.

In the present boy–girl example, then, we have matched seven of the eight girls, and found that the eighth girl is a Hungarian acorn. Here the algorithm stops. It stops because, as we have seen, it is impossible to match more than seven girls by any means. In general, the algorithm continues looking for augmenting paths, until *either* all A-vertices are matched, *or* all unmatched A-vertices are Hungarian acorns. When this happens, as many A-vertices as possible are matched, i.e., the matching is maximal. The algorithm is summarized in the flowchart and the pseudocode of Figure 4.24.

We will shortly prove that the Hungarian algorithm always produces a maximal matching, but first let's work another example.

EXAMPLE 4.8 Using the Hungarian algorithm of Figure 4.24, find a maximal matching in the bipartite graph described in Figure 4.25.

$$
\begin{array}{cc} & B \\ & \begin{array}{cccc} 1 & 2 & 3 & 4 \end{array} \\ A \begin{array}{c} 1 \\ 2 \\ 3 \\ 4 \\ 5 \end{array} & \begin{pmatrix} 1 & 0 & 1 & 0 \\ 1 & 1 & 0 & 1 \\ 1 & 0 & 0 & 0 \\ 0 & 1 & 0 & 0 \\ 1 & 1 & 0 & 0 \end{pmatrix} \end{array}
$$

(a)

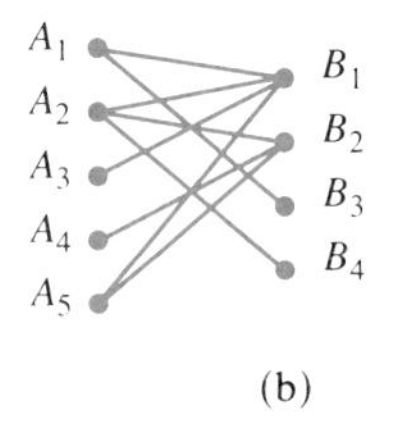

(b)

Figure 4.25 Graph for Example 4.8: (a) The adjacency matrix. (b) The graph.

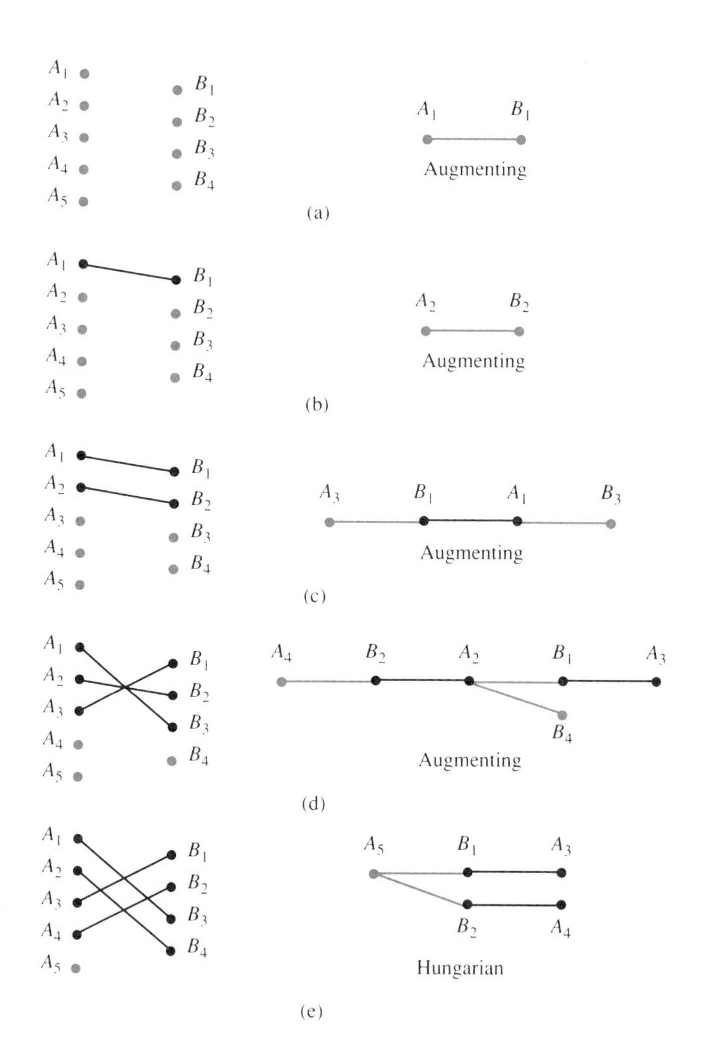

Figure 4.26 Solution to Example 4.8.

SOLUTION Recall that in Example 4.7 we began with a simple matching (the "greedy matching" of Figure 4.16) and proceeded from there. It is important to realize, however, that *this isn't necessary*, and we can begin the algorithm with any matching, even the *empty* matching, i.e., a matching in which nothing is yet matched. And although the algorithm can be sped up a little, perhaps, by starting with a tentative matching, the Hungarian algorithm as described in Figure 4.24 always begins with the empty matching.

Thus in the left half of Figure 4.26(a) we see the empty matching. If we grow an alternating tree, starting with the vertex A_1, we immediately find the augmenting tree shown in the right half of Figure 4.26(a), which produces the matching shown in Figure 4.26(b).

Given the matching in Figure 4.26(b), if we grow an alternating tree, starting with A_2, we get the augmenting tree shown in Figure 4.26(b), which produces the matching in Figure 4.26(c). Starting with A_3, we next get the augmenting tree in Figure 4.26(c), which leads to the matching in Figure 4.26(d). This matching can be augmented by the tree in Figure 4.26(d), resulting in the matching in Figure 4.26(e). However, if we grow an alternating tree from the last remaining unmatched A-vertex A_5, the result is the Hungarian tree of Figure 4.26(e). The algorithm now stops; the matching of Figure 4.26(e) is maximal. ■

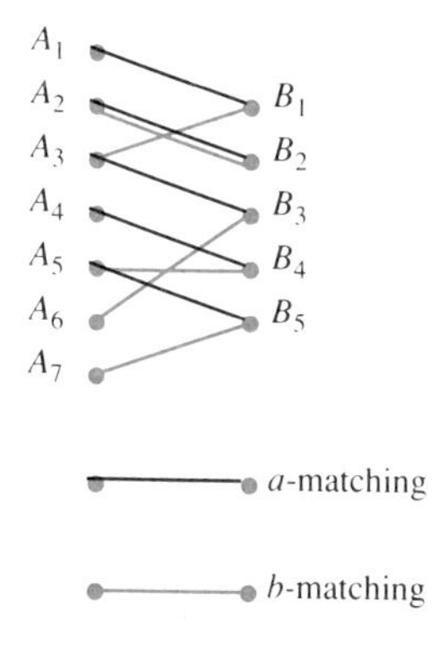

Figure 4.27 Is the b-matching really better than the a-matching?

We now return to the question of why, when the Hungarian algorithm terminates, the resulting matching is as large as possible.

Suppose then that the *algorithm* produces a matching (the *a-matching*) in which every unmatched A-vertex is a Hungarian acorn. Suppose also that somehow a *better* matching exists (the *b-matching*), which matches more A-vertices than the a-matching (see Figure 4.27). We wish to show that this is impossible, i.e., that there is no better matching.

Since the b-matching is alleged to be better than the a-matching, there must exist A-vertices which are *b-matched* but not *a-matched* (vertices A_6 and A_7 in Figure 4.27). Starting with such a vertex, we can build an *alternating path* (b-edge, a-edge, etc.). For example, in Figure 4.27, starting with vertex A_6, we get the alternating path shown in Figure 4.28(a).

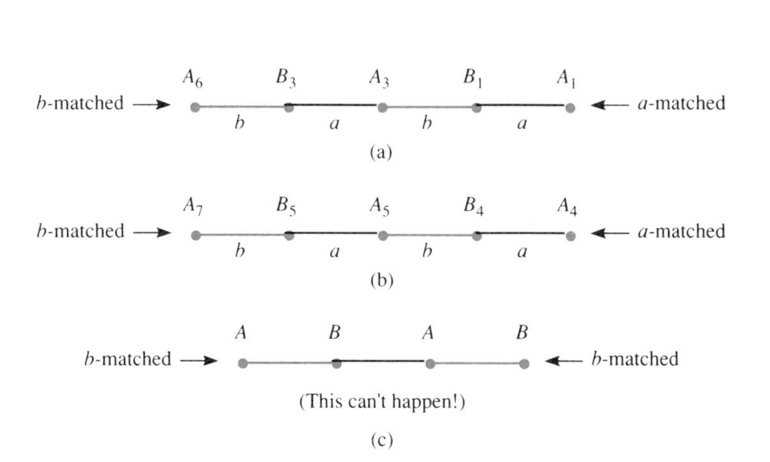

Figure 4.28 Some alternating paths in the bipartite graph of Figure 4.27.

In this case, we find at the end of the alternating path, a vertex (A_1) which is *a-matched* but not *b-matched.* Since all the intermediate vertices on the path are *both a-* and *b*-matched, the *b*-matching enjoys no net advantage on this particular path. Similarly, if we start the alternating path with A_7, we reach A_4 (Figure 4.28b), which is *a*-matched but not *b*-matched.

What happened in Figures 4.28(a) and 4.28(b) is no accident. Starting with any *b-matched, a-unmatched* vertex, we can construct a unique alternating path. Such a path cannot end with another *b-matched, a-unmatched* vertex, as depicted in Figure 4.28(c), since such a path would be an augmenting path for the *a*-matching; as we know that the Hungarian algorithm produces a matching in which there are no augmenting paths. Thus Figure 4.28(c) is impossible, and hence beginning with any *b-matched, a-unmatched* vertex we can find an alternating path which ends with a corresponding *a-matched, b-unmatched* vertex. Furthermore, this correspondence is one-to-one (see Problem 21), so that the *b*-matching does *not* match more vertices than the *a*-matching, after all. Thus the Hungarian algorithm produces the largest matching possible, as advertised!

We conclude this section with a discussion of the question of whether or not a given bipartite graph with vertex sets A and B has a **complete matching**, i.e., a matching which includes *all* of the vertices of A. We have already seen that the girl–boy bipartite graph of Figure 4.15 does *not* have a complete matching. Recall why this is so: Using the Hungarian tree of Figure 4.23, we saw that the five girls $\{B, D, E, F, H\}$ between them only like *four* boys: $\{b, c, d, h\}$—and this fact alone rules out the possibility of a complete matching. This simple idea can be generalized as follows: If in a given bipartite graph, there is a subset A_0 of A such that there are *fewer* than $|A_0|$ vertices in B which are compatible with vertices in A_0, then plainly a complete matching is impossible. Therefore the following condition is *necessary* for the existence of a complete matching:

(4.3) Each subset A_0 of A must be connected to at least $|A_0|$ elements of B.

Surprisingly, however, this condition is also *sufficient*! Why? Because, as we have seen, if a complete matching is impossible, the Hungarian algorithm will eventually produce a Hungarian tree, which always contains one more A-vertex than B-vertex, producing a violation of condition (4.3). This interesting and important result is called *Hall's theorem.**

HALL'S THEOREM Let G be a bipartite graph with vertex sets A and B. A complete matching exists for G if and only if condition (4.3) holds.

* Philip Hall (1904–1982), English mathematician. One of the most influential mathematicians of his generation, he spent his entire long career as a fellow of King's College in Cambridge. A lifelong bachelor, retiring and modest to a fault, he published relatively little, but scholars from all over the world made pilgrimages to Cambridge to hear his brilliant lectures on group theory and related topics.

Problems for Section 4.3

Find a maximal matching for the bipartite graph in each of Problems 1–6. In each case, if the matching is not complete, exhibit a subset A_0 of A which is connected to fewer than $|A_0|$ elements of B.

1.

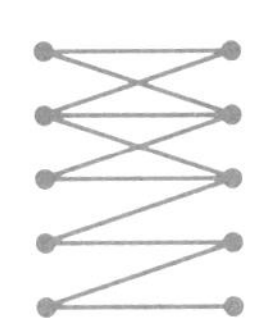

2.

	1	2	3	4	5
1	1				
2	1	1			1
3	1		1		
4		1	1	1	
5			1		

3.

	1	2	3	4
1	1	1		
2		1	1	
3	1	1	1	
4		1	1	1
5	1	1		
6			1	1

4.

	a	b	c	d	e	f
A	1	1	1			
B	1	1		1	1	
C	1	1		1	1	1
D	1				1	1
E		1				1
F	1		1	1	1	

5.

	A	B	C	D	E	F
1	1	1				
2		1	1			
3				1	1	
4					1	1
5	1					1

6.

	1	2	3	4	5	6	7	8
A	1				1			
B		1				1		
C			1				1	
D				1				1
E	1							
F		1						
G			1					
H				1				

7. We saw in the text that all eight girls in the bipartite graph of Figure 4.15 cannot be satisfied simultaneously, but that a set of *seven* girls (all but Hazel) can be. Can we satisfy Hazel (and all the other girls) if *Alice* sits out the dance?

8. In Figure 4.16 we saw that being "greedy" allowed us to match only four of the girls. Suppose, however, we had begun our "greedy" algorithm with *Hazel* and worked backwards. What happens now?

9. For each of the bipartite graphs in Figure 4.13, answer the following questions: How many matchings are there? How many *maximal* matchings?

10. In Figure 4.17, why do you suppose some of the vertices are marked in black and some in color?

In Problems 11–14, continue from the initial circled matchings (in Problem 14, start from scratch, i.e., from the empty matching) and find a maximal matching. If a matching is not complete, find a trouble spot (a set of girls who like too few boys).

11.

	B_1	B_2	B_3	B_4	B_5
G_1	①	1	0	0	0
G_2	0	1	0	①	0
G_3	0	0	①	0	1
G_4	1	1	1	0	0
G_5	1	0	0	0	0

12.

	a	b	c	d
A	①	0	1	0
B	1	①	0	1
C	1	0	0	0
D	0	1	0	0
E	1	1	0	0

13.

	a	b	c	d	e	f
A	0	0	①	1	1	0
B	0	①	0	1	0	1
C	0	0	1	0	①	0
D	0	1	0	0	1	①
E	0	0	0	①	0	0
F	0	0	1	1	0	1

14.

	B_1	B_2	B_3	B_4
G_1	1	1	0	0
G_2	1	1	0	0
G_3	1	1	0	0
G_4	1	1	0	0
G_5	1	1	1	1

15. If $S_1, S_2, \ldots, S_n$ are n subsets of a finite set X, a *system of distinct representatives* (SDR) for them is a list $(s_1, s_2, \ldots, s_n)$ of n distinct elements of X, with $s_i \in S_i$, $i = 1, 2, \ldots, n$. For example, if $S_1 = \{1, 2\}$, $S_2 = \{2, 3\}$, $S_3 = \{1, 2, 3\}$, then $(1, 3, 2)$ is an SDR. For each of the following families of sets, find (if possible) an SDR.
(a) $\{1, 2, 6\}, \{4, 5, 6\}, \{2, 6\}, \{1, 2, 4, 6\}, \{1, 6\}$.
(b) $\{a\}, \{a, b\}, \{a, b\}, \{a, b, c, d, e\}$.
(c) $\{a, e, f, g, h\}, \{b, c, h\}, \{c, e, f, g\}, \{d, h\}, \{b, c, d\}, \{b, d\}, \{a, b, c, d\}, \{b, c, d\}$.

16. Continuing Problem 15, show that a necessary and sufficient condition for the existence of an SDR for a collection of n sets $\{S_i\}$ is that $|\bigcup_{j=1}^{k} S_{i_j}| \geq k$, for any subcollection $S_{i_1}, \ldots, S_{i_k}$ of k sets, for $k = 1, 2, \ldots, n$. (*Hint*: Use Hall's theorem.)

17. If A and B are two finite sets, count the number of bipartite graphs with vertex sets A and B, in terms of $|A|$ and $|B|$.

18. The **complete bipartite graph** $K_{m,n}$ (see also Section 3.5) is defined to be a bipartite graph with $|A| = m$, $|B| = n$, with every A-vertex connected to every B-vertex. The accompanying figure shows $K_{2,3}$.
(a) How many edges does $K_{m,n}$ have?
(b) How many edges are in a maximal matching?
(c) When is there a *complete* matching?
(d) How *many* maximal matchings are there?

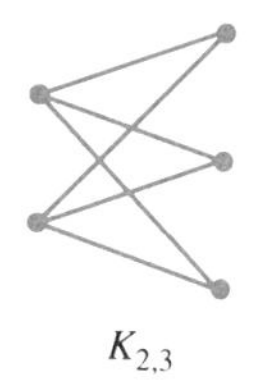

$K_{2,3}$

19. Suppose, in Example 4.7, that each girl likes *exactly four* boys, and that each boy is liked by exactly *four* girls. Without knowing anything else about the graph, show that a complete matching is possible.

20. Continuing Problem 19, suppose, in a bipartite graph, that each A-vertex is connected to exactly m B-vertices, and each B-vertex is connected to exactly m A-vertices. Show that a complete matching is possible.

21. In the course of our proof that the Hungarian algorithm always produces a maximal matching, we asserted that the correspondence between the set of *b-matched, a-unmatched* and the set of *a-matched, b-unmatched* vertices is one-to-one. Show that this is so, by explaining why, for example, it is not possible to have a pair of alternating paths like

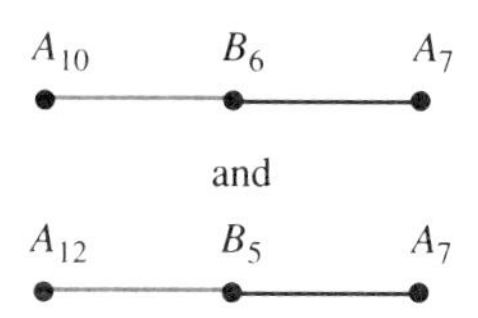

in which there are two advantages for b, viz. A_{10} and A_{12}, but only one for a, viz. A_7. Similarly explain why it isn't possible to have a situation like

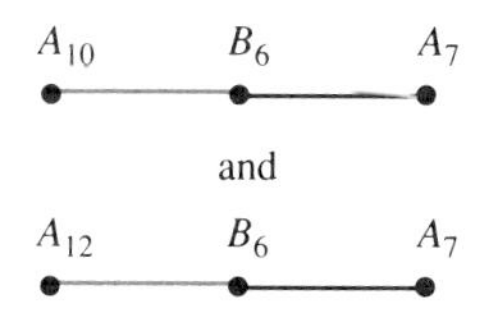

again with two advantages for b but only one for a.

4.4 Running Times of Algorithms

In this chapter, we have studied several different algorithms, and in the remainder of the book we will study many more. Many of the inventors of these algorithms are famous, and some of them are rich, because a good computer algorithm can be worth its weight in gold.*

But what exactly makes an algorithm good? One answer is that a good algorithm is a fast algorithm, i.e., one which performs its task rapidly. We admit that being the fastest doesn't always mean being the best; other considerations, like memory requirements and simplicity, are often important. But it's certainly

* Two recent examples of profoundly important algorithms are Andrew Viterbi's decoding algorithm (1969), which has revolutionized the telecommunications industry and which we will study in Chapter 6, and Narendra Karmarkar's linear programming algorithm (1982), which has made possible calculations in economics and other fields which were previously impossible.

no great honor to be slow! Anyway, in this section, we will naively assume that "good" means "fast," and try to find out just how fast a few simple algorithms are. We will find that the time an algorithm takes to perform a given task, i.e., its **running time**, can be expressed as a function of the *size* of the task the algorithm is asked to perform. Running times for algorithms are like snowflakes; no two are exactly the same. Still, we will find that running times can be classified into a few major categories, corresponding to their so-called "orders of magnitude." Algorithms whose running times have small orders of magnitude are considered fast algorithms, and those with large orders of magnitude are considered slow.

We will now explain briefly what we mean by the order of magnitude of the running time of an algorithm. Then we will consider a series of examples to illustrate this important concept.

Suppose we are given an algorithm which performs a specific task, and that associated with the task is an integer n which somehow measures the size of the task. (For example, in Prim's algorithm, n is usually taken to be the number of vertices in the graph.) We want to know how long the algorithm will take to run, as a function of n, especially for large values of n. (Large values of n correspond to large tasks, which is what we need computers for!) The exact running time will doubtless be a complicated function of n, say $t(n)$. However, it almost always happens that for large values of n, $t(n)$ can be closely approximated by a much simpler function, a function which will provide us with important insight about the algorithm.

For example, suppose we were studying an algorithm and we found its running time, expressed as a function of n, to be

(4.4) $$t(n) = 1.75n^3 - 0.33n^2 + 1.47n - 0.05 \text{ microseconds.}$$

The important thing to notice about (4.4) is that for large values of n, the term $1.75n^3$ will completely dominate the others. For example, if $n = 100$, the term $1.75n^3$ is 1,750,000, whereas the remaining terms $-0.33n^2 + 1.47n - 0.05$ equal only -3153.05. Thus for large values of n, it is reasonable to approximate $t(n)$ by its dominant term:

(4.5) $$t(n) \approx 1.75n^3 \text{ microseconds.}$$

With $n = 100$, the exact expression (4.4) gives 1.747 seconds, whereas the approximation (4.5) gives 1.750 seconds; and for larger values of n the approximation is even better.

The approximation in (4.5) is simple and accurate, but for some purposes it is still unnecessarily complicated! Indeed, a computer scientist interested in the speed of this particular algorithm would almost certainly disregard the constant "1.75" and say something like "Oh. I see. That algorithm's running time has order of magnitude n^3." To see why the constant 1.75 might not be important, suppose $n = 1000$. Using the exact formula (4.4) with $n = 1000$, we find that $t(1000) = 29.16119$ minutes, and the approximation (4.5) gives 29.16667 minutes. On the other hand, if we use the crude "order-of-magnitude" approximation $t(n) \approx n^3$, we get instead 16.66667 minutes. Of course, 16 minutes isn't the same as 29 minutes. But in this case the approximation n^3 nevertheless gives us a good feeling for how long the algorithm takes to run: not milliseconds, not hours, but a few tens of minutes. For many purposes this kind of information is adequate, and as n gets larger and larger, the order-of-magnitude approximation becomes even more valuable. For example, for $n = 100{,}000$, the order-of-magnitude approximation gives 31.71 years, whereas the true value is 55.49 years. You may think that these two numbers are different, but from a programmer's viewpoint, they are exactly the same: much too big!*

Order-of-magnitude statements about running times are so common that there is a special notation for them. The statement "$t(n)$ has order of magnitude at most n^3" is usually expressed by writing

(4.6)
$$t(n) = O(n^3),$$

where the symbol $O(n^3)$ is pronounced "**big-*O*** of n^3."†

The technical definition of big-O is as follows. If the actual running time of an algorithm is $t(n)$, and if $s(n)$ is a "simple" function, like n or n^2 or n^3 or 2^n, we are allowed to say that $t(n)$ has order of magnitude $s(n)$, or that $t(n)$ is big-O of $s(n)$, and write

$$t(n) = O(s(n)),$$

if $t(n)$ is less than or equal to a constant times $s(n)$, for large values of n.

For example, we have

$$3n + 2 = O(n),$$

because $3n + 2 \leq 4n$ for all $n \geq 2$. Similarly,

$$4n^2 + 5n - 6 = O(n^2),$$

because $4n^2 + 5n - 6 \leq 5n^2$ when n is large enough. [It is important to bear in mind that a statement like $t(n) = O(n^3)$ means only that the running time is *at most* a constant times n^3 for large n. In practice, the algorithm may run faster than this, possibly because the analysis that led to the estimate was imprecise, or possibly because the algorithm was given a particularly easy problem to solve. However, in the rest of this section we won't emphasize this point.]

We said at the beginning of this section that algorithms are often classified according to the order of magnitudes of their running times. For example, algorithms whose running times are $O(n)$ are often grouped together and called "linear-time" algorithms. Algorithms with $O(n^2)$ running times are called "quadratic-time" algorithms, those with $O(n^3)$ are called "cubic-time" algorithms, and so on. With these classifications it is possible to make generalized statements about whole families of algorithms. For example, $O(n)$ algorithms are regarded as very fast algorithms, $O(n^2)$ algorithms are fast, but not as fast as $O(n)$ algorithms, $O(n^3)$ are a bit slower, etc. There is a whole spectrum of speeds, as

* Another reason for disregarding specific constants, like those appearing in (4.4), is that they often depend on the particular computer, compiler, etc. being used, and are usually difficult to estimate accurately.

† The big-O notation was invented in 1892 by the German number theorist Paul Bachman; but modern computer scientists have found it to be essential for describing the running times of algorithms.

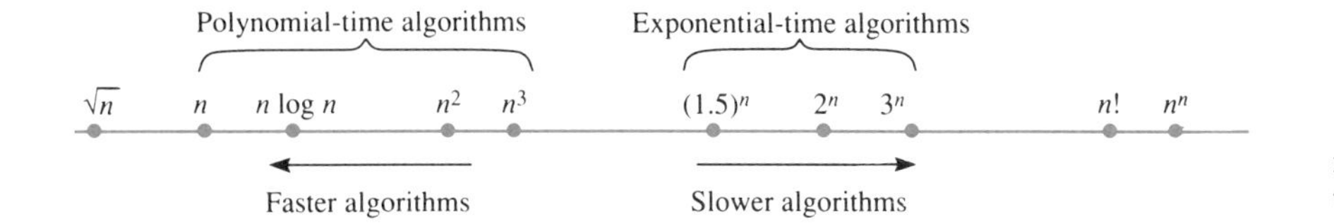

Figure 4.29 Order-of-magnitude chart for running times.

shown in Figure 4.29. The "powers-of-n" algorithms, i.e., the $O(n)$, $O(n^2)$, $O(n^3)$ algorithms, are at the "fast" end of the spectrum. These algorithms are commonly called "polynomial-time" algorithms. At the "slow" end of the spectrum, on the other hand, are those algorithms whose running times are *exponential* functions of n, like $(1.5)^n, 2^n, 3^n, \ldots$. These algorithms are commonly called "exponential-time" algorithms.

To get a feeling for the speeds of the running times in Figure 4.29, consider Table 4.4. There we have listed three polynomial functions and one exponential function, i.e., n, n^2, n^3, and 2^n, and have assumed that these four functions represent the running times, measured in microseconds, of four typical algorithms. The table shows how large a task can be performed by each of the algorithms in 1 second, 1 minute, 1 hour, and 1 day. For example, consider the "1 day" entry for the n^3 algorithm. One day consists of exactly 8.64×10^{10} microseconds. Therefore, the maximum value of n which can be handled by the n^3 algorithm in 1 day equals the largest value of n such that $n^3 \leq 8.64 \times 10^{10}$. This value is $n = 4420$, as indicated in the table. The other entries are computed similarly.

Table 4.4 makes it easier to understand the chart in Figure 4.29. We see, for example, that the $O(n)$ algorithm can perform a huge task in 1 second, let alone an hour or a day. The $O(n^2)$ algorithm can do a lot in an hour, but not nearly as much as the $O(n)$ algorithm can do in a second. The n^3 algorithm is very much slower, as you can see, although given a whole day to work, it can do a pretty big task. The $O(2^n)$ algorithm, on the other hand, presents a very different picture. The value of n it can handle grows extremely slowly as a function of the available computing time, and in a day it can only do $n = 36$, which even the $O(n^3)$ algorithm could do in less than a second. Considerations like this are why computer scientists say simply that polynomial-time algorithms are "fast" algorithms and that exponential-time algorithms are "slow" algorithms.

After this preliminary discussion, we are in a position to study the running times of some specific algorithms.

TABLE 4.4 Limits of task size as a function of running time

Running Time (Microseconds)	Maximum Task Size			
	1 second	1 minute	1 hour	1 day
n	1,000,000	60,000,000	3.6×10^9	8.64×10^{10}
n^2	1,000	7,745	60,000	293,938
n^3	100	391	1,532	4,420
2^n	19	25	31	36

```
/* Algorithm 1: Print the Integers from 1 to n */

{
  for (i = 1 to n)
      print i;
}
```

Figure 4.30 Algorithm 1 in pseudocode.

```
/* Algorithm 2: Find the Largest of a[1], . . . , a[n] */

{
  largest ← a[1];
  for (i = 1 to n)
      if (a[i] > largest)
         largest ← a[i];
  print largest;
}
```

Figure 4.31 Pseudocode for algorithm 2.

EXAMPLE 4.9 Find the running time for algorithm 1, which is described in pseudocode in Figure 4.30. Show that the running time is $O(n)$.

SOLUTION The basic operation performed by the algorithm is the printing of a single number, which is carried out n times. Therefore, the running time is $O(n)$. (But see Problem 14.)

EXAMPLE 4.10 Consider algorithm 2, shown in Figure 4.31, which performs the task of finding and then printing the largest of the n numbers $a[1], a[2], \ldots, a[n]$. Show that the running time is $O(n)$.

SOLUTION Almost all of the algorithm's work is done in the `for` loop. For each of the n executions of this loop, one comparison (`a[i] > largest`) and at most one assignment (`largest ← a[i]`) is made. Thus each execution of the loop requires a constant amount of time; and since the loop is executed n times, the running time is $O(n)$.

In the next example, we will analyze Prim's algorithm (Section 3.3), and find that while it is slower than algorithms 1 and 2, it is still a fast algorithm.

EXAMPLE 4.11 In Figure 4.32 we show a version of Prim's algorithm (cf. Figures 3.27 and 4.7.). Show that the running time for Prim's algorithm is $O(n^2)$.

SOLUTION In Prim's algorithm most of the work is inside the `while` loop, where we need to find the cheapest new vertex and update the current cost of the vertices which are still not in the growing tree. Suppose there are k vertices outside the growing tree. Then, the algorithm does two basic things:

```
/* Prim's algorithm without details */

{
   tree ← vertex v0;
   while (there are remaining ? vertices) {
        add cheapest ? vertex to tree;
        update costs of remaining vertices;
   }
}
```

Figure 4.32 Prim's algorithm revisited.

1. Finds the cheapest of the k remaining vertices. This amounts to finding the minimum of k numbers, which can be done in $O(k)$ units of time, if a modified version of algorithm 2 is used.
2. Updates the cost of each vertex which is still not in the growing tree. Since there are $k - 1$ such vertices, the time required for this step is also $O(k)$.

Thus, one execution of the `while` loop requires $O(k)$ units of time. Since as the algorithm progresses, k runs from $n - 1$ down to 1, the running time is big-O of $(n-1)+(n-2)+\cdots+1$. But since $1+2+\cdots+(n-1) = n(n-1)/2 = n^2/2 - n/2$, as can be proved by induction, it follows that the running time is $O(n^2)$. ■

In the next example, we will learn about the running time for Floyd's algorithm. It runs slower than any of the algorithms we have studied so far, although it is still a polynomial-time algorithm.

```
/* Floyd's algorithm */

{
  A0 ← graph's length
      matrix;
  for (i = 1 to n)
      for (all x and y)
        compute Ai[x, y];
}
```

Figure 4.33 Pseudocode for Floyd's algorithm.

EXAMPLE 4.12 In Figure 4.33 we see a pseudocode program for Floyd's algorithm, which is a slightly expanded version of the pseudocode in Figure 4.12. Show that the running time is $O(n^3)$, where n denotes the number of vertices in the graph.

SOLUTION In Floyd's algorithm the basic operation is the computation of the n^2 entries in the matrix A_i (see Equation 4.1, Section 4.2). For each of these entries the algorithm must perform an addition $(A_{i-1}[x, i] + A_{i-1}[i, y])$ and a comparison $(A_{i-1}[x, y]$ vs. $A_{i-1}[x, i] + A_{i-1}[i, y])$. Thus the time required to compute A_i is $O(n^2)$. Since this computation must be done for $i = 1, 2, \ldots, n$, the total running time is $O(n^3)$. ■

In the final example in this section we will study an algorithm which isn't named after anyone—which isn't surprising; no one would want to be remembered for an algorithm this slow!

EXAMPLE 4.13 A function $f(x_1, x_2, \ldots, x_n)$ of n variables is called a **Boolean function** if it can assume only the values 0 and 1, and if each of the variables $x_1, x_2, \ldots, x_n$ can also assume only the values 0 and 1. We will study Boolean functions in detail in Chapter 5. A Boolean function $f(x_1, x_2, \ldots, x_n)$ is said to

```
/* Satisfiability algorithm */

{
   answer ← NO;
   for (x = (0,0,...0) to (1,1,...,1))
      if (f(x) = 1)
         answer ← YES;
   print answer;
}
```

Figure 4.34 Pseudocode for the satisfiability algorithm.

be *satisfiable* if there is at least one assignment of the Boolean variables $x_1, x_2, \ldots, x_n$ for which $f(x_1, x_2, \ldots, x_n) = 1$. In Figure 4.34 we see, in pseudocode form, an algorithm which tests a given Boolean function $f(x_1, x_2, \ldots, x_n)$ for "satisfiability," by trying each of the 2^n possible values of $(x_1, x_2, \ldots, x_n)$, one at a time. Show that the running time of this algorithm is $O(2^n)$.

SOLUTION Most of the work in the satisfiability algorithm is done in the for loop which is executed 2^n times, once for each possible assignment of $x_1, x_2, \ldots, x_n$. Assuming that each execution of the loop takes a fixed amount of time, the running time is then $O(2^n)$. ■

Note: The problem solved by algorithm 5 is called the *satisfiability problem.* Algorithm 5 is not the fastest known algorithm for solving this problem (see Problem 22, for example). Still, no one has ever found a polynomial-time algorithm for it. Indeed, most computer scientists believe that there is no polynomial-time algorithm for the satisfiability problem or any of its many equivalent manifestations, although no one has been able to prove this. In a certain sense, the satisfiability problem is the archetype for all "hard" computational problems, and as of this writing (1988) the single most important research problem in theoretical computer science is whether or not there is a polynomial-time algorithm for the satisfiability problem.

Problems for Section 4.4

1. Compare (4.4) and (4.5) for $n = 1000, 10000$, and 100000.

2. Find the smallest constant K such that $3n + 2 \leq Kn$ for all $n = 1, 2, 3, \ldots$.

3. Find the smallest constant K such that $4n^2 - 5n - 6 \leq Kn^2$ for all n.

4. Using the definition given in the text, show the following:
(a) $n = O(n^2)$.
(b) $n^2 + 100000n = O(n^2)$.
(c) $2^n = O(n^n)$.

5. In Figure 4.29 we assert that a $(1.5)^n$ algorithm is slower than an n^3 algorithm. Verify this assertion by showing that for large enough n, $(1.5)^n > n^3$. (*Hint*: Use induction.)

6. Explain where the following functions should go in Figure 4.29.
(a) $\log n$.
(b) $n^{1/3}$.
(c) e^n, where $e = 2.718\ldots$ is the base of natural logarithms.

7. Verify the entries in Table 4.4.

8. Extend Table 4.4 by adding a column for "1 week."

9. Referring to Table 4.4, how much time would the $O(n^2)$ algorithm require to handle as large a task as the $O(n)$ algorithm handles in 1 second?

10. How long would it take to perform a task of size 100 with the $O(2^n)$ algorithm in Table 4.4? Do you expect to live long enough to see it completed?

11. In Table 4.4 we assumed that the basic time unit was 1 microsecond. Suppose that a computer 10 times as powerful were available, i.e., with a basic time unit of 100 nanoseconds. How would this increase in computing power affect the numbers in Table 4.4?

12. Suppose you have to solve a problem with $n = 1000$, using an $O(n^3)$ algorithm. Given a choice between a tenfold increase in computer speed or a more efficient algorithm, say a $O(n^2)$ algorithm, which would you prefer? Why?

13. In the solution to Example 4.11, we asserted that

$$1 + 2 + \cdots + (n - 1) = O(n^2).$$

Verify this.

14. In Example 4.9, we assumed that the time required to print the number i is a constant independent of i. However, the time required to print i depends on the number of digits in i, which is proportional to log i. Thus the time required to `print i` is $O(\log i)$. Bearing this fact in mind, reconsider algorithm 1. Is it still a $O(n)$ algorithm?

15. Here is an algorithm (Bubblesort) whose task is to take n numbers $x_1, x_2, \ldots, x_n$ and rearrange them into increasing order:

```
/* Bubblesort */
{
  for (MAX = n to 2)
       for (i = 1 to MAX - 1)
            SORT(x_i, x_{i+1});
}
```

In this algorithm the operation SORT(x_i, x_{i+1}) arranges the two numbers x_i and x_{i+1} into increasing order; i.e., if $x_i \le x_{i+1}$, it does nothing, but if $x_i > x_{i+1}$, it interchanges x_i and x_{i+1}. Find the order of magnitude of the running time.

16. Here is an algorithm ("Prime Testing") whose task is to take an integer n and to decide which of the numbers in the set $\{1, 2, \ldots, n\}$ are *not prime.* Estimate the running time.

```
/* Prime Testing */
{
  for (I = 1 to n)
       for (J = 2 to √I)
            if (I/J is an integer)
               print "I isn't prime";
}
```

17. Estimate the running time for the "sorting" algorithm in the accompanying algorithm. In this algorithm S is a set of n elements.

```
/* Sorting Algorithm */
{
  T ← S;
  while (T isn't empty) {
    x ← least element of T;
    print x;
    T ← T-{x};
  }
}
```

18. Estimate the running time for the "matrix multiplication" algorithm that follows.

```
/* Matrix Multiplication */
{
  for (i = 1 to n)
      for  (j = 1 to n) {
           c[i, j] ← 0;
           for (k = 1 to n)
               c[i, j] ← c[i, j]
                       +a[i, k]b[k, j];
      }
}
```

19. In the version of Prim's algorithm shown in Figure 4.32 we recommend using algorithm 2 to find the cheapest new vertex. Explain how to do this; and show that the running time of this modified algorithm is $O(k)$, where k denotes the number of vertices not in the growing tree.

20. Estimate the running time for Dijkstra's algorithm. (*Hint*: Use the remark at the end of Section 4.1 about the relationship between Dijkstra's algorithm and Prim's algorithm.)

21. Estimate the running time for Warshall's algorithm.

22. The algorithm in Figure 4.34 is somewhat inefficient in that if it finds a value of the Boolean variables $x_1, \ldots, x_n$ for which $f(x_1, \ldots, x_n) = 1$, it doesn't immediately print `YES` and stop. Devise a satisfiability algorithm that does print `YES` and stop; and estimate its running time. Is it faster than $O(2^n)$?

23. In the text we introduced you to the big-O notation. There is also a "little-o" notation. If $f(n)$ and $g(n)$ are functions of n, and if for large values of n, $f(n)$ is "much smaller" than $g(n)$, $f(n)$ is said to be "little-o" of $g(n)$, written $f(n) = o(g(n))$. The precise definition is that $f(n) = o(g(n))$ if and only if $\lim_{n\to\infty} f(n)/g(n) = 0$. Show the following:

(a) $n = o(n^2)$.
(b) $n^2 = o(n^3)$.
(c) $2^n = o(3^n)$.

Summary

We have described several important and practical graph algorithms.

- **Dijkstra's algorithm** constructs a spanning tree that gives shortest paths from a given vertex to all other vertices of a graph.
- **Floyd's algorithm** finds the shortest distances between all pairs of vertices in a directed graph.
- **Warshall's algorithm** determines **reachability**, i.e., for each pair of vertices in a directed graph, whether or not there is a path connecting them.
- The **Hungarian algorithm** finds a maximal matching in a bipartite graph and provides a proof of **Hall's theorem**, which gives necessary and sufficient conditions for the existence of a *complete matching*.

Finally, we introduced the **big-O** notation and learned how to use it to estimate the *running times* of various algorithms.

CHAPTER FIVE

PROPOSITIONAL CALCULUS AND BOOLEAN ALGEBRA

We will begin this chapter with a study of *propositional calculus*, which is a branch of pure mathematics straight out of the nineteenth century, when mathematicians hoped to reduce human thought to an exact science. This hope has proved to be in vain, but the "symbolic logic" invented by these nineteenth-century idealists, most prominently by George Boole, has unexpectedly proved to be invaluable in the study of computers! In Section 5.2 we will begin a serious study of *Boolean algebra*, as it has evolved as a tool for the design and analysis of digital electronic circuits. In the remaining sections of this chapter we will learn more about this important branch of applied mathematics, until in Section 5.5 we conclude by studying the *Karnaugh map*, which is a tool used daily by electrical engineers in the design of efficient digital electronic circuits.

5.1 Propositional Calculus

Consider the following five simple declarative sentences, or as we shall call them, **propositions**:

A: Art loves Betty.
B: Batman is left-handed.
C: Columbus discovered America.
D: Dynamite is dangerous.
E: Elephants have wings.

For the sake of argument, we shall assume that each of these propositions is either *true* or *false*. The **propositional calculus** has as its goal the discovery of *valid arguments about propositions*. For example, consider the following *compound proposition*:

(5.1) If either Art loves Betty or Batman is left-handed, and Batman is not left-handed, then Art loves Betty.

Granted, this is a silly proposition. But it is logically correct, and indeed it is a special case of the following more respectable-sounding *general proposition*:

(5.2) If either A or B is true, and B is not true, then A is true.

This proposition, which is true *independent* of the truth or falsity of the propositions A and B, is called a **theorem** of the propositional calculus. It represents a universally valid way of drawing a certain **conclusion** (A is true) from certain **premises** (either A or B is true; B is not true).

Of course, not every proposition is a theorem. For example, consider the following proposition.

(5.3) If P is true, then Q is true; if Q is false, then R is true; R is false; therefore P is true.

This argument may sound impressive, but we shall see later that it is *not* logically correct, i.e., proposition (5.3) isn't a theorem. The propositional calculus will give us a foolproof procedure for deciding whether or not any given proposition, however complicated, is a theorem and for producing an unlimited supply of new theorems. In fact, this decision procedure is completely mechanical, and could be stated in the form of a computer algorithm. Thus the propositional calculus is one branch of mathematics in which theorem proving need involve no creativity on the part of the mathematician. Unfortunately, this is the exception rather than the rule. For almost all branches of useful mathematics, no theorem-proving algorithm exists.*

Before we can start proving theorems, however, we need to introduce the standard formal notation which is used to express complex propositions like (5.3) precisely. We shall see that complex propositions are all built from simpler ones by using five kinds of **connectives**: *and*, *or*, *not*, *implies*, and *if and only if.* For example, the compound proposition "Art loves Betty and Columbus discovered America" is built from propositions A and C using the *and* connective. The *and* connective is represented by the symbol "$\wedge$," and in the notation of the propositional calculus this compound proposition is represented as follows:

(5.4) $(A \wedge C)$ ("*A and C*").

The parentheses in (5.4) are important, and it is not correct, for example, to write $A \wedge C$ instead of $(A \wedge C)$. Similarly, the compound proposition "Either A or B," which itself is short for "Either A is true, or B is true, or both," is represented by the following symbol:

(5.5) $(A \vee B)$ ("*A or B*").

The *negation* of a proposition is represented by prefixing the proposition with the symbol "$\sim$." Thus, the proposition "Dynamite is not dangerous" would be represented by

(5.6) $\sim D$ ("*not D*").

* This remarkable discovery was made in 1936 by Alonzo Church, and independently by Alan Turing, whose "Turing machines" we will meet in Chapter 6. In fact, Turing machines were invented for the explicit purpose of demonstrating the nonexistence of theorem-proving algorithms, or decision procedures, as they are usually called.

An *if-then* type of proposition uses the symbol "$\rightarrow$" (pronounced "implies"). Thus the proposition "If elephants have wings, then Batman is left-handed" would be represented by

(5.7) $$(E \rightarrow B) \quad \text{("}E\text{ implies }B\text{")}.$$

Finally, the connective *if and only if* is represented by the symbol "$\leftrightarrow$." The compound proposition "Dynamite is dangerous if and only if Art loves Betty," for example, is denoted by

(5.8) $$(D \leftrightarrow A) \quad \text{("}D\text{ if and only if }A\text{")}.$$

Starting with any given set of "atomic" propositions (like propositions A, B, C, D, E), and using the five rules (5.4)–(5.8), it is possible to form arbitrarily complex propositions. But the rules must be applied exactly; the propositional calculus allows for no variations. Propositions that are formed by successive application of the rules (5.4)–(5.8) are called **well-formed propositions** (WFPs), and for future reference we state the rules for forming WFPs in the box that follows. (The propositional calculus is an example of a **recursively defined formal language**. In a recursively defined formal language we begin with a set of "atomic" strings, and new strings are built from old ones by applying a basic set of rules. We will study some other languages of this type in Chapter 7.)

RULES FOR FORMING WELL-FORMED PROPOSITIONS

- *Rule 1*: Variable letters like A, B, C, ... are WFPs called *atoms*.

If x and y are WFPs, then so are each of the following:

- *Rule 2*: $(x \wedge y)$.
- *Rule 3*: $(x \vee y)$.
- *Rule 4*: $\sim x$.
- *Rule 5*: $(x \rightarrow y)$.
- *Rule 6*: $(x \leftrightarrow y)$.

For example, each of the following is a WFP:

(5.9)	P	(Rule 1)
(5.10)	Q	(Rule 1)
(5.11)	R	(Rule 1)
(5.12)	$(P \rightarrow Q)$	(Rule 5)
(5.13)	$\sim Q$	(Rule 4)
(5.14)	$(\sim Q \rightarrow R)$	(Rule 5)
(5.15)	$((P \rightarrow Q) \wedge (\sim Q \rightarrow R))$	(Rule 2)
(5.16)	$(Q \vee \sim Q)$	(Rule 3)
(5.17)	$(((P \rightarrow Q) \wedge (\sim Q \rightarrow R)) \rightarrow (Q \vee \sim Q))$	(Rule 5).

The WFP in (5.17) looks intimidating, but it has been formed in a straightforward manner by combining the WFPs in (5.15) and (5.16) with the *implies* connective, using Rule 5.

Now that we know how to *create* WFPs, we reverse the process and learn how to *recognize* them.

EXAMPLE 5.1. Which of the following propositions are WFPs?

(a) A
(b) (A)
(c) $\sim(A)$
(d) $(A \leftrightarrow \sim A)$
(e) $(A \wedge B \wedge C)$
(f) $((A \wedge B) \wedge C)$
(g) $((\sim(A \wedge B) \vee C) \rightarrow \sim A)$
(h) $((((A \rightarrow B) \rightarrow C) \rightarrow \sim(A \vee B) \leftrightarrow \sim\sim(C \wedge A)))$

SOLUTION

(a) This is an atom, so it is a WFP, by Rule 1.
(b) There is no rule that permits us to surround an atom with parentheses. This is *not* a WFP.
(c) This proposition appears to be an application of Rule 4 to the proposition (A); but in part (b) we saw that (A) isn't a WFP. So this is *not* a WFP.
(d) This is a WFP. Rule 6 has been applied to the WFPs A and $\sim A$.
(e) Not a WFP. Rule 2 doesn't allow us to combine *three* WFPs with the *and* connective!
(f) This is a WFP. Rule 2 has been applied to the WFPs $(A \wedge B)$ and C.
(g), (h) In the previous examples it was more or less possible to check for "well-formedness" by eye. However, given complicated propositions like (g) and (h) it will plainly be helpful to have a more systematic approach. In Figure 5.1, therefore, we give detailed instructions for dissecting, or **parsing**, to use the official term, propositions (g) and (h).

In Figure 5.1(a), then, we show how to parse proposition (g). We begin by writing the complete proposition as line 1. As a preliminary step, we replace each of the atoms A, B, and C by the symbol "$*$" in line 2. We do this because

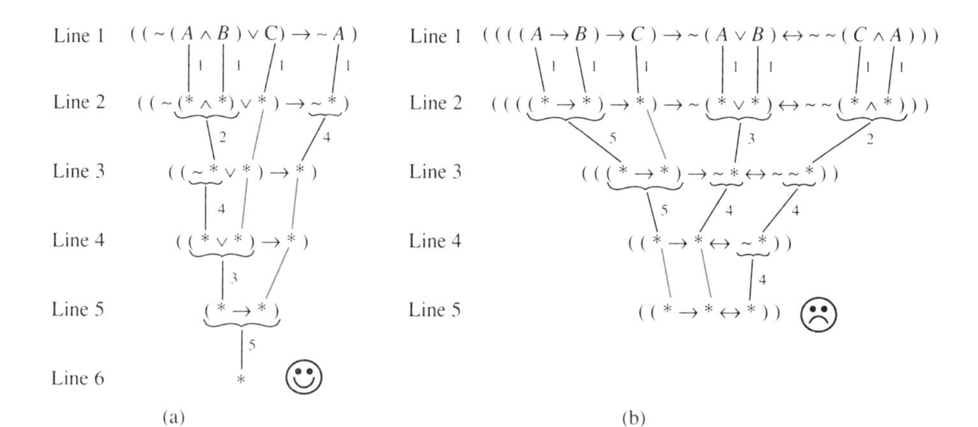

Figure 5.1 Solutions to Example 5.1: (a) Parsing proposition (g); (b) parsing proposition (h).

according to Rule 1, every atom is a WFP, and "$*$" is a symbol we shall use to represent "a well-formed proposition." If we now scan line 2, we see two "subpropositions" that are themselves WFPs, viz., "$(* \wedge *)$" and "$\sim *$". The first one is well formed by Rule 2, and the second one is well formed by Rule 4. Thus in line 3 we replace these two WFPs by the WFP symbol "$*$." Then in line 3 we find the WFP "$\sim *$," which is replaced in line 4 by "$*$." (You should resist the temptation of assuming that the subproposition $* \vee *$ in line 3 is well formed—remember that parentheses are needed!) Continuing on, we finally reach line 6, where we see that the entire complicated proposition has collapsed to merely "$*$," which means it is well-formed.

In Figure 5.1(b), we use the same procedure to parse proposition (h). After replacing the atoms with $*$, the idea is always to scan the current proposition for well-formed subpropositions, and to replace each one with $*$, until no further simplification is possible. In this case the final result is $((* \rightarrow * \leftrightarrow *))$, which cannot be further simplified; this means that proposition (h) is *not* well formed. (It is possible to use the ideas of Figure 5.1 to produce a formal parsing algorithm for the propositional calculus, but we will not do so.) ■

That concludes our discussion of well-formedness; from now on the only propositions we'll be dealing with will be WFPs. We now return to our main concern, the discovery of *valid arguments* and *theorems* in the propositional calculus.

We begin by giving the rules by which the truth or falsehood of a proposition formed by one of the rules in the box on page 179 can be determined from the truth or falsehood of the component propositions. For example, if x and y are WFPs, then $(x \wedge y)$ is true if and only if both x and y are true; and $(x \vee y)$ is true if and only if either x or y (or both) is true. Similarly, $\sim x$ is true if and only if x is false. These rules are summarized in the "truth tables" of Table 5.1,

TABLE 5.1 Truth tables for the five propositional connectives

x	y	$(x \wedge y)$
T	T	T
T	F	F
F	T	F
F	F	F

and

x	y	$(x \vee y)$
T	T	T
T	F	T
F	T	T
F	F	F

or

x	$\sim x$
T	F
F	T

not

x	y	$(x \rightarrow y)$
T	T	T
T	F	F
F	T	T
F	F	T

implies

x	y	$(x \leftrightarrow y)$
T	T	T
T	F	F
F	T	F
F	F	T

if and only if

TABLE 5.2 The truth tables for intersection, union, and complement are the same as those for *and, or,* and *not.*

A	B	$A \cap B$
T	T	T
T	F	F
F	T	F
F	F	F

Intersection

A	B	$A \cup B$
T	T	T
T	F	T
F	T	T
F	F	F

Union

A	A'
T	F
F	T

Complement

where "T" stands for "true" and "F" stands for "false." The truth tables for *and*, *or*, and *not* in Table 5.1 are essentially identical to the truth tables for intersection, union, and complement that we studied in Section 1.2; see Table 5.2. In fact, the term *truth table* which is used in set theory is inherited from the propositional calculus, where "truth" is studied.

The truth tables for $\rightarrow$ and $\leftrightarrow$, however, do not correspond to any common set operations and require further explanation. The $\leftrightarrow$ connective is fairly easy. The idea is that the proposition $(x \leftrightarrow y)$ is regarded as true if and only if x and y both have the *same* truth value. Thus (assuming that Columbus did in fact discover America and that dynamite is indeed dangerous), the compound propositions

Columbus discovered America if and only if dynamite is dangerous

and

Elephants have wings if and only if dynamite is not dangerous

are regarded as *true*, while

Dynamite is dangerous if and only if elephants have wings

is regarded as *false.*

The $\rightarrow$ connective is slightly more difficult to explain. The proposition $(x \rightarrow y)$ can be interpreted in words as

If x, then y,

where x is called the **premise** and y the **conclusion**. According to the truth table in Table 5.1, this proposition should be interpreted as *true* if x is true and y is true. Thus, the proposition

If dynamite is dangerous, then Columbus discovered America

is *true.* Similarly $(x \rightarrow y)$ is regarded as *false* if x is true but y is false. Thus

If Columbus discovered America, then elephants have wings

is a *false* proposition. This much is in accordance with common sense. The difficulty arises when we try to decide whether $(x \rightarrow y)$ is true or false when the *premise* x itself is false, as in

(5.18) If elephants have wings, then dynamite is dangerous

or

(5.19) If elephants have wings, then dynamite is not dangerous.

Strange propositions like (5.18) and (5.19) don't usually occur in polite conversation, and it would certainly be reasonable to regard them as meaningless. However, in the propositional calculus, everything must be either true or false; "meaningless" isn't allowed. In fact, according to the truth table in Table 5.1, both of the propositions must be regarded as *true.* On reflection, you may accept this as reasonable. In any event, it would be difficult to accuse someone who pronounces (5.18) or (5.19) of lying!

A more mathematical example should make things clearer. Consider a three-element set $U = \{1, 2, 3\}$ and the two subsets $A = \{1\}$ and $B = \{1, 2\}$. Then A is a subset of B, and in Section 1.1 we saw that this is the same as saying

(5.20) If $a \in A$, then $a \in B$.

Proposition (5.20) is regarded as a *true* statement for any a. As a assumes the three values 1, 2, and 3, the statements $a \in A$ and $a \in B$ assume all possible combinations of truth values *except* TF:

a	$a \in A$	$a \in B$
1	T	T
2	F	T
3	F	F

Since (5.20) is true for the three cases TT, FT, and FF, we can see why the proposition $(x \rightarrow y)$ must be true in these same three cases, as shown in Table 5.1. The point to remember is that the only circumstances under which $(x \rightarrow y)$ is regarded as *false* is when x *is true* and y *is false.* Incidentally, this convention forces us to accept the empty set $\varnothing$ as a subset of any set X, for in the proposition

(5.21) If $a \in \varnothing$, then $a \in X$,

the premise $a \in \varnothing$ is *always false*, and so by the truth table in Table 5.1, the proposition in (5.21) is true.

The simple truth tables in Table 5.1 enable us to calculate truth tables for any WFP, as we illustrate in the next example.

EXAMPLE 5.2 Calculate truth tables for each of the following WFPs:

(a) $(A \leftrightarrow \sim A)$
(b) $((A \rightarrow B) \wedge \sim A)$
(c) $(((A \vee B) \wedge \sim B) \rightarrow A)$
(d) $((\sim(A \wedge B) \vee C) \rightarrow \sim A)$

SOLUTION We show the required truth tables in Table 5.3. Detailed comments on the calculations follow.

TABLE 5.3 Solution to Example 5.2

A	$\sim A$	$(A \leftrightarrow \sim A)$
T	F	F
F	T	F

(a)

A	B	$(A \to B)$	$\sim A$	$((A \to B) \wedge \sim A)$
T	T	T	F	F
T	F	F	F	F
F	T	T	T	T
F	F	T	T	T

(b)

A	B	$(A \vee B)$	$\sim B$	$((A \vee B) \wedge \sim B)$	$(((A \vee B) \wedge \sim B) \to A)$	
T	T	T	F	F	T	
T	F	T	T	T	T	
F	T	T	F	F	T	(theorem)
F	F	F	T	F	T	

(c)

A	B	C	$(A \wedge B)$	$\sim(A \wedge B)$	$(\sim(A \wedge B) \vee C)$	$((\sim(A \wedge B) \vee C) \to \sim A)$
T	T	T	T	F	T	F
T	T	F	T	F	F	T
T	F	T	F	T	T	F
T	F	F	F	T	T	F
F	T	T	F	T	T	T
F	T	F	F	T	T	T
F	F	T	F	T	T	T
F	F	F	F	T	T	T

(d)

(a) This proposition is a function of the single atom A, and so in Table 5.3(a) we show in the left column of the truth table, the two possible values for A, viz. T and F. In the column headed "$\sim A$," we show the corresponding truth values for the proposition $\sim A$. Finally, in the column headed "$(A \leftrightarrow \sim A)$," we combine the truth values A and $\sim A$, using the truth table for *if and only if* given in Table 5.1.

(b) This proposition is a function of the two atoms A and B, and so in the two left columns of Table 5.3(b) we show all possible combinations for the truth values of A and B. Then we use the *implies* rule from Table 5.1

to compute the "$(A \to B)$" column, the *not* rule to compute the "$\sim A$" column, and finally combine those two columns using the *and* rule.

(c) This proposition is also a function of the two atoms A and B, and in Table 5.3(c) we have worked up to the final truth table by successively computing the truth table for $(A \vee B)$, $\sim B$, $((A \vee B) \wedge \sim B)$.

(d) This is a three-atom proposition, and in Table 5.3(d) we have worked up to the required truth table by successively computing the truth tables for $(A \wedge B)$, $\sim(A \wedge B)$, and $(\sim(A \wedge B) \vee C)$. The last column is the desired truth table, and has been obtained by combining the "$(\sim(A \wedge B) \vee C)$" column with the "$A$" column, using the shortcut that "$(x \to \sim y)$" is false if and only if x is true and y is true. ■

In Example 5.2(c), we discovered that the proposition $(((A \vee B) \wedge \sim B) \to A)$ assumed the value T for all possible truth values for A and B. A proposition that is *always true* is called a **theorem** of the propositional calculus.* In fact, this particular theorem is simply the formally precise way of stating the proposition (5.2) at the beginning of the section; see Figure 5.2.

There are many, in fact infinitely many, theorems in the propositional calculus, and many of them have been known (in principle) since ancient times. In the box that follows we have listed 13 of the most famous of these theorems. In each theorem x, y, z, and w can be replaced with any WFP. Each one of these theorems represents a logically valid "reasoning tool." We will leave as problems the verification that the WFPs in the box are indeed theorems, and content ourselves with an informal discussion of a few of them.

THIRTEEN FAMOUS THEOREMS OF THE PROPOSITIONAL CALCULUS

1. $(x \vee \sim x)$	Excluded middle
2. $(x \leftrightarrow \sim\sim x)$	Double negative
3. $((x \wedge y) \to x)$	Separation
4. $(x \to (x \vee y))$	Joining
5. $(((x \to y) \wedge x) \to y)$	Modus ponens or detachment
6. $((x \to y) \leftrightarrow (\sim y \to \sim x))$	Contrapositive
7. $(((\sim x \to y) \wedge (\sim x \to \sim y)) \to x)$	Proof by contradiction
8. $(((x \vee y) \wedge \sim y) \to x)$	Disjunctive syllogism
9. $((x \vee y) \leftrightarrow (\sim x \to y))$	Switcheroo
10. $(((x \to y) \wedge (y \to z)) \to (x \to z))$	Deduction
11. $((((x \to y) \wedge (z \to w)) \wedge (x \vee z)) \to (y \vee w))$	Constructive dilemma
12. $(\sim(x \vee y) \leftrightarrow (\sim x \wedge \sim y))$	DeMorgan's law I
13. $(\sim(x \wedge y) \leftrightarrow (\sim x \vee \sim y))$	DeMorgan's law II

* In many texts the word **tautology** is used. However, we feel that *tautology* carries a negative connotation which is inappropriate here.

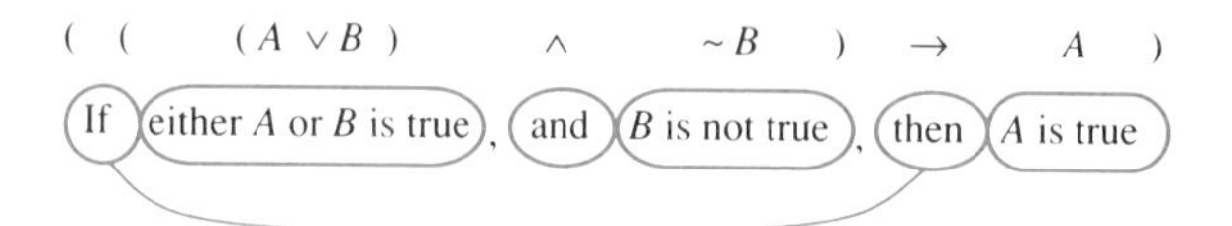

Figure 5.2 Proposition (5.2) is the same as the WFP in Example 5.2(c).

- *Law of the excluded middle*: This theorem say that every proposition is either *true* or *false*, there being no "middle ground." *Example*: "Either Art loves Betty or Art does not love Betty."
- *Law of double negative*: This theorem says that the negation of the negation of a proposition is the same as the original proposition. *Example*: "It is not the case that dynamite is not dangerous"—same as "Dynamite is dangerous."
- *Modus ponens*: This theorem is exemplified by the most famous logical argument of them all: "If Socrates is a man, then Socrates is mortal. Socrates is a man. Therefore Socrates is mortal." Here x is the proposition "Socrates is a man," and y is the proposition "Socrates is mortal."
- *Contrapositive*: This rule says that the propositions "If A, then B" and "If not B, then not A" are logically equivalent; that is, the propositions have identical truth tables. *Example*: "If you are studying discrete mathematics, then you are clever" is logically equivalent to "If you are not clever, then you are not studying discrete mathematics." (See Problems 7 10 for additional examples of logically equivalent.)

At the beginning of this section we said that the goal of the propositional calculus is the discovery of *valid arguments*, but since then we have made no further mention of valid arguments, only *theorems*. It turns out that each valid argument is a special kind of theorem. For example, recall the following proposition (see 5.2):

(5.22) If either A or B is true, and B is not true, then A is true.

We have seen that in the formal symbolism of the propositional calculus, this statement is equivalent to the theorem $(((A \vee B) \wedge \sim B) \rightarrow A)$. Another way of saying that (5.22) is logically sound is to say that the two **premises** $(A \vee B)$ and $\sim B$ lead to the **conclusion** A. Symbolically, we will say that

(5.23) $$(A \vee B), \sim B. \quad \therefore A$$

is a *valid argument*. (In 5.23 the symbol "∴" means "therefore.") As another example of a valid argument, consider the modus ponens theorem, which informally says

If x is true, then y is true. x is true.
Therefore y is true.

To put this in the form of a valid argument, we would write

$$(x \rightarrow y), x. \quad \therefore y.$$

More generally, if $x_1, x_2, \ldots, x_n$, and y are all WFPs, we shall call

(5.24) $$x_1, x_2, \ldots, x_n. \quad \therefore y$$

an **argument**. The WFPs $x_1, x_2, \ldots, x_n$ are called the **premises** of the argument, and the WFP y is called the **conclusion**. It is called a **valid argument** if the compound WFP

(5.25) $$(((\cdots((x_1 \wedge x_2) \wedge x_3) \wedge \cdots) \wedge x_n) \rightarrow y)$$

is a theorem. Informally, we can see that the argument (5.24) and the WFP (5.25) both assert the same thing, viz. "If x_1 and x_2 and $\cdots$ and x_n and are all true, then y is also true." In the box on page 185 there are several more theorems that can be turned into valid arguments. For example, the deduction theorem can be written as

$$(x \rightarrow y), (y \rightarrow z). \quad \therefore (x \rightarrow z),$$

and the constructive dilemma theorem is the same as

$$(x \rightarrow y), (z \rightarrow w), (x \vee z). \quad \therefore (y \vee w).$$

Similarly, the proof-by-contradiction theorem can be phrased as a valid argument, as follows:

$$(\sim x \rightarrow y), (\sim x \rightarrow \sim y). \quad \therefore x.$$

EXAMPLE 5.3 Which of the following are valid arguments?

(a) $(P \rightarrow Q), (\sim Q \rightarrow R), \sim R. \quad \therefore P.$
(b) $(A \rightarrow (B \rightarrow C)), B. \quad \therefore (A \rightarrow C).$
(c) If Betty has bought a fur coat, then either she has robbed a bank or her rich uncle had died. Betty has not bought a fur coat, or else she has not robbed a bank. Therefore her rich uncle has not died.
(d) Today is Sunday. Today is not Sunday. Therefore the moon is made of green cheese.

SOLUTION

(a) In order to check the validity of this argument, we need to see whether or not the WFP

(5.26) $$(x \rightarrow P)$$

is a theorem, where x denotes the WFP obtained by anding together the three premises, viz.

(5.27) $$x = (((P \rightarrow Q) \wedge (\sim Q \rightarrow R)) \wedge \sim R).$$

The argument will be valid if (5.26) is identically true. Now the only circumstances under which (5.26) can be *false* is when x *is true* and P *is false* (see Table 5.1). Since x is obtained by anding together the three premises, x will be true if and only if each of the premises is true. Therefore the argument will be invalid if and only if we can assign truth values to the atoms P, Q, and R in such a way that $(P \rightarrow Q)$, $(\sim Q \rightarrow R)$, and $\sim R$ are all true, but P is false. Stated another way, an argument is valid if for every assignment of truth values to the atoms which makes all the premises true, the conclusion is also true.

TABLE 5.4 Solution to Example 5.3: (a) The argument $(P \to Q)$, $(\sim Q \to R)$, $\sim R$. $\therefore P$ is not valid. (b) The argument $(A \to (B \to C))$, B. $\therefore (A \to C)$ is valid. (c) The argument $x \to (y \vee z))$, $(\sim x \vee \sim y)$. $\therefore \sim z$ is not valid.

			Premises			Conclusion
P	Q	R	$(P \to Q)$	$(\sim Q \to R)$	$\sim R$	P
T	T	T	T	T	F	T
T	T	F	T	T	T	T
T	F	T	F	T	F	T
T	F	F	F	F	T	T
F	T	T	T	T	F	F
→ F	T	F	T	T	T	F
F	F	T	T	T	F	F
F	F	F	T	F	T	F

(a)

			Premises		Conclusion
A	B	C	$(A \to (B \to C))$	B	$(A \to C)$
T	T	T	T	T	T
T	T	F	F	T	F
T	F	T	T	F	T
T	F	F	T	F	F
F	T	T	T	T	T
F	T	F	T	T	T
F	F	T	T	F	T
F	F	F	T	F	T

(b)

			Premises		Conclusion
x	y	z	$(x \to (y \vee z))$	$(\sim x \vee \sim y)$	$\sim z$
T	T	T	T	F	F
T	T	F	T	F	T
→ T	F	T	T	T	F
T	F	F	F	T	T
→ F	T	T	T	T	F
F	T	F	T	T	T
→ F	F	T	T	T	F
F	F	F	T	T	T

(c)

In Table 5.4(a) we show a truth table for the three premises $(P \to Q)$, $(\sim Q \to R)$, and $\sim R$; and for the conclusion P. We find that there is one choice of the truth values for P, Q, and R, viz. FTF, for which the premises are all true but the conclusion is false. Therefore the argument is invalid. [Incidentally, this particular argument is the one cited as proposition (5.3) at the beginning of the section.]

(b) In Table 5.4(b) we show truth tables for the two premises $(A \to (B \to C))$ and B, and for the conclusion $(A \to C)$. Here we find that whenever the premises are both true, the conclusion is also true. Therefore this argument is valid.

(c) Here we have a practical application of the propositional calculus! To solve the problem, we first must realize that there are three propositions involved, viz.

x: Betty has bought a fur coat.
y: Betty has robbed a bank.
z: Betty's rich uncle has died.

The given argument can thus be written symbolically as

$$(x \to (y \vee z)), (\sim x \vee \sim y). \quad \therefore \sim z.$$

As in parts (a) and (b), the way to check this argument is to see if there is a way to assign truth values to the atoms x, y, and z so that the two premises are true but the conclusion is false. In Table 5.4(c) we show the appropriate truth table, and find that the argument is invalid. There are *three* lines, viz. TFT, FTT, and FFT, for which the premises are both true but the conclusion is false. This is not an argument worthy of Sherlock Holmes!

(d) If we denote the proposition "Today is Sunday" by A, and the proposition "The moon is made of green cheese" by B, the given argument can be represented symbolically as follows:

$$A, \sim A. \quad \therefore B.$$

As we have seen before, the argument will be *invalid* if and only if for some assignment of truth values the premises are both true but the conclusion is false. In this case, however, there is no possible way for both premises to be true! If A is true, then $\sim A$ is false; and vice versa. Therefore, strange to say, this is a valid argument. More generally, we can say that any argument with *contradicting premises*, i.e., premises that cannot all be simultaneously true, must be valid. Sometimes this peculiar fact is expressed by saying "a contradiction implies anything." ■

Problems for Section 5.1

1. Which of the following propositions are WFPs?
(a) $X \to \sim Y$.
(b) $(Y \to \sim\sim Y)$.
(c) $(P \leftrightarrow (\wedge \sim P))$.
(d) $((\sim a \to b) \to ((\sim a \to \sim b) \to a))$.

2. Which of the following propositions are WFPs?
(a) $\sim((A \to B) \to \sim(B \to A))$.
(b) $(S \wedge (((P \to Q) \wedge (\sim Q \to R)) \to (Q \vee \sim Q)))$.
(c) $((((A \wedge \sim B) \vee (\sim A \wedge B)) \leftrightarrow \sim\sim\sim\sim C))$.
(d) $(((A \to B) \to C) \to \sim((A \vee B) \leftrightarrow \sim\sim(C \wedge A)))$.

3. In the text we stated that the $\rightarrow$ and $\leftrightarrow$ connectives of the propositional calculus do not correspond to any set operations in Section 1.2. Define set operations "$A \rightarrow B$" and "$A \leftrightarrow B$" that do correspond to these connectives, and draw the appropriate Venn diagrams.

4. Calculate truth tables for the following WFPs.
(a) $(\sim A \wedge \sim B)$. (b) $(x \rightarrow (y \rightarrow z))$.

5. Calculate truth tables for the following WFPs.
(a) $((P \leftrightarrow \sim Q) \vee Q)$. (b) $((\sim P \wedge Q) \rightarrow (\sim Q \wedge R))$.

6. Two WFPs x and y are said to be **logically equivalent** if the WFP $(x \leftrightarrow y)$ is a theorem. Show that this is the same as saying that x and y have identical truth tables.

In Problems 7–10, determine whether or not the given pairs of WFPs are logically equivalent.

7. (a) x and $(x \wedge (y \vee \sim y))$.
(b) x and $(x \wedge (y \rightarrow x))$.

8. (a) $(x \rightarrow y)$ and $(\sim y \rightarrow \sim x)$.
(b) $((A \rightarrow B) \rightarrow C)$ and $(A \rightarrow (B \rightarrow C))$.

9. (a) $(x \wedge y)$ and $(\sim x \rightarrow y)$.
(b) P and $(Q \leftrightarrow (P \rightarrow Q))$.

Note: The appropriate truth table for P is as follows:

P	Q	P
T	T	T
T	F	T
F	T	F
F	F	F

10. (a) $\sim(x \wedge y)$ and $(\sim x \wedge \sim y)$.
(b) $\sim(x \vee y)$ and $(\sim x \wedge \sim y)$.

11. Show that $((x_1 \wedge x_2) \wedge x_3)$ and $(x_1 \wedge (x_2 \wedge x_3))$ are logically equivalent.

12. Show that $((x_1 \vee x_2) \vee x_3)$ and $(x_1 \vee (x_2 \vee x_3))$ are logically equivalent.

13. Show that logical equivalence is an equivalence relation on the set of WFPs.

14. Here is a set of six WFPs:

$$\{(x \vee \sim x), (x \wedge \sim x), (x \vee y), (x \rightarrow x), (\sim x \wedge \sim y), ((x \wedge y) \rightarrow x)\}.$$

Find the logical equivalence classes.

15. In Example 5.2(d) we used the shortcut that the WFP $(x \rightarrow \sim y)$ is false if and only if x and y are both true. Explain why this shortcut is valid.

In Problems 16–28, use a truth table to verify that the indicated WFP from the box on page 185 is indeed a theorem.

16. Excluded middle.

17. Double negative.

18. Separation.

19. Joining.

20. Modus ponens, or detachment.

21. Contrapositive.

22. Proof by contradiction.

23. Disjunctive syllogism.

24. Switcheroo.

25. Deduction.

26. Constructive dilemma.

27. DeMorgan's law I.

28. DeMorgan's law II.

In Problems 29–31, determine which of the given WFPs are theorems.

29. (a) $(x \rightarrow x)$. (b) $\sim(x \leftrightarrow x)$.
(c) $(((P \rightarrow Q) \wedge (\sim P \rightarrow Q)) \rightarrow Q)$.

30. (a) $(\sim A \rightarrow (B \rightarrow A))$. (b) $((A \vee B) \rightarrow (\sim B \rightarrow A))$.

31. (a) $((\sim P \wedge Q) \wedge (Q \rightarrow P))$.
(b) $(((X \rightarrow Y) \rightarrow X) \rightarrow Y)$.

32. Suppose A is an atom. Define a sequence $A_0, A_1, A_2, \ldots$ of WFPs as follows:

$$A_0 = A \quad \text{and for} \quad n \geq 1, \quad A_n = \sim A_{n-1}.$$

For which values of n (if any) is $(A_n \leftrightarrow A)$ a theorem?

33. Suppose A is an atom. Define a sequence $B_0, B_1, B_2, \ldots$ of WFPs as follows:

$$B_0 = A \quad \text{and for} \quad n \geq 1, \quad B_n = (B_{n-1} \rightarrow A).$$

For which values of n (if any) is B_n a theorem?

In Problems 34–38, determine whether or not the given argument is valid.

34. $(x \rightarrow y), \sim y. \quad \therefore \sim x.$

35. $(P \vee Q), \sim P. \quad \therefore Q.$

36. $(x \wedge y), \sim x. \quad \therefore \sim y.$

37. $(A \rightarrow B), \sim(B \vee C), (\sim B \wedge \sim C), \sim B, \sim A. \quad \therefore \sim A.$

38. $(A \rightarrow B) \wedge (C \rightarrow D), (\sim B \vee \sim D). \quad \therefore (\sim A \vee \sim C).$

In Problems 39–46, decide whether or not the given argument is valid.

39. If I were a movie star then I would be famous. I am not a movie star. Therefore I am not famous.

40. My book is either on my desk or on the bookshelf. It is not on the bookshelf. Therefore it is on the desk.

41. If the function f is not continuous, then the function g is not differentiable. The function g is differentiable. Therefore the function f is continuous.

42. If Betty has bought a fur coat, then either she has robbed a bank or her rich uncle has died. Her rich uncle has not died. Therefore, if Betty has not robbed a bank, she has not bought a fur coat.

43. If there is life on Mars, then the experts are wrong and the government is lying. If the government is lying, then the experts are right or there is no life on Mars. The government is lying. Therefore there is life on Mars.

44. (Lewis Carroll.) Babies are illogical. Nobody who can manage a crocodile is despised. Illogical persons are despised. Therefore babies cannot manage crocodiles.

45. (Lewis Carroll.) No terriers wander among the Zodiac. Nothing that does not wander among the Zodiac is a comet. Nothing but a terrier has a curly tail. Therefore no comet has a curly tail.

46. (With Apologies to Lewis Carroll.) All unripe fruit is unwholesome. All these apples are unwholesome. No fruit, grown in the shade, is ripe. These apples are not grown in the sun. Therefore all ripe fruit is wholesome.

47. There is, beside the ones listed in Table 5.1, one other common propositional connective, called the *stroke*:

$$(x|y) \qquad (\text{"}x \text{ stroke } y\text{"}).$$

Its truth table is given below:

x	y	$(x\|y)$
T	T	F
T	F	T
F	T	T
F	F	T

(a) Show that $(x|y)$ has the same truth table as $\sim(x \wedge y)$.

(b) Show that the following pairs of WFPs have the same truth tables.

  (i) $\sim x$ and $(x|x)$.

  (ii) $(x \wedge y)$ and $((x|y)|(x|y))$.

  (iii) $(x \vee y)$ and $((x|x)|(y|y))$.

  (iv) $(x \rightarrow y)$ and $(x|(y|y))$.

(c) Using part (b), or otherwise, find a WFP which uses only the stroke connective and which has the same truth table as $(x \leftrightarrow y)$.

5.2 Basic Boolean Functions: Digital Logic Gates

In the previous section, we studied the branch of symbolic logic called propositional calculus. One of the earliest investigators of symbolic logic was George Boole,* who invented a systematic way of manipulating logic symbols which became known as *Boolean algebra.* The striking feature of Boolean algebra, which distinguishes it from all other branches of algebra, is that all the variables and constants must assume just one of two values, which are usually called *true* and *false.*

It is also true that computers and most other digital electronic devices do their work using only two symbols, which are usually called *one* and *zero.* In 1938 Claude Shannon† discovered that this is much more than just a coincidence, and that there is such a great similarity between the way logicians reason and the way electronic devices calculate, that Boolean algebra can be profitably applied to digital electronics! Indeed, since Shannon's discovery, what was previously an extremely "pure" branch of mathematics has become an indispensable tool in all branches of modern digital electronics. In this section, we will introduce you to the basics of Boolean algebra, as it has evolved as a tool for electrical engineers. The key idea is that of a *Boolean function.*

* George Boole (1815–1864), English mathematician, married to Mary Everest, niece of Sir George Everest, for whom Mt. Everest is named. He invented his eponymous algebra in the 1854 book *Investigation of the Laws of Thought.*

† Claude Shannon (1916–), American mathematician and engineer. His discovery of the relationship between Boolean algebra and digital electronics is now just a footnote to his career, since in 1948 he created *information theory*, the mathematical theory of communications, which is one of the most profound scientific accomplishments of the twentieth century.

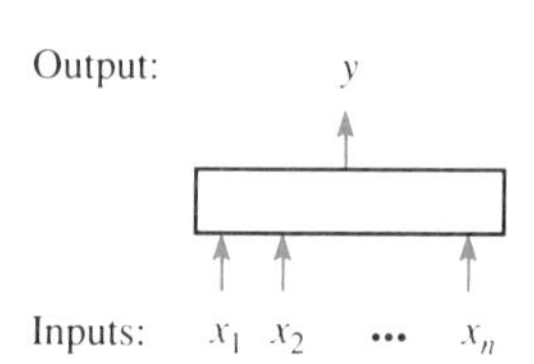

Figure 5.3 A generic Boolean function: $y = f(x_1, x_2, \ldots, x_n)$.

In Figure 5.3 we see a schematic representation of a generic **Boolean function** $y = f(x_1, x_2, \ldots, x_n)$. In Figure 5.3 each of the n **input variables** $x_1, x_2, \ldots, x_n$ can assume either of the values 0 or 1, and the **output variable** y, which is determined by $x_1, x_2, \ldots, x_n$, is also either 0 or 1. Thus, in the jargon of Chapter 1, a Boolean function is a function whose domain is the Cartesian product $\{0, 1\} \times \{0, 1\} \times \cdots \times \{0, 1\}$ and whose range is $\{0, 1\}$. The purpose of this introductory section is to describe some typical computer operations and their corresponding Boolean functions.

In Section 1.3, we saw that a function can always be represented in tabular form. For Boolean functions, such a table is called a **truth table**. (This terminology is a remnant of symbolic logic, in which 0 represents a "false" statement, and 1 represents a "true" statement.) Table 5.5 shows the truth table for one particular Boolean function of three variables.

Incidentally, a Boolean variable is also called a **bit** (short for *b*inary dig*it*). For example, one says that in the function of Table 5.5 the output bit y is determined by the three input bits x_1, x_2, and x_3.

TABLE 5.5 The truth table for one particular Boolean function of three variables

x_1	x_2	x_3	y
0	0	0	0
0	0	1	1
0	1	0	1
0	1	1	0
1	0	0	1
1	0	1	0
1	1	0	0
1	1	1	1

EXAMPLE 5.4 The "Half Adder." One of the simplest and most important arithmetic operations which can be done by a computer is **binary addition with carry**. The simplest case of this operation is the process of adding together two one-bit numbers x_1 and x_2:

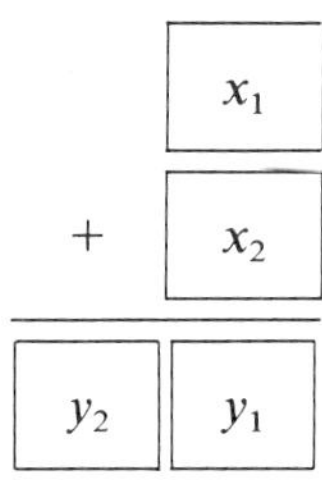

There are four possibilities, depending on the values of x_1 and x_2:

$$\begin{array}{cccc} 0 & 0 & 1 & 1 \\ +0 & +1 & +0 & +1 \\ \hline 0\,0 & 0\,1 & 0\,1 & 1\,0 \end{array}$$

Each of the values y_1 and y_2 is a Boolean function of the two variables x_1 and x_2. Calculate the truth tables for y_1 and y_2.

SOLUTION There are, as illustrated, only four possibilities for the "input pair" (x_1, x_2), and the corresponding values of y_1 and y_2 can be read off directly. The required truth tables are as follows:

Truth table for y_1

x_1	x_2	y_1
0	0	0
0	1	1
1	0	1
1	1	0

Truth table for y_2

x_1	x_2	y_2
0	0	0
0	1	0
1	0	0
1	1	1

TABLE 5.6 Seven important Boolean functions (digital logic gates)

Name	Symbol	Verbal Condition for $z = 1$	Truth Table	Boolean Formula
AND	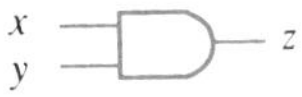 x y z	$x = 1$ and $y = 1$	x y z 0 0 0 0 1 0 1 0 0 1 1 1	$z = xy$
OR	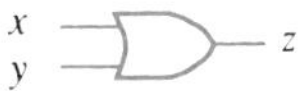 x y z	$x = 1$ or $y = 1$	x y z 0 0 0 0 1 1 1 0 1 1 1 1	$z = x + y$
NOT (inverter)	 x z	$x = 1$ is not true	x z 0 1 1 0	$z = \bar{x}$
NAND	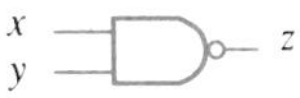 x y z	($x = 1$ and $y = 1$) is not true	x y z 0 0 1 0 1 1 1 0 1 1 1 0	$z = \overline{xy}$
NOR	 x y z	($x = 1$ or $y = 1$) is not true	x y z 0 0 1 0 1 0 1 0 0 1 1 0	$z = \overline{x + y}$
XOR (Exclusive-OR)	 x y z	($x = 1$ or $y = 1$) but not both	x y z 0 0 0 0 1 1 1 0 1 1 1 0	$z = x \oplus y$ $= \bar{x}y + x\bar{y}$
XNOR (Exclusive-NOR)	 x y z	($x = 1$ or $y = 1$ but not both) is not true	x y z 0 0 1 0 1 0 1 0 0 1 1 1	$z = \overline{x \oplus y}$ $= xy + \bar{x}\bar{y}$

Many of the functions performed by a modern computer are, of course, extremely complicated Boolean functions of many variables. However, in most cases such complicated functions are built up by combining simpler functions. There are, in fact, a handful of *very* simple one- or two-variable Boolean functions that are commonly used as building blocks for more complex functions. The seven most important such functions, or logic gates as they are sometimes called, are described in Table 5.6. These seven gates are all easily fabricated, readily available digital electronic devices.*

In Table 5.6 we have represented each function in five ways: by its usual name; by the symbol usually used to represent it in electronic circuit diagrams; by a possibly clumsy "verbal description," which helps explain the name; by the truth table, which describes the function precisely; and by a "Boolean formula," which will be explained fully in Section 5.3. The usefulness of these gates will appear slowly as the chapter develops; for now, let us just notice that the half adder of Example 5.4 can be represented by the functional description

$$y_1 = x_1 \text{ XOR } x_2, \qquad y_2 = x_1 \text{ AND } x_2,$$

and the logic circuit of Figure 5.4.

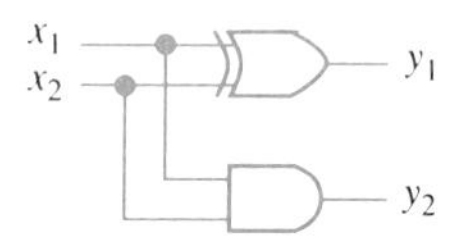

Figure 5.4 Building a half adder with an XOR gate and an AND gate.

The seven basic gates described in Table 5.6 can be combined in many ways. Example 5.5 is a simple illustration of how complex function can be built from simple ones.

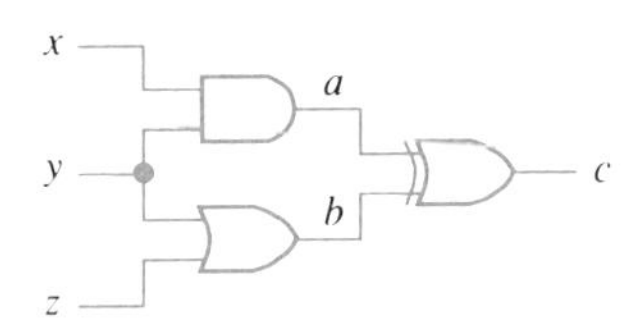

Figure 5.5 Logic diagram for Example 5.5.

EXAMPLE 5.5 In the logic diagram of Figure 5.5, c represents a Boolean function of the three Boolean variables x, y, and z. Calculate the truth table for c.

SOLUTION For us, the logic diagram is just a graphical way of describing the following relationships:

$$a = x \text{ AND } y, \qquad b = y \text{ OR } z,$$

$$c = a \text{ XOR } b = (x \text{ AND } y) \text{ XOR } (y \text{ OR } z).$$

The truth tables for a and b can be determined directly from Table 5.6. In terms of the three variables x, y, and z we have the following table:

x	y	z	a	b
0	0	0	0	0
0	0	1	0	1
0	1	0	0	1
0	1	1	0	1
1	0	0	0	0
1	0	1	0	1
1	1	0	1	1
1	1	1	1	1

$a = x \text{ AND } y$
$b = y \text{ OR } z$

* These devices are most often seen packaged in highly miniaturized integrated circuits (ICs). The NAND gate is the commonest of these gates, and the commonest form of the NAND gate is the model 7400 transistor-transistor logic (TTL) IC, which contains four NAND gates and costs about 25 cents!

Knowing a and b, we can compute $c = a$ XOR b from Table 5.6:

a	b	c
0	0	0
0	1	1
1	0	1
1	1	0

$c = a$ XOR b

Combining these two tables, we obtain the truth table for c:

x	y	z	(ab)	c
0	0	0	(00)	0
0	0	1	(01)	1
0	1	0	(01)	1
0	1	1	(01)	1
1	0	0	(00)	0
1	0	1	(01)	1
1	1	0	(11)	0
1	1	1	(11)	0

By combining the simple gates of Table 5.6 in various ways, we can obviously manufacture many complicated Boolean functions. In fact *any* Boolean function, of any number of variables, can be built from the basic gates! We will see why this remarkable fact is so in the next section. For now, we notice only that certain of the gates in Table 5.6 are redundant, in the sense that they can be built from the other six. For example, as the notation suggests, NAND can be built from NOT and AND (see Problem 15). The technique needed to verify this fact is given in the next example.

EXAMPLE 5.6 Show that the circuit on the left in Figure 5.6, which consists of three NOR gates, is equivalent to one AND gate. (So if NOR gates are available, we don't need AND gates!)

SOLUTION If a and b are intermediate functions as shown, the given logic diagram is equivalent to

$$a = x \text{ NOR } x, \qquad b = y \text{ NOR } y, \qquad z = a \text{ NOR } b.$$

The truth tables for a and b are therefore as follows (cf. Table 5.6):

x	a
0	1
1	0

y	b
0	1
1	0

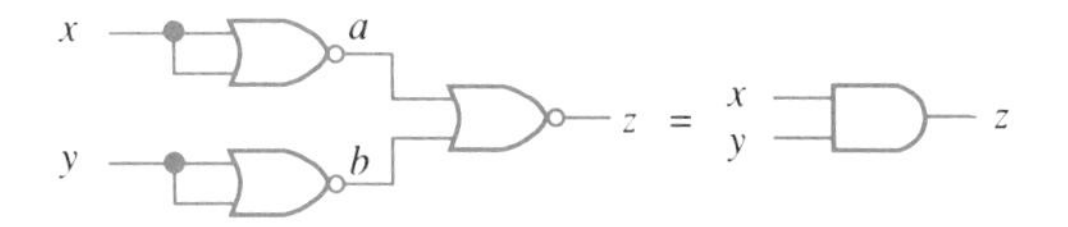

Figure 5.6 Circuit diagrams for Example 5.6.

It follows that the truth table for z is as follows:

x	y	(ab)	z
0	0	11	0
0	1	10	0
1	0	01	0
1	1	00	1

This is, however, exactly the same as the truth table for AND as given in Table 5.6.

More examples of the redundancy of the set of seven gates in Table 5.6 appear in Problems 15–20.

Problems for Section 5.2

1. (Limitations of the Half Adder.) Consider the addition (in binary) of the following four-bit numbers:

$$\begin{array}{rcccc} & 1 & \overset{\downarrow}{0} & 1 & 1 \\ + & 0 & 1 & 1 & 0 \\ \hline 1 & 0 & 0 & 0 & 1 \end{array}$$

Look carefully at the indicated column and verify that in fact *three* digits are being added: $0 + 1 + 1$ (a carry from the previous computation). Thus, a half adder is not sufficient for the addition of two arbitrary integers. However, see the next problem.

2. (The **Full Adder**.) Consider the process of adding *three* one-bit numbers:

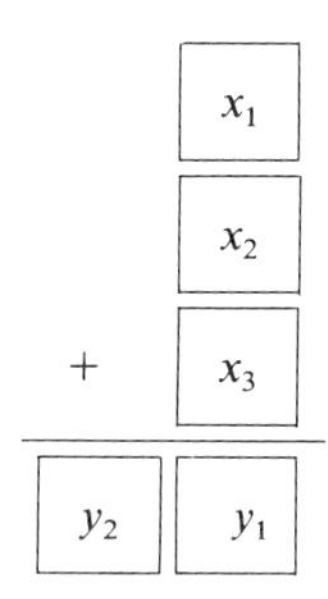

For example:

$$\begin{array}{r} 0 \\ 0 \\ +0 \\ \hline 00 \end{array}, \quad \begin{array}{r} 0 \\ 0 \\ +1 \\ \hline 01 \end{array}, \quad \begin{array}{r} 0 \\ 1 \\ +0 \\ \hline 01 \end{array}, \quad \ldots, \quad \begin{array}{r} 1 \\ 1 \\ +1 \\ \hline 11 \end{array}.$$

Write truth tables for the two Boolean functions y_1 and y_2. (*Note*: In circuitry capable of adding two integers written in binary notation, the full adder is the basic component, since it facilitates the addition of two bits from the two numbers plus the previous carry bit. A full adder can also be built from two half adders. See Problem 21.)

In Problems 3–7, calculate the truth table for the Boolean function c described in the given logic diagrams.

3.

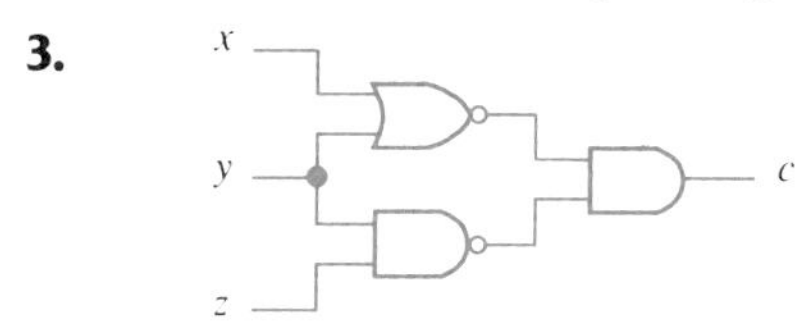

4.

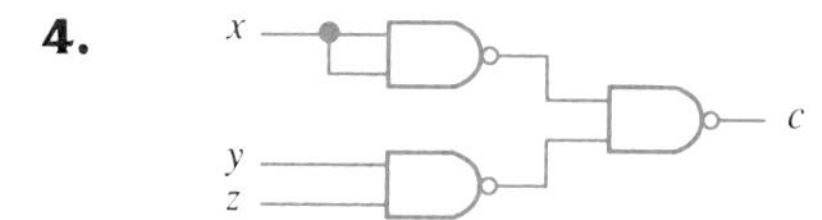

5.

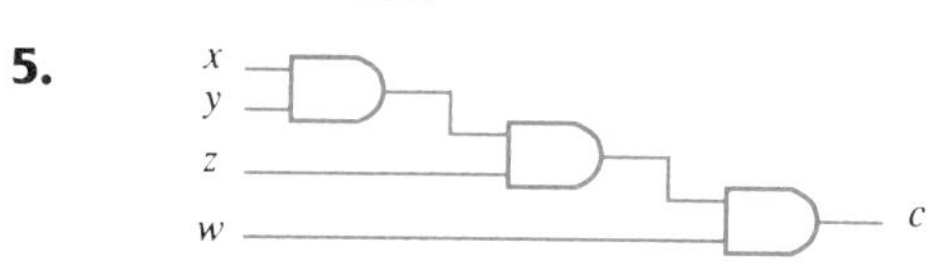

6.

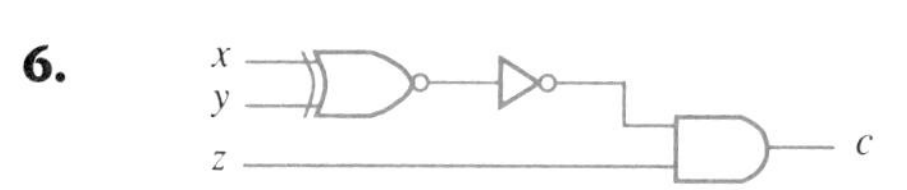

7.

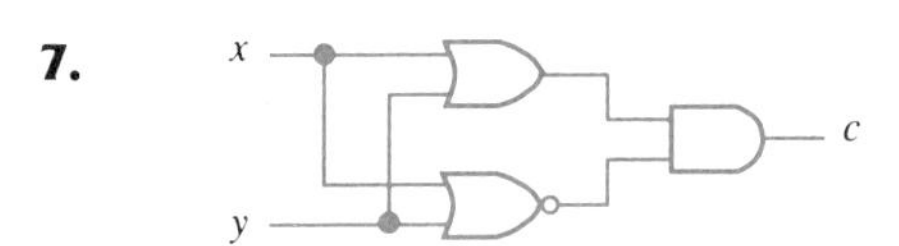

In Problems 8–14 a circuit is given which represents one of the gates in Table 5.6 in terms of the others. In each case identify the gate whose function is being simulated.

8.

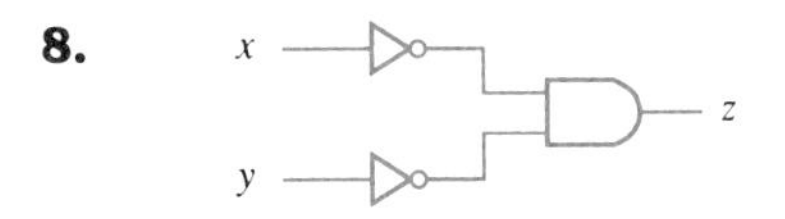

9.

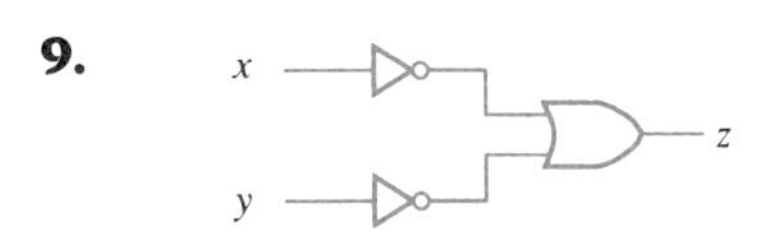

10.

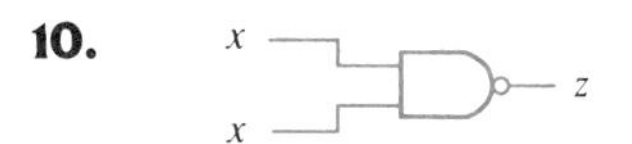

11.

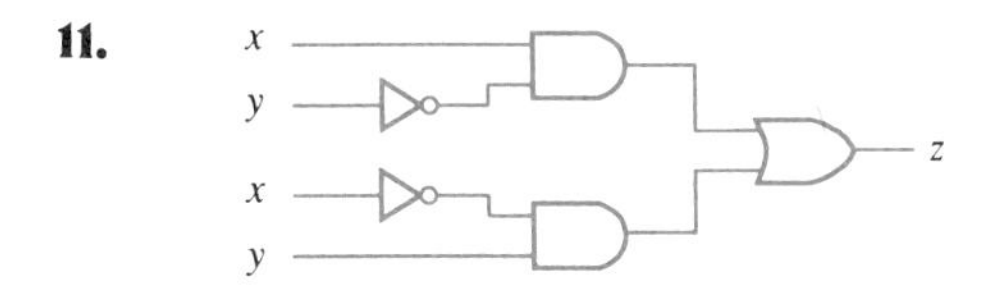

12.

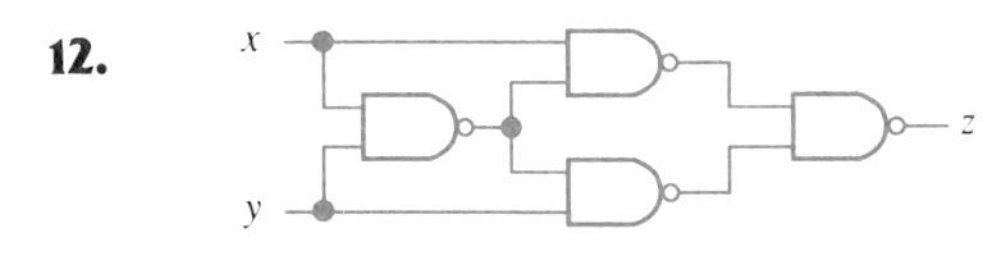

13.

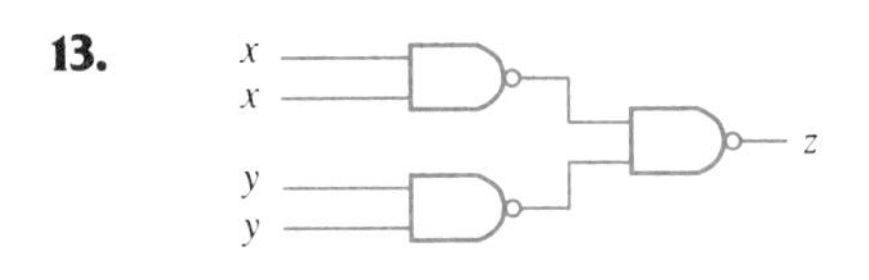

14. x, y → z

In Problems 15–20, try to build the given function by using the given collection of basic gates.

15. NAND, using NOT and AND.

16. NOR, using OR and NOT.

17. AND, using three NOTs and one OR.

18. OR, using three NOTs and one AND.

19. NOT, using one NAND gate.

20. The function of Table 5.5, using two XNOR gates.

21. Show that a full adder (Problem 2) can be built from two half adders, as follows:

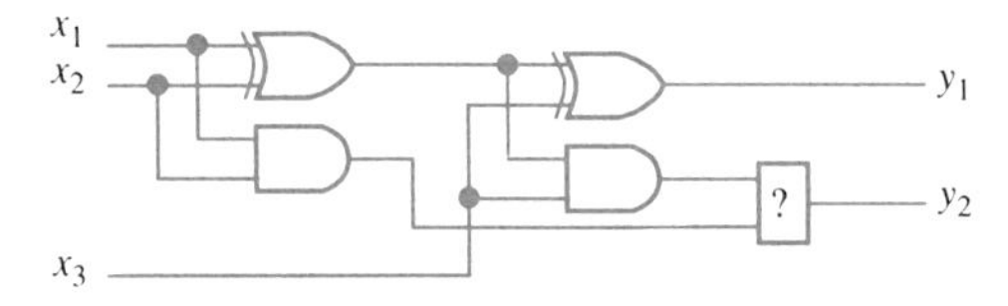

where the gate marked "?" is one of the gates from Table 5.6. (Of course, you must say which gate "?" is!)

22. Table 5.6 gives six Boolean functions of two variables (NOT only involves *one* variable). How many other Boolean functions of two variables are there? Write truth tables for them.

23. Two of the functions you found in Problem 22 are very easy, one whose truth table has all 1s in the right-hand column and one whose truth table has all 0s. Build these functions by using the gates of Table 5.6. (Use only a single gate in each case.)

24. Continuing Problem 23, build the remaining functions found in Problem 22 by using the gates of Table 5.6.

25. Continuing Problem 24, consider an arbitrary Boolean function f of two variables x and y:

x	y	$f(x, y)$
0	0	a
0	1	b
1	0	c
1	1	d

For example, if $a = b = c = 0$ and $d = 1$, we obtain the AND gate. How many ways are there of filling in the last column of the truth table? You have now found the total number of Boolean functions of two variables. (Return to Problem 22 and make sure you have not omitted any functions.)

26. Continuing Problem 25, find the total number of Boolean functions of n variables.

27. You are familiar with the *associative* property of ordinary addition: $(a + b) + c = a + (b + c)$. In other words, if we first add a to b, and then add c to the result, we get the same answer as if we had first added b to c and then added a to this result. Verify that ordinary multiplication is associative but that subtraction and division are not.

28. Continuing Problem 27, the associative property may be represented symbolically by the following diagram:

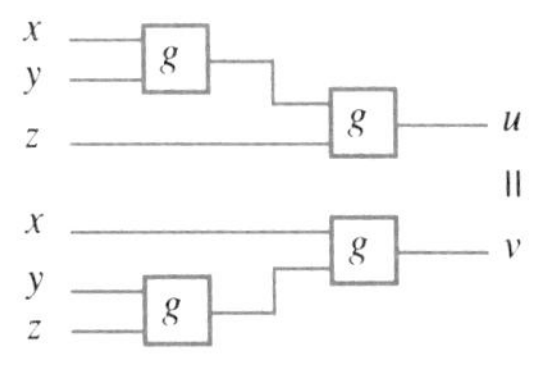

If g stands for ordinary addition, then $u = (x + y) + z$ and $v = x + (y + z)$; so $u = v$. Similarly, if g stands for ordinary multiplication, then $u = (xy)z$ and $v = x(yz)$; so $u = v$. The fact that $u = v$ means that the operation is associative. Now suppose that g is an AND gate, so that $g(x, y) = 1$ if

$x = y = 1$, and $g(x, y) = 0$ otherwise. Show that u still equals v, so that AND is associative.

29. Continuing Problem 28, show that OR is associative.

30. Continuing Problem 28, if g is an arbitrary Boolean function of two variables, g will be associative provided that

$$g(g(x, y), z) = g(x, g(y, z)).$$

Which of the remaining four two-variable Boolean functions (if any) of Table 5.6 are associative?

31. What does *eponymous* mean?

5.3 Minterm and Maxterm Expansions

Our goal in this section is to show that any Boolean function can be built from the basic gates of Table 5.6. Before doing this, however, we will introduce a famous shorthand notation that greatly simplifies the drawing of logic diagrams. This system of notation is called **Boolean algebra**.

Boolean algebra is a way of describing algebraically the operation of the AND, OR, and NOT gates of Table 5.6. For example, in Boolean algebra the simple AND gate circuit

x
y
z

is represented by the equation

(5.28)
$$z = (x \cdot y), \quad z = (xy) \qquad \text{(Boolean AND notation: } x \cdot y \text{ and } xy \text{ mean the same thing).}$$

Similarly, the OR gate notation

x
y
z

is represented in Boolean algebra by the equation

(5.29)
$$z = (x + y) \qquad \text{(Boolean OR notation).}$$

And finally the inverter

x
z

is represented by the notation

(5.30)
$$z = \bar{x} \qquad \text{(Boolean NOT).}$$

These three formulas appear under the heading "Boolean Formula" in Table 5.6, opposite the corresponding gate. The Boolean AND operation of Equation (5.28), which is often called **Boolean multiplication**, behaves just like ordinary multiplication, viz.

$$0 \cdot 0 = 0 \qquad 1 \cdot 0 = 0$$
$$0 \cdot 1 = 0 \qquad 1 \cdot 1 = 1.$$

However, the Boolean OR operation, also called **Boolean addition**, is slightly different from ordinary addition:

$$0 + 0 = 1 \qquad 1 + 0 = 1$$
$$0 + 1 = 1 \qquad 1 + 1 = 1(!).$$

Finally, Boolean NOT, also called **complementation**, simply reverses its arguments:

$$\bar{0} = 1, \qquad \bar{1} = 0.$$

Of course, this notation wouldn't be very useful if it could only be used to represent circuits containing one gate! However, *any* logic circuit involving only AND, OR, and NOT gates can be represented using Boolean algebra. For example, consider the two-gate circuit below:

(5.31)

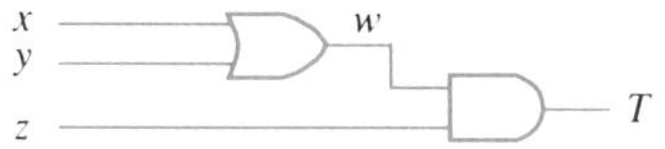

According to rule (5.29), we can replace w with $(x + y)$; this replacement takes care of the OR gate, and so the circuit can be simplified as follows:

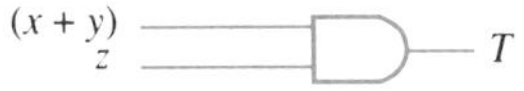

According to rule (5.28), this circuit is equivalent to the Boolean equation $T = ((x + y)z)$. Normally the outermost pair of parentheses in an expression like this is omitted, and the equation

(5.32) $$T = (x + y)z$$

is written instead. We emphasize that (5.31) and (5.32) are completely equivalent; they are simply two different ways of representing the same Boolean function. If we are given a circuit diagram, we can produce the corresponding Boolean equation, and vice versa. The next two examples illustrate this.

EXAMPLE 5.7 Write Boolean algebra expressions corresponding to the following two logic diagrams.

(a)

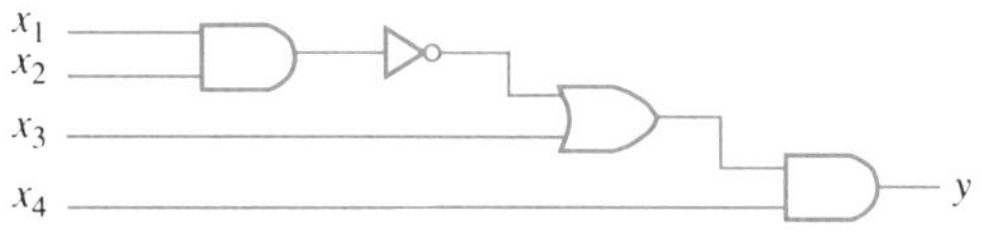

(b)

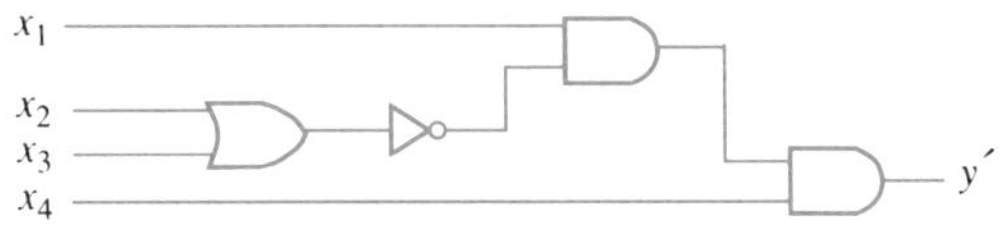

SOLUTION Both of these functions involve two ANDs, one OR, and one NOT, but, as we will see, they are quite different.

(a) Here the first operation is x_1x_2; this quantity is then complemented; then ORed with x_3; and finally ANDed with x_4. This sequence of operations is equivalent to the Boolean expression

$$y = (\overline{x_1x_2} + x_3)x_4.$$

(b) Here the first operation is $x_2 + x_3$, which is then complemented; then ANDed with x_1; then ANDed with x_4:

$$y' = ((\overline{x_2 + x_3})x_1)x_4.$$

EXAMPLE 5.8 Draw logic diagrams to represent the following two Boolean equations:

(a) $A = (x_1x_2)x_3$ (b) $B = \overline{(\bar{x}_1x_2)} + (x_1x_2)$.

SOLUTION

(a) In expression A, we see that first x_1 and x_2 are ANDed; then, the result is ANDed with x_3. The corresponding logic diagram is shown in Figure 5.7(a).
(b) Similarly, in expression B, x_1 is complemented; then ANDed with x_2; this result is again complemented; and this result is then ORed with the AND of x_1 and x_2. The corresponding logic diagram is shown in Figure 5.7(b).

We have now seen that the three Boolean algebra operations $\cdot$, $+$, and $\bar{\ }$ can be used to represent AND, OR, and NOT gates. You might think that four more Boolean symbols would be needed to represent the four remaining gates from Table 5.6 (NAND, NOR, XOR, and XNOR), but no! It turns out that each of these four gates can be represented in terms of Boolean multiplication, addition, and complementation. These representations are as follows:

(5.33) $z = \overline{(xy)}$ (Boolean NAND)

(5.34) $z = \overline{(x + y)}$ (Boolean NOR)

(5.35) $z = (\bar{x}y) + (x\bar{y})$ (Boolean XOR*)

(5.36) $z = (xy) + (\bar{x}\bar{y})$ (Boolean XNOR).

Of course, these four equations require explanation! We will now explain (5.36) in detail; the other three are left as problems at the end of the section.

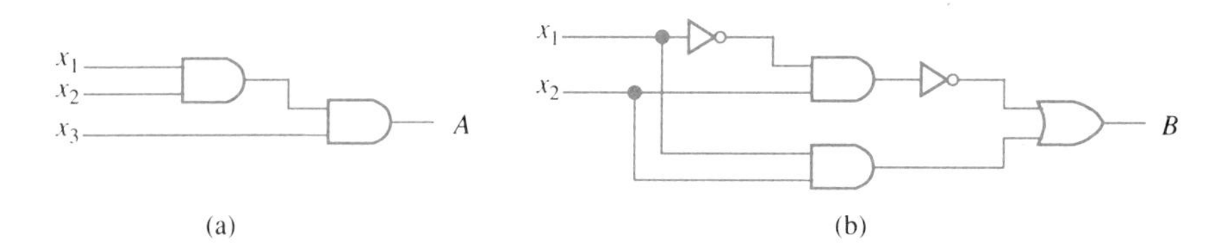

Figure 5.7 Solution to Example 5.8.

* Actually, there is a special Boolean algebra symbol for Exclusive-OR, viz. $z = x \oplus y$. We will have a little more to say about $\oplus$ in Problem 5.5.38 and later on in Chapter 6.

EXAMPLE 5.9 Verify that Equation (5.36) does indeed represent the XNOR gate.

SOLUTION The equation $z = (xy) + (\bar{x}\bar{y})$ is Boolean shorthand for the following logic diagram:

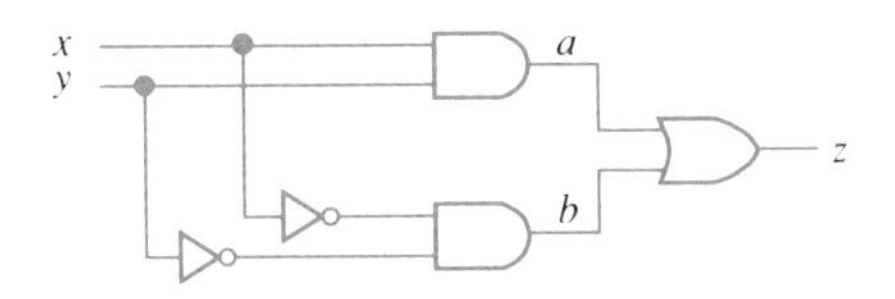

Plainly, this is not a picture of the XNOR gate of Table 5.6. However, it turns out that the *Boolean function* represented by this circuit is identical to the Boolean function XNOR as defined in Table 5.6. To see that this is so, consider the intermediate Boolean variables a and b shown in the figure: $a = (xy)$ and $b = (\bar{x}\bar{y})$. The Boolean function z is $z = a + b$, and the truth table for z is now easily computed as follows:

x	y	a	b	z
0	0	0	1	1
0	1	0	0	0
1	0	0	0	0
1	1	1	0	1

Comparing this truth table with the one for XNOR given in Table 5.6, we see that the two are the same; and so Equation (5.36) represents XNOR, as advertised.

The verifications that the Boolean equations (5.33), (5.34), and (5.35) represent NAND, NOR, and XOR gates are similar (see Problems 15–17).

It may seem surprising that all seven of the basic Boolean functions in Table 5.6 can be represented in terms of the three basic functions AND, OR, and NOT. But an even more surprising thing is also true: *Every* Boolean function, of any number of variables, can be built from ANDs, ORs, and NOTs. The key to this result is contained in the next example.

EXAMPLE 5.10 Describe the Boolean functions represented in Figure 5.8.

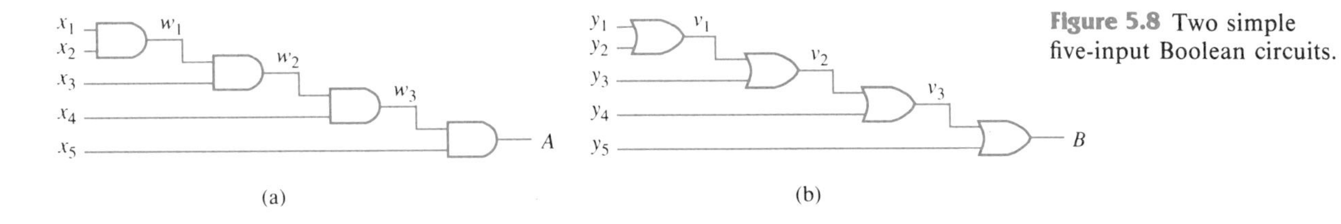

Figure 5.8 Two simple five-input Boolean circuits.

SOLUTION We could proceed by calculating the truth tables for the two functions, by assigning the Boolean variables their $2^5 = 32$ different values. But these two particular Boolean functions are, on reflection, so simple that most of the work can be avoided. In the AND tree of Figure 5.8(a), for example, the output A will be 1 if and only if $x_5 = 1$ and $w_3 = 1$; w_3, in turn, will be 1 if and only if $x_4 = 1$ and $w_2 = 1$; $w_2 = 1$ if and only if $x_3 = 1$ and $w_1 = 1$; finally, $w_1 = 1$ if and only if $x_2 = 1$ and $x_1 = 1$. Combining these simple observations, we conclude that

$$A = \begin{cases} 1, & \text{if } x_1 = x_2 = x_3 = x_4 = x_5 = 1 \\ 0, & \text{in all other cases.} \end{cases}$$

This A is a very simple function indeed—it is 0 for 31 of the 32 possible values of the variables x_1, x_2, x_3, x_4, x_5! In Boolean algebra notation we would represent the AND tree by the equation

$$A = (((x_1 x_2) x_3) x_4) x_5 .$$

However, we will see in the next section when we study the associative law that all these parentheses aren't really necessary, and so we can write instead

$$A = x_1 x_2 x_3 x_4 x_5 .$$

Similarly, the function B (the OR tree) is seen to be almost always equal to 1. In fact, $B = 1$ unless $y_5 = 0$ and $v_3 = 0$; but $v_3 = 1$ unless $y_4 = 0$ and $v_2 = 0$; etc. Continuing this way, we conclude that

$$B = \begin{cases} 0, & \text{if } y_1 = y_2 = y_3 = y_4 = y_5 = 0 \\ 1, & \text{otherwise.} \end{cases}$$

In Boolean algebra notation, we have

$$B = (((y_1 + y_2) + y_3) + y_4) + y_5 ,$$

or, since the parentheses again turn out to be redundant,

$$B = y_1 + y_2 + y_3 + y_4 + y_5 .$$

Of course the functions described in Example 5.10 can be generalized to more than five variables. By combining $(n - 1)$ AND gates in a tree like that in Figure 5.8(a), we can build a Boolean function of n variables $x_1, x_2, \ldots, x_n$, which is 1 if and only if $x_1 = x_2 = \cdots = x_n = 1$. Similarly, we can build an OR tree that will yield a Boolean function of n variables $y_1, y_2, \ldots, y_n$, which is 0 if and only if $y_1 = y_2 = \cdots = y_n = 0$. These functions are important enough to have been given logic symbols of their own, as shown in Figure 5.9.

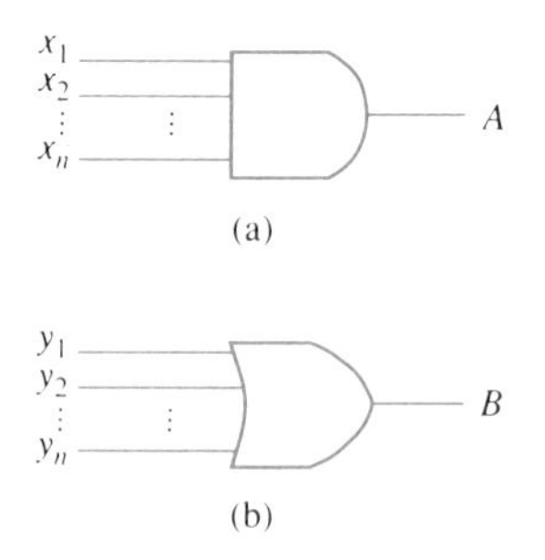

Figure 5.9 Generalized AND and OR gates.

Although it will be convenient to use the symbols in Figure 5.9 in what follows, we emphasize that these multi-input gates are actually built from the basic two-input AND and OR gates of Table 5.6.

In complicated Boolean circuits, it is helpful to think of the output 1 as "acceptance" and 0 as "rejection" of the current input. From this point of view, the big AND gate of Figure 5.9 *accepts* only the input $11 \cdots 1$, and the big OR gate *rejects* only $00 \cdots 0$. The next example shows that by bringing NOT gates (inverters) into the picture, we can build circuits to accept or reject *any* given input pattern, not just $11 \cdots 1$ or $00 \cdots 0$.

EXAMPLE 5.11 Describe the Boolean functions represented by the two circuits in Figure 5.10.

SOLUTION In Figure 5.10(a) the output of the big AND gate will be 1 if and only if all five of its inputs are 1; but because of the three inverters, this requires

$$x_1 = 0; \quad x_2 = 1; \quad x_3 = 0; \quad x_4 = 0; \quad x_5 = 1;$$

i.e., the *uninverted* inputs must all equal 1, and the *inverted* inputs must all equal 0. Thus

$$A = \begin{cases} 1, & \text{if } (x_1, x_2, x_3, x_4, x_5) = (0, 1, 0, 0, 1) \\ 0, & \text{otherwise.} \end{cases}$$

Thus the function A accepts only the input 01001. Using the parentheses-free notation for big AND gates mentioned in Example 5.10, we can write

$$A = \bar{x}_1 x_2 \bar{x}_3 \bar{x}_4 x_5.$$

In an exactly similar way, we find that

$$B = \begin{cases} 0, & \text{if } (y_1, y_2, y_3, y_4, y_5) = (0, 1, 0, 1, 1) \\ 1, & \text{otherwise.} \end{cases}$$

Thus B rejects only the input 01011. It can be represented by the equation

$$B = y_1 + \bar{y}_2 + y_3 + \bar{y}_4 + \bar{y}_5.$$

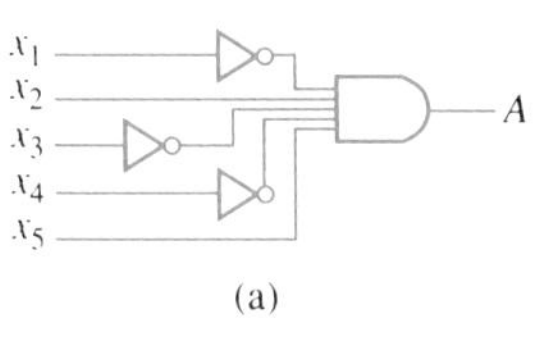

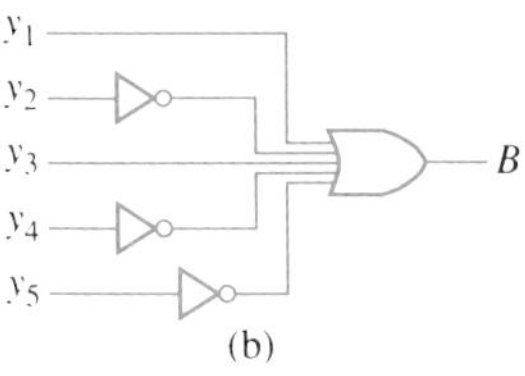

Figure 5.10 Two more Boolean functions: (a) accepts only 01001; (b) rejects only 01011.

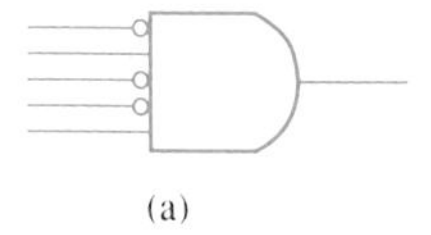

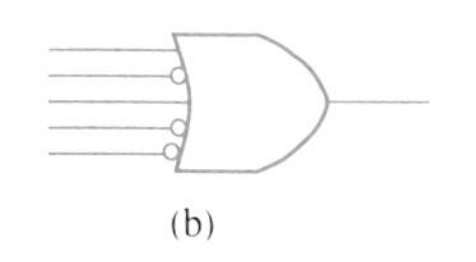

Figure 5.11 Simplified notation for the gates in Figure 5.10: (a) 01001 accepter; (b) 01011 rejecter.

The point of Example 5.11 is that given *any* pattern of 0s and 1s, it is easy to build a circuit that either accepts or rejects that sequence and no other. To build an "accepter," we use a multi-input AND gate, and invert the inputs to the gate corresponding to the 0s in the given pattern. To build the "rejecter," we use a big OR gate, and invert the inputs corresponding to the 1s in the given sequence. Circuits like this occur so often in practice that a simplified notation for them is often used. In Figure 5.11 we see this notation for the gates of Figure 5.10. The idea is simply that the inverters are represented by small circles.

Now we are ready to show how to build *any* Boolean function using only ANDs, ORs, and NOTs. We will illustrate the general technique by considering the specific Boolean function y of the three variables x_1, x_2, x_3 whown in Table 5.7. The function y in Table 5.7 accepts any one of the three input patterns 001, 100, 101. So to build y, we first construct separate accepters for these three inputs (see Figure 5.12).

Since we want y to be 1 when the input is 001 *or* 100 *or* 101, we connect the outputs of the three accepters in Figure 5.12 to one three-input OR gate, as shown in Figure 5.13. The circuit in Figure 5.13 accepts any input from the set {001, 100, 100} and rejects everything else. Since this is exactly what the function y of Table 5.7 does, we have succeeded in building y using only ANDs, ORs,

TABLE 5.7 Another truth table

x_1	x_2	x_3	y
0	0	0	0
0	0	1	1
0	1	0	0
0	1	1	0
1	0	0	1
1	0	1	1
1	1	0	0
1	1	1	0

and NOTs. In Boolean algebra notation, the circuit in Figure 5.13 is represented by the equation

$$y = \bar{x}_1\bar{x}_2x_3 + x_1\bar{x}_2\bar{x}_3 + x_1\bar{x}_2x_3. \tag{5.37}$$

This expression for y is called the **minterm expansion** for y.* By now, our various shorthand conventions may have obscured the important fact that Figure 5.13 and Equation (5.37) show how to build the function y defined in Table 5.7 by using only the basic AND, OR, and NOT gates of Table 5.6. The next example should clarify this.

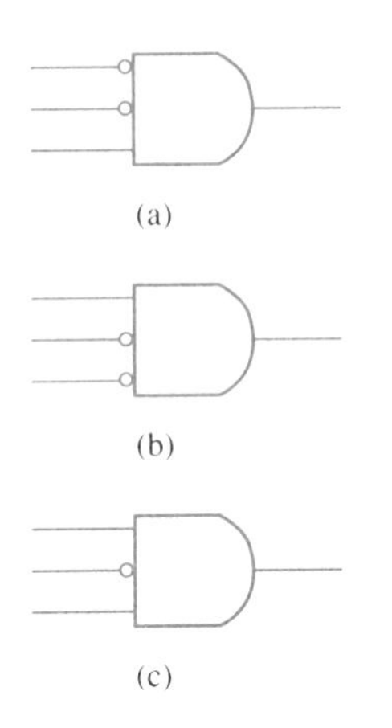

Figure 5.12 Accepters:
(a) 001 accepter, $\bar{x}_1\bar{x}_2x_3$;
(b) 100 accepter, $x_1\bar{x}_2\bar{x}_3$;
(c) 101 accepter, $x_1\bar{x}_2x_3$.

EXAMPLE 5.12 How many basic AND, OR, and NOT gates are needed to build the circuit of Figure 5.13?

SOLUTION Each of the big three-input AND gates represents a circuit using *two* basic two-input AND gates (see Figure 5.8); and the big three-input OR gate represents two basic two-input OR gates. The five small circles represent five NOTs. The count is therefore six ANDs, two ORs, and five NOTs.

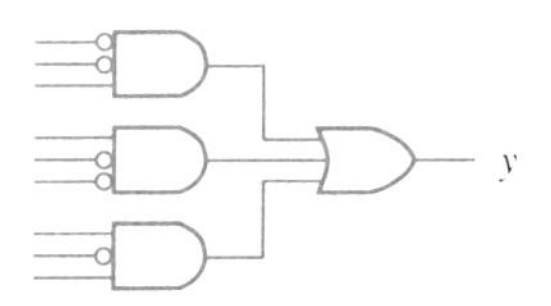

Figure 5.13 A circuit that accepts $\{001, 100, 101\}$.

Although we have chosen to view the function of Table 5.7 as one that accepts the set $\{001, 100, 101\}$, it is equally valid to view it as a function that *rejects* the complementary set, viz. $\{000, 010, 011, 110, 111\}$. This viewpoint leads to a completely different representation of y. To build a function that rejects 000 and 010 and 011 and 110 and 111, we need to connect the outputs of the five separate rejecters for these five inputs to one five-input AND gate, as shown in Figure 5.14.

The Boolean algebra description of the circuit in Figure 5.14 is

$$y = (x_1 + x_2 + x_3)(x_1 + \bar{x}_2 + x_3)(x_1 + \bar{x}_2 + \bar{x}_3)(\bar{x}_1 + \bar{x}_2 + x_3)(\bar{x}_1 + \bar{x}_2 + \bar{x}_3). \tag{5.38}$$

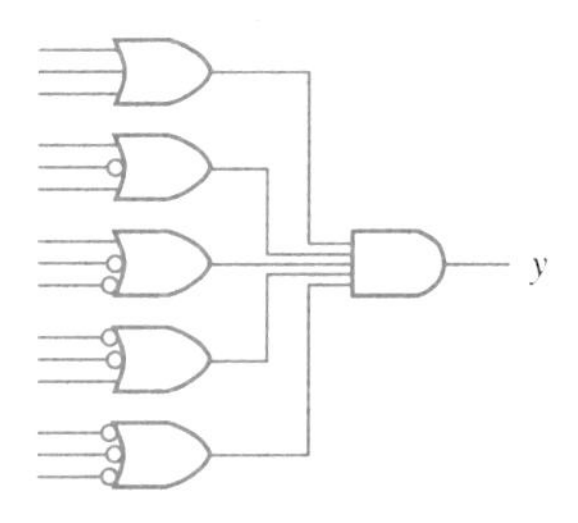

Figure 5.14 A circuit that rejects the set $\{000, 010, 011, 110, 111\}$.

This expression for y is called the **maxterm expansion** for y.† Reasoning like that in Example 5.12 shows that this expansion actually requires 4 ANDs, 10 ORs, and 8 NOTs, and so for this particular function the minterm expansion is more efficient than the maxterm expansion (assuming approximately equal cost for the various gates). Every Boolean function has both a minterm and a maxterm expansion; just as often as not, the maxterm expansion will be more efficient than the minterm expansion. However, usually *neither* expansion is the most efficient possible. For example, the circuit of Figure 5.15 also represents the function y of Table 5.7. Plainly, we have much to learn about simplifying Boolean functions!

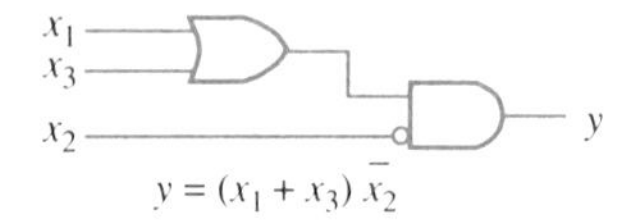

Figure 5.15 Another circuit for the function of Table 5.7.

* So-called because a term like $x_1\bar{x}_2x_3$ is traditionally called a **minterm**. The minterm expansion is also called the **sum-of-products expansion**, or the **disjunctive normal form**, for y.

† So-called because a term like $\bar{x}_1 + \bar{x}_2 + \bar{x}_3$ is traditionally called a **maxterm**. The maxterm expansion is also called the **product-of-sums expansion**, or the **conjunctive normal form**, for y.

Problems for Section 5.3

In Problems 1–5 you are asked to write a Boolean equation corresponding to the logic circuit given in one of the problems in Section 5.2.

1. Problem 5.2.3.

2. Problem 5.2.4.

3. Problem 5.2.5.

4. Problem 5.2.6.

5. Problem 5.2.7.

In Problems 6–14 you are asked to draw logic circuits corresponding to the given Boolean equations.

6. $a = x(y + \bar{z})$.

7. $b = \overline{xy} + xyz$.

8. $c = \overline{x_1(x_2(x_3 + x_4))}$.

9. $d = \overline{x_1 x_2 x_3}$.

10. $e = \overline{(x_1 x_2 x_3)(x_4 x_4)}$.

11. $y = (x_1 + x_2)x_3 + x_4$.

12. $y = (x_1 + x_2)(x_3 + x_4)$.

13. $y = x_1 + x_2 x_3 + x_4$.

14. $y = x_1 + x_2(x_3 + x_4)$.

15. Verify that Equation (5.33) does represent the NAND gate.

16. Verify that Equation (5.34) represents the NOR gate.

17. Verify that Equation (5.35) represents the XOR gate.

18. Show that both of the four-input AND trees in the accompanying figure produce the same Boolean function.

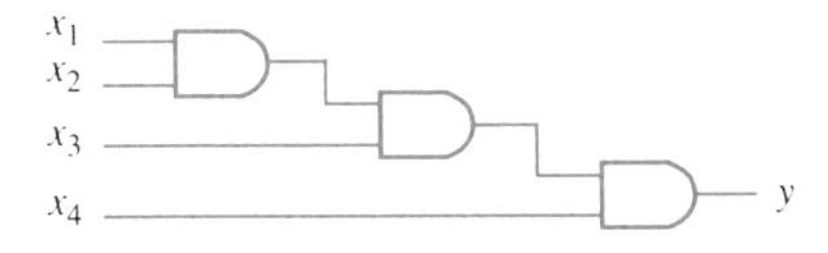

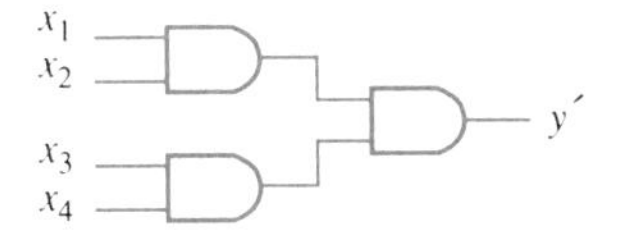

19. Write Boolean equations for each of the circuits in Problem 18.

20. Build a circuit that accepts the set
{000111, 100001, 010011}.

21. Build a circuit that rejects the set
{000111, 100001, 010011}.

22. Suppose you wanted to build a circuit that signalled "acceptance" with a 0 instead of a 1.

(a) Explain how you would (cf. Figure 5.10a) build a 01001 accepter.

(b) How would you build a circuit to accept an arbitrary set?

23. How would you build a circuit that accepted *all* inputs? Rejected all inputs?

24. How many basic AND, OR, and NOT gates are needed to build the functions in Figure 5.10?

25. Suppose you had a supply of three-input AND gates. How many of these would be needed to build a five-input AND gate like the one in Figure 5.9(a)?

26. Find the minterm expansion for the following two functions a and b:

x_1	x_2	x_3	a	b
0	0	0	0	0
0	0	1	1	0
0	1	0	1	0
0	1	1	0	1
1	0	0	1	0
1	0	1	0	1
1	1	0	0	1
1	1	1	1	1

27. Find the maxterm expansion for the functions a and b in Problem 26.

28. What would happen to the output y if the big OR gate in Figure 5.13 were replaced with a big AND?

29. What would happen to the output y if the big AND gate in Figure 5.14 were replaced with a big OR?

30. See if you can give a rule for predicting when the minterm expansion will require fewer basic AND/OR gates than the corresponding maxterm expression. (Don't worry about the NOTs.)

31. Verify that Figure 5.15 is a valid implementation of the function in Table 5.7.

32. Apply the minterm and maxterm constructions to the truth tables given in Table 5.6 for the NAND, NOR, XOR, and XNOR gates. Compare the results to the implementations in Equations (5.33)–(5.36).

33. See if you can think of a plausible explanation of why a term like $x_1\bar{x}_2x_3$ is called a *min*term and $\bar{x}_1 + \bar{x}_2 + \bar{x}_3$ is called a *max*term.

34. See how many different five-input AND trees you can find that implement a five-input AND gate like the one in Figure 5.9(a).

5.4 The Basic Theorems of Boolean Algebra

In Sections 5.2 and 5.3 we have seen several examples of different circuits that represent the same Boolean function. For example, we saw at the end of the previous section that the Boolean function y described in the truth table of Table 5.7 could be represented in the following three ways:

$$\begin{aligned} y &= \bar{x}_1\bar{x}_2x_3 + x_1\bar{x}_2\bar{x}_3 + x_1\bar{x}_2x_3 \qquad \text{(minterm expansion)} \\ &= (x_1 + x_2 + x_3)(x_1 + \bar{x}_2 + x_3)(x_1 + \bar{x}_2 + \bar{x}_3) \\ &\quad \cdot (\bar{x}_1 + \bar{x}_2 + x_3)(\bar{x}_1 + \bar{x}_2 + \bar{x}_3) \qquad \text{(maxterm expansion)} \\ &= (x_1 + x_3)\bar{x}_2 \qquad \text{(??? expansion).} \end{aligned}$$

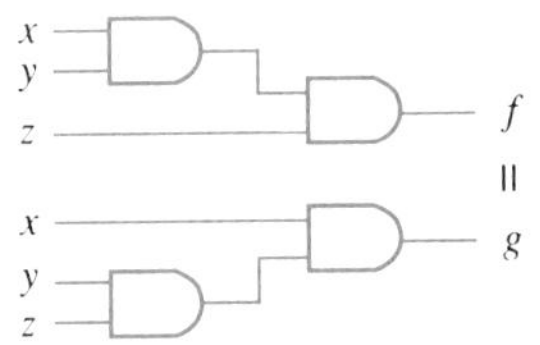

What this shows is that a given Boolean function can be built in many different ways, some complicated (and therefore expensive) and some simple (and therefore cheap). Our eventual goal (to be reached in Section 5.5) is to learn how to build Boolean functions cheaply by learning how to simplify complicated Boolean expressions. In this section we will take the preliminary step of at least learning how to tell when two apparently dissimilar Boolean expressions are in fact equal.

Figure 5.16 Illustrating the associative law.

As an example, we consider the so-called *associative law* mentioned briefly in the previous section, in connection with the AND and OR trees of Figure 5.8. Consider the two logic circuits of Figure 5.16. Plainly, the two circuits in Figure 5.16 are physically different. And yet f and g are the same *as Boolean functions*, since both circuits equal 1 if and only if the input is 111. We express this simple but important fact by writing

$$(xy)z = x(yz), \tag{5.39}$$

which is called the **associative law for Boolean multiplication**. What (5.39) implies is that the placing of the parentheses isn't important; in fact, it is usual to omit the parentheses altogether and write

$$f = g = xyz \tag{5.40}$$

instead. Similarly, although there are numerous ways to form the Boolean product of n variables $x_1, x_2, \ldots, x_n$, they all yield the same result, which is normally denoted by

$$x_1x_2x_3 \cdots x_n \tag{5.41}$$

rather than by $((x_1x_2)(x_3x_4)(x_5 \cdots))$, etc. (See Problem 1).*

Similarly, if we replaced the AND gates in Figure 5.16 with OR gates, we would obtain two different, but equivalent, circuits, and we would conclude that

$$(x + y) + z = x + (y + z), \tag{5.42}$$

which is called the **associative law for Boolean addition**.

* It is *not* safe to omit parentheses when both + and · are present in the same formula. For example, $(x + y)z$ and $x + (yz)$ are two quite different functions. However, there is a *convention* that says that if an ambiguous expression like $(x + yz)$ appears, multiplication is to be performed *before* addition. Thus in ordinary arithmetic, the expression $1 + 2 \times 3$ is 7, not 9. To be safe, however, we will use parentheses and not rely on this convention much.

EXAMPLE 5.13 Show that (5.42) is true.

SOLUTION We can do this by using a truth table:

xyz	$(x+y)$	$(y+z)$	$(x+y)+z$	$x+(y+z)$
000	0	0	0	0
001	0	1	1	1
010	1	1	1	1
011	1	1	1	1
100	1	0	1	1
101	1	1	1	1
110	1	1	1	1
111	1	1	1	1

Alternatively, as in the solution to Example 5.10, we can see that both expressions in Equation (5.42) are 1 unless $x = y = z = 0$.

This discussion of the associative law brings up a question you might be wondering about. We have *already* discussed the associative law, in Chapter 1 (Section 1.2, to be exact). Only there we were talking about operations on *sets*, rather than Boolean functions. Similarly, we have already mentioned truth tables as a tool for proving various things about sets. Is there a connection? The answer is very definitely yes, and in fact we have already seen the connection: In the previous section we showed that any Boolean function can be described in terms of the sequences it accepts. For example, the function given in Table 5.7 accepts the sequences in the subset $\{001, 100, 101\}$ of the universal set $\{000, 001, 010, 011, 100, 101, 110, 111\}$. Similarly, any Boolean function of n variables can be identified with a subset of a set with 2^n elements, viz. $\{00\cdots0, 00\cdots1, \ldots, 11\cdots1\}$. This subset is sometimes called the **truth set** for f.

EXAMPLE 5.14 Consider the function $A = \overline{(x_1 + x_2)\bar{x}_3 + x_4}$. Find its truth set.

SOLUTION The problem is to find out for which of the 16 possible values of x_1, x_2, x_3, x_4 the function A equals 1. Inspection of the form of A shows that $A = 1$ if $(x_1 + x_2)\bar{x}_3 + x_4 = 0$; this requires $(x_1 + x_2)\bar{x}_3 = 0$ *and* $x_4 = 0$. Also, $(x_1 + x_2)\bar{x}_3 = 0$ requires $x_1 + x_2 = 0$ *or* $x_3 = 1$. Finally, $x_1 + x_2 = 0$ requires $x_1 = x_2 = 0$. A little bookkeeping now shows that $A = 1$ only for $(x_1x_2x_3x_4) = (0000), (0010), (0110), (1010), (1110)$. Thus the truth set for A is the set $\{0000, 0010, 0110, 1010, 1110\}$.

Now consider the logic circuits in Figure 5.17. In Figure 5.17(a) we see a circuit which produces the product fg of f and g. Since $fg = 1$ if and only if $f = 1$ *and* $g = 1$, it follows that the truth set of the function fg must be the *intersection* of the truth sets for f and g. For example, suppose the truth set for

f, g → $h = fg$
(a)

f, g → $h = f + g$
(b)

f → $\bar{f}$
(c)

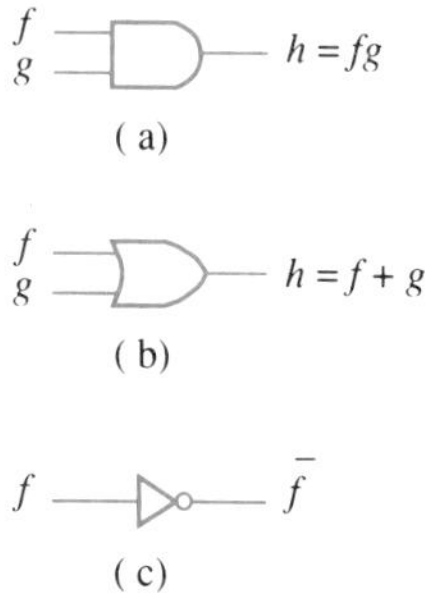

Figure 5.17 Boolean operations and their set-theoretic equivalents.
(a) AND = intersection.
(b) OR = union.
(c) NOT = complement.

f is {000, 011, 101, 110} and the truth set for g is {011, 101, 111}. Then the function h will equal 1 only if the input is in the set

$$\{000, 011, 101, 110\} \cap \{011, 101, 111\} = \{011, 101\}.$$

In the same way, we can see that ORing two Boolean functions together corresponds to taking the *union* of their truth sets, and NOTing a function corresponds to taking the *complement* of its truth set.

With this correspondence between Boolean functions and subsets of a set in mind, let's return to the two circuits of Figure 5.16 and imagine that x, y, and z are Boolean functions, the outputs, perhaps, of complicated logic circuits. Suppose that x, y, and z have truth sets A, B, and C, respectively. Then the truth set for the function f in Figure 5.16 is $(A \cap B) \cap C$ and the truth set for g is $A \cap (B \cap C)$. But we already know that these two sets are equal, by the associative law for *sets* (Section 1.2)! Since the functions f and g have the same truth sets, they are equal; and this gives another proof of the associative law (5.39) for Boolean multiplication.

Similarly, we can interpret the truth table in the solution to Example 5.13 as a truth table in the spirit of Section 1.2, as follows. Replace the Boolean functions x, y, and z by their truth sets A, B, and C, and replace 0 by F and 1 by T. Then we get the following table:

A	B	C	$A \cup B$	$B \cup C$	$(A \cup B) \cup C$	$A \cup (B \cup C)$
F	F	F	F	F	F	F
F	F	T	F	T	T	T
F	T	F	T	T	T	T
F	T	T	T	T	T	T
T	F	F	T	F	T	T
T	F	T	T	T	T	T
T	T	F	T	T	T	T
T	T	T	T	T	T	T

And this is just a "truth table" proof of the associative law for sets.

Thus every statement in Boolean algebra can be translated into set theory, and vice versa. Similarly, if we recall from Section 5.1 the relationship between the propositional connectives $\wedge$, $\vee$, and $\sim$ and the set theory connectives intersection, union, and complement, we can write a *third* version of every statement. For example, $A \cap B' \cap (A \cup C)$ corresponds to the well-formed proposition $((x \wedge \sim y) \wedge (x \vee z))$. Table 5.8 gives a short but illustrative list of statements in all three versions.

In the remainder of this section, we will see many theorems stated in Boolean algebra. All of these theorems, however, can be stated in both set theory and propositional calculus.

The associative law is only one of many known relationships among the operations of Boolean algebra. In Table 5.9, for example, we list 23 such laws—some important, and some not so important! (There are 24 entries in Table 5.9 but only 23 distinct laws; see Problem 19.)

TABLE 5.8 Some translations from Boolean algebra to set theory to propositional calculus

Boolean Algebra	Set Theory	Propositional Calculus
$f\bar{g}(f+h)$	$A \cap B' \cap (A \cup C)$	$((x \wedge \sim y) \wedge (x \vee z))$
$\bar{f} + (\overline{gh})$	$A' \cup (B \cap C)'$	$(\sim x \vee \sim(y \wedge z))$
$fgh + \overline{fgh}$	$(A \cap B \cap C) \cup (A \cap B \cap C)'$	$(((x \wedge y) \wedge z)$ $\vee \sim((x \wedge y) \wedge z))$
$(\bar{f}+g)(f+\bar{h})$	$(A' \cup B) \cap (A \cup C')$	$((\sim x \vee y) \wedge (x \vee \sim z))$

Note: f corresponds to A and x, g to B and y, and h to C and z.

The laws in Table 5.9 are listed in two columns, labelled "Primal Form" and "Dual Form." This format is motivated by the important *law of duality.*

LAW OF DUALITY

If, in a valid Boolean identity, you replace all ANDs by ORs, ORs by ANDs, 0s by 1s, and 1s by 0s, the result is also a valid Boolean identity.

In every case, the equation in the "Dual Form" column has been obtained from the corresponding equation in the "Primal Form" column by applying the law of duality. For example, consider the primal form of the distributive law:

$$x(y + z) = xy + xz.$$

TABLE 5.9 Twenty-three laws of Boolean algebra

Name of Law	Primal Form	Dual form
Associative	$x + (y + z) = (x + y) + z$	$x(yz) = (xy)z$
Commutative	$x + y = y + x$	$xy = yx$
Distributive	$x(y + z) = xy + xz$	$x + (yz) = (x + y)(x + z)$
DeMorgan	$\overline{x + y} = \bar{x}\bar{y}$	$\overline{xy} = \bar{x} + \bar{y}$
Weak absorption	$1 + x = 1$	$0 \cdot x = 0$
Strong absorption	$y + xy = y$	$y(x + y) = y$
Involution	$\bar{\bar{x}} = x$	$\bar{\bar{x}} = x$
Idempotency	$x + x = x$	$x \cdot x = x$
Complementarity	$x + \bar{x} = 1$	$x \cdot \bar{x} = 0$
Redundancy	$x + \bar{x}y = x + y$	$x(\bar{x} + y) = xy$
Consensus	$xy + \bar{x}z + yz = xy + \bar{x}z$	$(x + y)(\bar{x} + z)(y + z) = (x + y)(\bar{x} + z)$
Identity	$x + 0 = x$	$x \cdot 1 = x$

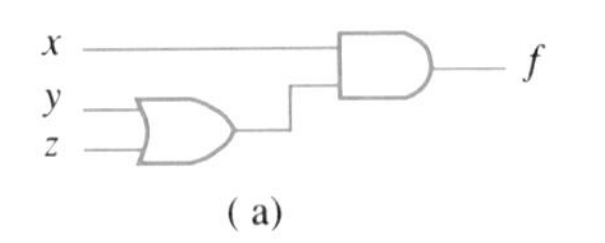

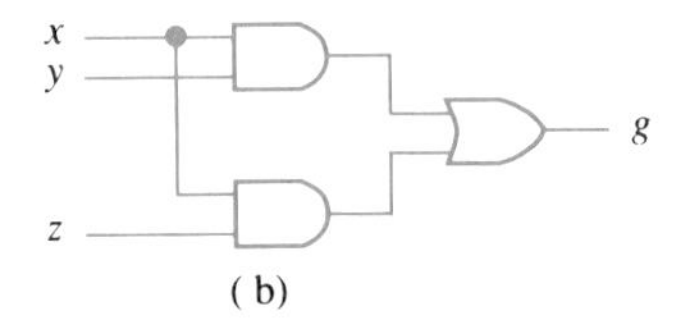

Figure 5.18 Illustrating the distributive law (primal form). (a) $f = x(y + z)$; (b) $g = (xy) + (xz)$.

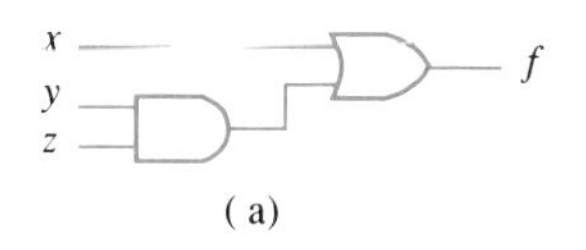

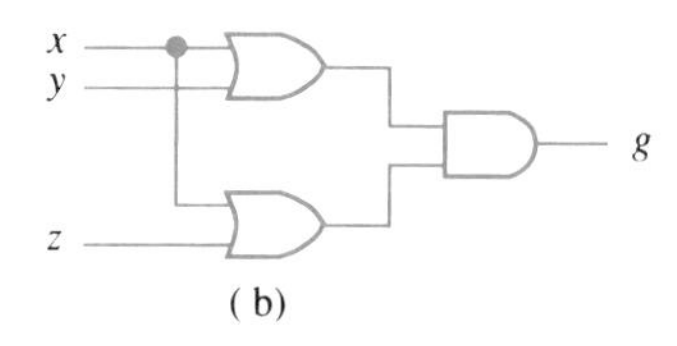

Figure 5.19 Illustrating the distributive law (dual form). (a) $f = x + (yz)$; (b) $g = (x + y)(x + z)$.

In the notation of logical circuits, this says that the two circuits of Figure 5.18 yield the same Boolean function.

The law of duality says to replace all AND gates with OR gates, and all OR gates with AND gates. In the case of the distributive law, the result is the pair of circuits in Figure 5.19. Thus the dual form of the distributive law is

$$x + (yz) = (x + y)(x + z),$$

as given in Figure 5.19.

EXAMPLE 5.15 The following is a valid Boolean identity:

$$x(y + \bar{y}z) + \bar{y}z = xy + \bar{y}z.$$

What is the dual form of this identity?

SOLUTION By interchanging $\cdot$ and $+$, we arrive at the following:

$$(x + (y \cdot (\bar{y} + z)))(\bar{y} + z) = (x + y) \cdot (\bar{y} + z).$$

This example should be studied carefully. The ability to go from primal to dual form is a valuable tool that often helps in simplifying complicated Boolean expressions.

By now, you should be wondering why the law of duality works. Actually, it's quite easy to see why it's true, if you look at it the way electrical engineers do. To an electrical engineer, an AND gate, for example, is a physical device that accepts as input two voltages, V_1 and V_2, each of which can be "high" or "low," and which, in response, produces another voltage, again either H or L. The truth table for this device is given below:

V_1	V_2	Output
H	H	H
H	L	L
L	H	L
L	L	L

If a high voltage is interpreted as 1 and a low voltage as 0, this truth table becomes, of course,

V_1	V_2	Output
1	1	1
1	0	0
0	1	0
0	0	0

which is the truth table for the AND gate of Table 5.6. The system of notation in which high voltage = 1 and low voltage = 0 is called the **positive logic** system. But there's no law that says H = 1 and L = 0! If the *other* correspondence, viz.

H = 0 and L = 1 is made—this is called the **negative logic** system—the truth table becomes instead

V_1	V_2	Output
0	0	0
0	1	1
1	0	1
1	1	1

which is the truth table for an OR gate! Thus the same *physical* device can serve either as an AND gate or an OR gate, depending on which system of notation is used. Similarly, a physical device whose H/L truth table is

V_1	V_2	Output
H	H	H
H	L	H
L	H	H
L	L	L

serves as an OR gate in positive logic, and is an AND gate in negative logic. Finally, a device whose truth table is

V_1	Output
H	L
L	H

is an inverter in either system.

Given these simple facts, we can see why the law of duality works. A law of Boolean algebra is just a statement that two different logic circuits perform the same function. For example, the distributive law says that the two circuits in Figure 5.18 perform identically. But the diagrams in Figure 5.18 aren't really the circuits, but merely *pictures* of the circuits, using one specific logic system, say positive logic. Now if we switch to negative logic—presto! The circuits don't change, but the pictures do. In fact, Figure 5.19 is just the *negative logic* representation of the circuits in Figure 5.18. Thus the law of duality is just an electrical engineer's way of saying that "a rose by any other name would smell as sweet."

We now return to the 23 laws of Boolean algebra in Table 5.9. As we have said before, merely *stating* a law doesn't make it true—proof is required for each of the 23 laws in Table 5.9. However, the law of duality shows that there are only 12 different laws—and we have already proved the associative law (Example 5.13) and the distributive law (Section 1.2). Two down, ten to go! In the remainder of this section we will prove three more, by three different methods. You will then be asked to prove the remaining seven in the problems at the end of this section.

EXAMPLE 5.16 Prove DeMorgan's law (primal form).

SOLUTION The simplest proof is via truth table:

x	y	$\bar{x}$	$\bar{y}$	$x + y$	$\overline{x + y}$	$\bar{x}\bar{y}$
0	0	1	1	0	1	1
0	1	1	0	1	0	0
1	0	0	1	1	0	0
1	1	0	0	1	0	0

We see that the $\overline{x + y}$ column is identical to the $\bar{x}\bar{y}$ column, and this proves DeMorgan's Law.

EXAMPLE 5.17 Prove the law of redundancy (dual form).

SOLUTION Once again, a proof via truth table is certainly possible; but it's also possible to give an algebraic proof, using two *other* laws, as follows:

$$\begin{aligned} x(\bar{x} + y) &= x\bar{x} + xy && \text{(distributive law)} \\ &= 0 + xy && \text{(complementarity)} \\ &= xy && \text{(identity and commutative laws).} \end{aligned}$$

This proof, besides being pleasantly short, also illustrates the fact that the 23 laws of Table 5.9 aren't independent of each other.

EXAMPLE 5.18 Prove the law of consensus (primal form).

SOLUTION In this case we give a short but slightly tricky proof, which is characteristic of proofs that are sometimes needed in advanced courses in Boolean algebra. We are to show that

$$xy + \bar{x}z + yz = xy + \bar{x}z.$$

Note first that either $yz = 0$ or $yz = 1$. In the first case, if $yz = 0$, then the two sides of the equation are identical, and there is nothing to prove. What if $yz = 1$? Then necessarily $y = 1$ and $z = 1$, and the equation reads

$$x + \bar{x} + 1 = x + \bar{x},$$

which is true since both sides are equal to 1, whether $x = 0$ or $x = 1$.

Problems for Section 5.4

1. Consider the product $y = x_1x_2x_3x_4$ of the four Boolean variables x_1, x_2, x_3, x_4. From our discussion of the associative law, we know that y is the same as the parenthesized expression $((x_1(x_2x_3))x_4)$. How many such equivalent parenthesized expressions are there? (Note that although all such expressions represent the same Boolean function, each expression has a different representation in terms of AND trees such as those in Figure 5.8a.)

2. Refer to Example 5.13. Use a truth table to show that AND is associative.

3. Find the truth sets for the following Boolean functions.
(a) $\overline{x + y + z}$.
(b) $x_1x_2x_3 + \overline{x_2x_3x_4}$.
(c) $(\bar{x}_1 + x_2)(1 + x_3)$.

4. Find the truth sets for the following Boolean functions.
(a) $(x + \bar{y})z$.
(b) $\overline{1 + x_1x_2\bar{x}_3}$.
(c) $x_1x_2x_3 \cdots x_n + \bar{x}_1\bar{x}_2 \cdots \bar{x}_n$.

5. Give the dual form of each of the following Boolean expressions.
(a) $1 + xy$.
(b) $0 + (\bar{x}_1 + x_2x_3)$.
(c) $\overline{x_1x_2x_3 \cdots x_n}$.

6. Give the dual form of each of the following Boolean expressions.
(a) $\overline{xy} + x$.
(b) 0.
(c) $x_1x_2 + x_3x_4 + x_5x_6$.

In Problems 7–13 you are asked to prove one of the laws in Table 5.9.

7. The commutative law.

8. Weak absorption.

9. Strong absorption. [Prove it in *two* ways: (a) Using truth tables; (b) using some other method.]

10. Involution.

11. Idempotency.

12. Complementarity.

13. Identity.

14. Suppose the Boolean functions f and g involve *different* sets of variables, e.g., $f = x_1 + x_2$, $g = x_2x_3\bar{x}_4$. How do you interpret the statement "The truth set for fg is the intersection of the sets f and g"?

15. What are the set theory and propositional calculus analogues for the Boolean functions represented by the minterm and maxterm gates in Figure 5.11?

16. We have seen in the text that the AND and OR gates of Table 5.6 satisfy both the associative and commutative laws. Investigate these laws for the other four two-input gates of Table 5.6 by completing the following table.

Gate	Associative?	Commutative?
AND	Yes	Yes
OR	Yes	Yes
NAND	?	?
NOR	?	?
XOR	?	?
XNOR	?	?

17. By consulting a dictionary if necessary, see if you can give plausible explanations for the *names* of the following laws from Table 5.9.
(a) Involution.
(b) Idempotency.
(c) Redundancy.
(d) Consensus.

18. In our discussion of the law of duality we saw that AND and OR gates are really the same physical devices, described in different notational systems. For each of the following four gates, determine what it becomes when the notational system is changed from postive to negative logic.
(a) NAND. (b) NOR. (c) XOR. (d) XNOR.

19. Why does the title of Table 5.9 say "Twenty-three" laws, when there are 24 entries?

In Problems 20–26 you are asked to prove the given Boolean identity *without* using truth tables (as in Examples 5.17 and 5.18).

20. $g(f + \bar{f}h) + \bar{f}h = fg + \bar{f}h$.

21. $(f + \bar{g})(f + \bar{h})(g + h) = f(g + h)$.

22. $(f + g + h)(f + g + \bar{h})(f + \bar{g} + h) = f + gh$.

23. $h(\overline{f + g}) + \bar{f}gh = \bar{f}h$.

24. $fg + (\overline{f\bar{g} + h}) + \bar{f}h = \bar{f} + g$.

25. $(f + g + h)\overline{fgh} = f\bar{g} + g\bar{h} + h\bar{f}$.

26. The identity given in Example 5.15.

27. Suppose that the Boolean functions $f_1, \ldots, f_k$ have truth sets $A_1, \ldots, A_k$. What statements about the f_i are equivalent to the following statements about the A_i?
(a) The sets $A_1, \ldots, A_n$ are pairwise disjoint.
(b) The sets $A_1, \ldots, A_n$ are pairwise disjoint and their union is the entire space.
(c) At least one A_i is nonempty.
(d) At least one A_i is not the entire space.
(e) The intersection of the A_i is empty.
(f) $A_i \subseteq A_j$.

28. If f and g are Boolean functions, let us say $f \leq g$ if $f(x_1, \ldots, x_n) \leq g(x_1, \ldots, x_n)$ for all choices of the Boolean variables $x_1, x_2, \ldots, x_n$. For example, $x_1x_2 \leq x_1 + x_2$.
(a) Show that $\leq$ is a partial order, as defined in Section 1.4.
(b) Draw the Hasse diagram for the partial ordering of the 16 Boolean functions of two variables. Where have you seen this diagram before?

29. Continuing Problem 28, if f and g are two Boolean functions, prove that $f \leq g$ if and only if $f + g = g$.

30. Consider the following statements for Boolean functions f and g:

$$P_1: fg = 0, \qquad P_2: f = \bar{g}.$$

(a) Does P_1 imply P_2? Prove or give an explicit counterexample.
(b) Does P_2 imply P_1? Prove or give an explicit counterexample.

31. We said in the text that every law of Boolean algebra has a corresponding version in set theory and in propositional calculus. How is the law of duality stated in set theory? In propositional calculus?

32. Consider the following high/low truth table:

x	y	z
H	H	H
H	L	H
L	H	H
L	L	L

In positive logic, which (if any) of the gates in Table 5.6 does this table represent? In negative logic?

33. Consider another high/low truth table:

x	y	z
H	H	L
H	L	H
L	H	H
L	L	L

See if you can design a circuit that realizes this function, using the basic gates of Table 5.6, in (a) positive logic and (b) negative logic.

34. What happens if we apply the law of duality to the 12 dual-form laws in Table 5.9? Do we get 12 new laws?

35. Suppose we have three Boolean variables x, y, and z, such that for particular values of x, y, and z we have $x + y = x + z$. Is it safe to conclude that $y = z$?

36. The *generalized* DeMorgan's law says that for any $n \geq 2$

$$\overline{x_1 + x_2 + \cdots + x_n} = \bar{x}_1\bar{x}_2 \cdots \bar{x}_n.$$

(a) Prove this law for all $n \geq 2$, using mathematical induction.
(b) What is the *dual* form of the generalized DeMorgan's law?

37. For each of the following three Boolean equations, say whether it is true or false. If it is true, give a proof. If false, give a counterexample.
(a) $\bar{x} + xy = \bar{x} + y$.
(b) $x\bar{y} + x\bar{z} + xy = x + yz$.
(c) $(x + \bar{y})(x + \bar{z})(x + y) = x(y + z)$.

38. Express each of the following laws of Boolean algebra (from Table 5.9) as a *theorem in the propositional calculus.*
(a) Associative law (dual form).
(b) Distributive law (primal form).
(c) Commutative law (primal form).
(d) Commutative law (dual form).
(e) Strong absorption (primal form).
(f) Consensus (dual form).
(g) Weak absorption (dual form).
(h) Complementary (primal form).

39. The following WFP is a theorem of the propositional calculus: $(\sim x \rightarrow \sim(x \wedge y))$. Which of the 23 laws of Boolean algebra in Table 5.9 does it most closely correspond to? Explain.

40. Which of the 23 laws of Boolean algebra (Table 5.9) does the theorem of the excluded middle in the box on page 185 most closely correspond to? Explain.

5.5 Simplifying Boolean Functions with Karnaugh Maps

In Section 5.3 we learned that any Boolean function can be built from ANDs, ORs, and NOTs, via the minterm expansion. However, a practicing engineer who needs to build a specific Boolean function will almost never be satisfied with the minterm expansion, because as a rule it requires many more gates than are necessary. The laws of Boolean algebra of Section 5.4 will almost always allow us to simplify the minterm expansion considerably. As an illustration, consider the Boolean function f of three variables whose truth set is

(5.43) $$A = \{000, 001, 010, 011, 110, 111\}$$

According to the rules of Section 5.3, the minterm expansion of this function is

(5.44) $$f = \bar{x}_1\bar{x}_2\bar{x}_3 + \bar{x}_1\bar{x}_2x_3 + \bar{x}_1x_2\bar{x}_3 + \bar{x}_1x_2x_3 + x_1x_2\bar{x}_3 + x_1x_2x_3.$$

This expression represents a circuit that requires 12 ANDs, 5 ORs, and 9 NOTS.

Now consider the following sequence of simplifications:

$$
\begin{aligned}
f &= \bar{x}_1\bar{x}_2(\bar{x}_3 + x_3) + \bar{x}_1 x_2(\bar{x}_3 + x_3) + x_1 x_2(\bar{x}_3 + x_3) && \text{(by the distributive law)} \\
&= \bar{x}_1\bar{x}_2 + \bar{x}_1 x_2 + x_1 x_2 && \text{(by the complementarity and identity laws)} \\
&= \bar{x}_1(\bar{x}_2 + x_2) + x_1 x_2 && \text{(distributive law, again)} \\
&= \bar{x}_1 + x_1 x_2 && \text{(complementarity and identity, again)} \\
&= \bar{x}_1 + x_2 && \text{(redundancy law).}
\end{aligned}
$$

Amazing! The expression $f = \bar{x}_1 + x_2$ represents a circuit requiring *no* ANDs, one OR, and one NOT, and computes exactly the same function as the long minterm expansion in Equation (5.44)! This is only one example, but it shows that it can be well worth our while to try to simplify the minterm expansion using the laws of Boolean algebra. However, knowing which laws to apply and in which order is a bit of a black art. Our goal in this section is to present a simple procedure which can be learned by anyone, and which usually leads to a significant simplification of the minterm expansion. And although it won't always produce the most efficient possible form of a given Boolean function, it is quite commonly used by practicing engineers anyway, since as a rule it is quite difficult to do much better. The procedure is called the *Karnaugh* map method.* It is normally applied only to Boolean functions of two, three, or four variables. It can, with difficulty, be applied to functions of five or more variables, but we will not discuss this (or any of several other known algorithms for simplifying Boolean functions) here.

A **Karnaugh map** is a two-dimensional representation of the truth table of a Boolean function. For example, in Figure 5.20 we show the ordinary truth table and the Karnaugh map for the function $f(x_1, x_2, x_3)$ given in (5.44). In Figure 5.20(a) we see that the function f has six 1s in its truth table. In Figure 5.20(b) these six 1s are placed in the appropriate six cells of the Karnaugh map. It would

x_1	x_2	x_3	$f(x_1, x_2, x_3)$
0	0	0	1
0	0	1	1
0	1	0	1
0	1	1	1
1	0	0	0
1	0	1	0
1	1	0	1
1	1	1	1

(a)

x_1 \ x_2x_3	00	01	11	10
0	1	1	1	1
1			1	1

(b)

Figure 5.20 (a) Truth table and (b) Karnaugh map for the Boolean function whose minterm expansion is given by (5.44).

* Maurice Karnaugh (1924–), American telecommunications engineer. He developed the Karnaugh map at Bell Laboratories in 1953 while studying the application of digital logic to the design of telephone switches.

	0	1
0	$\bar{x}_1\bar{x}_2$	$\bar{x}_1x_2$
1	$x_1\bar{x}_2$	x_1x_2

(a)

	00	01	11	10
0	$\bar{x}_1\bar{x}_2\bar{x}_3$	$\bar{x}_1\bar{x}_2x_3$	$\bar{x}_1x_2x_3$	$\bar{x}_1x_2\bar{x}_3$
1	$x_1\bar{x}_2\bar{x}_3$	$x_1\bar{x}_2x_3$	$x_1x_2x_3$	$x_1x_2\bar{x}_3$

(b)

	00	01	11	10
00	$\bar{x}_1\bar{x}_2\bar{x}_3\bar{x}_4$	$\bar{x}_1\bar{x}_2\bar{x}_3x_4$	$\bar{x}_1\bar{x}_2x_3x_4$	$\bar{x}_1\bar{x}_2x_3\bar{x}_4$
01	$\bar{x}_1x_2\bar{x}_3\bar{x}_4$	$\bar{x}_1x_2\bar{x}_3x_4$	$\bar{x}_1x_2x_3x_4$	$\bar{x}_1x_2x_3\bar{x}_4$
11	$x_1x_2\bar{x}_3\bar{x}_4$	$x_1x_2\bar{x}_3x_4$	$x_1x_2x_3x_4$	$x_1x_2x_3\bar{x}_4$
10	$x_1\bar{x}_2\bar{x}_3\bar{x}_4$	$x_1\bar{x}_2\bar{x}_3x_4$	$x_1\bar{x}_2x_3x_4$	$x_1\bar{x}_2x_3\bar{x}_4$

(c)

Figure 5.21 Karnaugh "templates" for two, three, and four variables; each cell corresponds to the indicated minterm.

be logical to place 0s in the remaining two cells, corresponding to the two 0s in the truth table, but it is customary not to do so. Each one of the eight cells in the Karnaugh map corresponds to one of the eight possible minterms. For example, the cell corresponding to $x_1 = 1$, $x_2 = 0$, $x_3 = 1$ (which is blank in Figure 5.20b) corresponds to the minterm $x_1\bar{x}_2x_3$. The complete correspondence between cells and minterms is shown in Figure 5.21(b), where we also show the "templates" for Karnaugh maps in two variables (Figure 5.21a) and four variables (Figure 5.21c).

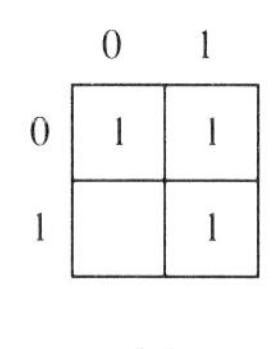

EXAMPLE 5.19 Use the templates in Figure 5.21 to find the Karnaugh maps for the functions whose minterm expansions are given below.

(a) $f(x_1, x_2) = \bar{x}_1\bar{x}_2 + \bar{x}_1x_2 + x_1x_2$.
(b) $f(x_1, x_2, x_3) = \bar{x}_1\bar{x}_2\bar{x}_3 + x_1\bar{x}_2\bar{x}_3 + x_1x_2x_3 + x_1x_2\bar{x}_3$.
(c) $f(x_1, x_2, x_3, x_4) = \bar{x}_1x_2\bar{x}_3x_4 + \bar{x}_1x_2x_3x_4 + x_1x_2\bar{x}_3x_4 + x_1x_2x_3x_4 + x_1\bar{x}_2\bar{x}_3x_4 + x_1\bar{x}_2x_3x_4$.

SOLUTION This is simply a matter of putting a 1 in the cells of the templates in Figure 5.21 corresponding to each of the minterms. The results are shown in Figure 5.22.

	0	1
0	1	1
1		1

(a)

	00	01	11	10
0	1			
1	1		1	1

(b)

	00	01	11	10
00				
01		1	1	
11		1	1	
10		1	1	

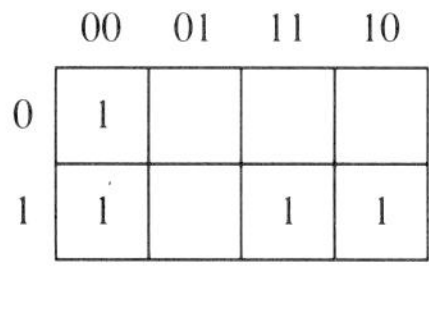

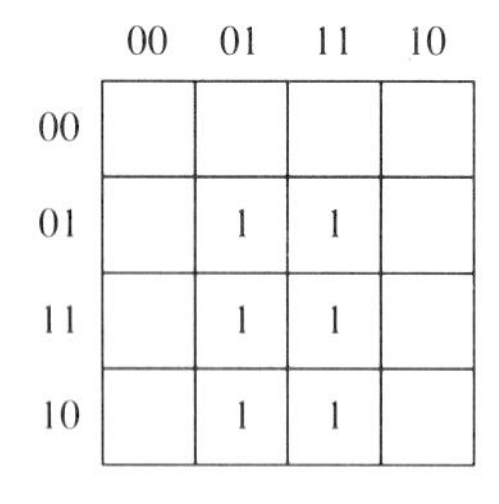

(c)

Figure 5.22 Solution to Example 5.19.

The basic reason Karnaugh maps are so useful is that certain very simple Boolean functions also have very simple Karnaugh maps. These simple functions are the **product functions**, which are simple products of some or all of the variables and their complements. For example, x_1, $\bar{x}_2x_4$, and xyz are product functions, but $x_1 + \bar{x}_2$ and $xy + zw$ are not.

As an example, consider the product function $f(x_1, x_2, x_3) = x_1x_3$. This function accepts the two inputs 101 and 111, and its Karnaugh map is shown in Figure 5.23(a). The thing to notice about Figure 5.23(a) is that the minterms involved (i.e., the 1s in the Karnaugh map) lie in a *rectangular block*, in this case a 1×2 block. As another example, consider the function $g(x_1, x_2, x_3) = \bar{x}_2$, which accepts the set {000, 001, 100, 101} and whose Karnaugh map appears in Figure 5.23(b). Again we see that the minterms are confined to a rectangular block, in this case a 2×2 block.

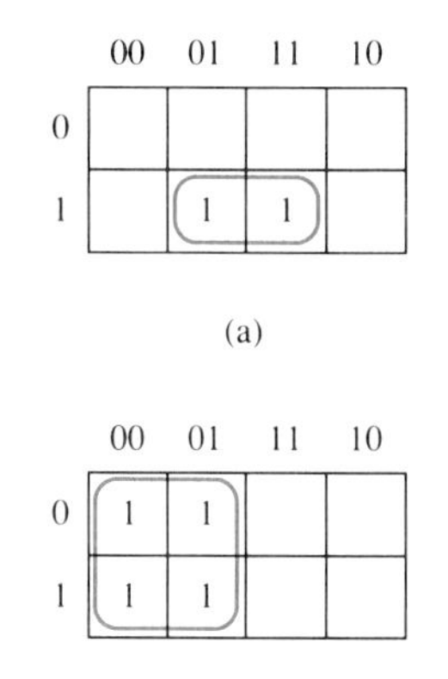

Figure 5.23 (a) The Karnaugh map for $f(x_1, x_2, x_3) = x_1x_3$. (b) The Karnaugh map for $g(x_1, x_2, x_3) = \bar{x}_2$.

It turns out that *every* product function has a Karnaugh map whose minterms are confined to a rectangular block whose sides are of size, 1, 2, or 4, and conversely, every such block of minterms corresponds to a product function. It is important to be able to identify a product function from its Karnaugh map. For example, Figure 5.24 shows the Karnaugh map for an unknown product function. Which function is it?

To identify the product function represented by the Karnaugh map in Figure 5.24, we reason as follows. The truth set for the given function is {011, 111} In this truth set, the first variable x_1 assumes the values 0 and 1, the second variable x_2 assumes only the value 1, and the third variable also assumes only the value 1. Thus we can represent the truth set as {*11}, where * is a *wildcard* symbol that can represent either 0 or 1. Now if the unknown product function involved the variable x_1 (as in $x_1\bar{x}_3$ or $\bar{x}_1\bar{x}_2x_3$), it would only accept inputs for which x_1 assumed one particular value ($x_1 = 1$ in the case of $x_1\bar{x}_3$, and $x_1 = 0$ in the case of $\bar{x}_1\bar{x}_2x_3$). But the truth set {*11} contains inputs with both $x_1 = 0$ *and* $x_1 = 1$. We conclude that x_1 is *not* involved. Is x_2 involved? Yes, it must be; since $x_2 = 1$ for all inputs in the truth set, x_2 (and not $\bar{x}_2$) is a term in the unknown product. Finally, we see that x_3 (and not $\bar{x}_3$) must also be involved, since $x_3 = 1$ for all inputs in the truth set. Therefore the unknown product function in Figure 5.24 is x_2x_3.

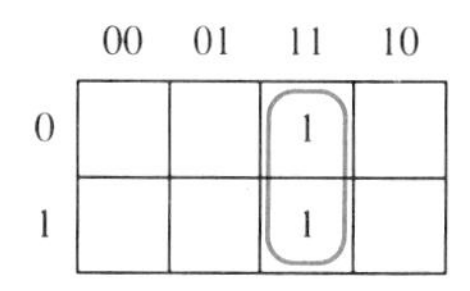

Figure 5.24 The Karnaugh map for an unknown product function.

With a little practice, you'll be able to identify the product function corresponding to a block of minterms almost immediately.

EXAMPLE 5.20 Identify each of the three-variable product functions represented by the Karnaugh maps of Figure 5.25.

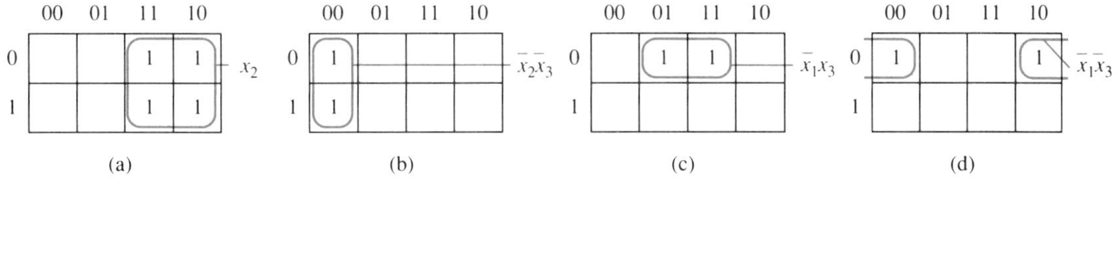

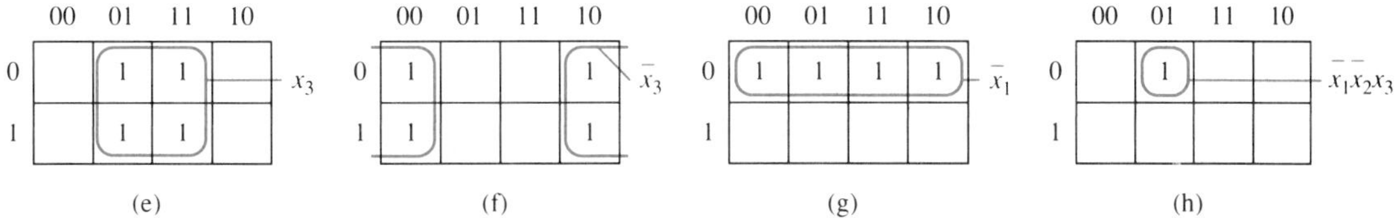

Figure 5.25 Karnaugh maps for Example 5.20.

SOLUTION

(a) Here the truth set is {011, 010, 111, 110}. Using the wildcard symbol *, we can represent the truth set as {*1*}, meaning that x_1 assumes the values 0 and 1, x_2 assumes only the value 1, and x_3 assumes the values 0 and 1. It follows that the product function does not include x_1, does include x_2, and does not include x_3. It is thus simply x_2.
(b) Here the truth set is {000, 100} = {*00}; it follows that the function is $\bar{x}_2\bar{x}_3$.
(c) Truth set = {001, 011} = {0*1}; function = $\bar{x}_1 x_3$.
(d) Notice that in this case the block of 1s in the Karnaugh map "goes around the corner." It is as if the Karnaugh map were pasted on the surface of a cylinder with the two vertical edges just touching. Anyway, the truth set is {000, 010} = {0*0}, and so the product function is $\bar{x}_1\bar{x}_3$.
(e) Truth set = {001, 011, 101, 111} = {**1}; function = x_3.
(f) Here the block goes around the corner again. Truth set = {000, 100, 010, 110} = {**0}; function = $\bar{x}_3$.
(g) Truth set = {000, 001, 011, 010} = {0**}; function = $\bar{x}_1$.
(h) Here the truth set is {001}, which cannot be simplified using the * notation. The product function is the single minterm $\bar{x}_1\bar{x}_2 x_3$. ■

In Example 5.20 we considered only three-variable Karnaugh maps; in the next example, we'll see how to identify product functions in two- and four-variable Karnaugh maps. In the two-variable map, any 1 × 2 or 2 × 1 block of minterms corresponds to a product function. In the four-variable map, product functions correspond to rectangular blocks with dimensions 1, 2, or 4.

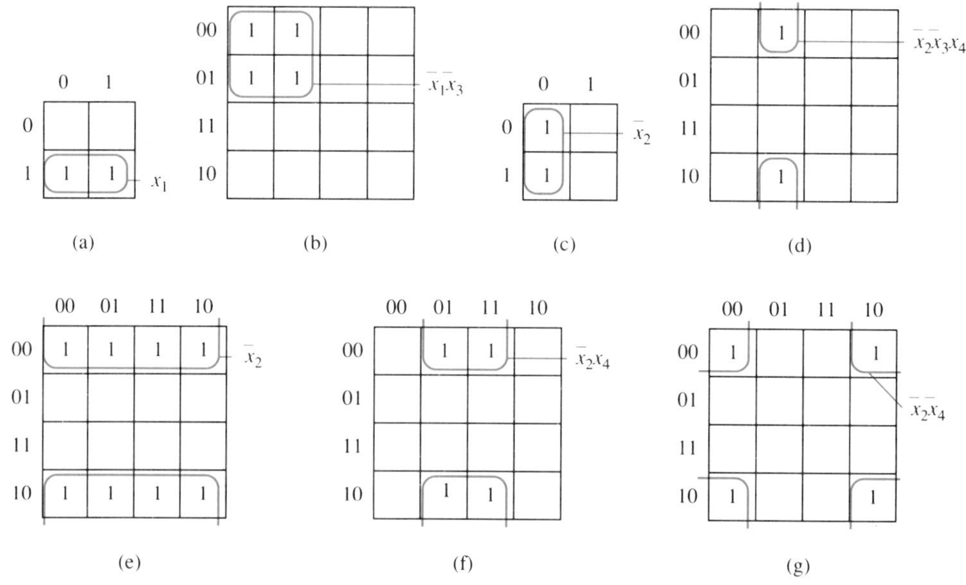

Figure 5.26 Karnaugh maps for Example 5.21.

EXAMPLE 5.21 Identify the product function corresponding to each of the Karnaugh maps in Figure 5.26.

SOLUTION

(a) Truth set = {10, 11} = {1*}; function = x_1.
(b) Truth set = {0000, 0001, 0100, 0101} = {0*0*}; function = $\bar{x}_1\bar{x}_3$.
(c) Truth set = {00, 10} = {*0}; function = $\bar{x}_2$.
(d) Truth set = {0001, 1001} = {*001}; function = $\bar{x}_2\bar{x}_3x_4$.
(e) Here the 2 × 4 block goes around the corner *vertically*. Truth set = {0000, 0001, 0011, 0010, 1000, 1001, 1011, 1010} = {*0**}; function = $\bar{x}_2$.
(f) Truth set = {0001, 0011, 1001, 1011} = {*0*1}; function = $\bar{x}_2x_4$.
(g) Here's a 2 × 2 block that goes around *two* corners! Truth set = {0000, 0010, 1000, 1010} = {*0*0}; function = $\bar{x}_2\bar{x}_4$.

In Examples 5.20 and 5.21 we have seen how to use the Karnaugh map to improve upon the minterm expansion for one important class of Boolean functions, the product functions. However, most Boolean functions aren't products. For example, the function described in Figure 5.20 isn't a product, since its minterms don't form a product block like the ones in Figure 5.25. Nevertheless, we can still use the Karnaugh map to simplify the function, as shown in Figure 5.27.

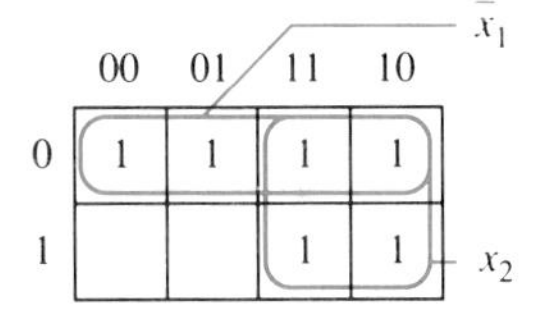

Figure 5.27 Why the function of Figure 5.20 equals $\bar{x}_1 + x_2$.

In Figure 5.27 we see that the minterms in the Karnaugh map, although not forming a *single* product block, can be decomposed into the *union of two* product blocks, one representing the function $\bar{x}_1$ and one representing the function x_2. Therefore the function obtained by ORing together these two product functions, viz. $\bar{x}_1 + x_2$, will have a Karnaugh map which is identical to the one in Figure 5.20, and so the function in Figure 5.20 (whose minterm expansion is given in Equation (5.44) must be equal to $\bar{x}_1 + x_2$. Of course we already proved this at the beginning of the section, using five laws of Boolean algebra—but now a quick glance at the Karnaugh map in Figure 5.27 tells us the same thing!

The technique illustrated in Figure 5.27 can, in principle, be applied to any Boolean function. The idea is to cover the minterms using as few product blocks as possible, and then to write the original functions as the sum of the corresponding product functions. The great advantage of this method is that it fully utilizes our human ability to recognize patterns. It is best learned by practice, but before we give another example, there is one rule of thumb worth mentioning immediately. *The product blocks used to cover the minterms should all be as large as possible.* In Figure 5.28 we see an example of what can happen if product blocks that are too small are chosen. The decomposition shown in Figure 5.28 leads to the expression

$$f(x_1, x_2, x_3) = \bar{x}_1\bar{x}_2 + x_2x_3 + x_2\bar{x}_3$$

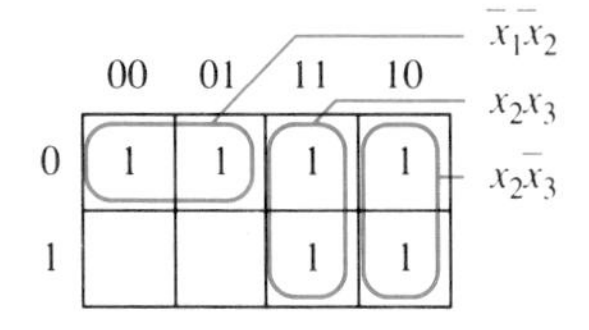

Figure 5.28 The function in Figure 5.20 can also be written as $\bar{x}_1\bar{x}_2 + x_2x_3 + x_2\bar{x}_3$.

for the function in Figure 5.20. This expression isn't wrong, but it is much inferior (in terms of complexity) to the expression $\bar{x}_1 + x_2$ obtained via the decomposition in Figure 5.27. The problem is, of course, that the product blocks in Figure 5.28

are too small: The $\bar{x}_1\bar{x}_2$ block of Figure 5.28 is contained in the $\bar{x}_1$ block of Figure 5.27, and the x_2x_3 and $x_2\bar{x}_3$ blocks in Figure 5.28 are both contained in the x_2 block of Figure 5.27. In general, the larger the blocks, the more $*$'s in the resulting truth set, and the simpler the resulting expression.

In general, any product function whose minterms form a subset of the minterms of a given Boolean function is called an **implicant** of the function. For example, the product $\bar{x}_1\bar{x}_2$ is an implicant of the function* whose Karnaugh map is shown in Figure 5.28. Implicants that cover as many cells as possible (i.e., whose minterms aren't contained in some larger implicant) are called **prime implicants**. For the function of Figures 5.20, 5.27, and 5.28, there are only two prime implicants, viz. $\bar{x}_1$ and x_2, as shown in Figure 5.27. As we have seen in Figure 5.28, it never pays to use a nonprime implicant, since a prime implicant can always be used to cover the same minterms, and will correspond to a less complex product function. Therefore the Karnaugh map method can be summarized as follows:

Cover the minterms with as few prime implicants as possible.

The next example will give us practice using this technique.

EXAMPLE 5.22 Use the Karnaugh map method to simplify the six Boolean functions in Figure 5.29.

SOLUTION We show the prime implicants for each of the six functions in Figure 5.30. In cases (a), (b), (c), (d), and (e), we must use *all* of the prime

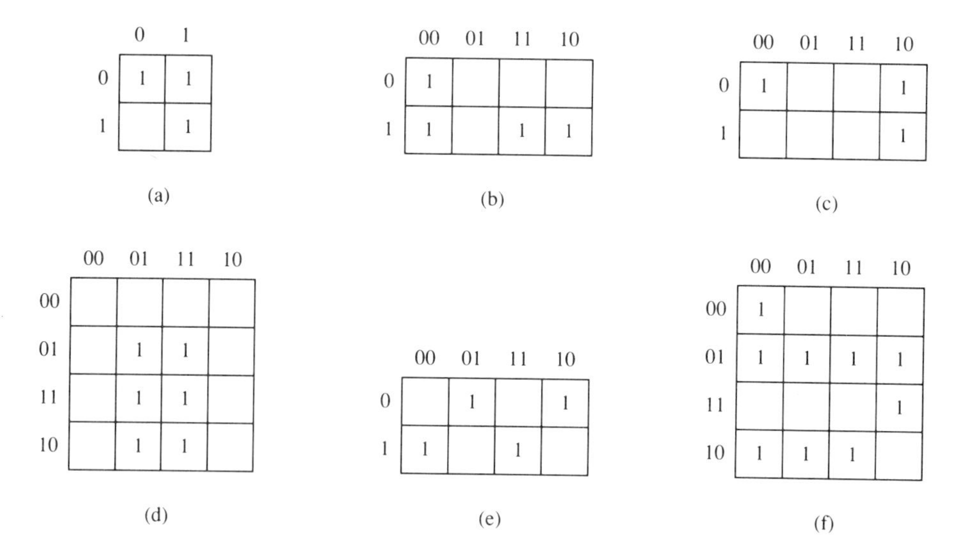

Figure 5.29 Karnaugh maps for Example 5.22.

* If we denote this function by $f(x_1, x_2, x_3)$, then $\bar{x}_1\bar{x}_2 = 1$ *implies* that $f(x_1, x_2, x_3) = 1$—hence the name *implicant*.

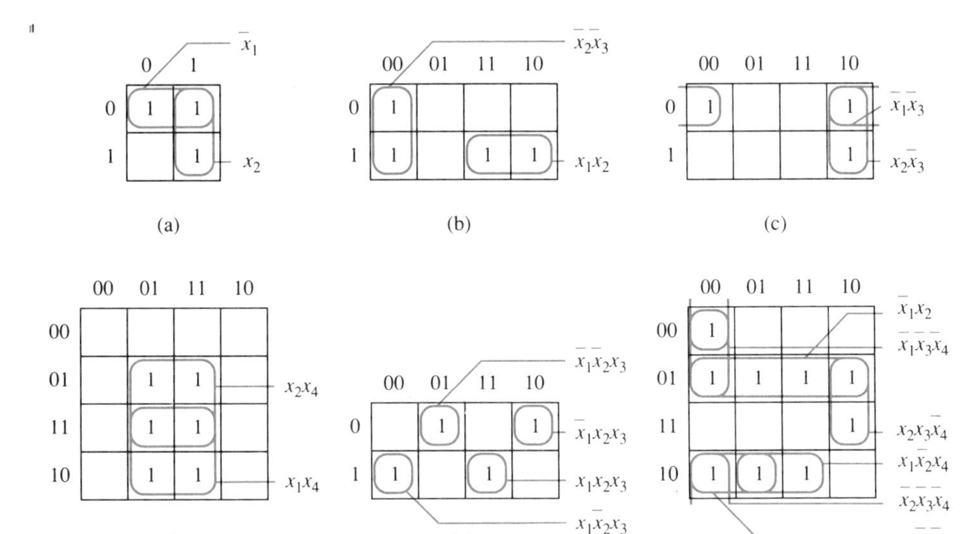

Figure 5.30 Solutions to Example 5.22.

implicants to cover the minterms, and so the resulting simplified expressions are as follows:

(a) $\bar{x}_1 + x_2$. (b) $\bar{x}_2\bar{x}_3 + x_1x_2$. (c) $\bar{x}_1\bar{x}_3 + x_2\bar{x}_3$.
(d) $x_2x_4 + x_1x_4$ (notice that in this case the minterms form a 3×2 rectangular block, but not a *product* block, since a product block must have sides of size 1, 2, or 4).
(e) $\bar{x}_1\bar{x}_2x_3 + \bar{x}_1x_2\bar{x}_3 + x_1\bar{x}_2\bar{x}_3 + x_1x_2x_3$.
(f) This example is a bit more difficult. The problem is that not all of the prime implicants are needed to cover the minterms. We can make some progress by noticing that the prime implicant $\bar{x}_1x_2$ is definitely needed, since it is the only prime implicant that covers the minterms 0101 and 0111. A prime implicant that covers minterms covered by no other prime implicant is called an **essential prime implicant**. We see that $x_2x_3\bar{x}_4$ is essential, since it is the only prime implicant that covers 1110; and $x_1\bar{x}_2x_4$ is also essential, since it is the only prime implicant that covers 1011. The other three prime implicants, viz. $\bar{x}_1\bar{x}_3\bar{x}_4$, $\bar{x}_2\bar{x}_3\bar{x}_4$, and $x_1\bar{x}_2\bar{x}_3$, are not essential, since each minterm covered by one of them is also covered by some other prime implicant. However, since we already know that $\bar{x}_1x_2$, $x_2x_3\bar{x}_4$, and $x_1\bar{x}_2x_4$ are essential, we may as well remove them, together with the minterms they cover, from the Karnaugh map of Figure 5.30(f). The result is shown in Figure 5.31. Now we can clearly see that the remaining minterms can be covered with just one more prime implicant, viz. $\bar{x}_2\bar{x}_3\bar{x}_4$. Therefore the smallest number of prime implicants needed to cover the minterms is *four*, and the corresponding expression for the function in Figure 5.29(f) as a sum of products is

$$\bar{x}_1x_2 + x_2x_3\bar{x}_4 + x_1\bar{x}_2x_4 + \bar{x}_2\bar{x}_3\bar{x}_4.$$

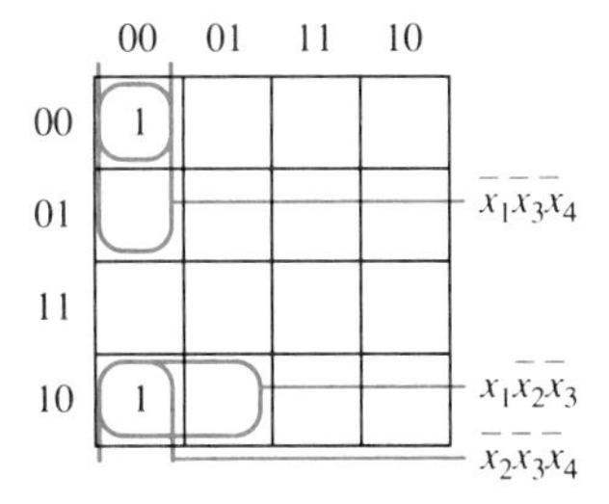

Figure 5.31 The Karnaugh map of Figure 5.30(f) after the essential prime implicants $\bar{x}_1x_2$, $x_2x_3\bar{x}_4$, and $x_1\bar{x}_2x_4$ have been removed.

In the next example, we'll see that sometimes there can be more than one way to cover the minterms with a minimum number of prime implicants.

EXAMPLE 5.23 Simplify the function whose Karnaugh map is shown in Figure 5.32(a).

SOLUTION In Figure 5.32(b) we see that this function has six prime implicants, which we have denoted by A, B, C, D, E, and F. The corresponding product functions are as follows:

$$A: \bar{x}_1 x_3 \quad C: x_1 \bar{x}_3 \quad E: \bar{x}_1 \bar{x}_2 \bar{x}_4$$
$$B: x_2 x_3 \quad D: x_1 x_2 \quad F: \bar{x}_2 \bar{x}_3 \bar{x}_4$$

Of these six prime implicants, only two are essential, A and C. A is essential, because it covers the minterm $\bar{x}_1\bar{x}_2x_3x_4$ (0011), which no other prime implicant covers. C is essential, because it covers $x_1\bar{x}_2\bar{x}_3x_4$ (1001), which no other prime implicant covers. The other four prime implicants are not essential, since every minterm covered by any one of them is covered by some other prime implicants as well. For example, B isn't essential since its four minterms (0111, 0110, 1111, 1110) are also covered by A or D.

It follows that any covering of the minterms by prime implicants must include A and C. In Figure 5.32(c) we see what is left of the Karnaugh map after the prime implicants A and C and the eight minterms they cover are removed. There are only three remaining uncovered minterms, viz. 0000, 1111, and 1110, but there are several possible ways of covering them with the remaining prime implicants B, D, E, and F.

In order to cover 0000 we need either E or F; to cover 1110, we need either B or D; and to cover 1111, we also need either B or D. Thus to cover the remaining three minterms, we need a set of prime implicants that includes

$$(E \text{ or } F) \text{ and } (B \text{ or } D),$$

or in Boolean algebra shorthand,

$$(E + F)(B + D).$$

Here is an opportunity to apply Boolean algebra to Boolean algebra! If we expand $(E + F)(B + D)$, using the distributive and commutative laws, we get

$$BE + DE + BF + DF,$$

which means that we need *two* prime implicants to cover the remaining minterms; either B *and* E, or D *and* E, or B *and* F, or D *and* F. There are thus *four* minimal

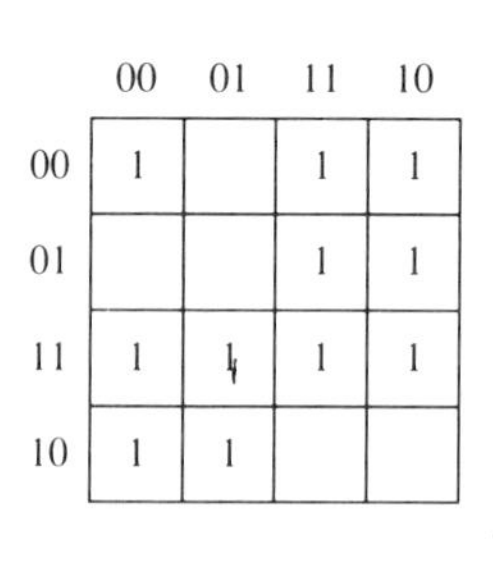

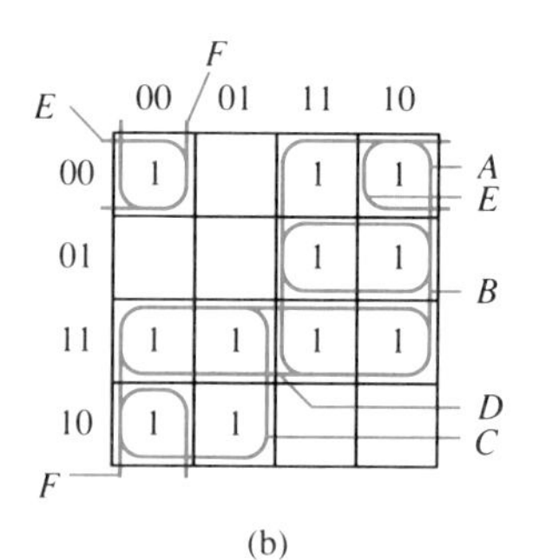

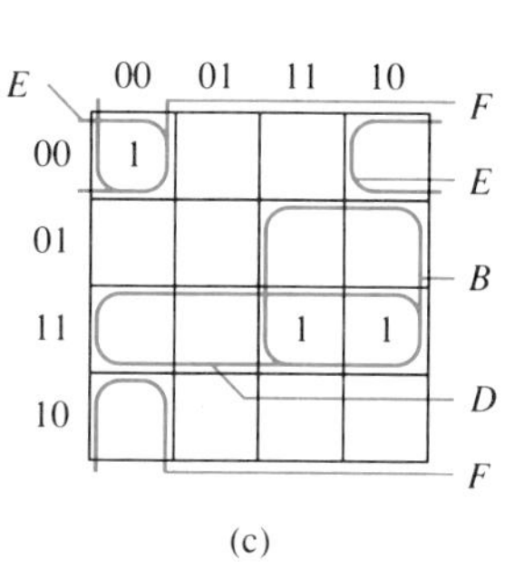

Figure 5.32 Karnaugh maps for Example 5.23.

expansions:

$$ABCE:\ \bar{x}_1x_3 + x_2x_3 + x_1\bar{x}_3 + \bar{x}_1\bar{x}_2\bar{x}_4$$

$$ACDE:\ \bar{x}_1x_3 + x_1\bar{x}_3 + x_1x_2 + \bar{x}_1\bar{x}_2\bar{x}_4$$

$$ABCF:\ \bar{x}_1x_3 + x_2x_3 + x_1\bar{x}_3 + \bar{x}_2\bar{x}_3\bar{x}_4$$

$$ACDF:\ \bar{x}_1x_3 + x_1\bar{x}_3 + x_1x_2 + \bar{x}_2\bar{x}_3\bar{x}_4.$$

That is nearly all we have to say about simplifying Boolean functions. The Karnaugh map method will usually, though not always, produce an expression that is much simpler than the minterm or maxterm expression. It is sometimes possible to do still better (see Problem 38), although there is no simple rule that tells how this can be done. There is one trick, however, that is usually worth a try and which is closely related to the notion of *duality* we discussed in the last section: *Complement* the function, use the Karnaugh map method on the complemented function, and then complement again, using DeMorgan's law.

EXAMPLE 5.24 Use this trick on the function whose minterm expansion is $f(x_1, x_2, x_3) = \bar{x}_1\bar{x}_2x_3 + x_1\bar{x}_2x_3 + x_1\bar{x}_2\bar{x}_3$.

SOLUTION In Figure 5.33(a) we see the Karnaugh map for the given function, and in Figure 5.33(b) we see that the three minterms can be covered with two prime implicants. The resulting expression for the original function, viz. $x_1\bar{x}_2 + \bar{x}_2x_3$, requires two ANDs, two NOTs, and one OR. But we can do better.

Figure 5.33(c) shows the Karnaugh map for the *complement* of the function in Figure 5.33(a), and in Figure 5.33(d) we see that the five minterms of the complemented function can be covered with two prime implicants, and so we have

(5.45) $$\bar{f} = \bar{x}_1\bar{x}_3 + x_2.$$

To obtain an expression for f itself, we complement both sides of Equation (5.45):

$$\bar{\bar{f}} = \overline{\bar{x}_1\bar{x}_3 + x_2}.$$

But $\bar{\bar{f}} = f$ by the law of involution; and $\overline{\bar{x}_1\bar{x}_3 + x_2} = \overline{\bar{x}_1\bar{x}_3} \cdot \bar{x}_2$ by DeMorgan's law. Hence

$$\begin{aligned} f &= \overline{\bar{x}_1\bar{x}_3} \cdot \bar{x}_2 \\ &= (\bar{\bar{x}}_1 + \bar{\bar{x}}_3)\bar{x}_2 \qquad \text{(DeMorgan again)} \\ &= (x_1 + x_3)\bar{x}_2 \qquad \text{(involution, again).} \end{aligned}$$

Thus we have arrived at an expression for f that uses only *one* AND, *one* OR, and *one* NOT. In fact, it is exactly the expression we pulled out of thin air in Figure 5.15! We have learned much.

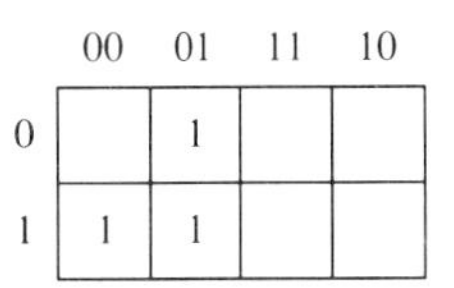

$\bar{x}_1\bar{x}_2x_3 + x_1\bar{x}_2x_3 + x_1\bar{x}_2\bar{x}_3$

(a)

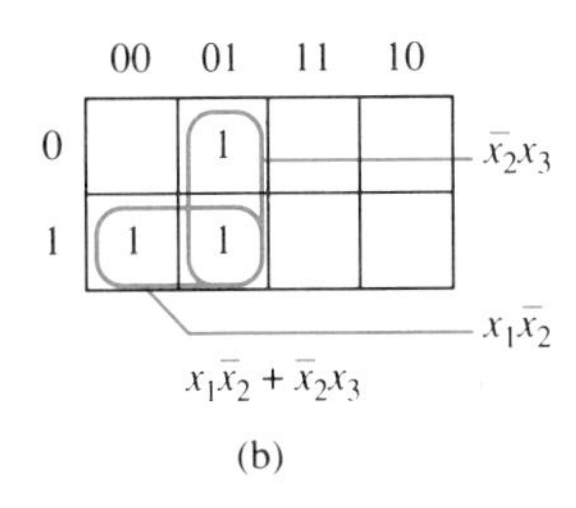

(b)

	00	01	11	10
0	1		1	1
1			1	1

$\bar{x}_1\bar{x}_2\bar{x}_3 + \bar{x}_1x_2x_3 + \bar{x}_1x_2\bar{x}_3 + x_1x_2x_3 + x_1x_2\bar{x}_3$

(c)

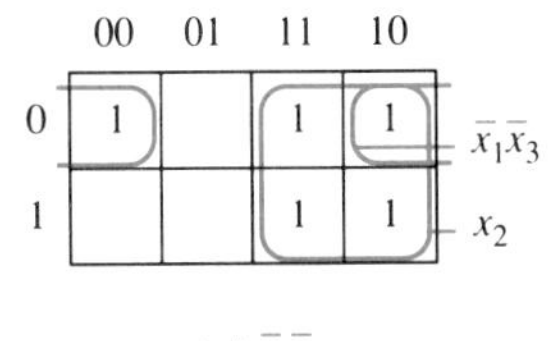

(d)

Figure 5.33 Solving Example 5.24.

Problems for Section 5.5

1. The following sequence of four steps shows how to use the laws of Table 5.9 to simplify one particular Boolean function. Explain which law is being used in each step (four questions).

(a) $\bar{x}_1\bar{x}_2x_3 + x_1\bar{x}_2x_3 + x_1\bar{x}_2\bar{x}_3 = \bar{x}_2(\bar{x}_1x_3 + x_1x_3 + x_1\bar{x}_3)$

(b) $= \bar{x}_2(\bar{x}_1x_3 + x_1(x_3 + \bar{x}_3))$

(c) $= \bar{x}_2(\bar{x}_1x_3 + x_1)$

(d) $= \bar{x}_2(x_3 + x_1)$.

2. Use the templates in Figure 5.21 to find the Karnaugh maps for the functions whose minterm expansions are given below.

(a) $g(x, y) = x\bar{y} + \bar{x}y + xy$.

(b) $g(x_1, x_2, x_3) = \bar{x}_1x_2\bar{x}_3 + x_1\bar{x}_2x_3 + x_1x_2\bar{x}_3$.

(c) $g(x_1, x_2, x_3, x_4) = \bar{x}_1\bar{x}_2\bar{x}_3\bar{x}_4 + \bar{x}_1x_2\bar{x}_3x_4 + x_1x_2x_3x_4 + x_1\bar{x}_2x_3\bar{x}_4$.

In Problems 3–7 you are asked to produce a Karnaugh map for the Boolean functions represented by the indicated logic diagrams.

3. Problem 5.2.3.

4. Problem 5.2.4.

5. Problem 5.2.5.

6. Problem 5.2.6.

7. Problem 5.2.7.

8. A Boolean function can be described with a truth table, a Karnaugh map, and a minterm expansion. In each case, supply the missing two descriptions.

(a) $xyzw + \bar{x}y\bar{z}w + x\bar{y}\bar{z}w$.

(b)

x	y	z	w	Out
0	0	1	0	1
1	1	1	1	1
0	0	1	1	1
Others				0

(c)

	xy	$x\bar{y}$	$\bar{x}\bar{y}$	$\bar{x}y$
z		1	1	
$\bar{z}$			1	1

9. Identify the product function represented by each of the three-variable Karnaugh maps in the figures.

(a)

1	1		

(b)

	1		
	1		

(c)

(d)

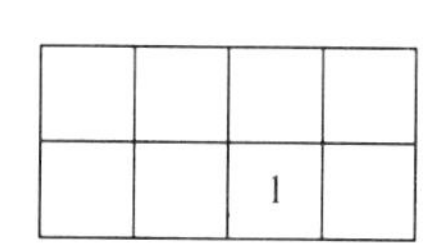

(e)

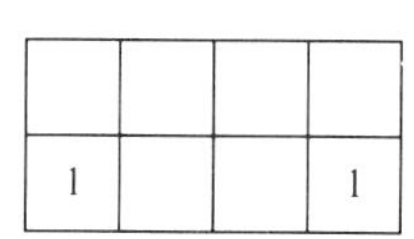

(f)

1	1	1	1

(g)

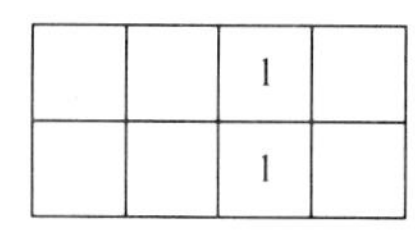

(h)

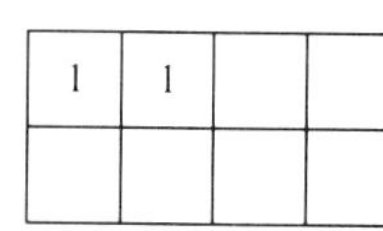

(i)

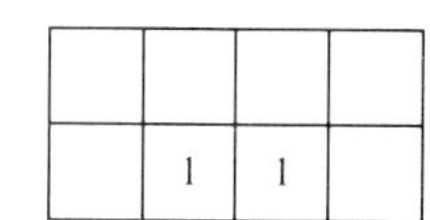

10. Identify the product function represented by each of the two-variable Karnaugh maps in the figures.

(a)

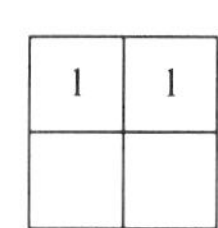

(b)

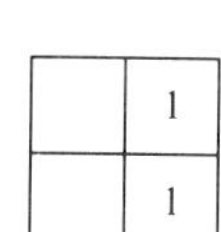

(c)

	1

(d)

1	

11. Identify the product function represented by each of the four-variable Karnaugh maps in the figures.

(a)

	1	1	
	1	1	

(b)

1			1
1			1
1			1
1			1

(c)

	1		
	1		
	1		
	1		

(f)

		1	1
		1	1

(d)

		1	

(g)

		1	1
		1	1

(e)

1			1

(h)

1			1
1			1

12. Give the truth table for each of the product functions described in Problem 9.

13. Give the minterm expansion for each of the product functions described in Problem 9.

14. Identify the Boolean functions represented by the following two Karnaugh maps. In both cases give the simplest possible expression.

(a)

1	1
1	1

(b)

15. (a) How many Boolean functions of the three variables x_1, x_2, x_3 are there?
(b) How many three-variable product functions are there?
(c) What percentage of the three-variable Boolean functions are product functions?

16. Simplify each of the following two Boolean functions, using two methods: the laws of Boolean algebra and the Karnaugh map.
(a) $\bar{x}y + xyz$. (b) $x + \bar{x}yz$.

17. Use the Karnaugh map method to simplify each of the following functions.

(a)

1			
1	1	1	
	1	1	

(b)

	1		1
1	1		

(c)

	1	1	1
1	1		

(d)

1	1		
1	1		1
	1	1	
	1	1	

(e)

1	1
	1

(f)

	1		1
1	1	1	1
1		1	

(g)

	1		
	1	1	1
	1	1	

(h)

1	1	1	
	1	1	
		1	

(i)

1	1		1
		1	1

(j)

1	1	1	
	1	1	1
1	1	1	1
1	1	1	

18. Find the simplest sum-of-products expression for the indicated function.

x	y	z	w	Out
0	0	0	0	1
0	0	0	1	0
0	0	1	0	0
0	1	0	0	0
1	0	0	0	1
1	1	0	0	0
1	0	1	0	0
1	0	0	1	0
0	1	0	1	1
0	1	1	0	1
0	0	1	1	0
1	1	1	0	1
1	1	0	1	1
1	0	1	1	0
0	1	1	1	1
1	1	1	1	1

19. Find the simplest sum-of-products expression for the following function:

$$xyz + \bar{x}y\bar{z} + xy\bar{z} + x\bar{y} + \bar{x}yz + \bar{x}\bar{y}z.$$

20. Use a Karnaugh map to find a simpler equivalent circuit for the indicated logic diagram.

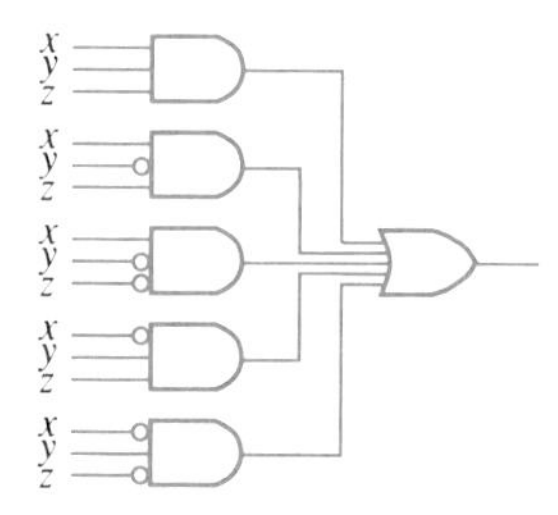

21. For each of the functions represented in Figure 5.29, do the following:

(a) Count the number of ANDs, ORs, and NOTs required by the minterm expansion.

(b) Count the number of ANDs, ORs, and NOTs required by the Karnaugh map method.

In Problems 22–28, use the Karnaugh map method to simplify the Boolean function whose truth set is given.

22. {000, 010, 100, 101, 110}.

23. {0000, 0001, 0010, 1000, 0101, 0110, 1001, 1100, 1101, 1110}.

24. {000, 001, 010, 101, 110, 111}.

25. {0000, 0001, 0100, 0101, 0110, 1000, 1010, 1101, 1110}.

26. {0001, 0010, 0100, 0101, 0111, 1000, 1001, 1100, 1110, 1111}.

27. {0000, 0001, 0010, 1000, 1010, 1011, 1110, 1111}.

28. {01101, 01111, 10001, 10010, 10011, 10100, 10101, 10111, 11001, 11011, 11101, 11111}. (Note: This problem will need a *five* variable Karnaugh map!)

In Problems 29–35, apply the Karnaugh map method to the complement of the functions described in Problems 22–28. Then apply DeMorgan's law to get a representation of the original function.

29. Problem 22.

30. Problem 23.

31. Problem 24.

32. Problem 25.

33. Problem 26.

34. Problem 27.

35. Problem 28.

36. Consider the function whose truth set is

{0001, 0100, 0101, 1010, 1011}.

Is it better to use the Karnaugh map method on f or $\bar{f}$? ("Better" means uses fewer ANDs, ORs, and NOTs.)

37. Among all Boolean functions of four variables, what is the largest possible number of minterms you can find for which the Karnaugh map method yields *no* simplification?

38. This problem concerns the two Boolean functions $f(x_1, x_2) = x_1 \oplus x_2$ and $g(x_1, x_2, x_3) = x_1 \oplus x_2 \oplus x_3$, where $\oplus$ denotes Exclusive-or (XOR), as in Table 5.6.

(a) Use a Karnaugh map to find a sum-of-products expression for f that uses as few ANDs, ORs, and NOTs as possible.

(b) Show how to design a logic circuit for g that uses only two XOR gates.

(c) Replace each XOR gate in the logic circuit in part (b) with the equivalent (AND, OR, NOT) circuit from part (a).

(d) Draw a Karnaugh map for g.

(e) Finally, explain how to build a circuit for the function in Figure 5.30(e) that uses fewer ANDs, ORs, or NOTs than the circuit produced by the Karnaugh map method.

39. If P and Q are two implicants of a given Boolean function, let us say that *P is covered by Q*, and write $P \leq Q$, if every minterm involved in P is also involved in Q. For example, comparing Figures 5.27 and 5.28, we can see that $\bar{x}_1\bar{x}_2 \leq \bar{x}_1$, $x_2x_3 \leq x_2$, etc.

(a) Show that the relation of covering is a *partial ordering* (Section 1.4), i.e., that $\leq$ is reflexive, transitive, and antisymmetric.

(b) Draw a Hasse diagram for the *implicant poset* for the function of Figure 5.20. (Don't forget that the minterms count as implicants!)

(c) Identify the prime implicants for the Hasse diagram in part (b).

Summary

Propositional calculus

- Connectives ($\wedge$, $\vee$, $\sim$, $\rightarrow$, $\leftrightarrow$); formation and recognition of well-formed propositions.
- Truth tables.
- Theorems in the propositional calculus; valid arguments.

Boolean algebra

- Boolean functions and their truth tables; basic digital logic gates—AND, OR, NOT, NAND, NOR, XOR, XNOR.
- Boolean multiplication, addition, and complementation.
- Representation of logic circuits; minterm and maxterm expansions.
- Basic laws of Boolean algebra; connection between Boolean algebra, set theory, and propositional calculus; duality.
- Simplification of Boolean functions by means of Karnaugh maps; prime implicants.

CHAPTER SIX

MATHEMATICAL MODELS FOR COMPUTING MACHINES

This chapter should be thought of as a continuation of Chapter 5 (Boolean algebra). There we discussed the basic mathematics involved in the efficient computation of Boolean functions, our motivation being that Boolean computations lie at the very heart of modern digital electronics. And so they do. But nearly every useful digital electronic device, including all general-purpose computers, requires one basic component nowhere mentioned in Chapter 5: *memory*. In this chapter, we will extend our earlier discussions, and introduce you to the basic mathematical tools needed to study computations involving memory. We will begin with a brief discussion of certain specific electronic parts with memory (*flip-flops*), continue with an abstract mathematical model that engineers use to design and analyze circuits with memory (*finite state machines* and *finite automata*), and conclude with an even more abstract model which mathematicians and computer scientists use to study the ultimate limits of machine computation (*Turing machines*). In the final section of this chapter, we describe an important specific application of these ideas: *convolutional codes with Viterbi decoding*, a technique which has been used by the *Voyager* spacecraft to aid in the transmission of photographs of distant planets back to earth.

6.1 Boolean Functions with Memory: Sequential Circuits

In Chapter 5 we learned how to construct circuits for computing Boolean functions, using only the basic AND, OR, NOT, etc., gates depicted in Table 5.6. For example, the Boolean function $y = x_1x_2x_3x_4x_5$ can be calculated with the circuit of Figure 6.1, using four two-input AND gates. There is nothing wrong with the circuit in Figure 6.1; it works fine, *provided the five inputs are all available at the same time.*

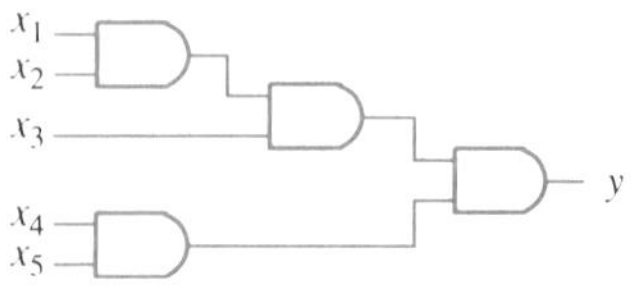

Figure 6.1 A circuit for computing $y = x_1x_2x_3x_4x_5$.

However, it is quite common in practice to encounter situations in which it is necessary to compute a Boolean function whose inputs do not arrive simultaneously, in *parallel*, but one at a time, or *serially* (see Figure 6.2). In a circuit with serial inputs, there is always a *master clock* present which produces clock pulses (CPs) at precisely regular intervals.* With each clock pulse, a new input signal appears on the input wire. Thus in the circuit of Figure 6.2(b), the input x_1 appears with the first CP, x_2 arrives at the second CP, etc.

The circuit of Figure 6.1 is inadequate (as it stands) if the five inputs x_1, x_2, x_3, x_4, and x_5 arrive at the circuit serially. What is needed is some way for the circuit to remember the inputs as they arrive so that inputs which appear at *different* times can be appropriately còmbined at some *later* time. In this section we will learn something about how circuits with memory can be designed.

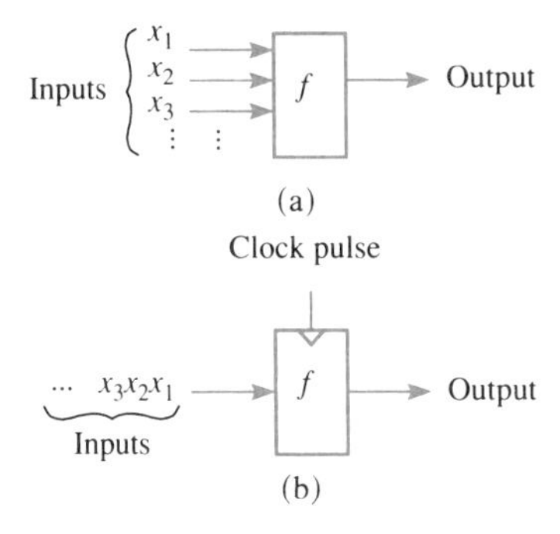

Figure 6.2 Two kinds of inputs to a Boolean function. (a) Parallel inputs; (b) serial input.

We saw in Chapter 5 that any Boolean function with parallel inputs can be built using only AND, OR, and NOT gates. In this chapter we will see that even if some or all of the function's inputs are serial, it can still be built with ANDs, ORs, and NOTs, together with just one further part called a **flip-flop**. There are several different kinds of flip-flops in common engineering use. However, since it is not our aim to learn the details of digital electronic design, we will only describe the simplest flip-flop, the *delay*, or *D flip-flop.* (Some of the other flip-flops are discussed in the problems at the end of the section.)

The D flip-flop (see Figure 6.3) is a primitive storage device that stores just one bit, either a 0 or a 1. The bit stored is equal to the most recent input. When a new input arrives, it is stored in the flip-flop, and the bit previously stored becomes the output. The effect of this is that the output sequence is equal to the input sequence *delayed by one CP.* For this reason, the D flip-flop is sometimes called a *unit delay.* Thus if (as in Figure 6.3) the input and output after the nth CP are denoted by x_n and y_n, respectively, after the $(n+1)$st CP, the new output y_{n+1} will be equal to the old input x_n:

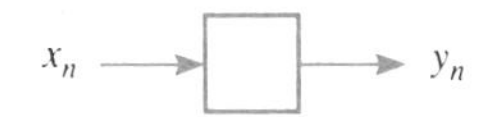

Figure 6.3 The D flip-flop: After the next CP, the new output (y_{n+1}) will be equal to the present input (x_n).

(6.1a)
$$y_{n+1} = x_n.$$

An equivalent way to say this is to say that the *present* output y_n is equal to the *previous* input x_{n-1}:

(6.1b)
$$y_n = x_{n-1}.$$

Equations (6.1a) and (6.1b) are the defining equations for the D flip-flop.

EXAMPLE 6.1 Suppose the sequence of inputs to the D flip-flop of Figure 6.3 is as given in the table below:

Clock pulse number (n)	1	2	3	4	5	6	7	8	9	...
Input (x_n)	1	0	0	1	0	1	1	1	0	...

Find the corresponding outputs y_n.

SOLUTION According to Equation (6.1a), y_1 is equal to x_0, which isn't specified, so we can't tell what y_1 is. However, again by (6.1a), $y_2 = x_1 = 1$, $y_3 = x_2 = 0, \ldots, y_{10} = x_9 = 0$. Thus the list of y_n's can be obtained just by shifting

* These intervals can be very tiny indeed. In current technology, clocks which produce tens of millions of clock pulses per second are commonplace.

Present State	Next State $x = 0$	Next State $x = 1$	Output $x = 0$	Output $x = 1$
0	0	1	0	0
1	0	1	1	1

(a)

0/0 1/0 1/1 0/1

(b)

Figure 6.4 (a) State table, and (b) state diagram, for the D flip-flop.

the list of x_n's one place to the right:

n	1	2	3	4	5	6	7	8	9	10	
x_n	1	0	0	1	0	1	1	1	0	...	
y_n	?	1	0	0	1	0	1	1	1	0	...

We can give a more formal description of the D flip-flop, based on the notion of its **state**. The state of the flip-flop is either 0 or 1, depending on what is being stored. When a new input arrives, two things happen: An output is produced, and a new state is determined. The output is equal to the old state, and the new state is equal to the new input. These facts are summarized in Figure 6.4(a), which is called the **state table** for the D flip-flop, and in Figure 6.4(b), which is called the **state diagram**. The state table describes how the flip-flop behaves in all possible circumstances. For example, if the flip-flop is in state 1 and receives an input of 0, we look in the table under "Present State 1" and "$x = 0$", and find that the new state is 0 and the output is 1. The state diagram gives exactly the same information in graphical form. In Figure 6.4(b) the state diagram is a labelled, directed graph with two vertices, one for each state. The edge labels show the input and the output (in the format "input/output") corresponding to a change of state. For example, the transition from state 1 to state 0 is labelled "0/1" to indicate that the input that causes this transition is 0 and the output is 1.

Thus if the input, output, and state after the nth CP are denoted by x_n, y_n, and s_n, respectively, the D flip-flop can be described by the following pair of equations:

(6.2a) $$s_{n+1} = x_{n+1} \quad \text{(next state = next input)}$$

(6.2b) $$y_{n+1} = s_n \quad \text{(next output = previous state).}$$

Note that if we combine (6.2b) and (6.2a), we get $y_{n+1} = s_n = x_n$, the same as (6.1a).

As a simple example to show the usefulness of the D flip-flop, let's see how it can be used to design a circuit to compute the Boolean function $y = x_1x_2x_3x_4x_5$, when the inputs arrive serially, rather than in parallel (see Figure 6.5). Figure 6.5 shows a Boolean circuit with one input line and one output line, whose inputs arrive serially. It is built from one two-input AND gate and one D flip-flop, connected as shown. Figure 6.5 shows the situation just after the nth clock pulse,

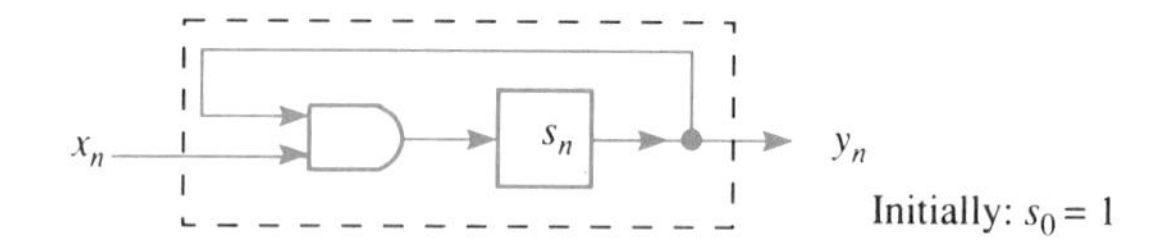

Figure 6.5 A serial ANDing circuit, just after the nth clock pulse.

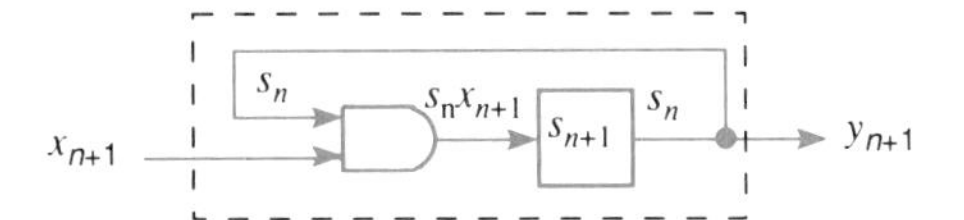

Figure 6.6 The same circuit, just after the $(n + 1)$st clock pulse.

with the input, output, and state of the flip-flop denoted by x_n, y_n, and s_n, respectively. If the circuit is initialized by forcing the flip-flop into state 1 before computation begins, i.e., by setting $s_0 = 1$, then the $(n + 1)$st output is equal to the AND of the first n inputs, i.e.,

(6.3) $$y_{n+1} = x_1 x_2 \cdots x_n, \qquad \text{for} \qquad n = 1, 2, 3, \ldots.$$

We can see why (6.3) is true by visualizing what happens after the $(n + 1)$st CP, as shown in Figure 6.6. The new input, output, and state will be x_{n+1}, y_{n+1}, and s_{n+1}. Furthermore, by Equation (6.2b), the output of the flip-flop will be its previous state, i.e., s_n. This tells us two important things. First, since the output of the flip-flop is also the output of the circuit, it tells us that $y_{n+1} = s_n$. Second, since the output of the flip-flop is also an input to the AND gate, it tells us that the output of the AND gate is $s_n x_{n+1}$; and since the output of the AND gate, in turn, is the input to the flip-flop, by (6.2) the new state of the flip-flop will be $s_n x_{n+1}$. In summary, the behavior of the circuit in Figure 6.5 is described by the following pair of equations:

(6.4) $$y_{n+1} = s_n$$

(6.5) $$s_{n+1} = s_n x_{n+1}.$$

Now we can prove (6.3), by proving that

(6.6) $$s_n = x_1 x_2 \cdots x_n$$

for all $n \geq 1$, by induction. With $n = 0$ in (6.5), we get $s_1 = s_0 x_1$. But $s_0 = 1$ is the initialization chosen for the circuit, and so $s_1 = x_1$. This proves (6.6) for $n = 1$. Next, we assume that (6.6) is true for a given value of n, and try to prove it for $n + 1$. Here goes:

$$\begin{aligned} s_{n+1} &= s_n x_{n+1} && \text{[by (6.5)]} \\ &= (x_1 x_2 \cdots x_n) x_{n+1} && \text{(induction assumption)} \\ &= x_1 x_2 \cdots x_{n+1} && \text{(associative law for AND).} \end{aligned}$$

Thus (6.6) is true for $n + 1$, and so by induction it is true for all values of n. Finally, $y_{n+1} = s_n$ by (6.4) and $s_n = x_1 \cdots x_n$ by (6.6), so that $y_{n+1} = x_1 x_2 \cdots x_n$, as advertised in (6.3).

Circuits like the one in Figure 6.5, containing flip-flops as well as the gates described in Chapter 5, are usually called **sequential circuits**. (Circuits using only the gates of Chapter 5 are sometimes called *combinational circuits.*) For many purposes the most convenient description of a sequential circuit is a state table or diagram like the ones in Figure 6.4.

EXAMPLE 6.2 Give a state table and diagram for the sequential circuit of Figure 6.5.

Present State	Next State		Output	
	$x = 0$	$x = 1$	$x = 0$	$x = 1$
0	0	0	0	0
1	0	1	1	1

(a)

0,1/0 0/1 1/1

(b)

Figure 6.7 (a) State table and (b) state diagram for the circuit of Figure 6.5.

S O L U T I O N The circuit has two possible states, 0 and 1. Using Equations (6.4) and (6.5), we can calculate the state table shown in Figure 6.7(a). For example, if the present state (s_n) is 1, and the next input (x_{n+1}) is 0, according to (6.4a), the next output is 1, and according to (6.5), the next state is 0. This gives the "Output" and "Next State" entries corresponding to "Present State 1," and "$x = 0$" in the state diagram. The other entries are calculated similarly. It is easy to use the state table to produce a state diagram, like the one shown in Figure 6.7(b). ■

Many of the most useful and interesting sequential circuits involve exclusive-OR (XOR) gates, which were mentioned but not emphasized in Chapter 5. For reference, we review the XOR gate in Figure 6.8.

The XOR operation, denoted by "$\oplus$," enjoys the following properties, which though innocuous looking, largely explain why it is so useful:

(a)

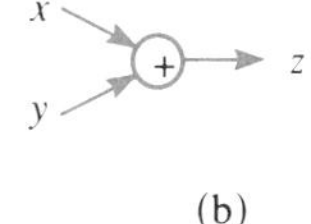

(b)

x	y	z
0	0	0
0	1	1
1	0	1
1	1	0

(c)

Figure 6.8 The XOR gate. (a) Standard symbol. (b) Simplified symbol. (c) Truth table.

(A) $x \oplus 0 = 0 \oplus x = x$ (zero element)

(B) $x \oplus y = y \oplus x$ (commutativity)

(C) $x \oplus (y \oplus z) = (x \oplus y) \oplus z$ (associativity)

(D) $a(x \oplus y) = ax \oplus ay$ (distributivity)

(E) $x \oplus x = 0$ (Nimishness*).

These five properties are easily verified, and we leave the verification as Problem 8. The next example investigates a typical circuit containing XOR gates.

EXAMPLE 6.3 Consider the sequential circuit shown in Figure 6.9(a), which consists of two D flip-flops and three XOR gates.† Define the *state* of the circuit

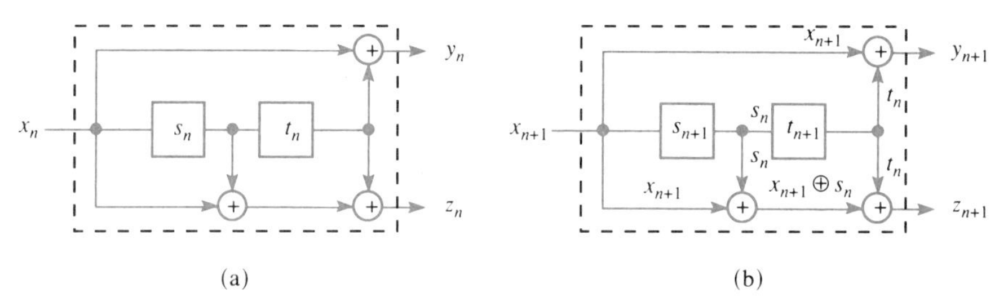

Figure 6.9 The circuit for Example 6.3: (a) Just after the nth clock pulse; (b) just after the $(n + 1)$st clock pulse.

* So-called because of the importance of this property to the game of Nim, which we will study in Chapter 8.

† This circuit is an example of a *convolutional encoder*, which we will discuss in more detail in Section 6.6.

after the nth clock pulse as the pair (s_n, t_n), where s_n is the state of the first flip-flop and t_n is the state of the second. Similarly, the *output* of the circuit is the pair (y_n, z_n).

(a) Express the next state (s_{n+1}, t_{n+1}) in terms of the present state (s_n, t_n) and the next input x_{n+1}.
(b) Express the next output (y_{n+1}, z_{n+1}) in terms of the present state and next input.
(c) Derive a state table and a state diagram.

SOLUTION (a) The key to this problem is Figure 6.9(b), which shows the situation just after the $(n+1)$st CP. The input, output, and state are x_{n+1}, (y_{n+1}, z_{n+1}), and (s_{n+1}, t_{n+1}), respectively; the output of the first flip-flop is s_n, and the output of the second flip-flop is t_n (see Equation 6.2b). Since the state of a D flip-flop is always its current input, it follows from Figure 6.9(b) that

(6.7a) $$s_{n+1} = x_{n+1};$$

(6.7b) $$t_{n+1} = s_n.$$

(b) Equations (6.7a) and (6.7b) express the circuit's next state in terms of the present state and the next input. Similarly, Figure 6.9(b) tells us that the first output y_{n+1} is the XOR of x_{n+1} and t_n; and that the second output z_{n+1} is the XOR of x_{n+1}, s_n and t_n. Therefore

(6.8a) $$y_{n+1} = x_{n+1} \oplus t_n;$$

(6.8b) $$z_{n+1} = x_{n+1} \oplus s_n \oplus t_n.$$

Equations (6.8a) and (6.8b) express the circuit's next output in terms of the present state and the next input.

(c) Finally, using Equations (6.7) and (6.8), it is a simple matter to obtain the state table, which is shown in Figure 6.10(a), and the state diagram, which is shown in Figure 6.10(b). For example, if the state is 10 and the input is 1, then $(s_n, t_n) = (1, 0)$, and $x_{n+1} = 1$. Then by (6.7) we have $(s_{n+1}, t_{n+1}) = (1, 1)$ and by (6.8) we have $(y_{n+1}, z_{n+1}) = (1, 0)$. Therefore in the state table of Figure 6.10(a), under "Present State 10," the "Next State" with $x = 1$ is 11, and the "Output" with $x = 1$ is 10. Similarly, in the state diagram of

Present State	Next State		Output	
	$x = 0$	$x = 1$	$x = 0$	$x = 1$
00	00	10	00	11
01	00	10	11	00
10	01	11	01	10
11	01	11	10	01

(a)

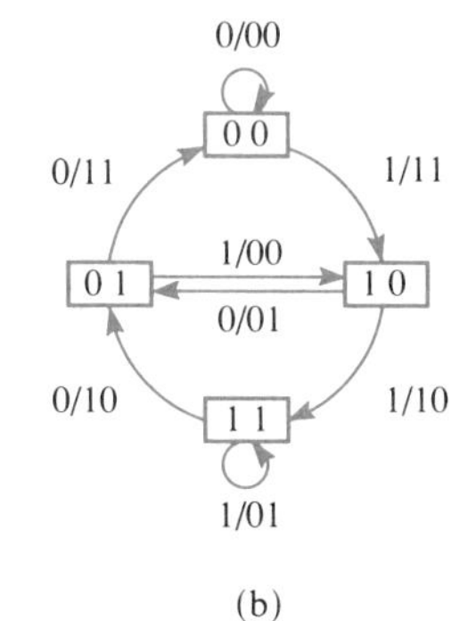

(b)

Figure 6.10 (a) State table for the circuit of Figure 6.9. (b) State diagram for the circuit of Figure 6.9.

Figure 6.10(b), there is a directed edge from state 10 to state 11 labelled 1/10.

In this section we have only scratched the surface of the subject of sequential circuits. It is not our intention to go farther along these lines, since the detailed study of the design of sequential circuits properly belongs to the electrical engineering curriculum. However, before leaving this topic, we wish to point out that all sequential circuits have many important features in common, features that do not depend on the engineering details. Each sequential circuit has inputs, outputs, and a finite number of internal states. Furthermore, whenever a new input arrives at the circuit, two things happen: The circuit moves to a new state, and an output is produced. In the next section we will study an abstract model for sequential circuits, based on these common features, called the **finite state machine**.

Problems for Section 6.1

1. Suppose the sequence of inputs to the D flip-flop of Figure 6.3 is as given in the table below.

n	0	1	2	3	4	5	6	7	8	9	. . .
x_n	0	1	1	0	1	1	0	1	1	0	. . .

Find the corresponding outputs y_n.

2. Suppose that the inputs from Problem 1 are used in the sequential circuit in the accompanying figure.

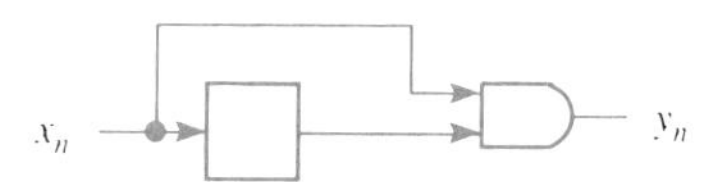

(a) Find the corresponding outputs.
(b) Assuming that the input sequence continues to repeat the pattern 011, for which values of n will the output y_n be equal to 1?

3. In the accompanying figure we show a simple sequential circuit.

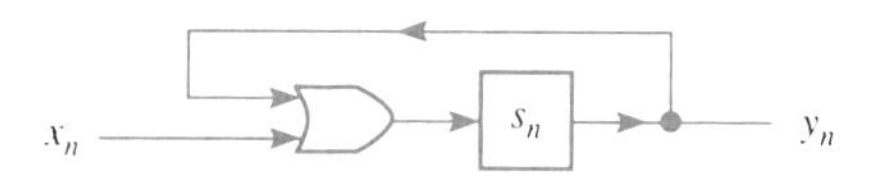

(a) Express the next output y_{n+1}, and the next state s_{n+1} in terms of the next input x_{n+1} the present state s_n.
(b) Assuming the circuit is initialized with $s_0 = 0$, calculate the output sequence corresponding to the following input sequence:

n	1	2	3	4	5	6	7	. . .
x_n	0	0	0	0	1	0	0	. . .

4. (a) Repeat Problem 3(b), but assume the initialization is $s_0 = 1$.
(b) Again assuming $s_0 = 0$, find a formula analogous to Equation (6.3) expressing y_{n+1} as a Boolean function of $x_1, x_2, \ldots, x_n$.

5. In the text we discussed the circuit in Figure 6.5, assuming the initialization $s_0 = 1$. What happens if $s_0 = 0$? Specifically:
(a) Suppose the input sequence is as follows:

n	1	2	3	4	5	6	7	. . .
x_n	1	1	1	0	1	1	1	. . .

Calculate the output sequence.
(b) Find a formula, analogous possibly to Equation (6.3), expressing y_{n+1} as a function of $x_1, x_2, \ldots, x_n$.
(c) Which initialization ($s_1 = 0$ or $s_1 = 1$) do you think is most "useful"? Explain.

6. In the accompanying figure we show another simple sequential circuit.

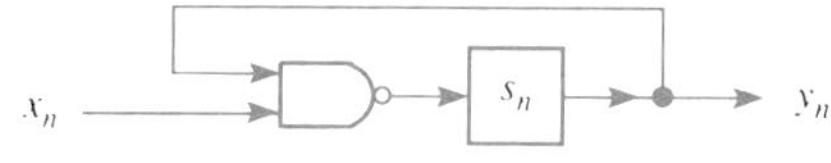

(a) Find a pair of formulas expressing s_{n+1} and y_{n+1} in terms of x_{n+1} and s_n.
(b) Assuming the initialization $s_0 = 1$, express y_4 as a Boolean function of x_1, x_2, x_3, and x_4.

7. Repeat Problem 6 for the accompanying sequential circuit (assume $s_0 = 0$).

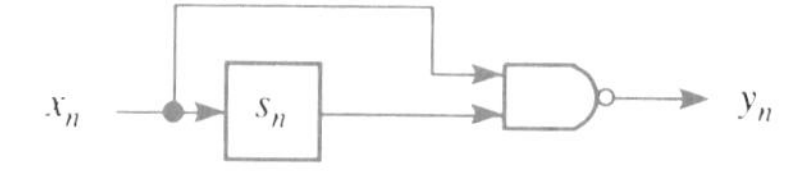

8. Verify by truth table properties (A)–(E) of "$\oplus$" cited in the text.

9. Here is another sequential circuit.

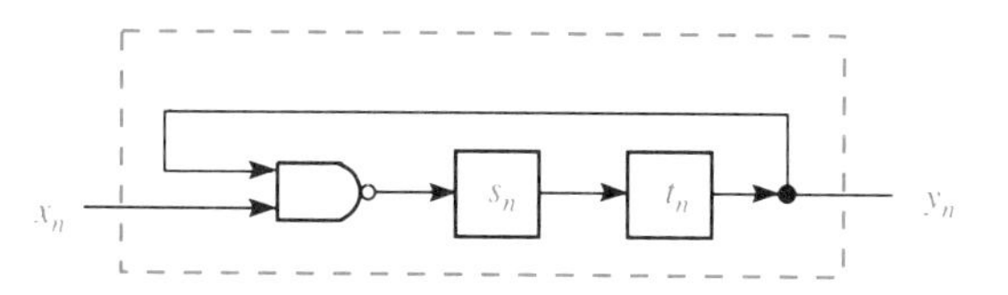

(a) Produce a state table.
(b) Produce a state diagram.
(c) Assuming that the initialization is $s_0 = t_0 = 0$, find the output, if the input is as given below.

n	1	2	3	4	5	6	...
x_n	1	1	0	1	0	0	...

10. Here is another sequential circuit.

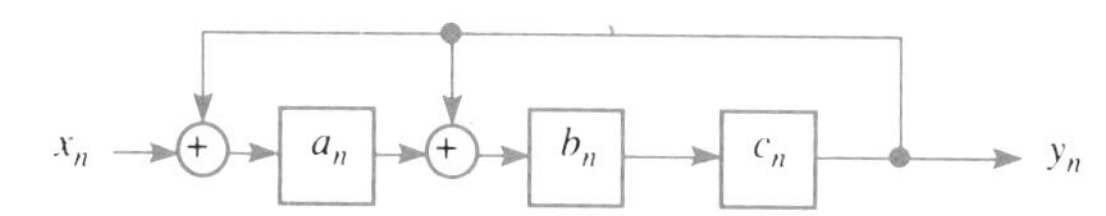

(a) Produce a state table.
(b) Produce a state diagram.
(c) Assuming that the initialization is $a_0 = b_0 = c_0 = 0$, find the output, if the input is as given below.

n	1	2	3	4	5	6	...
x_n	0	1	0	1	0	1	...

11. Here is another sequential circuit.

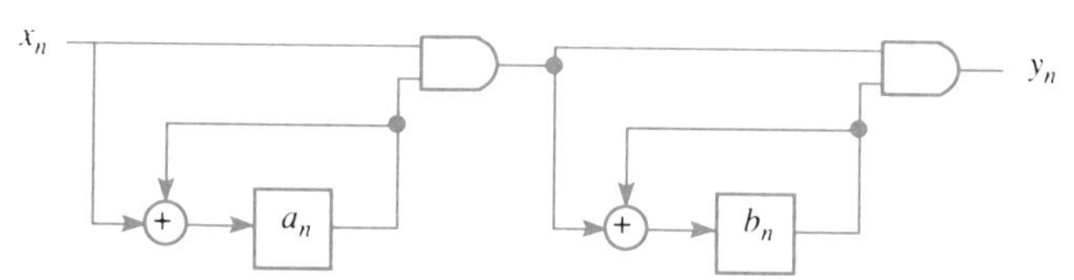

(a) Produce a state table.
(b) Produce a state diagram.
(c) Assuming the initialization is $a_0 = b_0 = 0$, find the output if the input is as given below.

n	1	2	3	4	5	6	...
x_n	1	1	1	1	1	1	...

In the next several problems we will be dealing with three different types of flip-flops; their state diagrams and tables are given below. Note that the state of each flip-flop is either 0 or 1, and the next output is always the present state. Note also that two of the flip-flops (the SR flip-flop and the JK flip-flop) have two inputs. In the case of the JK flip-flop, there are in effect *four* possible inputs (00, 01, 10, 11); in the case of the SR flip-flop, 11 is forbidden and so there are *three* possible inputs (00, 01, 10).

T (toggle) Flip-Flop

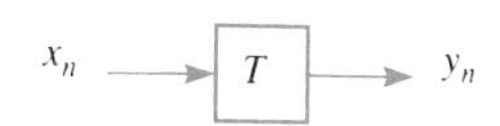

Present State	Next State		Output
	$x = 0$	$x = 1$	
0	0	1	0
1	1	0	1

SR (Set-Reset) Flip-Flop

(Input $x_n = 1$ and $x'_n = 1$ is not allowed)

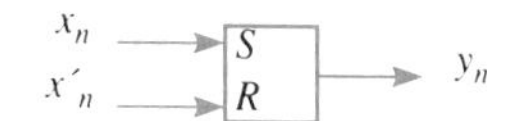

Present State	Next State			Output
	$xx' = 00$	$xx' = 01$	$xx' = 10$	
0	0	0	1	0
1	1	0	1	1

JK Flip-Flop

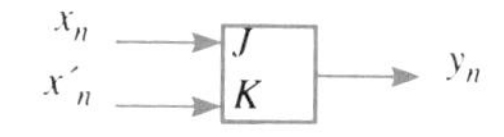

Present State	Next State				Output
	$xx' = 00$	$xx' = 01$	$xx' = 10$	$xx' = 11$	
0	0	0	1	1	0
1	1	0	1	0	1

12. Draw a state diagram for the T flip-flop.

13. Draw a state diagram for the SR flip-flop.

14. Draw a state diagram for the JK flip-flop.

For each flip-flop, derive a pair of equations expressing the next output y_{n+1} and the next state s_{n+1} in terms of the next input and the present state. (Hint: Use a truth table to express s_{n+1} as a Boolean function of s_n, x_{n+1}, and x'_{n+1}, and use a Karnaugh map. If a line of the truth table has an unspecified value for s_{n+1}—a so-called "don't care" condition—you may choose it to be either 0 or 1. Make your choice so as to simplify the Karnaugh map.)

15. The T flip-flop.

16. The SR flip-flop.

17. The JK flip-flop.

18. Show that the T flip-flop is equivalent to the circuit in the accompanying figure.

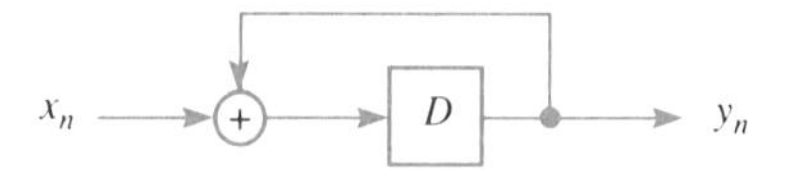

19. The sequential circuit in the accompanying figure consists of two AND gates, one inverter, and one SR flip-flop.

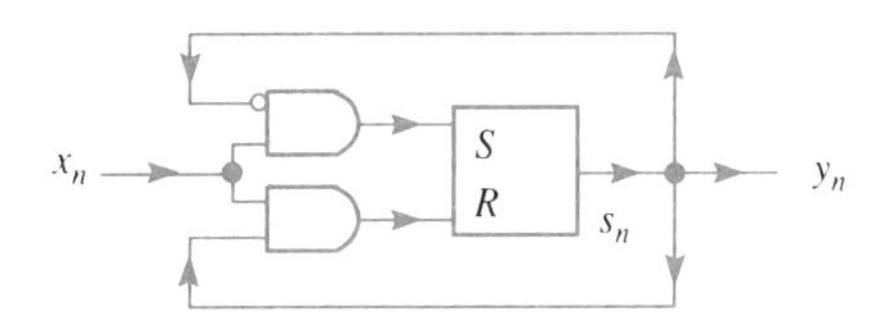

(a) Using the state table for the SR flip-flop given above, find the output sequence $y_1, y_2, \ldots$ corresponding to the input sequence 1100100011 (assume $s_0 = 1$).
(b) This circuit in fact behaves exactly like one of the flip-flops. Which one?

20. Consider the problem of building a circuit that will calculate the Boolean function $y = x_1 \oplus x_2 \oplus x_3 \oplus x_4$.
(a) Design such a circuit, assuming *parallel* inputs, using only ANDs, ORs, and NOTs.
(b) Design such a circuit, assuming *serial* inputs, using only ANDs, ORs, NOTs, and D flip-flops.
(c) Compare the number of parts required for each of the circuits in parts (a) and (b).

21. The sequential circuit in the accompanying figure consists of one AND gate, one inverter, and one JK flip-flop.

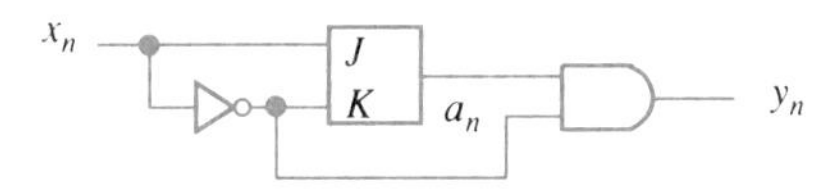

(a) Express y_{n+1} as a Boolean function of a_n and x_{n+1}.
(b) Express a_{n+1} as a Boolean function of x_{n+1}.
(c) Combine parts (a) and (b), and express y_{n+1} as a function of x_{n+1} and x_n.
(d) Calculate the *output* sequence $y_1, y_2, y_3, \ldots$ corresponding to the *input* sequence $(x_1, x_2, x_3, \ldots) = (100101101\ldots)$. Does your answer depend on the "initial value" a_0?

22. Replace the AND gate in Figure 6.5 with a NAND gate, and then calculate the output sequence corresponding to the input sequence 1011001111 .... Does your answer depend on the "initial value" s_0?

23. If several D flip-flops are connected in series, the resulting circuit is called a **shift register**. For example, the accompanying figure depicts a three-stage shift register.

(a) Express y_{n+1} as a Boolean function of x_{n+1}, x_n, x_{n-1}, and x_{n-2}.
(b) Calculate the state table.
(c) Draw the state diagram.

24. Continuing Problem 23, consider an M-stage shift register, as shown in the accompanying figure.

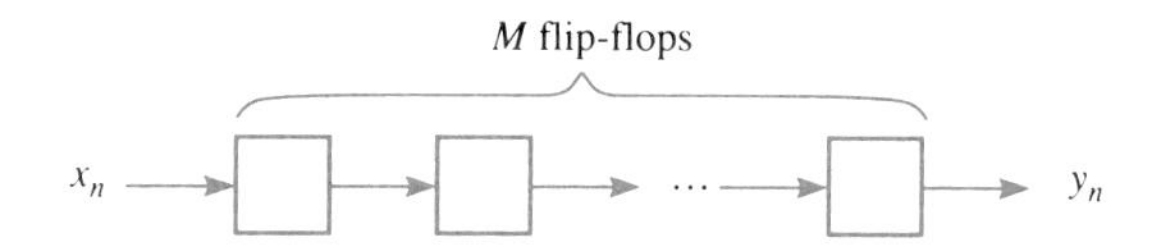

(a) How many vertices will the state diagram have?
(b) How many edges will be directed *out* of each vertex?
(c) How many edges will be directed *in*?
(d) Does the state diagram have an Euler cycle?

25. The sequential circuit in the accompanying figure has two input wires and three output wires.

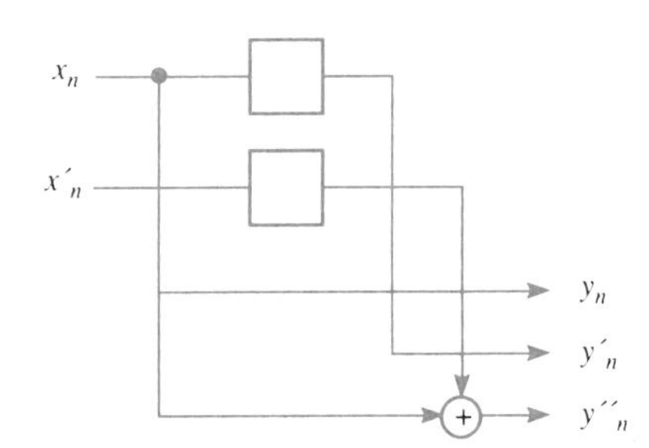

(a) Give the state table.
(b) Draw the state diagram.
(c) Assume the initial state is $(0, 0)$. Calculate the outputs (y_n, y'_n, y''_n) corresponding to the inputs $(x_n, x'_n) = (0, 1)$, $(1, 0)$, $(0, 0)$, $(1, 1)$, $(0, 1)$, $(1, 0)$, $(0, 0)$, $(1, 1), \ldots$ (repeats).

6.2 Abstract Sequential Circuits: Finite State Machines

In the previous section we learned something about sequential circuits, which are Boolean circuits built from the logic gates of Chapter 5 plus two further components, clocks and flip-flops. As we saw in Section 6.1, any sequential circuit can be functionally described by a state table (equivalently, a state diagram). In Section 6.1 we focused on the problem of *analyzing* a given sequential circuit. In this section, we will focus on the opposite problem, that of *designing* a sequential circuit that can do a specified job. Our approach will be to specify a state table or diagram for a sequential machine that will do the desired job. Electrical engineers know how to take a state table and build a corresponding sequential circuit using standard parts, and so we will stop with the state table, simply assuming that there is a sequential circuit that behaves as described. Abstract sequential circuits which are described only by their state tables are called **finite state machines** (FSMs), and so what we will be studying in this section is the design of finite state machines.

Although a finite state machine is an idealized abstraction, it can be thought of in very concrete terms. An FSM has a finite **set of states**, an **input alphabet**, and an **output alphabet**. The machine has an internal clock that "ticks" at regular intervals. Just prior to each tick, the machine will be in one of its states. (The machine is initialized by placing it in a fixed **initial state** before the clock is started.) At the tick, the machine will accept an input (a symbol from its input alphabet), and in response *move to another state*, and *produce an output* (a symbol from its output alphabet). In this respect, an FSM behaves just like a sequential circuit. The important difference is that with an FSM, we don't worry about the engineering details (which gates and flip-flops to use, and how to connect them) but focus instead on the underlying structure of the desired function.

We begin by considering a specific FSM design problem.

EXAMPLE 6.4 Design an FSM whose present output will be 1 if the present input is *the same* as the previous input, and 0 if the present input is *different* from the previous input, i.e.,

$$y_n = \begin{cases} 1 & \text{if } x_n = x_{n-1} \\ 0 & \text{if } x_n \neq x_{n-1}. \end{cases} \tag{6.9}$$

For example, if the input stream is 100111101001011 . . . , the output stream will be *01011100010001 . . . . (The "*" means that the first output is indeterminate, since there is no previous input.)

SOLUTION To design an FSM, we must start with three sets: an input alphabet, an output alphabet, and a set of states. In this example, we know that

$$\text{input alphabet} = \{0, 1\}, \qquad \text{output alphabet} = \{0, 1\},$$

and that the input-output relationship is given by (6.9).

A machine whose input-output function is given by (6.9) can be built with only two states, say A and B, to represent the two possible values of the previous

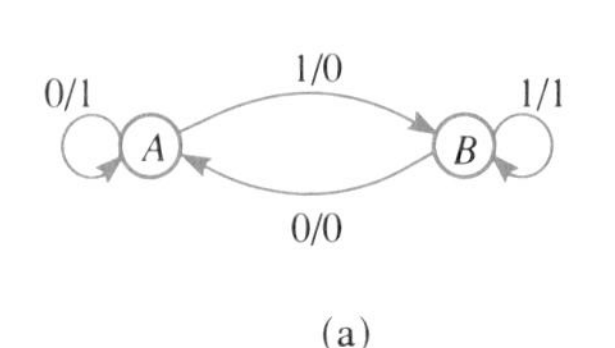

(a)

Present State	Next State		Output	
	$x = 0$	$x = 1$	$x = 0$	$x = 1$
A	*A*	*B*	1	0
B	*A*	*B*	0	1

(b)

Figure 6.11 Solution to Example 6.4. (State *A* represents a previous input of 0; state *B* represents a previous input of 1.)

input:

$$\text{state set} = \{A, B\}.$$

The machine will be in state *A* when the previous input was 0, and in state *B* when the previous input was 1. These two states will allow the machine to remember the value of the previous input, which can then be compared with the present input. For example, if the machine is in state *A*, and the present input is 1, the present input (1) is *not* the same as the previous input (0), and so by Equation (6.9) the present *output* should be 0. At the same time, the machine should move into state *B*, to indicate that the most recent input is now 1. What all this means is that in the state diagram for the machine, there should be an edge directed from vertex *A* to vertex *B* and labelled "1/0," as shown in Figure 6.11(a). The remaining edges and labels in the state diagram of Figure 6.11(a) can be computed similarly. The corresponding state table is shown in Figure 6.11(b). Figure 6.11 gives a complete description of the desired FSM. ■

EXAMPLE 6.5 Design a "mod 2 counter," i.e., an FSM whose input sequence is a sequence $(x_1, x_2, x_3, \ldots)$ of 0s and 1s, and whose output sequence $(y_1, y_2, y_3, \ldots)$ indicates the arrival of *every other* input 1, i.e.,

$$y_n = \begin{cases} 1 & \text{if } x_n = 1 \text{ and if there have been an } \textit{even} \text{ number of input 1s so far;} \\ 0 & \text{in all other cases.} \end{cases}$$

For example:

Input sequence	1	0	1	1	0	0	1	1	1	1	1	0	1	...
Output sequence	0	0	1	0	0	0	1	0	1	0	1	0	0	...

Thus the output $y_n = 1$ signals the arrival of the second, fourth, sixth, eighth, ..., 1 in the input stream.

SOLUTION Our FSM must have input and output alphabets both equal to $\{0, 1\}$. It will have two states, say *A* and *B*, to represent the following two conditions:

- *State A*: An *even* number of 1's have arrived so far.
- *State B*: An *odd* number of 1's have arrived so far.

The machine should be initialized by placing it in *A*, indicating that before the clock starts, an even number (zero) of 1s have arrived. When the present input x_n equals 0, no change in state should occur, and the output should be 0. When

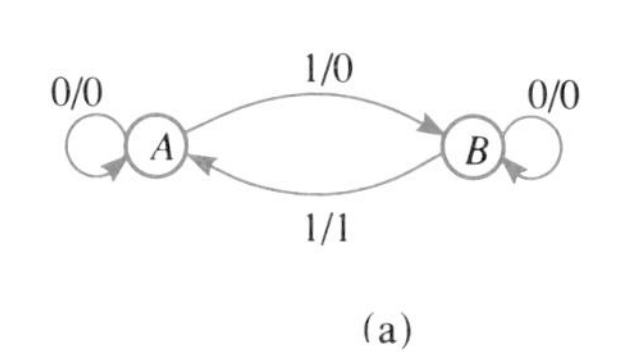

(a)

Present State	Next State $x = 0$	Next State $x = 1$	Output $x = 0$	Output $x = 1$
A	A	B	0	0
B	B	A	0	1

(b)

Figure 6.12 FSM for Example 6.5. (State A represents an even number of 1s; state B represents an odd number of 1s; the initial state is A.)

the present input x_n equals 1, the state should change; and if the change is from state B to state A, the output should be 1. The state diagram and the state table for this machine appear in Figure 6.12.

The next example illustrates one of the most important tasks that an FSM can perform, that of *recognizing the occurrence of a given pattern* in the input string. (We will return to pattern recognition in Section 6.4 and in Chapter 7.)

EXAMPLE 6.6 Design an FSM that recognizes the pattern 1111, i.e., one that accepts a serial stream of 0s and 1s as input, and signals the arrival of four 1s in a row by outputting a 1, otherwise outputting only 0s. In other words,

$$y_n = \begin{cases} 1 & \text{if } x_n = x_{n-1} = x_{n-2} = x_{n-3} = 1 \\ 0 & \text{in all other cases.} \end{cases}$$

SOLUTION In this example, we are told that both the input and the output alphabet are $\{0, 1\}$. One way to proceed is to design an FSM with these input/output alphabets and five states, S_0, S_1, S_2, S_3, and S_4, which will be used to remember how many "1s in a row" the machine is working on. The informal significance of each of the five states is as follows:

S_0: no 1s in a row
S_1: one 1 in a row
S_2: two 1s in a row
S_3: three 1s in a row
S_4: at least four 1s in a row

This machine should be started in state S_0, because initially no 1s have yet arrived. If a 1 input arrives, the machine will move "up" one state (unless it is already in the top state S_4, in which case it stays there), but when a 0 arrives, it will drop down to the bottom state S_0. The output always will be 0 unless the machine is in S_3 or S_4, and the input is 1, in which case the pattern 1111 is recognized and the output is 1.

Figure 6.13 shows an FSM that fits this description. Its operation will become clearer if we trace the sequence of states in response to a particular sequence of inputs:

x_n		0	0	1	1	1	1	0	1	1	1	1	1	1	0	1	...
State	(0)	S_0	S_0	S_1	S_2	S_3	S_4	S_0	S_1	S_2	S_3	S_4	S_4	S_4	S_0	S_1	...
y_n		0	0	0	0	0	1	0	0	0	0	1	1	1	0	0	...

We see that the arrival of a 0 always causes the state to be S_0; but as consecutive

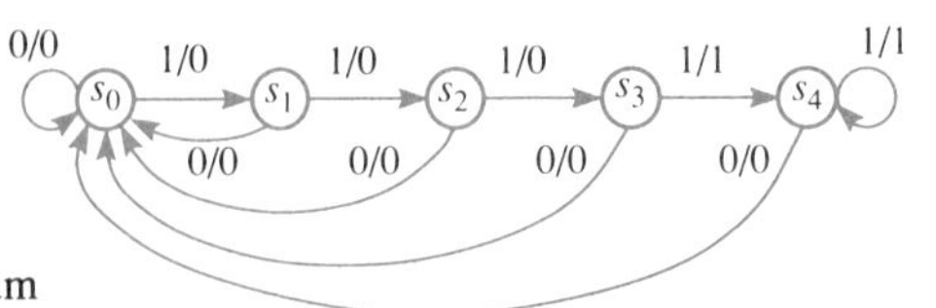

Present State	Next State $x = 0$	Next State $x = 1$	Output $x = 0$	Output $x = 1$
S_0	S_0	S_1	0	0
S_1	S_0	S_2	0	0
S_2	S_0	S_3	0	0
S_3	S_0	S_4	0	1
S_4	S_0	S_4	0	1

Figure 6.13 State diagram and state table for Example 6.6 (initial state = S_0).

1s arrive, the states "build up" toward state S_4, where the machine remains as long as 1s continue to arrive. Notice that the output is 1 if and only if the corresponding state is S_4, so that S_4 serves as a special "recognizer" state. In Section 6.4, when we consider the pattern recognition problem more seriously, we will introduce a special kind of FSM called a *finite automaton*, in which some of the states are identified as "recognizer," or "accepting," states, and for which the output is 1 if and only if the machine is in an accepting state.

In our final example, we'll design a "mod 3 counter." We remind you (review Example 1.19 if necessary) that two integers x and y are said to be *congruent mod 3* if the difference $x - y$ is a multiple of 3. In Example 1.19 we showed that every integer is congruent to either 0, 1, or 2 (mod 3), according to the following pattern:

n	0	1	2	3	4	5	6	7	8	...
n mod 3	0	1	2	0	1	2	0	1	2	...

In fact, n mod 3 is just the *remainder* obtained when n is divided by 3.

EXAMPLE 6.7 Design a mod 3 counter i.e., an FSM whose output at a given time equals the *total number of 1s* (mod 3) in the input stream, up to that time. As an example, here is one possible input sequence and the corresponding outputs:

(Inputs) x_n	0	1	1	0	1	1	1	1	0	1	1	0	0	1	...
(Outputs) y_n	0	1	2	2	0	1	2	0	0	1	2	2	2	0	...

SOLUTION Informally, we want the FSM to count the arriving 1s as follows: 0, 1, 2, 0, 1, 2, 0, 1, . . . , and so we will design our FSM with three states, S_0, S_1, S_2, representing the following conditions:

S_0: The number of input 1s is congruent to 0 (mod 3), i.e., a multiple of 3 (0, 3, 6, 9, 12, . . .).
S_1: The number of input 1s is congruent to 1 (mod 3), i.e., one more than a multiple of 3 (1, 4, 7, 10, . . .).
S_2: The number of input 1s is congruent to 2 (mod 3), i.e., two more than a multiple of 3 (2, 5, 8, 11, . . .).

The FSM should be started in state S_0 (because initially *no* 1s have arrived at the input), and as the 1s in the input stream arrive, the machine should move

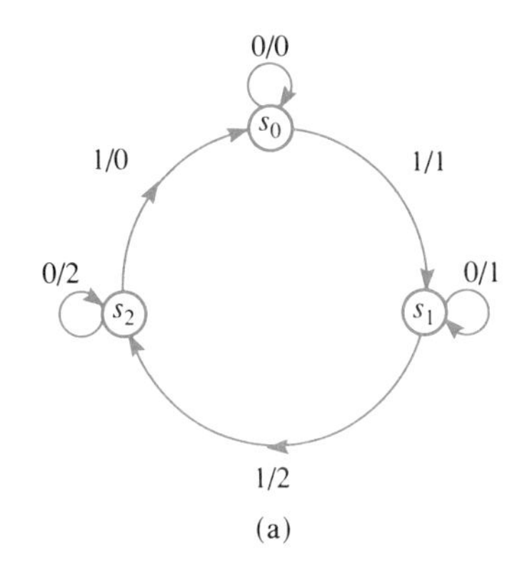

(a)

Present State	Next State $x = 0$	Next State $x = 1$	Output $x = 0$	Output $x = 1$
S_0	S_0	S_1	0	1
S_1	S_1	S_2	1	2
S_2	S_2	S_0	2	0

(b)

Figure 6.14 Solution to Example 6.7: (a) State diagram; (b) state table. (Initial state = S_0.)

periodically through the sequence of states S_0, S_1, S_2, S_0, S_1, S_2, S_0, S_1, etc. (An input 0 should not cause a change in states.) The *output* will always be the subscript on the next state. For example, if the machine is in state S_1 and the input is 1, the machine will move to S_2 and output 2. Similarly, if the machine is in S_2 and the input is 0, the machine will remain in S_2 and output 2. The state diagram and state table for this machine are shown in Figure 6.14. ■

Problems for Section 6.2

In Problems 1–3, give the state diagram and state table for a finite state machine with input alphabet = output alphabet = {0, 1} and for which the present output y_n equals 1 if and only if the indicated condition holds.

1. x_n and x_{n-1} are *different.*

2. x_n, x_{n-1}, and x_{n-2} are *all the same.*

3. x_n and x_{n-2} are *both zero.*

4. In the text, we noted that the first output y_1 of the FSM in Example 6.4 is indeterminate. Suppose we wished to specify that, in addition to (6.9), we also have $y_1 = 0$. Show how to design a *three-state* FSM to do this; its state diagram is partially given in the accompanying figure. Show how to connect the "start" state to the state diagram so that y_1 will always be correct.

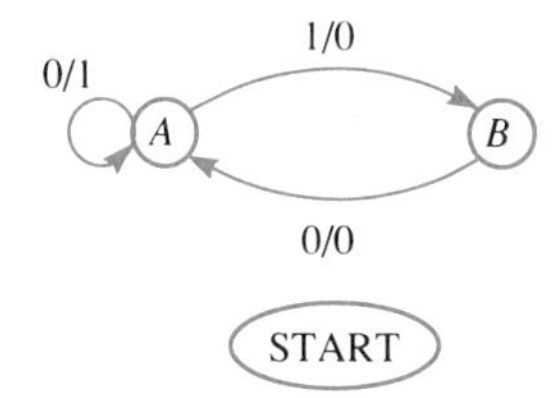

5. Repeat Problem 4 for $y_1 = 1$.

6. In Example 6.5, we designed an FSM which signalled the arrival of every *second* input 1 by outputting a 1. In this problem you are asked to design an FSM which similarly outputs a 1 to signal the arrival of every *fourth* input 1. Give the state table and the state diagram.

7. In Example 6.6, we designed a circuit which recognized the pattern 1111, i.e., which emitted a 1 whenever the input stream contained four 1s in a row. In this problem, design a 1101 recognizer.

8. Repeat Problem 7, but this time design a 1010 recognizer.

9. Design a 111 recognizer.

10. Design a mod 4 counter, analogous to the mod 3 counter in Example 6.7, i.e., an FSM whose output at a given time equals the total number of input 1s (mod 4) up to that time. For example:

x_n	0	1	1	0	1	1	1	1	0	1	1	0	0	1	...
y_n	0	1	2	2	3	0	1	2	2	3	0	0	0	1	...

11. Design an FSM counter whose output is 0 or 1, and such that a 1 indicates that the total number of input 1s so far is a multiple of 5. For example:

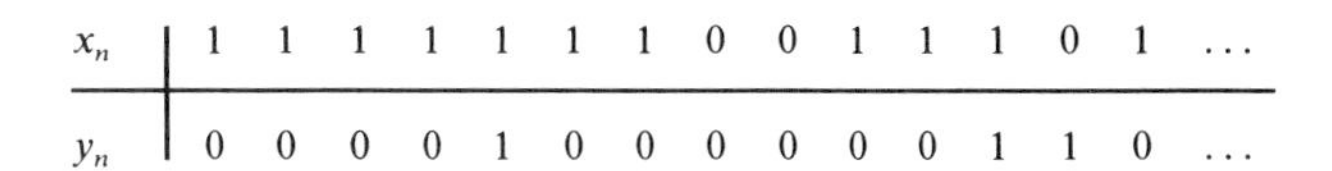

x_n	1	1	1	1	1	1	1	0	0	1	1	1	0	1	...
y_n	0	0	0	0	1	0	0	0	0	0	0	1	1	0	...

12. In Example 6.5 we specified that the initial state should be A. Suppose the machine were started instead in state B. What function would the FSM perform?

13. A "delay line" is a circuit for which the output sequence is a delayed version of the input sequence. For example, a two-unit delay line has $y_n = x_{n-2}$ for $n = 1, 2, \ldots$.

(a) Design a two-unit delay line FSM, with input and output alphabet $\{0, 1\}$.

(b) Design a two-unit delay line, with input and output alphabet $\{0, 1, 2\}$.

(c) If the input/output alphabet has N elements, how many states does a two-unit delay line have?

14. In Example 6.6, we specified S_0 as the initial state. What would happen if the machine were started in some other state? Be as specific as possible.

15. In Example 6.6, we used five states to build the FSM. It is actually possible to use only *four*. Verify this by appropriately labelling the edges in the accompanying figure.

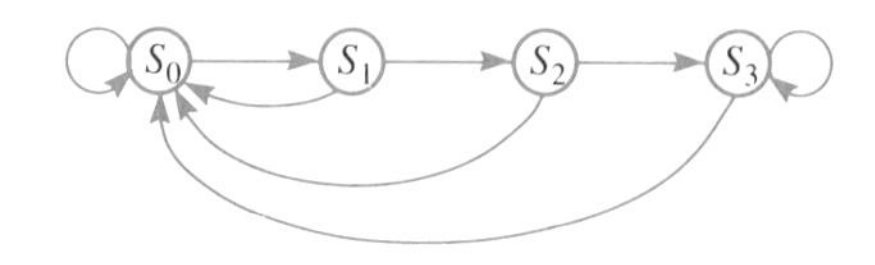

16. The sequential circuit in the accompanying figure is a realization of the FSM in Example 6.4, 6.5, 6.6, or 6.7. Which one?

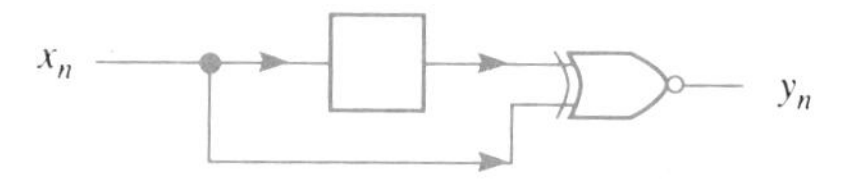

17. The sequential circuit in the accompanying figure is a realization of the FSM in Example 6.4, 6.5, 6.6, or 6.7. Which one?

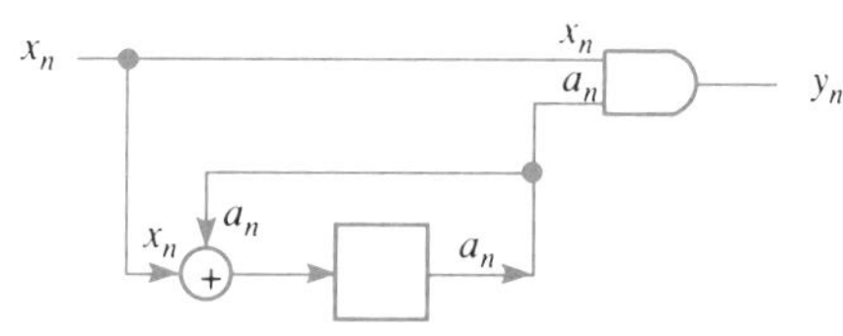

6.3 Reducing the Number of States with the State Equivalence Algorithm

In the first two sections of this chapter, we have seen how finite state machines and sequential circuits perform given tasks. The complexity of building a sequential circuit depends on the number of states in the FSM. It follows that when we design an FSM, we should use as few states as possible. In this section we will learn how to test a given FSM to see if there are any unnecessary states, and if there are, to get rid of them.

Let's begin with a somewhat absurd example. Consider the two sequential circuits in Figure 6.15. The circuit in Figure 6.15(a) consists of a single D flip-flop, and as we have seen, the relationship between the input x_n and the output y_n is $y_n = x_{n-1}$ (see Equation 6.1). Now consider the circuit in Figure 6.15(b). It is built from two flip-flops, and yet because of the way it is connected, it performs *exactly the same function* as the circuit in Figure 6.15(a). The second circuit contains a useless extra flip-flop!

However, before we judge the engineer who built the second circuit too harshly, let's have a look at the finite state machines that were used to design the two circuits (Figure 6.16). The FSM of Figure 6.16(a) corresponds to the circuit of Figure 6.15(a) and the FSM of Figure 6.16(b) corresponds to the circuit

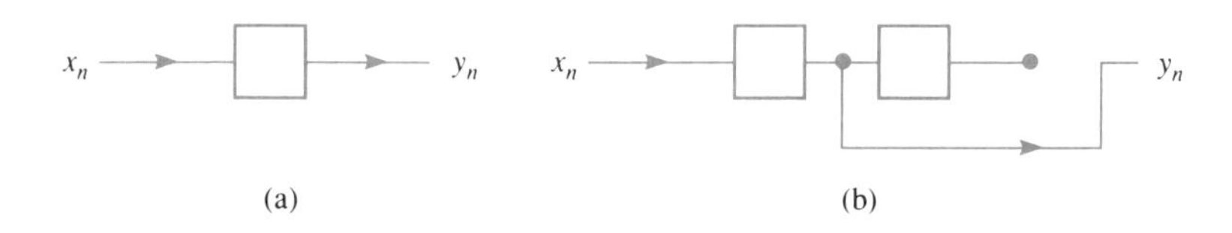

Figure 6.15 Two equivalent circuits(!).

of Figure 6.15(b). And although it is obvious how to simplify the circuit of Figure 6.15(b) (remove the useless flip-flop!), it is not so obvious how to simplify the corresponding FSM of Figure 6.16(b). Still, it is *possible* to simplify the FSM, and in this section we will learn a general technique for reducing the number of states in an FSM without disturbing the function it performs. When this technique is applied to the FSM of Figure 6.16(b), the FSM of Figure 6.16(a) results.

The key to our simplification procedure is the **state equivalence algorithm**, which we will soon describe completely. But before we plunge into the details of this algorithm, we'll give a somewhat informal discussion of how it works, and verify that the FSM in Figure 6.16(b) can safely be replaced by the simpler FSM of Figure 6.16(a).

The main idea involved is the notion of *state equivalence*. Informally, two states are said to be *equivalent* if the machine performs identically starting in either state. In other words, if S and S' are two different states, and if for every possible input sequence the corresponding output sequence is the same, whether the machine is started in S or S', then we say that S and S' are *equivalent states*. We shall use the notation $S \equiv S'$ to indicate that two states are equivalent.

For example, in the FSM of Figure 6.16(b) states A and C are equivalent, because A and C have *identical entries* in the state table:

	Next State		**Output**	
	$x = 0$	$x = 1$	$x = 0$	$x = 1$
A	A	B	0	0
C	A	B	0	0

The state table shows clearly that the machine must perform the same, whether started in A or C. Similarly, $B \equiv D$, again because the corresponding entries in the state table are identical.

Present State	**Next State**		**Output**	
	$x = 0$	$x = 1$	$x = 0$	$x = 1$
A	A	B	0	0
B	A	B	1	1

(a)

Present State	**Next State**		**Output**	
	$x = 0$	$x = 1$	$x = 0$	$x = 1$
A	A	B	0	0
B	C	D	1	1
C	A	B	0	0
D	C	D	1	1

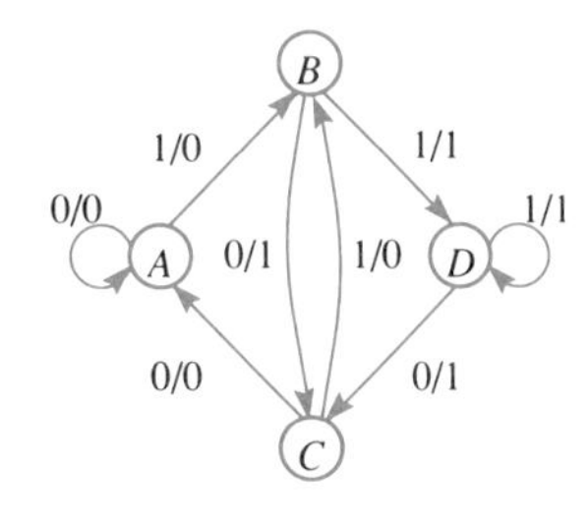

(b)

Figure 6.16 Two equivalent FSMs (?).

Equivalent states do not always have identical entries in the state table, as we shall soon see. However, equivalent states *do* always have identical entries in the "Output" part of the state table. In other words, if two states *differ* in the "Output" column, they cannot be equivalent. For example, consider states A and B in Figure 6.16(b). They differ in the state table's "Output" column:

	Output	
	$x = 0$	$x = 1$
A	0	0
B	1	1

If the machine is started in state A, and if the input sequence begins with 0, the output sequence will begin with 0. But if the machine is started instead in state B, and given the *same* input sequence starting with 0, the output sequence will begin with 1; and so A and B cannot be equivalent. More generally, if S and S' are two states, and if there is any "Output" entry for which S and S' differ, say for $x = a$, then the response of the machine with initial state S to an input stream with $x_1 = a$ will be different from the response to the same sequence if the initial state is S'. Thus S and S' cannot be equivalent.

The relationship of state equivalence is an *equivalence relation* as defined in Section 1.4 (see Problem 6), and so the set of states can be partitioned into a set of *equivalence classes*. If we choose one state from each equivalence class as a *representative* of that class, we can define a *reduced machine* whose states are the equivalence class representatives. For the FSM of Figure 6.16(b) we might have the following:

Equivalence Class	Representative
$\{A, C\}$	A
$\{B, D\}$	B

Here the reduced machine will have state set $\{A, B\}$. Its state table is derived from the state table of the original machine simply by deleting the rows not corresponding to equivalence class representatives, and replacing each remaining "Next State" entry with the representative of its equivalence class. This procedure is illustrated in Table 6.1. In Table 6.1 we begin with the state table from Figure 6.16(b) and delete rows C and D, since C and D are *not* the representatives of their equivalence classes. Then we replace C by A and D by B, since A is the representative of the equivalence class containing C and B is the representative of the equivalence class containing D. The resulting state table is shown in Table 6.1(b). It is the same as the state table in Figure 6.16(a), and so we have shown that the two FSMs in Figure 6.16 are indeed equivalent.

This simplification procedure can be applied to any FSM. One first finds the state equivalence classes; then a representative from each equivalence class is chosen; and finally a new machine is constructed using only the representatives as states. As we shall see, the first step (finding the equivalence classes) is the most difficult, but the following example will give us practice with the other two steps.

TABLE 6.1 Reducing the FSM of Figure 6.16(b), using the equivalence class representatives $\{A, C\} \rightarrow A$ and $\{B, D\} \rightarrow B$

Present State	Next State		Output	
	$x = 0$	$x = 1$	$x = 0$	$x = 1$
A	A	B	0	0
B	~~C~~ A	~~D~~ B	1	1
~~C~~	~~A~~	~~B~~	~~0~~	~~0~~
~~D~~	~~C~~	~~D~~	~~1~~	~~1~~

(a)

Present State	Next State		Output	
	$x = 0$	$x = 1$	$x = 0$	$x = 1$
A	A	B	0	0
B	A	B	1	1

(b)

EXAMPLE 6.8 Consider the FSM whose state table is given in Table 6.2. Given that the state equivalence classes are $\{A\}$, $\{B\}$, $\{C, D\}$, $\{E, F, G\}$, $\{H\}$, construct the reduced machine.

SOLUTION The first step is to select equivalence class representatives. In this case there are only two equivalence classes in which we have any choice, viz. $\{C, D\}$ and $\{E, F, G\}$, and if we choose C and E as the representatives of

TABLE 6.2 FSM for Example 6.8

Present State	Next State		Output	
	$x = 0$	$x = 1$	$x = 0$	$x = 1$
A	B	H	0	0
B	F	D	0	0
C	A	F	0	1
D	A	G	0	1
E	D	B	1	0
F	C	B	1	0
G	D	B	1	0
H	C	A	0	0

these classes, we have the following:

Equivalence Class	Representative
$\{A\}$	A
$\{B\}$	B
$\{C, D\}$	C
$\{E, F, G\}$	E
$\{H\}$	H

To produce the state table for the reduced machine, we start with the original state table, delete rows D, F, and G (states which are *not* representatives), and replace each state with the representative of its equivalence class; see Table 6.3(a). The final reduced machine is shown in Table 6.3(b). ■

Now we know what to do once the equivalence classes are given; but how do we find the equivalence classes in the first place? For the FSM described in Figure 6.16(b), finding equivalence classes was easy, since equivalent states have *identical* entries in the state table. However, in Example 6.8 the situation is more

TABLE 6.3 Solution to Example 6.8. (a) The original state table. (b) The reduced state table.

Present State	Next State		Output	
	$x = 0$	$x = 1$	$x = 0$	$x = 1$
A	B	H	0	0
B	~~F~~ ^{E}	~~D~~ ^{C}	0	0
C	A	~~F~~ ^{E}	0	1
~~D~~	~~A~~	~~G~~	~~0~~	~~1~~
E	~~D~~ ^{C}	B	1	0
~~F~~	~~C~~	~~B~~	~~1~~	~~0~~
~~G~~	~~D~~	~~B~~	~~1~~	~~0~~
H	C	A	0	0

(a)

Present State	Next State		Output	
	$x = 0$	$x = 1$	$x = 0$	$x = 1$
A	B	H	0	0
B	E	C	0	0
C	A	E	0	1
E	C	B	1	0
H	C	A	0	0

(b)

complicated. There, for example, we have $F \equiv G$, and yet the entries in the state table for F and G aren't identical:

	Next State		Output	
	$x = 0$	$x = 1$	$x = 0$	$x = 1$
F	C	B	1	0
G	D	B	1	0

We notice two important things about this abbreviated state table. First, the entries under "Output" *are* identical. This is necessarily so, as we observed earlier. Second, although the entries under "Next State" aren't *identical*, they *are equivalent*. That is, if we replace each entry in the state table with the representative of its equivalence class, rows corresponding to equivalent states will be equal:

	Next State		Output	
	$x = 0$	$x = 1$	$x = 0$	$x = 1$
F	C	B	1	0
G	C	B	1	0

(Here we have used C as the representative of the equivalence class $\{C, D\}$.) These two observations lead to a general algorithm for finding the state equivalence classes, which is described in Figure 6.17.

The first step in the state equivalence (SE) algorithm is to partition the set of states into subsets for which the "Output" entries in the state table are identical. This "first partition" we denote by P_1. The next step is to pretend that the subsets of P_1 are indeed the desired equivalence classes, by choosing representatives for them, and replacing each entry in the "Next State" part of the original state table by the representative of its subset. If all of the states in each subset now have identical entries in the state table, we say that the partition is "consistent"; the subsets really are the equivalence classes, and the algorithm stops. Otherwise, the partition is "refined" by breaking each subset into smaller subsets for which the entries in the modified state table *are* identical, and the procedure is continued, as shown in Figure 6.17. The following example will make this clearer.

EXAMPLE 6.9 Use the SE algorithm of Figure 6.17 to find the state equivalence classes for the FSM described in Figure 6.18.

SOLUTION The first step is to calculate the partition P_1, i.e., to divide the state set $\{A, B, C, D, E\}$ into subsets according to the "Output" entries in the state table. We find that states A, B, and C all have the same entries under "Output," and that D and E also have identical entries. So our *first partition* of the state set is as follows:

$$P_1: \{A, B, C\}, \{D, E\}.$$

To see whether P_1 is "consistent," we choose representatives for its subsets, say A to represent $\{A, B, C\}$ and D to represent $\{D, E\}$. Next, we replace each entry

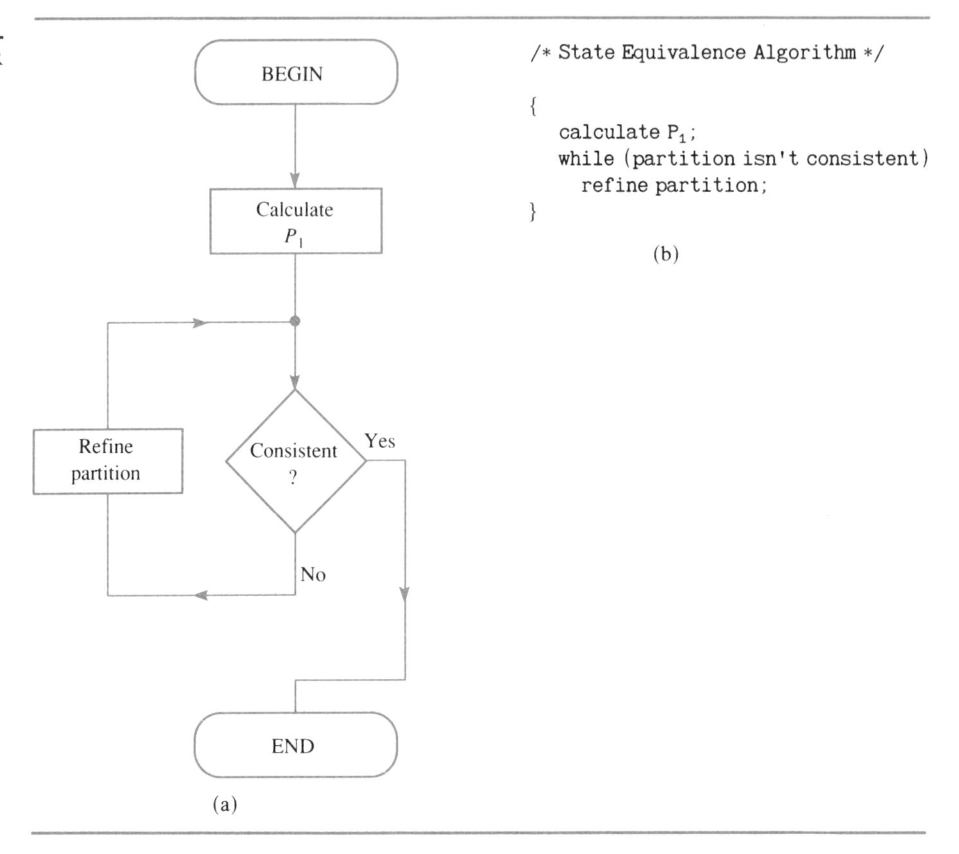
BEGIN
Calculate P_1
Consistent ?
Yes
No
Refine partition
END
(a)
/* State Equivalence Algorithm */
{
calculate P1;
while (partition isn't consistent)
refine partition;
}
(b)

Figure 6.17 The state equivalence algorithm: (a) Flowchart; (b) pseudocode.

	Next State		Output	
	$x = 0$	$x = 1$	$x = 0$	$x = 1$
A	B	C	1	0
B	C	D	1	0
C	B	E	1	0
D	E	B	0	1
E	D	C	0	1

Figure 6.18 The FSM for Example 6.9.

in the "Next State" part of the state table with the representative of its subset, i.e., A, B, and C are all replaced with A, and D and E are both replaced with D. The result of this substitution, which we shall call the *first next-state table*, or NST_1 for short, is given in abbreviated form below:

NST_1:

	Next State	
A	A	A
B	A	D
C	A	D
D	D	A
E	D	A

Now we check to see whether the states in each subset of the first partition have identical entries in NST_1. We find that of the three states in the subset $\{A, B, C\}$, two of the states (B and C) have identical entries, but state A is different. This won't do; the partition P_1 is *inconsistent*, and we must *refine* it. We must partition the subset $\{A, B, C\}$ into subsets for which the entries in NST_1 *are* identical, i.e., into $\{A\}$ and $\{B, C\}$. In the other subset of P_1, $\{D, E\}$, everything is okay, since both B and C have the entry DA in NST_1.

We have now refined the first partition P_1 into subsets, according to the entries in NST_1. This new partition we call the *second partition*, or P_2 for short:

$$P_2: \{A\}, \{B, C\}, \{D, E\}.$$

According to the algorithm described in Figure 6.17, we now must test P_2 for consistency. To do this we choose representatives for the subsets of P_2, say A, B, and D, and substitute these representatives in the original state table. The result, which we'll call NST_2, is as follows:

NST_2:

	Next State	
A	*B*	*B*
B	*B*	*D*
C	*B*	*D*
D	*D*	*B*
E	*D*	*B*

To see if P_2 is consistent, we check NST_2 to see whether or not the states in each subset of P_2 have identical entries. For the subset $\{A\}$, there's nothing to check, since there's only one state. For the subset $\{B, C\}$, we find that both states have the entry BD in NST_2. For $\{D, E\}$, both entries are DB. Thus P_2 *is* consistent, and so the subsets of P_2 are the state equivalence classes. The corresponding reduced machine is shown in Figure 6.19. ■

Example 6.9 gives us some useful practical experience with the SE algorithm. But as we've said previously, it's not enough to know *how* an algorithm works; we also need to know *why* it works. So in the next few pages, we'll give a *proof* that the state reduction procedure we have described always works.

First, we need to state carefully what it is we propose to prove. We begin by taking as *state equivalence classes* the subsets of the final partition produced by the SE algorithm. Notice that this is a *precise* definition and differs from our

	Next State		Output	
	$x = 0$	$x = 1$	$x = 0$	$x = 1$
A	*B*	*B*	1	0
B	*B*	*D*	1	0
D	*D*	*B*	0	1

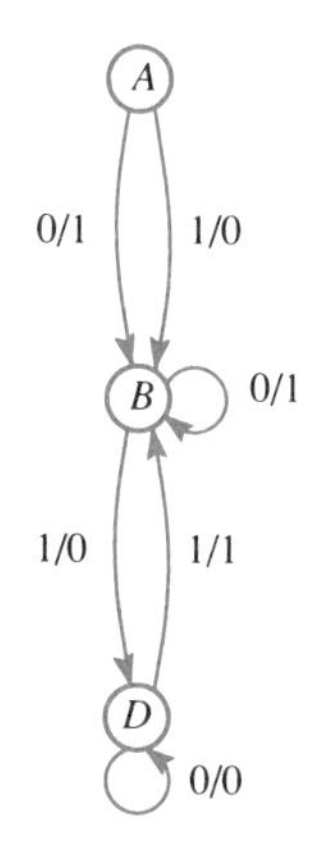

Figure 6.19 Reduced version of the FSM in Figure 6.18.

earlier *informal* definition ("the machine performs identically, starting in either state"). We then choose representatives for each equivalence class, and use them to construct a *new* FSM, as we did in Examples 6.8 and 6.9. The new FSM is called the **reduced** *machine.*

If the *original* machine is started in a certain state, say S_1, and given a certain input sequence, say $x_1, x_2, \ldots$, it will in response produce a certain output sequence, say $y_1, y_2, \ldots$. We think of this sequence of operations as a "computation" performed by the original machine on the input sequence. If now the *reduced* machine is started in a state which is *equivalent* to S_1, and given the *same* input sequence, we will show that it will produce the same output sequence as the original machine. What this means is that the reduced machine can perform any computation that the original machine can. The next example will illustrate this.

EXAMPLE 6.10 Consider the original machine in Table 6.2 and the corresponding reduced machine in Table 6.3(b). Suppose the original machine starts in state D, and is given the input sequence 00110001 . . . .

(a) Calculate the corresponding state sequence and output sequence.
(b) Start the reduced machine in a state equivalent to D, and give it the same input sequence. Compare the resulting state sequence and output sequence in part (a).

SOLUTION

(a) Using the state table in Table 6.2, we find that the *original* machine behaves as follows:

Input sequence		0	0	1	1	0	0	0	1
State sequence	$D \to A \to B \to D \to G \to D \to A \to B \to D$								
Output sequence		0	0	0	1	1	0	0	0

(b) As given in Example 6.8, the state equivalence classes are $\{A\}$, $\{B\}$, $\{C, D\}$, $\{E, F, G\}$, $\{H\}$. (Incidentally, in Problem 18 you will be asked to verify, using the algorithm in Figure 6.17, that they *are* actually the equivalence classes.) The state of the reduced machine in Table 6.3(b) which is equivalent to D is therefore C. Using the state table in Table 6.3(b), we now find that the *reduced* machine behaves as follows:

Input sequence		0	0	1	1	0	0	0	1
State sequence	$C \to A \to B \to C \to E \to C \to A \to B \to C$								
Output sequence		0	0	0	1	1	0	0	0

Comparing the behavior of the two machines, we find that the two *output sequences* are the same. Less evident, but equally important, is the fact that the *state sequences are equivalent*, i.e., that every state in the reduced state sequence is equivalent to the corresponding state in the original state sequence:

Original state sequence	D	A	B	D	G	D	A	B	D
	$\updownarrow$	$\updownarrow$	$\updownarrow$	$\updownarrow$	$\updownarrow$	$\updownarrow$	$\updownarrow$	$\updownarrow$	$\updownarrow$
Reduced state sequence	C	A	B	C	E	C	A	B	C

What happened in Example 6.10 happens in general. If the original machine and the reduced machine are started in equivalent states, and given identical input sequences, the resulting output sequences will be the same. Thus the reduced machine can perform the same computations as the original machine. We now intend to *prove* this important fact. The key to the proof is that the corresponding state sequences are equivalent.

The general situation is shown in Table 6.4. In Table 6.4, the initial states S_1 and S_1' are assumed to be equivalent. Both machines are given the same input sequence, viz. $x_1, x_2, x_3, \ldots$. The state sequence for the original machine is denoted by $S_1, S_2, S_3, \ldots$, and the corresponding output sequence by $y_1, y_2, y_3, \ldots$. Similarly the state sequence for the reduced machine is denoted by $S_1', S_2', S_3', \ldots$, and the output sequence by $y_1', y_2', y_3', \ldots$. Our goal is to show that the two output sequences are the same, i.e.,

(6.10) $$y_k' = y_k \quad \text{for} \quad k = 1, 2, 3, \ldots.$$

The way to do this is to show that the *state* sequences are equivalent, i.e.,

(6.11) $$S_k' \equiv S_k, \quad \text{for} \quad k = 1, 2, 3, \ldots.$$

Once we know that (6.11) is true, (6.10) follows immediately, since equivalent states give the same outputs.

It therefore remains to prove (6.11). We will use induction. The first step is to show that (6.11) is true for $k = 1$, i.e., that $S_1' \equiv S_1$. But this is true by assumption; we started both machines in equivalent states.

Next we assume that (6.11) is true for k, and try to prove it for $k + 1$. The diagram in Table 6.5 will be helpful here. We are assuming that S_k and S_k' are equivalent, and want to show that S_{k+1} and S_{k+1}' are equivalent. The key to

TABLE 6.4 Behavior of (a) the original machine and (b) the reduced machine, when the input is $x_1, x_2, \ldots$ (states S_1 and S_1' are assumed equivalent).

Input sequence		x_1		x_2		x_3	$\cdots$			x_k		x_{k+1}
State sequence	S_1	$\rightarrow$	S_2	$\rightarrow$	S_3	$\rightarrow$	$\cdots$	$\rightarrow$	S_k	$\rightarrow$	S_{k+1}	$\longrightarrow \cdots$
Output sequence		y_1		y_2		y_3				y_k		y_{k+1}

(a)

Input sequence		x_1		x_2		x_3	$\cdots$			x_k		x_{k+1}
State sequence	S_1'	$\rightarrow$	S_2'	$\rightarrow$	S_3'	$\rightarrow$	$\cdots$	$\rightarrow$	S_k'	$\rightarrow$	S_{k+1}'	$\longrightarrow \cdots$
Output sequence		y_1'		y_2'		y_3'				y_k'		y_{k+1}'

(b)

TABLE 6.5 Needed in the induction proof

Original machine	$S_k \xrightarrow{x_k} S_{k+1}$		
	$\parallel\!\!\mid \qquad \parallel\!\!\mid$	←	"Partition is consistent"
Original machine	$S'_k \xrightarrow{x_k} T$		
	$\qquad \parallel\!\!\mid$	←	"Definition of reduced machine"
Reduced machine	$S'_k \xrightarrow{x_k} S'_{k+1}$		

showing this is to consider the next state of the *original* machine when the present state is S'_k and the input is x_k. This state is called T in Table 6.5. Now S_{k+1} and T are equivalent, since equivalent states (S_k and S'_k) must have equivalent next states:

$$S_{k+1} \equiv T. \tag{6.12}$$

But also T must be equivalent to S'_{k+1}, since the next-state entries in the reduced machine are obtained by replacing the next-state entries in the original state table with the equivalence class representatives:

$$T \equiv S'_{k+1}. \tag{6.13}$$

And now it follows from (6.12) and (6.13) and the *transitivity property* of equivalence relations (Section 1.4), that $S_{k+1} \equiv S'_{k+1}$. This completes the induction proof of (6.11), and thus the reduced machine does perform the same function as the original machine, as advertised.

The reduced machine will almost always be preferred to the original machine, because it is less complex (has fewer states, anyway) and can do the same job. However, our final example shows that sometimes the reduced machine is the same as the original machine—the SE algorithm labors mightily but gets nowhere!

EXAMPLE 6.11 Apply the state equivalence algorithm to the FSM in Figure 6.20.

Present State	Next State		Output	
	$x = 0$	$x = 1$	$x = 0$	$x = 1$
A	A	B	0	0
B	B	C	0	0
C	C	D	0	0
D	D	E	0	0
E	E	F	0	0
F	F	A	0	1

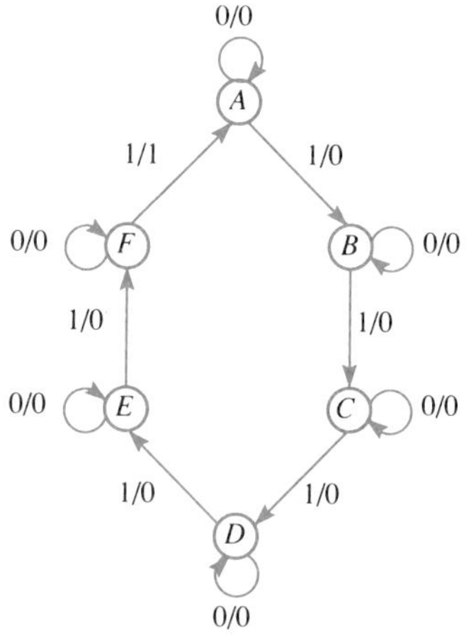

Figure 6.20 Can this mod 6 counter be simplified?

SOLUTION This FSM is similar to several we have considered earlier, for example, the mod 2 counter in Figure 6.12 and the mod 3 counter in Figure 6.14. What the machine in Figure 6.20 does is to signal the arrival of every sixth input 1 with an output 1 (otherwise the output is 0).

If we apply the algorithm as described in Figure 6.17, the first step is to calculate the first partition, P_1. Looking at the "Output" entries of the state table, we find that all of the states except state F have identical entries. Hence,

$$P_1: \{A, B, C, D, E\}, \{F\}.$$

Choosing A and F as subset representatives, the first next-state table is as follows:

NST$_1$:

	Next State
A	(A, A)
B	(A, A)
C	(A, A)
D	(A, A)
E	(A, F)
F	(F, A)

We see that in NST$_1$, all of the states in $\{A, B, C, D, E\}$ except E have identical entries, and so the second partition P_2 is given by

$$P_2: \{A, B, C, D\}, \{E\}, \{F\}.$$

Choosing A, E, and F as P_2's subset representatives, we calculate the second next-state table:

NST$_2$:

	Next State
A	(A, A)
B	(A, A)
C	(A, A)
D	(A, E)
E	(E, F)
F	(F, A)

We see that in NST$_2$, state D differs from the other states in the subset $\{A, B, C, D\}$, and so the third partition P_3 is given by

$$P_3: \{A, B, C\}, \{D\}, \{E\}, \{F\}.$$

Continuing, we calculate NST$_3$:

NST$_3$:

	Next State
A	(A, A)
B	(A, A)
C	(A, D)
D	(D, E)
E	(E, F)
F	(F, A)

Thus we have

$$P_4: \{A, B\}, \{C\}, \{D\}, \{E\}, \{F\},$$

and the next-state table is as follows:

NST$_4$:

	Next State
A	(A, A)
B	(A, C)
C	(C, D)
D	(D, E)
E	(E, F)
F	(F, A)

Too bad! In NST$_4$, even A and B differ, and so,

$$P_5: \{A\}, \{B\}, \{C\}, \{D\}, \{E\}, \{F\}.$$

And now the algorithm must stop, since no further refinement of the partition is possible. We conclude from all of this work that no two states are equivalent, and so no reduction in the number of states is possible. This conclusion is rather typical; if the machine is well-designed, it won't have any redundant states. But the state equivalence algorithm is usually worth trying, just in case. ■

Problems for Section 6.3

1. (Review Question.) How many partitions of the set $\{A, B, C, D\}$ are there?

2. In the course of our discussion of the FSM in Figure 6.16(b), we encountered the partition $\{A, C\}$, $\{B, D\}$ of the set $\{A, B, C, D\}$, and we chose as equivalence class representatives A and B.

(a) How many *other* choices of equivalence class representatives are there?

(b) Consider the partition $\{A, B, C\}, \{D, E\}$ that we encountered in Example 6.8. How many choices for equivalence class representatives are there?

3. Continuing Problem 2, in general, if $A = A_1 \cup A_2 \cup \cdots \cup A_r$ is a partition of the set A into r disjoint subsets, how many choices for equivalence class representatives are there?

4. Suppose we choose C and B as equivalence class representatives for the FSM of Figure 6.16(b). Construct the reduced machine (state table and state diagram).

5. In Example 6.8 we chose A, B, C, E, and H as equivalence class representatives. Suppose we choose A, B, D, G, and H, instead. Construct the state table for the reduced machine.

6. Show that state equivalence, as defined informally in the text ($S \equiv S'$ if and only if the machine performs identically starting in S or S'), is indeed an equivalence relation. That is, verify the following:

(a) $S \equiv S$ for all states S (reflexive)

(b) $S \equiv T$ if and only if $T \equiv S$ (symmetric)

(c) If $S \equiv T$ and if $T \equiv U$, then $S \equiv U$ (transitive).

7. Consider the FSM described in the accompanying table. Given that the state equivalence classes are $\{A, D\}$, $\{B, G\}$, $\{E\}$, $\{C\}$, $\{F\}$, construct the reduced machine.

	Next State		Output	
	$x = 0$	$x = 1$	$x = 0$	$x = 1$
A	B	E	0	0
B	F	C	1	1
C	A	C	0	1
D	G	E	0	0
E	C	F	1	1
F	F	D	1	0
G	F	C	1	1

8. Consider the FSM described in the accompanying table. Find the equivalence classes, and construct the reduced machine.

	Next State		Output	
	$x = 0$	$x = 1$	$x = 0$	$x = 1$
A	C	B	1	1
B	D	C	0	0
C	E	D	1	1
D	B	C	0	0
E	E	F	0	0
F	C	G	1	1
G	G	F	0	0
H	B	I	1	0
I	D	H	1	0

9. Consider the FSM described in the accompanying table. Find the equivalence classes, and construct the reduced machine.

Present State	Next State		Output	
	$x = 0$	$x = 1$	$x = 0$	$x = 1$
A	F	C	0	0
B	E	G	1	0
C	F	B	1	1
D	G	E	0	1
E	B	D	1	0
F	G	F	0	0
G	D	B	0	1
H	E	B	1	0

10. Consider the FSM described in the accompanying table. Find the equivalence classes, and construct the reduced machine.

	Next State		Output	
	$x = 0$	$x = 1$	$x = 0$	$x = 1$
A	F	C	0	0
B	E	G	1	0
C	F	B	1	1
D	G	E	0	1
E	B	D	1	0
F	G	F	0	0
G	D	B	0	0
H	E	B	1	0

11. Try to simplify the FSM in Figure 6.4.

12. Try to simplify the FSM in Figure 6.7.

13. Try to simplify the FSM in Figure 6.10.

14. Try to simplify the FSM in Figure 6.11.

15. Try to simplify the FSM in Figure 6.12.

16. Try to simplify the FSM in Figure 6.13.

17. Try to simplify the FSM in Figure 6.14.

18. Apply the SE algorithm to the FSM in Example 6.8 to verify that the given state equivalence classes are correct.

19. In Example 6.11 we discussed an FSM that signals the arrival of every sixth input 1 with an output 1.

(a) Design an analogous FSM that signals every seventh 1, i.e., exhibit the state table and state diagram.

(b) Apply the SE algorithm to the machine you designed in part (a).

(c) How many times did you have to `refine partition` in part (b)?

20. Continuing Problem 19, consider now the general problem of designing an FSM that signals the arrival of every nth input 1 with an output 1.

(a) How many states will such a machine have?

(b) If the SE algorithm is applied to it, how many times will the `refine partition` step be executed?

21. Consider an FSM for which you know that *all* the entries in the output portion of the state table are 0, but for which you know nothing at all about the next-state portion of the state table. Can you say anything about what will happen if the SE algorithm is applied?

Our proof of the correctness of the SE algorithm was not quite complete. We showed that, provided the answer to `consistent?` was eventually yes, the resulting partition could be used to build a reduced machine. But we *didn't* show that this would ever happen. To complete the proof, we need to show that the algorithm will *terminate*, i.e., will not loop endlessly around the `consistent?/refine partition` loop. Do this by solving Problems 22–25.

22. Suppose the state set contains n elements. What is the *largest possible* number of subsets in a partition? What is the *fewest*?

23. Continuing Problem 22, what can you say about the number of sets in the partition *before* and *after* the execution of a `refine partition` step? Does it increase? Decrease? Stay the same?

24. Based on the answers to Problems 22 and 23, what can you conclude about the number of times the `refine partition` step can be executed?

25. Continuing Problem 24, show that for any FSM, the SE algorithm will eventually terminate.

26. If you've done Problems 22–25, this one should be easy. Let n denote the number of states in a given FSM. In terms of n, what is the *largest possible* number of times the `consistent?` question will need to be answered?

27. In the text, we gave two *different* definitions of state equivalence. Our first, "informal," definition was this:

> Two states S and S' are *equivalent* if, *for every possible input sequence*, the corresponding output sequence is the same whether the initial state is S or S'.

Our second, "precise," definition was this:

> Two states S and S' are *equivalent* if they belong to the same subset of the final partition produced by the SE algorithm.

Show that two states which are equivalent according to the second definition are also equivalent according to the first. [*Hint*: Let S_1 and T_1 be two states which are in the same subset of the SE algorithm partition, and let $(x_1, x_2, \ldots)$ be an arbitrary input sequence. Show that the corresponding output S with S_1 and T as the starting states are both given by the output of the reduced machine as depicted in Table 6.4(b).]

28. In this problem we will consider a slightly modified version of the SE algorithm, which is described by the accompanying flowchart. The difference between this algorithm and the one in Figure 6.17 is that this algorithm never checks to see whether the current partition is "consistent"; it just happily continues around the loop forever, continuing to "refine" the current partition, even when this step isn't necessary! It therefore calculates an *endless sequence* of partitions $P_1, P_2, P_3, \ldots$, even though after a while the partitions will all be the same. (These partitions are the same as those produced by the algorithm in Figure 6.17, up to the point where that algorithm terminates.)

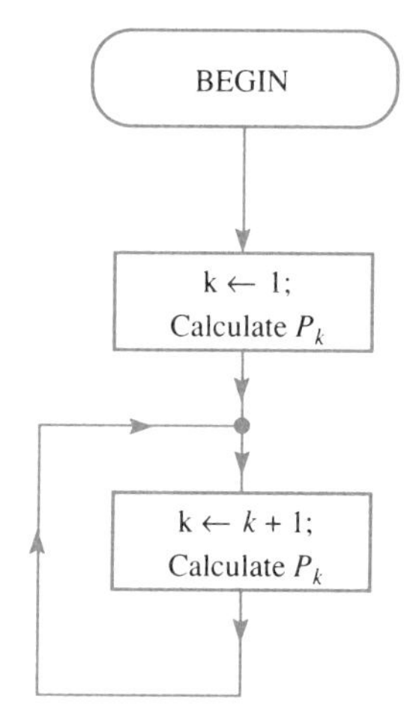

In this problem, we will discover an interesting property of these partitions P_k. Let's first define a new equivalence relation on the set of states:

> Two states S and S' are *k-equivalent*, if for *every possible input sequence of length* k, the corresponding output sequence is the same, whether the initial state is S or S'.

Prove that the following statement is true for all $k = 1, 2, 3, \ldots$, using induction:

> B_k: States which are in the same subset of P_k are k-equivalent.

6.4 Finite State Machines and Pattern Recognition; Finite Automata

In Example 6.6, we designed an FSM that recognizes the pattern 1111 by outputting a 1 whenever it encounters four 1s in a row in the input string. FSMs that recognize patterns turn out to be of considerable importance in many applications, especially in the design of system software like compilers, text editors, etc. In this section, we will continue our study of finite state machines, now emphasizing their pattern recognition capabilities. We will find that for purposes of pattern recognition, a somewhat simplified kind of FSM called a *deterministic finite automaton* can be used. We will also introduce a generalization called a *nondeterministic finite automaton*, which, despite its reputation for supernatural behavior, also proves to be a valuable tool for pattern recognition.

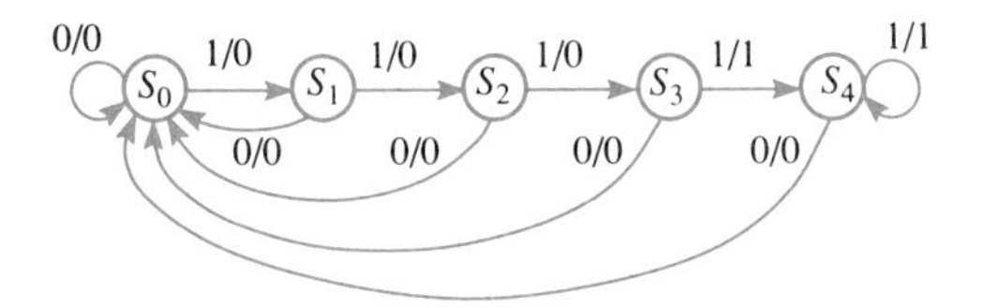

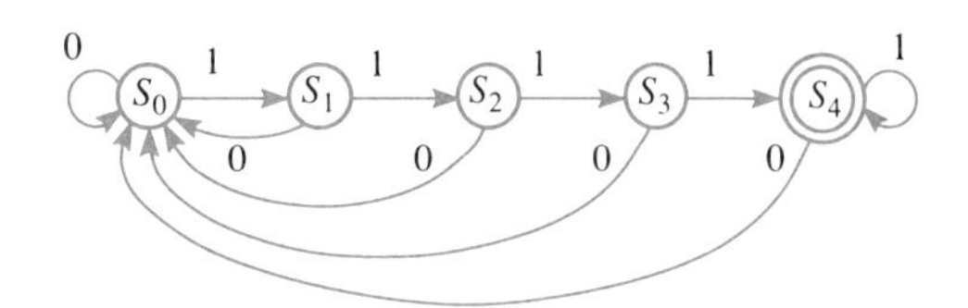

Figure 6.21 (a) State diagram for a 1111 recognizer. (b) Simplified state diagram, showing S_4 as an accepting state. (This is an example of a DFA.)

We begin by reviewing Example 6.6. Figure 6.21(a) shows the state diagram for the 1111 recognizer, which is identical to Figure 6.13. As we noted in the solution to Example 6.6, state S_4 is special; the output at a given point in time is 1 if and only if the machine enters S_4 at that point in time. Thus the outputs don't really have to be shown; we simply need to indicate that S_4 is a special, or accepting state, as we have done in Figure 6.21(b). In fact, we can think of Figure 6.21(b) as describing a special kind of FSM called a **deterministic finite automaton** (DFA), which differs from an ordinary FSM in two respects: It has no output, and some of its states are distinguished as **accepting states**. In Figure 6.22 we show several more DFAs, represented both by state diagrams and state tables. In each state diagram we have indicated the initial state with an arrow marked "start," and the accepting states with double circles. (Notice that several edges have *two* labels; for example, in Figure 6.22(a) the loop from state c back

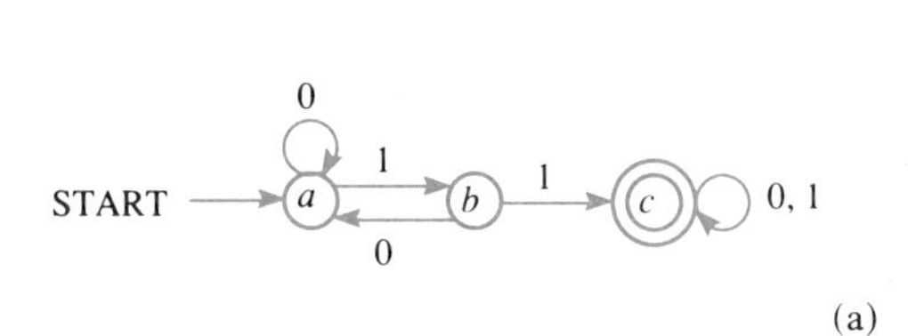

(a)

Present State	Next State $x = 0$	Next State $x = 1$
a	a	b
b	a	c
c	c	c

(b)

Present State	Next State $x = 0$	Next State $x = 1$
S_0	S_0	S_1
S_1	S_1	S_2
S_2	S_2	S_0

(c)

Present State	Next State $x = 0$	Next State $x = 1$
0	3	1
1	3	2
2	2	2
3	3	3

Figure 6.22 Three simple DFAs, represented by state diagrams and state tables: (a) Initial state = a and accepting state = c; (b) initial state = S_0 and accepting states = S_1, S_2; (c) initial state = 0 and accepting state = 2.

to state c is labelled "0, 1." This is just a shorthand way of indicating that there are really two loops at c, one labelled "0" and one labelled "1," so that if the DFA is in state c and the input is either 0 or 1, the next state will be state c.)

It may be helpful to think of a DFA as having an internal clock and an indicator light. Initially, the DFA is in its initial state and the light is off. Just prior to each clock tick, the DFA will be in one of its states, and the light will either be on or off. At the tick, the machine will read an input symbol, and in response move to a new state, which is *determined* by its old state and the input. (This explains the D in DFA.) If the new state is an accepting state, the indicator light is turned on; otherwise, it is turned off. For example, if the DFA in Figure 6.21(b) begins in state S_0 and is given the input string 01111100, the machine will behave as follows:

Tick number		1	2	3	4	5	6	7	8
Input		0	1	1	1	1	1	0	0
State	(0)	0	1	2	3	4	4	0	0
Light	(*Off*)	*Off*	*Off*	*Off*	*Off*	*On*	*On*	*Off*	*Off*

A given DFA is said to recognize—or more commonly, to **accept**—a string of symbols if, when started in its initial state and given the string as input, it ends up in an accepting state after the last symbol in the string is read. For example, the above computation shows that the DFA of Figure 6.21(b) does not accept 01111100, although it does accept the two "prefixes" 01111 and 011111. In fact, of course, the DFA of Figure 6.21(b) accepts a string of 0s and 1s if and only if its last four symbols are all 1s. The set of all strings accepted by a given DFA is called the **language** accepted by the DFA. We will have much more to say about languages in Chapter 7. For now, we will just examine the languages accepted by the DFAs in Figure 6.22.

EXAMPLE 6.12 Which of the following strings are accepted by the DFA of Figure 6.22(a)?

(a) 0101. (b) 0001100. (c) 10000010.

SOLUTION

(a) Starting in the initial state, which in this case is state a, and using the string 0101 as input, the given DFA travels through the following sequence of states: $(a)abab$. Since the final state, in this case state b, is not an accepting state, this string is not accepted.
(b) Again, starting at state a but this time with the input 0001100, the state sequence is $(a)aaabccc$. Since state c is an accepting state, this string is accepted.
(c) Here the state sequence is $(a)baaaaba$, so the string isn't accepted.

(Actually, this particular DFA accepts a given string of symbols if and only if the string contains two consecutive 1s. See Problem 2.)

EXAMPLE 6.13 Describe the language accepted by the DFA of Figure 6.22(b).

SOLUTION This DFA is much like the mod 3 counter in Example 6.6. In fact, we can see that if it is started in S_0, the state will tell us the mod 3 value of the number of 1s in the input stream. When the machine is in S_0, the number of 1s is congruent to 0 (mod 3). When it is in S_1, the number of 1s is congruent to 1 (mod 3); and when it is in S_2, the number of 1s is congruent to 2 (mod 3). Since only S_1 and S_2 are accepting states, it follows that this DFA will accept any string of 0s and 1s for which the number of 1s is not a multiple of 3. ■

EXAMPLE 6.14 Describe the language accepted by the DFA in Figure 6.22(c).

SOLUTION Notice that if the input string begins with 0, the machine moves to state 3, a nonaccepting state from which it can never exit. If the string begins with a 1, the machine moves to state 1. If the next symbol is a 0, the machine drops down to state 3, and can never leave. But if the next symbol is also a 1, the machine enters the accepting state 2, which it can never leave. The conclusion: This DFA accepts only those strings that begin with two 1s. ■

In Examples 6.12, 6.13, and 6.14, we have considered the problem of finding the set of strings accepted by a given DFA. However, the opposite problem, i.e., to find a DFA which accepts a given set of strings, is of much more practical importance. For example, in text editors it is useful to have a command that allows the user to search a file for all strings of a given type, say all words beginning with *po*, or all words whose letters are in alphabetical order.* As we will see, once we have a DFA capable of accepting a given set of strings, it's not hard to write a computer program to simulate the operation of the DFA, and then to search a file for all strings of the desired type. However, finding such a DFA may not be easy; indeed, it may not even be possible! (For example, it is known that there is no DFA that accepts the set {10, 11, 101, 111, 1011, 1101, 10001, . . .} of the binary expansions of the prime numbers.) The sets which can be accepted by DFAs are called **regular sets**. We will study regular sets in some detail in Chapter 7, and we will show that many interesting sets are regular. (However, we won't be able to show that any particular set *isn't* regular; this topic is best left to advanced texts on automata theory.) Often, however, even when the set in question is regular, the DFA that accepts it may be very complicated and difficult to find. Even for a set as simple as "strings ending with four 1s in a row," the DFA can be fairly complicated, as we have already seen in Figure 6.21.

Luckily, there is a more general type of device, called a **nondeterministic finite automaton**, or NFA for short, that can be used to accept sets of strings, and which

* The popular UNIX operating system has two commands, GREP and EGREP, that allow such searches, and these powerful utilities are based on the techniques discussed in this section and in Chapter 7.

is usually much easier to design than the corresponding DFA. An NFA, like a DFA, is described by a state diagram or a state table (see Figure 6.23). One state is distinguished as an initial state, one or more states are distinguished as accepting states, and there are labelled, directed edges, connecting the states. (Notice that some edges are labelled ϵ; we will explain this later.) A string is **accepted** by the NFA if and only if there is a path from the initial state to one of the accepting states, such that the edge labels on the path generate the string.

For example, consider the NFA of Figure 6.23(a); T_0 is the initial state, and T_4 is the only accepting state. A path from T_0 to T_4 must consist of a number (possibly zero) of trips around the loop at T_0, followed by the direct trip $T_0T_1T_2T_3T_4$ to the accepting state T_4. The loop at T_0 is labelled "0, 1", which, as we have seen, means that there are really two loops at T_0, one labelled 0 and one labelled 1. Thus while the path is travelling around the loop at T_0, the path label can be any combination of 0s and 1s. However, on the $T_0T_1T_2T_3T_4$ part of the path, the label must be 1111. Therefore, every path from T_0 to T_4 generates a string of 0s and 1s ending in four 1s. Conversely, any string ending with four 1s can be generated by a path from T_0 to T_4. For example, 0101001111 is generated by a path consisting of six loops at T_0 followed by $T_0T_1T_2T_3T_4$. In other words, the NFA of Figure 6.23(a) accepts exactly the same strings as the DFA of Figure 6.21. This is true despite the fact that the NFA is considerably simpler than the DFA.

If we view an NFA as a labelled, directed graph, there are no problems. However, if we try to interpret it as the state diagram of a real machine, we encounter serious difficulties. For example, suppose the NFA of Figure 6.23(a) is in T_1 and the input is 0. Since there is no edge out of T_1 labelled 0, there is no next state. This embarrassing situation is sometimes explained away by saying that the NFA simply "shuts down," and accepts no further inputs. But if the NFA is in state T_0 and the input is 1, things get worse. Since there is an edge from T_0 to T_0, and an edge from T_0 to T_1, both labelled 1, there appear to be two possibilities for the next state: T_0 and T_1; which is it? Some texts resolve this dilemma by asserting that when faced with this problem, an NFA somehow "reproduces itself," thereby producing a copy of itself in state T_0 and a copy in state T_1. This is called *nondeterministic* behavior, and explains the N in NFA. We feel, however, that such interpretations are best left to science fiction writers, and will stick to our unexciting but comprehensible directed-graph viewpoint.

In an NFA, one of the legal edge labels is ϵ, which indicates "no edge label." An ϵ edge label means that the corresponding state transition can occur even if there is no input, or, equivalently, if the input is the empty string containing no symbols. This possibility further simplifies the design of NFAs, as the next example shows.

EXAMPLE 6.15 Describe the language accepted by the NFA of Figure 6.23(b).

SOLUTION The initial state is I, and the states 0 and 1 are both accepting states. There are two kinds of "accepting" paths, those from I to 0, and those from I to 1. A path from I to 0 consists of the edge from I to 0, whose label is

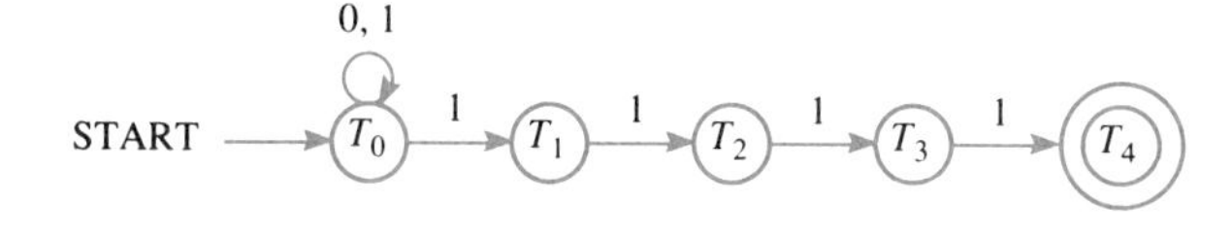

Present State	Next State	
	$x = 0$	$x = 1$
T_0	T_0	T_0, T_1
T_1	—	T_2
T_2	—	T_3
T_3	—	T_4
T_4	—	—

(a)

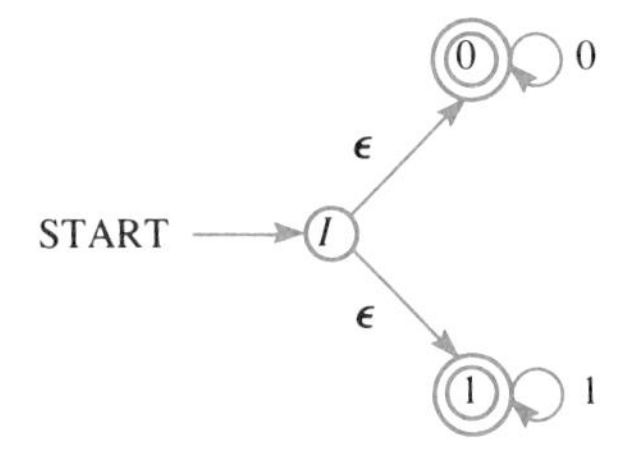

Present State	Next State		
	$x = 0$	$x = 1$	$x = \epsilon$
I	—	—	0, 1
0	0	—	—
1	—	1	—

(b)

Present State	Next State			
	$x = a$	$x = b$	$x = c$	$x = \epsilon$
0	0	—	—	1
1	—	1	—	2
2	—	—	2	—

(c)

Figure 6.23 Three simple NFAs, represented by state tables and state diagrams: (a) Initial state = T_0 and accepting state = T_4; (b) initial state = I and accepting states = 0, 1; (c) initial state = 0 and accepting state = 2.

ϵ, followed by a number of loops at 0, whose labels are all 0. It follows that the label on this type of path is just a run of 0s. Similarly, the paths from I to 1 produce strings consisting of a run of 1s. Thus the language accepted by the NFA of Figure 6.23(b) consists of those strings of 0s and 1s which are either all 0s or all 1s.

The next example further illustrates the use of ϵ's in designing NFAs.

EXAMPLE 6.16 Describe the language accepted by the NFA of Figure 6.23(c).

SOLUTION Here the alphabet is the three-element set $\{a, b, c\}$, the initial state is 0, and the accepting state is 2. The paths from state 0 to state 2 all consist of some (possibly no) loops at state 0 each labelled a, a (no-label) jump to state 1, some loops (labelled b) at state 1, a no-label jump to state 2, and finally some loops (labelled c) at state 2. Therefore the language accepted by this automaton consists of all strings formed from a run of a's followed by a run of b's followed by a run of c's. Note that some or all of these runs may be empty, so that ϵ, *acc*, and a are all accepted. (The label on the path 001222 is $a\epsilon\epsilon cc$, which is the same as *acc*, since ϵ is the empty string. ■

If we have a finite automaton, either deterministic or nondeterministic, which recognizes a certain set of strings, we can use that automaton to produce a computer program capable of recognizing the same set of strings. Such a program is said to "simulate" the automaton. In the case of a DFA, writing a simulation program is easy, as shown in Figure 6.24. This algorithm takes a string of symbols as input, reads them one at a time, and then either prints YES or NO, depending on whether or not the DFA accepts the string. In the algorithm, the variable s represents the current state of the DFA, and x represents the most recently read input symbol. After initializing s by setting it to the DFA's initial state, the program reads the input string, one symbol at a time. Each time a new symbol is read, a new state is calculated by the rule "$s \leftarrow \mathtt{next}(s, x)$," where the `next` function contains the next-state information given in the state table. This process continues until all of the symbols have been read, i.e., when `EOF` (end of file) is reached. Then if the final state s is an accepting state, the program prints `YES` and stops; otherwise, it prints `NO` and stops.

The algorithm in Figure 6.24 is a straightforward exercise in computer programming. A much more challenging task is to write a program that simulates the behavior of an NFA, because of the NFA's tendencies to "shut down" and to "reproduce itself," noted above. Nevertheless, such a program is possible, and we have outlined it in Figure 6.25. As with the DFA simulation algorithm, the idea is that we give the algorithm a string of symbols, and after reading the string, the algorithm prints `YES` if that string is accepted by the NFA, or `NO` if it isn't.

Here is a brief description of how the NFA simulation algorithm works. The algorithm is assumed to know the NFA's state table, and is given a string of n symbols, denoted by $a_1a_2 \ldots a_n$. Its job is to decide whether or not this string is accepted by the NFA, that is, whether it is the label on a path from the initial state to an accepting state, or as we shall say, is *produced* by such a path. To make this decision, the algorithm makes the optimistic assumption that the string *is* accepted, i.e., is produced by an accepting path. This path is not known to the algorithm, but the algorithm does its best to trace the unknown path as it reads the n symbols in the string, by keeping track of a certain set (S) of states. The initialization in step 1 is a little difficult to explain at this stage, but let's ignore this technicality for the moment. The algorithm reads the symbols $a_1, a_2, \ldots, a_n$, one at a time, at step 3. After the algorithm has read the first i symbols, S will be the set of all possible states where a path starting at the initial state and producing the string $a_1a_2 \ldots a_i$ could possibly be. The set S is computed by using the symbols $a_1, a_2, \ldots, a_i$ as clues. As new symbols are read, S is updated in step 4 in a way we will describe shortly. When all n symbols have been read, the algorithm takes the YES branch at step 2, and then checks (step 5) to see whether or not the set S contains an accepting state. If it does, the algorithm prints YES (step 6), indicating that the string $a_1a_2 \ldots a_n$ is accepted, and halts; if not, it prints NO (step 7), indicating that the string is not accepted, and halts. This would all be easy, except that some of the edges in the state diagram may be labelled ϵ, and whenever the unknown path takes an "ϵ edge," it will produce no label. It is the presence of ϵ edges that makes the `initialize` S and the `update` S steps of the algorithm a little tricky, but the diagram in Figure 6.26, together with the following discussion, should make things clearer.

In Figure 6.26 we see a part of a path, going from state s_1 to s_4, with many intermediate states. This long path produces only the string xy, since all but

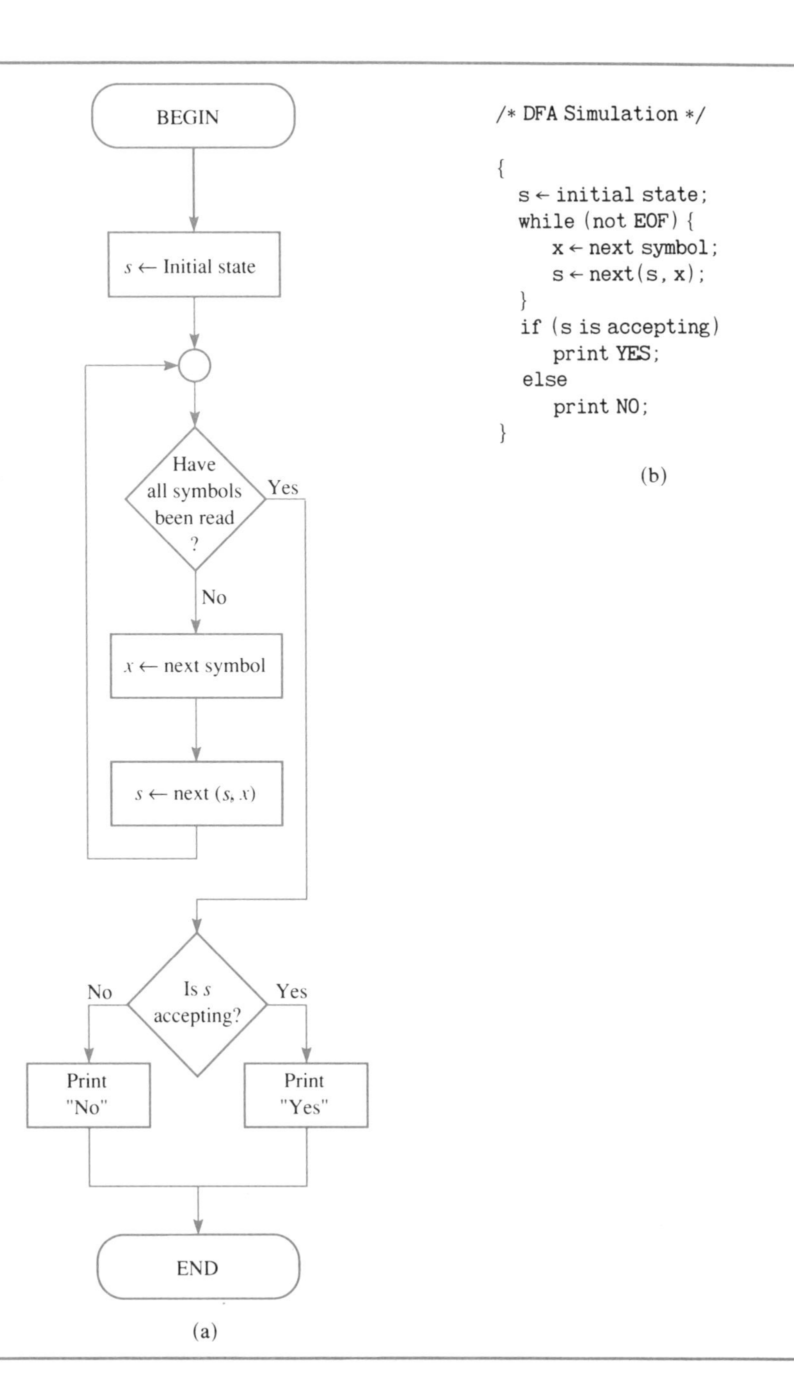


```
/* DFA Simulation */

{
  s ← initial state;
  while (not EOF) {
     x ← next symbol;
     s ← next(s, x);
  }
  if (s is accepting)
     print YES;
  else
     print NO;
}
```

(b)

Figure 6.24 A program to simulate a DFA.

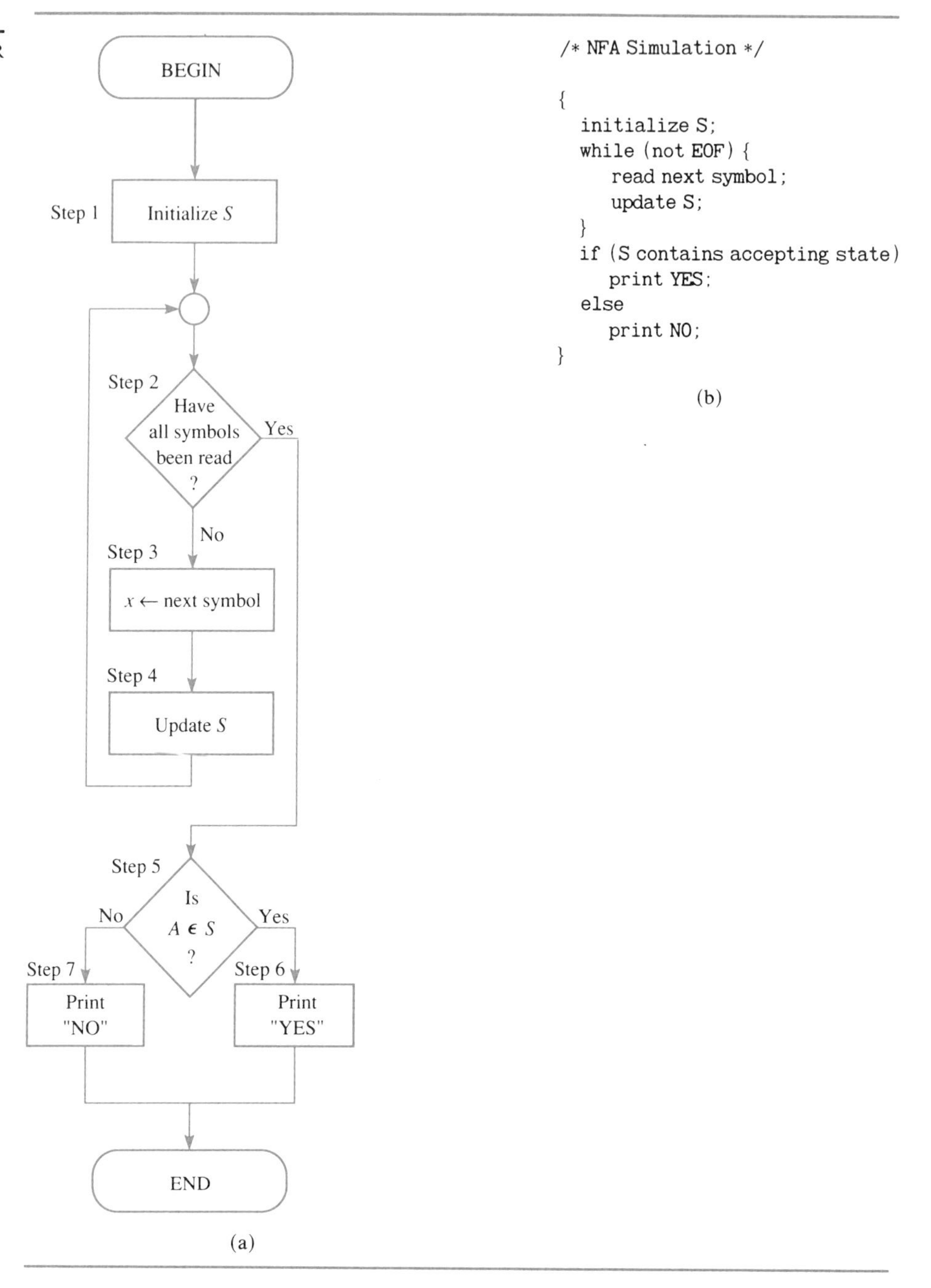

Figure 6.25 A program to simulate an NFA.

two of the edges in the path are ϵ edges. Now suppose that the NFA simulation algorithm is given the string "$\ldots xy \ldots$" produced by this path. Suppose further that the algorithm of Figure 6.25 has somehow correctly calculated, after reading all of the symbols in the input string up to, but not including, the x in the input string, a set S that includes the actual state s_1 that the path passed through just before producing the x. Let's see what it has to do in order to insure that after

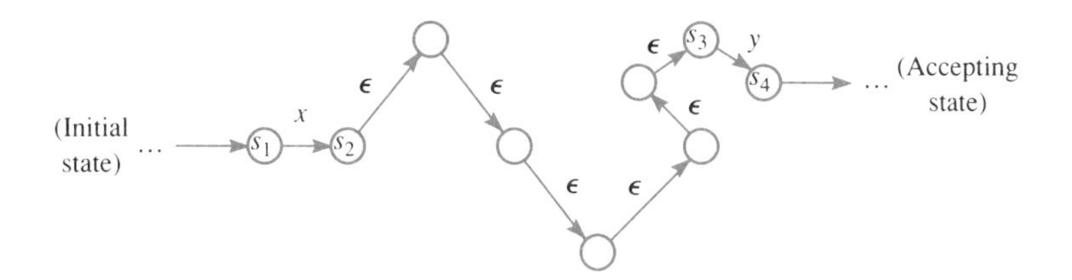

Figure 6.26 A portion of a path in an NFA's state diagram. State s_2 in "x-adjacent" to s_1; and s_3 is "ϵ-reachable" from s_2. Note that this portion of the path produces only the string "xy."

reading the x, S will be updated so that s_3 (the state reached just before the y was produced) will be in S.

When the algorithm reads the symbol x, it knows that the next state, whatever it is, must be connected to the present state by an edge labelled x. Let's agree to say that a state s' is "x-adjacent" to another state s if there is an edge from s to s' labelled x. It follows that if we replace every state in S with all of the states that are x-adjacent to it, we are sure to capture the actual next state, which in this case is s_2. This replacement forms the first part of the `update S` step in the NFA simulation algorithm (see Figure 6.27b).

However, there's more to the `update S` step than this, since after the path produces the label x, but before it produces y, it may include one or more ϵ edges, which produce no visible label. In Figure 6.26, for example, we see that the path goes from s_2 to s_3 without producing any label. In other words, it is possible to reach state s_3 from state s_2, using only ϵ edges, or as we will say, s_3 is "ϵ-reachable" from s_2. Therefore, in order to be sure that we haven't lost track of the correct state because of the ϵ edges, we need to replace each state in S with all states which are ϵ-reachable from it, i.e., with the state itself and all states which can be reached from it via a path all of whose edges are labelled ϵ.* (We always consider that a state is ϵ-reachable from itself.) If we make this

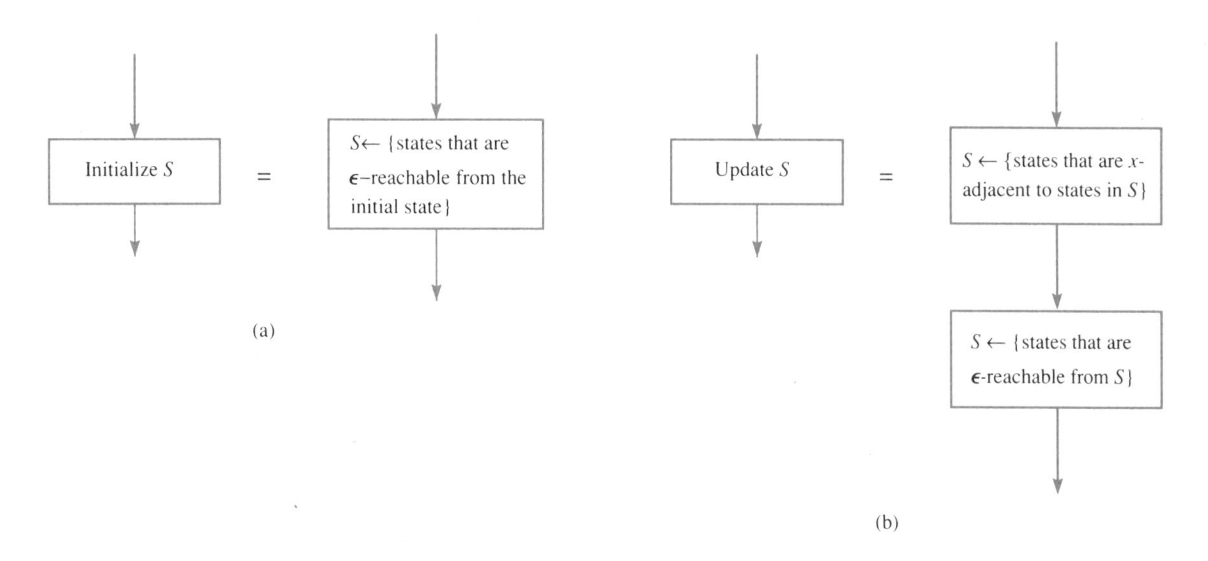

Figure 6.27 Details of the NFA simulation algorithm.

* We could, for example, modify Warshall's reachability algorithm (Section 4.2) to implement this part of the algorithm.

replacement, the resulting set S must include the state reached by the true path just before the next visible symbol (which is y in the case of the path in Figure 6.26) is produced.

To complete the discussion of the NFA simulation algorithm, we must return to the `initialize` S step. The initial value of S must include all possible states for the NFA just before the first visible symbol is produced. Since the path must start at I, S must be initialized as the set of all states which are ϵ-reachable from I, and in Figure 6.27(a) we have indicated this.

EXAMPLE 6.17 Use the NFA simulation algorithm to decide whether the following two strings are accepted by the NFA of Figure 6.23(c).

(a) *aab*. (b) *aba*.

SOLUTION

(a) We know from Figure 6.27(a) that S should be initialized as the set of states which are ϵ-reachable from the NFA's initial state, which in this case is state 0. All three states 0, 1, and 2 are ϵ-reachable from state 0, so that initially $S = \{0, 1, 2\}$. The algorithm now begins to read the symbols in the string *aab*. When the first a is read, each state in S must be replaced by the states which are a-adjacent to it. The only state a-adjacent to state 0 is state 0; and there are no states which are a-adjacent to either state 1 or state 2. Thus after the first half of the `update` S step is completed, $S = \{0\}$. The second half of the `update` S step says to replace each state in S with all those states which are ϵ-reachable from it. We have already seen that all three states are ϵ-reachable from state 0; so that after the first execution of the `update` S step, $S = \{0, 1, 2\}$. We might summarize the action so far by writing

$$\text{read } a\colon \{0, 1, 2\} \xrightarrow{a} \{0\} \xrightarrow{\epsilon} \{0, 1, 2\}.$$

Similarly, when the next "a" is read, S changes as follows:

$$\text{read } a\colon \{0, 1, 2\} \xrightarrow{a} \{0\} \xrightarrow{\epsilon} \{0, 1, 2\}.$$

Finally, when the "b" is read, we compute

$$\text{read } b\colon \{0, 1, 2\} \xrightarrow{b} \{1\} \xrightarrow{\epsilon} \{1, 2\}.$$

Thus the final value for S is $\{1, 2\}$, which contains the accepting state, state 2. Therefore the algorithm prints YES; this string is accepted by the NFA.

(b) Again the initial value of S is $\{0, 1, 2\}$. The set S progresses as follows:

$$\text{read } a\colon \{0, 1, 2\} \xrightarrow{a} \{0\} \xrightarrow{\epsilon} \{0, 1, 2\};$$

$$\text{read } b\colon \{0, 1, 2\} \xrightarrow{b} \{1\} \xrightarrow{\epsilon} \{1, 2\};$$

$$\text{read } a\colon \{1, 2\} \xrightarrow{a} \varnothing \xrightarrow{\epsilon} \varnothing.$$

Note that since there are no states which are a-adjacent to either state 1 or state 2, when the last "a" is read, S becomes empty, and stays that way. Since $\varnothing$ doesn't contain the accepting state, the algorithm prints NO; this string is rejected by the NFA. ■

We'll conclude this section with something that might surprise you. After all our discussion of the superiorities of NFAs over DFAs, we're going to show that any set of strings that can be accepted by an NFA can also be accepted by a DFA! In fact, we will show that given an NFA, it is always possible to construct a DFA that simulates the action of the NFA, i.e., which accepts exactly the same strings.

The DFA construction is based on the NFA simulation algorithm, and in particular on the `update` S step of the algorithm. Recall that S is a variable subset of the NFA's state set. Initially, S is the set of states which are ϵ-reachable from the NFA's initial state (we will call this the "initial subset"). After each new input symbol is read, the algorithm *determines* a new subset of states, and a string is accepted if and only if the final subset S contains at least one of the NFA's accepting states. We will call subsets containing at least one accepting state "accepting subsets." The action of the algorithm can therefore be completely described by a labelled, directed graph with one vertex for each possible subset of states. The vertex corresponding to the subset S is connected to the vertex corresponding to the subset T with an edge labelled x if and only if the `update` S step of the NFA simulation algorithm changes S into T when the input is x. This graph will serve as the state diagram for a DFA simulating the original NFA, provided we identify the vertex corresponding to the "initial subset" as the initial state, and the vertices corresponding to "accepting subsets" as accepting states. The next example will illustrate this construction.

EXAMPLE 6.18 Construct a DFA that simulates the NFA of Figure 6.23(b).

SOLUTION Since the NFA has three states, I, 0, and 1, the corresponding DFA will have eight states, corresponding to the eight subsets of $\{I, 0, 1\}$. One representation of such a DFA is shown in Figure 6.28. Since the NFA's initial state is I, and all three states I, 0, and 1 are ϵ-reachable from I, in Figure 6.28(a) the vertex corresponding to the subset $\{I, 0, 1\}$ is designated as the initial state. Since the accepting states of the NFA are 0 and 1, in Figure 6.28(a) each vertex corresponding to a subset containing either 0 or 1 is designated as an accepting state. The directed, labelled edges are determined by the `update` S step of the NFA simulation algorithm.

For example, consider the $\{I, 1\}$ vertex. With $S = \{I, 1\}$ and $x = 0$, the `update` S step of Figure 6.27 produces the empty set $\varnothing$, since there are no states which are 0-adjacent to either I or 1 in the NFA. On the other hand, with $S = \{I, 1\}$ and $x = 1$, the `update` S step produces the set $\{1\}$, since state 1 is 1-adjacent to state 1, there are no states which are 1-adjacent to I, and the only state which is ϵ-reachable from state 1 is state 1. This explains the edge from $\{I, 1\}$ to $\varnothing$ labelled 0 and the edge from $\{I, 1\}$ to $\{1\}$ labelled 1. The other labels in Figure 6.28(a) can be explained similarly.

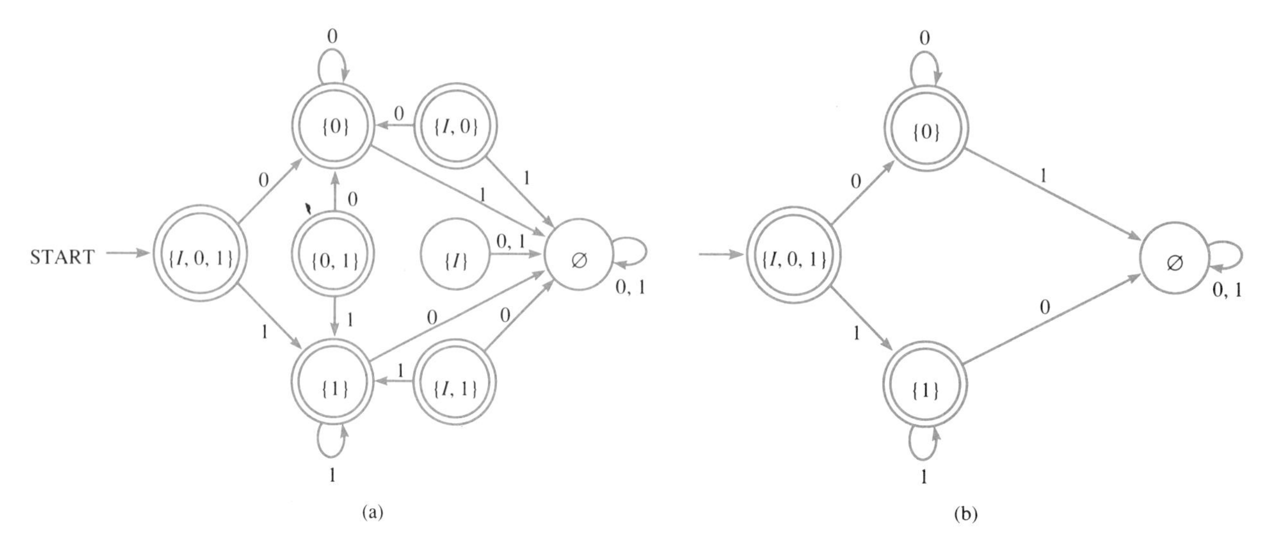

Figure 6.28 A DFA that simulates the NFA of Figure 6.23(b): (a) Before and (b) after extraneous states are removed.

Incidentally, the state diagram of Figure 6.28(a) can be simplified, if we notice that several of the states, viz. $\{I\}$, $\{I, 0\}$, and $\{0, 1\}$, can never be reached from the initial state $\{I, 0, 1\}$. We might call such states "extraneous" states. If we remove the extraneous states from the state diagram, the simpler state diagram of Figurc 6.28(b) results. There are various tricks that can be used to identify extraneous states. For example, if beginning with the initial state, we generate the states reachable from it under all possible inputs, and then generate more new states, etc., an extraneous state will never appear. (See Problem 32.) ■

Problems for Section 6.4

1. Produce a state table for the DFA of Figure 6.21(b).

2. Explain why the DFA of Figure 6.22(a) accepts a string if and only if the string contains two consecutive 1s.

3. Design a DFA which accepts a string of 0s and 1s if and only if the string does not contain two consecutive 1s.

4. How many strings of length 10 does the DFA of Figure 6.22(b) accept?

5. Find a formula for the number of strings of length n accepted by the DFA of Figure 6.22(c).

6. Describe in words the set of strings accepted by the DFA in the accompanying figure.

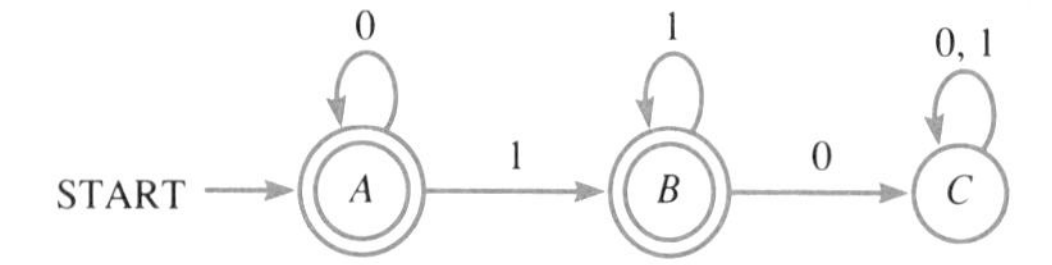

7. Describe in words the set of strings accepted by the DFA in the accompanying figure.

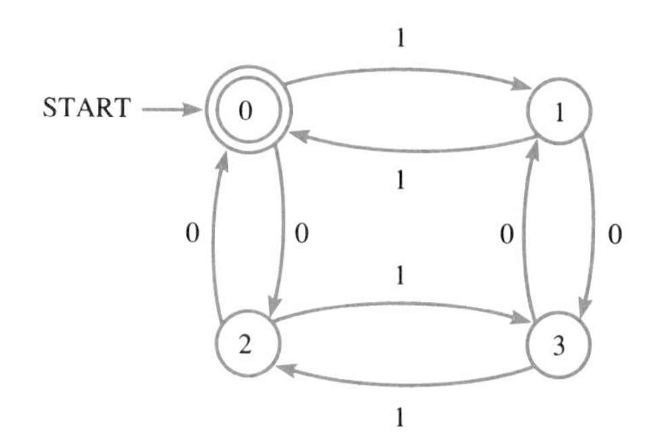

8. Describe in words the set of strings accepted by the DFA in the accompanying figure.

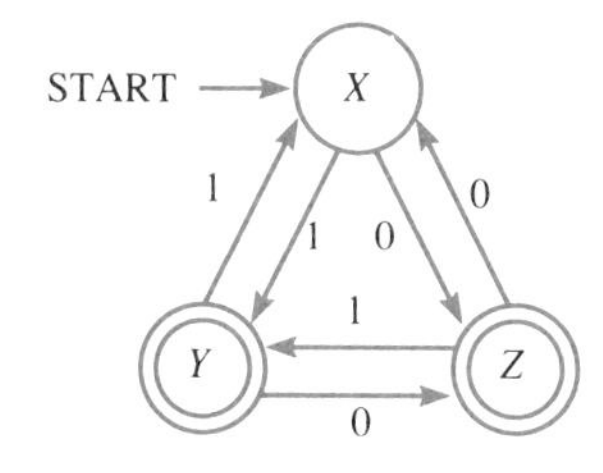

9. Design a DFA that accepts the same language as the NFA of Figure 6.23(a).

10. How many strings of length n does the NFA of Figure 6.23(b) accept?

11. Design a DFA that accepts the same language as the NFA of Figure 6.23(c).

12. Determine whether the NFA in the accompanying figure accepts (a) 000 and (b) 0110.

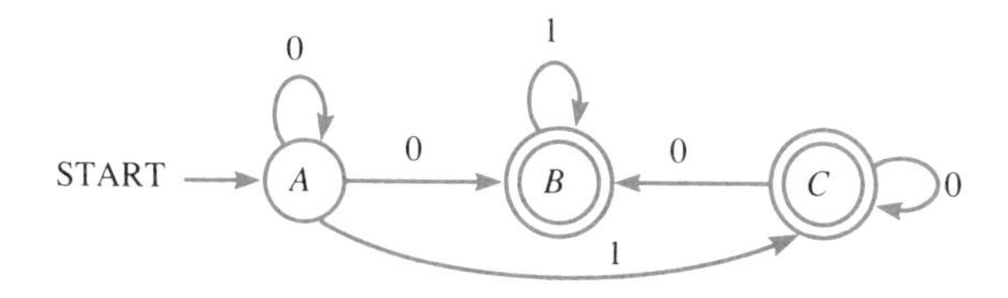

13. Consider the NFA in the accompanying figure.
(a) Is 11 accepted?
(b) Describe in words the set of accepted strings.

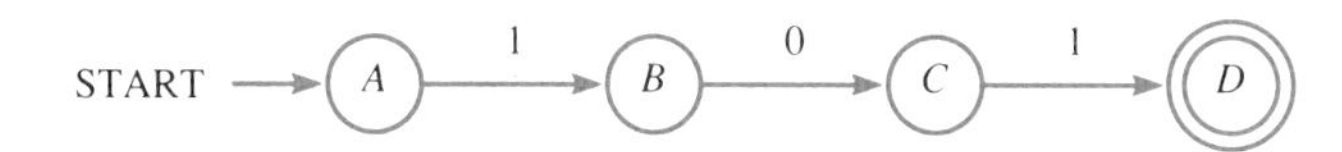

14. Consider the NFA in the accompanying figure.
(a) Is 11011 accepted?
(b) Describe in words the set of accepted strings.

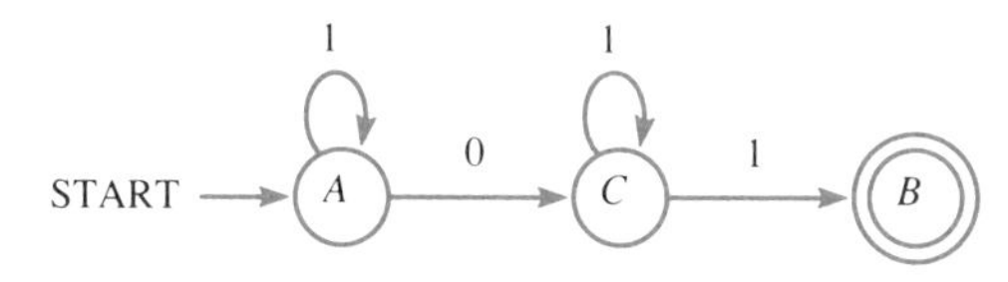

15. After reading the text, a student tried to design an NFA that would accept only those English words that begin with *po* and proposed the NFA in the accompanying figure. Show that this solution is incorrect. How can it be modified so that it is correct?

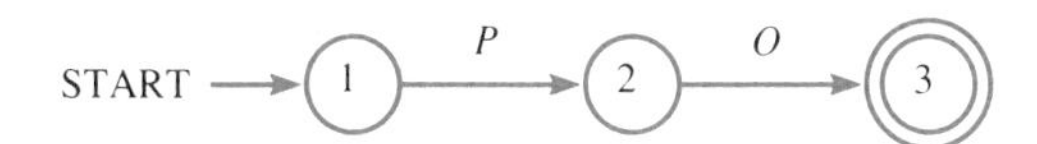

16. Design an NFA that will accept only those English words whose letters are in alphabetical order.

17. How many strings of length 6 are accepted by the NFA of Figure 6.23(c)?

18. Construct an NFA that accepts the set of strings of 0s and 1s ending in 000.

19. Construct an NFA that accepts the set of strings of 0s and 1s containing an even number of 1s.

20. Construct an NFA that accepts the set of strings of 0s and 1s containing no 1s.

21. Construct an NFA that accepts the set of strings of 0s and 1s containing exactly two 1s.

22. Construct an NFA that accepts the set of strings of 0s and 1s containing at least two 1s.

23. Construct an NFA that accepts the set of strings of 0s and 1s containing no two 1s in a row.

24. Construct an NFA that accepts the set of strings of a's, b's, and c's consisting of a nonempty run of a's followed by a nonempty run of b's followed by a nonempty run of c's.

25. Consider the set of strings representing *fixed-point numbers*, i.e., strings like 34.21 or 234 or 0.006. Each allowable string is either (i) a nonempty string of digits, or (ii) a nonempty string of digits followed by a decimal point followed by another nonempty string of digits. For example, "234" and "234.0" are legal, but "234" is not. Furthermore, a number between 0 and 1 should have exactly one 0 before the decimal point, so that 00.006 is illegal, but 0.006 is okay. Finally, strings with extra leading 0s, like 034.21, are also illegal. Construct an NFA that accepts this set of strings.

26. Use the algorithm in Figure 6.24 to decide whether or not the string 11101 is accepted by the DFA of Figure 6.22(b). Keep a careful list of the state sequence.

27. Explain how Warshall's algorithm can be used to implement the second half of the `updateS` step in the NFA simulation algorithm shown in Figure 6.27(b).

28. If the state diagram for an NFA has n states, m of which are accepting, how many states will the DFA that simulates the NFA have? How many of them will be accepting states?

29. Construct a DFA that simulates the NFA in the accompanying figure.

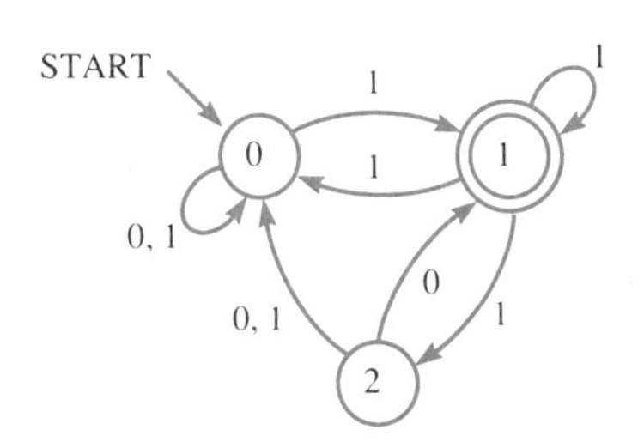

30. Construct a DFA that simulates the NFA in the accompanying figure.

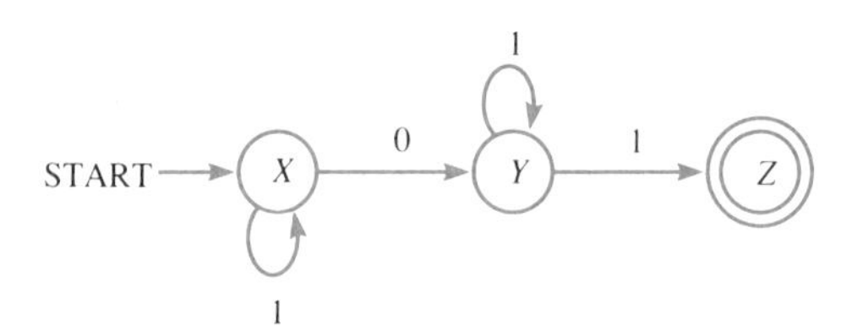

31. Construct a DFA that simulates the NFA in the accompanying figure.

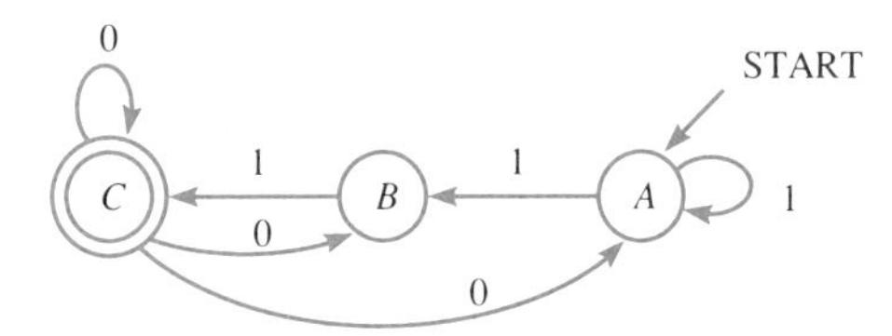

32. Construct a DFA that simulates the NFA of Figure 6.23(a). (See the remark at the end of Example 6.18.)

33. Construct a DFA that simulates the NFA of Figure 6.23(c).

34. Call a set S of states in an NFA "ϵ-closed" if every state which is ϵ-reachable from a state in S is also in S.

(a) How many ϵ-closed subsets does the NFA of Figure 6.23(c) have?

(b) Explain why, in the NFA simulator construction, any state corresponding to a subset which is not ϵ-closed must be extraneous.

35. Here is an old problem that can be solved with an NFA: A man with a fox, a goat, a cabbage, and a small boat is on the left bank of a river. The boat is only large enough to carry the man and one of the other three. The man wishes to move his goods to the other bank by ferrying them across the river, one at a time. However, if the fox and the goat are left unattended, the fox will eat the goat; and if the goat and the cabbage are left unattended, the goat will eat the cabbage. Can the man achieve his wish?

Solve this problem by constructing an NFA with a set of states ccorresponding to the location of the four objects: man (M), fox (F), goat (G), and cabbage (C). For example, when all four are on the left bank, the state is "MFGC—∅," and "C—MFG" means that the cabbage is alone on the left bank and the man, the fox, and the goat are on the right bank. The inputs to this NFA are the actions that the man can take. He can cross alone (call this input m), or with the fox (input f), or with the goat (input g), or with the cabbage (input c). For example, if the state is "C—MFG" and the input is g, the next state is "MCG—F."

(a) Draw the state diagram.

(b) Solve the problem by deciding whether there is a path from the initial state "MFGC—∅" to the accepting state "∅—MFGC."

6.5 Turing Machines

In the first four sections of this chapter, we've been studying finite state machines. FSMs are valuable models for many kinds of computation, and have important practical applications, for example, in the design of sequential circuits and pattern recognition algorithms. Nevertheless, there are many important computations that FSMs cannot do. For example, in a certain sense, no FSM can multiply by two! By this we mean that it is impossible to construct an FSM that takes as input a string of 1s and produce as output a string of 1s twice as long (see Figure 6.29). (The basic reason for this is that an FSM has, by definition, only finite memory, and if the input string has too many 1s, the machine will lose count, and forget how many 1s to output.)

The most serious problem with FSMs, however, is that they are inadequate models for the most important kind of computational machine, the computer.

Input (1111 . . . 10000 . . .) n → FSM ? → Output (1111 . . . 11111 . . . 10000 . . .) $2n$

Figure 6.29 An FSM that "doubles" a string of 1s (it doesn't exist!).

Again, the essential problem is the limited memory of the FSM. Any modern computer has so much memory that a FSM with the same number of states as the computer would be hopelessly unwieldy. (For example, a computer with a modest 1 megabyte of memory has, in principle, $2^{2^{23}}$ states!)

Fortunately, there is another class of models for computing machines, called **Turing machines,*** which have a potentially infinite amount of memory and yet are simple enough to be studied by mathematicians and computer scientists and to shed light on the capabilities and limitations of real computers. Turing machines are capable of performing any reasonable computation, even multiplying by two, as we will see at the end of this section.

A Turing machine consists of two parts: a "read/write" **tape head**, and a long† paper **tape** ruled into squares (see Figure 6.30). The tape head contains an FSM, and a **tape controller**, which we will describe shortly. On each square of the tape a symbol may be written; in Figure 6.30 the symbols are 0, 1, and *b*, with *b* denoting a *blank.* (In all of our examples, *b* will always be included as one of the symbols.)

The Turing machine is controlled by a master clock. At any given time, the tape head will be reading the symbol written on one square of the tape. For example, in Figure 6.30, the head is reading the 1 (shown in color) on the indicated square. The symbol being read serves as the input to the FSM inside the head. At the next clock pulse, the FSM will change states, and produce an output symbol, which is then written on the tape (this erases the old symbol). At the same time, the tape controller performs an operation, which can be one of three types:

- Move the head one square to the *right.*
- Move the head one square to the *left.*
- Cause the machine to *halt.* (This ends the computation.)

The important point here is that the tape head can move in *both directions.* If the tape head could only move in one direction, say only to the right, a Turing machine would actually be no different from an FSM. Indeed, a Turing machine that always moves right is simply reading the symbols written on the tape, one after another, and in response writing its output symbols on the tape, one after

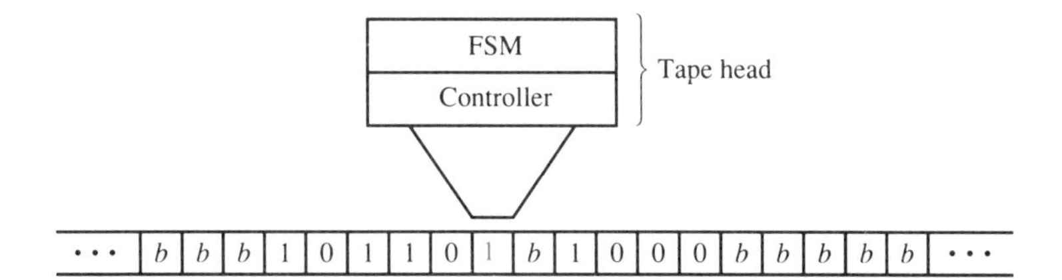

Figure 6.30 Finite state machine + two-way tape drive = Turing machine.

* Alan Turing (1912–1954), English mathematician. Arguably the world's first computer scientist, he invented the abstract "Turing machine" in 1936, and for the rest of his short life he was involved in the development of computers. For example, during World War II he helped design the Colossus, the computing machine that broke the famous German Enigma cypher. (Ahead of his time in other ways, he was a serious runner and posted a 2:46 marathon time in 1947!)

† Exactly how long, we won't say. But we'll assume it's so long that no calculation we need to make will ever run off the end.

the other. But a Turing machine can move in *either* direction, and as we shall see, this capability allows the machine to use the tape as a "scratch pad" to help it keep track of things.

Turing machines, like finite state machines, can be described by a **state table**, which tells how the machine will react to a given input symbol from a given state. In addition to the usual "Next State" and "Output" columns, however, there must also be an "Operation" column, to describe the action taken by the controller. For example, in Table 6.6(a) we have the state table description of a very simple Turing machine, which as we'll see, is capable of *erasing a string of* 1s.

The Turing machine described in Table 6.6(a) has only one state, called state A, and there are only two tape symbols, 1 and b (again, b denotes "blank"). If the machine *reads* a 1, it *writes* a b (thus erasing the 1), and *moves right*. The rightward motion is indicated by the letter R under "Operation, $x = 1$" in the state table. If, on the other hand, the machine reads a b, it writes a b (thus making no change on the tape), and the machine *halts*. The halting is indicated by the letter H under "Operation, $x = b$" in the state table.

To better understood how this machine works, let's see what happens if we give it a tape which is entirely blank except for four consecutive 1s, and initially position the tape head over the leftmost 1 (see Figure 6.31a). In Figure 6.31(a) we show the tape head reading a 1. According to the rules in the state table, the machine must print a blank on the current square and move right one square, leading to the situation in Figure 6.31(b). Again the machine is instructed to erase thc 1, and move right. This action continues (Figures 6.31b, 6.31c, 6.31d) until the last 1 is erased. At this point thc state table instructs the machine to halt, and the computation ends. Thc tape has been erased!

The state table is a good way to describe a Turing machine, but there is another way which is in some ways better, called the **program listing**. In Table 6.6(b) we give the program listing for the machine in Table 6.6(a). The idea is to list all possible combinations of *state* and *input*, followed by the corresponding *next state*, *output*, and *operation*. The format we use for program lines is

$$\text{(state)(input)}:\text{(next state)(output)(operation)}.$$

Thus for example, the first program line in Table 6.6(b) is

1. $A1:AbR$,

which means that if the machine is in state A and the input is 1, the machine will remain in state A, print a b (blank) on the tape, and move one square to the right.

TABLE 6.6 Description of a Turing machine that erases a string of 1s: (a) State table; (b) program listing

	Next State		Output		Operation	
	$x = 1$	$x = b$	$x = 1$	$x = b$	$x = 1$	$x = b$
A	A	A	b	b	R	H

(a)

1. $A1:AbR$
2. $Ab:AbH$

(b)

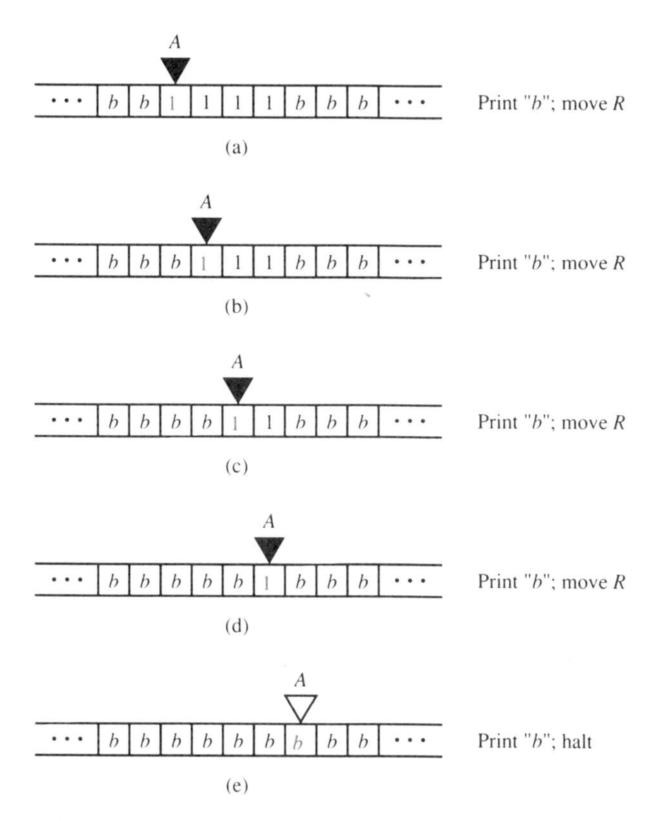

Figure 6.31 Operation of the Turing machine described in Table 6.6 when the input is a string of four 1s (▼ = active tape head; ▽ = halted tape head.

EXAMPLE 6.19 Figure 6.32 shows the program listing for a certain Turing machine. The machine is to be started in state Q_0, reading the leftmost 1 of a string of consecutive 1s written on an otherwise blank tape. Experiment with this machine, and find out what function it performs.

SOLUTION We note first that there are two states, Q_0 and Q_1, there are two input symbols, 1 and b, and there are *three* output symbols, 1, the word "Even," and the word "Odd." We also notice that after the machine prints either the word "Even" or "Odd," it halts. Hmm.... Maybe this machine decides whether the number of 1s is even or odd? Let's experiment and find out.

As a first experiment, we write only one 1 on the input tape (Figure 6.33a). The calculation proceeds as shown, first using program line 1 (because the state

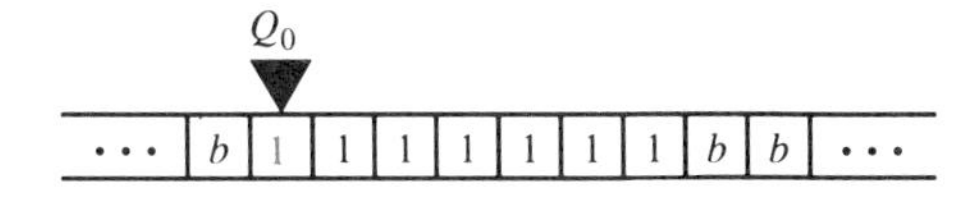

1. $Q_0 1 : Q_1 1 R$
2. $Q_1 1 : Q_0 1 R$
3. $Q_0 b : Q_0$ "Even" H
4. $Q_1 b : Q_1$ "Odd" H

Figure 6.32 Turing machine for Example 6.19.

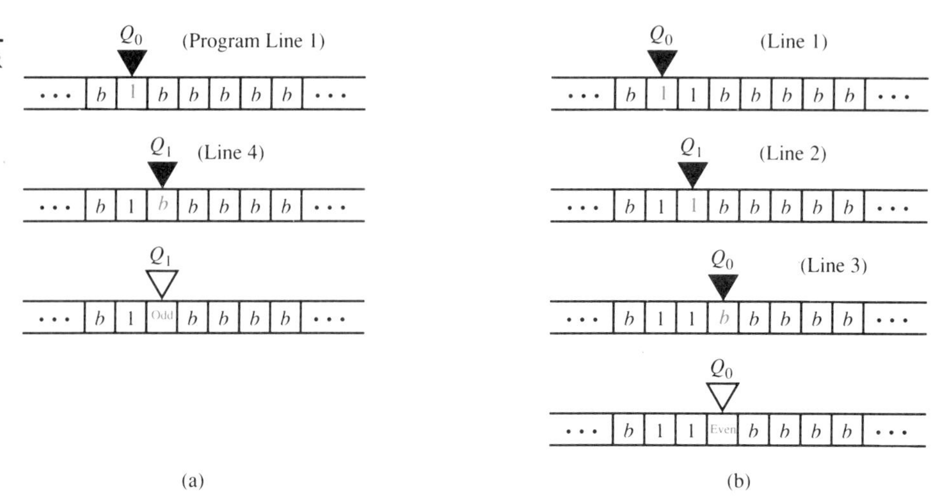

Figure 6.33 Experimenting with the Turing machine of Figure 6.32: (a) Tape has only one 1; (b) tape has two 1s.

is Q_0 and the input is 1), and then line 4 (state Q_1, input b). This causes the machine to halt after printing the word "Odd," as shown.

Next, let's try a tape with two 1s, as shown in Figure 6.33(b). The calculation proceeds as shown, but this time the machine prints the word "Even" and halts because the machine was in state Q_0 when it first encountered a blank square, which caused line 3 to be executed.

At this point it's pretty clear what's going on. As the machine reads the consecutive 1s on the tape, it alternates between states Q_0 and Q_1 (lines 1 and 2). After reading an *even* number of 1s, it'll be in Q_0, and after reading an *odd* number of 1s, it'll be in Q_1. When it reaches the end of the 1s, i.e., the first blank square, it will print the word "Even" if it is in state Q_0 (line 3), and "Odd" if it is in state Q_1 (line 4). Thus as we conjectured, this machine is designed to determine whether the number of 1s is even or odd. ■

In the two Turing machines we have now studied in detail (Table 6.6 and Figure 6.32), the tape head has only moved in one direction (right); but as we mentioned earlier, the real power of a Turing machine lies in its ability to move in *both* directions. In the next example, we'll need to use this power.

EXAMPLE 6.20 Design a Turing machine which, when given a tape which is completely blank except for a single 1 somewhere on it, finds the 1, and halts. (By "finding" the 1 we mean for the machine to position its tape head over the square containing the 1.)

SOLUTION The essential difficulty here is that we're not told whether the 1 is to the *right* or to the *left* of the tape head (see Figure 6.34). The Turing

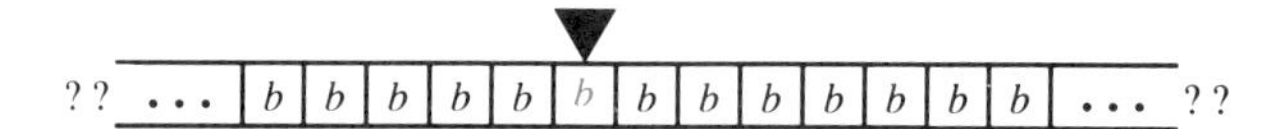

Figure 6.34 Where's the treasure?

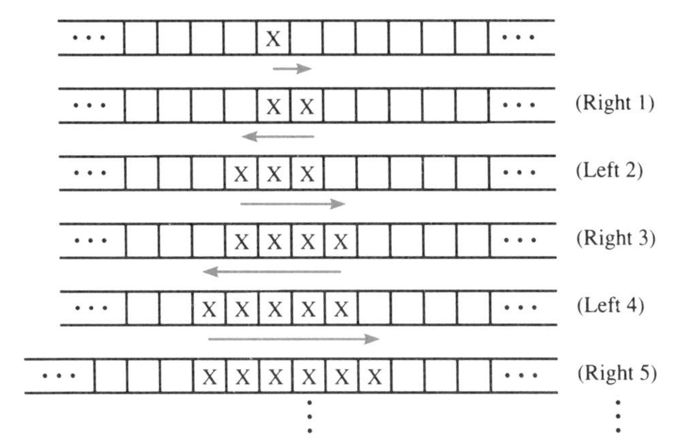

Figure 6.35 A search strategy for Example 6.20.

machine must discover this for itself. It might help to rephrase the problem. You are in a long narrow corridor which has been partitioned into many identical rooms. One of the rooms contains a treasure chest, but you can only examine one room at a time. How can you find the treasure?

After a little thought, you might decide to proceed as follows. You move to the *right* a few rooms, and if you don't find the treasure, you backtrack and move *left* a few rooms, and if you still don't find the treasure, you move right a few *more* rooms, etc. This procedure will work, but only if you make sure that every time you change directions you examine *more* rooms in that direction than you did the previous time. This search strategy is illustrated in Figure 6.35, with the rooms that have already been examined marked X. Starting from the initial square (or room), first you move *one* square right, then *two* squares *left*, then *three* squares *right*, etc. Eventually, you'll reach the treasure (or the 1), no matter how far away it is, or in which direction.

You might run into problems, however, if the treasure is very far away and you try to keep track of how far you need to travel. If you're a *human* examining rooms, you might forget how far you need to go on the next pass. If you're a Turing machine, you'll find that the FSM in your tape head might not be able to count high enough (for example if the FSM has only 100 states and the treasure is 101 squares away). In either case, it's best to play safe and *mark* the rooms (squares) that have already been examined. We could, for example, put a spot of paint on the floor of each room we examine, so that we will know when we enter a new room. In the case of blank squares on the tape, we could write a special symbol, say X, on each square we examine.

After these preliminary remarks, we're prepared to write a program for a Turing machine that does what's required, and in Table 6.7 we give such a program. This machine has two states, called state A and state B. When the machine is in state A, it will move *leftward* along the tape, passing all the squares previously marked X (program line 3), until it either finds the 1 and halts (line 5), or finds a blank square, marks it, and changes directions (line 1). Similarly, when the machine is in state B, it will move *rightward*, passing all the X's (line 4), until it either finds the 1 and halts (line 6), or finds a blank square, marks it, and changes direction (line 2).

TABLE 6.7 A Turing machine that finds the treasure (state A: searching left; state B: searching right

Marking a Blank Square, Changing Directions

1. $Ab : BXR$
2. $Bb : AXL$

Searching for an Unmarked Square

3. $AX : AXL$
4. $BX : BXR$

***Eureka*!**

5. $A1 : A1H$
6. $B1 : B1H$

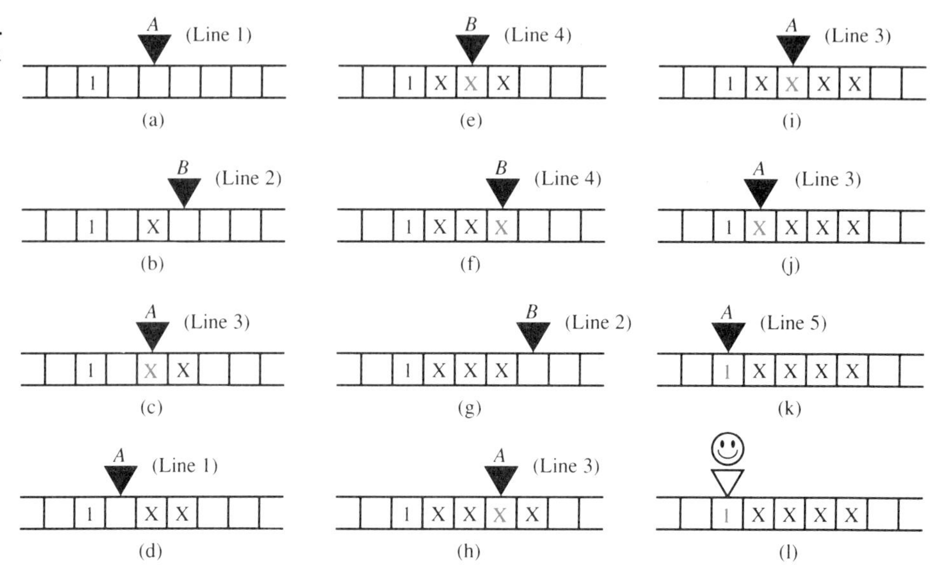

Figure 6.36 How the Turing machine of Table 6.7 works.

In Figure 6.36 we show this Turing machine in action, starting in state *A* on the blank square shown in Figure 6.36(a). Initially it finds itself in state *A* scanning a blank square, so (using program line 1), it marks the square X, goes into state *B*, and moves right one square. It then proceeds through the sequence of steps shown, until it finds the 1, and halts, as shown in Figure 6.36(l). (If you don't like all those X's left on the tape, do Problem 18.)

As our final example, we will design a Turing machine that can multiply by 2.

EXAMPLE 6.21 Design a Turing machine which when started at the leftmost 1 on a tape which is blank except for a single string of consecutive 1s, will *double* the length of the string, return to the leftmost 1, and halt (see Figure 6.37).

SOLUTION The first thing to do is to devise an overall strategy, and then translate the strategy into specific machine instructions. One possible strategy is shown in Figure 6.38. We use *two* extra symbols, X and Y. The symbol X will mark the location of a 1 that has already been copied, and the symbol Y will

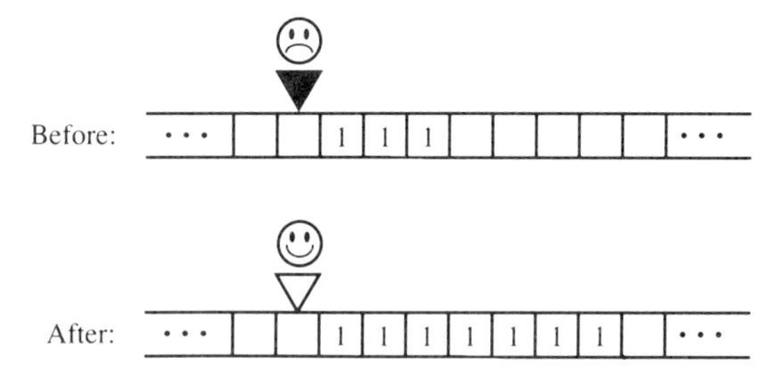

Figure 6.37 What the Turing machine in Example 6.21 is supposed to do. (Compare with Figure 6.29.)

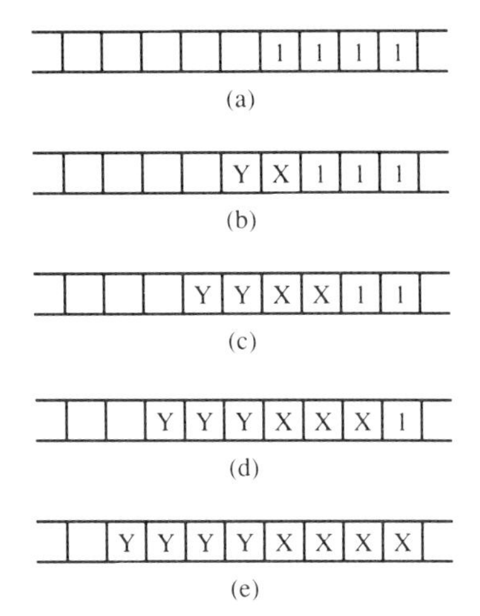

Figure 6.38 The strategy for Example 6.21.

mark the location of a copy of one of the original 1s. We'll first copy the leftmost 1, as shown in Figure 6.38(b). The *copy* is placed in the first available blank square to the *left* of the original block of 1s. Then we'll copy the second 1 (counting from the left), again using the first available blank square on the left (Figure 6.38c). We continue as shown, always marking with an X the first available 1 on the *right*, and placing a corresponding Y on the *left*, until there are no more 1s on the tape. Then it will be a simple matter to make one last run through the tape, changing the X's and Y's to 1s, finally arriving at a tape like the one depicted in the *after* part of Figure 6.37.

Table 6.8 describes a Turing machine that will implement the strategy depicted in Figure 6.38. There are two states, called Q_0 and Q_1, and the machine should be started in Q_0. When the machine is in state Q_0, it is looking for a 1 that has not yet been copied (marked X). If it finds a 1, it marks it X, enters state Q_1, and moves left (program line 1). Otherwise, if it sees either an X or a Y, it keeps looking (lines 2 and 3). Finally, if it encounters a blank square, there are no more 1s on the tape and the machine halts (line 4).

TABLE 6.8 A Turing machine that solves Example 6.21 (almost)

Looking for a New 1	
1. $Q_0 1 : Q_1 X L$	(Found it!)
2. $Q_0 X : Q_0 X R$	(Keep looking.)
3. $Q_0 Y : Q_0 Y R$	(Keep looking.)
4. $Q_0 b : Q_0 b H$	(Finished.)
Copying a 1	
5. $Q_1 b : Q_0 Y R$	(Copied it!)
6. $Q_1 X : Q_1 X L$	(Keep moving left.)
7. $Q_1 Y : Q_1 Y L$	(Keep moving left.)

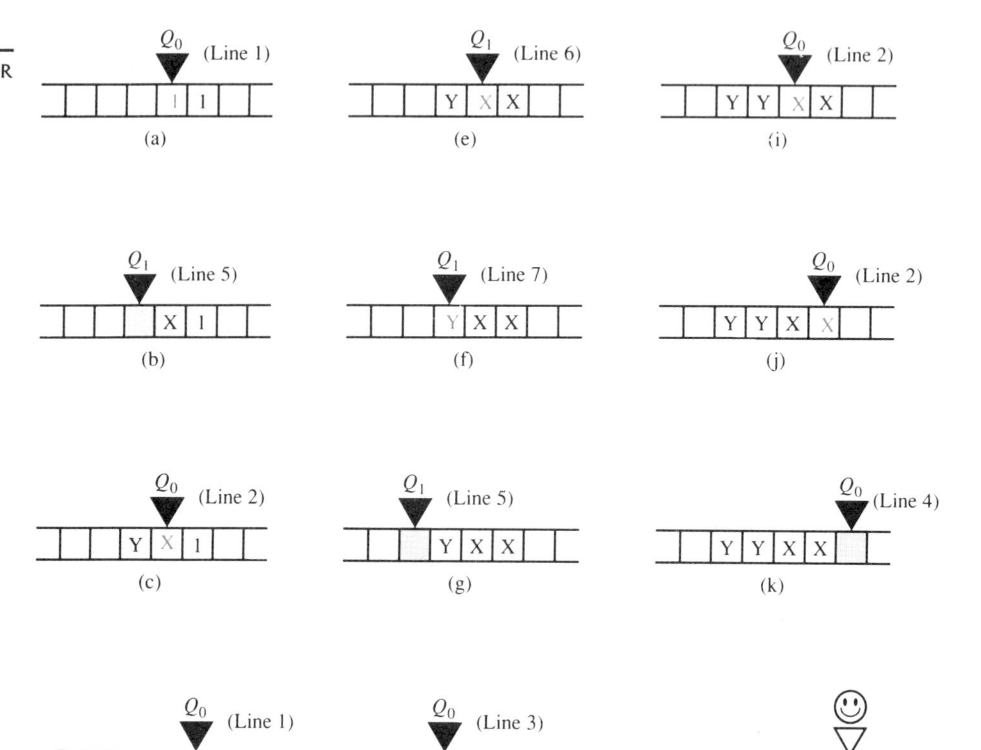

Figure 6.39 The Turing machine of Table 6.8 in action.

When the machine is in state Q_1, it has just found an uncopied 1 and is hurrying leftward to write a corresponding Y. As it passes the X's and Y's, it keeps going (lines 6 and 7), until it reaches a blank square, on which it writes Y. Then it enters state Q_0 and moves right (line 5).

In Figure 6.39 we show one example of how the Turing machine described in Table 6.8 works. As you can see, it does implement the strategy developed in Figure 6.38. However, it doesn't *quite* do what was asked for in Figure 6.37. We leave as Problem 22 the "cleanup patrol," i.e., the modifications needed so that instead of stopping with the tape in the unsightly condition shown in Figure 6.40, the machine makes one final sweep left, changing all the X's and Y's to 1s. ■

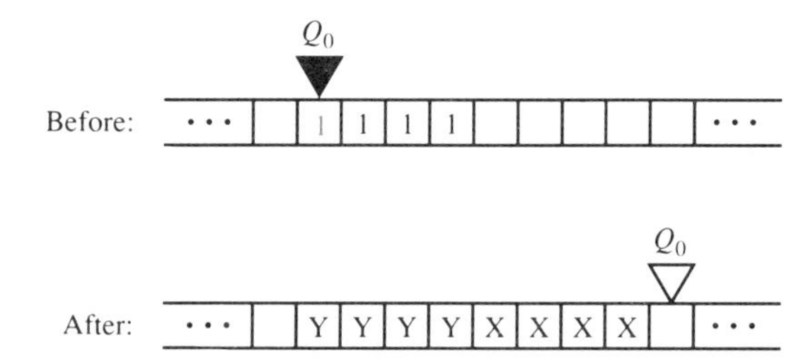

Figure 6.40 What the Turing machine in Table 6.8 really does. (Cleaning up is still required!)

Problems for Section 6.5

1. The accompanying figure shows the state diagram for an *infinite state machine.* Start the machine in state 0. What is the output if the input is

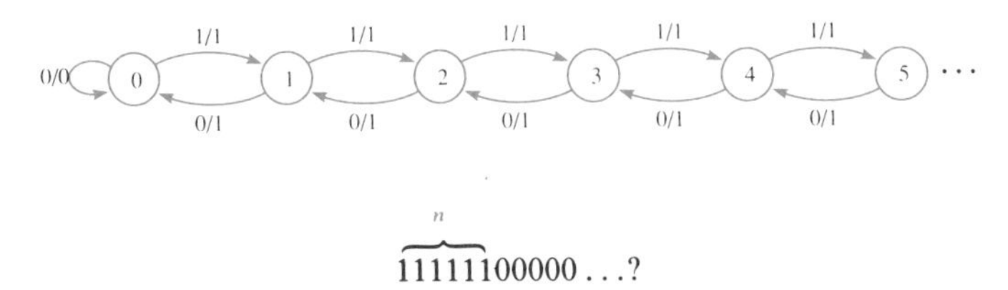

$$\overbrace{111111}^{n}00000\ldots?$$

2. Is it possible to design an FSM that *divides by* 2? That is, given an input sequence like

$$\overbrace{11111111}^{n}000\ldots,$$

where n is an even number, the output will be

$$\overbrace{1111}^{n/2}000\ldots0\ldots.$$

Explain your answer.

3. For a fixed integer M, design an FSM with M states that *can* double any string of $1, 2, \ldots,$ up to $(M-1)$ 1s. (*Hint*: Peek at Problem 1.)

4. Why do you suppose we specified that the tape for the Turing machine is a *paper* tape? Would some other material be better?

5. What would happen if we initialized the TM of Table 6.6 by placing it over the *rightmost* 1 on the tape in Figure 6.31(a)?

6. Design a Turing machine that erases *every other* 1 on a tape like the one in Figure 6.31(a) that contains a single string of 1s and is otherwise blank.

7. Make a state table (like the one in Table 6.6a) for the Turing machine in Figure 6.32.

8. Design a TM like the one in Example 6.19 but with this difference: The letters in the words "Even" and "Odd" are to be printed with only one letter per square. Thus, "Even" needs four squares and "Odd" needs three.

9. Prove by induction that the TM in Example 6.19 has the following property:

> P_n: After reading n 1s, the machine is in state Q_0 if n is even, and in state Q_1 if n is odd.

10. Design a TM to compute $n \pmod 3$. More precisely, if the TM is started on the leftmost 1 on a tape that contains a string of n 1s and is otherwise blank, it reads the n 1s and prints 0, 1, or 2, depending on whether n is congruent to 0, 1, or 2 (mod 3), and halts.

11. Design a Turing machine, which when given a sequence of 0s and 1s as input, will translate it into a sequence of X's and Y's—e.g., 101001 is translated XYXXYX—and halt.

12. Design a Turing machine that *adds.* That is, if the tape originally contains a string of m 1s, followed by a blank, followed by a string of n 1s, the machine should perform as shown in the accompanying figure.

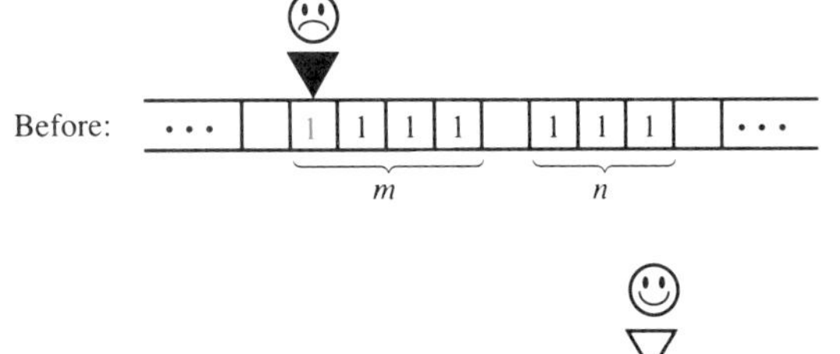

13. Here is a program listing for a certain Turing machine:

1. $Q_0 1 : Q_1 bR$
2. $Q_1 1 : Q_0 1R$
3. $Q_0 b : Q_0 bH$
4. $Q_1 b : Q_1 bH$

If the machine is started in state Q_0, reading the leftmost 1 on a tape that is blank except for a string of 1s, what will it do?

14. Make a flowchart for the algorithm performed by the TM in Example 6.20 (Table 6.7).

15. Make a flowchart for the algorithm performed by the TM in Example 6.21 (Table 6.8).

16. We said in the text that if a Turing machine only moved in one direction, its function could be performed by an FSM. The TMs in Table 6.6 and Figure 6.32 only move right.

(a) Design an FSM that performs the same function as the TM in Table 6.6.

(b) Design an FSM that performs the same function as the TM in Figure 6.32.

17. In the solution to Example 6.20 we said that if the TM only had 100 states and the treasure was 101 squares away, you'd be in trouble. Why?

18. Embellish the TM in Table 6.7 so that when the machine halts, all those unsightly X's have been erased, and only the 1 remains. Thus the tape should look like the diagram in the accompanying figure when the machine halts.

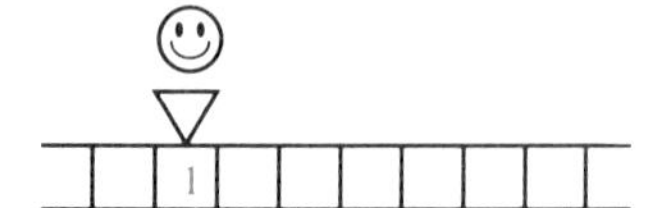

19. Solve Example 6.20, using a TM that only needs to write 1s and b's. (*Hint*: Use 1s to mark squares that have been reached, and look for the 1 *two* squares beyond the last marked square.)

20. Design a TM that solves the problem in Example 6.21, using the search strategy outlined in the accompanying figure (cf. Figure 6.38).

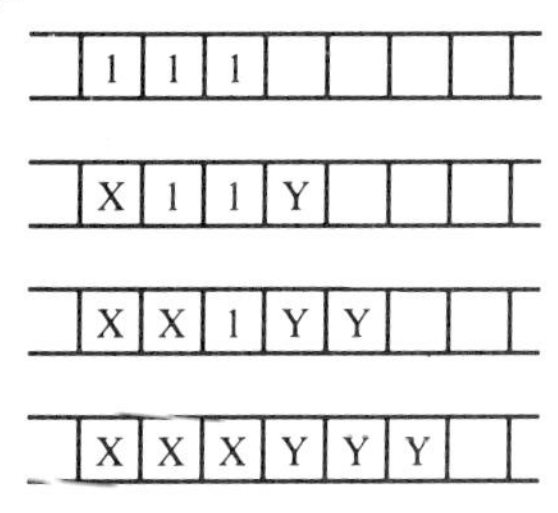

Compare this TM with the one given in Table 6.8. Which one do you prefer? Why?

21. In Table 6.8 there is no instruction that tells the machine how to behave when the state is Q_1 and the input is 1. Why do you suppose this is?

22. Embellish the TM in Table 6.8 so that when it finishes, all those unsightly X's and Y's have been replaced by 1s. That is, the machine should perform as indicated in the accompanying figure.

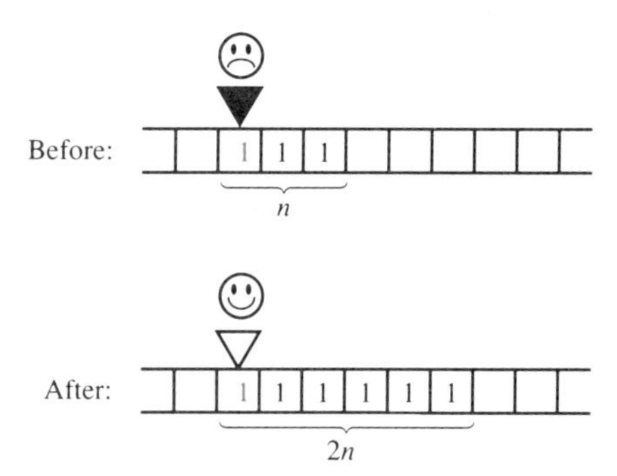

6.6 An Application: Convolutional Codes and Viterbi's Decoding Algorithm

In this final section we'll tackle a topic called *convolutional coding theory*, which has many important practical applications, including applications to deep-space telecommunications. This topic is a relatively new and advanced part of discrete mathematics, but don't worry. What we have already learned is more than enough to allow us to understand the essential facts about convolutional codes.

We begin by recalling one of the circuits from Section 6.1, viz. the circuit depicted in Figure 6.9. We repeat its circuit diagram (Figure 6.9), as well as its state table and diagram (Figure 6.10) as Figure 6.41. (Note that the state diagram has been slightly modified, so that the states are called A, B, C, and D, instead of 00, 01, 10, and 11.)

Recall that the *state* of the circuit in Figure 6.41(a) is defined to be the contents of the two D flip-flops. Thus for example if the first flip-flop contains a 1 and the second a 0, the circuit is in state 10, which is also called state C in Figure 6.41.

This machine has the property that at every clock pulse *one* bit goes in and *two* bits come out. A common use for such a machine is to *encode* a sequence of data bits; the machine is started in state A (00), and the data bits are used as input. The corresponding *output* bits are called the *encoded sequence.* This procedure is called a **convolutional encoding** of the data stream. Encoding is a way

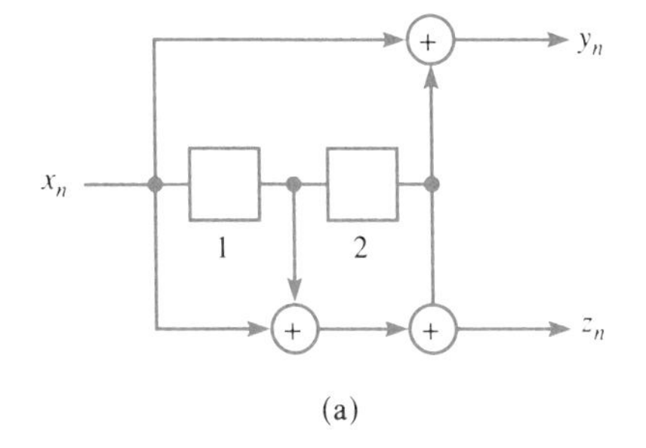

(a)

	Present State	Next State		Output	
		$x = 0$	$x = 1$	$x = 0$	$x = 1$
A	00	00	10	00	11
B	01	00	10	11	00
C	10	01	11	01	10
D	11	01	11	10	01

(b)

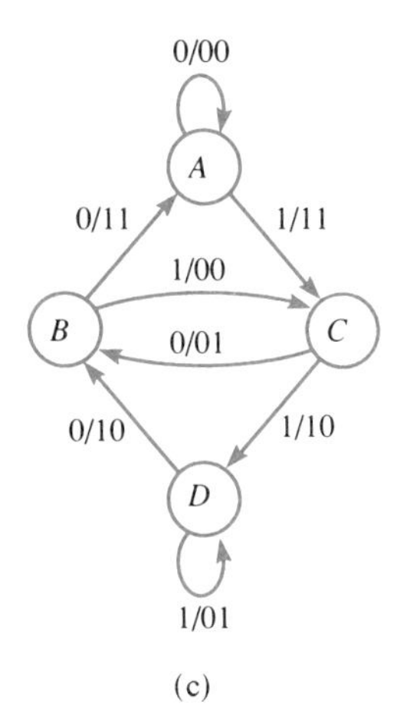

(c)

Figure 6.41 (a) Circuit diagram, (b) state table, and (c) state diagram for a simple *convolutional encoder.*

to add redundancy to data, prior to storage or transmission. When redundancy is present, it may be possible to correct errors which occur in the storage/retrieval or transmission/reception process. In Section 1.7 we saw another way to add redundancy, viz. the Hamming code. The Hamming code is used mostly to protect data from storage/retrieval errors, but convolutional codes are used mostly to protect data which is to be transmitted over long distances.

EXAMPLE 6.22 Use the FSM in Figure 6.41 to encode the data stream 11100100. . . .

SOLUTION Starting in state A, and given the input sequence 11100100 . . . , we can use the state diagram (Figure 6.41b) to calculate the corresponding state sequence and output sequence:

Clock pulse number		1	2	3	4	5	6	7	8	. . .
Input sequence		1	1	1	0	0	1	0	0	. . .
State sequence	[A]	C	D	D	B	A	C	B	A	. . .
Output sequence		11	10	01	10	11	11	01	11	. . .

Therefore the encoded sequence corresponding to the given data sequence is

11 10 01 10 11 11 01 11.

When the FSM of Figure 6.41 is used to encode data as in Example 6.22, it's called a **convolutional encoder.*** There are many other kinds of convolutional encoders, and some of them are shown in Figure 6.42.† However, since this section is only an introduction to the subject, for the most part we'll confine our discussion to the particular convolutional encoder shown in Figure 6.41. In any case, the encoder in Figure 6.41 is already complex enough to illustrate the most important features of the general theory.

* So called because, in the standard jargon of digital signal processing, the encoder *convolves* the input sequence with fixed sequences called the *encoding sequences.*

† The convolutional encoder in Figure 6.42(b) occupies a special place in the history of science. It was used to encode the historic photographs of Jupiter, Saturn, Uranus, and Neptune that the *Voyager* spacecraft transmitted to earth in the 1980s.

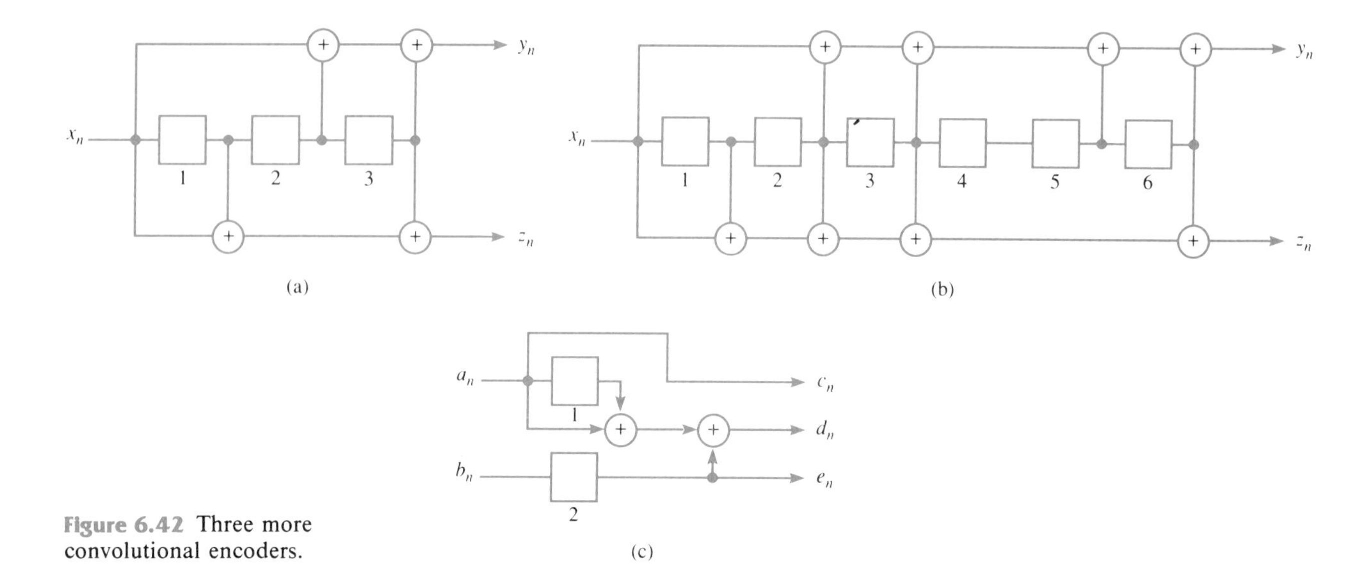

Figure 6.42 Three more convolutional encoders.

You may be thinking that all this "encoding" is used to prevent unauthorized persons from understanding the data stream; but no, that would be called "encryption." Actually, convolutional encoding is done for just the opposite reason: to make it *easier* to understand the data stream! Someone who was merely interested in recovering the original data sequence from the encoded stream would not have to work very hard, as Figure 6.43 shows. In Figure 6.43 we see the encoded stream (i.e., the two output sequences y_n and z_n) being transmitted to a remote location (the dots represent a long distance). At this remote location we find a simple circuit that performs an XOR operation on the two sequences, so that the output w_n is given by the Boolean equation

(6.14) $$w_n = y_n \oplus z_n.$$

This circuit enables the receiver to read the original input sequence x_n. Why? By Equations (6.7) and (6.8), or simply by inspection of the circuit diagram in Figure 6.41(a), we have

(6.15a) $$y_n = x_n \oplus x_{n-2}$$

(6.15b) $$z_n = x_n \oplus x_{n-1} \oplus x_{n-2}.$$

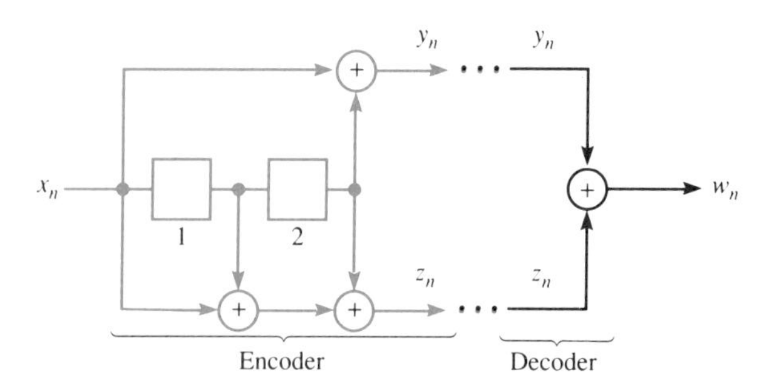

Figure 6.43 A simple decoder for the encoder in Figure 6.41 ($w_n = x_{n-1}$).

If we substitute these expressions for y_n and z_n into Equation (6.14), and use the known properties of "$\oplus$" (Equations A–E) on page 232, we obtain

$$\begin{aligned} w_n = y_n \oplus z_n &= (x_n \oplus x_{n-2}) \oplus (x_n \oplus x_{n-1} \oplus x_{n-2}) \\ &= (x_n \oplus x_n) \oplus x_{n-1} \oplus (x_{n-2} \oplus x_{n-2}) = x_{n-1}. \end{aligned} \tag{6.19}$$

EXAMPLE 6.23 Verify that Equation (6.14) does produce a one-clock-pulse-delayed version of the input sequence, using the sequences from Example 6.22.

SOLUTION As found in Example 6.22, the encoded stream is as follows:

n	1	2	3	4	5	6	7	8
(y_n, z_n)	(1, 1)	(1, 0)	(0, 1)	(1, 0)	(1, 1)	(1, 1)	(0, 1)	(1, 1)

To apply the decoding algorithm described in Figure 6.43 or Equation (6.14), all we do is XOR the corresponding (y_n, z_n) pairs:

n	1	2	3	4	5	6	7	8
(y_n, z_n)	(1, 1)	(1, 0)	(0, 1)	(1, 0)	(1, 1)	(1, 1)	(0, 1)	(1, 1)
w_n	0	1	1	1	0	0	1	0

We see that the w_n sequence calculated in this way is indeed the same as the input x_n sequence from Example 6.22, except that the w_n sequence has an initial 0.

In Figure 6.43 we have assumed that the values y_n and z_n which are received by the decoder are the same as those transmitted by the encoder. Unfortunately, however, real communication channels are often subject to *errors*. A transmitted 0 might be received as a 1, and a transmitted 1 might be received as a 0. The real value of convolution encoding, as we shall see, is that it allows the receiver to accurately read the original data stream *in spite of* occasional channel errors! With a more sophisticated decoding algorithm called *Viterbi's* algorithm*, it is possible to decode the data stream perfectly—provided there aren't *too* many errors.

In order to understand Viterbi's algorithm, we first need to take a closer look at the encoding process. For a convolutional encoder, as for any FSM, the key is the state sequence. In Example 6.22 the state sequence was $ACDDBACBA\ldots$, and in Figure 6.44 we have attempted to show the corresponding path through the state diagram. (For the next few pages, we won't worry about the encoder's output; we'll focus entirely on the state sequence.)

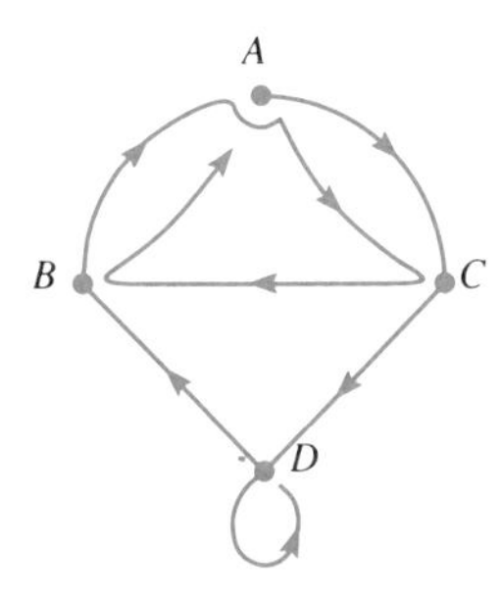

Figure 6.44 An attempt to trace the path $ACDDBACBA$ through the state diagram of Figure 6.41.

* Andrew Viterbi (1935–), Italian-born American engineer, industrialist, and entrepreneur. He invented "Viterbi's algorithm" in 1967 while working on deep-space communications problems for Caltech's Jet Propulsion Laboratory. Initially Viterbi believed his algorithm to be of no practical value, but now it plays an essential part in hundreds of high-performance telecommunications systems.

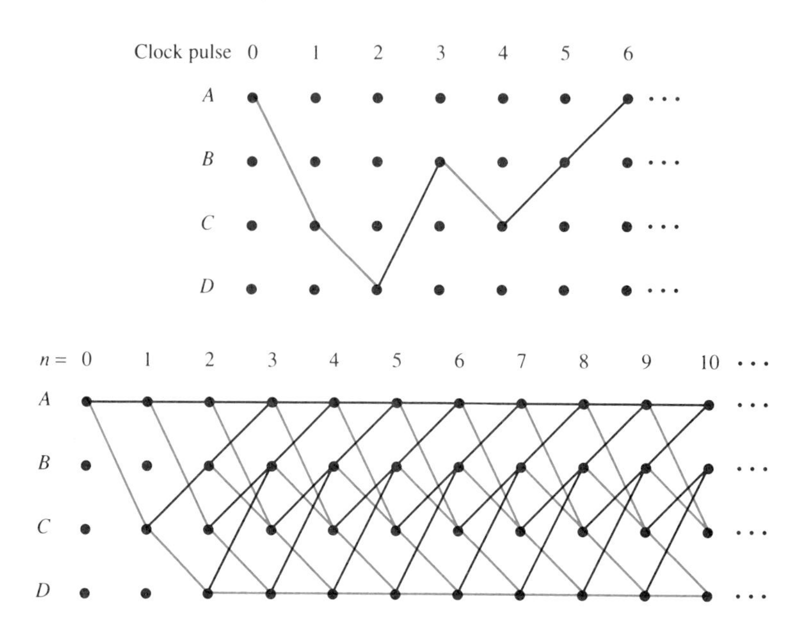

Figure 6.45 A trellis diagram, showing the path *ACDDBACBA* (black edge = input 0; color edge = input 1).

Figure 6.46 The trellis diagram can represent any path through the state diagram.

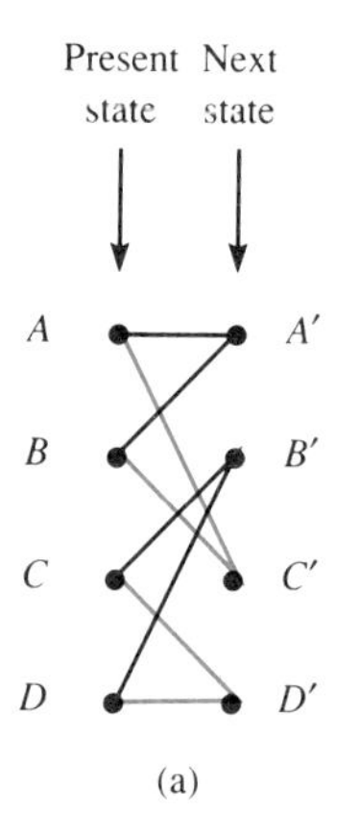

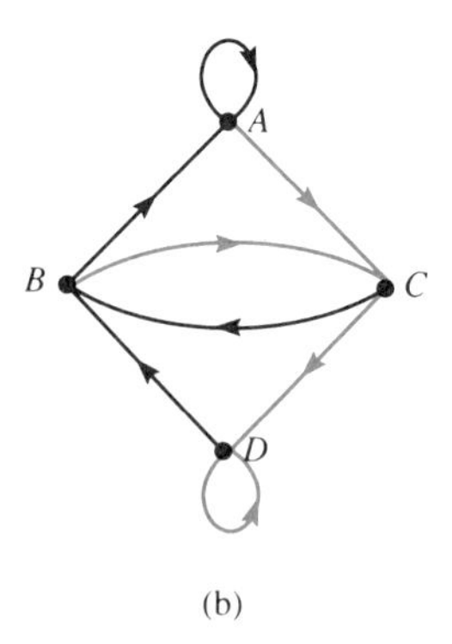

Figure 6.47 (a) The *trellis module*, compared with (b) the original state diagram.

Figure 6.44 isn't especially pretty; some edges are used twice, there are loops, etc. If the path had contained 16 edges instead of 8, the diagram would have been hopelessly messy. Still, it would be nice to have a clear graphic record of the encoder's travels through the state diagram. Fortunately, there is another diagram, called a **trellis diagram**, that is well suited for this purpose, and in Figure 6.45 we show how the path *ACDBCBA* is represented in a trellis diagram. The idea is simply to have a different copy of the state set (in this case $\{A, B, C, D\}$) corresponding to each clock pulse. In Figure 6.45 color is used to indicate the value of the input bits. A *black* edge represents a 0 input, and a *color* edge represents a 1 input. Thus the sequence "color, color, black, color, black, black" in Figure 6.45 corresponds to the input sequence 110100.

The trellis diagram in Figure 6.45 shows just *one* possible path through the state diagram. Actually, we can show *all* possible paths which start at state A with just a little more work (see Figure 6.46, which does actually look a little like a garden trellis!)

The trellis diagram has a regular repetitive structure; starting at the second clock pulse it consists of many identical copies of the **trellis module** in Figure 6.47(a), which is just an alternative form of the original state diagram (Figure 6.47b).

What we have done so far for the FSM in Figure 6.41 can be done for any FSM, as the next example shows.

EXAMPLE 6.24 Draw the trellis module, and a trellis diagram, for the mod 3 counter of Figure 6.14. Assume B is the initial state.

SOLUTION We repeat the state diagram of Figure 6.14 for reference here as Figure 6.48(a), again color-coded so that a black edge corresponds to a 0 input, and a color edge corresponds to a 1 input. Figure 6.48(b) shows the corresponding trellis module. There are two copies of each state, called A, A', B, B', C, C'; a directed edge in the state diagram, say from state X to state Y, is represented by an undirected edge in the trellis module from state X to state Y'.

In order to make the trellis *diagram*, we need to glue together many copies of the trellis module. However, since we're asked to assume that the initial state is B, the first two modules in the trellis diagram will be different from the others (see Figure 6.49). In the $n = 0$ to $n = 1$ module, for example, only the edges BB (black) and BC (color) are represented, since the paths must all start at B. At $n = 1$ the path could be at either B or C, and so all edges in the state diagram starting at either B or C are represented in the $n = 1$ to $n = 2$ module. At $n = 2$, the path could be at any one of the three states A, B, or C, and so from $n = 2$ onward, all the modules are the same as the "full" trellis module in Figure 6.48(b). ■

The paths in the trellises in Figures 6.46 and 6.49 appear to go on forever. In practice, of course, things can't go on forever, and so we'll need a graceful way to bring the paths to an end. One way to do this is to decree that all of the paths have the same length, and that they all end up back in state A. In Figure 6.50 we show a finite version of the trellis in Figure 6.46, in which all the paths are of length 6 and all end up in the same state. This is an example of a **truncated**

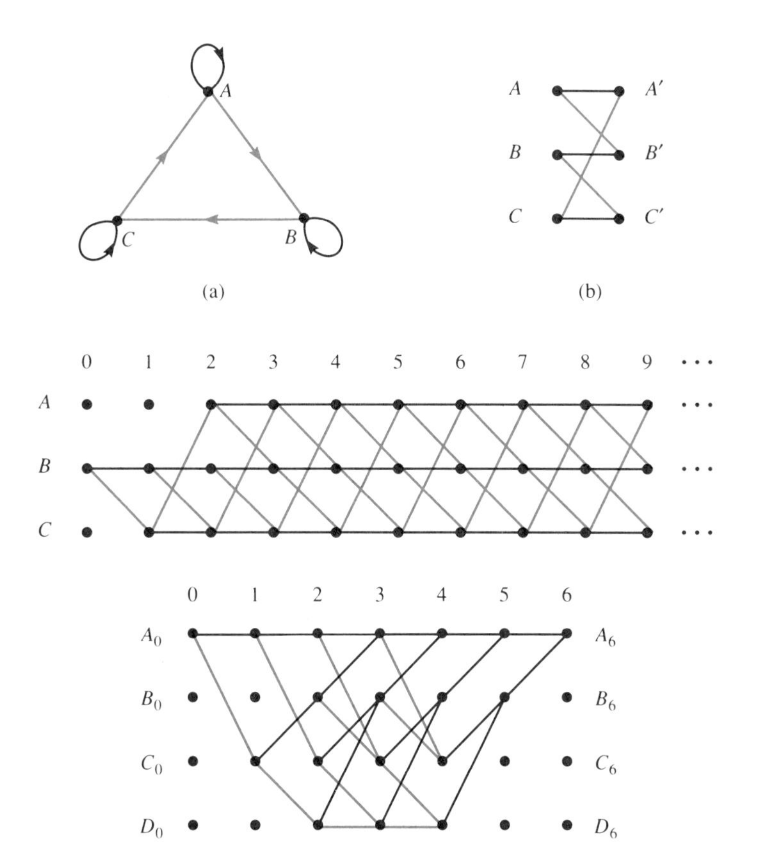

Figure 6.48 (a) The original state diagram; (b) the corresponding trellis module.

Figure 6.49 The trellis diagram for Example 6.24.

Figure 6.50 A truncated version of the trellis diagram in Figure 6.46.

trellis diagram. Notice that after the fourth clock pulse all of the paths "head for home" along *black* edges.

The truncated trellis in Figure 6.50 gives us a good opportunity to practice our counting skills.

EXAMPLE 6.25 How many paths from A_0 to A_6 does the trellis of Figure 6.50 contain?

SOLUTION As we have seen, each such path can be characterized uniquely by a sequence of *six input bits* (black or color). For example, the path *ACDBCBA* in Figure 6.45 is characterized by the sequence "color, color, black, color, color, black, black." Any other sequence of six bits will represent a path from A_0 to A_6, *provided that the last two bits are black.* (This condition ensures that the path ends up at A_6.) Therefore the number of paths from A_0 to A_6 equals the number of sequences of six bits, in which there are *two* possibilities for each of the first four bits (color or black), but only *one* possibility for the last two bits (black). By the multiplication rule (Section 2.1), then, the total number of possibilities is

$$2 \times 2 \times 2 \times 2 \times 1 \times 1 = 16.$$

In the last few pages we've been focusing our attention on the *state sequence* of the encoder in Figure 6.41 and haven't been paying attention to the *output.* Now we'll bring the output back into the picture by labelling each edge in the trellis diagram with the appropriate pair of output bits; see Figure 6.51. In Figure 6.51 each edge corresponds to one of the edges in the state diagram (Figure 6.41c), and the labels for the trellis edges are just copied from the corresponding edges in the state diagram.

Figure 6.51 is, in essence, Viterbi's view of the convolutional encoding process.* The truncated trellis diagram can be used to convert any sequence of 4 input bits into a 12-bit **codeword**. The idea is to start at A_0 and trace a path to A_6, with the input bits used for guidance. A 0 input bit directs us to follow a black edge, and a 1 input directs us to follow a color edge. After the 4 input bits

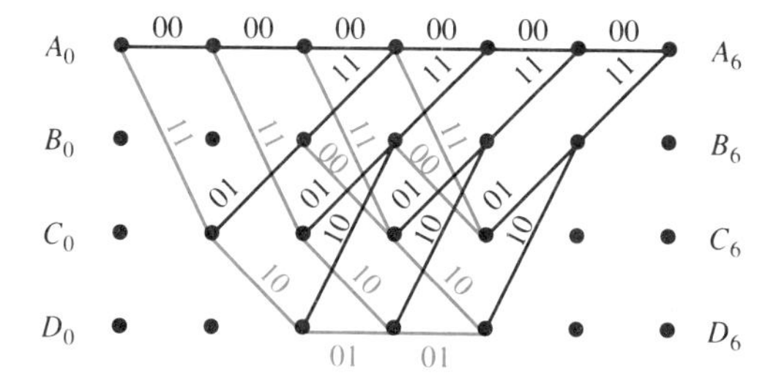

Figure 6.51 The completely labelled truncation of the trellis diagram for the convolutional encoder of Figure 6.41.

* Bear in mind that the truncated trellis in Figure 6.51 is only one example of a trellis diagram. There are many other kinds of convolutional codes, and each has a different trellis diagram. Moreover, each trellis diagram can be truncated at any depth. But what we shall say about this one *particular* trellis applies, in principle, to all convolutional codes.

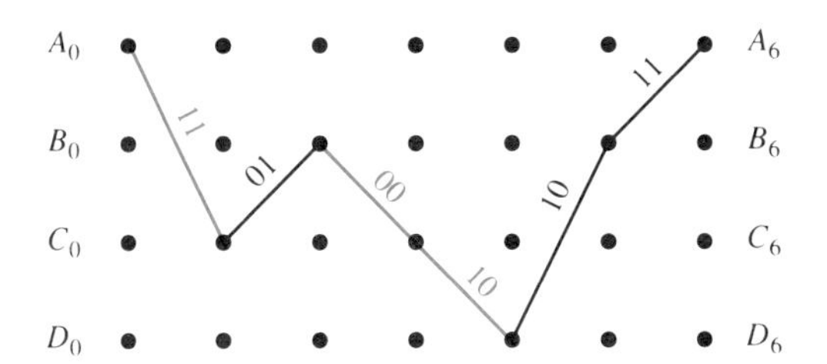

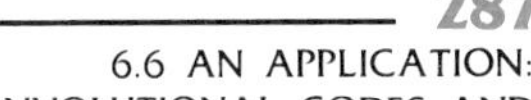
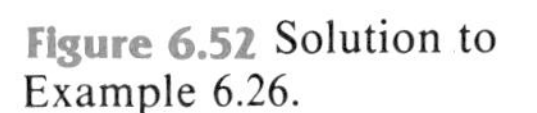

Figure 6.52 Solution to Example 6.26.

are used up, the path continues along two additional black edges to A_6. This path we shall call the **codepath**; the codeword is simply the labels on the edges of the codepath. Note that according to Example 6.25, there are 16 codewords.

EXAMPLE 6.26 Suppose the four input bits are 1011. Find the corresponding codepath and codeword.

SOLUTION We start at A_0. The given input bits correspond to the path whose edge colors are "color, black, color, color, (black), (black)." This codepath is shown in Figure 6.52, with the corresponding output bits attached to each of the edges. The codeword is seen to be 11 01 00 10 10 11.

Now we are ready to study the decoding problem. Here is the situation: There are two individuals, the *Sender* and the *Receiver*, who are physically separated by a considerable distance, and who can only communicate with each other over a *noisy channel.* The Sender has some information to send to the Receiver, information which is in the form of 4-bit "packets." The Sender and Receiver have previously agreed that the Sender will *encode* each 4-bit packet into a 12-bit codeword, using the trellis of Figure 6.51, and will then send the 12-bit codeword on to the Receiver via the noisy channel. When the 12-bit codeword arrives at its destination, it may be somewhat *garbled*; some of the transmitted 0s may be received as 1s, and vice versa. The Receiver's *decoding problem* is to make an intelligent guess as to what the original 4-bit packet was, basing this guess only on the received garbled codeword.

Guess really is the appropriate word here, since any one of the 16 codewords *could* have been garbled by the channel into any given received word, if the channel noise was bad enough. The best the Receiver can do is to search among the codewords and find the one that *most nearly resembles* what was received. As we shall see, the trellis structure in Figure 6.51 makes this search much easier than it would otherwise be.

Let's do an example. Suppose that the garbled codeword is

$$R = 10\ 10\ 10\ 10\ 01\ 11.$$

Which codeword most clearly resembles R? One approach to this problem would be to list all 16 codewords, and compare them, one by one, with R. For example,

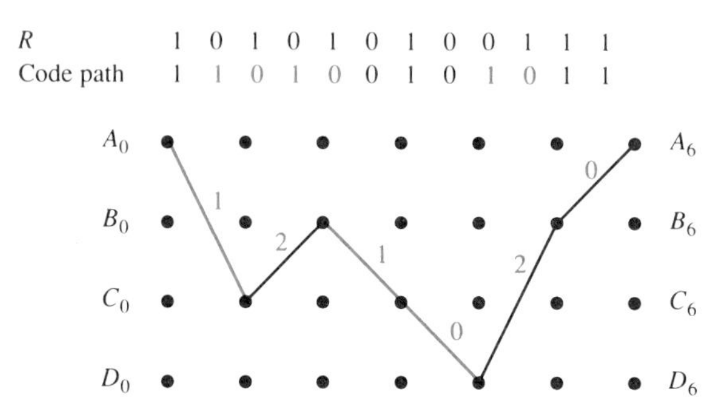

Figure 6.53 The codepath from Figure 6.52, with the edges labelled to show the number of disagreements with R.

the codeword in Figure 6.52 differs from R in six positions:

Garbled codeword R	10	10	10	10	01	11
Codeword from Figure 6.52	11	01	00	10	10	11
Disagreements	1	2	1	0	2	0

We can illustrate the relationship between R and this particular codepath by labelling each edge with 0, 1, or 2 to indicate the number of disagreements between the corresponding pairs of bits (see Figure 6.53). We could continue comparing R with the other 63 codepaths, but we won't, because there's a nice shortcut that takes advantage of the fact that each edge in the trellis is part of many different codepaths.

In Figure 6.53 we labelled 6 of the trellis edges to show the number of disagreements between one particular codepath and R; in Figure 6.54 we have finished the job and labelled all 28 edges similarly. Each edge label in Figure 6.54 was obtained simply by comparing the 2-bit edge labels in Figure 6.51 with the corresponding pair of bits in R. For example, consider the $B_3 - C_4$ edge, circled in Figure 6.54. In Figure 6.51 this edge is labelled 00; the corresponding bits of R are 10; and so in Figure 6.54 this edge is labelled 1, to indicate a disagreement of 1 bit.

Figure 6.54 makes it easy to compare codepaths with R. The number of disagreements between a given codepath and R is just the sum of the edge labels on the codepath. For example, Figure 6.54 tells us that the number of disagreements between R and the codepath corresponding to the input bits 0000(00) (i.e., the all-black codepath) is $1 + 1 + 1 + 1 + 1 + 2 = 7$. Our goal, remember, is to find the codepath that differs from R in the *fewest positions*, and in Figure 6.54 this corresponds to the codepath for which the sum of the edge labels is as

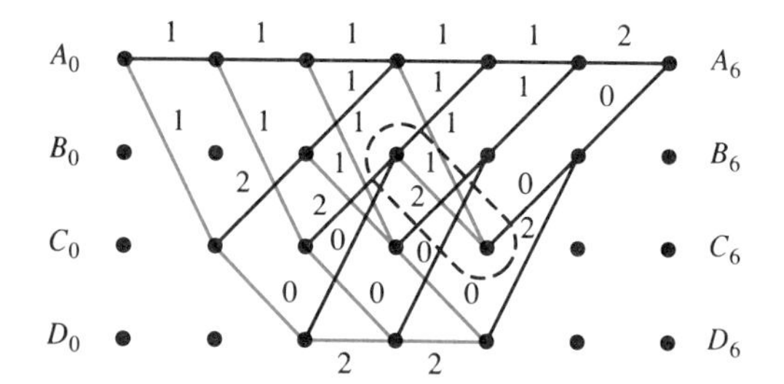

Figure 6.54 The entire trellis relabelled to show the number of disagreements between R and any codepath.

small as possible. If we now think of each edge label as the length of that edge, we see that the decoding problem boils down to *finding the shortest path from A_0 to A_6* in Figure 6.54.

Aha! If the decoding problem is nothing but a shortest-path problem in disguise, we already know how to solve it! As we learned in Chapter 4, we can use *Dijkstra's* algorithm to solve shortest-path problems, and indeed Dijkstra's algorithm can be used to solve the decoding problem for convolutional codes. However, the section title promises *Viterbi's* algorithm, not Dijkstra's. The reason for this is that Viterbi discovered his algorithm quite independently from Dijkstra; but subsequent research has shown that Viterbi's algorithm is only a slight modification of Dijkstra's, a modification which takes advantage of the extremely regular structure of the trellis diagram, and which lends itself more readily to practical implementation. Rather than give a detailed description of Viterbi's algorithm, we will conclude this section by explaining how Viterbi's algorithm finds the shortest path from A_0 to A_6 in Figure 6.54, which is the same as finding the codepath closest to the received sequence R in Figure 6.53.

We show the progress of Viterbi's algorithm on the trellis of Figure 6.54 in Figures 6.55(a)–6.55(f). We'll get to the details shortly, but for now you should

Figure 6.55 The progress of Viterbi's algorithm on the trellis of Figure 6.54.

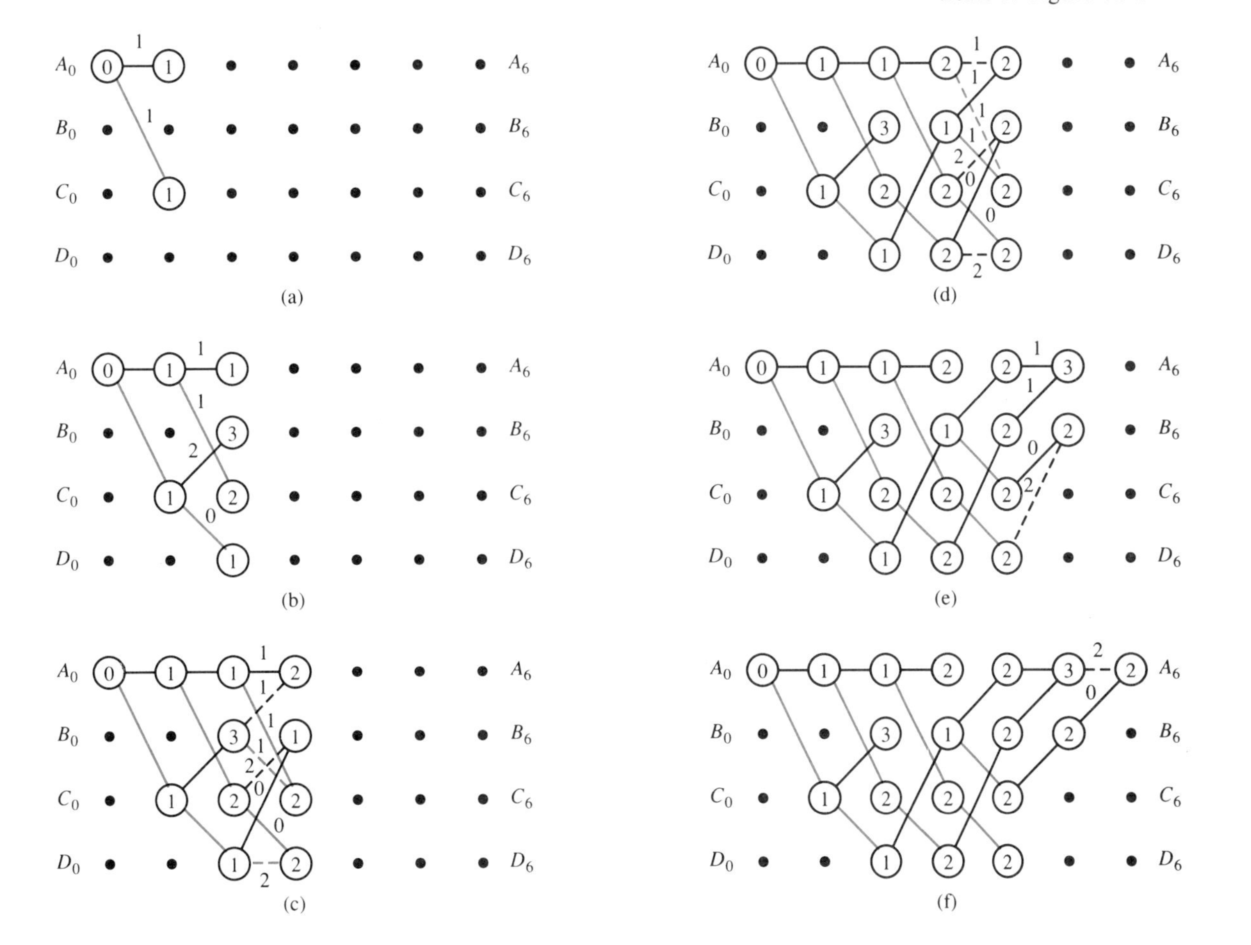

notice how the algorithm grows a tree, starting from A_0, and at each stage the tree is exactly one unit longer than it was before, until A_6 is reached. This orderly growth is the key to Viterbi's algorithm. After each iteration, Viterbi's algorithm will have found all of the shortest paths out to some fixed depth in the trellis. the next iteration, these shortest paths will be extended one unit deeper into the trellis.

For example, in Figure 6.55(d), all the shortest paths out to depth 3 have already been found, and are indicated by solid edges:

Vertex	Shortest Path	Length of Shortest Path
A_3	$A_0A_1A_2A_3$	2
B_3	$A_0C_1D_2B_3$	1
C_3	$A_0A_1A_2C_3$	2
D_3	$A_0A_1C_2D_3$	2

In Figure 6.55(d) we can also see how Viterbi's algorithm extends the shortest paths out to depth 4. For example, in the computation of the shortest path out to B_4, the reasoning is as follows: The shortest path to B_4 must pass through either C_3 or D_3 (see Figure 6.56). If the shortest path passes through C_3, the *portion* of that path from A_0 to C_3 must also be the shortest path to C_3, by the commonsense "principle of optimality" mentioned in Section 4.1. Thus, if it passes through C_3, its length must be

$$(\text{length of shortest path to } C_3) + (\text{length of edge } C_3B_4) = 2 + 2 = 4.$$

On the other hand, if the shortest path passes through D_3, its length must be

$$(\text{length of shortest path to } D_3) + (\text{length of edge } D_3B_4) = 2 + 0 = 2.$$

We conclude that the shortest path passes through D_3 and so in Figure 6.55(d), we make the edge D_3B_4 a solid edge.

In exactly the same way the shortest paths to A_4, C_4, and D_4 have been found, and are shown, in Figure 6.55(d). At the next iteration, viz., Figure 6.55(e), the paths are extended out to depth 5; then in Figure 6.55(f), out to depth 6. (Notice that in one case, a *tie* occurs: in Figure 6.55e, the two possible paths to A_5 both have length 3, and so *both edges* A_4A_5 and B_4A_5 are solid.

Thus Viterbi's algorithm proceeds happily along until it reaches the situation in Figure 6.55(f), at which point the shortest path from A_0 to A_6 has been found. In Figure 6.57 we isolate this path, and compare it with the received codeword R. Our conclusion is that R differs from the closest codepath in exactly two

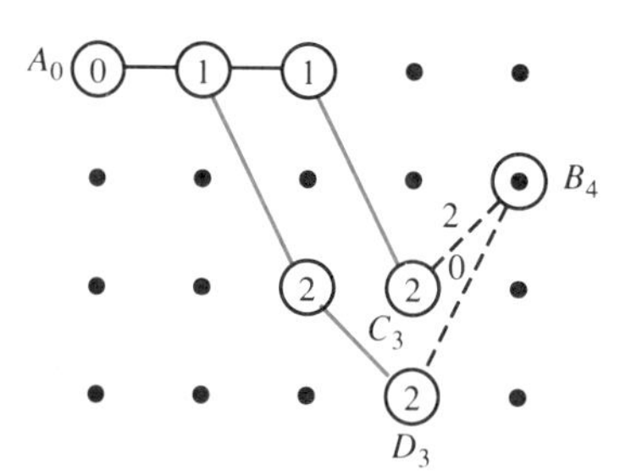

Figure 6.56 Finding the shortest path to B_4.

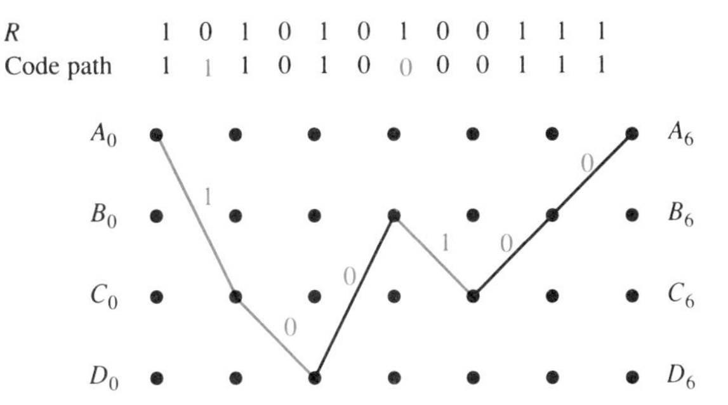

Figure 6.57 The codepath closest to R.

positions. Thus despite the presence of channel errors, the decoder can be reasonably certain that the transmitted four-bit packet was 1101.

Problems for Section 6.6

1. Use the FSM in Figure 6.41 to encode each of the following data streams. (Start in state A.)
(a) 00011011 . . . .
(b) 1010101010 . . . .
(c) 1111111111 . . . .

2. For the convolutional encoders in Figures 6.42(a) and 6.42(c), indicate how many states will be in the corresponding state diagram.

3. Draw state diagrams for the convolutional encoders in Figures 6.42(a) and 6.42(c).

4. *Don't* draw a state diagram for the convolutional encoder in Figure 6.42(b), but answer the following questions.
(a) How many states will there be?
(b) How many edges will there be?

5. Find a simple decoder for (a) the encoder of Figure 6.42(a); (b) the encoder of Figure 6.42(c) (cf. Figure 6.43).

6. Verify (i.e., give reasons for) each of the four steps leading to Equation (6.15).

7. Verify that Equation (6.14) does produce a one-clock-pulse-delayed version of the input sequence, for each of the input sequences from Problem 1.

8. Notice that Equation (6.15) doesn't make sense if $n = 1$, since x_{-1} and x_0 aren't defined. Still, in Example 6.23, we calculated $w_1 = 0$. How do you explain this? How should w_1 be defined in general?

9. Not every possible sequence of 0s and 1s occurs as the output of the convolutional encoder of Figure 6.41. Which of the following sequences of (y_n, z_n)'s could actually be an encoded stream?
(a) $(1, 1), (1, 0), (0, 1), (1, 0), (1, 1), \ldots.$
(b) $(1, 1), (1, 1), (0, 1), (1, 0), (1, 1), \ldots.$
(c) $(1, 0), (0, 1), (1, 1), (1, 0), (1, 1), \ldots.$
(d) $(0, 0), (0, 0), (0, 0), (1, 0), (0, 0), \ldots.$
(e) $(1, 1), (1, 0), (0, 1), (0, 1), (0, 1), \ldots.$

10. Draw trellis modules (cf. Figure 6.47a) for the convolutional encoder in (a) Figure 6.42(a) and (b) Figure 6.42(c).

11. Draw a truncated trellis of depth 3 (i.e., only three data bits) for the convolutional encoder in (a) Figure 6.42(a) and (b) Figure 6.42(c). (*Note*: The trellis in Figure 6.50 is truncated at depth 4.)

12. *Do not* draw the trellis module for the convolutional encoder of Figure 6.42(b) but *do* answer the following questions.
(a) How many vertices will there be?
(b) How many *color* edges?
(c) How many *black* edges?

In problems 13–17, draw trellis modules for each indicated FSM.

13. The FSM in Figure 6.11.

14. The FSM in Figure 6.12.

15. The FSM in Figure 6.13.

16. The FSM in Figure 6.14.

17. The FSM in Figure 6.16b.

In Problems 18–22, draw a trellis diagram for each of the indicated FSMs, assuming the given initial state.

18. The FSM in Figure 6.11 (initial state: A).

19. The FSM in Figure 6.12 (initial state: B).

20. The FSM in Figure 6.13 (initial state: S_1).

21. The FSM in Figure 6.14 (initial state: S_0).

22. The FSM in Figure 6.16b (initial state: B).

23. The "simple decoder" of Figure 6.43 won't work if there are *errors* in the received (y_n, z_n) sequence, or will it? There are certain patterns of *two* errors that the simple decoder *will* correct. See if you can find such patterns.

24. Consider the trellis diagram of Figure 6.46. There are *two* edges out to depth 1, *six* edges out to depth 2, 14 out to depth 3, etc. Find a general formula for the number of edges out to depth n. (*Hint*: Guess the formula and prove its correctness by induction.)

25. Repeat Problem 24, but use the trellis of Figure 6.49.

26. Consider the trellis of Figure 6.49. Count the number of paths

(a) From B_0 to A_6.
(b) From B_0 to A_9.
(c) From C_1 to A_5.

27. (a) Count the total number of edges in the trellis of Figure 6.50.

(b) Figure 6.50 is the "$L = 4$" truncation of the infinite trellis in Figure 6.46. Find a formula for the total number of edges in the Lth truncation.

28. Count the number of paths from A_0 to A_n, as a function of n, in the trellis of Figure 6.46.

29. Use Viterbi's algorithm, and Figure 6.51, to decode each of the following garbled codepaths.

(a) 10 01 11 01 00 00.
(b) 11 11 00 11 11 00.
(c) 11 10 01 01 10 11.

30. In Example 6.25 we counted 16 paths from A_0 to A_6 in the trellis of Figure 6.50. How many of these paths pass through (a) the edge D_3B_4 and (b) the edge D_2D_3?

31. In the text (cf. Figure 6.57) we found the codepath in Figure 6.54 which was *most* like $R = 10\ 10\ 10\ 10\ 01\ 11$. Which codepath is *least* like R?

32. Exactly one of the vertices in Figure 6.51 has more than one shortest path to A_0. Which one is it? (*Hint*: Examine Figure 6.55.)

Summary

If Boolean logic gates are combined with **flip-flops**, **sequential circuits** can be constructed. These circuits have **memory** and can handle inputs that occur *serially* rather than in parallel. A sequential circuit can be represented by the **state diagram** of a **finite state machine**; the present state and the next input determine the next state and next output. Finite state machines can be designed to solve many problems that involve counting or pattern recognition. The efficiency of the design can sometimes be improved by the **state equivalence algorithm**.

A simplified finite state machine with no outputs and with designated **accepting states** is called a **deterministic finite automaton** (DFA). If more than one transition (or no transition) from a particular state is allowed for a given input, a **nondeterministic finite automaton** (NFA) results. NFAs are often more convenient than DFAs for the problem of designing a machine to accept a particular set of strings. An efficient algorithm (the **NFA simulation algorithm**) can be used to determine whether a particular string is accepted by a given NFA.

Finite state machines are fundamentally limited by their finite memory. The **Turing machine** is an infinite-memory device that can in principle perform any reasonable computational task.

Sequential circuits can be used to build **convolutional encoders** and **decoders** designed to correct errors in the transmission of information. The problem of determining, for a given received sequence, which transmitted sequence is most likely, is shown to be equivalent to a shortest-path problem, which is most efficiently solved by **Viterbi's algorithm**.

CHAPTER SEVEN

FORMAL LANGUAGES AND DECISION ALGORITHMS

In Chapter 6 we mentioned briefly the subject of *formal languages*, and in this chapter we will go into this subject in more detail. Formal languages were originally introduced as a part of mathematical logic, as simplified models for complex human languages like English or Chinese. However, no formal language ever proposed seems close to capturing the essence of even the most primitive human language, so language theory has traditionally been considered a rather impractical branch of pure mathematics.

Much to everyone's surprise, however, formal languages have proved to be of tremendous help in the design of modern programming languages, compilers, operating systems, and other software! For this reason language theory is an important tool for modern computer scientists, especially software specialists.

In this chapter, our goal is to introduce you to some of the basics of formal language theory, focusing in particular on the important *decision problem*. (For a given formal language, the decision problem is to decide whether a given string is in the language or not. The goal is always to find an efficient algorithm to do the job.) In Section 7.1 we will give three specific examples of formal languages, and discuss the decision problem for each one. Section 7.2 is a digression, concerning one of the most famous of all mathematical counting problems, the problem of counting the number of strings of each length in one of the languages, the "balanced parenthesis language," introduced in Section 7.1. In Section 7.3 we will introduce the important class of *regular* languages, and show, among other things, that every language accepted by a finite automaton is a regular language. Then in Section 7.4 we will show that conversely, for every regular language, it is possible to construct a (nondeterministic) finite automaton that accepts it. This fact, combined with the NFA simulation algorithm we studied in Section 6.4, shows how to produce a decision algorithm for any regular language.

7.1 Three Examples of Formal Languages

In Chapter 5 we learned how "propositions" in the propositional calculus can be represented as *strings of symbols*. The symbols allowed are the seven symbols in the set $\{(,), \sim, \wedge, \vee, \rightarrow, \leftrightarrow\}$, plus the atoms. We learned how to decide whether or not a given string of symbols represented a well-formed proposition; and how to decide whether or not a well-formed proposition was a theorem. Although we didn't emphasize the point, these decision procedures can be formulated as *algorithms.* We could program a computer to tell us whether a given string of symbols is a WFP or a theorem. Similarly, in Chapter 6 we studied sets of strings accepted by finite automata (deterministic or nondeterministic) and devised algorithms capable of deciding whether or not a given string is in such a set.

It turns out that there are other important situations in which we are given a set of strings built from a finite set of symbols, and are asked to find an algorithm for deciding whether or not a given string is in the set. In problems like this, the underlying set is called the **alphabet**, the set of strings is called a **formal language**, or just a **language**, and the algorithm is called a **decision algorithm**. In this section, we will consider three representative formal languages, called languages A, B, and C, and in each case try to find a decision algorithm.*

Our first example is language A, which is a language built from the three-letter alphabet $\{a, b, c\}$. Language A is similar to the propositional calculus, in that it is defined *recursively*, i.e. a set of "atomic" strings is given, and then new strings are built from old ones by applying a basic set of rules. Here are the rules defining language A (in these rules, x denotes an arbitrary string of a's, b's, and c's.

RULES DEFINING LANGUAGE A

- Rule $(A0)$: The empty string ϵ is in Language A.
- Rule $(A1)$: If x is in Language A, so is xa.
- Rule $(A2)$: If x is in Language A, so is xab.
- Rule $(A3)$: If x is in Language A, so is xbc.

In $(A0)$, we refer to the empty string ϵ, which we encountered earlier in Chapter 6. It is the string formed from no symbols: a very small string, but also very important. Its role in the theory of languages is analogous to the role played by the empty set in the theory of sets. The rules $(A1)$–$(A3)$ are sometimes called **production rules** for the language, since they can be used to *produce* new strings from old ones.† In what follows, we'll call the strings of a's, b's, and c's that are in language A *well-formed strings* (WFSs).

Starting with the empty string, and applying the production rules $(A1)$, $(A2)$, and $(A3)$ in any order, we can produce as many WFSs as we wish. For example, the following strings are all in language A:

* Actually, there are languages for which it is known to be *impossible* to find a decision algorithm! The most famous example of this is the formal language that is used to express the theorems of *number theory.* In 1936 Alonzo Chuch and Alan Turing (see Section 6.5) showed that there is no algorithm which will automatically decide whether or not a "well-formed" statement in number theory (like "every even number is the sum of two primes") is true.

† Production rules like these are similar to, but not quite the same as, the "productions" that computer scientists use to give formal descriptions of computer languages.

ϵ	(rule $A0$)
a	(rule $A1$ applied to ϵ)
aab	(rule $A2$ applied to the previous string)
$aabbc$	(rule $A3$)
$aabbcab$	(rule $A2$ again),

etc. Of course there are many more, in fact infinitely many more, strings that can be produced, and language A is the set of *all* such strings. In the next example, we will consider the decision problem for language A.

EXAMPLE 7.1 Which of the following strings belong to language A?

(a) *abaaabca.* (b) *bcaababa.* (c) *abbaabca.*

SOLUTION

(a) Let's begin by assuming that the given string *abaaabca* is indeed in language A, and see if we can determine the sequence of rules that was used to produce it. It's actually easier to work *backwards*, and to begin by determining the *last* rule applied. The given string ends in a; and the only one of the rules that produces a string ending in a is rule ($A1$), which simply appends a as a "suffix" to a previously generated WFS. Therefore if we *delete* the suffix a from the given string, the result must be a WFS:

$$abaaabc\not{a} \xrightarrow{(A1)} abaaabc.$$

This new string ends in the suffix bc, which means that rule ($A3$) must have been used to produce it. If we delete the suffix bc, we get a still shorter string:

$$abaaa\not{b}\not{c} \xrightarrow{(A3)} abaaa.$$

This string ends in a, which means that ($A1$) was used to produce it:

$$abaa\not{a} \xrightarrow{(A1)} abaa.$$

Continuing in this way, we can shrink the string three more times,

$$aba\not{a} \xrightarrow{(A1)} aba, \qquad ab\not{a} \xrightarrow{(A1)} ab, \qquad \not{a}\not{b} \xrightarrow{(A2)} \epsilon,$$

finally reaching the empty string ϵ, which is by ($A0$) well formed. Therefore the original string is in language A; and the sequence of rules used to produce it was ($A2$), ($A1$), ($A1$), ($A1$), ($A3$), ($A1$).

(b) We'll tackle this string in just the same way we handled the string in part (a). Omitting further discussion, the computation proceeds as follows:

$$bcaabab\not{a} \rightarrow bcaabab, \qquad bcaab\not{a}\not{b} \rightarrow bcaab,$$

$$bca\not{a}\not{b} \rightarrow bca, \qquad bc\not{a} \rightarrow bc, \qquad \not{b}\not{c} \rightarrow \epsilon.$$

Therefore the original string is in language A, and was produced by the sequence of rules ($A3$), ($A1$), ($A2$), ($A2$), ($A1$).

(c) Here we go again:

abbaabc~~a~~ → *abbaabc*, *abbaa~~bc~~* → *abbaa*,

abba~~a~~ → *abba*, *abb~~a~~* → *abb*, *abb* → ?

Oops! We have arrived at a string (*abb*) which is not empty, and which cannot have been produced by an application of ($A1$), ($A2$), or ($A3$)! [A string produced by ($A1$) ends in *a*; a string produced by ($A2$) ends in *ab*; and a string produced by ($A3$) ends in *bc*.] Therefore the given string is *not* in language A. ■

The technique introduced in Example 7.1 for deciding whether or not a given string is in language A can be given in the form of an *algorithm*; and in Figure 7.1 we show this algorithm. It is called a *decision algorithm* for language A, because it takes as input an arbitrary string formed from the letters *a*, *b*, and *c*, and *decides* whether that string is in language A. If the string is in language A, the algorithm will print YES, and stop. If the string is not in language A, the algorithm will print NO, and stop. (We will return to language A in Section 10.1, where we will consider the problem of counting the number of strings of length n in the language.)

Our next example of a formal language will be called language B. The strings in language B will be formed entirely from the two symbols "(" (left parenthesis) and ")" (right parenthesis). This example is of some practical importance because of the way parentheses are used in mathematics to ensure clarity. (Of course, parentheses have other uses as well.) For example, the four expressions

(7.1) $$a + b^2 + c + d^2$$

(7.2) $$(a + b)^2 + c + d^2$$

(7.3) $$((a + b)^2 + c + d)^2$$

(7.4) $$(a + b)^2 + (c + d)^2$$

are identical, apart from the placement of the parentheses, yet represent four entirely different things.

The use of parentheses in mathematics follows certain simple but important rules that make it possible to find typographic errors in mathematical manuscripts *without* having to understand the mathematics. One rule is that every left parenthesis must have a matching right parenthesis, and vice versa. For example, an expression like

(7.5) $$\ell_i^*(c) \le \ell_i^*(c - 1))^{c-1}$$

(whatever it is supposed to mean) *cannot possibly be correct*, because it has fewer left parentheses (two) than right parentheses (three). Furthermore, for each left parenthesis, the matching right parenthesis must come *later* in the formula, so that an expression like

(7.6) $$(x + 1)) + ((y - z)$$

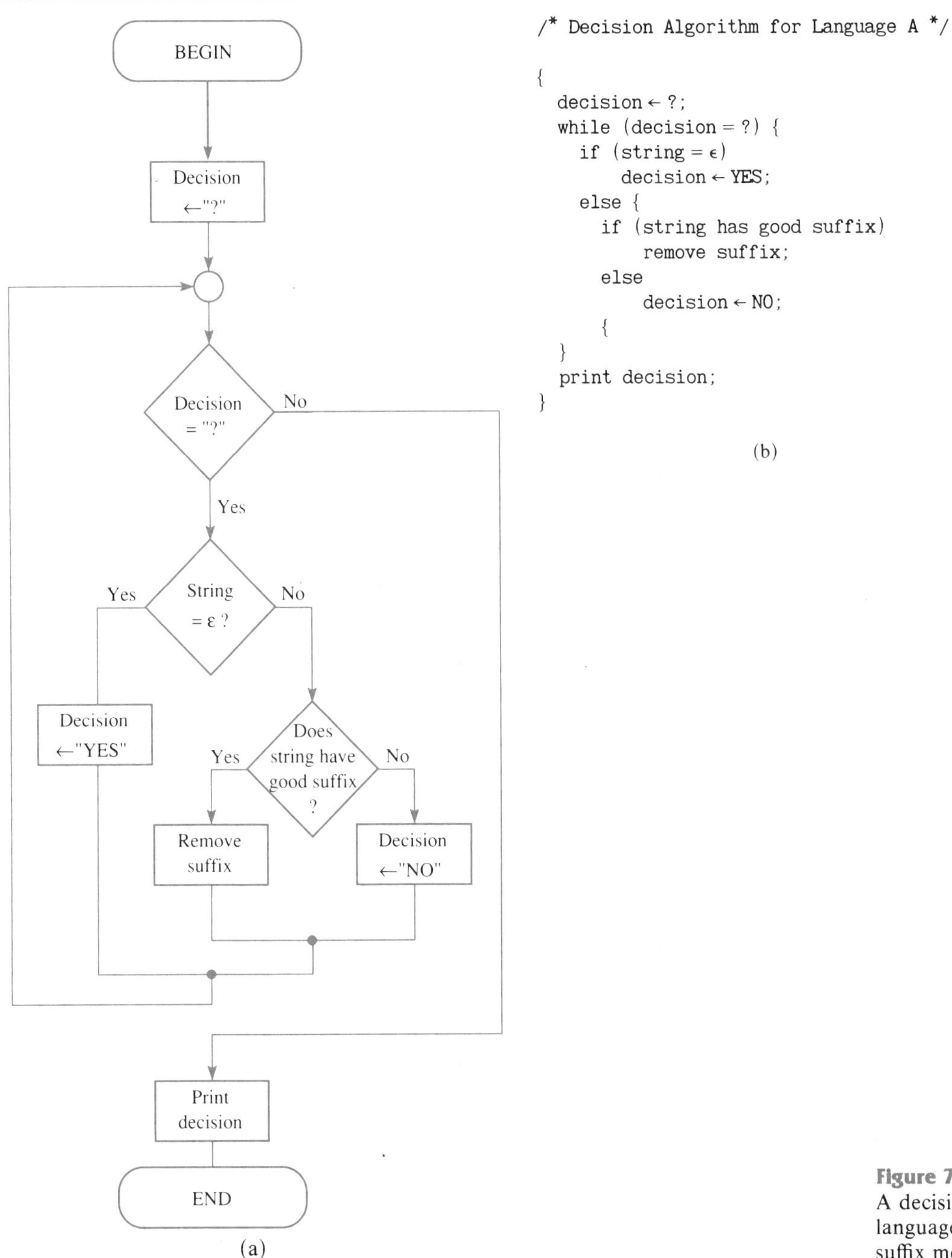

Figure 7.1 Algorithm A: A decision algorithm for language A. (*Note*: A "good" suffix means *a*, *ab*, or *bc*.)

also cannot be correct, even though it has an equal number of left and right parentheses, because its second right parenthesis comes before its second left parenthesis. With more complicated formulas, the business of parenthesis checking can become more complicated, as Table 7.1 shows.

TABLE 7.1 (a) An expression with balanced parentheses. (b) An expression with unbalanced parentheses

$$\log((a+b)^2+\cos((x+y)^2+z)) = (a-b)^{1/2} - \sin((x+y)^2+z)$$
$$((\quad) \qquad ((\quad) \quad)) \quad (\quad) \qquad ((\quad) \quad)$$
(a)

$$\log((a+b)^2+\cos((x+y)^2+z) = (a-b)^{1/2} - \sin((x+y)^2+z)$$
$$((\quad) \qquad ((\quad) \quad) \quad (\quad) \qquad ((\quad) \quad)$$
(b)

In Table 7.1(a) we see an algebraic expression in which the parentheses are all nicely matched, or, as we shall say from now on, *balanced.* In Table 7.1(b), however, we see an almost identical expression in which the parentheses are *unbalanced.*

In our study of language B (the language of "balanced parentheses"), we shall be interested only in the pattern of left and right parentheses, and not in the intervening mathematics. For example, for the expressions (7.1)-(7.6), the corresponding *parenthesis strings* are as follows:

(7.1′) ϵ (the empty string)

(7.2′) ( )

(7.3′) (( ))

(7.4′) ( )( )

(7.5′) ( )( ))

(7.6′) ( ))(( ).

The strings (7.1′)-(7.4′) are well-formed parenthesis strings, and are in language B. The strings (7.5′) and (7.6′), on the other hand, are not well formed, and are not in language B. The parenthesis strings corresponding to the formulas in Table 7.1 are shown in the table. The upper string is in language B, but the lower one is not.

It is time to give the official rules for language B. (In these rules, x denotes an arbitrary string of ('s and)'s).

RULES DEFINING LANGUAGE B

- Rule ($B0$): The empty string ϵ is in language B.
- Rule ($B1$): If x is in language B, so is (x).
- Rule ($B2$): If x and y are in language B, so is xy.

These rules correspond to the way parentheses are built up in practice, and using them we can build up as many well-formed parenthesis strings as we wish. For example, the parenthesis string corresponding to the formula in Table 7.1(a) can be built up as follows:

(a) ϵ	(rule *B*0)
(b) ( )	(rule *B*1 applied to a)
(c) (( ))	(rule *B*1 applied to b)
(d) ( )(( ))	(rule *B*2 applied to b and c)
(e) (( )(( )))	(rule *B*1 applied to d)
(f) (( )(( )))( )(( ))	(rule *B*2 applied to e and d).

We'll now describe a simple decision algorithm for language B, called algorithm B. (See Figure 7.2.) If algorithm B is given a string of left and right parentheses, it will "read" the string from left to right, one symbol at a time. After reading the last symbol (and sometimes even sooner), the algorithm will print either YES or NO, indicating that the given string is either in or not in language B, and halt.

The key to the operation of algorithm B is the variable `count`, which is an integer indicating the current imbalance between left and right parentheses. Initially (step 1) `count` is set to 0. When a left parenthesis is read, `count` is increased by 1 (step 3), and when a right parenthesis is read, `count` is decreased by 1 (step 4). Thus we have at all times

`count` = no. of left parens read − no. of right parens read.

If the string is in language B, each left parenthesis is matched by a right parenthesis that comes later, so as the string is read, `count` will always be greater than or equal to 0. Thus if `count` ever goes negative, the string cannot be in language B. The algorithm always checks for a negative `count` (step 2), and if it finds one, it prints NO (step 7), and halts.

However, if `count` never goes negative, the algorithm will need to read all the symbols in the input string before making its decision. After all the symbols have been read (this fact is determined by the "end of string" query in step 2), the algorithm again checks the `count` (step 5). If `count` is not 0, the total number of left and right parentheses aren't equal, which means that the string isn't in language B, and so the algorithm prints NO (step 7), and halts. On the other hand, if `count` does equal 0, then the number of left and right parentheses are equal, and moreover at no time did a right parenthesis come before its matching left parenthesis. When this happens, the algorithm prints YES (step 6), and halts.

We leave as an exercise the formal proof that algorithm B works (Problem 16). The following example will give us some practice in using it.

EXAMPLE 7.2 Apply algorithm B to the following three strings.

(a) ( )( )). (b) ( ))(( ). (c) (( ))(( ))( ).

SOLUTION

(a) The following table summarizes the behavior of the algorithm when the input string is "( )( ))."

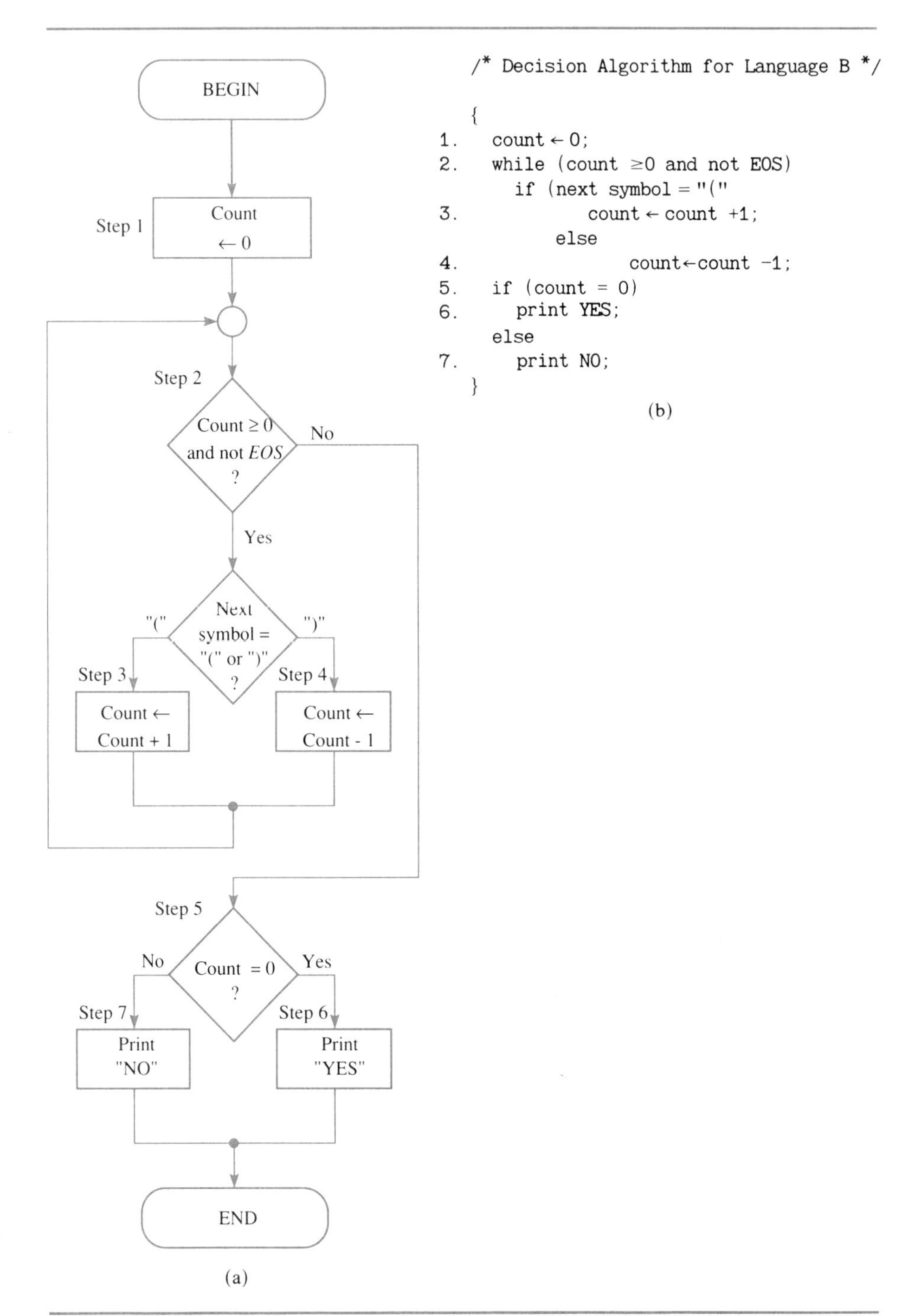

Figure 7.2 (a) Flowchart for Algorithm B. (b) Pseudocode for algorithm B. (*Note*: "EOS" means "end of string.")

String Symbols	Count
	0
(	1
)	0
(	1
)	0
)	−1; print NO

As you see, count remains greater than or equal to 0 until after the last symbol is read, at which point it becomes −1. Therefore when the algorithm exits from the while loop, it prints NO, and halts.

(b) The following table summarizes the behavior of the algorithm when the input string is "( ))(( )."

String Symbols	Count
	0
(	1
)	0
)	−1; print NO
(	
(	
)	

As you see, after the third symbol is read, the count goes negative, at which point the algorithm exits from the while loop, prints NO, and halts.

(c) The following table summarizes the behavior of the algorithm when the input string is "(( ))(( ))( )."

String Symbols	Count
	0
(	1
(	2
)	1
)	0
(	1
(	2
)	1
)	0
(	1
)	0; print YES

As you see, after all 10 symbols are read, count is 0, so the algorithm takes the YES branch at step 5, prints YES, and halts. ■

We will conclude this section with a discussion of a third formal language, which we will call language C, a language that was invented by Douglas Hofstadter in his Pulitzer Prize–winning book *Gödel, Escher, Bach: An Eternal Golden Braid* (New York: Basic Books, 1979). This language is an example of a language for which it is quite difficult to find a decision algorithm. In fact, apparently no decision algorithm for this language was known until Laif Swanson,* a colleague of ours, read an early draft of this chapter and discovered one.

Language C consists of certain strings formed from the three letters M, I, and U. Like languages A and B, it is defined by certain rules. The rules for language C follow. (In these rules, x and y denote arbitrary strings of M's, I's and U's.)

RULES DEFINING LANGUAGE C

- Rule $(C0)$: The string MI is in language C.
- Rule $(C1)$: If the string xI is in language C, so is the string xIU.
- Rule $(C2)$: If the string Mx is in language C, so is the string Mxx.
- Rule $(C3)$: If the string $xIIIy$ is in language C, so is the string xUy.
- Rule $(C4)$: If the string $xUUy$ is in language C, so is the string xy.

These five rules are summarized in the following box.

A SUMMARY OF THE RULES OF LANGUAGE C

$(C0)$	MI
$(C1)$	$xI \to xIU$
$(C2)$	$Mx \to Mxx$
$(C3)$	$xIIIy \to xUy$
$(C4)$	$xUUy \to xy$

The production rules for language C are a bit more complicated than those for languages A and B; let's get some practice using them. Here is a sequence of legal applications of the rules $(C0)$–$(C4)$:

MI	(rule $C0$)
MII	(rule $C2$, applied to $M\overbrace{I}^{x}$)
$MIIII$	(rule $C2$, applied to $M\overbrace{II}^{x}$)
MUI	(rule $C3$, applied to $\overbrace{M}^{x}III\overbrace{I}^{y}$)
$MUIU$	(rule $C1$, applied to $\overbrace{MU}^{x}I$)
$MUIUUIU$	(rule $C2$, applied to $M\overbrace{UIU}^{x}$)
$MUIIU$	(rule $C4$, applied to $\overbrace{MUI}^{x}UU\overbrace{IU}^{y}$).

* Laif Swanson (1950–), American mathematician and engineer. Formerly a mathematics professor at Texas A&M, in 1981 she joined the research staff at Caltech's Jet Propulsion Laboratory, where she led the team that discovered the convolutional code (see Chapter 6) used by *Galileo*, a NASA spacecraft scheduled to explore Jupiter and its satellites in the 1990s.

TABLE 7.2 The birthdays of some strings in language C

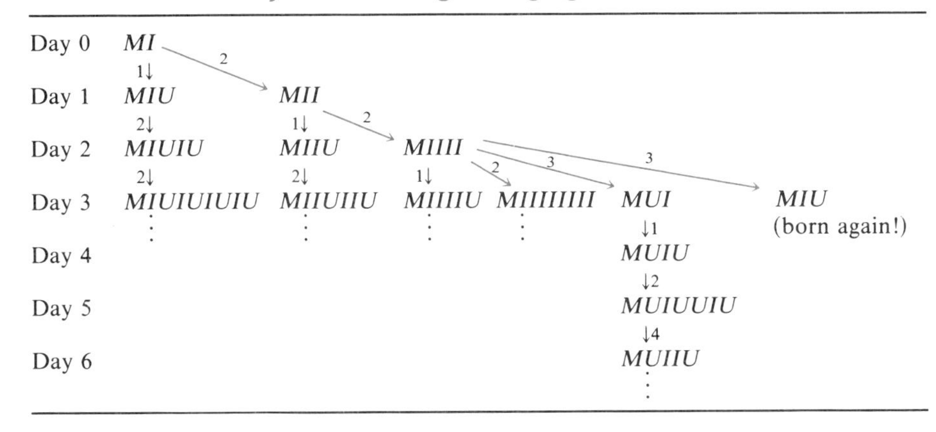

We conclude that each of the above-listed strings is in language C, and in particular, that the string *MUIIU* is in language C.

The strings in language C can in principle be displayed in a *tree* structure, as shown in Table 7.2. At the top of the table we show the string *MI*, which is given as belonging to language C by rule (*C*0). We might say that *MI* is *born on day 0.* Once *MI* is given, we may apply rule (*C*1), and obtain *MIU*, or rule (*C*2), and obtain *MII*. We say that these two strings are born on day 1. In Table 7.2, we see that three new strings are born on day 2, and five further new strings are born on day 3. Notice that if rule (*C*3) is applied to the string *MIIII*, which is born on day 2, we can get *either MUI* or *MIU*, depending on which group of three *I*'s is changed to *U*. Notice also the string *MIU* was *already born* on day 1, so we might say that it is *born again* on day 3! We have not shown all strings born on day 4, but we do show the part of the tree leading to the birth of *MUIIU* on day 6.

In principle, Table 7.2 can be extended indefinitely to show all of the strings in language C. In practice, of course, this isn't satisfactory, since the tree of Table 7.2 gets very complicated very fast, as the next example illustrates.

EXAMPLE 7.3 Table 7.2 shows that the string *MIIIIIIII* is born on day 3. Find all of its descendents which are born on day 4.

SOLUTION There are only four rules which can be applied to the given string, viz. rules (*C*1), (*C*2), (*C*3), and (*C*4). Let's see what happens when each of these rules is applied.

- *Rule (C1)*: This rule says that a *U* can be added at the end of any string in language C that ends in *I*. When this rule is applied to the given string, it produces *MIIIIIIIIU*.
- *Rule (C2)*: This rule says that the string appearing after the *M* in any string in language C can be *doubled*. Applying this rule to the given string, we get *MIIIIIIIIIIIIIIII*.

- *Rule (C3)*: This rule says that any three I's in a row can be replaced by U. In the given string, viz. $MIIIIIIII$, three I's in a row appear in six places, and so rule $(C3)$ can be applied in six different ways:

$$M\overbrace{III}IIIII \to MUIIIII \qquad MIII\overbrace{III}II \to MIIIUII$$

$$MI\overbrace{III}IIII \to MIUIIII \qquad MIIII\overbrace{III}I \to MIIIIUI$$

$$MII\overbrace{III}III \to MIIUIII \qquad MIIIII\overbrace{III} \to MIIIIIU.$$

- *Rule (C4)*: This rule says that two U's in a row can be deleted from any string in language C. But the given string does not have two U's in a row (indeed, it has no U's at all), and so rule $(C4)$ doesn't apply.

In summary, we have shown that there are exactly eight descendents of *MIIIIIIII* born on day 4, and they are *MIIIIIIIIU*, *MIIIIIIIIIIIIIIII*, *MUIIIII*, *MIUIIII*, *MIIUIII*, *MIIIUII*, *MIIIIUI*, and *MIIIIIU*. ■

Given a string of M's, I's, and U's, it can be very difficult to decide whether or not that string is in language C. One approach is to extend the tree of Table 7.2, hoping that the given string will appear. If the string does appear, well and good. But if the string refuses to appear, is it because we haven't extended the tree far enough, or is it perhaps because the string isn't in the language at all? For example, consider the string I. It doesn't appear in Table 7.2, but is it perhaps born on day 4 or later? With a little thought, you can see that the answer is no, because every string in language C must begin with an M. Indeed, the following theorem is true.

THEOREM 7.1 Every string in language C must begin with M and have no other M's in it.

The evidence of Table 7.2 certainly supports Theorem 7.1; but we leave its formal proof as Problem 25. Theorem 7.1 allows us to eliminate not only I, but also *IM*, *MIM*, *MM*, and many more strings from language C.

However, there are many other strings that are much harder to decide. For example, consider the string *MU*. Is *MU* in language C? On one hand, according to Table 7.2, *MU* has not yet been born by day 3; on the other hand, we can see no obvious reason why it can't be born on some later day. For example, if say the string *MIII* were born on day 15, then *MU* would be born on day 16 via the rule $(C3)$.

In fact, the string *MU* is *not* in language C, because of the following rather peculiar theorem.

THEOREM 7.2 In any string in language C, the number of I's cannot be a multiple of 3.

We will give a proof of Theorem 7.2 below. But before doing so, we note that Theorem 7.2 does eliminate *MU* from language C, since the number of I's in *MU* is *zero*, which is a multiple of 3. (Zero equals 0 times 3!) Theorem 7.2 also eliminates *MIII*, *MUU*, *MIIUI*, and many other strings not already eliminated by Theorem 7.1.

To prove Theorem 7.2, we will use induction on the *birthday number* of the strings in language C. The exact induction statement we shall use is the following (from now on we will call the number of *I*'s in a string of *M*'s, *I*'s, and *U*'s the *I-count* of the string).

> C_n: In any string born on day n, the I-count is not a multiple of 3.

We can see by examining Table 7.2 that the statements C_0, C_1, C_2, and C_3 are all true. For example, the three strings born on day 2, viz. *MIUIU*, *MIIU*, and *MIIII*, have I-counts 2, 2, and 4, respectively, and none of these numbers is a multiple of 3. Our induction proof is off to a good start!

Next, we make the *strong induction assumption* that $C_0, C_1, \ldots, C_n$ have all been proved, i.e., that every string in language C with birthday less than or equal to n has an I-count which is not a multiple of 3, and try to prove C_{n+1}. To do this, we must show that a string born on day $n + 1$ cannot have an I-count which is a multiple of 3.

Thus let z be a string born on day $n + 1$. It must be descended from a string born on day n via an application of one of the four rules $(C1)$, $(C2)$, $(C3)$, and $(C4)$. We need to consider these four possibilities separately.

- *Rule (C1)*: If z is born via rule $(C1)$, then z is of the form xIU, where xI is a string in language C born on day n. But the I-count of xIU is *exactly the same* as the I-count of xI, and since xI was born on day n, our induction assumption tells us that its I-count is not a multiple of 3. Therefore the I-count of z is also not a multiple of 3, in this case.
- *Rule (C2)*: In this case z is of the form Mxx, where Mx is a string born on day n. The I-count of Mxx is *twice* the I-count of Mx; and the I-count of Mx is not a multiple of 3, by the induction assumption. But multiplying a number which is not a multiple of 3 by 2 will never produce a multiple of 3. Therefore the I-count of z is not a multiple of 3, in this case.
- *Rule (C3)*: In this case z is of the form xUy, where $xIIIy$ is a string born on day n. The I-count of xUy is *three less* than the I-count of $xIIIy$; and the I-count of $xIIIy$ is not a multiple of 3, by the induction assumption. But subtracting 3 from a number which is not a multiple of 3 will never produce a multiple of 3. Therefore the I-count of z isn't a multiple of 3, in this case either.
- *Rule (C4)*: In this case z is of the form xy, where $xUUy$ is a string born on day n. But the I-count of xy is *exactly the same* as the I-count of $xUUy$, whose I-count is not a multiple of 3, by the induction assumption. So in this case, too, the I-count of z isn't a multiple of 3.

These four cases exhaust the possibilities for parentage of z; and so our proof of C_n by induction is complete. ■

We now give an example showing how Theorems 7.1 and 7.2, together with the tree of Table 7.2, can help us decide membership in language C.

EXAMPLE 7.4 How many strings of length 3 can be formed from the letters *M*, *I*, and *U*? How many of these are in language C?

SOLUTION The multiplication rule (Section 2.1) tells us that there are $3 \times 3 \times 3 = 27$ possible strings of *M*'s, *I*'s, and *U*'s of length 3. However, Theorem 7.1 says that a string in language C must begin with *M*, and have no further *M*'s. This means that for a string of length 3 in language C there is only *one* possibility for the first letter (*M*), *two* possibilities for the second letter (*I* or *U*), and *two* possibilities for the third letter (*I* or *U*). By the multiplication rule again, this means that there are at most $1 \times 2 \times 2 = 4$ length-3 strings in language C: *MII*, *MIU*, *MUI*, and *MUU*. Examining Table 7.2, we see that *MII* and *MIU* are born on day 1, and *MUI* is born on day 3. The string *MUU* does not appear in Table 7.2, and in fact, since its *I*-count is zero, Theorem 7.2 tells us that it is *not* in language C. Therefore, of the 27 possible strings of three *M*'s, *I*'s, and *U*'s, only 3 are in language C. ■

Theorem 7.2 gives an interesting *necessary* condition on the strings in language C, but it leaves us to wonder whether or not it is also *sufficient*. For example, is *MIUI* in the language? The *I*-count of *MIUI* is two, which is not a multiple of 3, and so it isn't ruled out by Theorem 7.2. But *MIUI* doesn't appear in Table 7.2, and in fact a computer search shows that it isn't born by day 6. The authors of this book couldn't decide whether or not *MIUI* was in language C, until Laif Swanson proved the following remarkable theorem.

THEOREM 7.3 A string of *M*'s, *I*'s and *U*'s is in language C if and only if it begins with *M* and is followed by a string of *I*'s and *U*'s for which the *I*-count is not a multiple of 3.

The proof of Theorem 7.3 isn't too hard, and it is outlined in the problems at the end of this section. But we can illustrate it by showing that indeed *MIUI* is in language C:

MI	(rule *C*0)
MII	(rule *C*2)
MIIII	(rule *C*2)
MIIIIIIII	(rule *C*2)
MIIIIIIIIU	(rule *C*1)
MIIIIIUU	(rule *C*3)
MIIIII	(rule *C*4)
MIUI	(rule *C*3).

Therefore *MIUI* is born on day 7! If you're interested in the proof of theorem 7.3, turn to Problems 31–34.

Problems for Section 7.1

1. Which of the following strings belongs to language A?
(a) *aaaaaaaaaaaa*. (c) *bcbaaaab*.
(b) *babababababa*.

2. How many strings of length 6 are in language A?

3. Suppose, in the definition of language A, that rule ($A0$) were changed to read "The string *bbb* is well formed." How many strings of length 5 would language A now contain? Length 6?

4. Consider a new language, language A′, which is defined as follows:

(A′0) The strings a, ab, and bc are in language A′.

(A′1) If the strings x and y are in language A′, then so is the string xy.

(a) Which of the following strings belong to language A′: *abaaabca*, *bcaababa*, *abbaabca*?
(b) Design a decision algorithm for language A′.
(c) What is the relationship between languages A and A′?

5. Suppose language A had been defined as follows:

(A0) ϵ is in language A.
(A1) If x is in language A, then so is ax.
(A2) If x is in language A, then so is abx.
(A3) If x is in language A, then so is bcx.

Design a decision algorithm based on this definition.

6. Language A″ consists of certain strings of 0s and 1s. It is defined as follows:

(A″0) 0, 10, 110, and 111 are all in language A″.
(A″1) If x and y are in language A″, then so is xy.

(a) Which of the following strings is in language A″: (0, 1, 10111011, 101110110)?
(b) Let K_n denote the number of strings of length n in language A''_n. Calculate $K_0, K_1, K_2, \ldots, K_5$.
(c) Design a decision algorithm for language A″.

7. Suppose that the production rule ($A2$) for language A were changed to read "If x is a WFS, then so is xba," and suppose we changed the decision algorithm in Figure 7.1 accordingly, (now the good suffixes are a, ba, and bc). This decision algorithm would unfortunately not work, since some strings (e.g., *aba* and *bcaaba*) end in *both* a and ba! Show how to modify the decision algorithm so that it does work.

8. Prove by induction that algorithm A works. The appropriate induction statement is:

A_n: Algorithm A correctly decides membership/nonmembership in language A for all strings of length n.

9. For each of the following mathematical expressions, find the corresponding parenthesis string.
(a) $(x+1)^2 + (y+2)^2 + \cdots + (z+n)^2$.
(b) $(\cos^2(a+b) + \cos^2(a+c) + \cos^2(b+c))^{1/2}$.
(c) $a_0 + x(a_1 + x(a_2 + x(\cdots + x(a_{n-1} + a_n x))))$.
(d) $((((x \to y) \wedge (z \to w)) \wedge (x \vee z)) \to (y \vee w))$.

10. Which of the following parenthesis strings is in language B?
(a) ((((( ))))). (c) (((((( ))))). (e) (((((( ))))(( ))).
(b) ((((( )))))). (d) ((( )( )( )( ))( )).

11. Suppose you are given a string of left and right parentheses of length n. What is the largest possible value, as a function of n, that the variable `count` can assume in algorithm B? If the string is in language B, how large can `count` be?

12. The string ((( )(( ))( ) contains six left parentheses but only four right parentheses, so it cannot be in language B. However, if we change one of the left parentheses to a right parenthesis, the string might be in language B. For which of the six left parentheses in the string ((( )(( ))( ) is it true that changing it to a right parenthesis will produce a string accepted by language B?

13. Prove that algorithm B will always reject a string containing an odd number of symbols.

14. Find numbers a, b, c, and d such that the four expressions (7.1)–(7.4) all yield *different* numbers.

15. Prove the following theorem about language B.

Theorem. A string of left and right parentheses is in language B if and only if the following two conditions are satisfied.

(i) There are an equal number of left and right parentheses.
(ii) If we start counting the symbols from the left, the number of *left* parentheses is always at least as large as the number of *right* parentheses.

16. Prove by induction that algorithm B works. The appropriate induction statement is as follows:

B_n: Algorithm B correctly decides membership/nonmembership in language B for all strings of length n.

17. Suppose we define language B′ to be identical to language B, *except* that the production rule ($B2$) is changed as follows:

(B2′) If x and y are in language B′, then so is (xy).

(a) Which of the three strings listed in Example 7.2 is in language B′?
(b) Design a decision algorithm for language B′.

18. In the rules for language C, suppose we changed rule ($C0$) to "*MU* is in language C." Draw a tree like the one in Table 7.2, showing all strings born on days 0, 1, 2, and 3.

19. Consider the language defined in the previous problem. Is *MI* in this language?

20. List all strings born on day 4 in language C. Which, if any, of these strings are actually born again on day 4?

21. Prove, by induction, that the string $MII \cdots I$ (2^n I's) is in language C. On which day is it born?

22. Find the day 4 descendants of each of the following strings born on day 3 in language C.
(a) *MIUIUIUIU.* (b) *MIIUIIU.* (c) *MIIIIU.* (d) *MUI.* (e) *MIU.*

23. Is the string "*M*" in language C?

24. Given the string $MIII \cdots I$ (with n I's). How many *different* strings can be produced from it by using the following?
(a) Rule ($C1$).
(b) Rule ($C2$).
(c) Rule ($C3$).
(d) Rule ($C4$).

25. Use induction on the "birthday number" to prove Theorem 7.1 about language C.

26. How many strings of length 1 are in language C?

27. How many strings of length 2 are in language C?

28. Which of the following strings is in language C: (*MUU, MII, MIM, MIII, MUUU*)? In each case, if the string *is* in language C, give its birthday. If it is *not* in language C, explain why. Do *not* use Theorem 7.3.

29. Using Theorem 7.3, count the number of strings of length 4 in language C.

30. Using Theorem 7.3, find a general formula for the number of strings of length n which are in language C.

In Problems 31–34, we will outline Swanson's proof of Theorem 7.3. Because of Theorem 7.2, all we need to show is that every string of M's, I's, and U's that begins with M and is followed by a string of I's and U's for which the I count is not a multiple of 3 is in language C. As you will see, the crux of the proof deals with "pure I strings," i.e., strings of the form $MIIII \cdots I$. If the number of I's in such a string is N, we will denote the string by MI^N. Thus MI^4 denotes *MIIII*, for example.

31. Show that MI, MI^2, MI^4, MI^8, ... are all in language C.

32. Show that if $xIII$ is in language C, so is x. [*Hint*: First adjoin U to $xIII$, using ($C1$).]

33. (a) Using the results of Problems 31 and 32, show that MI^{11} is in language C. (*Hint*: Use Problem 31 to generate MI^{32}, and then remove I's, three at a time, using Problem 32.)
(b) Now show that any string of the form MI^N, where N isn't a multiple of 3, is in language C.

34. (a) Using the result of Problem 33(b), show how to generate *MIUIUU*. [*Hint*: Generate MI^{11} as in Problem 33, and then apply ($C3$) appropriately.]
(b) Using part (a) as a model, show that every string of M's, I's, and U's that begins with M and is followed by a string of I's and U's for which the I count is not a multiple of 3 is in language C.

7.2 Counting Strings in Language B

In this chapter we are mainly interested in finding decision algorithms for formal languages. However, in discrete mathematics it is always good to ask "How many?", and for a language, the appropriate form of this question is "How many strings of length n does the language have?" For language A we shall answer the question in Chapter 10, as part of our study of difference equations. For language C, we have left the question as Problem 7.1.30. For language B, however, this question leads to one of the most famous counting problems in all of discrete mathematics, and we can't resist the temptation to devote this short section to its solution.

Language B contains no strings of odd length, since any string in language B must contain an equal number of left and right parentheses. The challenge is to count the number of strings of even length, and in this section we will prove the following theorem.

THEOREM 7.4 Language B contains no strings of odd length, and exactly

$$\frac{1}{n+1}\binom{2n}{n}$$

strings of length $2n$.

Before we give the proof of Theorem 7.4, we illustrate it in Table 7.3. For example, Table 7.3 asserts that language B contains five strings of length 6. This is the case $n = 3$ of Theorem 7.4, which indeed predicts that there are

$$\frac{1}{4}\binom{6}{3} = \frac{1}{4} \cdot \frac{6 \cdot 5 \cdot 4}{3 \cdot 2 \cdot 1} = 5$$

strings of length 6 in language B. In the third column of Table 7.3 we have explicitly listed each string of length less than or equal to 6 in language B. You can see that it's much easier to compute the *number* of strings of a given length (thanks to Theorem 7.1) than to actually *list* them!

Now we will give the proof of Theorem 7.4. It depends on a famous trick, which is to translate strings of left and right parentheses into certain geometric figures called **lattice paths**. This translation is illustrated in Figure 7.3, where we show the lattice paths corresponding to six particular strings of left and right parentheses. The idea is to start the path at a fixed point A, and for each parenthesis in the string to add one diagonal edge to the path. In the case of a *left* parenthesis, the edge should go one unit diagonally in a *northeast* direction; for a *right* parenthesis, the edge should go *southeast*. In this way, after all of the symbols in the string of parentheses have been read, the lattice path will have reached another point, say point B.

It will be easier to discuss these lattice paths if we introduce the *Cartesian coordinates* of the points involved. Let us agree that the point A is always the

TABLE 7.3 Illustrating Theorem 7.4

String Length	Number of Strings	List of Strings
0	1	ϵ
2	1	( )
4	2	( )( ), (( ))
6	5	( )( )( ), (( ))( ), ( )(( )), (( )( )), ((( )))
8	14	(Do Problem 3!)
10	42	⋮
12	132	⋮

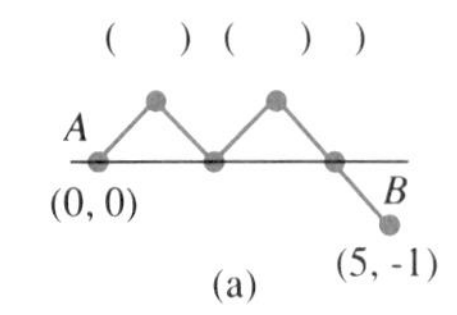

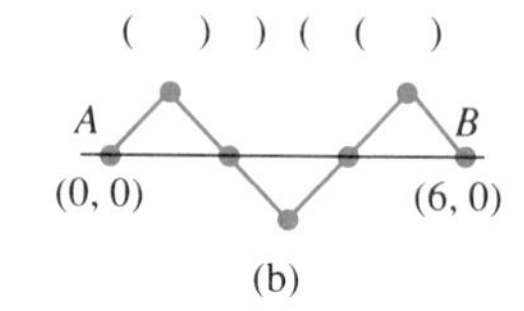

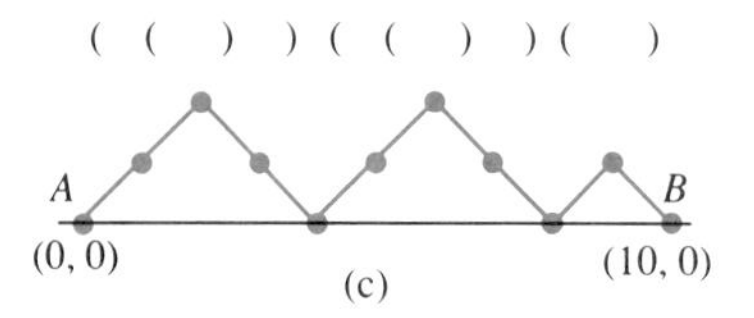

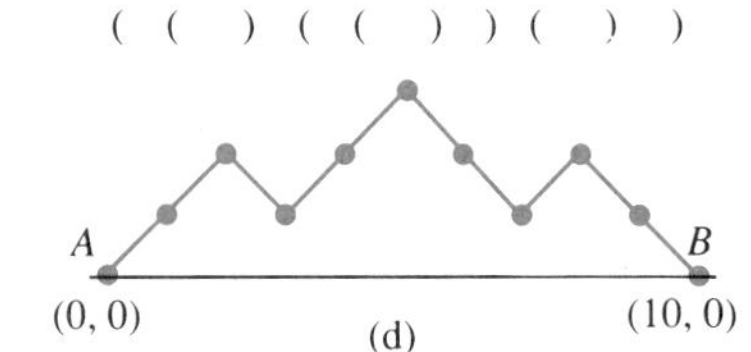

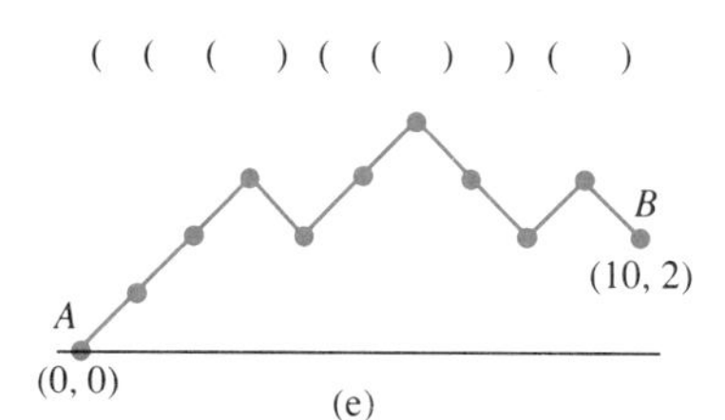

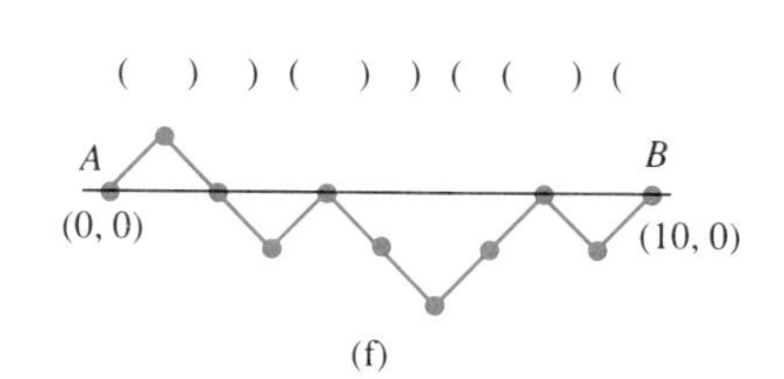

Figure 7.3 The lattice paths corresponding to six strings of left and right parentheses [only (c) and (d) are in language B].

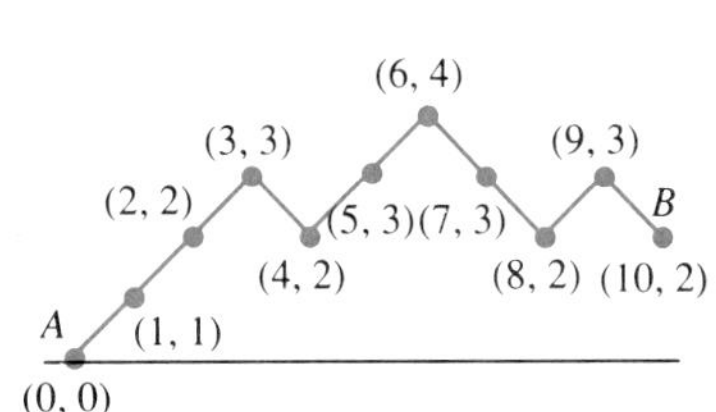

Figure 7.4 The cartesian coordinates of the points on the lattice path of Figure 7.3(e). The corresponding parentheses string ((( )(( ))( ) has six left and four right parentheses.

origin, i.e., the point with coordinates (0, 0). A *left* parenthesis, i.e., a northeast move, corresponds to increasing the x coordinate by 1 and increasing the y coordinate by 1. A *right* parenthesis, i.e., a southeast move, corresponds to increasing the x coordinate by 1 but *decreasing* the y coordinate by 1. It follows that the x coordinate of the final point B will be the total number of symbols in the string, and the y coordinate will be the difference between the number of left and right parentheses.

We illustrate this in Figure 7.4, where we show the coordinates of the points of the lattice path of Figure 7.3(e), corresponding to the string ((( )(( ))( ). This string consists of six left and four right parentheses, and indeed the x coordinate of the final point B is $10 = 6 + 4$, and the y coordinate is $2 = 6 - 4$.

The point of introducing lattice paths is that it is possible to tell by *visual inspection* of the lattice path whether or not the corresponding parentheses string is in language B. The rule is this: A string is in language B if and only if its lattice path *terminates on the x axis*, and *never goes below the x axis*. This is so because a lattice path is just a geometric description of how the variable `count` in algorithm B behaves. The x coordinate of the lattice path represents the number of symbols that have been read, and the y coordinate represents the `count`. Thus, for example, the lattice path in Figure 7.3(a) represents the behavior of algorithm B on the string ( )( )), because as we saw in Example 7.2(a), as the string is read, `count` is $0, 1, 0, 1, 0, -1$. Similarly, the lattice paths of Figures 7.3(b) and 7.3(c) correspond to the calculations in Examples 7.2(b) and 7.2(c). Thus saying that a lattice path terminates on the x axis and never goes below the x axis is the same as saying that the string has an equal number of left and right parentheses and that as the string is scanned, the number of left parentheses is always at least as large as the number of right parentheses.

If we apply this rule to the six lattice paths in Figure 7.3, we find that only paths (c) and (d) pass the test. String (a) doesn't terminate on the x axis *and* goes below the x axis; string (b) terminates on the x axis but goes below it; string (e) stays above the x axis but doesn't terminate on it; and string (f) terminates on the x axis but goes below it. And yes, if we apply algorithm B to the corresponding parentheses strings, we discover that only strings (c) and (d) belong to language B.

Our goal is to prove Theorem 7.4, which is stated in terms of strings in language B. If we translate Theorem 7.4 into the language of lattice paths, we obtain Theorem 7.5.

THEOREM 7.5 The number of lattice paths which begin at the point $(0, 0)$, and terminate at the point $(2n, 0)$ without ever going below the x axis, is

$$\frac{1}{n+1}\binom{2n}{n}.$$

It is Theorem 7.5 that we will now prove.

Saying "a lattice path which begins at $(0, 0)$ and terminates at $(2n, 0)$ without ever going below the x axis" is a mouthful, so from now on we'll simply say "a *legal path* from A to B." A path from A to B which is not a legal path will be called an *illegal path.* In Figure 7.5(a) we show one legal and one illegal path from A to B. We want to count the number of legal paths, and to do so we will use the elementary fact that *every* path from A to B is either legal or illegal, so that

(7.7) total number of paths = number of legal paths + number of illegal paths.

We will now count the *total number of paths* and the *number of illegal paths*; the formula for the number of legal paths will follow by subtraction.

Counting the *total* number of paths from A to B is relatively easy. Every path from A to B must contain $2n$ edges, n upward edges and n downward edges. The path is uniquely specified by the position of the n upward edges. The number of ways of selecting n positions for the upward edges from the $2n$ available locations is by the formula for unordered samples without repetition (Section 2.3) exactly $\binom{2n}{n}$. Therefore,

(7.8) $$\text{total number of paths from } A \text{ to } B = \binom{2n}{n}.$$

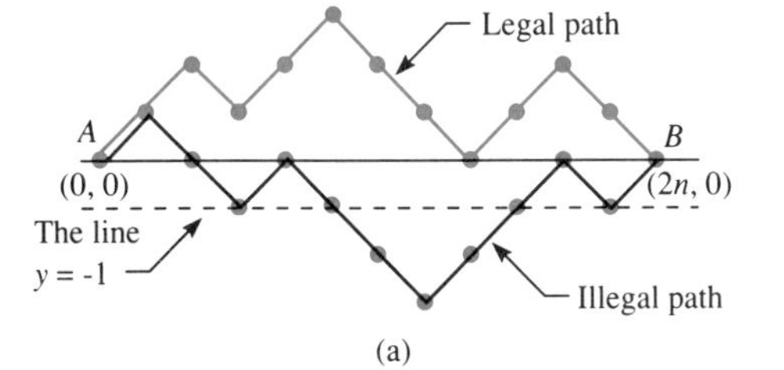

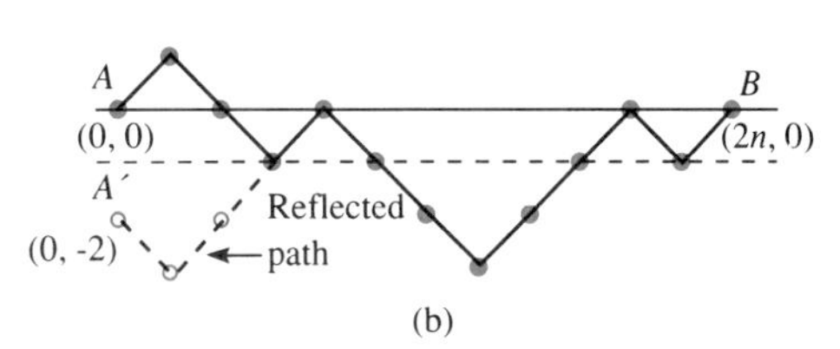

Figure 7.5 (a) A legal and an illegal path from A to B. (b) Every illegal path corresponds to a reflected path from A' to B.

Counting the *illegal* paths from A to B looks like a much harder problem, but fortunately there is an amazing trick that greatly simplifies things. As Figure 7.5(a) shows, every illegal path from A to B must either *touch* or *cross* the line $y = -1$, which is the horizontal line exactly one unit below the x axis. This means that every illegal path will have a *first point of contact* with the line $y = -1$. In Figure 7.5(a), for example, the first point of contact is the point $(3, -1)$. If we then *reflect* the first part of the illegal path with respect to the line $y = -1$, as illustrated in Figure 7.5(b), we will obtain a path from A' to B, where A' is the point $(0, -2)$. Similarly, any path from A' to B must pass through the line $y = -1$, and so if its first part is reflected, we will obtain an illegal path from A to B.

This argument shows that there is a one-to-one correspondence between the set of *illegal paths* from A to B and the set of *all paths* from A' to B. It follows that the number of illegal paths from A to B is the same as the total number of paths from A' to B:

(7.9)
$$\text{number of illegal paths from } A \text{ to } B = \text{total number of paths from } A' \text{ to } B.$$

This curious fact is very helpful, because counting the total number of paths from A' to B is relatively easy, and in fact is quite similar to counting the total number of paths from A to B. Every path from A' to B must contain $2n$ edges, but since the total "rise" of such a path is 2 (it starts at $y = -2$ and terminates at $y = 0$), there must be $n + 1$ upward edges and $n - 1$ downward edges. The path is uniquely specified by the position of its $n + 1$ upward edges, and so again by the rule in Section 2.3, there are exactly $\binom{2n}{n+1}$ such paths:

(7.10)
$$\text{total number of paths from } A' \text{ to } B = \binom{2n}{n+1}.$$

Finally, then, we can count the number of *legal* paths from A to B:

$$\begin{aligned} \text{number of legal paths} &= \text{total paths} - \text{illegal paths} && \text{(from 7.7)} \\ &= \binom{2n}{n} - \text{illegal paths} && \text{(from 7.8)} \\ &= \binom{2n}{n} - \text{paths from } A' \text{ to } B && \text{(from 7.9)} \\ &= \binom{2n}{n} - \binom{2n}{n+1} && \text{(from 7.10).} \end{aligned}$$

Therefore the total number of paths from A to B, i.e., the number of strings of length $2n$ in language B, is $\binom{2n}{n} - \binom{2n}{n+1}$. To see that this expression is also equal to $\binom{2n}{n}/(n + 1)$ requires a short exercise in algebra:

$$\begin{aligned} \binom{2n}{n} - \binom{2n}{n+1} &= \frac{(2n)!}{n!n!} - \frac{(2n)!}{(n+1)!(n-1)!} = \frac{(2n)!}{n!(n-1)!}\left(\frac{1}{n} - \frac{1}{n+1}\right) \\ &= \frac{(2n)!}{n!(n-1)!} \cdot \frac{1}{n(n+1)} = \frac{1}{n+1} \cdot \frac{(2n!)}{n!n!} = \frac{1}{n+1}\binom{2n}{n}. \end{aligned}$$

Problems for Section 7.2

1. How many strings of length 10 does language B contain? How many strings of length 12?

2. How many strings of length 14 does language B contain? How many strings of length 16?

3. According to Table 7.3, language B contains 14 strings of length 8. List each of these strings.

4. Draw lattice paths corresponding to each of the following strings of left and right parentheses.

(a) ( ). (b) (( )). (c) ( )( )). (d) (( )( )( )( )). (e) (((((. (f)))))). (g))( )( )(. (h) ϵ.

5. Apply algorithm B to each of the six strings of left and right parentheses shown in Figure 7.3.

6. Count the total number of lattice paths.

(a) From $(1,3)$ to $(5,9)$.
(b) From $(-3,1)$ to $(10,0)$.
(c) From $(0,0)$ to $(13,0)$.
(d) From $(-1,-3)$ to $(1,10)$.

7. Find a *general formula* for the number of lattice paths from (x, y) to the point (x', y').

8. This problem requires some knowledge of probability. Suppose a string of 12 left and right parentheses is selected at random, with each symbol in the string equally likely to be "(" or ")." What is the probability that the resulting string is in language B?

9. This problem requires some knowledge of probability. Suppose seven right and seven left parentheses are placed in an urn, and withdrawn, one after another. What is the probability that the resulting string of "("'s and ")"'s is in language B?

10. In the following diagrams we show three illegal paths from *A* to *B*. In each case find the corresponding reflected path from *A'* to *B*.

(a)

(b)

(c)

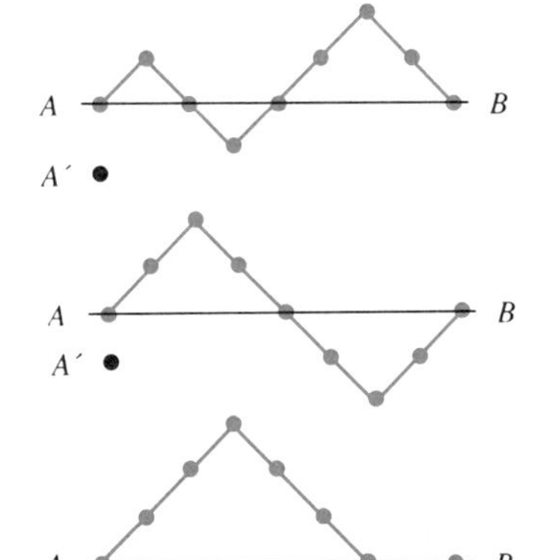

11. In the following diagrams we show three reflected paths from *A'* to *B*. In each case find the corresponding illegal path from *A* to *B*.

(a)

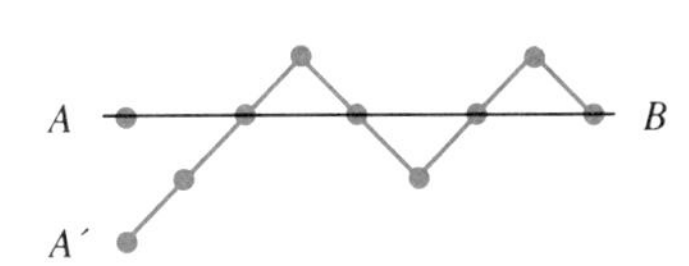

(b)

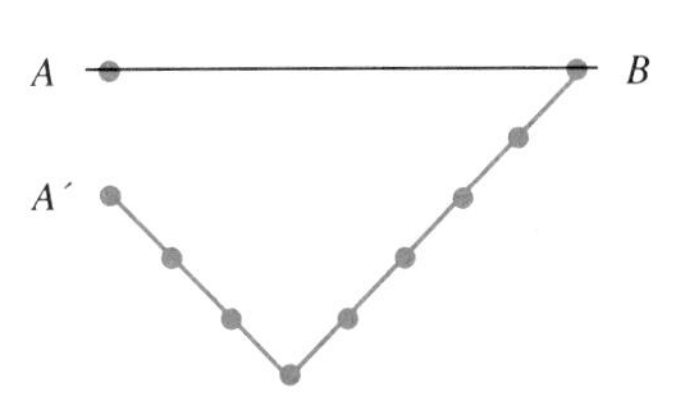

(c)

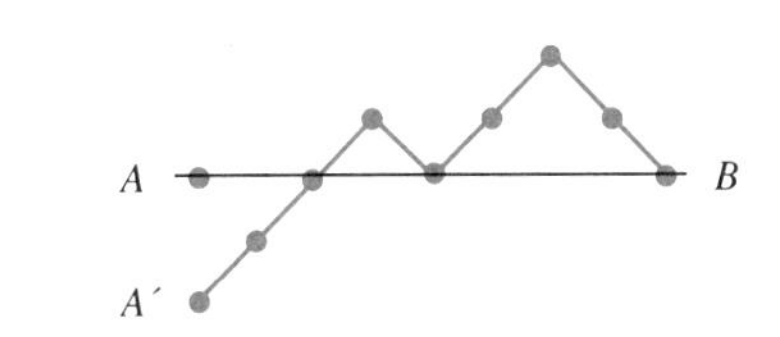

12. Let *A* denote the point $(0,1)$ and *B* denote the point $(10,3)$.

(a) How many lattice paths are there connecting *A* and *B*?
(b) How many such paths do not touch or cross the *x* axis? (*Hint*: See the accompanying figure.)

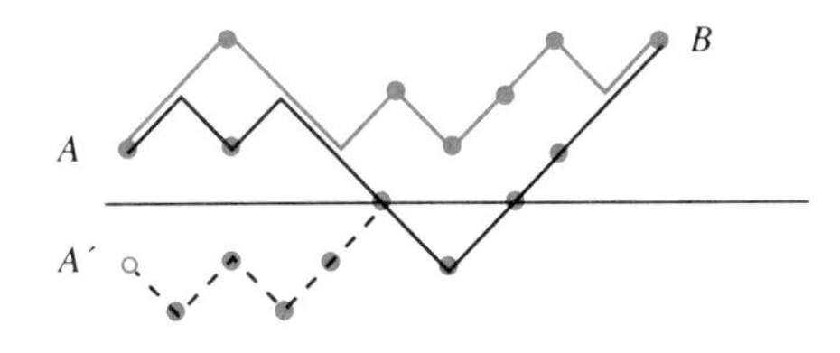

13. Let *A* denote the point (x, y), and *B* denote the point (x', y'). How many lattice paths are there from *A* to *B* which do not touch or cross the *x* axis? (You may assume x, y, x', and y' are all positive integers.)

14. Suppose we define language B' to be identical to language B, except that the production rule ($B2$) is changed to

(B2') If x and y are in language B, then so is (xy).

How many strings of length n does language B' contain?

7.3 Regular Languages

In Section 7.1 we learned what formal languages are, and worried about finding decision algorithms for them. We saw that for some languages, it's easy to find a decision algorithm, but for others it can be difficult or even impossible. In this section and the next, we'll study a very large and important class of languages, called *regular languages*, which arise in text-editing, data retrieval, and symbol manipulation applications. In this section, we will define regular languages, and find that every language accepted by a deterministic finite automaton is a regular language. To prove this important fact, we will use an algorithm much like the algorithms of Floyd and Warshall that we studied in Section 4.2. Then in Section 7.4, we will turn the tables and show that every regular language can be accepted by a nondeterministic finite automaton. Since we showed in Section 6.4 that every language accepted by an NFA has a decision algorithm (the NFA simulation algorithm of Figure 6.25), this will show that every regular language has an efficient decision algorithm. Also, since we saw in Section 6.4 that every set of strings accepted by an NFA can also be accepted by a DFA, when we finish Section 7.4, we will have proved the fact that the regular languages are exactly the same as the languages accepted by DFAs, which is one of the most important results in language theory.

A regular language, like any language, is a set of strings built from a basic underlying set A, called the *alphabet*. What makes a regular language special is that it is defined by something called a **regular expression**. We now define regular expressions.

There are two kinds of regular expressions: *atomic* and *compound*. The **atomic regular expressions** (we will often just call them **atoms**) are simply the letters in the alphabet A, plus the special symbols $\varnothing$ and ϵ, whose significance we will explain shortly. For example, if the alphabet is the three-element set $\{x, y, z\}$, there are exactly five atomic regular expressions: x, y, z, $\varnothing$, and ϵ. The atoms are the simplest regular expressions. **Compound regular expressions** are built up from the atoms, using the symbols (,), +, and $*$, by following a set of three rules. (In these rules, r_1 and r_2 denote previously defined regular expressions.)

- *Rule 1*: $(r_1 + r_2)$ is a regular expression.
- *Rule 2*: $(r_1 r_2)$ is a regular expression.
- *Rule 3*: $(r_1)^*$ is a regular expression.

For example, if the underlying alphabet is $\{0, 1\}$, the following seven expressions are all regular expressions:

(7.11) ϵ

(7.12) 0

(7.13) (01)

(7.14) $((00) + (11))$

(7.15) $((01)((00) + (11)))$

(7.16) $((0 + 1))^*$

(7.17) $(((0 + 1))^*(00))$.

The regular expressions (7.11) and (7.12) are atoms; the other five are all compound. For example, regular expression (7.13) is built by combining the atoms 0 and 1 using Rule 2. The regular expression in (7.14) is built from the regular expressions (00) and (11), using Rule 1. The regular expression in (7.16) is built from $(0+1)$, using Rule 3. We will leave the further details of the construction of these regular expressions as Problem 1 at the end of the section.

As we have said, every regular expression represents a certain language; the languages represented by regular expressions are called **regular languages**. It is time to describe the correspondence between regular expressions and regular languages. The **atomic regular languages**, i.e., the languages corresponding to atomic regular expressions, are the simplest. Indeed, the language corresponding to the atomic regular expression a is just the language $\{a\}$, consisting of the single string a. Similarly, the regular language corresponding to the atom ϵ is the small language $\{\epsilon\}$, which contains only the empty string, and finally, the language corresponding to the atom $\varnothing$ is the empty set itself. Thus, for example, if the underlying alphabet is $\{0, 1, 2\}$, there are five atomic regular expressions and corresponding regular languages:

Regular Expression	Regular Language
$\varnothing$	$\varnothing$
ϵ	$\{\epsilon\}$
0	$\{0\}$
1	$\{1\}$
2	$\{2\}$

We come now to **compound regular languages**, i.e., regular languages corresponding to compound regular expressions. Just as a compound regular expression is built from simpler regular expressions using Rules 1, 2, and 3, so compound regular languages are built from simpler regular languages using a corresponding set of rules. In order to understand these rules, you will need to know what the term *concatenation* means. If x and y are two strings, the **concatenation** of x and y, which is denoted by xy, is the string formed by "gluing" x and y together, one after the other, with no spaces. Thus if x is the string 0001, and y is the string 1101, then xy is the string 00011101, and yx is the string 11010001. Any number of strings can be concatenated. For example, if we concatenate the three English strings TO, READ, and OR, the result is TOREADOR.

Here are the three rules which tell us how to form regular languages from the corresponding regular expressions. As before, r_1 and r_2 denote regular expressions; and now R_1 and R_2 denote the corresponding regular languages.

- *Rule 1*: The language corresponding to $(r_1 + r_2)$ is called the *sum* of the languages R_1 and R_2, and is denoted by $(R_1 + R_2)$. It is just the union of the two languages.
- *Rule 2*: The language corresponding to $(r_1 r_2)$ is called the *product* of the languages R_1 and R_2, and is denoted by $(R_1 R_2)$. It consists of all strings that can be formed by taking a string from R_1 and concatenating it with a string from R_2.

- *Rule 3*: The language corresponding to $(r_1)^*$ is called the **closure** of the language R_1, and is denoted by $(R_1)^*$. It consists of all strings that can be built by concatenating together any number of strings from R_1, plus the empty string ϵ.

Using these rules, we can describe the regular languages corresponding to each of the regular expressions in (7.11)-(7.17). The regular expressions (7.11) and (7.12) are atoms, and so the corresponding regular languages both contain just one short string:

Regular Expression	Regular Language
ϵ	$\{\epsilon\}$
0	$\{0\}$

The regular language corresponding to the regular expression (7.13), viz. (01), is by Rule 2 the product of the regular languages {0} and {1}. But the regular language {0} contains only the string 0, and the regular language {1} contains only the string 1, and so the regular language corresponding to (01) consists of the string which is the concatenation of the strings 0 and 1. Hence:

Regular Expression	Regular Language
(01)	{01}

The regular language corresponding to the regular expression ((00) + (11)) in (7.14) is by Rule 1 the union of the regular languages corresponding to (00) and (11). These two regular languages, in turn, consist of one string each, viz. 00 and 11, and so:

Regular Expression	Regular Language
((00) + (11))	{00, 11}

The regular expression in (7.15) is built by applying Rule 2 to the regular expressions (01) and ((00) + (11)). The regular languages corresponding to (01) and ((00) + (11)) are {01} and {00, 11}, respectively, and so by Rule 2 the language corresponding to ((01)((00) + (11))) is obtained by concatenating the strings in these two languages. Thus we find the following:

Regular Expression	Regular Language
((01)((00) + (11)))	{0100, 0111}

The regular expression in (7.16) is formed by applying Rule 3 to the regular expression (0 + 1). According to Rule 3, this language consists of all strings which can be built by concatenating together any number of strings from the language corresponding to (0 + 1). Now the language corresponding to (0 + 1) is simply the set {0, 1}. Thus $((0+1))^*$ represents the language built by concatenating together any number of 0s and 1s. In other words, $(0+1)^*$ simply represents the set of *all possible* strings of 0s and 1s! Note that Rule 3 also specifies that ϵ is in the closure, so that we have:

Regular Expression	Regular Language
$((0+1))^*$	All strings of 0s and 1s, including ϵ

We will leave the description of the regular language corresponding to the regular expression in (7.17) as Problem 6.

If we adhere strictly to the rules for generating regular expressions, we can produce monstrosities like $(((((01)1)(01)) + (((01)1)(01))))^*$, in which the parentheses obscure what's going on. It is possible to write regular expressions without using so many parentheses, if we agree to observe certain notational rules. When these rules are observed, the regular expressions in (7.11)-(7.17) can safely be rewritten as follows:

(7.11′) ϵ

(7.12′) 0

(7.13′) 01

(7.14′) $00+11$

(7.15′) $01(00+11)$

(7.16′) $(0+1)^*$

(7.17′) $(0+1)^*00$

In order for "$00+11$" to mean "$((00)+(11))$," and not "$(0(0+(11)))$," for example, we have to agree that the product operation takes priority over the addition operation. This is just like the rule in ordinary arithmetic that says that $2+3\cdot 4$ means $(2+12)$ rather than $(5\cdot 4)$. Similarly $*$ has higher priority than product, so that 01^* represents the language consisting of strings like 0111111111, not like 0101010101. Similarly, if we use the associative law for regular expressions, which says that $((r_1r_2)r_3) = (r_1(r_2r_3))$, and $((r_1+r_2)+r_3) = (r_1+(r_2+r_3))$, we can write simply $r_1r_2r_3$ and $r_1+r_2+r_3$ instead, without fear of confusion. In the rest of this section we will use this streamlined notation for regular expressions and regular languages.

The next example will give us more practice with regular expressions and the languages they represent.

EXAMPLE 7.5 In each of the following four cases, describe the languages represented by the given regular expressions, and decide whether or not the given strings are in the language.

(a) $(0+1)^*001$; 0101010010, 001.
(b) $0^*1^*2^*$; 001120, 112.
(c) $a^*(b+c)^*d$; *abcd*, *aad*.
(d) $(0+10+110)^*+(1+01+001)^*$; 01110101010, 00111111.

SOLUTION

(a) This regular expression is the product of the regular expressions $(0+1)^*$ and 001. We have seen that $(0+1)^*$ represents the set of all strings over the alphabet {0, 1}. On the other hand, the regular expression 001 represents the single string 001. Therefore the regular expression $(0 + 1)^*001$ represents the strings which can be built by concatenating an arbitrary string with the string 001. In other words, this language consists of all strings ending with 001. It follows that the string 0101010010 is not in the language, but 001 is in the language (it is the concatenation of ϵ with 001).

(b) This regular expression is the product of the three regular expressions 0^*, 1^*, and 2^*. The regular expression 0^* represents all strings consisting of only 0s; 1^* represents all strings consisting of only 1s; and 2^* represents all strings consisting of only 2s. Therefore $0^*1^*2^*$ represents all strings consisting of any number of 0s followed by any number of 1s followed by any number of 2s. Hence 001120 is not in the language, but 112 is in the language (no 0s followed by two 1s followed by one 2).

(c) This regular expression is the product of the three regular expressions a^*, $(b + c)^*$, and d. It represents the language consisting of any number of a's followed by any number of b's and c's, followed by the single letter d. Thus both of the strings *abcd* and *aad* are in the language.

(d) This regular expression is the sum of the two regular expressions $(0 + 10 + 110)^*$ and $(1 + 01 + 001)^*$. The language represented by $(0 + 10 + 110)^*$ consists of all strings built by concatenating together any number of 0s, 10s, and 110s. Any such string (except ϵ) will end in 0, and can have at most two 1s in a row. Similarly, $(1 + 01 + 001)^*$ represents ϵ and all strings ending in 1 with no run of more than two 0s. Therefore the language represented by $(0 + 10 + 110)^* + (1 + 01 + 001)^*$ consists of ϵ and all the strings of both these types. Now the string 01110101010 ends in 0 and so if it is in the language, it can have no more than two 1s in a row; but there are three 1s in a row, so it is not in the language. On the other hand, 00111111 ends in 1 and has no run of three or more 0s, so it is in the language. ■

Figure 7.6 A simple DFA.

We will conclude this section by showing, as promised, that every language accepted by a DFA is a regular language. We will do this by exhibiting an algorithm which, when given a DFA, will produce a regular expression that represents the language accepted by that machine. This algorithm is similar to the "all-pairs" algorithms of Warshall and Floyd in Section 4.2.

Here is the idea. If the DFA has n states, labelled $\{1, 2, \ldots, n\}$, the algorithm produces a sequence of $n \times n$ matrices, $R_0, R_1, \ldots, R_n$. Each entry in each of these matrices is a regular expression. The initial matrix R_0 is defined as follows:

$$\textbf{(7.18)} \quad \begin{aligned} R_0[x, y] &= \text{the sum of all labels on the directed edges from } x \text{ to } y, \text{ if } x \neq y, \\ &= \epsilon + \text{the sum of all labels on the directed edges from } x \text{ to } y, \text{ if } x = y. \end{aligned}$$

Equation (7.18) tells us that each of the entries in R_0 is a regular expression. For example, starting with the DFA of Figure 7.6, Equation (7.18) yields the R_0 matrix shown in Table 7.4. For example, the $\varnothing$ entry in row 1, column 3, indicates that there are no directed edges from state 1 to state 3.

TABLE 7.4 The initial matrix R_0 for the DFA of Figure 7.6

	1	2	3
1	$\epsilon + 0$	1	$\varnothing$
2	0	ϵ	1
3	$\varnothing$	$\varnothing$	$\epsilon + 0 + 1$

For $i \geq 1$, the (x, y)th entry in the matrix R_i is a regular expression that represents the set of strings produced by all i-paths from state x to state y, i.e., paths from x to y which are restricted to pass only through intermediate states in the set $\{1, 2, \ldots, i\}$. The rule for computing R_i from R_{i-1} is as follows:

(7.19) $$R_i[x, y] = R_{i-1}[x, y] + R_{i-1}[x, i]R_{i-1}[i, i]^*R_{i-1}[i, y].$$

What Equation (7.19) says is that an i-path from x to y must be either an $(i-1)$-path from x to y, or else an $(i-1)$-path from x to i, followed by a number (possibly zero) of $(i-1)$-paths from i back to i, followed by an $(i-1)$-path from i to y. This rule is illustrated in Figure 7.7, which you should compare with Figure 4.11. For example, consider the state diagram in Figure 7.6. The state sequence [12112121] is a 2-path from state 1 to state 1. It can be broken up into the four segments [12], [2112], [212], and [21]. The segment [12] is a 1-path from 1 to 2, [2112] and [212] are 1-paths from 2 to 2, and [21] is a 1-path from 2 to 1.

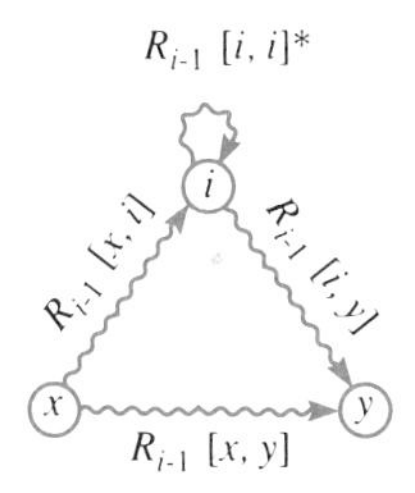

$R_i\ [x, y] = R_{i-1}\ [x, y] +$
$R_{i-1}\ [x, i]R_{i-1}[i, i]^*R_{i-1}[i, y]$

Figure 7.7 Calculating the i-paths from the $(i-1)$-paths (cf. Figure 4.11).

Because of Equation (7.19), if each entry in R_{i-1} is a regular expression, then so is each entry in R_i. Therefore when the algorithm terminates with the matrix R_n, its entries will be regular expressions that represent the strings generated by the n-paths, i.e., the unrestricted paths between the states. The language accepted by the DFA is therefore represented by the sum of the entries in R_n corresponding to the paths from the initial state to each of the accepting states. It follows that the set of strings accepted by the DFA is a regular language, although the regular expression resulting from this algorithm can be very complex, as the following example illustrates.

EXAMPLE 7.6 Find a regular expression that represents the language accepted by the DFA of Figure 7.6.

SOLUTION The initial matrix R_0 has already been given, in Table 7.4. The next step is to compute the matrix R_1. This matrix contains nine entries, and we won't go through the details of them all; instead we'll just verify the entries $R_1[1, 1]$, $R_1[1, 2]$, and $R_1[2, 2]$. If we to compute $R_1[1, 1]$, for example, using Equation (7.19), we get the following formidable regular expression:

$$\begin{aligned} R_1[1, 1] &= R_0[1, 1] + R_0[1, 1]R_0[1, 1]^*R_0[1, 1] \\ &= (\epsilon + 0) + (\epsilon + 0)(\epsilon + 0)^*(\epsilon + 0). \end{aligned}$$

Luckily, this expression can be simplified. Indeed, it is just a long-winded way of representing all strings consisting of a run of zero or more 0s, so that it is equivalent to the simpler regular expression 0^*, which we have entered in the matrix R_1 in Table 7.5. (A look at Figure 7.6 tells us that this is correct. The only way to get from state 1 to state 1 with state 1 as the only intermediate state is to spin around the loop at state 1 a number of times.)

Here is the calculation for $R_1[1, 2]$:

$$\begin{aligned} R_1[1, 2] &= R_0[1, 2] + R_0[1, 1]R_0[1, 1]^*R_0[1, 2] \\ &= 1 + (\varepsilon + 0)(\varepsilon + 0)^*1. \end{aligned}$$

Here again, simplification is possible. The regular expression $(\varepsilon + 0)(\varepsilon + 0)^*$ is

TABLE 7.5 Solving Example 7.6

$$R_1 = \begin{array}{c|ccc} & 1 & 2 & 3 \\ \hline 1 & 0^* & 0^*1 & \varnothing \\ 2 & 00^* & \epsilon + 00^*1 & 1 \\ 3 & \varnothing & \varnothing & \epsilon + 0 + 1 \end{array} \qquad R_2 = \begin{array}{c|ccc} & 1 & 2 & 3 \\ \hline 1 & ? & ? & 0^*1(00^*1)^*1 \\ 2 & ? & ? & ? \\ 3 & ? & ? & \epsilon + 0 + 1 \end{array} \qquad R_3 = \begin{array}{c|ccc} & 1 & 2 & 3 \\ \hline 1 & ? & ? & 0^*1(00^*1)^*1(0+1)^* \\ 2 & ? & ? & ? \\ 3 & ? & ? & ? \end{array}$$

equivalent to 0^*, so that the above expression for $R_1[1,2]$ simplifies to $1 + 0^*1$. However, since the language represented by 0^*1 includes 1, $1 + 0^*1$ boils down simply to 0^*1, as shown in Table 7.5.

Here is the calculation for $R_1[2,2]$:

$$R_1[2,2] = R_0[2,2] + R_0[2,1]R_0[1,1]^*R_0[1,2] = \epsilon + 0(\epsilon + 0)^*1, = \epsilon + 00^*1,$$

as indicated in Table 7.5. The entire matrix R_1 is displayed in Table 7.5.

Our next task is to calculate the entries in the matrix R_2. However, we can save a lot of work if we anticipate that to obtain the regular expression for the language accepted by the DFA of Figure 7.6, we will only need one of the entries in R_3, viz. $R_3[1,3]$ which gives the regular expression for all the strings from the initial state (state 1) to the accepting state (state 3). The calculation of $R_3[1,3]$, viz.

$$R_3[1,3] = R_2[1,3] + R_2[1,3]R_2[3,3]^*R_2[3,3],$$

involves only $R_2[1,3]$ and $R_2[3,3]$, and so in Table 7.5 we have only shown these two regular expressions, whose calculation via Equation (7.19) we leave as Problem 24. Using these values, we can calculate

$$\begin{aligned} R_3[1,3] &= R_2[1,3] + R_2[1,3]R_2[3,3]^*R_2[3,3] \\ &= 0^*1(00^*1)^*1 + (0^*1(00^*1)^*1)(\epsilon + 0 + 1)^*(\epsilon + 0 + 1). \end{aligned}$$

This expression simplifies to $0^*1(00^*1)^*1(0+1)^*$, a regular expression that represents the language accepted by the DFA of Figure 7.6.

Actually, the DFA in this example is simple enough so that it's possible to find the regular expression "by inspection." Indeed, note that starting in state 1, any path to state 3 consists of three parts:

1. Any number of loops at 1, together with any number of jumps back and forth between 1 and 2. The regular expression for this part of the path is then $(0 + 10)^*$.
2. A direct path from 1 to 2 to 3, with corresponding regular expression 11.
3. Finally, any number of loops at 3, with regular expression $(0 + 1)^*$.

Therefore the regular expression $(0 + 10)^*11(0 + 1)^*$ represents the language accepted by the DFA of Figure 7.6. This is not the same as the answer obtained previously, but both answers are nevertheless correct. There can be many regular expressions that represent a given regular language. ■

We hope that the messy details of the computation in Example 7.6 haven't obscured the main point: Any language accepted by a DFA is a regular language. In the next section, we'll show the surprising fact that the converse is also true, i.e., that every regular language can be accepted by a DFA.

Problems for Section 7.3

1. Explain in detail how the compound regular expressions in (7.15) and (7.17) were constructed from Rules 1, 2, and 3. Which atoms are involved? Which of the three rules was used?

2. Which of the following 12 expressions are regular expressions? (Assume that a, b, c, and d are all atoms.)

(a) (a). (b) (ab). (c) (abc). (d) $((a)(bc))$. (e) $((ab)c)$. (f) a^*b^*. (g) $(ab)^*$. (h) $(a + (b)^*)$. (i) $(a + b + c)$. (j) $((a + b) + c)^*$. (k) $(ab + aa)$. (l) $(a + (b + c)d)$.

3. Is the language $\{\epsilon\}$ the same as the language $\varnothing$?

4. Let x, y, and z be the strings over the alphabet $\{a, b, c, d, e, f\}$ defined as follows: $x = abc$, $y = def$, $z = \epsilon$. What are the following strings?

(a) xy. (b) xyz. (c) xzy. (d) yyz. (e) zzz.

5. If x and y are strings over some alphabet, it is usually not the case that $xy = yx$. However, if x and y are equal, then of course $xy = yx$. Now suppose x and y are strings, and $xy = yx$. Does it necessarily follow that $x = y$? If so, explain carefully why. If not, give an example of two different strings x and y such that $xy = yx$.

6. Describe in words the regular language corresponding to the regular expression (7.17).

7. Rewrite the monstrosity

$$(((((01)1)(01)) + (((01)1)(01))))^*,$$

without using so many parentheses. Describe the language it represents.

8. Find a regular expression that represents the language consisting of the strings $\epsilon, 01, 0101, 010101, 01010101, \ldots$.

9. Let R and S be regular languages. Answer the following.

(a) Is RS = SR?
(b) Is R + S = S + R?
(c) Is R(S + T) = RS + RT?

10. Explain how to build each of the four regular expressions in Example 7.5, using atoms and Rules 1, 2, and 3.

11. Two of the following three regular expressions are equivalent. Which two?

(a) $(00)^*(\epsilon + 0)$. (b) $(00)^*$. (c) 0^*.

12. Give regular expressions for the following languages over the alphabet $\{a, b\}$.

(a) All strings beginning and ending in b.
(b) All strings without two consecutive a's.
(c) All strings with an odd number of a's.
(d) All strings *not* containing the substring aba.
(e) All strings consisting of at least one a and at least one b.

13. Prove or disprove the following identities for regular expressions.

(a) $(rs + r)^*r = r(sr + r)^*$.
(b) $(r + s)^* = r^* + s^*$.
(c) $s(rs + s)^*r = rr^*s(rr^*s)^*$.

14. Describe in words the language represented by the following regular expression:

$$MU^*(IU^*IU^*I)^*U^*I(\epsilon + U^*I)U^*.$$

15. Explain why in the initial matrix R_0 defined in (7.18), the rule is different, depending on whether or not $x = y$.

16. Decompose the 3-path 112112333 from state 1 to state 3 in Figure 7.6 into a number of 2-paths.

17. Verify in detail that the regular expression

$$(\epsilon + 0) + (\epsilon + 0)(\epsilon + 0)^*(\epsilon + 0)$$

represents the same language as the regular expression 0^*.

18. Verify in detail that the regular expression

$$1 + (\epsilon + 0)(\epsilon + 0)^*1$$

is equivalent to 0^*1.

19. In Example 7.6, we verified only three of the nine entries in the matrix R_1. Verify the other six.

(a) $R_1[1, 3]$. (b) $R_1[2, 1]$. (c) $R_1[2, 3]$. (d) $R_1[3, 1]$. (e) $R_1[3, 2]$. (f) $R_1[3, 3]$.

20. Suppose that the accepting state in Figure 7.6 is changed to state 2. In order to compute a regular expression that represents the corresponding language, which entries in the matrix R_2 in Table 7.5 would be needed?

21. Calculate $R_2[1, 1]$ in Table 7.5.

22. Calculate $R_2[1, 2]$ in Table 7.5.

23. Calculate $R_2[2, 3]$ in Table 7.5.

24. Calculate $R_2[1, 3]$ and $R_2[3, 3]$ in Table 7.5.

25. In Table 7.5, most of the entries in R_3 are left blank. However, two of the entries turn out to be $\varnothing$. Which two?

26. Find a regular expression that represents the language accepted by the DFA described in the accompanying figure.

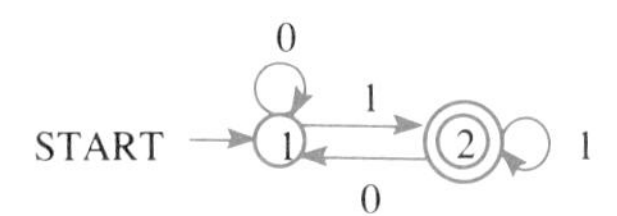

27. Find a regular expression that represents the language accepted by the DFA described in the accompanying figure.

START A 0, 1 B 0, 1 C 0, 1

28. By inspection, find a regular expression that describes the language generated by the following NFA.

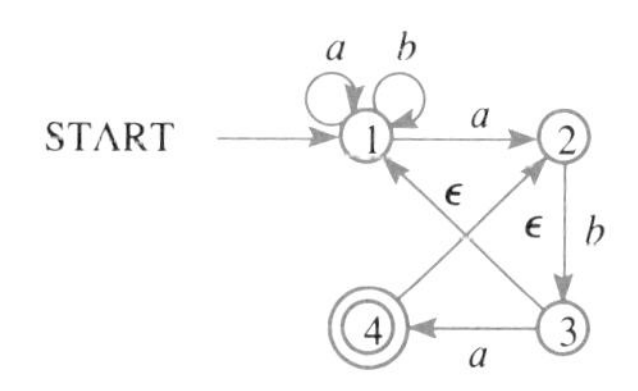

29. By inspection, find a regular expression that describes the language generated by the following NFA.

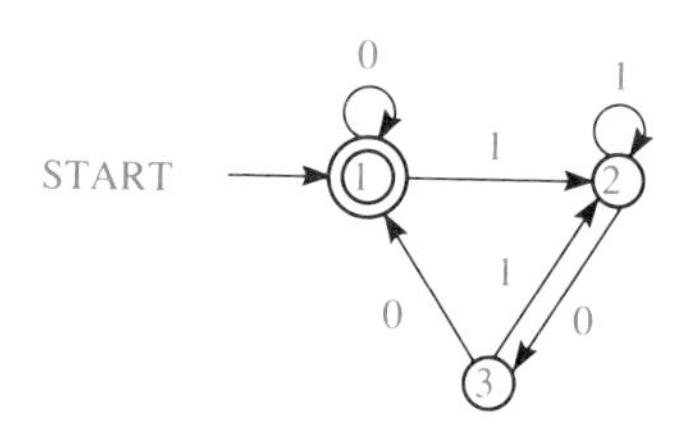

30. By inspection, find a regular expression that describes the language generated by the following NFA.

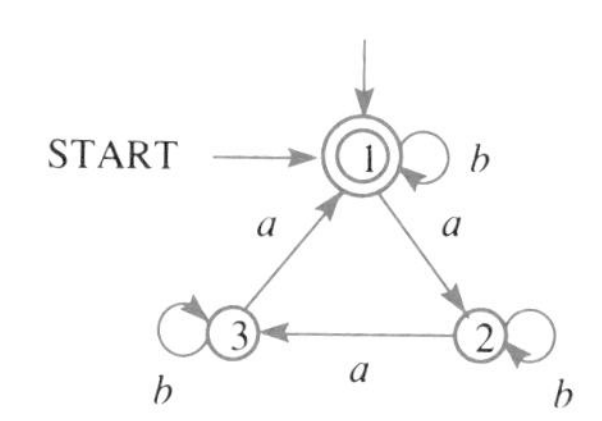

31. Given a DFA with three states, with initial state 1 and accepting states 1 and 3, decide which entries in R_0, R_1, R_2, and R_3 you actually need to get the regular expression for the corresponding language.

7.4 A Decision Algorithm for Regular Languages

In the previous section, we introduced the important class of regular languages, but said nothing about decision algorithms for them. In this section, we will remedy this by showing that indeed every regular language does have an efficient decision algorithm.

For many regular languages, it is easy to provide a decision algorithm. For example, consider the language represented by the regular expression $(0 + 1)^*001$ (Example 7.5a). This language can be decided by an algorithm that essentially asks the question "Does the string end in 001?" Similarly, the language of Example 7.5(b), represented by the regular expression $0^*1^*2^*$, can be decided by an algorithm that asks the question "Is the string made up of a run of 0s followed by a run of 1s followed by a run of 2s?" The language in Example 7.5(c), represented by the regular expression $a^*(b + c)^*d$, can be decided by the question "Does the string start with a run of a's, end with a single d, and otherwise contain only b's and c's?" However, these informal decision algorithms are quite different from each other, and it's not clear how one might produce a decision algorithm for a more complicated regular language. However, there is a way to take a regular expression and to automatically produce a decision algorithm for the corresponding regular language, and we will explain this procedure in this section.

The key fact here is that for a given regular language, as represented by a regular expression, it is always possible to construct a nondeterministic finite automaton (NFA) that represents the language. (By the results in Section 6.4, once we have an NFA, we can also construct a DFA that represents the language. Since by Section 7.3 we know that every language accepted by a DFA is regular, it follows that a language is regular if and only if it is accepted by a DFA. This is a famous result known as *Kleene's theorem.**) Once we have such an NFA, the NFA simulation algorithm of Section 6.4 will serve as a decision algorithm. All we need to do in this section is to show how to take a regular expression and produce a corresponding NFA.

Before we prove that *any* regular language can be generated by an NFA, let's warm up by considering the following seven regular expressions:

(7.20) ϵ

(7.21) 0

(7.22) 01

(7.23) $00 + 11$

(7.24) $01(00 + 11)$

(7.25) $(0 + 1)^*$

(7.26) $(0 + 1)^*00.$

EXAMPLE 7.7 Show that the seven NFAs in Figure 7.8 generate the seven regular languages represented by the regular expressions (7.20)–(7.26).

SOLUTION In Figure 7.8(a) we see an NFA with just two states. State 1 is the initial state, and state 2 is the accepting state. The edge from state 1 to state 2 is labelled ϵ, and that is the only edge in the state diagram. What language is generated by this state diagram? There is only one possible path from the initial state to the accepting state, viz. the direct jump on the edge from state 1 to state 2, which produces the string ϵ. Therefore, the language generated by this particular state diagram contains only the empty string, and this is the language represented by the regular expression ϵ in (7.20). A similar analysis shows that the state diagram in Figure 7.8(b) generates the language $\{0\}$, which is the language represented by the regular expression 0 in (7.21). In the state diagram of Figure 7.8(c), there is again only one path from the initial state (state 1) to the accepting state (state 3), which produces the string 01; and so this state diagram generates the language corresponding to the regular expression 01 in (7.22).

* Stephen C. Kleene (1909–), American mathematician. He has made important contributions to many branches of mathematical logic, and what we have called Kleene's theorem resulted from his work for the RAND Corporation in 1951. Also an avid lepidopterist, he discovered the variety *Beloria toddi ammiralis ob.* Kleene (Watson) 1921.

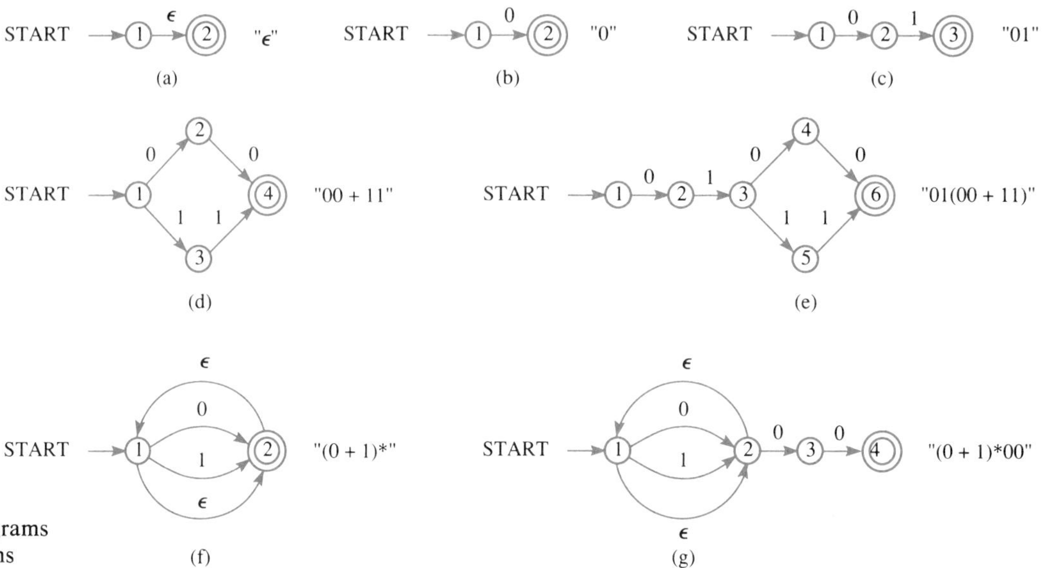

Figure 7.8 State diagrams for regular expressions (7.20)-(7.26).

In the state diagram of Figure 7.8(d), there are exactly two different paths from the initial state (state 1) to the accepting state (state 3): the path $1 \to 2 \to 4$, and the path $1 \to 3 \to 4$. These paths produce the strings 00 and 11, respectively, and so this state diagram generates the language {00, 11}, which is the same as the regular language represented by the regular expression 00 + 11 in (7.23). Similarly, in the state diagram of Figure 7.8(e), there are two paths from the initial state (state 1) to the accepting state (state 6), viz. $1 \to 2 \to 3 \to 4 \to 6$ and $1 \to 2 \to 3 \to 5 \to 6$, which produce 0100 and 0111, respectively. Thus this state diagram generates the language {0100, 0111}, which is also the language represented by the regular expression 01(00 + 11) in (7.24).

The state diagram of Figure 7.8(f), which corresponds to the regular language $(0 + 1)^*$ in (7.25), is more interesting. It has only two states: an initial state (state 1) and an accepting state (state 2). There are three edges going from state 1 to state 2, labelled 0, 1, and ϵ, respectively, and one edge going from state 2 to state 1, labelled ϵ. If we start at state 1, we can go back and forth between state 1 and state 2 as often as we wish, thus producing any possible string of 0s and 1s. For example, if we wished to produce the string 001 starting at the initial state and ending at the accepting state, we would begin at state 1 and travel to state 2 along the edge labelled 0, return to state 1 along the edge labelled ϵ, then travel back to state 2 along the 0 edge, then back to state 1 along the ϵ edge, and finally back to the accepting state 2 along the 1 edge. Notice that even the empty string ϵ is accepted by this state diagram, since we can travel from state 1 to state 2 via the edge labelled ϵ. Therefore, the language generated by the state diagram in Figure 7.8(f) consists of all possible strings of 0s and 1s, including the empty string, and so it represents the same language as the regular expression of (7.25).

Finally, it is not hard to see that the state diagram of Figure 7.8(g) produces the language generated by the regular expression in (7.26), viz. $(0 + 1)^*00$. This is because any path from the initial state 1 to the accepting state 4 must consist of two parts: a path from state 1 to state 2 (possibly traveling back and forth many times), followed by a path from state 2 to state 4. As we have just seen in our analysis of the state diagram in Figure 7.8(f), the first part of such a path can generate any string from the language $(0+1)^*$. The second part of the path will generate 00. Hence the entire path will generate a string from $(0+1)^*$ concatenated with the string 00, i.e., a string from the language $(0+1)^*00$. ■

Now that we have shown that the seven regular languages in (7.20)–(7.26) can all be generated by state diagrams, we must get down to the more serious business of showing that *any* regular language can be generated by a state diagram. We will do this in two stages. First, we will show that any *atomic* regular expression can be represented by an NFA. This is easy; indeed, in Figure 7.9, we show NFAs corresponding to the three atomic regular expressions $\varnothing$, ϵ, and a, where a is any letter in the underlying alphabet. The second stage, which is the hard part, is to show that if r_1 and r_2 are regular expressions which are representable by NFAs, then each of the compound regular expressions formed from r_1 and r_2 using Rules 1 through 3, viz. $(r_1 + r_2)$, $(r_1 r_2)$, and $(r_1)^*$, is also representable by an NFA. Since every regular expression is built up from atoms, using these three rules, this will complete the proof that every regular expression is representable by an NFA.

Figure 7.9 NFAs for the regular expressions $\varnothing$, ϵ, and a.

We begin with Rule 1. It says that if R_1 and R_2 are both regular languages, then so is $R_1 + R_2$. What we need to do is to show that if R_1 and R_2 can both be accepted by NFAs, then so can $R_1 + R_2$. We will do this by showing exactly how to construct such an NFA, starting with the NFAs for R_1 and R_2. This construction is described in Figures 7.10 and 7.11. (All of the NFAs we will construct will have just one accepting state.)

In Figure 7.10 we see abstract representations of the NFAs for the regular languages R_1 and R_2. We will call the corresponding state diagrams SD_1 and SD_2. The initial and accepting states in SD_1 are denoted by I_1 and A_1, and the corresponding states in SD_2 are denoted I_2 and A_2. In Figure 7.11 we see these two state diagrams combined into a single state diagram, a diagram which generates $R_1 + R_2$, as it turns out. In Figure 7.11 we have created a new initial state I and a new accepting state A. The new initial state I is connected with a directed edge to both of the old initial states I_1 and I_2; these edges are both labelled "ϵ." Both of the old accepting states A_1 and A_2 are connected with a directed edge to the new accepting state A; both of these edges are also labelled "ϵ." We might summarize this construction by saying that the state diagrams for R_1 and R_2 are connected "in parallel."

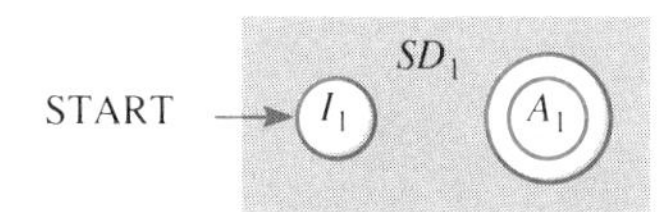

State diagram for R_1

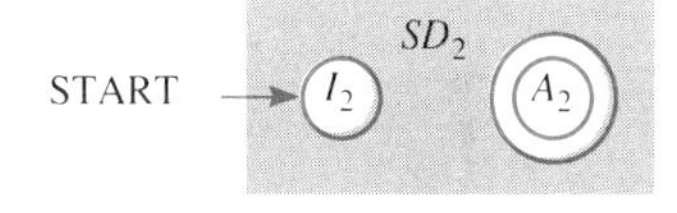

State diagram for R_2

Figure 7.10 NFAs for regular languages R_1 and R_2.

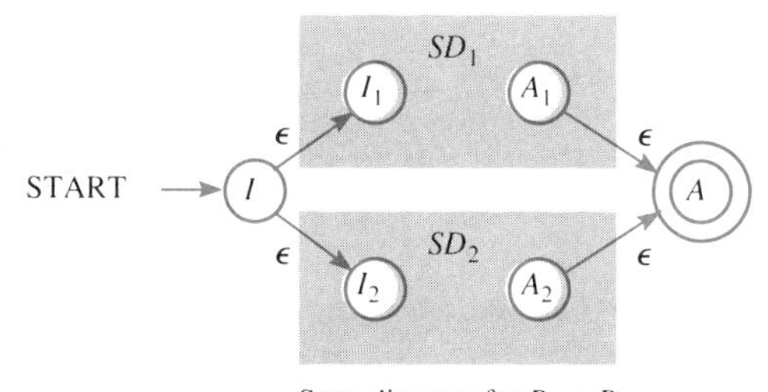

State diagram for $R_1 + R_2$

Figure 7.11 An NFA for the regular language $R_1 + R_2$ (Rule 1).

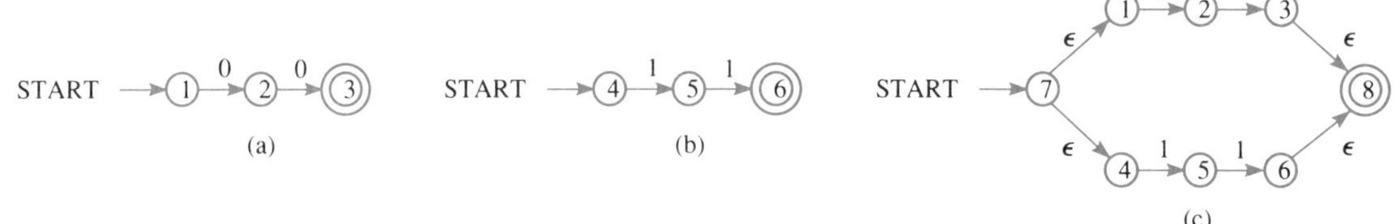

Figure 7.12 (a) An NFA for the language 00. (b) An NFA for 11. (c) An NFA for the sum 00 + 11, constructed according to the instructions given in Figure 7.11.

We claim that the NFA constructed as shown in Figure 7.11 generates the language $R_1 + R_2$. To see why, let's see what strings are generated by paths from I to A is the state diagram of Figure 7.11.

Starting at I, we can either go north, and enter SD_1, or south, and enter SD_2. This initial step produces nothing, since the edges $I \to I_1$ and $I \to I_2$ are both labelled "ϵ." If we enter SD_1, the only way we can reach A is to wander around in SD_1 for a while but eventually reach A_1, where we can exit directly to A. The string generated by such a path will be the same as that generated by the path from I_1 to A_1, since the final step $A_1 \to A$ is labelled ϵ. But the path from I_1 to A_1 will generate a string from the language R_1, since SD_1 generates R_1. Similarly, a path from I to A which begins by entering SD_2 will generate a string from R_2. In summary, any path from I to A in the state diagram constructed as shown in Figure 7.11 will generate either a string from R_1, or a string from R_2; in other words, this state diagram generates the language $R_1 + R_2$, as advertised.

In Figure 7.12, we see a simple illustration of this construction. In Figure 7.12(a) we see an NFA for the regular expression 00, and in Figure 7.12(b) we see an NFA for the regular expression 11. In Figure 7.12(c) we have created an NFA for the regular expression 00 + 11 by gluing a new initial state (state 7) and a new accepting state (state 8) to the old state diagrams, according to the instructions given in Figure 7.11. The resulting NFA accepts the language corresponding to the regular expression 00 + 11. Note that this NFA is not the same as the one in Figure 7.8(d), although they both accept the same language. Also note that the old initial states 1 and 4, and the old accepting states 3 and 6, are demoted to ordinary statehood in the new state diagram; there is only one initial state and one accepting state in the NFAs we're constructing.

Next we consider how to build an NFA that accepts a regular language which is constructed by Rule 2, i.e., which is the product of two other regular languages. Such an NFA is shown in Figure 7.13. The initial state I_1 for SD_1 becomes the initial state for the new state diagram; the accepting state A_1 for SD_1 is connected by a directed edge labelled ϵ to the initial state I_2 of SD_2; and the accepting state A_2 for SD_2 becomes the accepting state for the new state diagram. We might summarize this construction by saying that SD_1 and SD_2 are connected "in series," to produce the new state diagram.

The NFA constructed as shown in Figure 7.13 accepts the product language R_1R_2. Let us see why this is so, by examining the strings produced by paths from the initial state I_1 to the accepting state A_2 in the state diagram of Figure 7.13.

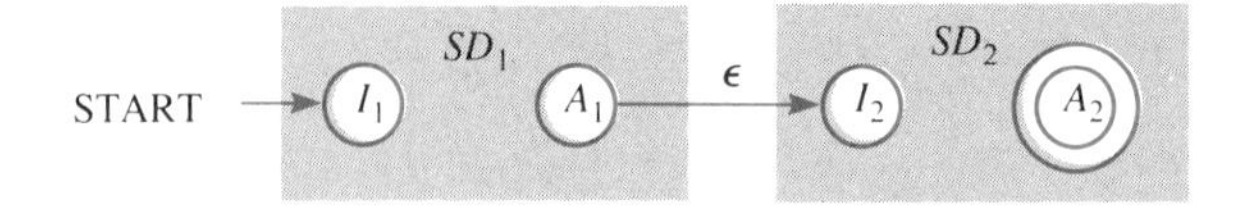

Figure 7.13 NFA for the regular language R_1R_2 (Rule 2).

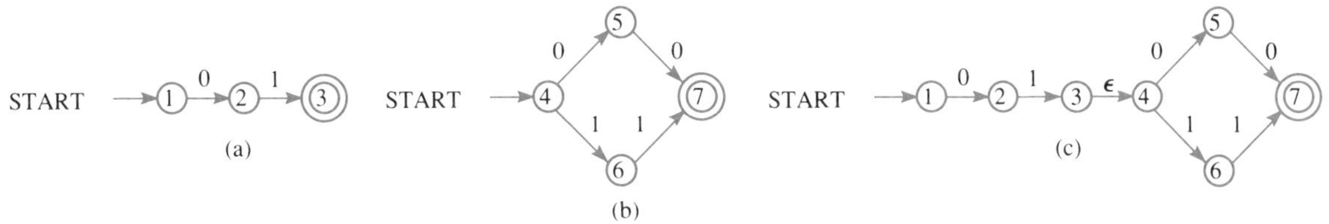

Figure 7.14 (a) An NFA for the language 01. (b) An NFA for 00 + 11. (c) An NFA for the product 01(00 + 11), constructed according to the instructions given in Figure 7.13.

Any path from I_1 to A_2 must consist of three parts: (1) a path from I_1 to A_1; (2) a direct jump from A_1 to I_2; and (3) a path from I_2 to A_2. The first part of this path will produce a string from R_1, since SD_1 generates R_1; the second part will produce nothing, since the edge from A_1 to I_2 is labelled ϵ; and the third part will produce a string from R_2, since SD_2 generates R_2. Therefore the entire path will produce a string from R_1 followed by a string from R_2, i.e., a string from R_1R_2. Thus any string accepted by the NFA of Figure 7.13 is in the language R_1R_2. Conversely, any string in R_1R_2 is accepted by the NFA of Figure 7.13 (see Problem 6). Hence the NFA in Figure 7.13 does accept the language R_1R_2, again as advertised.

In Figure 7.14, we see a concrete illustration of the general construction in Figure 7.13. In Figure 7.14(a), we have an NFA that accepts the regular language $R_1 = \{01\}$, and in Figure 7.14(b) we have an NFA that accepts $R_2 = \{00, 11\}$. In Figure 7.14(c) we have connected these two NFAs in series according to the directions given in Figure 7.13, and the resulting NFA accepts the regular language $R_1R_2 = \{0100, 0111\}$. Note that this NFA is slightly different from the state diagram in Figure 7.8(e), which accepts the same language.

Finally, we consider Rule 3. Here the problem is to construct an NFA for the language R_1^*, starting with an NFA for R_1. Such a construction is shown in Figure 7.15. The idea is to start with the state diagram for R_1, as shown in Figure 7.10, and to add a new initial state I and accepting state A, with new labelled edges as shown. Note especially the new edge from A_1 back to I_1 labelled ϵ.

The state diagram constructed as shown in Figure 7.15 generates the regular language R_1^*. To see why, consider a path from I to A in the state diagram of Figure 7.15. Such a path might be a direct jump from I to A, which produces ϵ. Otherwise, the path must consist of three parts: a jump from I to I_1 followed by a path from I_1 to A_1, followed by a jump from A_1 to A. Since $I \to I_1$ and $A_1 \to A$ both produce ϵ, the string produced by such a path is the same as the string produced by the subpath from I_1 to A_1. This subpath, in turn, must consist of one or more sub-subpaths from I_1 to A_1 in SD_1 connected by returns to I_1 along the ϵ edge. Each of these sub-subpaths will generate a string from R_1, since SD_1 generates R_1, but the return jumps back to I_1 generate nothing. Therefore any path from I to A will generate a string consisting of some number (possibly zero)

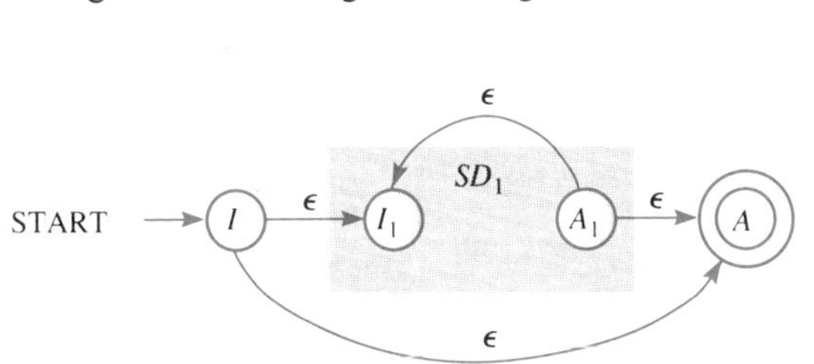

Figure 7.15 State diagram for the regular language R_1^* (Rule 3).

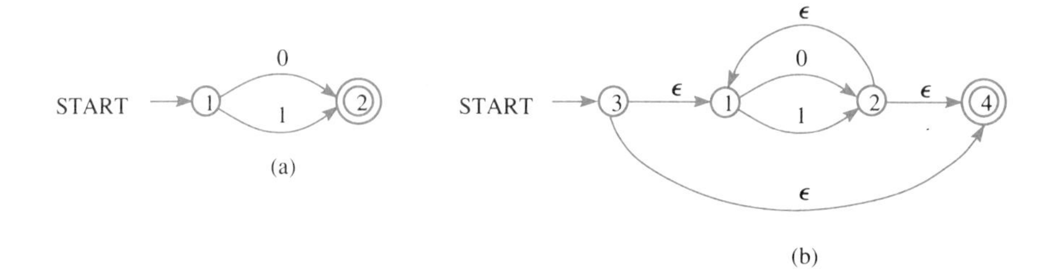

Figure 7.16 (a) An NFA for the language $(0 + 1)$. (b) A state diagram for the language $(0 + 1)^*$, constructed according to the instructions given in Figure 7.15.

of strings from R_1 concatenated together, and so the state diagram of Figure 7.15 generates the language R_1^*, as advertised.

In Figure 7.16 we see a specific application of the construction described in Figure 7.15. In Figure 7.16(a), we see a simple NFA that accepts the regular language $0 + 1$, and in Figure 7.16(b) we see this NFA modified according to the general instructions given in Figure 7.15, resulting in an NFA for the language $(0 + 1)^*$. Note that the new initial state is state 3, and the new accepting state is state 4. Also note that this NFA is different from the NFA in Figure 7.8(f), which accepts the same language.

By combining the atomic NFAs in Figure 7.9 with the constructions shown in Figures 7.11, 7.13, and 7.15, we can produce NFAs that accept any desired regular language. In the next example, we will show how this grand amalgamated construction works.

EXAMPLE 7.8 For each of the following four regular expressions, construct an NFA that accepts the corresponding regular language.

(a) 01. (b) $(0 + 1)^*$. (c) $a^*(b + \epsilon)$. (d) $a^*b^* + c^*$.

SOLUTION

(a) We have already seen one NFA that accepts this language (Figure 7.8c), but never mind that, let's start from scratch, using the construction techniques just described. The regular expression 01 is the product of the atomic regular expressions 0 and 1. Figure 7.9(c) shows how to construct NFAs for these two expressions, and we show such diagrams in Figure 7.17(a). In order to accept the product 01, we must connect these two NFAs in series, according to the instructions given in Figure 7.13. The result is shown in Figure 7.17(b). Notice that this NFA has one more state than the NFA in Figure 7.8(c). This inefficiency is expected: it almost always happens that if we start from scratch and construct a state diagram using only the constructions allowed in Figures 7.9, 7.11, 7.13, and 7.15, the resulting state diagram will have more, and often many more, states than necessary. However, starting from scratch has the advantage that it always produces a valid state diagram, and requires no cleverness on our part!

(b) This regular expression is the closure of the regular expression $0 + 1$, which in turn is the sum of the two primitive regular expressions 0 and 1. Starting with NFAs for 0 and 1 (Figure 7.17a), we can connect them in parallel according to the instructions in Figure 7.11 to produce a state

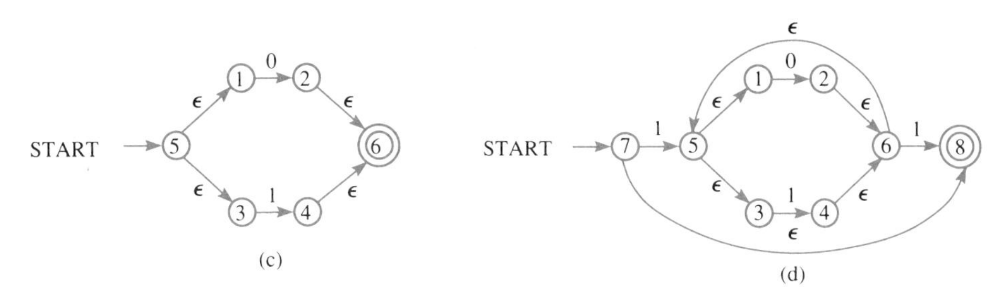

Figure 7.17 NFAs: (a) For 0 and 1; (b) for 01; (c) for $0 + 1$; (d) for $(0 + 1)^*$.

diagram for $0 + 1$ (Figure 7.17c). We can then apply the closure construction described in Figure 7.15 to this NFA to produce an NFA for $(0 + 1)^*$. Notice that this state diagram has eight states, whereas the state diagram of Figure 7.8(f) needs only two states to generate the same language. Sad. But at least we managed to construct a state diagram!

(c) This construction, starting from scratch, is summarized in Figure 7.18. Notice that we needed to use all three of our rules to construct this state diagram!

(d) NFAs for a^*, b^*, and c^* are shown in Figure 7.19(a). (These state diagrams have *not* been constructed from scratch, but you should be able to see that they work.) In Figure 7.19(b), we have connected the NFAs for a^* and b^* in series to produce an NFA for a^*b^*. Finally, in Figure 7.19(c), we have connected the NFAs for a^*b^* and c^* in parallel (with initial state 7 and accepting state 8) to produce an NFA for $a^*b^* + c^*$. ■

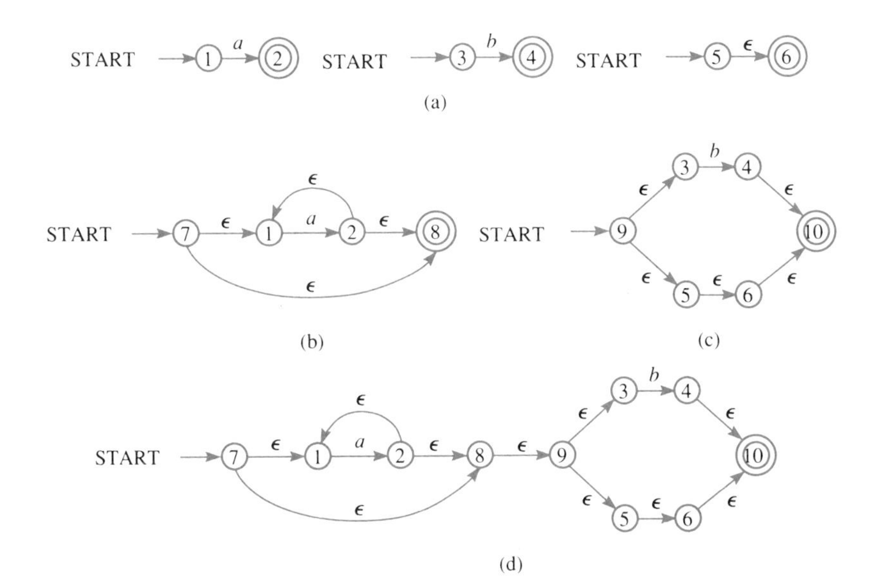

Figure 7.18 (a) NFAs for a, b, and ϵ. (b) NFA for a^*. (c) NFA for $b + \epsilon$. (d) NFA for $a^*(b + \epsilon)$.

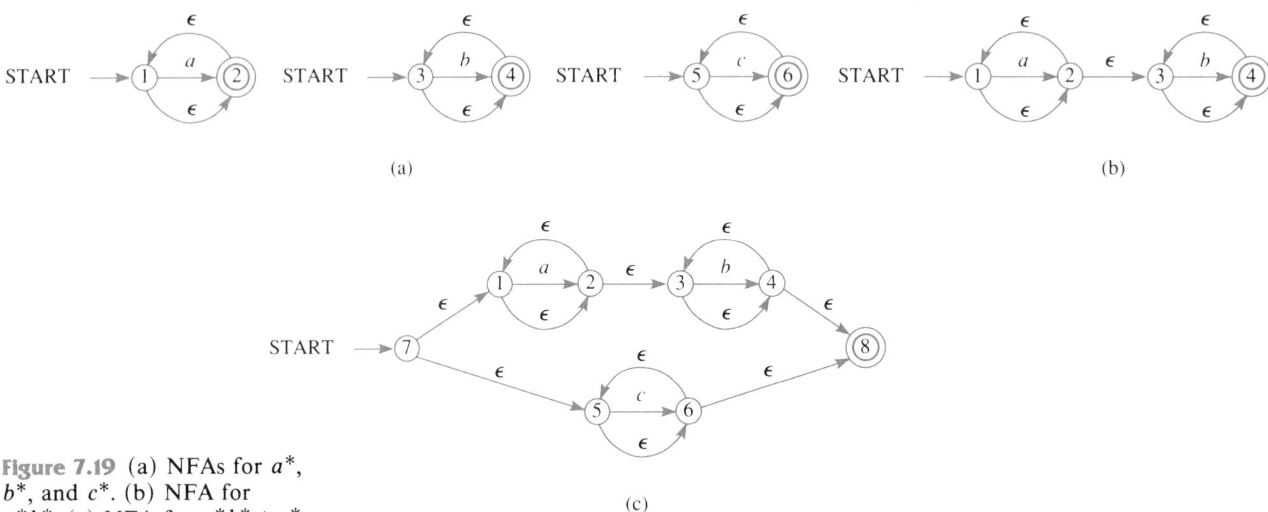

Figure 7.19 (a) NFAs for a^*, b^*, and c^*. (b) NFA for a^*b^*. (c) NFA for $a^*b^* + c^*$.

Problems for Section 7.4

1. Find *informal* decision algorithms for the seven regular languages in the expressions (7.20)–(7.26).

2. Construct a state diagram for the language represented by the regular expression $(00 + 11)01$.

3. Construct NFAs which accept the regular languages corresponding to the following regular expressions.

(a) $11 + 00$.
(b) $(00 + 11)01$.
(c) $(0 + 1)^*(000 + 11)$.
(d) $a^*b^* + b^*a^*$.
(e) $(x + \epsilon)(y + \epsilon)(z + \epsilon)$.
(f) $xy + (x + yy)z$.
(g) $xy(((yz)^* + zzz)^* + x)^*z$.
(h) $((x + y)(x + y))^* + ((x + y)(x + y)(x + y))^*$.

4. Let R be a regular language, over the alphabet A. Is the *complement* of R a regular language? Explain fully.

5. In Figure 7.9(b) we see a state diagram with two states that generates the language $\{\epsilon\}$. Find a state diagram with four states that does the same job.

6. Show that any string from R_1R_2 is accepted by the state diagram of Figure 7.13.

7. Find a different state diagram that generates the same language as the state diagram in Figure 7.14(c).

8. The language R_1^* is defined to consist of all strings that can be formed by concatenating together any number of strings from R_1. *Any number* includes *zero.* Does the state diagram of Figure 7.15 generate the strings that can be formed by concatenating zero strings from R_1? Explain.

9. Explain why the NFAs in Figure 7.19(a) correspond to the regular expressions a^*, b^*, and c^*.

10. Professor M once suggested an improved construction for producing a state diagram for the closure of a regular language. This construction is shown in the accompanying figure. However, a clever student politely suggested that Professor M was wrong! Who was right?

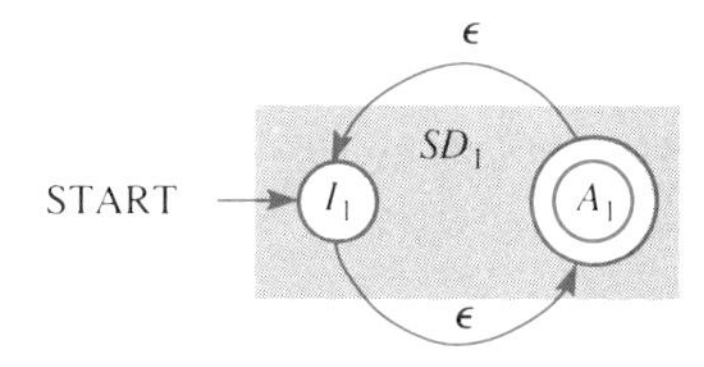

11. Professor M also once suggested an improved construction for producing a state diagram for the product of two regular languages. This construction is shown in the accompanying figure. The idea is to combine the accepting state for language 1 with the initial state for language 2. However, once again a clever student politely suggested that the learned professor was wrong! Who was right this time?

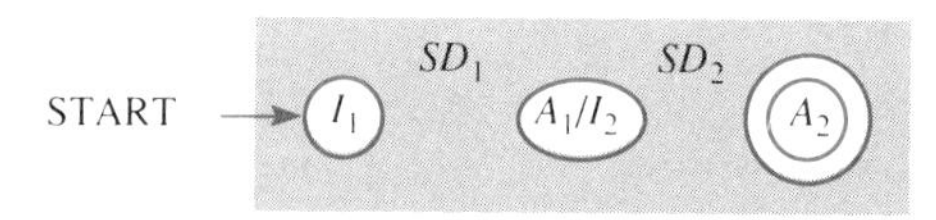

Improved (?) construction for R_1R_1

12. Construct an NFA for a^* "from scratch" (using only the constructions allowed in Figures 7.9 and 7.15), and compare it with the NFA given in Figure 7.18(b).

13. (a) Construct an NFA for $(0 + 1)^{**}$, using two applications of the construction in Figure 7.15.
(b) Find an NFA with as few states as possible that generates the same language.

14. Construct an NFA that accepts language C, as described in Section 7.1.

In Problems 15–19, give NFAs corresponding to the given regular languages, over the alphabet $\{a, b\}$. (Do them "by inspection.")

15. All strings beginning and ending in b.

16. All strings without two consecutive a's.

17. All strings with an odd number of a's.

18. All strings not containing the substring aba.

19. All strings containing at least one a and at least one b.

In Problems 20–22, give an NFA for the following languages over the alphabet $\{0, 1\}$. (Do them "by inspection.")

20. All strings ending in 00.

21. All strings containing three consecutive 0s.

22. All strings such that the fifth symbol from the right end is 1.

Summary

In this Chapter, we have discussed the following things:

- **Formal languages**—sets of strings over a given alphabet.
- **Recursively defined formal languages** such as languages A, B, and C.
- **Decision algorithms** for formal languages.
- Counting the strings in language B via legal and illegal paths.
- **Regular expressions** and **regular languages**.
- Every language accepted by a DFA is regular. (cf. Floyd-Warshall algorithm.)
- Every regular language can be accepted by an NFA.
- Every regular language has an efficient decision algorithm.
- **Kleene's theorem:** A language is regular if and only if it is accepted by a DFA.

CHAPTER EIGHT

MATHEMATICAL GAMES

In this chapter, we'll show you how to beat your friends at many kinds of mathematical games. We think this is a lot of fun, and at the same time, it will give us a chance to apply some of the things we've already learned.

In Section 8.1, we'll learn all about one very special game, the game of *Nim*. Then in Section 8.2, we will consider a large class of Nim-like games (called *impartial games*), of which Nim is the prototype. We will see that there is a general algorithm, the Sprague-Grundy algorithm, for finding winning strategies for impartial games. Finally, in Section 8.3, we will illustrate the general theory of Section 8.2 by describing winning strategies for many specific games, including that popular favorite, Hackenbush. Incidentally, nearly all of the material in this chapter, and much, much more about mathematical games, can be found in E. R. Berlekamp, John H. Conway, and Richard K. Guy's *Winning Ways* (London: Academic Press, 1982; two volumes), which is easily the best book ever written (or likely to be written) on the subject.

8.1 The Game of Nim

The game of Nim* is a game played between two players, using piles of counters, as in Figure 8.1. (We like to use checkers for counters, but almost anything will do, e.g., coins or matchsticks.) In Nim, the players alternate turns. When your turn comes, you must remove some (or all) of the counters from just *one* of the piles. The object is to be the one who removes the last counter from the table.†

Before going further, let's watch an actual game of Nim, played between Art and Betty, starting with the configuration in Figure 8.1. Art moves first, and the

* The mathematical theory presented in this chapter was discovered by Charles Bouton (American mathematician, 1869–1922) of Harvard University in 1901. Bouton also coined the name *Nim* apparently after the archaic English verb *nim*, meaning to take away, or steal.

† In the so-called *misère* version of Nim, the object is *not* to remove the last counter. We discuss this game in Problem 37.

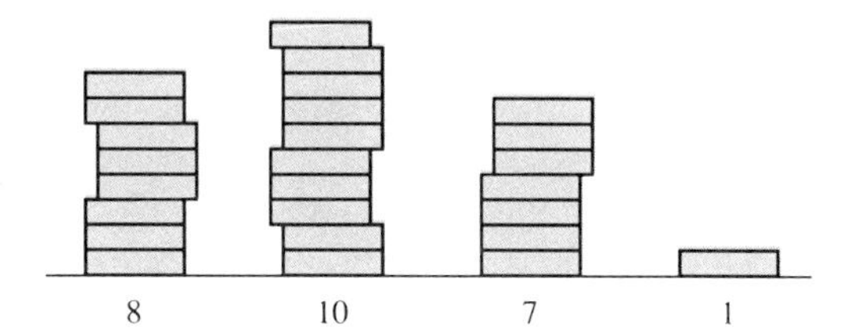

Figure 8.1 Anyone for Nim?

TABLE 8.1 How Betty beat Art

Move	Art Moves to	Betty Moves To
1	[8, 10, 3, 1]	[8, 7, 3, 1]
2	[7, 7, 3, 1]	[7, 5, 3, 1]
3	[6, 5, 3, 1]	[6, 4, 3, 1]
4	[5, 4, 3, 1]	[5, 4, 0, 1]
5	[5, 3, 0, 1]	[2, 3, 0, 1]
6	[0, 3, 0, 1]	[0, 1, 0, 1]
7	[0, 0, 0, 1]	[0, 0, 0, 0]

exciting action is shown in Table 8.1. In Table 8.1, we see that Art's first move is to remove four counters from the third pile, leaving Betty with the configuration [8, 10, 3, 1]. Betty then removes three counters from the second pile, leaving [8, 7, 3, 1]. The game continues as shown, and finally ends on Betty's seventh move, in which she clears the table and wins. For the moment, we will not comment on whether this particular game was well or poorly played, but simply use it as a fairly typical example of how a game of Nim might go in practice. Notice that Nim differs from games like chess or checkers in two important respects. Either player may move any one of the "pieces," and the game cannot possibly go on indefinitely, no matter how skilled (or inept) the players are.

It turns out that there is a complete mathematical theory for Nim, and when you learn it you will be able to play perfectly. You will always beat an opponent who doesn't know the theory, and you will be able to play on equal terms with anyone. The key to this theory is a strange new way of adding numbers, called **Nim-addition**. We therefore must interrupt our discussion of the game of Nim to discuss Nim-addition.

If a and b are two integers, their **Nim-sum** is another integer, denoted by $a \overset{*}{+} b$. Nim-addition shares some of the properties of ordinary addition, viz.:

(8.1) $$0 \overset{*}{+} a = a \overset{*}{+} 0 = a \qquad \text{(zero element)}$$

(8.2) $$a \overset{*}{+} b = b \overset{*}{+} a \qquad \text{(commutativity)}$$

(8.3) $$(a \overset{*}{+} b) \overset{*}{+} c = a \overset{*}{+} (b \overset{*}{+} c) \qquad \text{(associativity)}$$

but in other respects it is startlingly different. In Table 8.2, we present a table showing the Nim-sums of the numbers 0 through 15. For example, in Table 8.2

we see that

$$4 \overset{*}{+} 8 = 12, \qquad 3 \overset{*}{+} 4 = 7, \qquad 2 \overset{*}{+} 2 = 0\,(!), \qquad 5 \overset{*}{+} 3 = 6\,(!), \qquad \text{etc.}$$

Sometimes the Nim-sum is the same as the ordinary sum, but usually it's quite different! Of course, if you're not the curious type and you only need to Nim-add numbers which are less than 16, Table 8.2 is all you'll need. But to compute larger Nim-sums, e.g., $51 \overset{*}{+} 47$, you'll need to know two more facts about Nim-addition, which tell how to Nim-add powers of 2 (i.e., the numbers $1, 2, 4, 8, \ldots$):

(8.4) The Nim-sum of two *different* powers of two is the same as their *ordinary* sum. Thus for example $4 \overset{*}{+} 16 = 20$.

(8.5) The Nim-sum of two *equal* powers of two is *zero*. Thus for example $8 \overset{*}{+} 8 = 0$.

From the five rules (8.1)–(8.5), any Nim-sum can be computed, without reference to Table 8.2. The idea is to write the numbers as sums of distinct powers of two [ordinary sums and Nim-sums of distinct powers of two are the same, by (8.4)], and then to "cancel out" like powers of two, using (8.5). Properties (8.2) and (8.3) allow us to freely rearrange and ignore parentheses. For example, to compute $5 \overset{*}{+} 3$, we proceed as follows:

$$\begin{aligned} 5 \overset{*}{+} 3 &= (4 + 1) \overset{*}{+} (2 + 1) \\ &= (4 \overset{*}{+} 1) \overset{*}{+} (2 \overset{*}{+} 1) && \text{(by 8.4)} \\ &= 4 \overset{*}{+} \cancel{1} \overset{*}{+} 2 \overset{*}{+} \cancel{1} && \text{(by 8.5)} \\ &= 4 \overset{*}{+} 2 && \text{(by 8.5)} \\ &= 6 && \text{(by 8.4).} \end{aligned}$$

EXAMPLE 8.1 Compute the following Nim-sums:

(a) $51 \overset{*}{+} 47$; (b) $8 \overset{*}{+} 10 \overset{*}{+} 7 \overset{*}{+} 1$; (c) $5 \overset{*}{+} 1 \overset{*}{+} 3 \overset{*}{+} 7$.

SOLUTION

(a) $$\begin{aligned} 51 \overset{*}{+} 47 &= (\cancel{32} \overset{*}{+} 16 \overset{*}{+} \cancel{2} \overset{*}{+} \cancel{1}) \overset{*}{+} (\cancel{32} \overset{*}{+} 8 \overset{*}{+} 4 \overset{*}{+} \cancel{2} \overset{*}{+} \cancel{1}) \\ &= 16 \overset{*}{+} 8 \overset{*}{+} 4 = 28. \end{aligned}$$

(b) $$8 \overset{*}{+} 10 \overset{*}{+} 7 \overset{*}{+} 1 = \cancel{8} \overset{*}{+} (\cancel{8} \overset{*}{+} \cancel{2}) \overset{*}{+} (4 \overset{*}{+} \cancel{2} \overset{*}{+} \cancel{1}) \overset{*}{+} \cancel{1} = 4.$$

(c) $$5 \overset{*}{+} 1 \overset{*}{+} 3 \overset{*}{+} 7 = (\cancel{4} \overset{*}{+} \cancel{1}) \overset{*}{+} \cancel{1} \overset{*}{+} (\cancel{2} \overset{*}{+} \cancel{1}) \overset{*}{+} (\cancel{4} \overset{*}{+} \cancel{2} \overset{*}{+} \cancel{1}) = 0.$$

There is another popular way to compute Nim-sums, which is based on the *binary expansion* of the numbers involved. This method always gives the same result as the other method, and in fact it is the *same* method, slightly disguised. For example, to compute $8 \overset{*}{+} 3 \overset{*}{+} 6 \overset{*}{+} 4$ using this method, we first write the

TABLE 8.2 Nim-addition table

$\overset{*}{+}$	0	1	2	3	4	5	6	7	8	9	10	11	12	13	14	15
0	0	1	2	3	4	5	6	7	8	9	10	11	12	13	14	15
1	1	0	3	2	5	4	7	6	9	8	11	10	13	12	15	14
2	2	3	0	1	6	7	4	5	10	11	8	9	14	15	12	13
3	3	2	1	0	7	6	5	4	11	10	9	8	15	14	13	12
4	4	5	6	7	0	1	2	3	12	13	14	15	8	9	10	11
5	5	4	7	6	1	0	3	2	13	12	15	14	9	8	11	10
6	6	7	4	5	2	3	0	1	14	15	12	13	10	11	8	9
7	7	6	5	4	3	2	1	0	15	14	13	12	11	10	9	8
8	8	9	10	11	12	13	14	15	0	1	2	3	4	5	6	7
9	9	8	11	10	13	12	15	14	1	0	3	2	5	4	7	6
10	10	11	8	9	14	15	12	13	2	3	0	1	6	7	4	5
11	11	10	9	8	15	14	13	12	3	2	1	0	7	6	5	4
12	12	13	14	15	8	9	10	11	4	5	6	7	0	1	2	3
13	13	12	15	14	9	8	11	10	5	4	7	6	1	0	3	2
14	14	15	12	13	10	11	8	9	6	7	4	5	2	3	0	1
15	15	14	13	12	11	10	9	8	7	6	5	4	3	2	1	0

corresponding binary expansions:

		8	4	2	1
8	=	1	0	0	0
3	=			1	1
6	=		1	1	0
4	=		1	0	0

To compute the Nim-sum, we count the number of 1s in each column. If this number is *odd*, we put a 1 in that column; if it is *even*, we put a 0:

		8	4	2	1	
8	=	1	0	0	0	
3	=			1	1	
6	=		1	1	0	
4	=		1	0	0	
9	=	1	0	0	1	← Nim-sum

The Nim-sum is then the number represented (in binary) by this pattern of 0s and 1s. Thus we see that $8 \overset{*}{+} 3 \overset{*}{+} 6 \overset{*}{+} 4 = 9$. Computers especially like this definition of Nim-addition, since computers normally represent integers using the binary expansion anyway, and with respect to the binary expansion, Nim-addition is just addition *without carries*, or alternatively, componentwise XORing. However, you will need to perform Nim-additions mentally while playing human opponents, and we recommend you use the first method.

EXAMPLE 8.2 Use the second method to recompute the Nim-sums

(a) $51 \overset{*}{+} 47$. (b) $8 \overset{*}{+} 10 \overset{*}{+} 7 \overset{*}{+} 1$. (c) $5 \overset{*}{+} 1 \overset{*}{+} 3 \overset{*}{+} 7$.

SOLUTION

(a)

		32	16	8	4	2	1
51	=	1	1	0	0	1	1
47	=	1	0	1	1	1	1
28	=	0	1	1	1	0	0

(b)

		8	4	2	1
8	=	1	0	0	0
10	=	1	0	1	0
7	=	0	1	1	1
1	=	0	0	0	1
4	=	0	1	0	0

(c)

		4	2	1
5	=	1	0	1
1	=	0	0	1
3	=	0	1	1
7	=	1	1	1
0	=	0	0	0

Now we can return to our real interest, a winning strategy for the game of Nim. The key to analyzing a given Nim position is the *Nim-sum* of the number of counters in each pile. For example, the Nim-sum of the position in Figure 8.1 is $8 \overset{*}{+} 10 \overset{*}{+} 7 \overset{*}{+} 1 = 4$, as we calculated in Examples 8.1(b) and 8.2(b). Similarly, in the game depicted in Table 8.1, after Betty's second move, the Nim-sum is $7 \overset{*}{+} 5 \overset{*}{+} 3 \overset{*}{+} 1 = 0$, as we saw in Examples 8.1(c) and 8.2(c).

It turns out that if the game is played properly, the outcome can be predicted merely by knowing whether or not the Nim-sum of the position is 0. If the Nim-sum *isn't* 0, there is a winning strategy for the *first player* to move; but if the Nim-sum *is* 0, the *second player* can win. In the standard jargon of game theory, a position with a nonzero Nim-sum is called a **Fuzzy position**, and a position with a zero Nim-sum is called a **Zero position**. These facts are summarized in Table 8.3.

TABLE 8.3 Who should win a game of Nim

Nim-sum	Winner
$\neq 0$ (Fuzzy)	First player
0 (Zero)	Second player

The facts in Table 8.3 will allow you to *predict* the outcome of a well-played game of Nim, but won't help you *play* a well-played game of Nim! Don't worry, we'll explain the winning strategy in detail shortly. However, before we do that, let's check a few simple special cases of the rule given in Table 8.3.

- $[0, 0, 0, \ldots]$: Here there are *no* counters at all, and so the first player called upon to move will be unable to do so, and will immediately lose! This "empty position" is called the **Endgame**. The Nim-sum is zero, and indeed the second player will win, in agreement with Table 8.3.
- $[1, 0, 0, \ldots]$: Here there is only *one* counter, which will be taken by the first player, who therefore wins. The Nim-sum is 1, so this is a Fuzzy position, and indeed the first player wins.
- $[n, 0, 0, \ldots]$: Here all n of the counters are in one pile. The first player will want to take all these counters at once, and win. The Nim-sum is n, the position is Fuzzy, and the first player should win. But notice also that if the first player plays stupidly and does anything *other* than take all n counters, he will lose. The rule of Table 8.3 applies only to intelligent players!
- $[n, n, 0, 0, \ldots]$: Here there are two *equal* piles, the Nim-sum is 0, and so according to Table 8.3, the second player should be able to win. And in fact Betty (playing second) will be able to beat Art (playing first) simply by *copying* his moves. For example, if Art takes two counters from the first pile, Betty should take two from the second. In this way Betty will never be at a loss for a move, and so will win.
- $[1, 1, 1, \ldots, 1]$: Here there are n piles, each containing a single counter. This will be a boring game, with Art and Betty alternately removing single counters until none are left. If n is *odd*, Art will win, and if n is *even*, Betty will win. But the Nim-sum is 1 if n is odd, and 0 if n is even, once again in agreement with Table 8.3.

We hope our discussion of these special cases makes the rule of Table 8.3 seem plausible, but of course it does not *prove* that the rule is correct in general. In the next few pages, we will give such a proof; not the shortest proof possible, but one that will shed considerable light not only on Nim but on many of the other games we'll consider later in the chapter.

Everything depends on knowing what can happen to the Nim-sum when a player makes a legal move. The next example will give us some numerical experience in this matter.

EXAMPLE 8.3 For each of the following three Nim positions, calculate the Nim-sum, then calculate all possible Nim-sums that can result from a legal move from the given position.

(a) $[2, 2, 3]$. (b) $[5, 1]$. (c) $[3, 5, 2]$.

SOLUTION

(a) The Nim-sum is $2 \overset{*}{+} 2 \overset{*}{+} 3 = 3$. The legal moves are to $[1, 2, 3]$, $[0, 2, 3]$, $[2, 1, 3]$, $[2, 0, 3]$, $[2, 2, 2]$, $[2, 2, 1]$, and $[2, 2, 0]$, with corresponding

Nim-sums 0, 1, 0, 1, 2, 1, and 0. We summarize this list of Nim-sums as $(0^3, 1^3, 2^1)$, meaning that there are three Nim-sums of value 0, three of value 1, and one of value 2.

(b) The Nim-sum $5 \overset{*}{+} 1 = 4$. The legal moves are to [4, 1], [3, 1], [2, 1], [1, 1], [0, 1], and [5, 0], with corresponding Nim-sums 5, 2, 3, 0, 1, and 5. Thus the list of Nim-sums is $(0^1, 1^1, 2^1, 3^1, 5^2)$.

(c) The Nim-sum is $3 \overset{*}{+} 5 \overset{*}{+} 2 = 4$. The legal moves are to [2, 5, 2], [1, 5, 2], [0, 5, 2], [3, 4, 2], [3, 3, 2], [3, 2, 2], [3, 1, 2], [3, 0, 2], [3, 5, 1], and [3, 5, 0], with corresponding Nim-sums 5, 6, 7, 5, 2, 3, 0, 1, 7, and 6. The list of Nim-sums is therefore $(0^1, 1^1, 2^1, 3^1, 5^2, 6^2, 7^2)$. ■

If we examine the data from Example 8.3, we can see some interesting things:

(8.6) The *original* Nim-sum is *never* among the new Nim-sums.

(8.7) All values *less* than the original Nim-sum *are* among the new Nim-sums.

These two facts, which are at least true for the three Nim-positions in Example 8.3, turn out *always* to be true, and form the basis for the mathematical theory of Nim. Let's now see why they're always true.

The proof that (8.6) is true is relatively easy. Any legal move changes exactly one of the summands, as indicated schematically in Table 8.4. Changing one of the summands will, in turn, change some or all of the bits in the binary representation of that summand. (In Table 8.4, for example, the third bit of the changed summand changes from 1 to 0.) Any column where such a change occurs must *change parity*, i.e., the number of ones in that column changes from odd to even, or from even to odd, so that the new Nim-sum will differ from the old in this position at least. (In Table 8.4, the third bit in the Nim-sum will change.) Notice incidentally that this argument only uses the fact that one of the summands *changes*, and it doesn't matter whether the summand decreases (a legal move) or increases (an illegal move). So we can state a slightly stronger version of (8.6), which we will need in Section 8.3:

(8.6′) *Any* change in one of the numbers in a Nim-position (either an *increase* or a *decrease*) will result in a *different* Nim-sum.

TABLE 8.4 Why (8.6), or (8.6′), is true

Binary Representations of Original Numbers						**Binary Representations of New Numbers**				
x	x	x	x	x		x	x	x	x	x
x	x	x	x	x		x	x	x	x	x
x	x	1	x	x	— Summand changes →	y	y	0	y	y
x	x	x	x	x		x	x	x	x	x
x	x	x	x	x		x	x	x	x	x
		↑			Parity changes			↑		

The proof of (8.7) is a little harder, and we'll begin with an illustration, in Table 8.5, of how to change the Nim-sum of [8, 7, 5, 12] from 6 to 3 by decreasing just one of the summands. In Table 8.5(a), we show the binary representation of the summands 8, 7, 5, and 12, the actual Nim-sum 6, and the desired Nim-sum 3. The desired Nim-sum differs from the actual Nim-sum in two places, the 4's place and the 1's place. To achieve the desired Nim-sum, then, we must select one of the summands and change it in both of these places. But which summand should we select? To answer this, in Table 8.5(b), we show the result of changing each of the numbers 8, 7, 5, and 12 in the 4's place and the 1's place. We see that 8 changes to 13, 7 changes to 2, 5 changes to 0, and 12 changes to 9. The important thing is that the summand 8 has *increased*, while each of the other three numbers has *decreased*. What this means is that there are three legal moves from [8, 7, 5, 12] to a position with Nim-sum 3, viz. to [8, 2, 5, 12], [8, 7, 0, 12], or [8, 7, 5, 9]. (Additionally, there is one illegal move, viz. to [13, 7, 5, 12].)

In order to see that what happens in Table 8.5 is typical, we will need to know how to tell which of two distinct numbers, represented in binary, is larger. The rule is this: *Locate the leftmost place in which the two numbers differ*; *the number with a 1 in this place is larger.* For example, 1000 (8) is larger than 0111 (7), since the leftmost disagreement appears in the 8's place, where 1000 has a 1 and 0111 has a 0; and 11011 (27) is smaller than 11101 (29), since the leftmost disagreement appears in the 4's place, where 27 has a 0 and 29 has a 1.

Now we can complete the proof of (8.7). To change from one Nim-sum to another, smaller, Nim-sum, we first compare the binary representations of the

TABLE 8.5 (a) To change the Nim-sum from 6 to 3, change is required in the 4's place and in the 1's place. (b) Result of changing the 4's place and the 1's place of 8, 7, 5, and 12

	8	4	2	1	
8	1	0	0	0	
7	0	1	1	1	
5	0	1	0	1	
‡ 12	‡ 1	1	0	0	
6	0	1	1	0	← Actual Nim-sum
3	0	0	1	1	← Desired Nim-sum
		↑		↑	

Positions where change is needed

(a)

		↓		↓							
8	1	0	0	0	→	1	1	0	1	13	(increase)
7	0	1	1	1	→	0	0	1	0	2	(decrease)
5	0	1	0	1	→	0	0	0	0	0	(decrease)
12	1	1	0	0	→	1	0	0	1	9	(decrease)

(b)

two Nim-sums, and determine the places in which they differ, the "D-places" for short. If any one of the summands is changed in each of the D-places, the desired Nim-sum will result. If the summand has a 0 in the leftmost D-place, this will change to a 1, and the summand will increase; but if it has a 1 in the leftmost D-place, it will change to a 0, and the summand will decrease. Therefore, to find a legal move that produces the desired change in Nim-sum, one selects a summand with a 1 in the leftmost D-place, and changes it in each of the D-places. [For example, in Table 8.5(b), the D-places are the 4's place and the 1's place. The summand 12 has a 1 in the leftmost D-place (4's place), and changing it in the D-places does produce a smaller number.] To see that there will always be a summand with a 1 in the leftmost D-place, recall that the desired Nim-sum is *less* than the actual Nim-sum, which means that the actual Nim-sum must have a 1, and the desired Nim-sum must have a 0, in leftmost D-place. [For example, in Table 8.5(a), in the 4's place the actual Nim-sum has a 1, and the desired Nim-sum has a 0.] Since the actual Nim-sum has a 1 in the leftmost D-place, there must be an odd number of summands with a 1 in the leftmost D-place. Since zero isn't an odd number, there must be at least one summand with a 1 in the leftmost D-place, and so changing it in the D-places is a legal move that produces the desired Nim-sum.

This completes our proof of the important property (8.7) of Nim-addition. The next example will sharpen our numerical skills.

EXAMPLE 8.4 Start with the Nim-position [11, 22, 23], whose Nim-sum is 60. Show how to make a legal move which changes the Nim-sum to (a) 55; (b) 0.

SOLUTION First we check the Nim-sum:

		32	16	8	4	2	1
11	=	0	0	1	0	1	1
22	=	0	1	0	1	1	0
33	=	1	0	0	0	0	1
60	=	1	1	1	1	0	0

(a) We compare the binary representations of the actual Nim-sum and the desired Nim-sum, to determine the disagreement set:

		32	16	8	4	2	1
60	=	1	1	1	1	0	0
55	=	1	1	0	1	1	1
				↑		↑	↑

disagreements

The leftmost disagreement is thus in the 8's position. Of the summands 11, 22, and 33, only 11 has a 1 in this position, and so the only legal move which produces the Nim-sum 55 is to change 11 in the 8's, 2's, and 1's places, i.e., to reduce 11 (001011) to 0 (000000).

(b) Here we must compare the binary representatives of 60 and 0:

$$\begin{array}{rccccccc} & & 32 & 16 & 8 & 4 & 2 & 1 \\ \hline 60 & = & 1 & 1 & 1 & 1 & 0 & 0 \\ 0 & = & 0 & 0 & 0 & 0 & 0 & 0 \\ & & \multicolumn{4}{c}{\underbrace{\uparrow \qquad \uparrow \qquad \uparrow \qquad \uparrow}_{\text{disagreements}}} & & \end{array}$$

The leftmost disagreement is in the 32's position. Of the summands 11, 22, and 33, only 33 has a 1 in this position, and if we change it in the 32's, 16's, 8's, and 4's positions we obtain 011101 = 29. Therefore, the only legal move that produces a Nim-sum of 0 is to reduce the 33 to 29.

After this long but necessary discussion of the properties of Nim-addition, we will now return to our study of Nim-strategy. Recall that in Table 8.3 we asserted that the outcome of a well-played game of Nim depends only on whether or not the Nim-sum of the original position is zero. We will now see why this is true. Everything depends on the following three assertions:

(8.8) Any move from a *Zero* position will produce a *Fuzzy* position.

(8.9) From any *Fuzzy* position, it is always *possible* to move to a *Zero* position.

(8.10) The *Endgame* (from which no move is possible) is a Zero position.

We have already (p. 337) discussed (8.10), and assertions (8.8) and (8.9) follow easily from the basic facts (8.6) and (8.7) about Nim-sums. If we have a Zero position the Nim-sum is 0, and so by (8.6) after a legal move the Nim-sum cannot be 0, i.e., the new position will be Fuzzy. On the other hand, if the position is Fuzzy, the Nim-sum will be greater than 0, and so by (8.7) it will be possible to make a legal move to a Zero position.

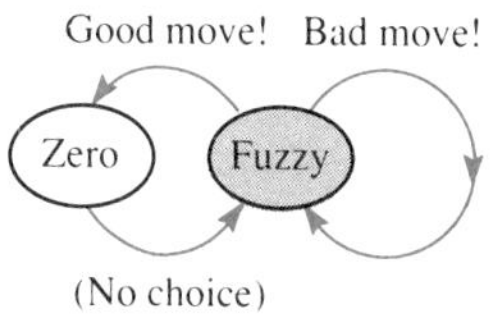

Figure 8.2 The secret of Nim.

Given the facts (8.8) and (8.9), it's easy to understand the rule of Table 8.3. For example, suppose the initial position is Fuzzy, and that Art plays first. Then according to (8.9) he can (and should) move to a Zero position; we call this a *good move* (see Figure 8.2). Then if Betty can move at all (i.e., if the position isn't the Endgame), she will by (8.8) be forced to move back to a Fuzzy position. Then Art will make another good move, Betty will move back to Fuzzy, etc. If Art plays this way, Betty will find herself faced with a series of Zero positions, each involving fewer counters than the one before. Eventually then, she will be faced with the Endgame, and lose.

Similarly, if the initial position is *Zero*, Art's first move will make it Fuzzy, and then Betty will be able to present Art with a series of Zero positions; eventually Art will be faced with the Endgame, and Betty will win.

In summary, if Nim is well played (i.e., if there are no *bad* moves as defined in Figure 8.2), the position will alternate between Fuzzy and Zero, until finally the Endgame is reached. What this means is that if the initial position is Fuzzy, the first player will win, and if the initial position is Zero, the second player will win, as promised in Table 8.3.

We close this section with two examples which illustrate the mathematical theory of Nim.

EXAMPLE 8.5 From each of the following Fuzzy positions, find all the good moves.

(a) [3, 4, 8, 9]. (b) [8, 10, 7, 1]. (c) [9, 25, 49]. (d) [6, 5, 3, 1].

SOLUTION

(a) The Nim-sum is $3 \overset{*}{+} 4 \overset{*}{+} 8 \overset{*}{+} 9 = 6$, as verified below:

		8	4	2	1
3	=	0	0	1	1
4	=	0	1	0	0
8	=	1	0	0	0
9	=	1	0	0	1
6	=	0	1	1	0

To make a good move, we must change the Nim-sum to zero, which means we must change the 4's and 2's position of one of the summands. The critical position is the 4's position, and the only summand with a 1 in this position is 4. Therefore the only good move is to change the 4 (= 0100) to 2 (= 0010), i.e., to remove two counters from the second pile. Incidentally, in practical play, you won't want to be secretly scribbling on bits of paper before each move. It is better to try to *visualize* the counters in each pile as partitioned into distinct subpiles whose sizes are powers of two, and then to mentally pair up these subpiles. For example, in Figure 8.3 we show how to visualize the position [3, 4, 8, 9] as subpiles of sizes 1, 2, 4, 8, 1, and 8. The two piles of size 1 and the two piles of size 8 "cancel out" leaving as unmatched only the pile of size 2 and the pile of size 4. From this it is plain that reducing the pile of size 4 to a pile of size 2 will produce a Zero position. It's true that this mental calculation is difficult if there are a lot of piles with many counters; but if you're playing a human opponent who doesn't know the theory, it's quite safe to make random moves until the position is simple enough for you to analyze mentally.

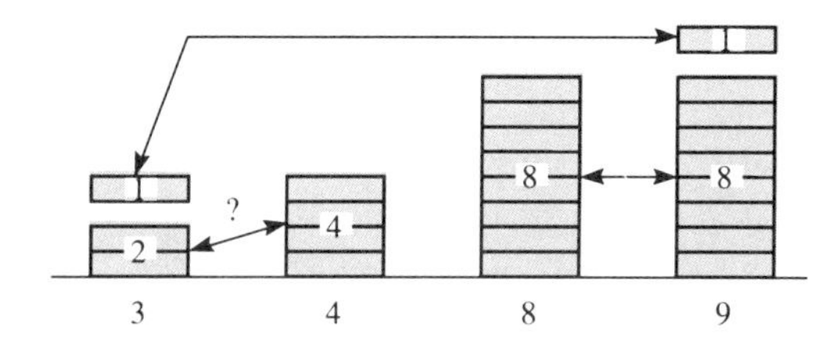

Figure 8.3 Finding a good move from [3, 4, 8, 9] mentally.

(b) The Nim-sum is 4:

$$\begin{array}{rcllll} 8 & = & 1 & 0 & 0 & 0 \\ 10 & = & 1 & 0 & 1 & 0 \\ 7 & = & 0 & 1 & 1 & 1 \\ 1 & = & 0 & 0 & 0 & 1 \\ \hline 4 & = & 0 & 1 & 0 & 0 \end{array}$$

To produce a Nim-sum of 0, we must change one of the summands in the 4's position. Only 7 has a 1 in this position, so that the only good move is to reduce 7 (0111) to 3 (0011), i.e., to remove four counters from the third pile.

(c) The Nim-sum is 33:

$$\begin{array}{rcllllll} 9 & = & 0 & 0 & 1 & 0 & 0 & 1 \\ 25 & = & 0 & 1 & 1 & 0 & 0 & 1 \\ 49 & = & 1 & 1 & 0 & 0 & 0 & 1 \\ \hline 33 & = & 1 & 0 & 0 & 0 & 0 & 1 \end{array}$$

To produce a Nim-sum of 0, we must change the 32's position and the 1's position of one of the summands. The critical position is the 32's position, and only 49 has a 1 in this position. Hence, the only good move is to reduce 49 (110001) to 16 (010000), i.e., to remove 33 counters from the third pile.

(d) The Nim-sum is 1:

$$\begin{array}{rclll} 6 & = & 1 & 1 & 0 \\ 5 & = & 1 & 0 & 1 \\ 3 & = & 0 & 1 & 1 \\ 1 & = & 0 & 0 & 1 \\ \hline 1 & = & 0 & 0 & 1 \end{array}$$

To produce a Nim-sum of 0, we must change the 1's position of one of the summands. Three of the summands (5, 3, 1) have a 1 in this position, and so in this case there are *three* good moves, i.e., removing exactly one counter from either the second, third, or fourth pile. ■

We conclude this section as we began, by considering the game between Art and Betty described in Table 8.1.

EXAMPLE 8.6 Make a critical analysis of the game between Art and Betty, i.e., identify all the *good* and *bad* moves.

SOLUTION The game consists of 14 moves, 7 by Art and 7 by Betty. In what follows, we number these moves $A1, B1, A2, \ldots, B7$.

- $A1$ (Art's first move): Here Art moves from the initial position [8, 10, 7, 1] to [8, 10, 3, 1]. But we saw in Example 8.5(b) that this is a good move, indeed the *only* good move from the initial position. Art is off to a good start!

- *B*1: Betty, faced with a Zero position, is helpless, and moves to the Fuzzy position [8, 7, 3, 1]. (When *you* are faced with a Zero position, perhaps the best strategy is to remove just one counter, leaving the position as complicated as possible, and hope that your opponent will fail to analyze it.)
- *A*2: The position [8, 7, 3, 1] has Nim-sum 13, and the only good move is to [5, 7, 3, 1]. But Art inexplicably misses this move, and moves to the Fuzzy position [7, 7, 3, 1] instead. A *bad* move, and now Betty has the advantage!
- *B*2: Betty's move to [7, 5, 3, 1] is good; but perhaps a *better* move would have been to [7, 7, 1, 1], from which position it would be slightly easier to find subsequent good moves.
- *A*3: Art helplessly moves back to a Fuzzy position.
- *B*3: Good move ([6, 5, 2, 1] and [6, 5, 3, 0] are also good).
- *A*4: Helpless move.
- *B*4: Betty finds the only good move.
- *A*5: Helpless.
- *B*5: Good move.
- *A*6: Helpless.
- *B*6: Good.
- *A*7: Helpless.
- *B*7: Betty takes the last counter and wins!

Problems for Section 8.1

1. Compute the following Nim-sums, using the first method given in the text (based on powers of 2).
(a) $17 \dotplus 5 \dotplus 13 \dotplus 4$. (c) $10000 \dotplus 100000$.
(b) $11 \dotplus 9 \dotplus 10 \dotplus 8$.

2. Repeat Problem 1 for the following Nim-sums.
(a) $11 \dotplus 22 \dotplus 33$. (c) $1 \dotplus 2 \dotplus 87 \dotplus 87 \dotplus 87$.
(b) $1 \dotplus 2 \dotplus 4 \dotplus 8 \dotplus 16 \dotplus 32$.

3. Compute the Nim-sums of Problem 1, using the second method given in the text (based on binary expansions).

4. Compute the Nim-sums of Problem 2, using the second method given in the text.

5. For the Nim-position [3, 1, 4], find the Nim-sum, and calculate all possible Nim-sums that can result from a legal move from the given position.

6. Repeat Problem 5 for the position [3, 2].

7. The Nim-sum of the position [8, 3, 17] is 26. Show how to change the sum to 8 by changing one of the numbers.

8. The Nim-sum of the position [10, 12] is 6. Show how to change the sum to 4 by changing one of the numbers.

9. If possible, find a legal move from the given Nim-position which produces the given Nim-sum s. If it *is* possible, exhibit all such moves.
(a) [17, 5, 13, 4], $s = 15$.
(b) [11, 9, 10, 8], $s = 9$.
(c) [10000, 100000], $s = 99999$.

10. Repeat Problem 9 for the following positions.
(a) [11, 22, 33], $s = 1$.
(b) [1, 2, 4, 8, 16, 32], $s = 63$.
(c) [1, 2, 87, 87, 87], $s = 128$.

11. For each of the positions in Problem 9, find all possible good moves.

12. For each of the positions in Problem 10, find all possible good moves.

13. Here is the box score of a game of Nim between Cora and Dave. Study the game and identify each good and bad move. The initial position is [17, 5, 13, 4].

Move	Cora	Dave
1	[12, 5, 13, 4]	[11, 5, 13, 4]
2	[11, 5, 13, 3]	[11, 5, 12, 3]
3	[11, 5, 11, 3]	[11, 3, 11, 3]
4	(Resigns)	

Why did Cora resign?

14. In Table 8.4, the implication is that from a Fuzzy position, it is always possible to find a *bad* move.

(a) Can you find a Fuzzy position from which it is *impossible* to make a bad move?

(b) Find *all* Fuzzy positions from which it is impossible to find a bad move.

15. Consider three-pile Nim positions of the form $[n_1, n_2, n_3]$ with each $n_i \leq 7$. There are by the multiplication rule $8 \times 8 \times 8 = 512$ such positions. How many of them are zero positions?

16. Consider the game of two-pile Nim, in which there are only two piles. Explain how the theory simplifies, and give a simple rule for identifying zero positions.

17. Consider the Nim-position $[1, 2, 4, \ldots, 2^n]$. Show that it is Fuzzy, and find a good move. Is there more than one good move?

18. Can you find a Fuzzy Nim-position for which there are exactly *two* good moves?

19. Explain why the two definitions of Nim-addition given in the text (as described in Example 8.1 and Example 8.2) are equivalent.

20. For which numbers a is it true that $a \overset{*}{+} 7 = a - 1$?

21. Show that $a \overset{*}{+} b = 0$ if and only if $a = b$.

22. For which pairs of numbers (a, b) is it true that $a \overset{*}{+} b = a + 1$?

23. Show that $a \overset{*}{+} b = c$ if and only if $a = b \overset{*}{+} c$.

24. If $b \neq c$, is it necessarily true that $a \overset{*}{+} b \neq a \overset{*}{+} c$?

25. Show that the Nim-sum of two numbers is always less than or equal to the ordinary sum, i.e., that $a \overset{*}{+} b \leq a + b$. When does equality hold?

26. The calculations in Example 8.3 support the following conjecture:

> Every number *less* than the original Nim-sum appears an *odd* number of times as a new Nim-sum; and every number *greater* than the original Nim-sum appears an *even* number of times as a new Nim-sum.

Is this conjecture true? If it is, give a proof; if not, give a counterexample, i.e., a specific instance in which it is false.

27. Consider a game between Betty and Art, with Betty going first. Assume the initial position is zero, and that Betty and Art each make two bad moves. Who wins the game?

28. We have seen that from any Zero position, any move which *decreases* the size of some pile will make the position fuzzy. Is it possible to *increase* the size of some pile and keep the position zero?

29. Consider the Boolean function $f(x_1, x_2) = x_1 \overset{*}{+} x_2$ (each x_i is 0 or 1). Identify this function as one of those in Table 5.6.

30. Use the techniques of Chapter 5 to design a circuit that computes the Nim-sum of a and b, where a and b are numbers between 0 and 15, represented by their four-bit binary expansions.

31. Show that if the highest power of two involved in b is not involved in a, then $a \overset{*}{+} b \geq a$. When does equality hold?

32. This problem requires some knowledge of probability. Suppose you are playing three-pile Nim, and that the initial position $[n_1, n_2, n_3]$ is chosen randomly, with each n_i independent of the others and uniformly distributed on $\{0, 1, 2, \ldots, 7\}$. What is the probability that the position will be Zero? (*Hint*: Do Problem 15.)

33. Generalize Problem 32 to a k-pile initial position $[n_1, n_2, \ldots, n_k]$ with each n_i selected independently and uniformly distributed in the range $\{0, 1, \ldots, 2^m - 1\}$. What is the probability that the position will be Zero?

In Problems 34–37, we will describe several *variants* on the game of Nim. In each case you are supposed to discover the relationship between the given game and Nim, and find a good move from the given position(s).

34. **Strip Nim** is played on a long strip of paper which is partitioned into squares labelled 0, 1, 2, ... On each square may be placed one or several counters. The legal move is to move just one of the counters *leftward* (toward 0). The winner is the player who moves the last counter "home," i.e., to square 0. Explain the relationship to Nim, and find a good move from the given position. (See the accompanying figure.)

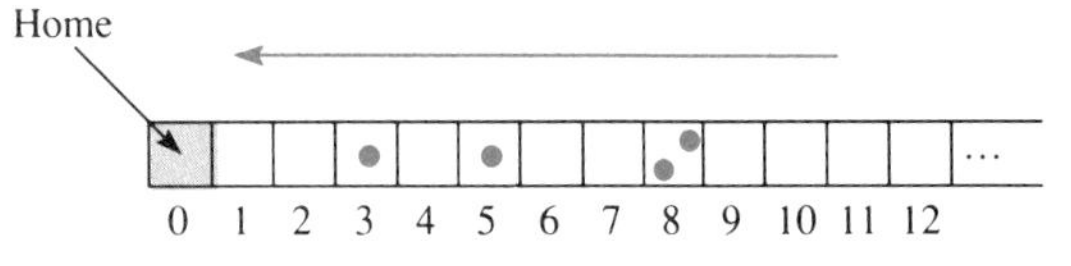

35. **Rook Nim** is played on a chessboard. On each square may be placed one or several *rooks*. The legal move is to move just one of the rooks any number of squares *upward* or *leftward* (but not at both). The object is to get all the rooks to the square marked "home," and the winner is the player who moves the last rook "home." Explain the relationship to Nim, and find a good move from the given position. (See the accompanying figures.)

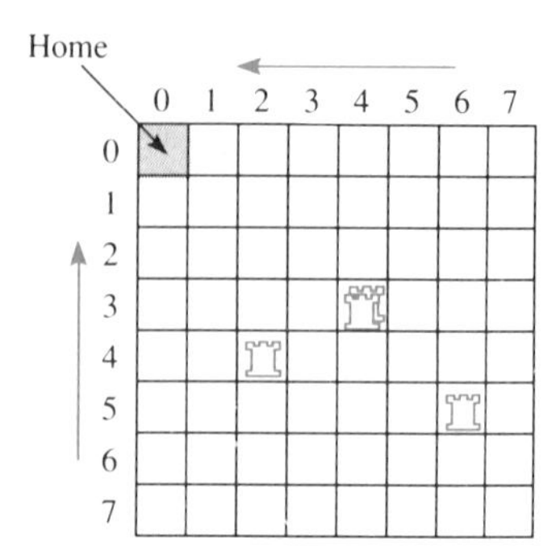

36. Poker Nim is the same as ordinary Nim, except that either player is given the alarming option of *putting counters back on the table.* These counters must be counters the player has previously removed from the table, and must all be put onto the same pile. For example, in the position shown in the accompanying figure, Betty, whose turn it is to move, might decide to put 10 extra counters on the second pile, thus changing the position from [5, 4, 1] to [5, 14, 1]. Describe a winning strategy for Poker Nim, and illustrate it by finding a good response for Art if Betty plays as suggested above.

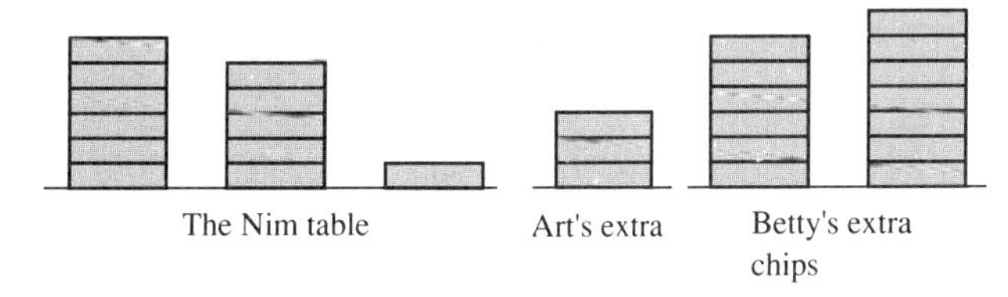

37. Misère Nim (Rough translation from the French: "Nim for losers") is played exactly like ordinary Nim, *except* that the object is to force *your opponent* to take the last counter. The winning strategy of Misère Nim can be stated as follows:

> Play just as you would in ordinary Nim, *except* that if your good move would leave a position of the form [1, 1, 1, . . . , 1] with an *even* number of counters, modify your move so that it leaves an *odd* number of one-counter piles instead.

(a) Show that this rule does indeed represent a winning strategy for Misère Nim.
(b) Find (if possible) a good Misère Nim move from each of the following Misère positions: [0, 3, 0, 1], [1, 2, 1, 1], [1, 1, 1], and [2, 2, 1, 1, 1].
(c) Review the game in Table 8.1. Assuming that Art plays as before, at which point should Betty make a different move, if she wants to win in Misère Nim?

8.2 General Impartial Games; the Sprague-Grundy Algorithm

In Section 8.1 we learned in detail how to play a winning game of Nim. But Nim, it turns out, is just one member of a large family of similar games, called *impartial games*, and each game in this family has a mathematical theory, similar to that of Nim. In this section we'll build on the theory of Section 8.1, and learn how to find winning moves in any impartial game.

Before we give the formal definition of an impartial game, we consider two more examples, **Wythoff's game** and **Strip Dominos**.

Wythoff's game is played on an ordinary chessboard using just one piece, a queen; see Figure 8.4. The queen is initially placed on one of the squares. There are two players, and they alternate turns, as in Nim. A legal move is to move the queen any number of squares *north*, *west*, or *northwest*. The player who moves the queen to the shaded square ("home") is the winner.

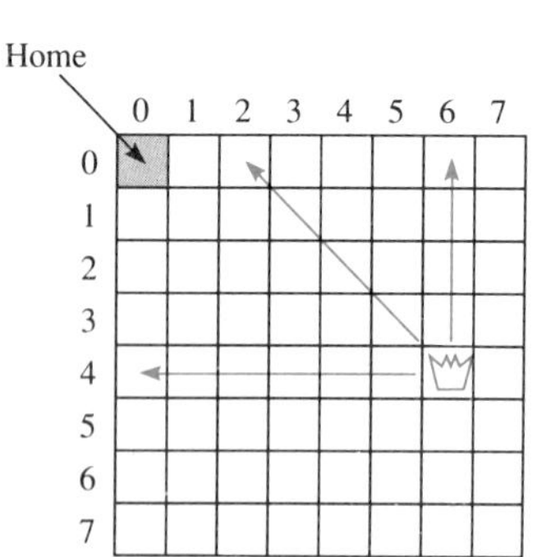

Figure 8.4 Anyone for Wythoff's game?

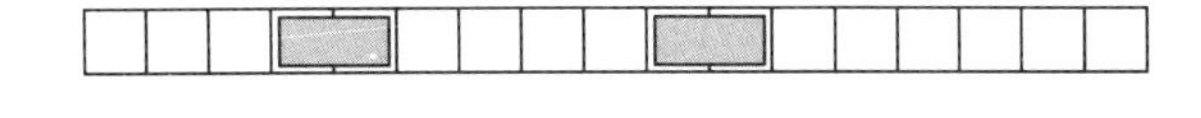

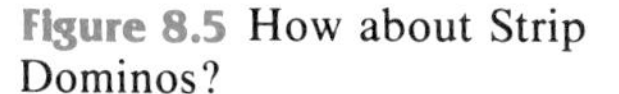

Figure 8.5 How about Strip Dominos?

Strip Dominos* is played on a long strip of paper which is ruled into squares. Each player is given a large supply of dominos, each of which will cover exactly two of the squares on the strip. The players take turns placing dominos on the strip (dominos may not overlap), until one of the players finds no room to place a domino. This player loses. In Figure 8.5 we see a game of Strip Dominos being played on a strip of length 17, with two dominos having already been played.

Nim, Wythoff's game, and Strip Dominos, despite their obvious dissimilarities, have much in common. For the record, we now list their common features:

(8.11) The game has a certain set of **positions**, and play begins from a particular **starting position**.

(8.12) Each position has associated with it a set of **options**, which are the positions to which it is legal to move from the given position. If the set of options is *empty*, no move is possible from the position, which is then called an **end position**.

(8.13) There are two players, who alternate turns. Each player, when it is his or her turn, must move from the current position to one of its options.

(8.14) A player who is unable to move (because the current position is an end position) *loses*.†

To illustrate some of these concepts, we note that in Wythoff's game there are 64 positions, and in the game of Figure 8.4 the starting position is the square in the fourth row and sixth column. The options of this starting position are the squares covered by the arrows; there are 14 options. For the Strip Dominos position shown in Figure 8.5, there are 10 options, corresponding to the 10 different places the next domino could go. In Nim, there is only one end position, viz. the *Endgame* position with no counters at all. In Wythoff's game there is also just one end position, viz. the square marked "home." However, in Strip Dominos, there are many end positions; see Figure 8.6.

Figure 8.6 Some end positions in Strip Dominos ($n = 7$).

Games that satisfy conditions (8.11)–(8.14) are called **impartial‡ games**. Most, but not all, of the impartial games we shall consider also satisfy one further

* In the book *Winning Ways* cited at the beginning of this chapter, this game appears under the name *Dawson's Kayles.*

† This is often called the *normal play* convention for impartial games. If we decree instead that the first player unable to move *wins*, we have what is called *misère play.* The theory for misère play is much harder than for normal play, and except in several problems, we won't consider it.

‡ This terminology comes from the fact that from any given position, exactly the same moves are available to each player. Thus chess, for example, isn't impartial, since from a given position each player will have a different set of options.

important condition, called the **ending condition**:

(8.15) No matter how the game is played, it must finally end with one player unable to move. Equivalently, there can be no *infinite sequence* of positions P, P', P'', P''', etc., each being an option of the previous one.

Nim, Wythoff's game, and Strip Dominos all satisfy the ending condition. As an example of an impartial game which *does not* satisfy the ending condition, consider the well-known child's game "Yes you are, No I'm not." In this game there are only two positions; one is called "Yes you are," and the other is called "No I'm not." From the position "Yes you are" the only option is "No I'm not," and from the position "No I'm not," the only option is "Yes you are." As childhood experience shows, this game can continue for quite a long time without anyone winning; still, it does qualify as an impartial game not satisfying the ending condition. There are actually some very interesting impartial games that don't satisfy the ending condition, and we shall meet some of them later on. However, in the rest of this section, we'll always be assuming that the ending condition holds.

Now that we have defined, and given several examples of, impartial games, we will describe their general mathematical theory.

We saw in Section 8.1 that in Nim, each position can be assigned a value (the Nim-sum of the number of counters in each pile), and that the good moves are just those that go from a position with a *nonzero* value (Fuzzy position) to a position with a *zero* value (Zero position). Surprisingly, it is just the same for any other impartial game. Each position can be assigned a value from $\{0, 1, 2, \ldots\}$ in such a way that the good moves are again those that go from a Fuzzy position (value not 0) to a Zero position (value equal to 0). In other words, the secret to winning *any* impartial game is revealed in Figure 8.2, which we repeat here as Figure 8.7, *once the values of the positions are known.*

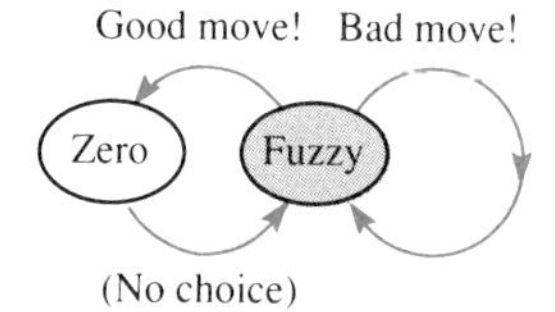

Figure 8.7 The secret of impartial game.

Everything therefore depends on being able to compute the position values, which are also called the **Nim-values**, of the game. In Nim, there is a simple explicit rule for computing the values: You just compute the Nim-sum of the number of counters in each pile. For other impartial games, the computation of the values is (as a rule) not as simple as for Nim. However, there is a general *algorithm* for computing the Nim-values in any impartial game. Before we can describe this algorithm, we need to introduce a new concept, the *minimal excludent.*

If S is a set of nonnegative integers (i.e., a set of numbers from $0, 1, 2, \ldots$), the **minimal excludent** of S, mex(S) for short, is defined to be the least number from $0, 1, 2, \ldots$ which is *not* in S. For example, if $S = \{0, 1, 2\}$, then $\text{mex}(S) = 3$, and $\text{mex}\{1, 2, 3\} = 0$.

EXAMPLE 8.7 Find the mex of each of the following three sets.

(a) $A = \{0, 1, 3, 4, 10\}$. (b) $B = \varnothing$. (c) $C = \{5, 6, 0, 3, 2\}$.

SOLUTION

(a) $\text{mex}(A) = 2$.

```
    /* Sprague-Grundy Algorithm */
    {
1.    while (not all positions have values) {
2.      P ← any unnumbered position,
        all of whose options are numbered;
3.      g(P) ← mex of values of P's options;
      }
    }
```

Figure 8.8 How to compute Nim-values (the Sprague-Grundy algorithm).

(b) This is a little tricky, as are most calculations involving the empty set. The reasoning here is that since $\varnothing$ contains *no* elements, the smallest number *not* in $\varnothing$ must be the smallest *possible* number, viz., 0.

(c) $\text{mex}(C) = 1$. ■

Now that we know about mex's, we can give a description of the algorithm for computing Nim-values for a general impartial game. We call this algorithm the **Sprague-Grundy* (SG) algorithm**, after its discoverers; see Figure 8.8. The description of the SG algorithm in Figure 8.8 is a little cryptic, and so before further theoretical discussion, we'll try to clarify things by using the algorithm to compute the Nim-values for Wythoff's game (Figure 8.4).

For clarity, we'll refer to the 64 positions in Wythoff's game by their *row/column addresses*; for example, the home square has address $(0, 0)$, and the position occupied by the queen in Figure 8.4 has address $(4, 6)$. As we go, when a position is assigned a value, we'll say that the position has been *numbered.* In honor of Grundy, we will denote the Nim-value of the position (x, y) by $g(x, y)$. (The Nim-values of a game are sometimes called the **Grundy numbers** for the game.)

Here we go: Line 1 just tests to see if we are finished, so we won't mention it further. Now at the first execution of line 2, the only possible choice for P is the home square $(0, 0)$, *which has no options at all* (and so in particular no *unnumbered* options!). Now we can execute the key line 3, which says:

(8.16) Nim-value of a position = mex{the Nim-values of its options}.

In this case $P = (0, 0)$ is therefore assigned a value equal to the mex of the *empty set*, which, as we have seen in Example 8.7(b), equals 0. Therefore the home square can be assigned Nim-value 0, and we denote this fact by writing $g(0, 0) = 0$ (see Table 8.6a). Once $(0, 0)$ has been numbered, we can number the positions $(1, 0)$ and $(0, 1)$, since both of these positions have only $(0, 0)$ as an option, and $(0, 0)$ has already been numbered. Thus the Nim-value for $(1, 0)$ is

$$g(1, 0) = \text{mex}\{g(0, 0)\} = \text{mex}\{0\} = 1;$$

* Patrick M. Grundy (English mathematician, 1917–1959) and Roland P. Sprague (German mathematician, 1894–1967). Grundy wrote an influential article in 1939, and since that time game theorists have spoken of Grundy numbers, Grundy functions, Grundy's theorem, etc. However, it was discovered somewhat belatedly that Sprague independently discovered and published essentially all of Grundy's results in a 1935 article! Grundy and Sprague are now usually given equal cobilling as the founders of modern impartial game theory. This is an interesting example of a remarkably common occurrence in science—the independent and nearly simultaneous discovery of an important new result.

TABLE 8.6 Computing the Nim-values for Wythoff's Game

	0	1	2	3	4	5	6
0	0						
1							
2							
3							
4							
5							
6							

(a)

	0	1	2	3	4	5	6
0	0	1					
1	1						
2							
3							
4							
5							
6							

(b)

	0	1	2	3	4	5	6
0	0	1	2				
1	1	2					
2	2						
3							
4							
5							
6							

(c)

	0	1	2	3	4	5	6
0	0	1	2	3			
1	1	2	0				
2	2	0					
3	3						
4							
5							
6							

(d)

	0	1	2	3	4	5	6
0	0	1	2	3	4		
1	1	2	0	4			
2	2	0	?				
3	3	4					
4	4						
5							
6							

(e)

	0	1	2	3	4	5	6
0	0	1	2	3	4	5	
1	1	2	0	4	5		
2	2	0	1	5			
3	3	4	5				
4	4	5					
5	5						
6							

(f)

similarly, $g(0, 1) = 1$; see Table 8.6(b). Continuing as shown in Table 8.6, we can systematically compute all the Nim-values for Wythoff's game. For example, in Table 8.6(e), we see how to compute $g(2, 2)$. If we use the notation $\text{Opt}(P)$ to denote the options of the position P, then we have

$$\text{Opt}(2, 2) = \{(2, 1), (1, 2), (2, 0), (1, 1), (0, 2), (0, 0)\}.$$

Therefore by line 3 of the SG algorithm,

$$\begin{aligned} g(2, 2) &= \text{mex}\{g(2, 1), g(1, 2), g(2, 0), g(1, 1), g(0, 2), g(0, 0)\} \\ &= \text{mex}\{0, 0, 2, 2, 2, 0\} = 1. \end{aligned}$$

	0	1	2	3	4	5	6	7
0	0	1	2	3	4	5	6	7
1	1	2	0	4	5	3	7	8
2	2	0	1	5	3	4	8	6
3	3	4	5	6	2	0	1	9
4	4	5	3	2	7	6	9	0
5	5	3	4	0	6	8	10	1
6	6	7	8	1	9	10	3	4
7	7	8	6	9	0	1	4	5

Figure 8.9 The Nim-values for Wythoff's game, and the winning move from the position of Figure 8.4.

The complete set of Nim-values for Wythoff's game is shown in Figure 8.9. From Figure 8.9 we can see that the starting position $(4, 6)$ shown in Figure 8.4 has Nim-value 9, and so is a Fuzzy position, i.e., a win for the first player. We can also see that there is only one good move from this position, viz., a one-square diagonal move to the Zero position $(3, 5)$.

Now that we have a little experience with the SG algorithm, we are in a position to see why the "Fuzzy-Zero" strategy of Figure 8.7 works. We will need to show three things:

(8.17) Any end position has Nim-value zero.

(8.18) Any option of P has a Nim-value different from that of P.

(8.19) All values less than the Nim-value of P occur as Nim-values of options of P.

We saw in Section 8.1 that these three facts underlie the theory of Nim; similarly, they underlie the theory of any impartial game. To see why (8.17) is true, note that by definition an end position has no options, and so its Nim-value is the mex of the empty set, which is zero. To see (8.18) and (8.19), suppose that the Nim-value of a position P is N. Then the mex of the Nim-values of its options must be N, so the Nim-values of its options must include $0, 1, \ldots, N-1$, but not N. This is just what (8.18) and (8.19) say!

Because of (8.17), (8.18), and (8.19), the "Fuzzy-Zero" strategy works, for any impartial game satisfying the ending condition. Indeed, (8.18) implies that any move from a Zero position will be to a Fuzzy position, and (8.19) implies that from any Fuzzy position there will be a legal move to a Zero position. Thus starting from a Fuzzy position, Art can present Betty with a series of Zero positions, and Betty will be forced to present Art with a series of Fuzzy positions. If the game satisfies the ending condition, eventually the players will reach an end position, which by (8.17) must be a Zero position, and so Betty will lose.

The next example will give us some more practice in computing Nim-values.

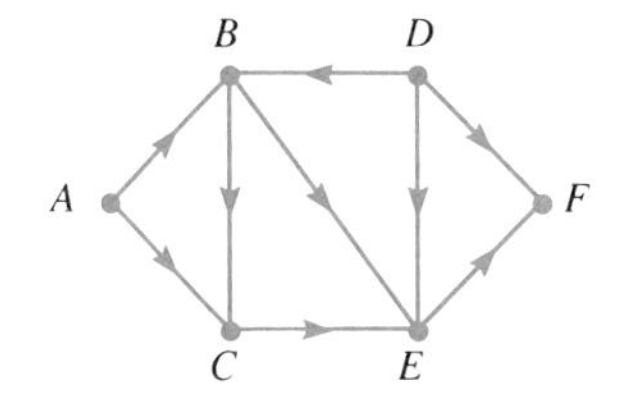

Figure 8.10 A graph for a game.

EXAMPLE 8.8 We define an impartial game, using the directed graph of Figure 8.10, as follows: The *positions* are the six vertices of the graph, and the *options* of a given position are given by the directed edges of the graph. Thus

$$\text{Opt}(A) = \{B, C\}, \qquad \text{Opt}(C) = \{E\}, \qquad \text{Opt}(E) = \{F\},$$
$$\text{Opt}(B) = \{C, E\}, \qquad \text{Opt}(D) = \{B, E, F\}, \qquad \text{Opt}(F) = \varnothing.$$

Find the Nim-values for this game.

SOLUTION We begin with the end position F, which must be numbered 0. Next, we can number E since its only option is F, and F has already been numbered. Thus the Nim-value for E is $\text{mex}\{0\} = 1$. If we again denote the Nim-value of a position P by $g(P)$, we thus have $g(F) = 0$ and $g(E) = 1$. Next, we can compute $g(C)$:

$$g(C) = \text{mex}\{g(E)\} = \text{mex}\{1\} = 0.$$

We continue as follows:

$$g(B) = \text{mex}\{g(C), g(E)\} = \text{mex}\{0, 1\} = 2.$$

Now, *either* A or D can be numbered. We choose D:

$$g(D) = \text{mex}\{g(B), g(E), g(F)\} = \text{mex}\{2, 1, 0\} = 3.$$

Finally,

$$g(A) = \text{mex}\{g(B), g(C)\} = \text{mex}\{2, 0\} = 1.$$

Thus the complete set of Nim-values is as follows:

Position	A	B	C	D	E	F
Nim-value	1	2	0	3	1	0

Now that we have had some practical experience computing Nim-values, we conclude this section by raising some important, though rather technical,

```
2. P ← any unnumbered position,
       all of whose options are numbered;
```

↓

```
2a. P ← any unnumbered position;
2b. while (P has unnumbered options)
2c.       P ← one of P's unnumbered options;
```

Figure 8.11 An expanded version of line 2 in the SG algorithm.

questions about the SG algorithm. There is a potential difficulty in line 2, where in order to proceed it is necessary to find an unnumbered position, *all of whose options have already been numbered.* The specific questions we wish to raise are these:

- *Question 1*: Will there always *be* such a position?
- *Question 2*: If there is *more than one* such position, does it matter which one we choose?

We shall see that the answers to these questions are yes and no, respectively, which means that the algorithm always succeeds in assigning a Nim-value to every position, and that a given position always gets the *same* Nim-value, regardless of the order in which the positions are numbered.

To answer Question 1, we will need to expand line 2 in the SG algorithm, as shown in Figure 8.11. The idea is simple; if a given unnumbered position P cannot yet be numbered because one of its options, say P', is unnumbered, we try to number P'. If P' cannot be numbered because one of *its* options, say P'', is unnumbered, we try to number P'', etc. In this way we get a sequence of positions $P, P', P'', P''', \ldots$, each of which is an option of the one before. But we know, because of the *ending condition* (8.15), that this sequence cannot go on forever, and so we *must* eventually reach an unnumbered position with no unnumbered options (note that at the *first* execution of line 2, this position will necessarily be an *end* position, with no options at all!). It follows then, with line 2 implemented as shown in Figure 8.11, the answer to Question 1 is yes. In Figure 8.12, we show the SG algorithm with the expanded line 2.

EXAMPLE 8.9 Use the Figure 8.12 version of the SG algorithm to once again find the Nim-values for the game of Figure 8.10.

SOLUTION At two places in the algorithm (steps 2a and 2c) we are asked to select a position with a given property, and in general there might be several possibilities. Let us agree beforehand that when we have a choice, we will always select the position that comes first *alphabetically*, i.e., A will be preferred to B, B to C, etc. (If you were programming a computer to implement the SG algorithm, you would have to adopt some such convention, though not necessarily this one.)

Thus we begin with the graph of Figure 8.10 with no Nim-values assigned. The answer to the question posed at step 1 is yes; we therefore proceed to step 2a, where (according to the convention agreed upon) position A is selected. Since

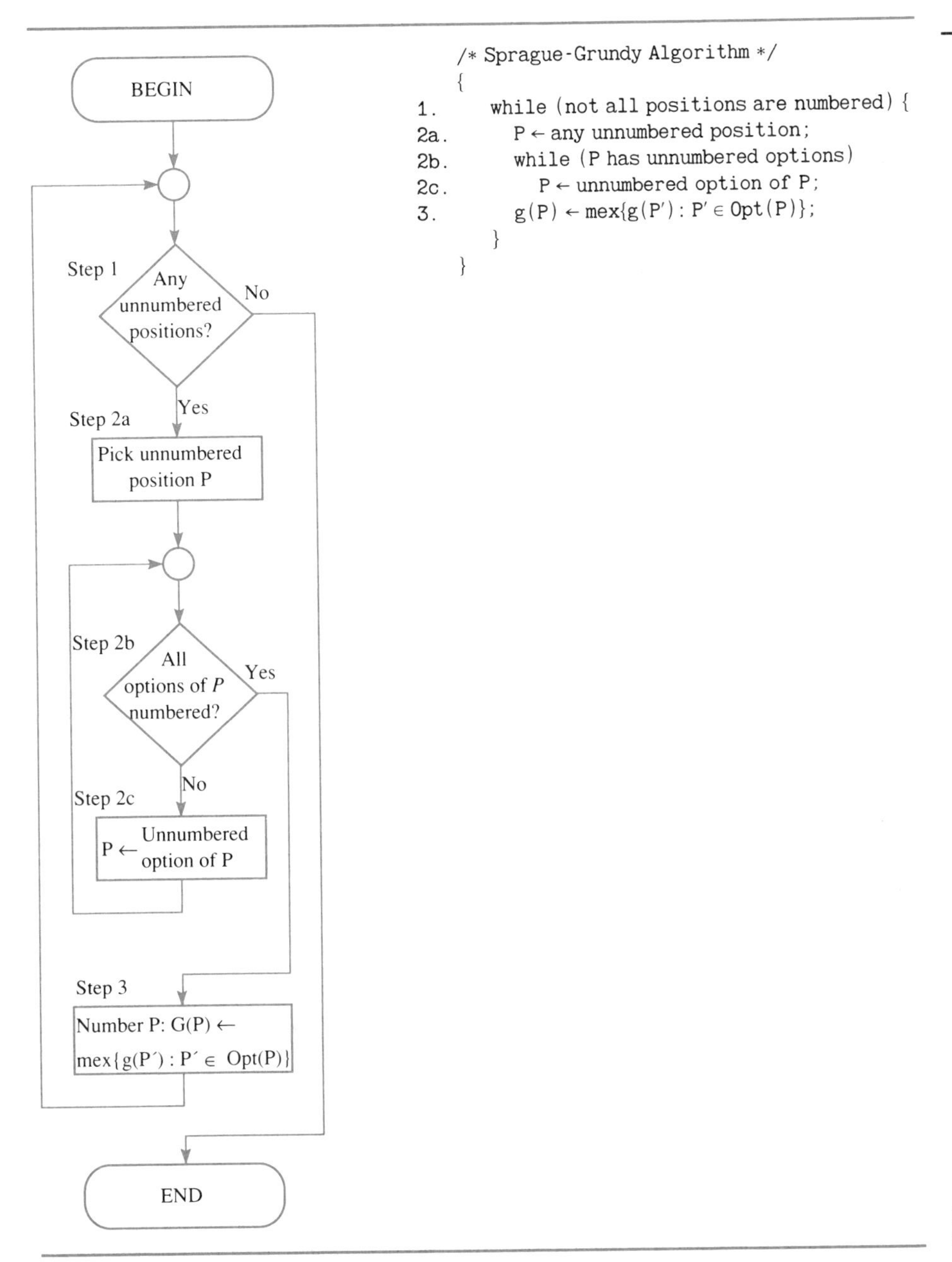

Figure 8.12 A flowchart and pseudocode for the SG algorithm.

$\mathrm{Opt}(A) = \{B, C\}$, we must take the "No" branch at step 2b, and then select B at step 2c. We continue around the `while` loop at steps 2b and 2c, successively choosing positions A, B, C, E, and finally F. When F has been selected, we must exit from the loop, since $\mathrm{Opt}(F) = \varnothing$, which means that F has no unnumbered options. Then at step 3 we calculate $g(F) = 0$, just as in Example 8.8.

Having completed our first execution of step 3, we return to step 1 where position A, which is still the (alphabetically) first unnumbered position, is again selected. On the second pass through the loop, the sequence of positions examined is A, B, C, E, each position being an option of the previous one, until we exit to step 3 and compute

$$g(E) = \text{mex}\{g(F)\} = \text{mex}\{0\} = 1.$$

Continuing in this way, we once again compute the Nim-values for the game of Figure 8.10, with the sequence of positions examined in the loop given below:

1st pass	A, B, C, E, F	$g(F) = 0$
2nd pass	A, B, C, E	$g(E) = 1$
3rd pass	A, B, C	$g(C) = 0$
4th pass	A, B	$g(B) = 2$
5th pass	A	$g(A) = 1$
6th pass	D	$g(D) = 3$

Notice that in Example 8.8, the positions were assigned Nim-values in the order F, E, C, B, D, A; whereas in Example 8.9 the order was slightly different (F, E, C, B, A, D). Still, the Nim-values assigned were the same in both cases. This brings us back to Question 2 about the SG algorithm raised earlier. It turns out that *in whatever order the positions are assigned Nim-values by the* SG *algorithm, the results are always the same*, so we are justified in calling the results of any version of the SG algorithm *the* Nim-values for the game. To prove this vitally important fact, we will actually prove a somewhat stronger result, the *uniqueness theorem for Nim-values.*

THE UNIQUENESS THEOREM FOR NIM-VALUES In any impartial game that satisfies the ending condition, there is one and only one way to assign a nonnegative integer $g(P)$ to each position P so that the following three properties hold:

(a) $g(P) = 0$ for each end position.
(b) If P' is an option of P, then $g(P') \neq g(P)$.
(c) Given a position P with $g(P) > 0$, and a nonnegative integer $n < g(P)$, P has an option P' with $g(P') = n$.

We already know that there is at least one way to assign such values, viz. via the SG algorithm. The trick is to show that there is no other way. To do this, let $g_0(P)$ denote the Nim-values assigned by the SG algorithm, and let $g(P)$ be any other function satisfying (a), (b), and (c). We will show that $g_0(P) = g(P)$ for all positions P, using induction.

Let's assume that the SG algorithm assigns values to the positions in the order $P_1, P_2, \ldots$. We will show that $g_0(P_n) = g(P_n)$ for $n = 1, 2, \ldots$, by induction on n. For $n = 1$, this follows from the fact that P_1, as the first numbered position, must be an end position, and (a), which says that the Nim-value of any end position must be 0. Thus $g_0(P_1) = g(P) = 0$. Now assume that g_0 and g agree on the n positions $P_1, P_2, \ldots, P_n$, and consider P_{n+1}. Then by line 3 of the SG algorithm, we know that

(8.20) $$g_0(P_{n+1}) = \text{mex}\{g_0(P'): P' \in \text{Opt}(P_{n+1})\}.$$

But since each option of P was numbered earlier than P_{n+1}, by induction we know that $g_0(P') = g(P')$, for all $P' \in \mathrm{Opt}(P_{n+1})$, so that in fact

(8.21) $$g_0(P_{n+1}) = \mathrm{mex}\{g(P')\colon P' \in \mathrm{Opt}(P_{n+1})\}.$$

But by (c) we know that every nonnegative integer smaller than $g(P_{n+1})$ occurs as a Nim-value of some option of P_{n+1}, and by (b) that $g(P_{n+1})$ does not occur, so that

(8.22) $$\mathrm{mex}\{g(P')\colon P' \in \mathrm{Opt}(P_{n+1})\} = g(P_{n+1}).$$

Combining (8.21) and (8.22), we have

$$g_0(P_{n+1}) = g(P_{n+1}),$$

which completes the induction proof of the Nim-value uniqueness theorem. ∎

Of course, Nim *itself* is an impartial game, and so it must have Nim-values associated with its positions. What are these values? You will not, we hope, be surprised to learn that

> the Nim-value associated with the Nim-position $[n_1, n_2, \ldots, n_m]$ is just the Nim-sum $n_1 \stackrel{*}{+} n_2 \stackrel{*}{+} \cdots \stackrel{*}{+} n_m$.

To prove this, all we have to do is show that the Nim-sums satisfy properties (a), (b), and (c). Property (a) holds because in Nim, the only end position is the Nim-position with zero counters, whose Nim-sum is certainly zero; property (b) holds because of property (8.6) proved in Section 8.1; and property (c) holds because of property (8.7) proved in Section 8.1.

With this, we end the section. However, we remind the reader that we have not yet found a good move in the Strip Dominos position of Figure 8.5. We will do this, and learn about many more impartial games, in Section 8.3. Stay tuned!

Problems for Section 8.2

1. Calculate the following minimal excludents.
(a) $\mathrm{mex}\{1, 2, 3, 4, 5\}$. (c) $\mathrm{mex}\{1 \stackrel{*}{+} 3, 1 \stackrel{*}{+} 3\}$.
(b) $\mathrm{mex}\{1, 1, 2\}$.

2. Calculate the following minimal excludents.
(a) $\mathrm{mex}\{0, 2, 4, 8\}$.
(b) $\mathrm{mex}\{3 \stackrel{*}{+} 5, 2 \stackrel{*}{+} 5, 1 \stackrel{*}{+} 5, 0 \stackrel{*}{+} 5, 4 \stackrel{*}{+} 4, 4 \stackrel{*}{+} 3, 4 \stackrel{*}{+} 2, 4 \stackrel{*}{+} 1, 4 \stackrel{*}{+} 0\}$.

3. Find the Nim-values for the game whose graph is given below. (Cf. Example 8.8.)

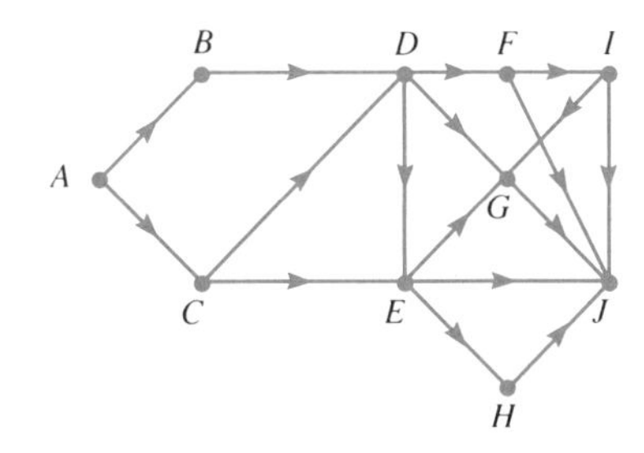

4. In Problem 3, if the starting position is A, will the first player win or lose?

5. Use the Figure 8.12 version of the Sprague-Grundy algorithm to once again find the Nim-values for the game of Problem 3.

6. In Example 8.8, the positions in the game of Figure 8.10 were numbered in the order F, E, C, B, D, A, whereas in Example 8.9, they were numbered in the order F, E, C, B, A, D. Are there any *other* possible orderings of the positions that might be produced by the SG algorithm? (*Suggestion*: Consider changing the convention given at the beginning of the solution to Example 8.9.)

7. Starting from the position shown in Figure 8.4, how many moves (maximum) can the game last?

8. From the general Nim-position $[n_1, n_2, \ldots, n_m]$, how many options are there?

9. Given the general position $[x, y]$ in Wythoff's game, how many options are there?

10. Given the following Strip Dominos position, how many options are there?

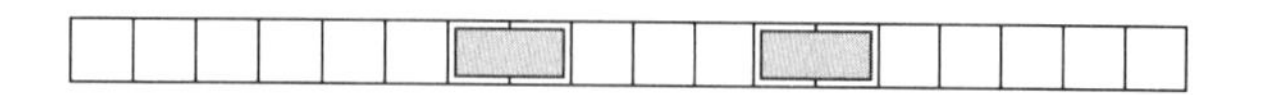

11. Suppose G is an impartial game with exactly three positions, A, B, and C, with $\text{Opt}(A) = \{B, C\}$, $\text{Opt}(B) = \{C\}$, and $\text{Opt}(C) = \varnothing$. Find the Nim-values of A, B, and C.

12. Repeat Problem 11, except now use nine positions $\{A, B, C, D, E, F, G, H, I\}$, with

$$\begin{aligned} &\text{Opt}(A) = \{B, C, E, F\}, && \text{Opt}(F) = \{H, I\}, \\ &\text{Opt}(B) = \{D, E, H\}, && \text{Opt}(G) = \{I\}, \\ &\text{Opt}(C) = \{D, F, G\}, && \text{Opt}(H) = \{E\}, \\ &\text{Opt}(D) = \{G, H\}, && \text{Opt}(I) = \varnothing. \\ &\text{Opt}(E) = \{G, I\}, && \end{aligned}$$

13. Review Problem 8.1.36, in which *Poker Nim* is described. Is Poker Nim an impartial game? Does it satisfy the ending condition?

14. Draw a graph for the game "Yes you are, No I'm not" described in the text.

15. Extend Wythoff's game to a 10×10 board, and find the Nim-values.

16. Consider the following impartial game, which is played with exactly *two* piles of counters. The legal moves are to remove any number of counters from *either* pile, or an equal number of counters from *both* piles. For example, from the position $[2, 3]$, the options are $[1, 3]$, $[0, 3]$, $[2, 2]$, $[2, 1]$, $[2, 0]$, $[1, 2]$, $[0, 1]$.

(a) Find the Nim-values for this game for all positions of the form $[a, b]$ with $a \leq 7$ and $b \leq 7$.

(b) Find a good move from the position $[4, 6]$.

17. Modify the rules of the game in Problem 3 so that the first player unable to move *wins*. Calculate the new Nim-values. (*Hint*: You will have to make a slight modification of the game's graph.)

18. Give an example of an impartial game, and a starting position, such that whether the first player unable to move is declared the *loser* or the *winner*, in either case the first player *loses*.

19. Repeat Problem 18, but now the first player *wins*.

20. Consider the graph of Figure 8.10, again defining an impartial game, but now allowing *double moves*. For example, it is now possible to move from vertex A to B, C, or E. Find the new Nim-values.

21. The **White Knight's game** is, like Wythoff's game, played on a chessboard using only one piece, in this case a knight. From a given position, the knight is permitted in general any one of four moves, as shown in the accompanying figure. The player who finds himself unable to move (because the knight is on one of the four shaded squares) loses. Find the Nim-values for each of the possible 64 positions.

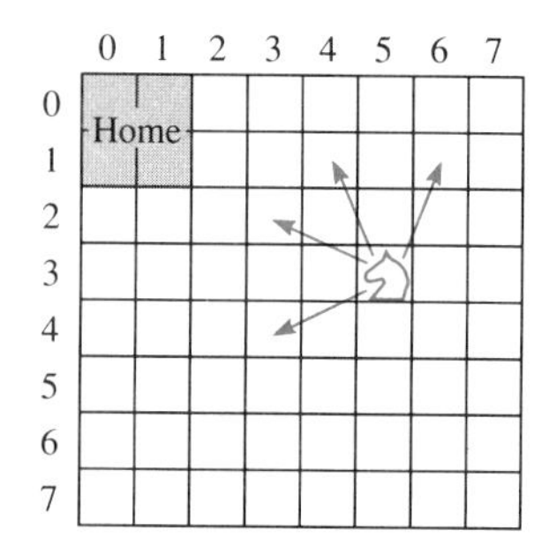

22. In Problem 21, find a good move (if there is one) from the position shown in the figure.

23. The **White Rook's game** is, like Wythoff's game, played on a chessboard using only one piece, in this case a rook. From a given position, the rook is permitted to move any number of squares *north* or *west* (but not both), as indicated in the accompanying figure. The player who finds himself unable to move (because the rook is on the shaded square) loses.

(a) Find the Nim-values for each of the 64 possible positions.

(b) Find (if possible) a good move from the position shown in the figure.

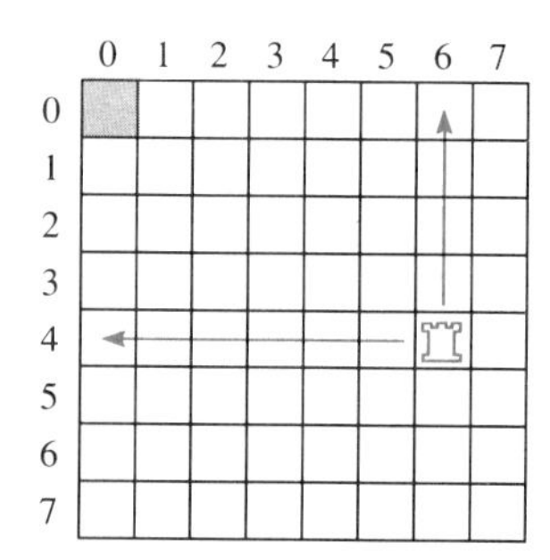

24. The White Rook's game is very closely related to *two-pile Nim*. Explain this relationship—and thereby find a short way to compute the Nim-values.

25. Consider the accompanying directed graph, which is almost identical to the graph of Figure 8.10. However, when the SG algorithm of Figure 8.12 is applied to it, the algorithm *fails*. Demonstrate this, and explain what has happened.

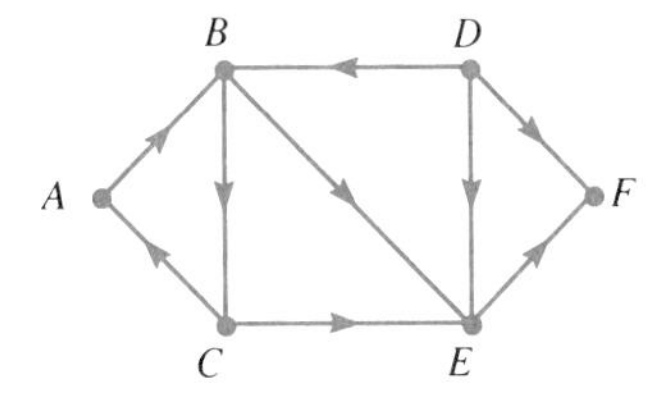

26. If P is a position in an impartial game, and $g(P)$ denotes its Nim-value, show that $g(P) \leq |\text{Opt}(P)|$.

27. Consider the graph in the accompanying figure. Show that *both* of the indicated assignments of "Nim-values" satisfy the defining property of Equation (8.16).

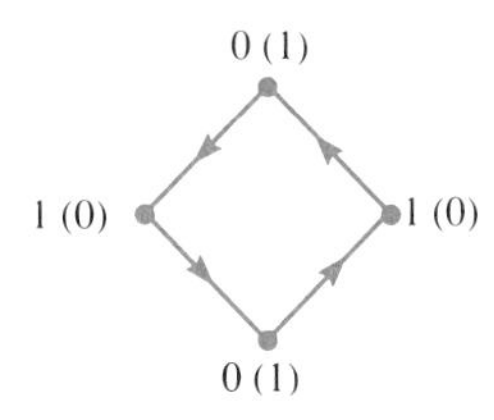

28. We proved in the text that no matter which choices are made in step 2 of the SG algorithm (Figure 8.8), the same Nim-values are obtained; in other words, Nim-values are unique. Do the results of Problem 27 contradict this assertion? Explain.

29. Apply the SG algorithm to *Nim itself*, as follows: Begin with the *endgame*, which, of course, has Nim-value 0, and using the SG algorithm as described in Figure 8.8 (not Figure 8.12!) work backward until you have found the Nim-value for all three-pile Nim-games $[n_1, n_2, n_3]$ with $n_1 + n_2 + n_3 \leq 5$.

30. Modify the SG algorithm, as depicted in Figure 8.12, so that, if it is applied to a game which is impartial but *does not* satisfy the ending condition (8.15), this fact will be detected. Apply your modified algorithm to the following game on a graph, and thereby demonstrate that it doesn't satisfy (8.15).

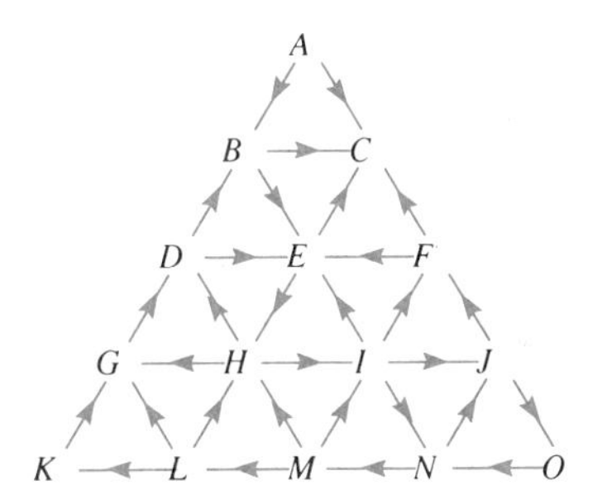

8.3 Some More Nim-like Games

Let us summarize what we have done so far in this chapter. In Section 8.1, we introduced the game of Nim, and described a winning strategy in terms of the Nim-sums of the number of counters in each pile. In Section 8.2, we introduced the family of impartial games (including Nim as a special case), and described a winning strategy for a general impartial game in terms of the Nim-values of the positions. These Nim-values can be calculated using the Sprague-Grundy algorithm. Finally, at the end of Section 8.2 we showed that the Nim-values for Nim itself are in fact the Nim-sums introduced in Section 8.1, so that a circle is closed.

In principle, the theory presented in Sections 8.1 and 8.2 will enable us to find good moves in any impartial game. In practice, however, we have much to learn. (For example, what is a good move from the Strip Dominos position in Figure 8.5?) In this section, we will refine our game-playing skills considerably. The main tool we shall use is the notion of a *sum* of impartial games, which we now describe.

Suppose we are given two different impartial games, for example, the two games shown in Figure 8.13, i.e., a one-pile Nim game with three counters and Wythoff's game with the queen in position (4, 6). We already know how to play both of these games *separately*: In the one-pile Nim game the only good move

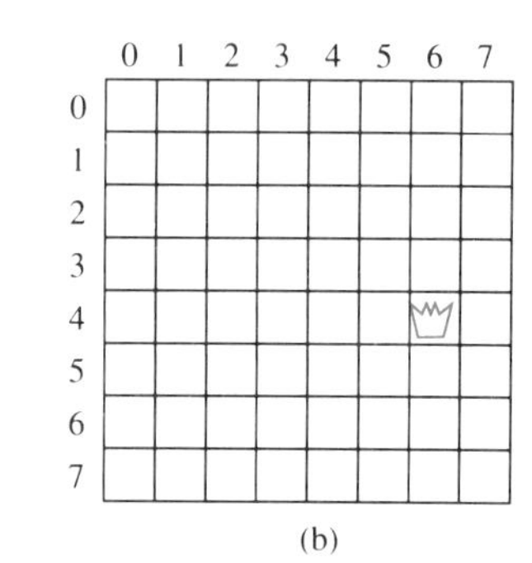

Figure 8.13 A sum of two games. When called on to move a player must move in (a) or (b), but not both.

is to remove all three counters, and in Wythoff's game the only good move is to the position (3, 5). But now suppose we are asked to play the games not *separately* but *simultaneously*, in the sense that when it is our turn to move we must make a move in one or the other of the two games, but not both, it being understood that the game ends when it is no longer possible to move in either component game. The game formed in this way is called the **sum*** of the two original games.

This new game is plainly an impartial game in the sense defined in Section 8.2, and so we could in principle find a good move by using the Sprague-Grundy algorithm to find the Nim-value of the given position, etc. But it's really not necessary to go to all that trouble, if we use the handy **Bogus Nim principle**. What this says is that since we already know that the Nim-value of the position in Figure 8.13(b) is 9 (from Figure 8.9), we should simply treat this position in Wythoff's game *as if it were a Nim-pile with nine counters.* If we do this, the unfamiliar and apparently difficult game of Figure 8.13 becomes the familiar and easy two-pile Nim game of Figure 8.14.

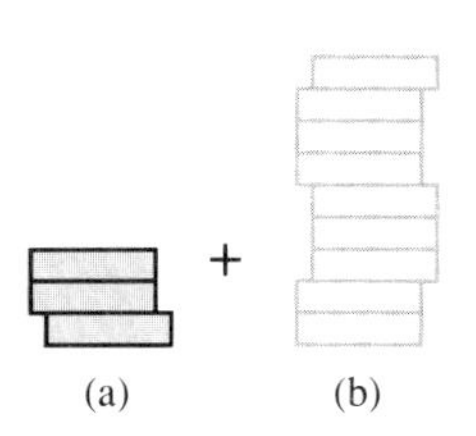

Figure 8.14 A simplified view of the game in Figure 8.13 (thanks to the Bogus Nim principle).

Now to find a good move in the game in Figure 8.13, we reason as follows: The only good move in the game of Nim depicted in Figure 8.14 is to reduce the 9-counter pile to a 3-counter pile. In order to implement this move in the original game of Figure 8.13, we need to find an option of the position in Figure 8.13(b) with Nim-value 3. We are *guaranteed* to be able to find such an option, since the Nim-value 9 is the mex of the Nim-values of the options of the given position, and so all numbers smaller than 9 must appear as option Nim-values. And indeed if we examine Figure 8.9, we see that by moving the queen from the given position (4, 6) to either of the positions (4, 2) or (2, 4), the Nim-value will be reduced to 3. Hence if we move the queen to say (4, 2), the effect is the same as if we had removed 6 counters from the Bogus Nim-pile of Figure 8.14(b) (see Figure 8.15). The resulting position is equivalent to the Nim-position [3, 3], which we know is a Zero position ($3 \stackrel{*}{+} 3 = 0$) (Figure 8.16).

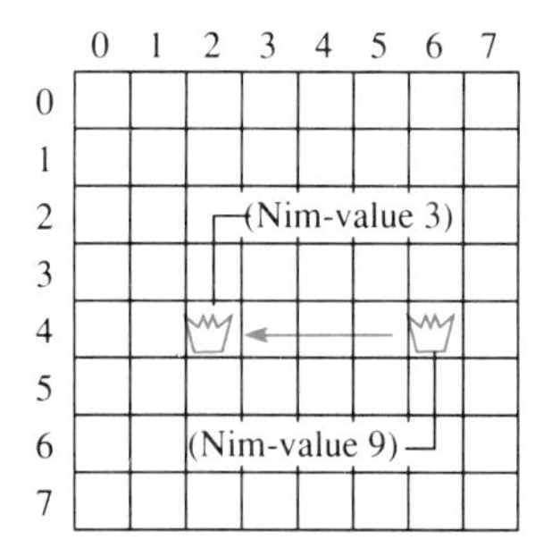

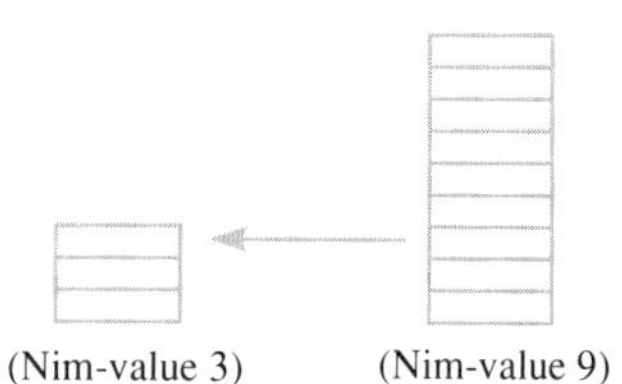

Figure 8.15 Moving the queen from (4, 6) to (4, 2) is the same as reducing the Bogus Nim-pile from 9 to 3 counters.

Just to be sure that the position in Figure 8.16 *is* a Zero position, though, let's check that all its options are Fuzzy. If we remove counters from the *real* Nim-pile of size 3, the new Bogus Nim-position will be one of [2, 3], [1, 3], or [0, 3], all of which are Fuzzy. On the other hand, if we move in the *Bogus* Nim-pile (i.e., if we move the queen), the Nim-value of the queen will be changed from 3 to one of the numbers in the set {0, 1, 2, 4, 5} (refer to Figure 8.9 again), and

* In some texts, the term *disjunctive compound* is used instead of *sum*.

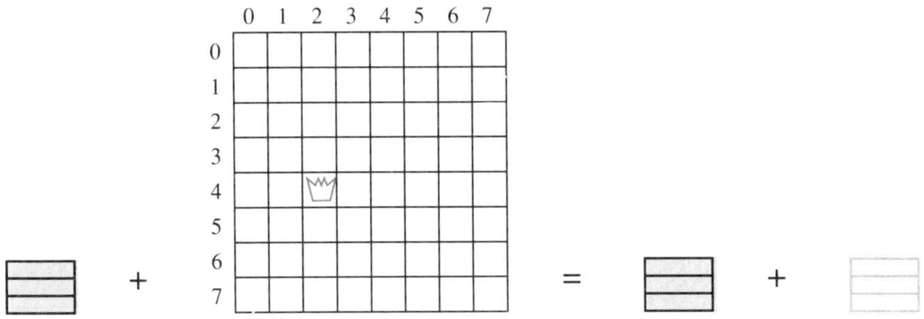

Figure 8.16 After the move depicted in Figure 8.15, the position is equivalent to the Zero Nim-position [3, 3].

so the new Bogus Nim-position will be one of [3, 0], [3, 1], [3, 2], [3, 4], or [3, 5], all of which are again Fuzzy positions.

Notice, however, that in one sense the game of Bogus Nim differs startlingly from ordinary Nim, in that a legal move may actually *increase* the size of one of the Bogus Nim-piles! For example, moving the queen in Figure 8.16 from (4, 2) to (4, 1) changes the Bogus Nim-position from [3, 3] to [3, 5]. Still, this move does the player who makes it no good, since the resulting position is fuzzy, and moving the queen from (4, 1) to (3, 0) will make the position Zero again (equivalent to the Nim-position [3, 3]).

The Bogus Nim principle (BNP) can be applied to the sum of any number of impartial games. Here by the *sum* of a number of *component* games, say $G_1, G_2, \ldots$, we mean a compound game, denoted by $G_1 + G_2 + \cdots$, which is played as follows: There are two players, who alternate turns. The player whose turn it is to move must choose one of the component games, and make a legal move in that component. The game ends when one of the players cannot find a move because an end position has been reached in each component game. This player is the loser. The following example will give us more practice in applying the BNP to the sum of a number of impartial games.

EXAMPLE 8.10 **White Queens** is a game played with several queens on an ordinary chessboard. Any number of queens can be on the same square, and the player whose turn it is must move one of the queens as in Wythoff's game. When all of the queens have reached the home square, the game ends with the player whose turn it is losing. Analyze the position in Figure 8.17, and find (if possible) a good move.

SOLUTION Because the queens move independently, this game is the sum of eight copies of Wythoff's game. We therefore apply the BNP and view each queen as a Nim-pile whose size equals the Nim-value of the queen's position. We have referred to Figure 8.9 again and written the Nim-value on each of the queens. The Nim-value of the position in Figure 8.17 is thus

$$3 \overset{*}{+} 3 \overset{*}{+} 4 \overset{*}{+} 2 \overset{*}{+} 6 \overset{*}{+} 0 \overset{*}{+} 0 \overset{*}{+} 5 = 5.$$

(We could do this calculation mentally by noticing that the two 3s cancel and

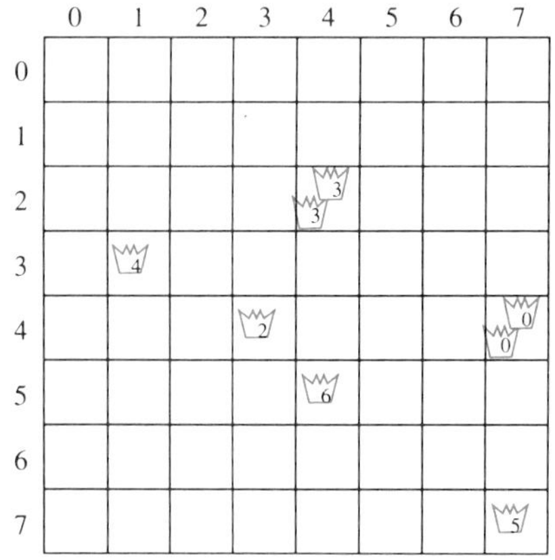

Figure 8.17 Eight White Queens.

that $4 \stackrel{*}{+} 2 \stackrel{*}{+} 6 = 0$.) It follows that if we reduce the Bogus Nim-pile of size 5 to 0, the resulting position will be a Zero position. We can plainly accomplish this reduction by moving the queen at $(7, 7)$ to $(4, 7)$. (There are a number of other good moves; see Problem 1.) ■

We have seen that the Bogus Nim principle can be very useful when we are playing a sum of impartial games, but until now our treatment of it has been rather informal. In fact, though, the BNP can be stated as a precise mathematical theorem about the Nim-values of a sum of games. This theorem says, for example, that the Nim-value of the compound game depicted in Figure 8.13 is $3 \stackrel{*}{+} 9 = 10$, exactly as suggested in Figure 8.14. Similarly, the BNP says that the Nim-value of the position in Figure 8.17 is

$$3 \stackrel{*}{+} 3 \stackrel{*}{+} 4 \stackrel{*}{+} 2 \stackrel{*}{+} 6 \stackrel{*}{+} 0 \stackrel{*}{+} 0 \stackrel{*}{+} 5 = 5,$$

as we assumed it was in our solution to Example 8.10. Before we discuss any more specific games, let's give the statement and proof of this theorem. We begin by reviewing (somewhat formally) the definition of the sum of a number of impartial games.

If $\{G_1, G_2, \ldots, G_m\}$ is any set of impartial games, the **sum** of these games, denoted by $G = G_1 + G_2 + \cdots + G_m$, is another impartial game defined as follows: The **positions** in G are the lists of the form $P = (P_1, P_2, \ldots, P_m)$, where each P_i is a position in the corresponding game G_i. The **options** of such a position P are the positions of the form $(P_1', P_2, \ldots, P_m)$, $(P_1, P_2', \ldots, P_m), \ldots, (P_1, P_2, \ldots, P_m')$ where each P_i' is an option of the corresponding position P_i.

We know from the theory developed in Section 8.2 that each of the component games $G_1, G_2, \ldots, G_m$ has a unique set of Nim-values; we denote the function which produces the Nim-values in G_i by g_i. Thus for example the Nim-value of the position P_2 in G_2 is $g_2(P_2)$, the Nim-value of the position P_3' in G_3 is $g_3(P_3')$, and so on.

It follows from this definition, and our earlier informal discussions, that the new game $G_1 + G_2 + \cdots + G_m$ is also an impartial game, and so it too will have a unique set of Nim-values. We will denote the Nim-value of the position $P = (P_1, P_2, \ldots, P_m)$ by $g(P_1, P_2, \ldots, P_m)$. The BNP identifies these Nim-values as the Nim-sums of the component Nim-values:

(8.23) $$g(P_1, P_2, \ldots, P_m) = g_1(P_1) \overset{*}{+} g_2(P_2) \overset{*}{+} \cdots \overset{*}{+} g_m(P_m).$$

To prove the BNP it is sufficient to show that the numbers $g_1(P_1) \overset{*}{+} g_2(P_2) \overset{*}{+} \cdots \overset{*}{+} g_m(P_m)$ satisfy properties (a), (b), and (c) of the uniqueness theorem for Nim-values in Section 8.2. This isn't hard, if we keep in mind the fact that each of the component "Grundy functions" $g_1, g_2, \ldots, g_m$ satisfies these conditions also. Here we go:

(a) If P is an end position, then so is each P_i; therefore $g_i(P_i) = 0$ for all i, so that the Nim-sum of the values $g_i(P_i)$ is certainly zero.
(b) A move from P to P' involves choosing one of the component games G_i and moving from P_i to P'_i in that game. But then $g_i(P'_i) \neq g_i(P_i)$, and it follows from property (8.6′) in Section 8.1 that the Nim-sum must change.
(c) If n is a nonnegative number less than the Nim-sum $g_1(P_1) \overset{*}{+} g_2(P_2) \overset{*}{+} \cdots \overset{*}{+} g_m(P_m)$, then by property (8.7) of Section 8.1, the Nim-sum can be changed to n by reducing one of the summands, say by reducing $g_1(P_1)$ to n_1. But since the component Nim-value g_1 satisfies condition (c), P_1 will have an option P'_1 whose Nim-value is n_1, and so there will be a legal move in the sum game that changes the Nim-sum to n.

That concludes our discussion of the *theory* of impartial games. In the rest of this section, we will *apply* the theory to a few more specific games. The first game we will consider is the game of Strip Dominos which we introduced in Section 8.2.

Recall that Strip Dominos is played on a long strip of paper which has been ruled into a number of squares, and that the two players alternately place dominos on the strip until no more can be placed. In Figure 8.5 we saw a position in Strip Dominos in which the original strip was of length 17, after two dominos had been played. We repeat that figure as Figure 8.18(a) here. In order to make a good move from the position in Figure 8.18(a), we need to compute its Nim-value. How can we do this? The trick is to notice that the somewhat complicated-looking position in Figure 8.18(a) is actually the *sum* of the three simpler positions shown in Figure 8.18(b). It follows then by the BNP that the Nim-value of the position in Figure 8.18(a) is the Nim-sum of the Nim-values of the three positions in Figure 8.18(b). In the same way, we can see that *any* position in Strip Dominos can be decomposed into a sum of blank strips of various lengths, so that we can completely solve the game if only we can find the Nim-values of blank strips of lengths $1, 2, 3, \ldots$. This we shall now do.

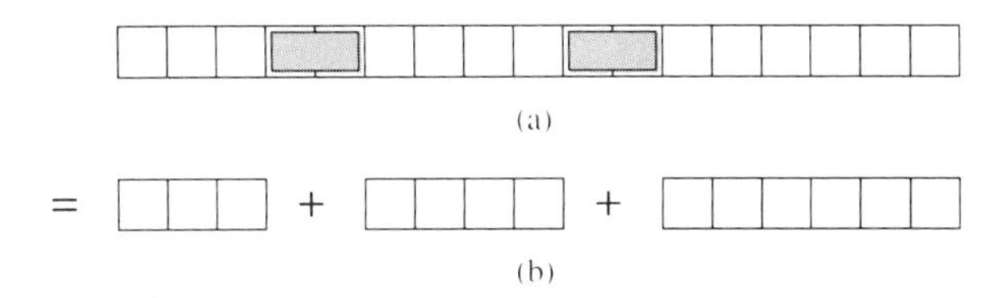

Figure 8.18 Two views of a position in Strip Dominos.

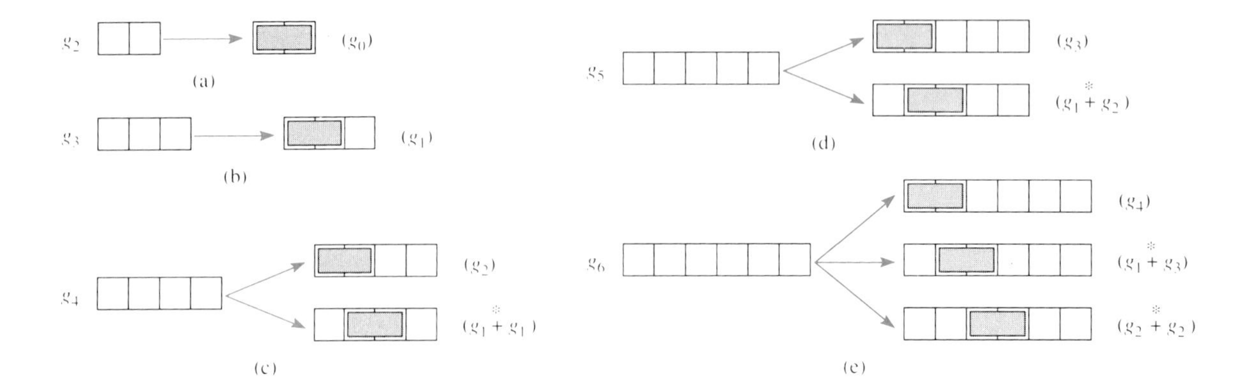

Figure 8.19 Computing (a) g_2, (b) g_3, (c) g_4, (d) g_5, and (e) g_6.

So let's denote the Nim-value of a blank strip of length n by g_n, and try to find the first few values of g_n, using the fundamental property of Nim-values (Equation 8.16), viz.

(8.24) $$g_n = \text{mex}\{\text{Nim-values of the options of } \overbrace{\square\square\square\square\square\square\square\square}^{n}\}.$$

Here we go: The first value, or rather the *zeroth* value, g_0, is plainly 0, since from a strip of length 0 (i.e., an empty strip) no moves are possible, so from (8.24)

$$g_0 = \text{mex}\{\varnothing\} = 0.$$

Similarly g_1 is zero since also from a strip of length 1 no moves are possible:

$$g_1 = \text{mex}\{\varnothing\} = 0.$$

Next, to calculate g_2 we notice that from a strip of length 2, there is only one possible move—to a strip of length 0 (Figure 8.19a), and so

$$g_2 = \text{mex}\{g_0\} = \text{mex}\{0\} = 1.$$

From a strip of length 3, the only move is to a strip of length 1 (Figure 8.19b), and so

$$g_3 = \text{mex}\{g_1\} = \text{mex}\{0\} = 1.$$

From a strip of length 4, there are *two* distinct options: a single strip of length 2, or two separate strips of size 1 (Figure 8.19c), and so

$$g_4 = \text{mex}\{g_2, g_1 \overset{*}{+} g_1\} = \text{mex}\{1, 0 \overset{*}{+} 0\} = \text{mex}\{1, 0\} = 2.$$

Continuing this way, and using Figure 8.19 for guidance, we compute

$$g_5 = \text{mex}\{g_3, g_1 \overset{*}{+} g_2\} = \text{mex}\{1, 0 \overset{*}{+} 1\} = 0$$

and

$$g_6 = \text{mex}\{g_4, g_1 \overset{*}{+} g_3, g_2 \overset{*}{+} g_2\} = \text{mex}\{2, 1, 0\} = 3.$$

Continuing in this way (a computer helps!), we arrive at Table 8.7.

There is not much of a pattern to the numbers in Table 8.7* but with its help you will be able to beat any opponent (who hasn't read this book!).

* Actually, there is a pattern. The Nim-values *repeat themselves* starting with g_{35}. Thus $(g_1, g_2, g_3, \ldots) = (g_{35}, g_{36}, g_{37}, \ldots) = (0, 1, 1, 2, 0, \ldots)$.

TABLE 8.7 The first 48 Nim-values for Strip Dominos

n	g_n	n	g_n	n	g_n
1	0	17	2	33	7
2	1	18	2	34	4
3	1	19	3	35	0
4	2	20	3	36	1
5	0	21	0	37	1
6	3	22	1	38	2
7	1	23	1	39	0
8	1	24	3	40	3
9	0	25	0	41	1
10	3	26	2	42	1
11	3	27	1	43	0
12	2	28	1	44	3
13	2	29	0	45	3
14	4	30	4	46	2
15	0	31	5	47	2
16	5	32	2	48	4

EXAMPLE 8.11 Find (if possible) good moves from each of the following Strip Dominos positions.

(a)

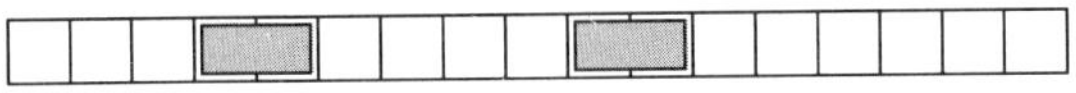

(b)

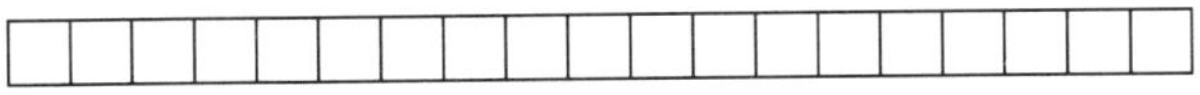

(c)

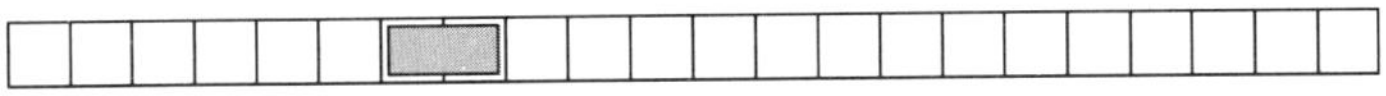

SOLUTION

(a) (Note that this is the position from Figures 8.5 and 8.18.) The Nim-value for this position is

$$g_3 \stackrel{*}{+} g_4 \stackrel{*}{+} g_6 = \text{(using Table 8.7) } 1 \stackrel{*}{+} 2 \stackrel{*}{+} 3 = 0.$$

Oops! This is a *Zero* position; our best strategy is to courteously offer our opponent the first move!

(b) From Table 8.7, we see that the Nim-value of this position is $g_{19} = 3$, so the position is Fuzzy, and there ought to be a good move. From the given position, the options have Nim-values g_{17} (if we place a domino at the extreme end of the strip), $g_1 \stackrel{*}{+} g_{16}$ (if we place the domino one square away from the end of the strip), etc. Here are the Nim-values of all possible options:

$$g_{17} = 2, \qquad g_5 \overset{*}{+} g_{12} = 0 \overset{*}{+} 2 = 2,$$
$$g_1 \overset{*}{+} g_{16} = 0 \overset{*}{+} 5 = 5, \qquad g_6 \overset{*}{+} g_{11} = 3 \overset{*}{+} 3 = 0\ (!),$$
$$g_2 \overset{*}{+} g_{15} = 1 \overset{*}{+} 0 = 1, \qquad g_7 \overset{*}{+} g_{10} = 1 \overset{*}{+} 3 = 2,$$
$$g_3 \overset{*}{+} g_{14} = 1 \overset{*}{+} 4 = 5, \qquad g_8 \overset{*}{+} g_9 = 1 \overset{*}{+} 0 = 1.$$
$$g_4 \overset{*}{+} g_{13} = 2 \overset{*}{+} 2 = 0\ (!),$$

It follows that there are two good moves—place the domino either four squares from either end, or six squares from either end.

(c) The position here has Nim-value $g_6 \overset{*}{+} g_{14} = 3 \overset{*}{+} 4 = 7$. If we think of this as the Bogus Nim-position [3, 4], we see that the only good move is to reduce the 4 to a 3. So we must find an option of the strip of length 14 with Nim-value 3. These options have the following Nim-values:

$$g_{12} = 2, \qquad g_4 \overset{*}{+} g_8 = 2 \overset{*}{+} 1 = 3\ (!),$$
$$g_1 \overset{*}{+} g_{11} = 0 \overset{*}{+} 3 = 3\ (!), \qquad g_5 \overset{*}{+} g_7 = 0 \overset{*}{+} 1 = 1,$$
$$g_2 \overset{*}{+} g_{10} = 1 \overset{*}{+} 3 = 2, \qquad g_6 \overset{*}{+} g_6 = 3 \overset{*}{+} 3 = 0.$$
$$g_3 \overset{*}{+} g_9 = 1 \overset{*}{+} 0 = 1,$$

So once again there are two good moves—either place a domino *one* square to the right of the original domino, or *four* squares to the right of it. ■

The next example will give us more practice in computing Nim-values for another typical impartial game.

EXAMPLE 8.12 **Grundy's game** is (like Nim) played with several piles of counters, but in this game the legal move is to select one pile and *divide it into two unequal piles* (Figure 8.20). The game ends when all the piles are of size 1 or 2, since no further division will then be possible. Denote by G_n the Nim-value of a single pile of size n in Grundy's game.

(a) Calculate $G_1, G_2, \ldots, G_{20}$.
(b) Use the results of part (a) to find a good move from the position with pile sizes (4, 19, 12, 10).

SOLUTION

(a) This problem is much like computing the Nim-values for Strip Dominos. Both G_1 and G_2 are 0, since as mentioned no moves are possible from piles of size 1 or 2. The only move from a pile of size 3 is to split it into a pile of size 1 and one of size 2, so

$$G_3 = \text{mex}\{G_1 \overset{*}{+} G_2\} = \text{mex}\{0 \overset{*}{+} 0\} = 1.$$

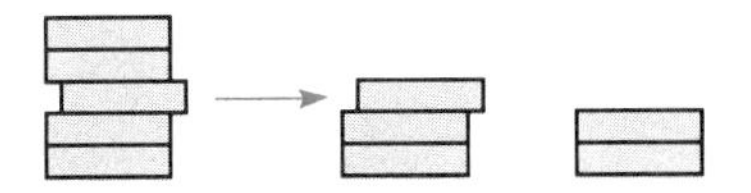

Figure 8.20 A legal move in Grundy's game: Splitting a pile of size 5 into a pile of size 3 and one of size 2.

A pile of size 4 can only be split into (3, 1) [(2, 2) is forbidden since the piles must be of unequal size], and so

$$G_4 = \text{mex}\{G_3 \overset{*}{+} G_1\} = \text{mex}\{1 \overset{*}{+} 0\} = 0.$$

Similarly,

$$G_5 = \text{mex}\{G_4 \overset{*}{+} G_1, G_3 \overset{*}{+} G_2\} = \text{mex}\{0 \overset{*}{+} 0, 1 \overset{*}{+} 0\}$$
$$= \text{mex}\{0, 1\} = 2$$

and

$$G_6 = \text{mex}\{G_5 \overset{*}{+} G_1, G_4 \overset{*}{+} G_2\} = \text{mex}\{2 \overset{*}{+} 0, 0 \overset{*}{+} 0\}$$
$$= \text{mex}\{2, 0\} = 1$$

and

$$G_7 = \text{mex}\{G_6 \overset{*}{+} G_1, G_5 \overset{*}{+} G_2, G_4 \overset{*}{+} G_3\} = \text{mex}\{1 \overset{*}{+} 0, 2 \overset{*}{+} 0, 0 \overset{*}{+} 1\}$$
$$= \text{mex}\{1, 2, 1\} = 0.$$

Continuing in this way, we obtain Table 8.8.

TABLE 8.8 The first 20 Nim-values for Grundy's game

n	G_n	n	G_n
1	0	11	2
2	0	12	1
3	1	13	3
4	0	14	2
5	2	15	1
6	1	16	3
7	0	17	2
8	2	18	4
9	1	19	3
10	0	20	0

(b) From Table 8.8, we calculate the Nim-value of the Grundy game position (4, 19, 12, 10) as

$$G_4 \overset{*}{+} G_{19} \overset{*}{+} G_{12} \overset{*}{+} G_{10} = 0 \overset{*}{+} 3 \overset{*}{+} 1 \overset{*}{+} 0 = 2.$$

The only good move from the Bogus Nim-position [0, 3, 1, 0] is to [0, 1, 1, 0], and so we must find an option of the Grundy pile of size 19 with Nim-value 1. Here are the options from a single pile of size 19, and their Nim-values:

(18, 1): $G_{18} \overset{*}{+} G_1 = 4 \overset{*}{+} 0 = 4$,

(17, 2): $G_{17} \overset{*}{+} G_2 = 2 \overset{*}{+} 0 = 2$,

(16, 3): $G_{16} \overset{*}{+} G_3 = 3 \overset{*}{+} 1 = 2$,

(15, 4): $G_{15} \overset{*}{+} G_4 = 1 \overset{*}{+} 0 = 1$ (!),

(14, 5): $G_{14} \overset{*}{+} G_5 = 2 \overset{*}{+} 2 = 0$,

(13, 6): $G_{13} \overset{*}{+} G_6 = 3 \overset{*}{+} 1 = 2$,

(12, 7): $G_{12} \overset{*}{+} G_7 = 1 \overset{*}{+} 0 = 1$ (!),

(11, 8): $G_{11} \overset{*}{+} G_8 = 2 \overset{*}{+} 2 = 0$,

(10, 9): $G_{10} \overset{*}{+} G_9 = 0 \overset{*}{+} 1 = 1$ (!).

This calculation shows that there are *three* good moves, viz. to split the 19-pile into (15, 4), (12, 7), or (10, 9). ■

We conclude this section by discussing one last relative of Nim, the popular game of **Hackenbush**. Ideally, Hackenbush should be played in your neighbor's rose garden using a pair of pruning shears, but it can, if necessary, be played using pencil and paper with a drawing like the one in Figure 8.21. The drawing

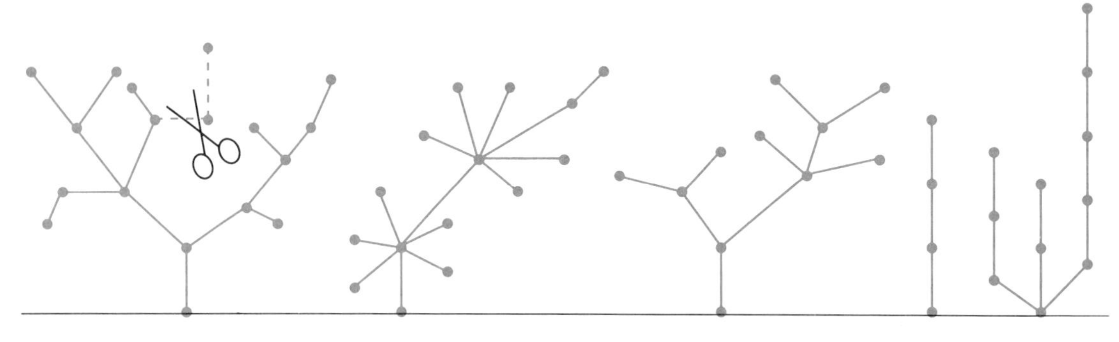

Figure 8.21 Anyone for Hackenbush?

consists of a dotted horizontal line (called the **ground**), and a number of **trees** (in the graph-theoretic sense of Chapter 3), each connected to the ground by a single **ground vertex**. The legal move is to *chop* one edge of one of the trees, which then causes that edge and all other edges no longer connected to the ground to disappear. In the position of Figure 8.21, for example, if the indicated edge is chopped, the two dotted edges will disappear. The game ends as usual when some player is unable to move because all of the edges have disappeared.

How can we find a good move from the position of Figure 8.21? Since the game is plainly the *sum* of the games played with each of the five individual trees, the BNP tells us that the Nim-value of the position in Figure 8.21 is the Nim-sum of the Nim-values of each of the component trees. So to compute the overall Nim-value, we must first be able to compute the five component Nim-values.

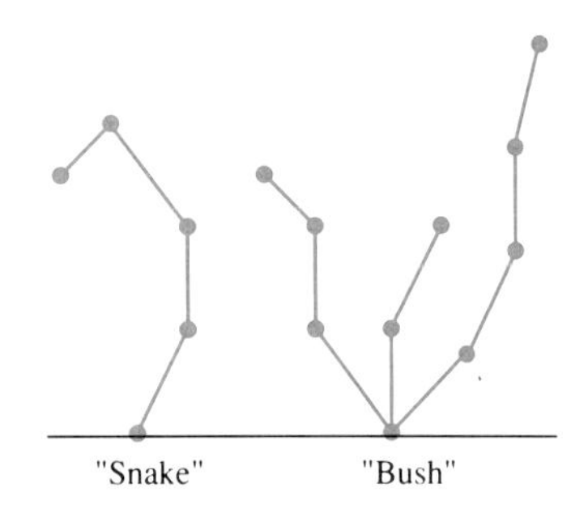

Figure 8.22 Two simple Hackenbush trees.

There is a simple rule for computing the Nim-value of a Hackenbush tree, which can best be explained by first considering two very simple kinds of trees, called **snakes** and **bushes** (see Figure 8.22). A snake is just a chain of edges with one end touching the ground, and a bush is just a number of snakes joined together at the ground. A Hackenbush snake is really just the same as a Nim-pile of the same size, and indeed if snakes were the only kind of tree available, Hackenbush would be no different from Nim (see Figure 8.23). Bushes aren't much harder, since despite the fact that the snakes are connected at the ground, players chop the snakes independently, so we can disconnect them, once again seeing an immediate equivalence to Nim (Figure 8.24). A bush which consists

Figure 8.23 Snake Hackenbush = Nim.

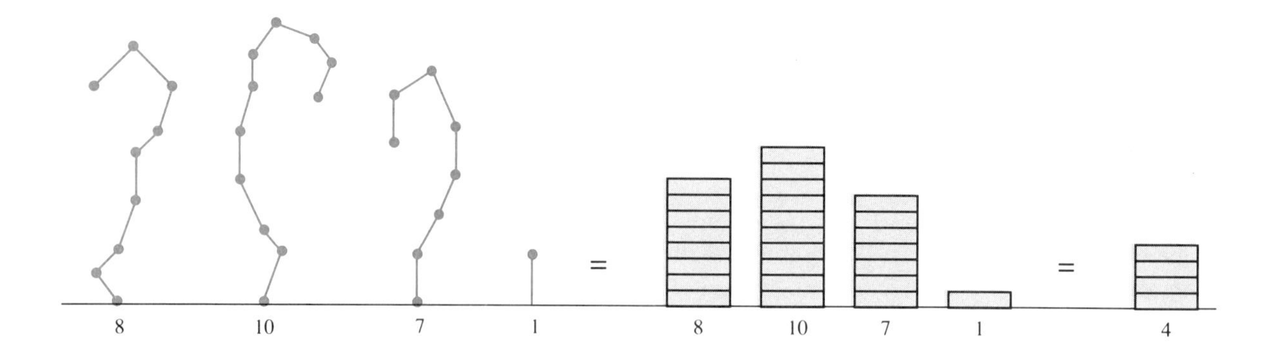

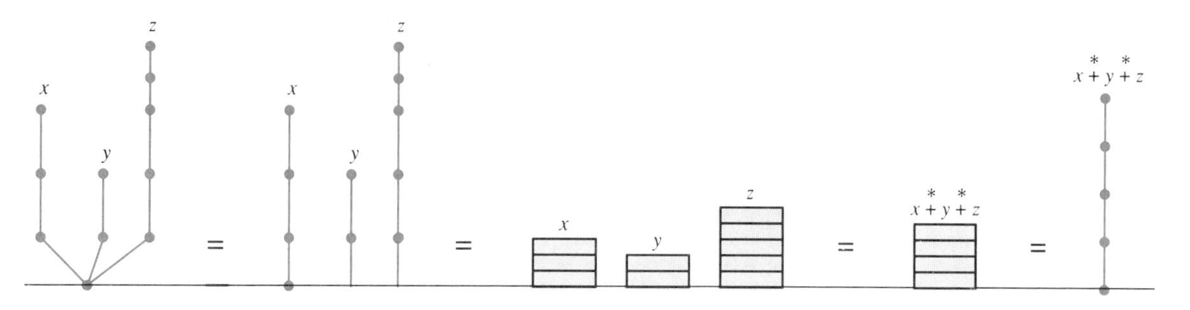

Figure 8.24 How to simplify a Hackenbush bush.

of snakes of lengths $n_1, n_2, \ldots$ is equivalent to a single snake of length $n_1 \overset{*}{+} n_2 \overset{*}{+} \cdots$.

It turns out that knowing only about snakes and bushes, we can compute the Nim-value of any tree. Consider, for example, the tree in Figure 8.25(a). Inspecting this tree carefully, we see that it consists of several bushes "glued together" appropriately. For example, there is a bush consisting of two snakes of length 1 growing from vertex A. According to the principle illustrated in Figure 8.24, this bush is equivalent to a snake of length $1 \overset{*}{+} 1 = 0$, and in Figure 8.25(b) we have replaced it with this (harmless) snake. Similarly, the bush growing from vertex B is equivalent to a snake of length $3 \overset{*}{+} 2 = 1$, and the bush growing from vertex C is equivalent to a snake of length $2 \overset{*}{+} 1 = 3$. When these three bushes have all been replaced by the corresponding snakes, the tree of Figure 8.25(b) results. In this tree, we can see a bush growing from vertex D, which is equivalent to a snake of length $2 \overset{*}{+} 2 \overset{*}{+} 4 = 4$. Replacing the bush with the equivalent snake, we get the tree of Figure 8.25(c). But this tree is just a snake of length 5, which has Nim-value 5. It follows then that the original tree, too, has Nim-value 5.

When computing Nim-values for Hackenbush trees, it's a nuisance to have to keep redrawing the pictures, and so in Figure 8.25(d) we have indicated a way to do the bookkeeping without having to erase anything. The idea is to attach a number to each edge which represents the Nim-value of the part of the tree consisting of the given edge, and all edges *above it.* We call these numbers the *edge labels*, and they can be calculated from the top down. We start by labelling the edges in the three bushes growing at vertices A, B, and C. Each of these

Figure 8.25 Computing the Nim-value of a tree, using the bush principle.

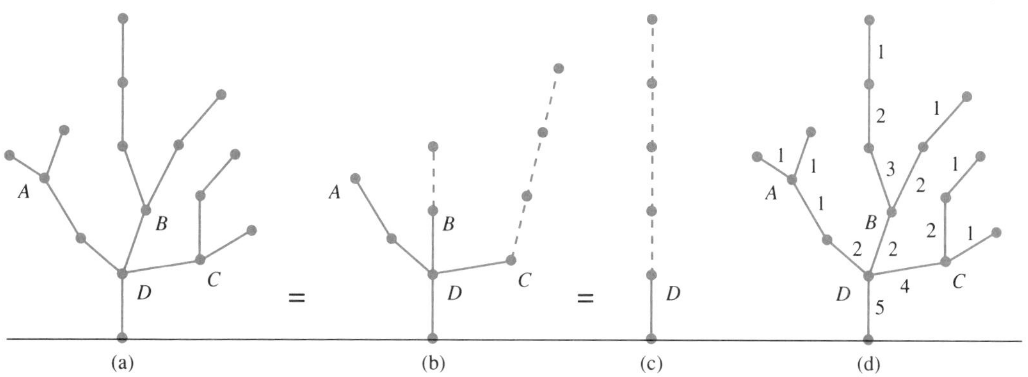

bushes consists of a number of snakes, and the edges in a snake are labelled $1, 2, 3, \ldots$, starting at the top.

Next, we assign labels to the edges just below the vertices A, B, and C. These values are each equal to *one more* than the length of the snake above the corresponding vertex in Figure 8.25(b); alternatively, and more conveniently, they are equal to one more than the Nim-sum of the edge labels just above. (The general rule for computing edge labels is illustrated in Figure 8.26.) Now we label the edge two below vertex A with 2 (the snake is getting longer). Finally, we use the rule of Figure 8.26 again to compute the label of the trunk of the tree: $(2 \overset{*}{+} 2 \overset{*}{+} 4) + 1 = 5$. This final label is the Nim-value of the original tree, as we saw before.

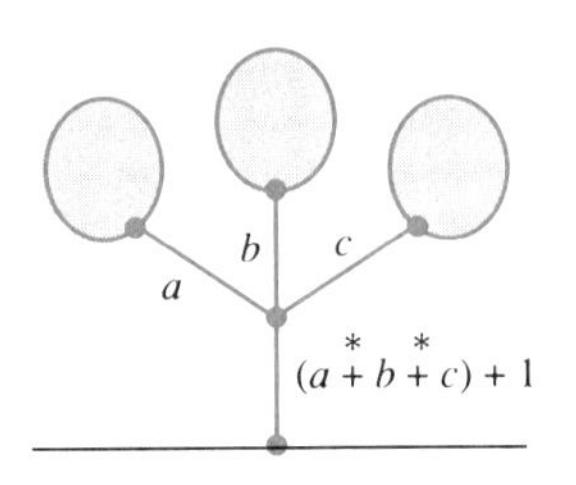

Figure 8.26 The basic rule for computing Hackenbush edge labels.

The final example in this chapter will give us some more practice in computing Nim-values for Hackenbush trees.

EXAMPLE 8.13 Find a good move (if there is one) from the Hackenbush position of Figure 8.21.

SOLUTION Using the techniques just developed, we find that the Nim-values of the component trees are 15, 6, 4, 3, and 4, respectively (see Figure 8.27).

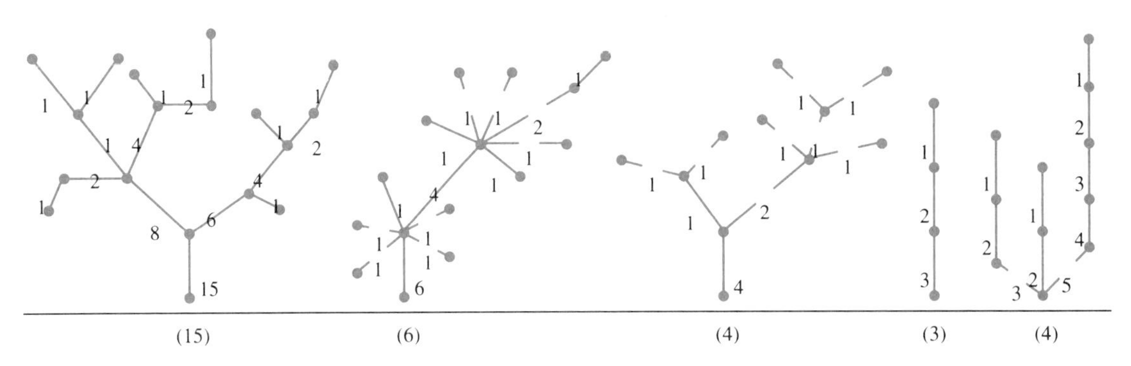

Figure 8.27 The Nim-values of the trees in Figure 8.21.

The overall Nim-value of the position is therefore $15 \overset{*}{+} 6 \overset{*}{+} 4 \overset{*}{+} 3 \overset{*}{+} 4 = 10$. Using the BNP we find that the only good move will be one that replaces the Nim-value 15 with the Nim-value 5. How can we find such a move? If we change the edge label on the trunk of the tree of Nim-value 15 to the desired 5 (see Figure 8.28), we then are forced to change the edge labels just above so that their Nim-sum is 4. Since a single edge chop can only affect one of these labels, what we need to find is a legal move from the Bogus Nim position [8, 6] which changes the Nim-value to 4. The only such move is to reduce the 8 to 2, again as shown in Figure 8.28. But to change the 8 to 2 requires moving from the Bogus Nim-position [2, 1, 4] so that the Nim-value changes to 1. The only such move is to reduce the 4 to 2, as indicated. Finally, to change the label 4 to 2 requires changing the Bogus Nim-value of the position [1, 2] from 3 to 1. This requires changing the 2 to 0, and this, finally, can be accomplished by chopping the edge indicated in Figure 8.28.

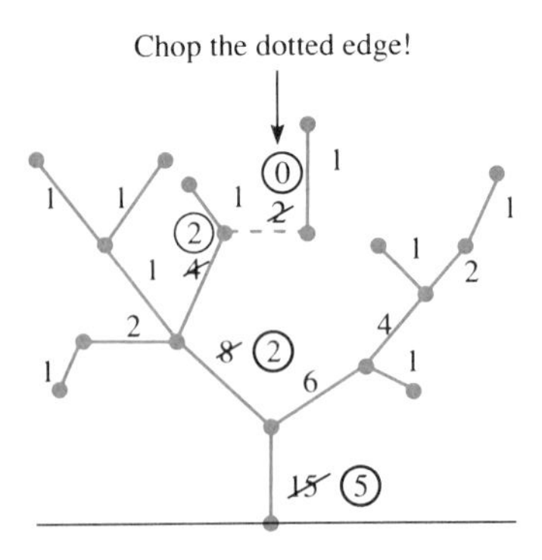

Figure 8.28 Climbing the tree to find the winning move.

So, the *only* good move from the Hackenbush position of Figure 8.21 is to chop the edge shown in Figure 8.28!

Problems for Section 8.3

1. In Example 8.10, we found *a* good move from the White Queens position in Figure 8.17. Find *all* the possible good moves.

In Problems 2–5 find (if possible) a good move from each of the White Queens positions described (we give the row-column addresses of the queens).

2. (2, 2), (2, 3), (2, 4).

3. (5, 5), (6, 6), (7, 7).

4. (7, 5), (4, 3), (5, 2), (4, 4).

5. (3, 3), (3, 3), (3, 3).

6. Consider the game which is the *sum* of the Nim-position in Figure 8.1 and the White Queens position in Figure 8.17. Find (if possible) a good move.

In Problems 7–10, find a good move from the illustrated Strip Dominos positions.

7.

8.

9.

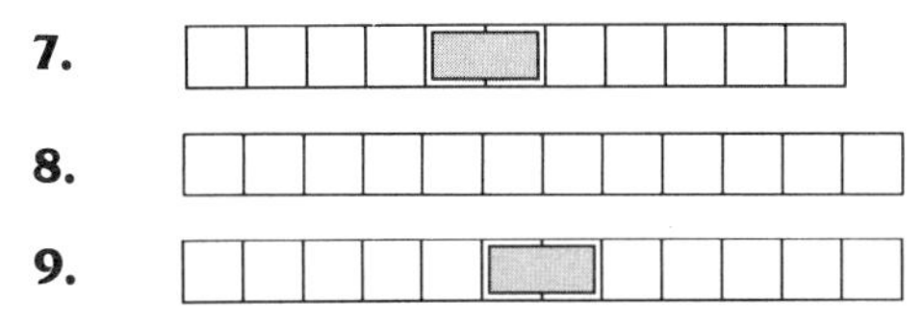

10.

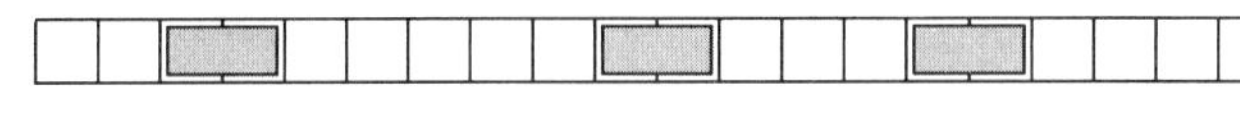

In Problems 11–13, find (if possible) good moves from the indicated positions in Grundy's game.

11. [4, 5, 3, 3]. **12.** [8, 10, 7, 1]. **13.** [17, 18, 19, 20].

14. Verify the values G_8–G_{20} given in Table 8.8.

15. Calculate the Nim-value of the following Hackenbush trees.

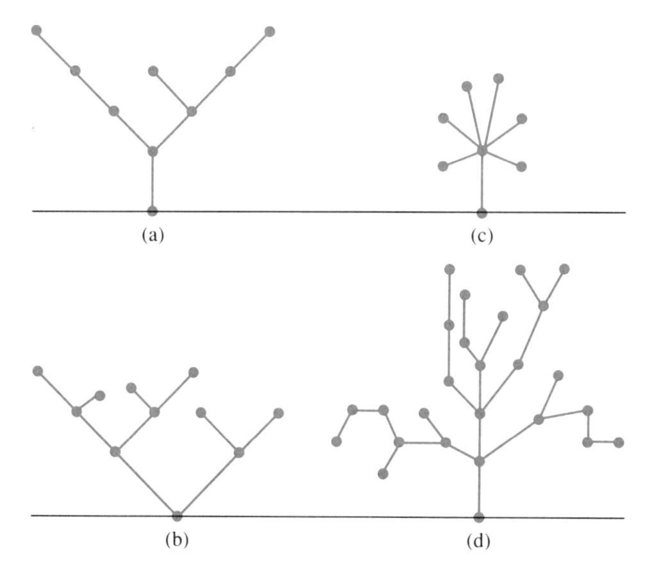

16. Find a good move from the following Hackenbush position.

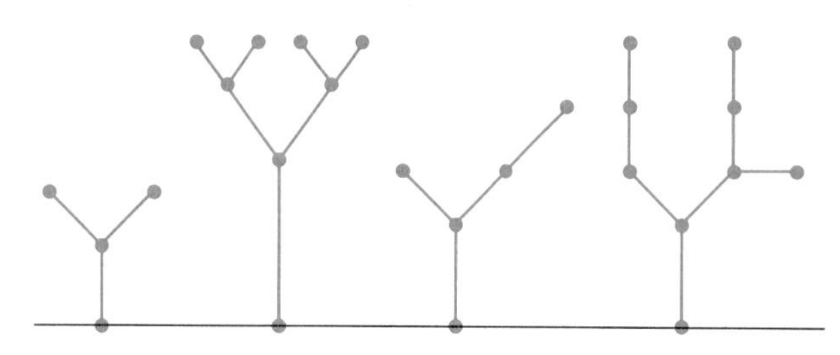

17. In the game of **White Knights** (cf. Problem 8.2.21), a number of knights are placed on a chessboard, with any number of knights being allowed on the same square, and the player whose turn it is to move must move just one of the knights as shown in the figure accompanying Problem 8.2.21. If the knights are on squares (3, 3), (3, 4), (4, 5), and (7, 7), find a good move.

18. In the game of **White Rooks** (cf. Problem 8.2.23), a number of rooks are placed on a chessboard, with any number of rooks being allowed on the same square. The player whose turn it is to move must move one of the rooks any number of squares *northward*, or *westward*, as shown in the figure accompanying Problem 8.2.23. If the rooks are located on the squares (3, 3), (3, 4), (4, 5), and (7, 7), find a good move.

19. Verify the values g_7–g_{20} in Table 8.7.

20. **Strip Trominos** is played just like Strip Dominos, except that the players must use *trominos*, which are 3 × 1 rectangles.

(a) Denoting the Nim-value of a blank Strip Trominos strip of length n by T_n, find $T_0, T_1, \ldots, T_{10}$.

(b) Find a good move from the following position.

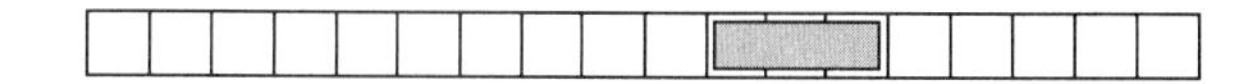

21. Find the next five Nim-values in Grundy's game, i.e., G_{21}–G_{25}.

22. Given a pile of n counters, find a formula for the number of options in Grundy's game.

23. **Dumbo's game** is just like Grundy's game, except that a given pile may be split into two nonempty piles of *any* size.

(a) Let D_n denote the Nim-value of a pile of n counters in Dumbo's game. Find a general formula for D_n.

(b) Where does this game get its name?

24. We might modify the game of Nim by requiring that in any move the number of counters taken from a pile is *at most three*. Let G_n denote the Nim-value of a pile of size n in this game, and compute G_0–G_{10}.

25. Continuing Problem 24, find a general formula for G_n.

26. Continuing Problem 25, find a good move from the position [11, 12, 13, 14, 15].

27. **Northcott's game** is played on a checkerboard with one black and one white piece in each row, as shown in the accompanying figure. There are two players, black and white. When it is your turn, you may move any piece of your own color to any other empty square in the same row, provided you don't jump over your opponent's piece. If you can't move (because all your pieces are pinned at the side of the board), you lose. This game is just Nim in disguise, and the important numbers are the number of spaces *between* the two pieces in each row, which we have written to the left of the board in the figure.

(a) Find a good move in the position shown, assuming you are black.
(b) Repeat part (a), but now you are white.
(c) Explain how you would respond if your opponent (from a Zero position) moved his piece so as to *increase* the distance between the pieces in that row.

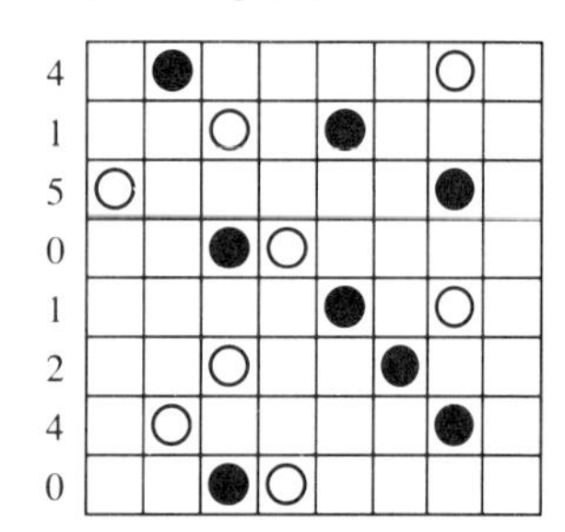

28. The **Silver Dollar game** is played on a long strip of squares with a number of silver dollars, placed on different squares (see the accompanying figure). A legal move is to push one coin leftward, not passing any other coin, to an unoccupied square. The game ends when all the coins are in a traffic jam at the end of the strip. This game is Nim in disguise, and the important numbers are the number of spaces between each pair of coins (the coins are paired up starting from the rightmost coin). We have written these numbers above the strip in the example shown.

(a) Find a good move from the position shown.
(b) Explain how you would respond if your opponent (from a zero position) moved in a way that *increased* one of the numbers.

0 2 3 2

29. The old English game of **Kayles** is played with a number of rows of ninepins (see the accompanying figure). The players are assumed to be so skillful that they can roll a ball so as to knock down any single pin or any two adjacent ones. However, it is physically impossible to knock down pins separated by any greater distance. The players bowl alternately, and the player who knocks down the last pin wins.

(a) Denote by K_n the Nim-value of a single row of n ninepins. Calculate K_1-K_{12}.
(b) Find (if possible) a good move from the given position.

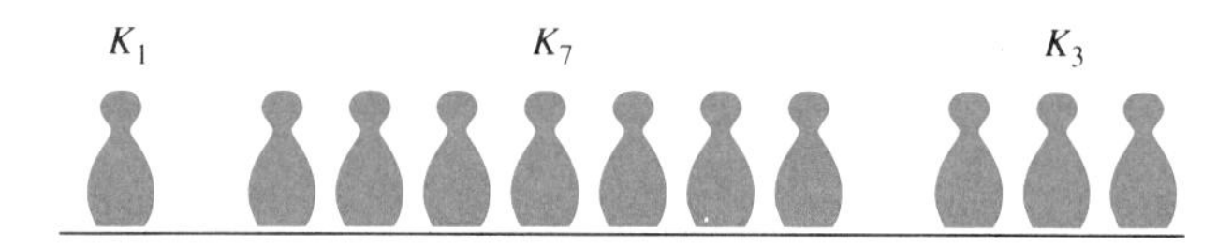

30. The *height* of a Hackenbush tree is the length (measured in edges) of the longest path from the ground to the extremities of the tree. For example, the leftmost tree in Figure 8.27 has height 5. Can you find a shorter tree of Nim-value 15?

31. Write a computer program to verify for $n \leq 100$ the assertion, made in the text, that the Nim-values g_n in Table 8.7 repeat periodically, starting with g_{35}.

32. **Board Dominos** is played on an ordinary chessboard. The player whose turn it is must place a domino horizontally on two adjacent empty squares. As usual, the player unable to move loses. Find a good move from the following Board Dominos position.

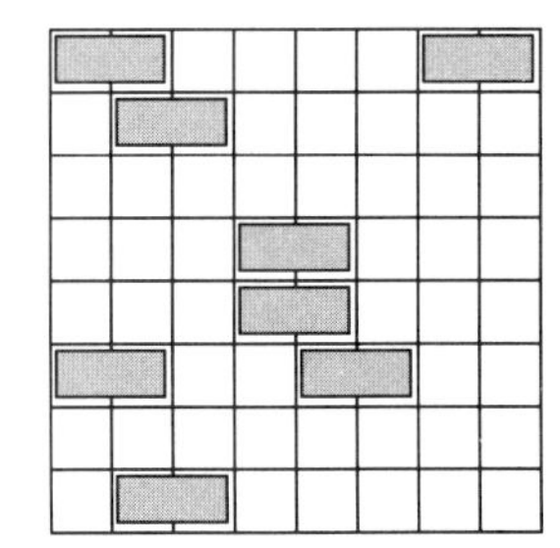

33. **Rims** is played with pencil and paper. The initial position is a figure consisting of numerous *dots* and closed *curves*; dots may appear on the curves. The legal move is to draw a new closed curve which contains at least one of the dots, which does not intersect any of the other curves, and which may contain some of the dots inside it. (See the accompanying figure.) Explain the relationship to Nim, and find a good move from the given position.

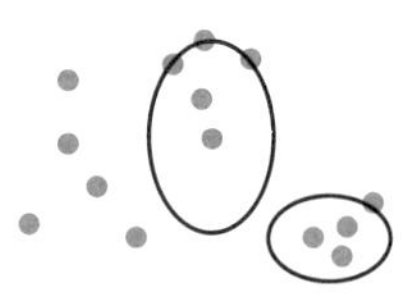

Summary

A position in the game of **Nim**, with piles of sizes $n_1, n_2, \ldots, n_k$, may be analyzed by computing the **Nim-sum** $n_1 \overset{*}{+} n_2 \overset{*}{+} \cdots \overset{*}{+} n_k$. A player faced with a nonzero Nim-sum (a **Fuzzy position**) can make a good move by removing counters from a pile so that the Nim-sum becomes 0 (a **Zero position**). A player faced with a Zero position is forced to create a Fuzzy position, and will lose if the opponent plays correctly. This key idea occurs more generally in any **impartial game**. In such a game, a Nim-value $g(P)$ can be defined for each position P; the basic property of Nim-values is

$$g(P) = \text{mex}\{g(P'): P' \in \text{Opt}(P)\},$$

where Opt (P) is the set of **options** of P and mex stands for **minimal excludent**. The $g(P)$'s may be found systematically using the **Sprague-Grundy algorithm**.

Sums of impartial games may be analyzed by the **Bogus Nim principle**, in which each component game is treated as if it were a single Nim-pile of size equal to the Nim-value of that game.

CHAPTER NINE

DISCRETE PROBABILITY THEORY

In this chapter, we will apply the ideas from some of the earlier chapters to the *theory of probability*, which is the branch of mathematics devoted to the study of uncertain events. Many important phenomena in the real world involve at least some degree of uncertainty (for example, football games, the weather, economics, quantum physics, telecommunications, life expectancy, etc.). The theory of probability allows us to make intelligent decisions in the face of uncertainty, and has become an indispensable tool in modern applied mathematics. Many books have been devoted exclusively to probability theory, and in this chapter we will barely scratch the surface of this important subject. Still, we will introduce the most important concepts in the subject: *probability spaces*, *independent events*, and *random variables*. If you pursue a career in engineering or science, you will surely need to study more about probability, but this chapter will give you a good start.

9.1 Discrete Probability Spaces

If a given event may or may not occur, the *probability* of that event is defined to be the fraction of time it is expected to occur. For example, if we flip a coin, we expect *heads* and *tails* to occur about equally often, and so the events "heads" and "tails" are assigned probability $\frac{1}{2}$ each. Similarly, if you roll one die, each of six "events"

is assigned probability $\frac{1}{6}$. Of course, many objections can now be raised: The coin *could* land on edge; it *might* be unbalanced; the die *could* be loaded; the person rolling the die *might* be cheating; etc. The answer to all such objections is the same: Mathematics cannot deal with a *real* coin or a *real* die at all! It can only deal with an idealized *model* of a coin or die. And to the extent that the idealized model fails to capture the essence of the real object (for example by ignoring the probability that the coin might land on edge), exactly to that extent will the mathematics fail to give us insight about the real object. It is no exaggeration to say that the most important and elusive of all scientific skills is the ability to produce realistic and tractable mathematical models of the real world.

Returning to the coin-flipping experiment, we notice that a real coin has many properties which are wholly irrelevant if we are *flipping* it and not *spending* it! Its denomination, size, color, weight, composition, and design are all unimportant. All that really *is* important, in fact, is that it has two sides (called "heads" and "tails," whatever the pictures really are), and that both occur about equally often. We thus produce a mathematical model for "essence of coin," as shown in Figure 9.1.

(a)

$\{H, T\}$

$\Pr\{H\} = \Pr\{T\} = \frac{1}{2}$

(b)

Figure 9.1 Two kinds of coins: (a) A real coin; (b) a mathematical coin.

In Figure 9.1(b) we see that the model for the coin (or more exactly, the model for *the experiment of flipping a fair coin*) consists of two things: a set with two elements, viz. {H, T}, representing the possible **outcomes** of the experiment (of course H = heads, T = tails), and an assignment of *probabilities* to the elements of the set.

Similarly, in Figure 9.2, we see a mathematical model for the experiment of rolling a fair die. In Figure 9.2(b), again we see that the model consists of a *set*, viz., {1, 2, 3, 4, 5, 6}, and a *probability assignment* on the set; in this case $\Pr\{a\} = \frac{1}{6}$, for $a = 1, 2, 3, 4, 5$, and 6. In Figures 9.1 and 9.2, the models are very simple; but sometimes it takes some serious thinking to make a good model for a probabilistic experiment, as the following example shows.

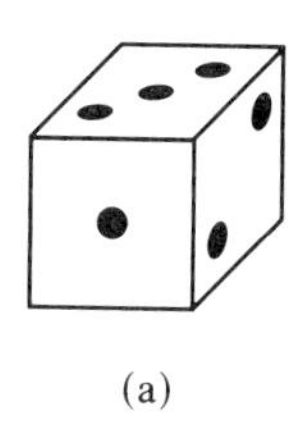

(a)

$\{1, 2, 3, 4, 5, 6\}$

$$\Pr\{1\} = \Pr\{2\} = \Pr\{3\} = \Pr\{4\} = \Pr\{5\} = \Pr\{6\} = \tfrac{1}{6}$$

(b)

Figure 9.2 Two kinds of dice: (a) A real die; (b) a model for a real die.

EXAMPLE 9.1 A *pair* of dice are thrown. Set up an appropriate model for this experiment, and calculate the probabilities of the "events" 2, 3, 4, 5, 6, 7, 8, 9, 10, 11, and 12, i.e., the possible value for the sum of the two numbers appearing on the dice.

SOLUTION Since there are 11 possibilities, we might argue that the appropriate set should be $\{2, 3, 4, \ldots, 12\}$, and that each event should be assigned probability $\frac{1}{11}$. This *is* a model for the given experiment, to be sure; but to test its accuracy, we might wish to perform the experiment, and see what happens. The authors actually did this (or rather our computer did) and the results of 1000 rolls of a pair of dice are shown in Table 9.1.

We see from Table 9.1 that the results of the experiment are not in close agreement with the predictions of the model—the model assigns probability $\frac{1}{11} = .091$ to each of the events! Of course the outcome of this one experiment *proves* nothing, but it does strongly *suggest* that the reasoning that led us to the answer $\frac{1}{11}$ was faulty. In fact, $\frac{1}{11}$ *is* wrong. Let us see why.

The problem is that (as Table 9.1 suggests) the 11 possible outcomes aren't *equally likely*. If you have any practical experience with dice, you already know this. For example, 7 is much more likely than 2, because 7 can occur in *several* ways (1 and 6, 2 and 5, 3 and 4, etc.), while 2 can occur in *only one* way (both dice show 1). Thus the event "7" is actually a *compound event*, composed of

TABLE 9.1 Result of 1000 rolls of a pair of dice (an actual experiment)

Outcome	Number of Occurrences	Relative Frequency
2	31	.031
3	69	.069
4	92	.092
5	101	.101
6	131	.131
7	146	.146
8	137	.137
9	137	.137
10	76	.076
11	50	.050
12	30	.030
	1000	1.000

First Die	Second Die	Sum	First Die	Second Die	Sum
1	1	2	4	1	5
1	2	3	4	2	6
1	3	4	4	3	7
1	4	5	4	4	8
1	5	6	4	5	9
1	6	7	4	6	10
2	1	3	5	1	6
2	2	4	5	2	7
2	3	5	5	3	8
2	4	6	5	4	9
2	5	7	5	5	10
2	6	8	5	6	11
3	1	4	6	1	7
3	2	5	6	2	8
3	3	6	6	3	9
3	4	7	6	4	10
3	5	8	6	5	11
3	6	9	6	6	12

Figure 9.3 The 36 atomic events when two dice are rolled.

several simpler events, whereas "2" is an indivisible, or *atomic* event. In Figure 9.3 we show the 36 atomic events in this experiment.

Why are there 36 events depicted in Figure 9.3? The multiplication rule (Section 2.1) tells us why. There are 6 possibilities for the first die and 6 for the second: $6 \times 6 = 36$. It is safe to assume that these 36 events are equally likely, since if the dice aren't loaded, no face is more likely to appear than any other. We therefore assign each of the 36 atomic events in Figure 9.3 probability $\frac{1}{36}$.

Now we can solve the given problem. To compute the probability of 8, for example, we just count the number of events in Figure 9.3 with sum 8. There are five such: $\{(2,6), (3,5), (4,4), (5,3), (6,2)\}$. The compound event "sum = 8," then, is composed of five atomic *events*, each of which has probability $\frac{1}{36}$. The probability of the event "sum 8" is therefore $\frac{1}{36} + \frac{1}{36} + \frac{1}{36} + \frac{1}{36} + \frac{1}{36} = .139$. Similar calculations for the other 10 possibilities lead to Table 9.2.

Notice that the theoretical probabilities in Table 9.2 are now in good agreement with the experimental results in Table 9.1. While we cannot "prove" that the model of Figure 9.3 is the right one for the experiment of rolling two dice, it is the model universally adopted. And if a real pair of dice gave experimental results significantly different from the predictions of Table 9.2, the model would not be suspected: The dice would be declared "loaded"! ■

Although we shall soon tackle problems much harder than Example 9.1, all such problems will have much in common. There will always be a set of **atomic events**, called the **sample space**.* The sample space is usually denoted by the

* In this text we only deal with finite sample spaces; but infinite sample spaces are allowed and important in the general theory of discrete probability.

TABLE 9.2 Probabilities when two ideal dice are rolled

Outcome	Probability	Probability (in Decimal)
2	$\frac{1}{36}$	.028
3	$\frac{2}{36}$	.056
4	$\frac{3}{36}$	.083
5	$\frac{4}{36}$	.111
6	$\frac{5}{36}$	.139
7	$\frac{6}{36}$	.167
8	$\frac{5}{36}$	.139
9	$\frac{4}{36}$	.111
10	$\frac{3}{36}$	.083
11	$\frac{2}{36}$	.056
12	$\frac{1}{36}$	.028

capital Greek letter omega (Ω), and the atoms by Greek lowercase omegas (ω):

(9.1)
$$\Omega = \{\omega_1, \omega_2, \ldots, \omega_n\}.$$

With each atom ω_i is associated its **probability**, denoted by $\Pr\{\omega_i\}$, which is a number between 0 and 1. The probabilities of the atoms must total 1. The sample space, together with its probability assignment, is called a **probability space**.

In a given probability space, a **compound event**, or **event** for short, is just a subset of Ω. Its probability is defined to be the sum of the probabilities of its component atoms. Thus if $A = \{\omega_1, \ldots, \omega_k\}$ is an event, then using the "$\sum$" notation introduced in Section 1.6, we have

(9.2)
$$\Pr\{A\} = \Pr\{\omega_1\} + \Pr\{\omega_2\} + \cdots + \Pr\{\omega_k\},$$
$$\Pr\{A\} = \sum_{\omega \in A} \Pr\{\omega\}.$$

For example, the empty set $\varnothing$ (sometimes called the "impossible event" in this context) has probability 0; and the whole sample space Ω (sometimes called the "certain event") has probability 1. Normally these two events will be the only ones with probability 0 or 1, but if the sample space contains atoms with probability 0, there will be other such events.

EXAMPLE 9.2 Consider the following five-point probability space: $\Omega = \{a, b, c, d, e\}$, with probabilities as follows:

ω	a	b	c	d	e
$\Pr\{\omega\}$	$\frac{1}{2}$	0	$\frac{1}{4}$	0	$\frac{1}{4}$

Enumerate all events of probability 0 and probability 1.

SOLUTION Since the atoms b and d both have probability 0, the event $\{b, d\}$ and all of its subsets will have probability 0. There are therefore four events of probability 0: $\varnothing$, $\{b\}$, $\{d\}$, and $\{b, d\}$. Similarly, any event which contains $\{a, c, e\}$ will have a probability 1; and there are four such sets: $\{a, c, e\}$, $\{a, b, c, e\}$, $\{a, c, d, e\}$, and $\{a, b, c, d, e\}$.

The sample spaces of Figures 9.1, 9.2(b), and 9.3 share one important property—within each sample space, the atoms all have the same probability. Not all probability spaces have this property (the sample space in Example 9.2 does not, for example), but those that do are called **uniform probability spaces**. In a uniform space with n atoms, each atom will have probability $1/n$. If the event A contains k atoms, then by (9.2), $\Pr\{A\} = 1/n + 1/n + \cdots + 1/n$ (k terms) $= k/n$. Thus we have the important fact that in a uniform space, *probabilities can be obtained by counting*:

(9.3) $$\Pr\{A\} = \frac{|A|}{|\Omega|} \quad \text{(uniform spaces only!)}.$$

We learned a lot about counting in Chapter 2. The next two examples give probabilistic applications of some of these techniques, based on Equation (9.3).

EXAMPLE 9.3 In a poker hand of 5 cards, what is the probability of being dealt "4 of a kind"?

SOLUTION Since there are 52 cards in a deck, there are $\binom{52}{5} = 2598960$ possible poker hands (the subset rule, Section 2.3). Having no reason to believe that any one of these hands is any more or less probable than any other, we take as our model a sample space with $\binom{52}{5}$ atoms, each with probability $1/\binom{52}{5}$. This is a uniform probability space, and so by (9.3) to answer the question, all we need to do is count the number of poker hands ranked as "4 of a kind." This, in turn, can be done using the multiplication rule, since each such hand can be classified according to (a) the card which appears 4 times, and (b) the "other card." There are 13 choices for (a), viz. {A, 2, 3, ..., Q, K}, and, for each such choice, there are $52 - 4 = 48$ choices for (b). Thus there are $13 \times 48 = 624$ poker hands rated "4 of a kind," and so the *probability* of this hand is

$$\Pr\{4 \text{ of a kind}\} = \frac{624}{2598960} = .00024.$$

EXAMPLE 9.4 If three dice are rolled, what is the probability that *exactly two* of the faces show a number greater than or equal to 5?

SOLUTION Since each of the three dice can show any number from 1 to 6, in this problem an appropriate sample space is the set of all ordered samples of size 3 from the set $\{1, 2, 3, 4, 5, 6\}$. By the rule for ordered samples with

TABLE 9.3 An 11-atom nonuniform probability space

ω	a	b	c	d	e	f	g	h	i	j	k
$\Pr\{\omega\}$	$\frac{1}{36}$	$\frac{2}{36}$	$\frac{3}{36}$	$\frac{4}{36}$	$\frac{5}{36}$	$\frac{6}{36}$	$\frac{5}{36}$	$\frac{4}{36}$	$\frac{3}{36}$	$\frac{2}{36}$	$\frac{1}{36}$

repetition, Section 2.2, there are $6 \times 6 \times 6 = 216$ such samples, ranging from $(1, 1, 1)$ to $(6, 6, 6)$, and we assign each of them probability $1/216$. The question is, How many of them have exactly two of their components greater than or equal to 5? Again, the multiplication rule gives the answer. The "small number" (the number which is less than 5) can occur in any one of the *three* positions; and given the location of the small number, there are four possibilities for the small number: 1, 2, 3, or 4, and *two* possibilities for each of the "big numbers," viz. 5 or 6. Hence there are $3 \times 4 \times 2 \times 2 = 48$ possible outcomes with two big and one small number. Since we are dealing with a uniform space, the probability of this event is by (9.3)

$$\Pr\{A\} = \tfrac{48}{216} = \tfrac{2}{9}.$$

We next consider another simple example of a *non*uniform probability space.

EXAMPLE 9.5 Consider the 11-point probability space of Table 9.3, and calculate the probabilities of the events $A = \{a, b, c, d\}$, $B = \{b, j\}$, and $C = \{h, i, j\}$.

SOLUTION This is just a simple matter of adding the probabilities of the atoms in three events:

$$\Pr\{A\} = \tfrac{1}{36} + \tfrac{2}{36} + \tfrac{3}{36} + \tfrac{4}{36} = \tfrac{5}{18},$$

$$\Pr\{B\} = \tfrac{2}{36} + \tfrac{2}{36} = \tfrac{1}{9},$$

$$\Pr\{C\} = \tfrac{4}{36} + \tfrac{3}{36} + \tfrac{2}{36} = \tfrac{1}{4}.$$

We hope you have noticed the connection between the 36-atom *uniform* probability space of Figure 9.3 and the 11-atom *nonuniform* probability space of Table 9.3 (if not, do Problem 10). What this connection illustrates is that it is often possible to model a given real situation in several different ways.

Next we discuss how to calculate, or at least estimate, the probability of a *union* of events. If A and B are disjoint events, i.e., if $A \cap B = \varnothing$, then we have

(9.4) $$\Pr\{A \cup B\} = \Pr\{A\} + \Pr\{B\}.$$

This important fact follows from (9.2). The probability of $A \cup B$ is defined to be the sum of the probabilities of the atoms in $A \cup B$. Since each such atom must lie in either A or B but not both (since A and B are disjoint), (9.4) follows. (See Figure 9.4.)

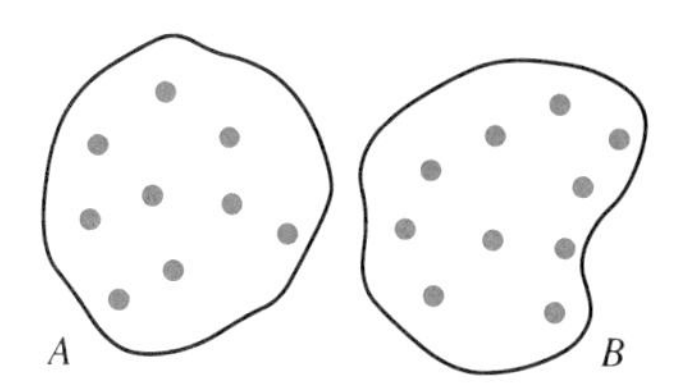

Figure 9.4 Every atom in $A \cup B$ lies in A or B, but not both.

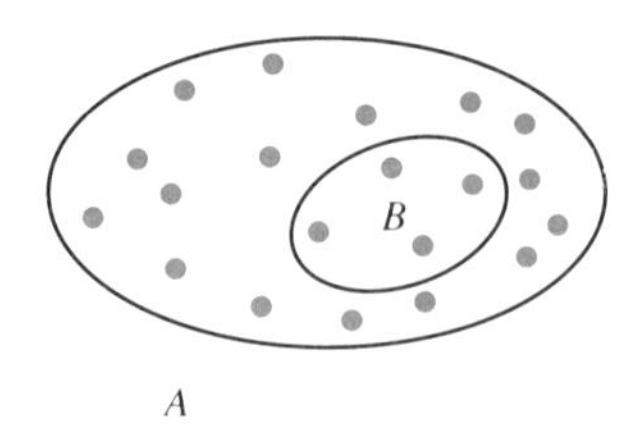

Figure 9.5 If $B \subseteq A$, then $A = (A - B) \cup B$, and so (9.6) follows from (9.4).

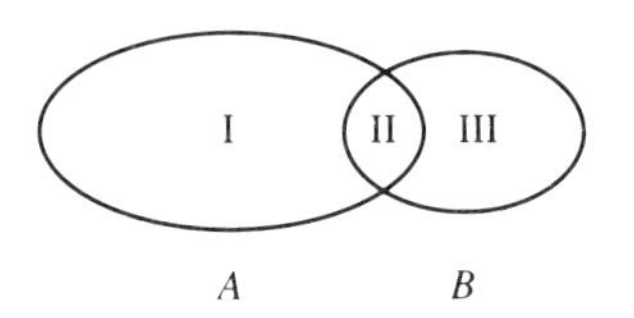

Figure 9.6 Every atom in $A \cup B$ must lie in exactly one of the regions I, II, or III.

Similarly, if the event A can be written as the *disjoint union* of n events,

$$A = A_1 \cup A_2 \cup \cdots \cup A_n \qquad \text{(pairwise disjoint sets)},$$

then we have

(9.5) $$\Pr\{A\} = \Pr\{A_1\} + \Pr\{A_2\} + \cdots + \Pr\{A_n\}.$$

Another simple, but useful, result concerns subsets:

(9.6) $$\Pr\{A - B\} = \Pr\{A\} - \Pr\{B\}, \qquad \text{if } B \subseteq A,$$

(See Figure 9.5.)

There is also a formula, corresponding to (9.4), when A and B are *not* disjoint:

(9.7) $$\Pr\{A \cup B\} = \Pr\{A\} + \Pr\{B\} - \Pr\{A \cap B\}.$$

You should recognize Equation (9.7) as a thinly disguised form of the principle of inclusion and exclusion (Section 2.6). But the underlying idea is so important, we'll review it here, with a slightly different proof.

To see why (9.7) is true, look at Figure 9.6 and notice that $A \cup B$ can be written as the disjoint union of three sets:

$$A \cup B = \underbrace{(A - A \cap B)}_{\text{I}} \cup \underbrace{(A \cap B)}_{\text{II}} \cup \underbrace{(B - A \cap B)}_{\text{III}}.$$

Thus by (9.5), $\Pr\{A \cup B\}$ is equal to the sum of the probabilities of these three sets:

$$\Pr\{A \cup B\} = \Pr\{A - A \cap B\} + \Pr\{A \cap B\} + \Pr\{B - A \cap B\}.$$

But since $A \cap B$ is a subset of both A and B, we have by (9.6) that

$$\Pr\{A - A \cap B\} = \Pr\{A\} - \Pr\{A \cap B\},$$

and

$$\Pr\{B - A \cap B\} = \Pr\{B\} - \Pr\{A \cap B\}.$$

Thus we have finally

$$\begin{aligned}\Pr\{A \cup B\} &= \Pr\{A\} - \Pr\{A \cap B\} + \Pr\{A \cap B\} + \Pr\{B\} - \Pr\{A \cap B\} \\ &= \Pr\{A\} + \Pr\{B\} - \Pr\{A \cap B\},\end{aligned}$$

as advertised. There is a complicated generalization of (9.7) for the union of n sets, corresponding to the principle of inclusion and exclusion (PIE_n) of Section 2.6 (see Problem 35). But sometimes the following simpler *inequality* is useful:

(9.8) $$\Pr\{A \cup B\} \leq \Pr\{A\} + \Pr\{B\}.$$

This follows immediately from (9.7). And it is not hard to prove by induction (see Problem 34) that in general,

(9.9) $$\Pr\{A_1 \cup A_2 \cup \cdots \cup A_n\} \leq \Pr\{A_1\} + \cdots + \Pr\{A_n\}.$$

This inequality is often called the **union bound** for probabilities.

EXAMPLE 9.6 As in Example 9.4, three dice are rolled. Let A be the event "at least two 6s appear." Use the union bound to estimate $\Pr\{A\}$.

SOLUTION The event A is the union of the following three events:

$$A_{12} = \{\text{dice 1 and 2 both show 6}\},$$
$$A_{13} = \{\text{dice 1 and 3 both show 6}\},$$
$$A_{23} = \{\text{dice 2 and 3 both show 6}\}.$$

The number of atoms in A_{12} is 6, since there are 6 choices for the third die. Thus

$$A_{12} = \{(6, 6, 1), (6, 6, 2), (6, 6, 3), (6, 6, 4), (6, 6, 5), (6, 6, 6)\},$$

and $\Pr\{A_{12}\} = \frac{6}{216} = \frac{1}{36}$. Similarly $\Pr\{A_{13}\} = \Pr\{A_{23}\} = \frac{1}{36}$. Thus by the union bound (9.9), we have

$$\begin{aligned}\Pr\{A\} &= \Pr\{A_{12} \cup A_{13} \cup A_{23}\} \leq \Pr\{A_{12}\} + \Pr\{A_{13}\} + \Pr\{A_{23}\} \\ &= \tfrac{3}{36} = \tfrac{1}{12}.\end{aligned}$$

In this case the exact answer is fairly easy to calculate (see Problem 36), but in more complicated problems, the union bound is a very good way to estimate probabilities that are otherwise quite difficult to find exactly (see Problem 37). ■

The last example in this section is one version of a famous old problem in discrete mathematics. It is sometimes called the **problem of derangements**. It is a very good exercise in the important skill of counting.

EXAMPLE 9.7 Consider the set of all *one-to-one and onto* functions from the set $S = \{1, 2, \ldots, n\}$ to itself. If f is one of these functions, a *fixed point* of f is an element $x \in S$ such that $f(x) = x$. If one of these functions is chosen at random, what is the probability that it has *no* fixed points?

SOLUTION Before launching into the full solution, let's consider the special case $n = 4$ to clarify things. As we saw in Section 1.3, the one-to-one and onto functions are the same as the rearrangements, i.e., the *permutations* of the set S. In Table 9.4, we show all 24 permutations of the set $\{1, 2, 3, 4\}$, with

TABLE 9.4 The 24 permutations of $\{1, 2, 3, 4\}$

Permutation number	1	2	3	4	5	6	7	8	9	10	11	12	13	14	15	16	17	18	19	20	21	22	23	24
1	1	1	1	1	1	1	2	2	2	2	2	2	3	3	3	3	3	3	4	4	4	4	4	4
2	2	2	3	3	4	4	1	1	3	3	4	4	1	1	2	2	4	4	1	1	2	2	3	3
3	3	4	2	4	2	3	3	4	1	4	1	3	2	4	1	4	1	2	2	3	1	3	1	2
4	4	3	4	2	3	2	4	3	4	1	3	1	4	2	4	1	2	1	3	2	3	1	2	1
Number of fixed points	4	2	2	1	1	2	2	0	1	0	0	1	1	0	2	1	0	0	0	1	1	2	0	0

the fixed points of each permutation underlined. For example, permutation 7 is the function given by

$$f(1) = 2, \qquad f(2) = 1, \qquad f(3) = \underline{3}, \qquad f(4) = \underline{4},$$

which has 2 fixed points. We see that there are exactly 9 permutations without fixed points. Thus if we model the experiment of choosing a function at random with a uniform probability space having as atoms the 24 permutations, we find that the probability of no fixed points is $\frac{9}{24}$.

To solve the given problem in general, we need to be able to count the number of permutations of $\{1, 2, \ldots, n\}$ with no fixed points. This turns out to be a job for the principle of inclusion and exclusion, PIE$'_n$, from Section 2.6. The idea is to let A_i represent the set of functions which have i as a fixed point, for $i = 1, 2, \ldots, n$. For example, looking at Table 9.4, we see that A_1 consists of permutations numbered 1, 2, 3, 4, 5, 6, A_2 consists of numbers 1, 2, 15, 16, 21, 22, etc. The set of functions with no fixed points is then the set $A'_1 \cap A'_2 \cap \cdots \cap A'_n$, i.e., the set of functions for which 1 is not a fixed point, 2 is not a fixed point, $\ldots$, n is not a fixed point.

To count $|A'_1 \cap A'_2 \cap \cdots \cap A'_n|$, we now use PIE$'_n$. The term $|S|$ is equal to the total number of permutations, which is, by the permutation rule, $n!$. The next term is A_1, which is the number of permutations which have 1 as a fixed point. This number is $(n-1)!$, since if 1 is fixed, the rest of the function is just a permutation of the $(n-1)$-element set $\{2, 3, \ldots, n\}$. Exactly similar reasoning shows that all n terms of the form $|A_i|$ are equal to $(n-1)!$. Next, there are $\binom{n}{2}$ terms of the form $|A_i \cap A_j|$. For example, consider the term $|A_1 \cap A_2|$, which is the number of permutations with both 1 and 2 as fixed points. This number is equal to $(n-2)!$, since if both 1 and 2 are fixed, the rest of the function is just a permutation of the $(n-2)$-element set $\{3, 4, \ldots, n\}$. Similarly, we can see that all $\binom{n}{2}$ of the terms $|A_i \cap A_j|$ are equal to $(n-2)!$.

Continuing this line of reasoning, we see that the $\binom{n}{3}$ terms of the form $|A_i \cap A_j \cap A_k|$ are all equal to $(n-3)!$, and so on. We conclude at last that the number of permutations with *no* fixed points is, by PIE$'_n$, equal to

$$\textbf{(9.10)} \qquad n! - n(n-1)! + \binom{n}{2}(n-2)! - \binom{n}{3}(n-3)! + \cdots,$$

the terms continuing until the term $\binom{n}{n} \cdot 0!$ is reached. (This final term is equal to 1, which makes sense, since it represents the number of permutations for which *every* element of the set $\{1, 2, \ldots, n\}$ is fixed, and there is only one such permutation.) For example, applying (9.10) to the case $n = 4$, we get for the number of permutations with no fixed points

$$4! - 4 \cdot 3! + 6 \cdot 2! - 4 \cdot 1! + 1 \cdot 0! = 24 - 24 + 12 - 4 + 1 = 9,$$

as we saw before by looking at Table 9.4.

Formula (9.10) can be simplified if we notice that the second term $n(n-1)!$ equals $n!$, that the third term $\binom{n}{2}(n-2)!$ equals $n(n-1)(n-2)!/2! = n!/2!$, and so on. Thus $n!$ can be factored out of each term in (9.10), revealing that in fact the number of permutations without fixed points is equal to

$$\textbf{(9.11)} \qquad n!\left(1 - 1 + \frac{1}{2!} - \frac{1}{3!} + \cdots \pm \frac{1}{n!}\right).$$

Finally, remembering that our goal is to compute the *probability* that a randomly selected permutation has no fixed points, we divide the expression in (9.11) by the total number of permutations, which is $n!$, and obtain the following formula:

$$\textbf{(9.12)} \quad \text{probability of having no fixed points} = 1 - 1 + \frac{1}{2!} - \frac{1}{3!} + \cdots \pm \frac{1}{n!}$$

$$= \sum_{k=0}^{n} (-1)^k \frac{1}{k!}.$$

In Table 9.5 we give a short table of these probabilities.

In Table 9.5, we can see that the probability of having no fixed points is not changing very much after $n = 6$ or thereabouts. And in fact, with the help of calculus, it can be shown that for larger values of n, the probability P_n is rapidly approaching the constant $e^{-1} = .36787944\ldots$, where $e = 2.71828\ldots$ is the base of natural logarithms. ■

TABLE 9.5 A short table of P_n, the probability of no fixed points in a randomly chosen permutation on $\{1, 2, \ldots, n\}$

n	P_n
1	.0000
2	.5000
3	.3333
4	.3750
5	.3667
6	.3681
7	.3679
8	.3679

Problems for Section 9.1

1. Suppose you discovered a rare coin which was found to land "on edge" 1% of the time. Make a model for the experiment of flipping this coin.

2. Make a model for the experiment of rolling a pair of four-sided dice (the sides are labelled 1, 2, 3, 4).

3. In Example 9.2, enumerate all events of probability $\frac{1}{2}$.

4. In Example 9.2, enumerate all events of probability $\frac{1}{4}$.

5. In a poker hand of five cards, find the probability of a royal flush (AKQJ10 of the same suit).

6. In a poker hand of five cards, find the probability of a full house (three of one kind and two of another kind, e.g., three kings and two 4s).

7. If three dice are rolled, find the probability that exactly one face shows a number less than or equal to 4.

8. If three dice are rolled, find the probability that the sum is 6.

9. In the probability space of Table 9.3, calculate the probabilities of the events $D = \{d, g, i\}$ and $E = \{b, j, k\}$.

10. What is the relationship between the 36-point probability space of Figure 9.3 and the 11-point probability space of Table 9.3? (*Hint*: Notice that in Table 9.3, $\Pr\{g\} = \frac{5}{36}$, which is the answer to one of the problems posed in Example 9.1.)

11. If three dice are rolled, use the union bound to estimate the probability of obtaining at least one 6.

12. If four dice are rolled, use the union bound to estimate the probability of obtaining at least three 6s.

13. An absent-minded mathematics professor is paying her bills. She has 10 bills to pay, so she writes 10 checks and addresses 10 corresponding envelopes. Unfortunately she pays no attention to which checks are supposed to go into which envelopes, but just randomly puts a check in each envelope. What is the probability that she gets *every* check in the wrong envelope?

14. Four bald men wearing hats go to a party, and when they arrive they put their hats in a dark closet. During the party someone yells, "Fire!" and the men rush to the closet, grab a hat at random, and leave. What is the probability that all four get the wrong hats?

15. Continuing Problem 14, find the probability of each of the following sets.
(a) All four men get the right hats.
(b) Exactly three get the right hats.
(c) Only two get the right hats.
(d) Exactly one gets the right hat.

16. If three dice are rolled, what is the probability that each die shows an odd number?

17. Consider the experiment of choosing an ordered sample of size 3 (without repetition) from the set $\{1, 2, 3, 4, 5\}$.
(a) How many such samples are there?
(b) Assuming that all samples are equally likely, find the probability that (i) the first number in the sample is even; (ii) the second number is even; (iii) the third number is even.

18. A single die is rolled six times. What is the probability that the six outcomes are all different?

19. An absentminded mathematics professor is rushing off to class. His dresser drawer contains four black socks, six blue socks, and five brown socks. He grabs two socks at random and puts them on without looking. What is the probability that his socks match?

20. A box contains 100 light bulbs, 5 of which are defective. If 20 are selected, what is the probability that they all work?

In Problems 21–25, you are asked to compute the probabilities of various kinds of poker hands; see Example 9.3 for guidance. (*Note*: "One pair" does *not* include "three of a kind," or "two pair," etc. See Problems 2.3.17 and 2.3.18, for terminology.)

21. Straight.

22. Three of a kind.

23. Two pair.

24. One pair.

25. "Nothing."

26. Conduct your own computer experiment in dice rolling, and give your results in a table like Table 9.1.

27. Suppose three dice are rolled. Calculate the probability that the sum is 10.

28. Consider the experiment of flipping four unbiased coins. (On each coin the outcome must be "heads" or "tails.") Describe a uniform probability space which models this experiment. Compute the probability of "at least two heads."

29. Look at Table 9.4 and notice that the total number of fixed points for the 24 permutations is $4 + 2 + 2 + \cdots + 0 + 0 = 24$. Is this an accident, or an instance of a general rule? Explain fully. (*Suggestion*: Look at the number of 1s in the first row, the number of 2s in the second row, etc.)

30. In a probability space with n atoms, k of which have probability 0, how many events of probability 0 are there? How many events of probability 1?

31. At a certain party, there are a married couples, b single men, and c single women. A man and a woman are selected at random. What is the probability that (a) both are married; (b) only one is married; (c) they are married to *each other*?

32. (With our apologies to Douglas Adams and Jacob Bernoulli.) Two prisoners, Arthur and Ford, are ordered to each roll one die, with the understanding that the one who gets the smaller number will be forced to listen to Vogon poetry, and the one with the larger number will be released. Both will be spared if the numbers are the same. Compute the following probabilities.

(a) Arthur will be spared.
(b) Ford will be spared.
(c) Both will be spared.
(d) Neither will be spared.

33. Prove that Formula (9.5) is true, using mathematical induction.

34. Use mathematical induction to prove the general union bound (9.9).

35. Develop a formula similar to (9.7), for the exact probability $\Pr\{A_1 \cup A_2 \cup \cdots \cup A_n\}$.

36. Find the exact value of $\Pr\{A\}$ in Example 9.6

37. Suppose n dice are rolled. Use the union bound to estimate the probability that at least k 6s appear.

38. Consider an experiment in which there are n possible outcomes, labelled $1, 2, \ldots, n$. The experiment is repeated r times. The result of the r trials is recorded in the form of a *histogram*, i.e., a list which tells how many times outcome 1 occurred, 2 occurred, ..., n occurred.

(a) How many possible histograms are there? (See Problem 2.4.18.)
(b) If a histogram is selected at random, what is the probability that outcome 1 will occur *exactly once*?

39. See if you can generalize the result of Example 9.7, and find a formula for the number of permutations on the set $\{1, 2, \ldots, n\}$ having exactly r fixed points.

40. (S. W. Golomb) Twelve people put their umbrellas in an umbrella stand. It starts to rain, and each person hurriedly grabs an umbrella at random. What is the *least likely* number of people who get their own umbrella?

9.2 Conditional Probabilities

Consider again the experiment of throwing a pair of dice. Here are four events:

A: "sum = 10," $\Pr\{A\} = \frac{1}{12}$,

B: "sum is even," $\Pr\{B\} = \frac{1}{2}$,

C: "first die = 6," $\Pr\{C\} = \frac{1}{6}$,

D: "second die = 3," $\Pr\{D\} = \frac{1}{6}$.

These probabilities can be computed using the ideas of the previous section, and Figure 9.3. The point we wish to make here is that some of these events are related to the others. For example, if A occurs, then B *must also occur*; in other words, the occurrence of the event A *changes* the probability of B from $\frac{1}{2}$ to 1. We say "the probability of B, given A, is 1," and write

(9.13) $$\Pr\{B|A\} = 1.$$

Similarly, if D occurs, A is *impossible*, and since an impossible event is always assigned probability 0, we write

(9.14) $$\Pr\{A|D\} = 0.$$

These two simple examples are special cases of the notion of *conditional probability*, which is a way of measuring how the occurrence of one event changes the probability of another.

Before going on to the formal definition of conditional probability, let's consider a more difficult example.

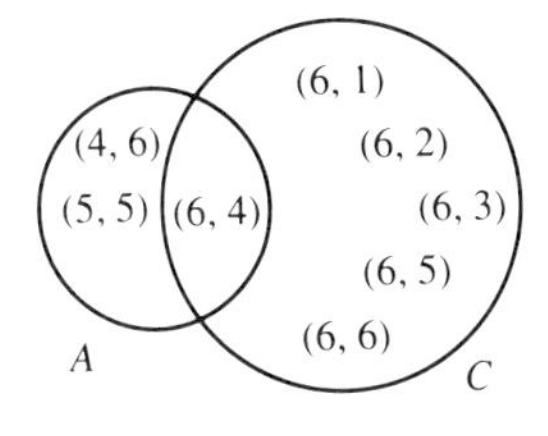

Figure 9.7 The events A and C for Example 9.8.

EXAMPLE 9.8 Calculate $\Pr\{A|C\}$, "the probability of A, given C."

SOLUTION We display the events A and C in a simple Venn diagram (Figure 9.7). The occurrence of C (first die = 6) makes A (sum = 10) neither *certain* nor *impossible*. Still, the occurrence of C does have a considerable effect on A. To measure this effect, note that A is composed of three atoms:

$$A = \{(4,6), (5,5), (6,4)\}; \qquad \Pr\{A\} = \tfrac{3}{36}.$$

The event C, on the other hand, is composed of six atoms:

$$C = \{(6,1), (6,2), (6,3), (6,4), (6,5), (6,6)\}; \qquad \Pr\{C\} = \tfrac{6}{36}.$$

If, however, we *know* that C has occurred (for example by rolling the dice one at a time, and noticing that 6 occurs on the first roll), the 30 points in the sample space of Figure 9.3 *not* in C are eliminated from consideration; in particular two of the three atoms in A [(4, 6) and (5, 5)] are eliminated. Thus if C has occurred, the only way for A to occur also is if the event

$$A \cap C = \{(6,4)\}$$

occurs. Since C has six atoms, and only *one* of these is in A, the probability of A, given C, is

$$\Pr\{A|C\} = \tfrac{1}{6}.$$

We might summarize this calculation as follows:

Before knowing C has occurred

$$\Pr\{A\} = \frac{|A|}{|\Omega|} = \frac{3}{36} = \frac{1}{12}.$$

After knowing C has occurred*

$$\Pr\{A|C\} = \frac{|A \cap C|}{|C|} = \frac{1}{6}.$$

In a sense, of course, the answer to Example 9.8 is obvious, since if the first die is 6, the only way the sum can be 10 is if the second die is 4; and we know that the probability of rolling a 4 is $\frac{1}{6}$. However, our long-winded solution does have a purpose; it leads us to the general definition of conditional probability on a uniform probability space.

DEFINITION If A and B are two events in a uniform probability space, the **conditional probability** of A, given B, is defined as follows:

(9.15) $$\Pr\{A|B\} = \frac{|A \cap B|}{|B|} \qquad \text{(assuming } |B| > 0\text{)}.$$

This definition is almost the same as (9.3) of Section 9.1, the definition of probabilities in a uniform space. The difference is that when B has occurred, B *itself* becomes in effect the new sample space, and all probabilities are calculated with respect to B.

In a nonuniform space the definition is almost the same, except the *sizes* of the sets A and B arc replaced by their *probabilities*:

(9.16) $$\Pr\{A|B\} = \frac{\Pr\{A \cap B\}}{\Pr\{B\}} \qquad \text{(if } \Pr\{B\} > 0\text{)}.$$

[In (9.16), we divide by $\Pr\{B\}$; if $\Pr\{B\} = 0$ this is illegal, and the conditional probabilities aren't defined (see Problem 24).]

EXAMPLE 9.9 Refer to the sample space of Table 9.3. Let $A = \{i\}$ and $B = \{a, c, e, g, i, k\}$. Compute $\Pr\{A|B\}$ and $\Pr\{B|A\}$.

SOLUTION We have $A \cap B = \{i\}$, and so from (9.16),

$$\Pr\{A|B\} = \frac{\Pr\{A \cap B\}}{\Pr\{B\}} = \frac{\frac{3}{36}}{\frac{1}{36} + \frac{3}{36} + \frac{5}{36} + \frac{5}{36} + \frac{3}{36} + \frac{1}{36}} = \frac{1}{6};$$

$$\Pr\{B|A\} = \frac{\Pr\{B \cap A\}}{\Pr\{A\}} = \frac{\frac{3}{36}}{\frac{3}{36}} = 1.$$

Example 9.9 shows that as a rule, $\Pr\{A|B\}$ and $\Pr\{B|A\}$ aren't the same. Still, there is a very close connection between these two conditional probabilities,

* In some texts on probability, the "before" and "after" probabilities are called by their Latin names: before—*a priori* probabilities; after—*a posteriori* probabilities.

which can be seen as follows: We have from Equation (9.16) that

$$\Pr\{A|B\} = \frac{\Pr\{A \cap B\}}{\Pr\{B\}}$$

and [by interchanging the roles of A and B in (9.16)],

$$\Pr\{B|A\} = \frac{\Pr\{B \cap A\}}{\Pr\{A\}};$$

it follows that

$$\Pr\{B|A\} = \frac{\Pr\{A|B\}\Pr\{B\}}{\Pr\{A\}}. \tag{9.17}$$

The relationship in Equation (9.17) is a handy way to compute $\Pr\{B|A\}$ when $\Pr\{A|B\}$ is given, or is easy to compute. The following example illustrates this, and also shows how probabilities can often be calculated when the underlying sample space isn't completely specified.

EXAMPLE 9.10 Suppose there are three urns, each containing two balls. The first urn contains two black balls; the second contains a black ball and a white ball; the third contains two white balls (see Figure 9.8). Suppose an urn is selected at random, and one ball from it is removed and found to be *white*. What is the probability that the urn selected is urn 3?

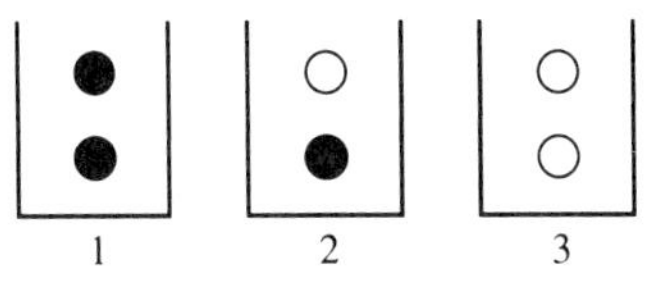

Figure 9.8 The three urns.

SOLUTION Without needing to construct the sample space explicitly, we know that it ought to contain events called U_1, U_2, and U_3, to represent the selection of the urns numbered 1, 2, and 3, respectively. These events form a partition of the sample space, and we interpret the phrase "an urn is selected at random" to mean

$$\Pr\{U_1\} = \Pr\{U_2\} = \Pr\{U_3\} = \tfrac{1}{3}.$$

There should also be events labelled B and W, to represent the selection of a black or white ball. Since there are three balls of each color in Figure 9.8, we should have

$$\Pr\{B\} = \Pr\{W\} = \tfrac{1}{2}.$$

The requested "probability that the urn selected is urn 3" is the conditional probability $\Pr\{U_3|W\}$. This probability can't be calculated directly (although it might be guessed by inspection of Figure 9.8), but, as often is the case, the *reverse* probability $\Pr\{W|U_3\}$ is easy to compute, and then formula (9.17) can be used to compute $\Pr\{U_3|W\}$. Indeed, because urn 3 contains *only* white balls, $\Pr\{W|U_3\} = 1$. Similarly (though we don't need these facts to solve the given problem), $\Pr\{W|U_1\} = 0$, and $\Pr\{W|U_2\} = \tfrac{1}{2}$. To find $\Pr\{U_3|W\}$, we now apply the "inversion formula" (9.17), with $A = W$ and $B = U_3$:

$$\Pr\{U_3|W\} = \frac{\Pr\{W|U_3\}\Pr\{U_3\}}{\Pr\{W\}} = \frac{1 \cdot \frac{1}{3}}{\frac{1}{2}} = \tfrac{2}{3}.$$

Thus if a white ball is drawn, it is twice as likely that the urn selected is urn 3 rather than urn 2. ∎

The calculation in Example 9.10 is an unusually simple application of (9.17). This technique can be souped-up considerably, but we postpone a more serious discussion of it until we discuss dependent trials in Section 9.4. In the next section we consider the simpler (and in many ways, more important) notion of *independent trials.*

Problems for Section 9.2

In Problems 1–9, refer to the events A, B, C, D defined at the beginning of the section, and calculate the indicated probabilities.

1. $\Pr\{A|B\}$. **4.** $\Pr\{B|C\}$. **7.** $\Pr\{C|B\}$.

2. $\Pr\{A|D\}$. **5.** $\Pr\{B|D\}$. **8.** $\Pr\{C|D\}$.

3. $\Pr\{B|A\}$. **6.** $\Pr\{C|A\}$. **9.** $\Pr\{A|A\}$.

10. Consider the four events A, B, C, D, listed at the beginning of this section. There are 16 possible conditional probabilities that can be computed, and they can be arranged in a table (the entry in row A, column C is $\Pr\{A|C\}$, for example):

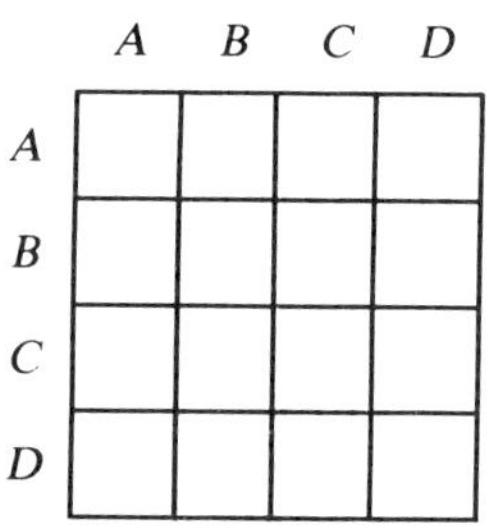

In Example 9.8 and Problems 1-9 we have computed most of these. Find the remaining conditional possibilities, and fill in the entire table.

In Problems 11–14, refer to Example 9.5 (Table 9.3) and let $C = \{c, e, g\}$, $D = \{e, g, h, i\}$, and $E = \{i, j, k\}$. Compute the indicated probabilities.

11. $\Pr\{C|D\}$. **13.** $\Pr\{D|E\}$.

12. $\Pr\{D|C\}$. **14.** $\Pr\{E|D\}$.

15. Construct a complete sample space for the solution of Example 9.10, using atoms (U_i, W) and (U_i, B), $i = 1, 2, 3$. Assign an appropriate probability to each atom.

16. Use the sample space constructed in Problem 15 to calculate the probability $\Pr\{U_3|W\}$ of Example 9.10.

In Problems 17–19, refer to the sample space of Figure 9.3, and compute the indicated probabilities.

17. Pr{sum = 8|at least one 4 showing}.

18. Pr{at least one 5|sum = 8}.

19. Pr{sum is prime|sum is even}.

In Problems 20–22, consider the experiment of rolling three unbiased three-sided dice. (Each die will show one of the numbers 1, 2, or 3.) Compute the indicated probabilities.

20. Pr{each face is odd}.

21. Pr{sum exceeds 6|at least one "1"}.

22. Pr{sum is prime|exactly two faces are the same}.

23. Show that the definition (9.16), when applied to a *uniform* space, specializes to the earlier definition (9.15).

24. Sometimes it is said that "something divided by 0 is infinity." Why then don't we define the conditional probability $\Pr\{A|B\}$, when $\Pr\{B\} = 0$, to be "infinity"?

In Problems 25 and 26, refer to Examples 2.24 and 2.25. (We have a room with 100 people, of whom 60 are men, 30 are young, 10 are young men, 40 are Republicans, 20 are Republican men, 15 are young Republicans, and 5 are young Republican men.)

25. Suppose one person is selected at random and discovered to be a man. What is the probability that the person is an old Republican?

26. Now suppose that *two* people are chosen, and it turns out that both are Democrats. What is the probability that one is a man and the other is a woman?

27. Let $\Omega = \{1, 2, 3, 4, 5, 6, 7, 8\}$ be a uniform probability space, i.e., each of the eight atoms has probability $\frac{1}{8}$. Let $A = \{1, 2, 3, 4\}$. Find all events B such that $\Pr\{A|B\} = \Pr\{A\}$.

28. The spinner shown in the accompanying figure is divided into four equal *colored* regions and three equal *numbered* regions. The outcome of a spin is a pair (number, color), e.g. (2, yellow). The probability that the spinner will land in a given region is the total angle spanned by the region, divided by 360°. Thus the probability of landing in each numbered region is $\frac{120}{360} = \frac{1}{3}$, and the probability of landing in each colored region is $\frac{90}{360} = \frac{1}{4}$. Construct a sample space that models this spinner, and assign an appropriate probability to each atom.

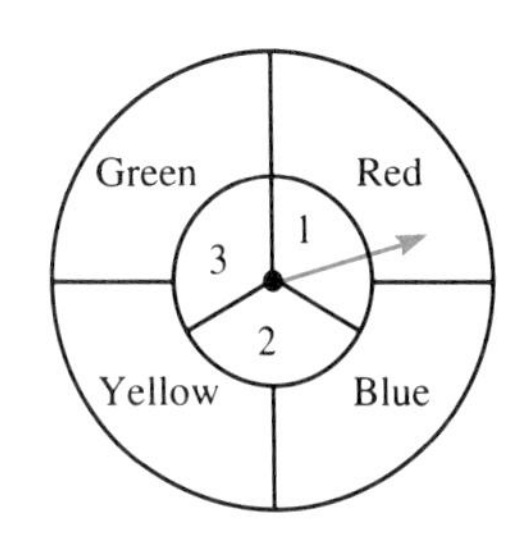

In Problems 29–32, refer to the spinner of Problem 28 and calculate the indicated probabilities.

29. Pr{yellow|2}.

30. Pr{2|yellow}.

31. Pr{red or yellow|1 or 2}.

32. Pr{2 or 3|not green}.

9.3 Independent Events, Product Spaces, and the Binomial Density

Consider the simple experiment of flipping a coin. In Section 9.1, we made a mathematical model of this experiment, with sample space

$$\Omega = \{\text{H}, \text{T}\},$$

and probabilities

$$\Pr\{\text{H}\} = \Pr\{\text{T}\} = \tfrac{1}{2}.$$

We assign the events H and T probabilities $\frac{1}{2}$ each, because practical experience tells us that if we flip a coin *a large number of times*, heads and tails will each occur about 50% of the time. However, the sample space $\Omega = \{\text{H}, \text{T}\}$ is a model for the experiment of flipping a coin *once*, not "a large number of times." If the theory of probability is to be of any practical value, it should be possible to construct a model for *repeated trials* of this (or of any other) experiment. In this section we will learn exactly how to do this.

In the previous section we saw that the occurrence of one event can change the probability of another event. However, in the real world, some events have no effect on others. The weather in Boston has no effect on the winner of a horse race in Hollywood Park; the outcome of the seventh flip of your coin has no effect on the eleventh flip, etc. Events that do not affect each other are called *independent events.* Our first goal in this section is to express the notion of independence mathematically.

Let us agree to call two events A and B **independent** if the occurrence of B has no effect on the probability of A. That is, events A and B are independent if

(9.18) $$\Pr\{A|B\} = \Pr\{A\}.$$

Since by (9.16), $\Pr\{A|B\} = \Pr\{A \cap B\}/\Pr\{B\}$, (9.18) can also be written as

(9.19) $$\Pr\{A \cap B\} = \Pr\{A\} \cdot \Pr\{B\}.$$

If we divide both sides of Equation (9.19) by $\Pr\{A\}$ we get

(9.20) $$\Pr\{B|A\} = \Pr\{B\}.$$

Thus if B has no effect on A (Equation 9.18), then necessarily also A has no effect on B (Equation 9.20). Any of the three equations (9.18), (9.19), or (9.20) can be used as a definition of independence. However, (9.19) is most common, because it is symmetric in A and B, and it applies even when $\Pr\{A\} = 0$ or $\Pr\{B\} = 0$, in which case $\Pr\{B|A\}$ or $\Pr\{A|B\}$ cannot be defined.

EXAMPLE 9.11 Consider the experiment of flipping a fair coin twice, and the three events

$$\begin{aligned} A &= \text{heads on the first flip}, \\ B &= \text{heads on the second flip}, \\ C &= \text{both flips are the same}. \end{aligned}$$

Show that each of the three events A, B, and C is independent of the other two.

SOLUTION The appropriate probability space is

$$\Omega = \{(\text{H}, \text{H}), (\text{H}, \text{T}), (\text{T}, \text{H}), (\text{T}, \text{T})\},$$

where each of the four possibilities is assigned probability $\frac{1}{4}$. In terms of this notation, we have

$$\begin{aligned} A &= \{(\text{H}, \text{H}), (\text{H}, \text{T})\}; & \Pr\{A\} &= \tfrac{1}{2}. \\ B &= \{(\text{H}, \text{H}), (\text{T}, \text{H})\}; & \Pr\{B\} &= \tfrac{1}{2}. \\ C &= \{(\text{H}, \text{H}), (\text{T}, \text{T})\}; & \Pr\{C\} &= \tfrac{1}{2}. \end{aligned}$$

To show that the events A and B are independent, we check to see if Equation (9.19) is true:

$$\begin{aligned} \Pr\{A \cap B\} &= \Pr\{(\text{H}, \text{H})\} = \tfrac{1}{4}, \\ \Pr\{A\} \cdot \Pr\{B\} &= \tfrac{1}{2} \cdot \tfrac{1}{2} = \tfrac{1}{4}. \end{aligned}$$

Similarly

$$\begin{aligned} \Pr\{A \cap C\} &= \Pr\{(\text{H}, \text{H})\} = \tfrac{1}{4}, \\ \Pr\{A\} \cdot \Pr\{C\} &= \tfrac{1}{2} \cdot \tfrac{1}{2} = \tfrac{1}{4}, \end{aligned}$$

and

$$\begin{aligned} \Pr\{B \cap C\} &= \Pr\{(\text{H}, \text{H})\} = \tfrac{1}{4}, \\ \Pr\{B\} \cdot \Pr\{C\} &= \tfrac{1}{2} \cdot \tfrac{1}{2} = \tfrac{1}{4}. \end{aligned}$$

■

Independence can also be defined for more than two events. For example, *three* events A, B, and C are said to be **pairwise independent** if any pair of them is independent of each of the other two, i.e., if

(9.21)
$$\begin{aligned} \Pr\{A \cap B\} &= \Pr\{A\} \Pr\{B\} \\ \Pr\{A \cap C\} &= \Pr\{A\} \Pr\{C\} \\ \Pr\{B \cap C\} &= \Pr\{B\} \Pr\{C\}. \end{aligned}$$

Furthermore, these same events are said to be **mutually independent** if, in addition to (9.21) we also have

(9.22) $$\Pr\{A \cap B \cap C\} = \Pr\{A\} \Pr\{B\} \Pr\{C\}.$$

Similarly, n events $A_1, A_2, \ldots, A_n$ are called *mutually independent* if

(9.23) $$\Pr\{A_1 \cap A_2 \cap \cdots \cap A_n\} = \Pr\{A_1\} \Pr\{A_2\} \cdots \Pr\{A_n\},$$

and if each collection of fewer than n of the sets is *itself* mutually independent. In other words, if we select any subset of $\{A_1, A_2, \ldots, A_n\}$, say $\{A_{i_1}, A_{i_2}, \ldots, A_{i_k}\}$, where k is an integer between 2 and n, then

(9.24) $$\Pr\{A_{i_1} \cap A_{i_2} \cap \cdots \cap A_{i_k}\} = \Pr\{A_{i_1}\} \Pr\{A_{i_2}\} \cdots \Pr\{A_{i_k}\}.$$

It is possible (but luckily unusual) for three events to be pairwise independent but not mutually independent, as the next example shows.

EXAMPLE 9.12 Show that the three events A, B, and C of Example 9.11 are not mutually independent.

SOLUTION We have already seen that the three events are *pairwise* independent, in the solution to Example 9.11. To check for *mutual* independence, we need to see whether (9.22) is true:

$$\Pr\{A \cap B \cap C\} = \Pr\{(\text{H}, \text{H})\} = \tfrac{1}{4},$$

but

$$\Pr\{A\} \cdot \Pr\{B\} \cdot \Pr\{C\} = \tfrac{1}{2} \cdot \tfrac{1}{2} \cdot \tfrac{1}{2} = \tfrac{1}{8}.$$

It follows that the events A, B, and C are not mutually independent. ■

The definition of mutual independence is rather complicated; and in a given situation it may be difficult to tell whether or not a given set of events is pairwise or mutually independent. However, as a practical matter, it is rarely necessary to check whether a given set of events is independent, because probability spaces are usually constructed in such a way that certain events are *forced* to be mutually independent. For the rest of this section, we will study this important construction. We begin with a simple example.

EXAMPLE 9.13 Consider these two experiments:

- *Experiment 1*: Flip a coin.
- *Experiment 2*: Roll a die.

Construct a single probability space to model the compound Experiment 3:

- *Experiment 3*: Independently flip a coin *and* roll a die.

SOLUTION We can begin by giving separate probability spaces for Experiments 1 and 2. They are, of course,

$$\Omega_1 = \{H, T\}; \qquad \Pr\{H\} = \Pr\{T\} = \tfrac{1}{2}$$

$$\Omega_2 = \{1, 2, 3, 4, 5, 6\}; \qquad \Pr = \tfrac{1}{6} \text{ each.}$$

Since an outcome of Experiment 3 is nothing more than a pair of outcomes, one from Experiment 1 and one from Experiment 2, it is natural to take as the sample space Ω_3 for Experiment 3 the *Cartesian product* of Ω_1 and Ω_2, which consists of the set of ordered pairs of outcomes from Experiments 1 and 2:

$$\Omega_3 = \{(H, 1), (H, 2), (H, 3), (H, 4), (H, 5), (H, 6), (T, 1), (T, 2), (T, 3), (T, 4), (T, 5), (T, 6)\}.$$

For example the event (T, 4) in Ω_3 represents the outcome "coin shows tails, and die shows 4."

To complete the construction of a probability space for Experiment 3, we must assign probabilities to each of the 12 atoms in Ω_3. But this assignment is forced on us by the definition (9.19) of independence. For example, the atom (T, 4) in Ω_3 represents the simultaneous occurrence of the events "coin shows tails" *and* "die shows 4." By the *definition* of Experiment 3, these events are independent, and so we have, from (9.19),

$$\Pr\{(T, 4)\} = \Pr\{T\} \Pr\{4\} = \tfrac{1}{2} \cdot \tfrac{1}{6} = \tfrac{1}{12}.$$

Similarly each of the other 11 atoms in Ω_3 must be assigned probability $\frac{1}{12}$. Thus our solution is $\Omega_3 = \Omega_1 \times \Omega_2$, with each atom in Ω_3 having probability $\frac{1}{12}$. ■

Using exactly the same ideas as in the solution to Example 9.13, if we are given any two probability spaces, we can construct a third probability space, which is a model for **independent trials**, i.e., an independent performance of the experiments corresponding to the two original spaces. Here is the construction: If Ω_1 and Ω_2 are the original two sample spaces, the new sample space is

(9.25) $$\Omega_3 = \Omega_1 \times \Omega_2.$$

The atoms of Ω_3 are the ordered pairs (ω_1, ω_2), where ω_1 and ω_2 are atoms in Ω_1 and Ω_2, respectively. The probability of the atom (ω_1, ω_2) is defined by

(9.26) $$\Pr\{(\omega_1, \omega_2)\} = \Pr_1\{\omega_1\} \Pr_2\{\omega_2\},$$

where $\Pr_1$ and $\Pr_2$ are the probability functions on Ω_1 and Ω_2. The probability space constructed in this way is called the **product** of the two original spaces.

Notice that when we take the product of two probability spaces, every event in the original two spaces Ω_1 and Ω_2 has a counterpart in Ω_3. Consider again Example 9.13. There the event "coin shows tails," which is the *atom* {T} in Ω_1, is represented by the *compound event*

$$\{(T, 1), (T, 2), (T, 3), (T, 4), (T, 5), (T, 6)\}$$

in Ω_3. Although these two versions of "tails" are different as *sets*, the probability of "tails" is the same in both spaces:

Space	Event	Probability
Ω_1	{T}	$\frac{1}{2}$
Ω_3	{(T, 1), (T, 2), (T, 3), (T, 4), (T, 5), (T, 6)}	$\frac{1}{12}+\frac{1}{12}+\frac{1}{12}+\frac{1}{12}+\frac{1}{12}+\frac{1}{12}=\frac{1}{2}$

Similarly, the event "die shows a multiple of 3" is represented as follows:

Space	Event	Probability
Ω_2	{3, 6}	$\frac{1}{6}+\frac{1}{6}=\frac{1}{3}$
Ω_3	{(H, 3), (T, 3), (H, 6), (T, 6)}	$\frac{1}{12}+\frac{1}{12}+\frac{1}{12}+\frac{1}{12}=\frac{1}{3}$

It *always* turns out, fortunately, that the probability of an event that depends only on the outcome of one of the experiments is the same, whether the probability is computed in the original space or the product space (see Problems 23 and 24). Much more important, though, is the following fact, which explains why the product space construction is exactly the right way to study independent experiments:

(9.27) If A is an event that depends only on the outcome of the first experiment, and B is an event that depends only on the outcome of the second experiment, then A and B are independent.

For example, let $A = \{T\} \subseteq \Omega_1$ and $B = \{3, 6\} \subseteq \Omega_2$. We have just seen how to view A and B as events in $\Omega_3 = \Omega_1 \times \Omega_2$:

$$A = \{(T, 1), (T, 2), (T, 3), (T, 4), (T, 5), (T, 6)\}; \qquad \Pr\{A\} = \tfrac{1}{2}.$$

$$B = \{(H, 3), (T, 3), (H, 6), (T, 6)\}; \qquad \Pr\{B\} = \tfrac{1}{3}.$$

To check that A and B are independent, we note that

$$A \cap B = \{(T, 3), (T, 6)\}; \qquad \Pr\{A \cap B\} = \tfrac{1}{6}.$$

Thus $\Pr\{A \cap B\} = \Pr\{A\} \cdot \Pr\{B\}$, and so A and B are independent by (9.19). Although this calculation only illustrates a particular case of (9.27), a careful study of it shows how (9.27) can be proved in general. (See Problems 23–25.)

It is possible to take the product of more than two probability spaces. If $\Omega_1, \Omega_2, \ldots, \Omega_n$ the n sample spaces, the product has sample space

$$\Omega = \Omega_1 \times \Omega_2 \times \cdots \times \Omega_n.$$

The atoms of Ω are the ordered samples $(\omega_1, \omega_2, \ldots, \omega_n)$, where each ω_i is an atom from Ω_i; the probability of such an atom is defined by

(9.28) $$\Pr\{\omega\} = \Pr_1\{\omega_1\} \Pr_2\{\omega_2\} \cdots \Pr_n\{\omega_n\},$$

where $\Pr_1, \Pr_2, \ldots, \Pr_n$ are the probability functions on $\Omega_1, \Omega_2, \ldots, \Omega_n$, respectively. In this situation, the statement corresponding to (9.27) is this (see Problem 26):

(9.29) If $A_1, A_2, \ldots, A_n$ are events in Ω such that A_i depends only on the ith experiment, then $A_1, A_2, \ldots, A_n$ are mutually independent.

This construction allows us to model the independent performance of any number of experiments.

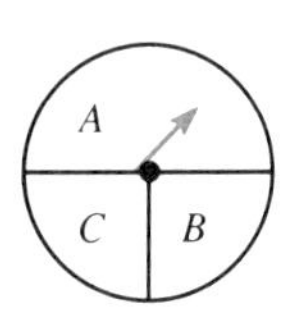

Figure 9.9 The spinner for Example 9.14.

EXAMPLE 9.14 Construct a probability space corresponding to the compound experiment of flipping a coin, rolling a die, and spinning the spinner of Figure 9.9. (The probability that the spinner will land in a given region is the total angle spanned by the region, divided by 360°. See Problems 9.2.28–9.2.32 for other examples.)

SOLUTION The component probability spaces are

$$\Omega_1 = \{\text{H}, \text{T}\}; \qquad \Pr\{\text{H}\} = \Pr\{\text{T}\} = \tfrac{1}{2}$$

$$\Omega_2 = \{1, 2, 3, 4, 5, 6\}; \qquad \Pr = \tfrac{1}{6} \text{ each}$$

$$\Omega_3 = \{A, B, C\}; \qquad \Pr\{A\} = \tfrac{1}{2}; \Pr\{B\} = \Pr\{C\} = \tfrac{1}{4}.$$

Thus the product space $\Omega = \Omega_1 \times \Omega_2 \times \Omega_3$ has $2 \times 6 \times 3 = 36$ atoms: $(\text{H}, 1, A), (\text{H}, 1, B), \ldots, (\text{T}, 6, C)$. The probabilities of the atoms are obtained by multiplying the probabilities of the components. Thus, $\Pr\{(\text{H}, 1, A)\} = \frac{1}{2} \cdot \frac{1}{6} \cdot \frac{1}{2} = \frac{1}{24}, \ldots, \Pr\{(\text{T}, 6, C)\} = \frac{1}{2} \cdot \frac{1}{6} \cdot \frac{1}{4} = \frac{1}{48}$. ∎

If we want a model for *n independent repetitions of the same experiment*, the component spaces Ω_i will all be the same. This is by far the most important application of the product construction.

EXAMPLE 9.15 Consider an experiment in which there are only two possible outcomes, "success" and "failure," which occur with probability p and q, respectively. (Of course $p + q = 1$.) Construct a probability space corresponding to n independent repetitions of this experiment.

SOLUTION The basic probability space here is

$$\Omega = \{\text{S}, \text{F}\}; \qquad \Pr\{\text{S}\} = p, \Pr\{\text{F}\} = q.$$

The sample space for n independent repetitions of this experiment is $\Omega \times \Omega \times \cdots \times \Omega$, which has 2^n atoms, i.e., all possible sequences of S's and F's. If the sequence has k S's and $(n - k)$ F's, its probability will be $p^k q^{n-k}$. For example with $n = 3$ we have the following table:

ω	$\Pr\{\omega\}$
SSS	p^3
SSF	p^2q
SFS	p^2q
SFF	pq^2
FSS	p^2q
FSF	pq^2
FFS	pq^2
FFF	q^3

Note that the sum of the probabilities is

$$p^3 + 3p^2q + 3pq^2 + q^3 = (p + q)^3 \quad \text{(by the binomial theorem)}$$
$$= 1.$$

EXAMPLE 9.16 In the probability space of Example 9.15, calculate the probability of having *at least one* success.

SOLUTION This problem is an example of a situation in which it's easier to compute the probability of the *complement* of the given event, and use the relationship

$$\text{(9.30)} \qquad \Pr\{A\} = 1 - \Pr\{A'\},$$

where A' is the complement of A. [Note that (9.30) is a special case of (9.6): $\Pr\{A\} = \Pr\{\Omega - A'\} = \Pr\{\Omega\} - \Pr\{A'\} = 1 - \Pr\{A'\}$.] In this case, the event A is "at least one success." The complementary event is "*not* at least one success," which is the same as "all failures." In terms of the sample space of Example 9.15, this, in turn, is the event FFFF · · · F, which has probability $q \cdot q \cdots q = q^n$. Hence by Equation (9.30), we have

$$\text{Probability of at least one success} = 1 - q^n.$$

Repeated independent identical trials, like those in Example 9.15, are called **Bernoulli* trials**. Often, when Bernoulli trials are involved, all that's important, as in Example 9.16, is simply the *number of successes* in n trials, rather than the exact sequence of S's and F's involved. Our last example deals with this situation in a general setting. It is so important that we will repeat it (in slightly different language) in Section 9.5.

EXAMPLE 9.17 In the probability space of Example 9.15, calculate the probability of having exactly k successes in n trials.

SOLUTION Let's begin by doing a special case, say $n = 4$ and $k = 2$. There are $2^4 = 16$ atoms in the given probability space, but only those with 2 S's and 2 F's will contribute to the desired probability:

$$\{\text{SSFF, SFSF, SFFS, FSSF, FSFS, FFSS}\}.$$

These 6 favorable cases correspond to the 6 ways of choosing 2 of the 4 possible positions for the S's: $\binom{4}{2} = 6$. Each of these 6 atoms has probability p^2q^2, and so the desired probability is $6p^2q^2$.

* Jacob Bernoulli (1654–1705), Swiss mathematician and physicist. He was the oldest of eight Bernoullis who made important scientific contributions in the seventeenth and eighteenth centuries. His intellectual battles with his younger brother Johann would make good material for a television soap opera. His book *Ars Conjectandi* published eight years after his death contains the first systematic treatment of combinatorics and its relationship to probability theory.

In general, to specify a sequence with exactly k S's and $(n-k)$ F's, we choose k of the n possible positions for the S's, and put F's in the remaining $(n-k)$ positions. Thus there will be $\binom{n}{k}$ atoms consisting of k S's and $(n-k)$ F's (the subset rule), and each will have probability $p^k q^{n-k}$. Hence the probability of exactly k successes in n trials is

$$\Pr\{k \text{ successes}\} = \binom{n}{k} p^k q^{n-k}. \tag{9.31}$$

For example with $n = 4$, these probabilities are given below:

k	$\Pr\{k\}$
0	q^4
1	$4pq^3$
2	$6p^2q^2$
3	$4p^3q$
4	p^4

Notice that $q^4 + 4pq^3 + 6p^2q^2 + 4p^3q + p^4 = (p+q)^4 = 1$ by the binomial theorem. In Section 9.5, Example 9.21, we will meet this set of numbers again, under the name *binomial density*. ■

Repeated independent identical trials in which each trial has more than two possible outcomes are called **generalized Bernoulli trials**. The analysis we have done here carries over to the more general case, except that *multinomial* rather than *binomial* coefficients appear. See Problems 36–38.

Problems for Section 9.3

1. In Example 9.11, let $A = \{\text{heads on the first flip}\}$, $B = \{\text{heads on the second flip}\}$ (as in the text), and $D = \{\text{exactly one head}\}$. Show that A, B, and D are pairwise independent.

2. In Problem 1, give an intuitive argument that the occurrence of A should not affect the probability of D.

3. Let A and B be as in Example 9.11, and D as in Problem 1. Show that A, B, and D are not mutually independent.

4. Let C be as in Example 9.11, and D as in Problem 1. Are C and D independent? Justify your answer.

5. Let A, B, C, and D be four *arbitrary* events in the same probability space. Write down the equations (9.24), which are needed to guarantee that A, B, C, and D are mutually independent.

6. How many equations are needed to guarantee that n sets $A_1, A_2, \ldots, A_n$ are mutually independent? [*Hint*: For $n = 2$, the answer is *one*, viz., Equation (9.19); for $n = 3$, the answer is *four*, viz., the equations in (9.21) and (9.22); for $n = 4$, see Problem 5.]

7. Consider the following experiments:

- *Experiment 1*: Pick a card from an ordinary deck.
- *Experiment 2*: Roll two fair dice.

Construct a single probability space to model the compound experiment consisting of Experiment 1 followed by Experiment 2.

8. In Problem 7, calculate the probability of the event

$$\{\text{queen of hearts, sum of dice} = 10\}.$$

9. In Example 9.14, find the probability of the event

{coin comes up heads, face of die is even, spinner lands in A}.

10. In Example 9.14, find the probability of the event

{coin comes up tails, face of die is less than 3, spinner lands in A or B}.

Problems 11–14 refer to a sequence of n Bernoulli trials; see Examples 9.15, 9.16, and 9.17.

11. Find the probability of exactly two successes in five trials.

12. Find the probability of at least two successes in five trials.

13. Find the probability of at most three successes in six trials.

14. Find the probability of an even number of successes in six trials.

15. Consider families with three children. If b denotes "boy" and g denotes "girl," assume each of the eight possibilities ($bbb, bbg, \ldots, ggg$) has probability $\frac{1}{8}$. Consider the three events A, B, and C:

A: Family has children of both sexes.
B: Family has exactly two boys.
C: Family has at most one girl.

Which (if any) of the three pairs of these events are independent?

16. Consider the probability space $\Omega = \{a, b, c, d, e\}$ with $\Pr\{a\} = \frac{2}{5}$, $\Pr\{b\} = \frac{4}{15}$, $\Pr\{c\} = \frac{2}{15}$, $\Pr\{d\} = \frac{2}{15}$, and $\Pr\{e\} = \frac{1}{15}$. Find all subsets of Ω which are independent of the event $A = \{a, b\}$.

17. Consider a normal six-sided die, on which, in addition to the usual numbering, each face is also marked either H or T, as follows:

1	2	3	4	5	6
H	T	H	H	T	T

(a) Construct a probability space corresponding to the experiment of rolling this die.
(b) Is the event H independent of the event $\{3, 6\}$? Is H independent of $\{2, 4, 6\}$?

18. Consider an urn that contains b black balls and w white balls.
(a) Describe a sample space for the experiment of selecting n balls (with replacement) from this urn.
(b) What is the probability that all n balls are white?
(c) In the special case $b = w = 1$ and $n = 2$, what is the answer to part (b)?

19. In a sequence of eight Bernoulli trials, compute the probability of $0, 1, \ldots, 8$ successes in terms of p.

20. In Problem 19, for which values of p is 4 the most probable number of successes?

21. A famous basketball player is shooting free throws. He will make the shot with probability .90, and will miss with probability .10. Suppose he attempts 10 shots in a row.
(a) What is the probability that he makes all 10 shots?
(b) What is the probability that he misses all 10?
(c) What is the probability that he misses at least 1?

22. Consider a product space $\Omega = \Omega_1 \times \Omega_2$. Is it possible for Ω to be a *uniform* space if Ω_1 and/or Ω_2 are *not* uniform? If your answer is yes, give an example of such an Ω_1 and Ω_2. If your answer is no, explain clearly why it isn't possible.

23. If $A \subseteq \Omega_1$ and $B \subseteq \Omega_2$, show that in the product space $\Omega_1 \times \Omega_2$, $\Pr\{A \times B\} = \Pr_1\{A\} \cdot \Pr_2\{B\}$.

24. Use the result of Problem 23 to show that an event in the product space which depends only on the first experiment has the same probability in Ω_1 as in $\Omega_1 \times \Omega_2$. (*Hint*: If $A \subseteq \Omega_1$ is the event in question, its representation in $\Omega_1 \times \Omega_2$ is $A \times \Omega_2$.)

25. Use the result of Problem 24 to prove that (9.27) is generally true.

26. Generalize Problem 24, and use the result to prove (9.29).

27. Show that the probability assignments (9.26) and more generally, (9.28), are legal, i.e., that the probabilities add to 1.

28. Refer to Example 9.12 and construct a probability space with *four* events A, B, C, D which are pairwise but not mutually independent.

29. Is it possible to find events A, B, and C such that (9.22) holds, but *none* of the three equations in (9.21) hold? Explain fully.

30. We have seen that $\Pr\{A \cap B\} = \Pr\{A\} \cdot \Pr\{B\}$ if the events A and B are *independent*. This is similar to (9.4), which says that $\Pr\{A \cup B\} = \Pr\{A\} + \Pr\{B\}$ if A and B are *disjoint*. However (9.8) says that $\Pr\{A \cup B\} \leq \Pr\{A\} + \Pr\{B\}$ in any case. Is it perhaps true that $\Pr\{A \cap B\} \leq \Pr\{A\} \Pr\{B\}$ for all pairs of events A and B? Explain fully.

31. A spinner is in the form of a circle divided into four equal sectors. If the spinner is spun n times, what is the probability that at least one sector has not yet occurred?

32. Consider the four spinners* depicted in the accompanying figure. Each of the obtuse angles is 120°, and the numbers inside the circles indicate the score given to the player who spins. For example, if the C spinner lands in the lower region, the player receives 6 points. Assume that each of the spinners is spun randomly and independently.
(a) Compute each of the six probabilities: Pr{A beats B}, Pr{A beats C}, ..., Pr{C beats D}.
(b) Which spinner would you prefer to have?

* The spinners in Problem 32 are due to Bradley Efron; those in Problem 33 are due to Colin Blyth. Both were reported by Martin Gardner in *Scientific American* (December 1970 and March 1976; "Mathematical Games" column. Efron's spinners were actually presented as dice.).

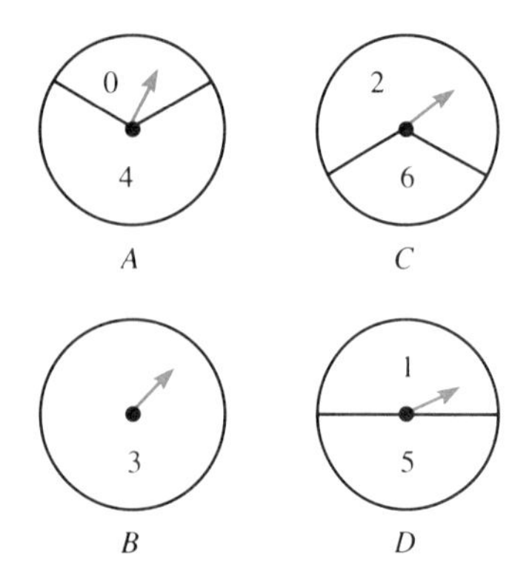

33. Consider the three spinners in the accompanying figure. (The probabilities of the various sectors are given on the drawings. Thus in spinner B, the probability of landing in the upper right sector is .22, and in this case the player scores 4 points.) The spinners are spun randomly and independently.

(a) Compute Pr{A beats B}, Pr{A beats C}, and Pr{B beats C}.

(b) Compute Pr{A beats B and C}, Pr{B beats A and C}, and Pr{C beats A and B}.

(c) Which spinner would you prefer to have?

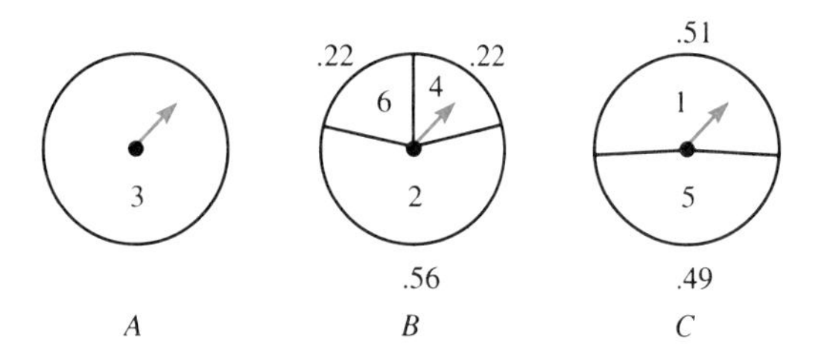

34. As in Equation (9.29), suppose $A_1, A_2, \ldots, A_n$ are events in a product space $\Omega = \Omega_1 \times \cdots \times \Omega_n$ such that A_i depends only on the ith experiment. Let A be the event "at least one of the A_i's occurs." Compute $\Pr\{A\}$ in terms of $\Pr\{A_1\}, \Pr\{A_2\}, \ldots, \Pr\{A_n\}$. (*Hint*: Look to Example 9.16 for guidance.)

35. (The Birthday Surprise.) A large urn contains n balls labelled $1, 2, \ldots, n$. A fixed number, say k, of these balls are selected with replacement.

(a) What is the probability that all k balls are different?

(b) For $n = 365$, find the largest value of k such that the probability found in part (a) is greater than $\frac{1}{2}$.

(c) If 30 people are selected at random, what is the probability that at least 2 will have the same birthday? [A calculator will be useful in parts (b) and (c).]

36. (Generalized Bernoulli Trials.) Consider an experiment in which there are m possible outcomes, labelled $b_1, \ldots, b_m$, and b_i occurs with probability p_i, $i = 1, \ldots, m$ (where all p_i are nonnegative and $p_1 + \cdots + p_m = 1$). Construct a probability space corresponding to n independent repetitions of this experiment. (The atoms will be all possible sequences of length n with components $b_1, \ldots, b_m$. When $m = 2, p_1 = p$, and $p_2 = q$, we obtain ordinary Bernoulli trials as in Example 9.15.)

37. Continue Problem 36 with $m = 3$, $p_1 = \frac{1}{2}$, $p_2 = \frac{1}{3}$, $p_3 = \frac{1}{6}$, and $n = 5$.

(a) Show that the probability of the sequence $b_1 b_3 b_3 b_2 b_1$ is $p_1^2 p_2 p_3^2$.

(b) Show that the number of sequences in which b_1 occurs exactly twice, b_2 once, and b_3 twice, is $5!/2!1!2!$.

(c) Show that the probability that in five trials, b_1 will occur exactly twice, b_2 once, and b_3 twice, is $(5!/2!1!2!)p_1^2 p_2 p_3^2$.

38. Consider a sequence of n generalized Bernoulli trials, as in Problem 37. Show that the probability that the event {outcome b_1 will occur exactly n_1 times and outcome b_2 will occur exactly n_2 times and ... outcome b_m will occur exactly n_m times} (where $n_1 + n_2 + \cdots + n_m = n$) is

$$\frac{n!}{n_1!n_2!\cdots n_m!} p_1^{n_1} p_2^{n_2} \cdots p_m^{n_m}.$$

9.4 Dependent Trials and Tree Diagrams

In Section 9.2, we introduced the notion of conditional probabilities, and promised a serious study of sequences of *dependent* events after we had studied the simpler (and more important) notion of *independent* events, in Section 9.3. So now we will resume our study of dependent events, and look at conditional probability calculations that are more complicated than those in Section 9.2. The general situation we wish to model here is a sequence of experiments, or *trials*, where the probabilities of the various possible outcomes depend on the earlier results.

We have already seen one simple pair of dependent trials, Example 9.10, which involved the three urns depicted in Figure 9.10. In the example, the experiments were as follows:

- *Experiment 1*: Select an urn.
- *Experiment 2*: Choose a ball from the urn.

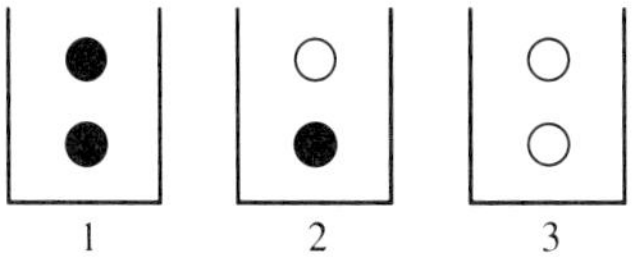

1 2 3

Figure 9.10 The three urns from Example 9.10.

Plainly, the outcome of Experiment 2 depends very much on the outcome of Experiment 1! We can illustrate the *exact* dependence using a **tree diagram** (see Figure 9.11). In this tree diagram, each of the branches is labelled with an appropriate probability. For example, the branch from the root to U_3 is labelled $\frac{1}{3}$, because $\Pr\{U_3\} = \frac{1}{3}$; the branch from U_2 to B is labelled $\frac{1}{2}$ because $\Pr\{B|U_2\} = \frac{1}{2}$; and so on. Furthermore, it is possible to compute compound probabilities like $\Pr\{U_2 \cap W\}$ using the tree diagram by *multiplying* the appropriate probabilities. In the case of the event $\{U_2 \cap W\}$, the corresponding path in the tree is root $\rightarrow$ $U_2 \rightarrow W$, and hence $\Pr\{U_2 \cap W\} = \frac{1}{3} \cdot \frac{1}{2} = \frac{1}{6}$. This is just another way of looking at Equation (9.16), which tells us that $\Pr\{U_2 \cap W\} = \Pr\{U_2\} \cdot \Pr\{W|U_2\}$. A tree diagram can sometimes be very helpful in visualizing a sequence of dependent trials; in general there will be a node corresponding to each possible intermediate result, and branches emanating from each node, one for each possible subsequent outcome.

EXAMPLE 9.18 A box contains six scrabble chips, one each for the letters in the word *BANANA*. A chip is withdrawn, examined, and *not* replaced (trial 1). A second chip is now examined (trial 2). Draw a tree diagram for this sequence of trials.

SOLUTION There are three possible outcomes for the first trial (A, B, or N), and because of the spelling of *BANANA*, the probabilities are

$$\Pr\{A\} = \tfrac{3}{6}, \qquad \Pr\{B\} = \tfrac{1}{6}, \qquad \Pr\{N\} = \tfrac{2}{6}.$$

There are also three possible outcomes for the second trial; but now the probabilities depend on the outcome of the first trial. For example, if the first letter chosen is A, then among the remaining five, two are A's, one is B, and two are

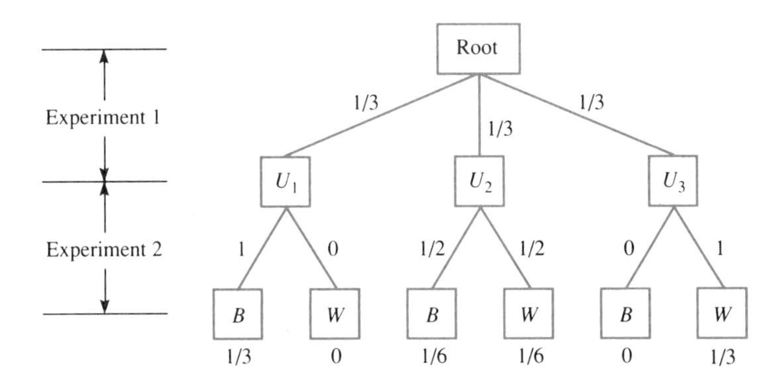

Figure 9.11 A tree diagram for Example 9.10.

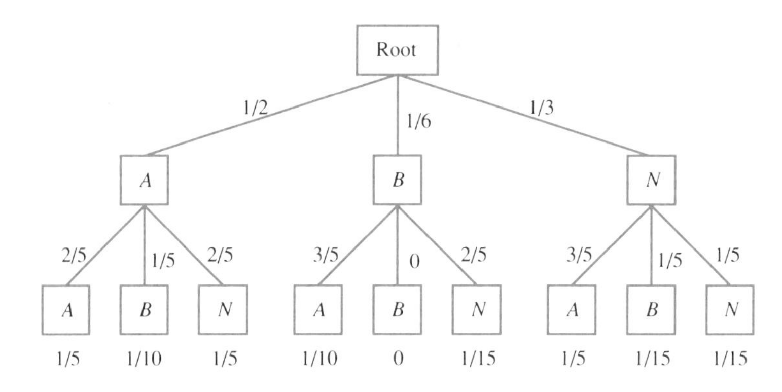

Figure 9.12 A tree diagram for Example 9.18.

N's. Thus

$$\Pr\{A|A\} = \tfrac{2}{5}, \qquad \Pr\{B|A\} = \tfrac{1}{5}, \qquad \Pr\{N|A\} = \tfrac{2}{5}.$$

Continuing in this way, we arrive at the tree diagram of Figure 9.12. ■

In applications, it is common to encounter a sequence of dependent trials in which only the outcome of the *last* trial is observable, and to be asked to infer something about the outcomes of earlier trials. In Example 9.10, for instance, we were told that the outcome of Experiment 2 was *white* and (essentially) asked to make an intelligent guess about the outcome of the earlier Experiment 1. Our primary goal in this section is to extend the ideas that led to the solution of Example 9.10; for while one rarely encounters urns filled with balls nowadays, it is often necessary to compute $\Pr\{B|A\}$, when $\Pr\{A|B\}$ is known (see Problem 24 for an almost realistic application). Here the main tool is (9.17), which says that $\Pr\{B|A\} = \Pr\{A|B\} \cdot \Pr\{B\}/\Pr\{A\}$. The calculation in Example 9.10, however, was a misleadingly simple application of (9.17). More often than not, the calculation of the denominator $\Pr\{A\}$ is somewhat involved, and one more idea is needed, the *theorem of total probability.*

Suppose the sample space Ω can be partitioned into several pairwise disjoint events $B_1, B_2, \ldots, B_n$, i.e. (see Section 1.1)

$$\Omega = B_1 \cup B_2 \cup \cdots \cup B_n,$$

where the B_i's are pairwise disjoint. In the language of probability, this says that *exactly one* of the events B_i will occur when the experiment is performed. In Example 9.18, for instance, we could take $B_1 = \{\text{first letter selected is } A\}$, $B_2 = \{\text{first letter is } B\}$, $B_3 = \{\text{first letter is } N\}$. Then if A is any other event,

$$A = (A \cap B_1) \cup (A \cap B_2) \cup \cdots \cup (A \cap B_n),$$

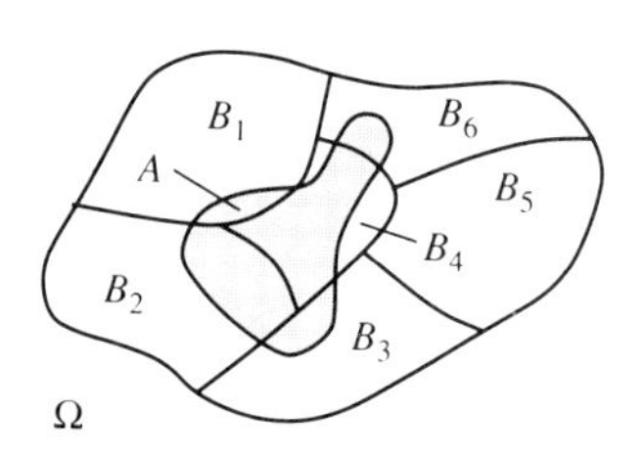

Figure 9.13 $A = (A \cap B_1) \cup \cdots \cup (A \cap B_n)$.

which is just a mathematical way of saying "the whole is equal to the sum of its parts" (see Figure 9.13).

In Example 9.18, for instance, if $A = \{\text{second letter is } N\}$, and if B_1, B_2, and B_3 are as above, then

$$A \cap B_1 = \{\text{first letter is } A, \text{ second letter is } N\}$$

$$A \cap B_2 = \{\text{first letter is } B, \text{ second letter is } N\}$$

$$A \cap B_3 = \{\text{first letter is } N, \text{ second letter is } N\}.$$

Since the events $(A \cap B_1), (A \cap B_2), \ldots, (A \cap B_n)$ are pairwise disjoint because the B_i are pairwise disjoint, we have, by (9.5),

$$\Pr\{A\} = \Pr\{A \cap B_1\} + \Pr\{A \cap B_2\} + \cdots + \Pr\{A \cap B_n\}. \tag{9.32}$$

But by (9.16), $\Pr\{A \cap B_i\} = \Pr\{A|B_i\}\Pr\{B_i\}$. Combining this fact with Equation (9.32), we get the promised **theorem of total probability**:

$$\begin{aligned}\Pr\{A\} &= \Pr\{B_1\}\Pr\{A|B_1\} + \cdots + \Pr\{B_n\}\Pr\{A|B_n\} \\ &= \sum_{i=1}^{n} \Pr\{B_i\}\Pr\{A|B_i\}.\end{aligned} \tag{9.33}$$

The following example is a typical application of these ideas.

EXAMPLE 9.19 Refer again to Figure 9.10; again an urn is selected at random. But this time, a ball is selected, found to be white, and *replaced.* Another ball is now selected from the urn, and *it too is white.* What is the probability that the urn selected was urn 3?

SOLUTION We proceed as before, and have

$$\Pr\{U_1\} = \Pr\{U_2\} = \Pr\{U_3\} = \tfrac{1}{3}.$$

If we denote the event {two white balls are selected} by WW, we can see by examining Figure 9.10 that

$$\Pr\{WW|U_1\} = 0, \qquad \Pr\{WW|U_2\} = \tfrac{1}{4}, \qquad \Pr\{WW|U_3\} = 1.$$

(Notice that once the urn is selected, the experiment of "selecting two balls, with replacement" is just an instance of Bernoulli trials with $n = 2$ and $p = 0, \frac{1}{2}$, or 1, depending on which urn is selected.) A tree diagram for this sequence of two dependent trials (select an urn; then select two balls with replacement) is shown in Figure 9.14.

The given problem is to find $\Pr\{U_3|WW\}$; by (9.17), this is

$$\Pr\{U_3|WW\} = \frac{\Pr\{WW|U_3\}\Pr\{U_3\}}{\Pr\{WW\}}. \tag{9.34}$$

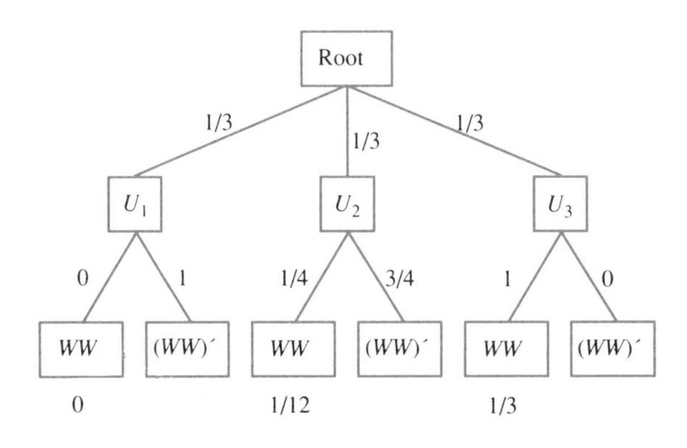

Figure 9.14 Tree diagram for Example 9.19.

We know that $\Pr\{WW|U_3\} = 1$, and $\Pr\{U_3\} = \frac{1}{3}$. All that remains is $\Pr\{WW\}$. To compute $\Pr\{WW\}$, we note that U_1, U_2, and U_3 partition the sample space (one and only one of the three urns is selected), and so by (9.33)

$$\begin{aligned}\Pr\{WW\} &= \Pr\{U_1\}\Pr\{WW|U_1\} + \Pr\{U_2\}\Pr\{WW|U_2\} \\ &\quad + \Pr\{U_3\}\Pr\{WW|U_3\} \\ &= \tfrac{1}{3}\cdot 0 + \tfrac{1}{3}\cdot\tfrac{1}{4} + \tfrac{1}{3}\cdot 1 = \tfrac{5}{12}.\end{aligned}$$

Notice that each term in this sum can be found from Figure 9.14 by tracing the paths in the tree from the root to the nodes labelled WW, and multiplying the corresponding branch labels. Hence by (9.34)

$$\Pr\{U_3|WW\} = \frac{1\cdot\frac{1}{3}}{\frac{5}{12}} = \tfrac{4}{5}.$$

■

Problems involving calculations like those in Example 9.19 are quite common, and often Formulas (9.17) and (9.33) are combined into one:

$$\Pr\{B|A\} = \frac{\Pr\{A|B\}\Pr\{B\}}{\sum_{i=1}^{n}\Pr\{B_i\}\Pr\{A|B_i\}},$$

which is called *Bayes' rule.** However, we recommend that you not memorize this complicated formula; it is much better just to bear the fundamental relationships (9.17) and (9.33) in mind, and combine them only when necessary.

* Thomas Bayes (1702–1761), English mathematician and clergyman. Bayes was one of the founders of modern statistical analysis, and even today there are Bayesian and non-Bayesian statisticians. However, "Bayes' rule" is not to be found (at least not explicitly) in his works!

Problems for Section 9.4

1. In Example 9.18, draw a tree diagram if *three* instead of two chips are drawn without replacement.

2. Suppose that in Example 9.18 the first chip is replaced before the second chip is drawn. Draw a tree diagram.

3. In Example 9.19, find $\Pr\{U_1|WW\}$ and $\Pr\{U_2|WW\}$.

4. Continuing Example 9.19, suppose *three* balls are examined (each ball is replaced before the next one is examined). Suppose all three balls are found to be *black.* Find the probabilities that the urn selected is U_1, U_2, or U_3.

5. A fair coin is flipped three times. Consider the following sequence of dependent trials:

- *Trial 1*: Count the number of heads on the first flip.
- *Trial 2*: Count the number of heads on the first two flips.
- *Trial 3*: Count the number of heads on the first three flips.

Draw a tree diagram for this situation.

6. (Tree Diagram for *In*dependent Trials.) Repeat Problem 5, but now the trials are as follows:

- *Trial 1*: Observe the outcome (i.e. H or T) of the first flip.
- *Trial 2*: Observe the outcome of the second flip.
- *Trial 3*: Observe the outcome of the third flip.

Draw a tree diagram.

7. Suppose an urn from Figure 9.10 is selected at random, three balls are selected (with replacement), and *all are white.* What is the probability that the next ball selected will be black?

8. Consider an urn containing b black and w white balls. Balls are selected, one at a time, from the urn, without replacement.
(a) What is the probability that the first ball selected will be white?
(b) What is the probability that the first two balls selected will be white?
(c) What is the probability that the first n balls selected will be white?

9. In a certain tire factory, tires are produced by three machines: M_1, M_2, and M_3. Machine M_1 produces 50% of the tires, M_2 produces $33\frac{1}{3}$%, and M_3 produces $16\frac{2}{3}$% of the total. Of the tires manufactured, 2% made by M_1 are defective, 4% made by M_2 are defective, and 3% made by M_3 are defective. A tire is selected at random and found to be defective. What is the probability that it was manufactured by M_1? By M_2? By M_3?

10. In Problem 9 a tire is selected at random and found *not* to be defective. What is the probability that it was manufactured by M_1? By M_2? By M_3?

11. Urn A has 10 white and 20 red balls, and urn B has 10 white and 50 red balls. An urn is picked at random and a ball is drawn. If the ball turns out to be white, find the probability that urn A was picked.

12. In a certain region, liberal arts colleges enroll 60% of the college population and technical schools get the rest. Liberal arts schools have a 20% dropout rate, while technical schools have a 30% dropout rate. A student picked at random is found to be a dropout. What is the probability that the student attended a technical school?

13. If a fair coin is tossed twice, find the conditional probability of two heads, given that at least one head appears.

14. In a certain city, registered Democrats outnumber registered Republicans by 4 to 1. In the last election, 90% of registered Republicans voted Republican, and 60% of registered Democrats voted Democratic. If a voter is known to have voted Republican, find the probability that she is a registered Democrat.

15. A biased penny has Pr{H} = .1, a biased nickel has Pr{H} = .2, and a biased dime has Pr{H} = .3. A coin is picked at random and tossed twice. Given that the first toss results in heads, find the probability of heads on the second toss.

16. If two cards are drawn without replacement from an ordinary deck, find the following probabilities:
(a) The first is red, given that the second is black;
(b) The first is red, given that there is at least one red.

17. If 10 tosses of a fair coin result in 6 heads and 4 tails, find the probability that one of those 6 heads occurred on the eighth toss.

18. Two cards are drawn without replacement from an ordinary deck.
(a) Find Pr{second card is a queen|first card is an ace} and Pr{first card is a queen|second card is an ace}.
(b) Give an intuitive argument to justify that the two probabilities computed in (a) are the same.

19. Consider the five urns shown in the accompanying figure. An urn is selected at random, and three balls are selected (with replacement); two are black and one is white. Calculate the probability of urns 1, 2, 3, 4, and 5.

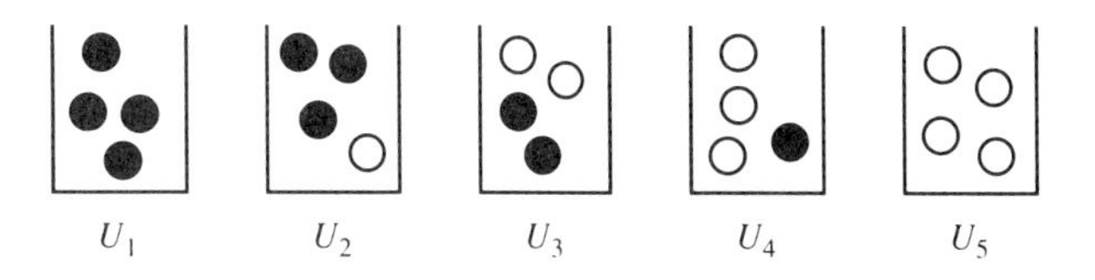

20. Repeat Problem 19, but now the balls are *not* replaced when they are withdrawn.

21. Suppose two dice are rolled and it is found that the two values showing differ by at most two. What is the most probable sum?

22. Suppose three dice are rolled and the sum is found to be greater than 12. What is the probability that no 6s are showing?

23. An urn from Figure 9.10 is selected according to the outcome of one spin of the depicted spinner. A ball is selected and found to be white. What is the probability that the urn is urn 2?

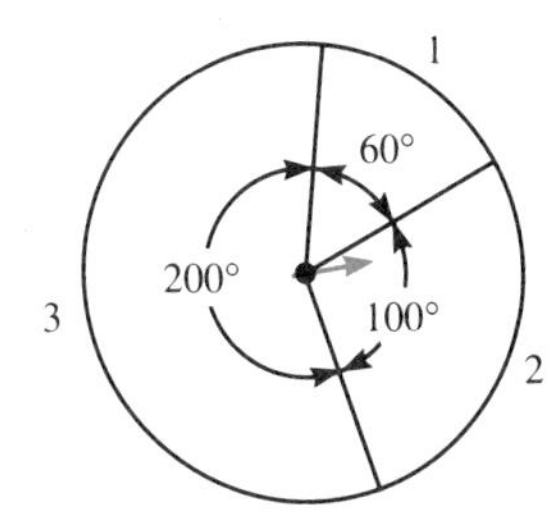

24. A message bit, i.e., a 0 or a 1, is selected at point A. The message 0 is selected with probability $\frac{3}{4}$, and 1 is selected with probability $\frac{1}{4}$. This message is then transmitted three times independently over an unreliable channel. This channel is described as follows: With probability $\frac{2}{3}$ the bit transmitted is correctly received, but with probability $\frac{1}{3}$ the bit received is the *opposite* of the bit transmitted. At point B the receiver gets three noisy versions of the transmitted message bit, i.e., one of the eight patterns 000, 001, 010, 100, 110, 101, 011, 111. For each of these eight possibilities, what is the receiver's best guess as to which message was selected at point A? (For example, if 000 is received, the best guess turns out to be that 0 was selected.)

25. There are n cars in a parking lot, with license plates numbered $1, 2, \ldots, n$. They leave, one at a time, at random.

(a) What is the probability that the second car will have a higher number than the first?

(b) What is the probability that the second *and* third cars will have numbers higher than the first?

(c) What is the probability that the first three cars will be in ascending order?

26. A certain fast-food chain once offered the following promotional game: Tickets were issued, and each ticket contained seven "spots," coated with opaque paint. Two of the spots contained the words *YOU WIN*; two contained the word *ZAP*! (The other three spots contained things irrelevant to this discussion.) The customer was supposed to scrape the paint off of the spots, one at a time, the object being to uncover both *YOU WIN* spots before either *ZAP*! What was the probability of winning?

9.5 Random Variables and Their Density Functions; Expectations

Consider yet again the experiment of rolling a pair of dice. We have now learned how to construct a mathematical model (i.e., a probability space) for this experiment (Section 9.1), or indeed for a series of independent repetitions of this experiment (Section 9.3). However, in our discussions so far we have been somewhat handicapped, because we have only been able to ask indirect yes-or-no questions of the form "Has event A occurred?"—for example, "Was the sum less than 9?" It is much more natural to ask more direct and quantitative questions, like "What *was* the sum?" In this section we will learn how to do this by introducing the notion of a *random variable.*

Roughly speaking, a random variable is the *numerical outcome* of an experiment. Tomorrow's high temperature, the number of babies to be born next month, the birthday of the next person you meet, etc., are all examples of random variables in the real world. Probability theory, however, cannot deal directly with the real world; it can deal only with *models* of the real world—probability spaces. And in a probability space, a random variable is not "uncertain" at all. The definition follows.

(9.35) A **random variable** is a function defined on the sample space of a probability space.*

For example, consider the 36-point probability space of Figure 9.3, which is a model for the dice-rolling experiment. Associated with each of the 36 atoms is an integer called SUM. This association can be thought of as a *function*, and indeed SUM is by the definition (9.35) a random variable. This particular random variable assumes only the values $2, 3, \ldots, 12$, with the by-now-familiar probabilities $\frac{1}{36}, \frac{2}{36}, \ldots$.

In general, the set of values assumed by a random variable is called the **range** of the random variable, and is often denoted by the letter R. The range of SUM in the example above is $R = \{2, 3, \ldots, 12\}$. If $R = \{x_1, x_2, \ldots, x_n\}$ is the range

* Thus a random variable isn't random, and isn't a variable!

of the random variable X, then the n events $\{X = x_i\}$, defined by

$$\{X = x_i\} = \{\omega : X(\omega) = x_i\},$$

form a partition of the sample space Ω, and the function f defined by

(9.36) $$f(x_i) = \Pr\{X = x_i\}, \qquad i = 1, 2, \ldots, n,$$

is called the **probability density function**, or more often just the **density function**, for the random variable X. For example, in the 36-point sample space for dice-rolling (Figure 9.3), we have

$$\begin{aligned}
&\{\text{SUM} = 2\} = \{(1,1)\}, && \Pr = \tfrac{1}{36},\\
&\{\text{SUM} = 3\} = \{(1,2),(2,1)\}, && \Pr = \tfrac{2}{36},\\
&\{\text{SUM} = 4\} = \{(1,3),(2,2),(3,1)\}, && \Pr = \tfrac{3}{36},\\
&\quad\vdots && \quad\vdots\\
&\{\text{SUM} = 12\} = \{(6,6)\}, && \Pr = \tfrac{1}{36}.
\end{aligned}$$

The density function for SUM is therefore given by the following table:

k	2	3	4	5	6	7	8	9	10	11	12
$f(k)$	$\frac{1}{36}$	$\frac{2}{36}$	$\frac{3}{36}$	$\frac{4}{36}$	$\frac{5}{36}$	$\frac{6}{36}$	$\frac{5}{36}$	$\frac{4}{36}$	$\frac{3}{36}$	$\frac{2}{36}$	$\frac{1}{36}$

EXAMPLE 9.20 Consider the experiment of flipping a biased coin three times; heads has probability $\frac{1}{3}$ and tails has probability $\frac{2}{3}$. Set up the appropriate probability space, and let X be the random variable that counts the total number of heads that occurs. Find the density function of X.

SOLUTION This experiment is an example of *Bernoulli trials* (Example 9.15), with $n = 3$, $p = \frac{1}{3}$, and $q = \frac{2}{3}$. Thus the appropriate probability space has eight atoms and is given below:

ω	$\Pr\{\omega\}$	X
HHH	$\frac{1}{27}$	3
HHT	$\frac{2}{27}$	2
HTH	$\frac{2}{27}$	2
HTT	$\frac{4}{27}$	1
THH	$\frac{2}{27}$	2
THT	$\frac{4}{27}$	1
TTH	$\frac{4}{27}$	1
TTT	$\frac{8}{27}$	0

The random variable X which counts the number of heads is given in the last column. The range of X is $\{0, 1, 2, 3\}$, and the density function is given below:

x	$f(x)$
0	$\frac{8}{27}$
1	$\frac{12}{27}$
2	$\frac{6}{27}$
3	$\frac{1}{27}$

For example,

$$\begin{aligned} f(1) &= \Pr\{X = 1\} = \Pr\{\text{HTT, THT, TTH}\} \\ &= \Pr\{\text{HTT}\} + \Pr\{\text{THT}\} + \Pr\{\text{TTH}\} \\ &= \tfrac{4}{27} + \tfrac{4}{27} + \tfrac{4}{27} = \tfrac{12}{27}. \end{aligned}$$

The other values of $f(x)$ are computed similarly. ■

Example 9.20 illustrates a special case of the *binomial density*, which is easily the most important density function in discrete probability. The general case is given in the next example, which is essentially identical to Example 9.17.

EXAMPLE 9.21 (The binomial density.) In a sequence of n Bernoulli trials (see Example 9.15), let X denote the number of successes. Find the density function of X.

SOLUTION Note that Example 9.20 is the special case $n = 3$, $p = \frac{1}{3}$, and $q = \frac{2}{3}$. In general, the random variable X can assume the $(n + 1)$ values $0, 1, \ldots, n$, and so its density function is $f(k)$, where

$$f(k) = \Pr\{X = k\}, \qquad \text{for} \qquad k = 0, 1, \ldots, n.$$

As we saw in Example 9.15, there are 2^n atoms in the underlying sample space, consisting of all n-letter words that can be made from the two letters S and F. Of these, exactly $\binom{n}{k}$ atoms have k S's and $(n - k)$ F's, and each such atom has probability $p^k q^{n-k}$. Hence the event $\{X = k\}$ is composed of $\binom{n}{k}$ atoms, each of probability $p^k q^{n-k}$, and so the required density function, which is called the **binomial density**, is

$$f(k) = \binom{n}{k} p^k q^{n-k}.$$

■

The density function $f(x)$ of the random variable X gives a complete description of X, for all practical purposes, and almost any question that can be asked about X can be answered, if $f(x)$ is known. For example, the question mentioned at the beginning of this section, "Has event A occurred?" is equivalent to "Is the value of X an element of A?" where A is some subset of the range R of X. The *probability* of the event $\{X \in A\}$ can be calculated by reasoning as follows:

Suppose A consists of m atoms: $A = \{x_1, x_2, \ldots, x_m\}$. Then the event $\{X \in A\}$ can be written as the disjoint union

$$\{X \in A\} = \{X = x_1\} \cup \{X = x_2\} \cup \cdots \cup \{X = x_m\},$$

and so by the rule for computing the probability of a disjoint union (see Equation 9.5), it follows that

(9.37)
$$\Pr\{X \in A\} = \Pr\{X = x_1\} + \Pr\{X = x_2\} + \cdots + \Pr\{X = x_m\}$$
$$= \sum_{k=1}^{m} \Pr\{X = x_k\}.$$

But since by the definition in Equation (9.36) $\Pr\{X = x_k\} = f(x_k)$, it follows from (9.37) that

(9.38)
$$\Pr\{X \in A\} = \sum_{x \in A} f(x).$$

Formula (9.38) is just a fancy way of expressing a very simple concept, as the next example shows.

EXAMPLE 9.22 Use the Formula (9.38) to calculate the probability "SUM is less than 9" in the experiment of rolling two dice.

SOLUTION For this application, $A = \{2, 3, 4, 5, 6, 7, 8\}$, and using the density for SUM given above, we have, applying (9.38),

$$\Pr\{\text{SUM} < 9\} = f(2) + f(3) + f(4) + \cdots + f(8)$$
$$= \tfrac{1}{36} + \tfrac{2}{36} + \tfrac{3}{36} + \cdots + \tfrac{5}{36} = \tfrac{26}{36}.$$

Alternatively, we could use the fact that $\Pr\{\text{SUM} < 9\} = 1 - \Pr\{\text{SUM} \geq 9\}$. (As we said in Example 9.16, it's sometimes easier to compute the probability of the complement of the event in question.)

$$\Pr\{\text{SUM} \geq 9\} = \Pr\{\text{SUM} = 9\} + \Pr\{\text{SUM} = 10\}$$
$$+ \Pr\{\text{SUM} = 11\} + \Pr\{\text{SUM} = 12\}$$
$$= \tfrac{4}{36} + \tfrac{3}{36} + \tfrac{2}{36} + \tfrac{1}{36} = \tfrac{10}{36},$$

which again leads to $\Pr\{\text{SUM} < 9\} = \frac{26}{36}$, but with a little less work. ■

Frequently more than one random variable is defined on a given probability space. In such a case it will usually be important to know how these random variables are related to each other. The *joint density function* of the random variables gives this information. If, for example, X and Y are two random variables, both defined on the same probability space, and if R_X and R_Y denote the ranges of X and Y, the **joint density function** of X and Y, denoted by $f(x, y)$, is defined, for all $x \in R_X$ and $y \in R_Y$, by

(9.39)
$$f(x, y) = \Pr\{X = x, Y = y\}.$$

(In (9.39), the notation $\{X = x,\ Y = y\}$ is short for $\{X = x \textit{ and } Y = y\}$.) More generally, if $X_1, X_2, \ldots, X_n$ are n random variables with ranges $R_1, R_2, \ldots, R_n$, their joint density function $f(x_1, x_2, \ldots, x_n)$ is defined by

$$f(x_1, x_2, \ldots, x_n) = \Pr\{X_1 = x_1, X_2 = x_2, \ldots, X_n = x_n\}, \tag{9.40}$$

for all ordered n-tuples $(x_1, \ldots, x_n)$ from the set $R_1 \times R_2 \times \cdots \times R_n$.

EXAMPLE 9.23 Continuing Example 9.20, let Y denote the number of *tails* in the *last two* flips. Calculate the joint density of X and Y.

SOLUTION We extend the table in Example 9.20 to include Y:

ω	$\Pr\{\omega\}$	X	Y
HHH	$\frac{1}{27}$	3	0
HHT	$\frac{2}{27}$	2	1
HTH	$\frac{2}{27}$	2	1
HTT	$\frac{4}{27}$	1	2
THH	$\frac{2}{27}$	2	0
THT	$\frac{4}{27}$	1	1
TTH	$\frac{4}{27}$	1	1
TTT	$\frac{8}{27}$	0	2

The range of X is as before $R_X = \{0, 1, 2, 3\}$. The range of Y is $R_Y = \{0, 1, 2\}$. The joint density function of X and Y is thus given by the following table:

(x, y)	$f(x, y)$	Contributing Atoms
(0, 0)	0	
(0, 1)	0	
(0, 2)	$\frac{8}{27}$	TTT
(1, 0)	0	
(1, 1)	$\frac{8}{27}$	THT, TTH
(1, 2)	$\frac{4}{27}$	HTT
(2, 0)	$\frac{2}{27}$	THH
(2, 1)	$\frac{4}{27}$	HHT, HTH
(2, 2)	0	
(3, 0)	$\frac{1}{27}$	HHH
(3, 1)	0	
(3, 2)	0	

Notice that each of the eight atoms in the sample space contributes to just one of the pairs (x, y). For example the atom HTH has $X = 2$ and $Y = 1$, and so it contributes its probability ($\frac{2}{27}$) to the (2, 1) entry in the joint density table. Notice also that some pairs $(x, y) \in R_X \times R_Y$ *cannot occur.* For example, even though $\{X = 0\}$ can occur (its probability is $\frac{8}{27}$), and $\{Y = 0\}$ can occur (its probability is $\frac{3}{27}$), these two events cannot occur *simultaneously*, and so $f(0, 0) = \Pr\{X = 0, Y = 0\} = 0$. Finally notice that the joint density function can be displayed in a *rectangular array*, as follows:

		X = 0	X = 1	X = 2	X = 3	Density of Y
Y	0	0	0	$\frac{2}{27}$	$\frac{1}{27}$	$\frac{3}{27}$
	1	0	$\frac{8}{27}$	$\frac{4}{27}$	0	$\frac{12}{27}$
	2	$\frac{8}{27}$	$\frac{4}{27}$	0	0	$\frac{12}{27}$
Density of X		$\frac{8}{27}$	$\frac{12}{27}$	$\frac{6}{27}$	$\frac{1}{27}$	

In this array the *columns* are indexed by the values assumed by X; i.e., 0, 1, 2, and 3; and the *rows* are indexed by the values assumed by Y, i.e., 0, 1, and 2. For example, the $X = 2$, $Y = 1$ entry is $\frac{4}{27}$, which means that $\Pr\{X = 2, Y = 1\} = \frac{4}{27}$. Notice that the *column sums*, i.e., $\frac{8}{27}, \frac{12}{27}, \frac{6}{27}, \frac{1}{27}$, give the density function of X; similarly the *row sums* give the density function of Y (see Problem 7). ■

Very often, in problems dealing with one or several random variables, only the density functions (and not the underlying probability space) will be given. With a little practice, you can learn how to solve the given problem without reference to the probability space. However, it is always *possible*, if necessary, to construct a probability space that will yield the given density function. The next example illustrates one way to do this.

EXAMPLE 9.24 Two random variables X and Y have the following joint density function:

		X = −1	X = +1
Y	1	0	$\frac{1}{4}$
	2	$\frac{1}{8}$	$\frac{1}{8}$
	3	$\frac{1}{2}$	0

Construct a probability space which yields the given joint density function.

SOLUTION The idea is to use a construction very closely related to the product space construction of Section 9.3. The *sample space* will have as atoms the six ordered pairs (x, y), where x is one of the possibilities for X, and y is one of the possibilities for Y. The *probability* of such an atom will be the corresponding entry in the joint density table. Finally, the random variables, X

and Y will be defined as the x and y components of the corresponding atoms:

ω	$\Pr\{\omega\}$	X	Y
$(-1, 1)$	0	-1	1
$(-1, 2)$	$\frac{1}{8}$	-1	2
$(-1, 3)$	$\frac{1}{2}$	-1	3
$(1, 1)$	$\frac{1}{4}$	1	1
$(1, 2)$	$\frac{1}{8}$	1	2
$(1, 3)$	0	1	3

We leave as Problem 9 the verification that the probability space above does yield the given density function for X and Y. ■

The procedure just illustrated can be used to produce a probability space corresponding to *any* given density function, as follows: If $f(x_1, x_2, \ldots, x_n)$ is the joint density function for the random variables $X_1, X_2, \ldots, X_n$, the desired probability space will have as a sample space Ω the *domain* of f, i.e., the set of all ordered lists $(x_1, x_2, \ldots, x_n)$ on which f is defined. The *probability* of the atom $(x_1, \ldots, x_n)$ is given by

$$\Pr\{(x_1, \ldots, x_n)\} = f(x_1, \ldots, x_n).$$

Finally, the random variable X_i is defined as the ith coordinate of the atom, i.e.,

$$X_i(x_1, x_2, \ldots, x_n) = x_i.$$

This construction, which we admit is a bit slippery, nevertheless shows why it is legal to talk about random variables when only their density functions, and not the underlying samples spaces, are given.

We will now discuss the problem of computing the *average value* of a given random variable. In probability texts the average value is traditionally called the *expectation*, or the *expected value* of the random variable, and we will use this terminology, too.

You already know how to compute the average value of a set of n numbers—you just add them up and divide by n. For example, the average of the six numbers $1, 2, 3, 4, 5, 6$ is

$$\frac{1+2+3+4+5+6}{6} = 3.5.$$

This simple rule is all you need to compute the expectation of a random variable defined on a *uniform* probability space. Thus if $\Omega = \{\omega_1, \omega_2, \ldots, \omega_M\}$ is a uniform probability space, i.e., if $\Pr\{\omega_i\} = 1/M$ for $i = 1, 2, \ldots, M$, and if X is a random variable defined on Ω, then the **expectation** of X, which is denoted by $E(X)$, is defined by

(9.41)
$$E(X) = \frac{1}{M} \sum_{i=1}^{M} X(\omega_i).$$

For example, let X represent the outcome of one roll of a fair die. Then X can be modelled using the six-element sample space shown in Figure 9.2(b), as follows:

ω	$\Pr\{\omega\}$	$X(\omega)$
1	$\frac{1}{6}$	1
2	$\frac{1}{6}$	2
3	$\frac{1}{6}$	3
4	$\frac{1}{6}$	4
5	$\frac{1}{6}$	5
6	$\frac{1}{6}$	6

Thus by Formula (9.41), the expectation of X is

$$E(X) = \tfrac{1}{6}(1 + 2 + 3 + 4 + 5 + 6) = 3.5.$$

This is no surprise!

EXAMPLE 9.25 Compute the expectation of the random variable SUM (Figure 9.3) which represents the sum of the values of two fair dice rolled simultaneously.

SOLUTION As Figure 9.3 shows, the underlying sample space contains 36 elements, each of which has probability $\frac{1}{36}$. This is a uniform space, and so we can apply Formula (9.41):

$$E(X) = \tfrac{1}{36}(2 + 3 + 4 + 5 + 6 + 7 + 3 + \cdots + 12) = \tfrac{252}{36} = 7.00.$$

Therefore, the average value of a roll of a pair of dice is 7.

When the underlying probability space isn't uniform, the definition of expectation has to be modified. For example, suppose Y represents the outcome of one roll of a dishonest die, as described by the following probability space:

i	$\Pr\{\omega_i\}$	$Y(\omega_i)$
1	.05	1
2	.10	2
3	.15	3
4	.20	4
5	.25	5
6	.25	6

This particular die will show "1" 5% of the time, "2" 10% of the time, etc. In a case like this, the expectation is defined to be the *weighted sum* of the values assumed by the random variable, where the weights are the given probabilities. For the random variable Y defined in the table above, we have

$$\begin{aligned} E(Y) &= (.05 \times 1) + (.10 \times 2) + (.15 \times 3) + (.20 \times 4) + (.25 \times 5) + (.25 \times 6) \\ &= 4.25. \end{aligned}$$

The average outcome of this die is 4.25, as compared to only 3.5 for an honest die, as we saw above.

In general, the expectation of a random variable X defined on an M-element sample space $\Omega = \{\omega_1, \omega_2, \ldots, \omega_M\}$ is given by the following formula:

(9.42)
$$E(X) = \sum_{i=1}^{M} \Pr\{\omega_i\} X(\omega_i).$$

We saw in Section 9.1 that the experiment of rolling a pair of dice can be modelled either by a 36-element uniform space or by an 11-element nonuniform space (Table 9.3). The next example shows that the expected value of SUM doesn't depend on which of these spaces is chosen!

EXAMPLE 9.26 Refer to Table 9.3, and consider the following probability space $\Omega = \{a, b, c, d, e, f, g, h, i, j, k\}$ and random variable S:

ω	$\Pr\{\omega\}$	$S(\omega)$
a	$\frac{1}{36}$	2
b	$\frac{2}{36}$	3
c	$\frac{3}{36}$	4
d	$\frac{4}{36}$	5
e	$\frac{5}{36}$	6
f	$\frac{6}{36}$	7
g	$\frac{5}{36}$	8
h	$\frac{4}{36}$	9
i	$\frac{3}{36}$	10
j	$\frac{2}{36}$	11
k	$\frac{1}{36}$	12

Calculate the expectation $E(S)$.

SOLUTION Using the definition (9.42), we have

$$E(S) = (\tfrac{1}{36} \cdot 2) + (\tfrac{2}{36} \cdot 3) + \cdots + (\tfrac{1}{36} \cdot 12) = 7.00.$$

Thus just as in Example 9.25, we find that the average value of one roll of a pair of dice is 7.00.

In this section we have seen that a random variable can be described just as well by its density function as by its probability space. And, as you might guess, there is a way to compute the expectation of a random variable directly from its density function. For example, consider the density function for the number of heads in three flips of a biased coin that we found in Example 9.20:

x	$f(x)$
0	$\frac{8}{27}$
1	$\frac{12}{27}$
2	$\frac{6}{27}$
3	$\frac{1}{27}$

To find the *average* number of heads in three flips, we form the *sum of the products* of the entries in the two columns of this table:

$$E(X) = (0 \cdot \tfrac{8}{27}) + (1 \cdot \tfrac{12}{27}) + (2 \cdot \tfrac{6}{27}) + (3 \cdot \tfrac{1}{27}) = 1.00.$$

Thus if the probability of heads on one flip is $\frac{1}{3}$, the *expected* number of heads in three flips is 1. In general, if X is a random variable with range $\{x_1, x_2, \ldots, x_n\}$, and density function $f(x_i)$, the expectation of X is given by the following formula:

(9.43)
$$E(X) = \sum_{i=1}^{n} x_i f(x_i).$$

[Equation (9.43) simply amounts to collecting the terms in Equation (9.42) according to the value of $X(\omega_i)$.]

EXAMPLE 9.27 Calculate the density function for the random variable Y in Example 9.23. Use this density function to calculate $E(Y)$.

SOLUTION Recall that Y represents the number of tails in the last two flips of three flips of a biased coin. Thus Y assumes the three values 0, 1, and 2. According to the table in Example 9.23, the probabilities of these three values are

$$\Pr\{Y = 0\} = \Pr\{\text{HHH, THH}\} = \tfrac{1}{27} + \tfrac{2}{27} = \tfrac{1}{9},$$

$$\begin{aligned}\Pr\{Y = 1\} &= \Pr\{\text{HHT, HTH, THT, TTH}\}\\ &= \tfrac{2}{27} + \tfrac{2}{27} + \tfrac{4}{27} + \tfrac{4}{27} = \tfrac{4}{9},\end{aligned}$$

$$\Pr\{Y = 2\} = \Pr\{\text{HTT, TTT}\} = \tfrac{4}{27} + \tfrac{8}{27} = \tfrac{4}{9}.$$

Therefore if we denote the density function for Y by f_Y, its values are given in the following table:

y	$f_Y(y)$
0	$\frac{1}{9}$
1	$\frac{4}{9}$
2	$\frac{4}{9}$

Using the sum-of-products rule (9.43), then, the expectation of Y is

$$E(Y) = (0 \cdot \tfrac{1}{9}) + (1 \cdot \tfrac{4}{9}) + (2 \cdot \tfrac{4}{9}) = \tfrac{4}{3}.$$

It is worth noting here that $E(Y)$ can also be calculated directly from the eight-element sample space whose table is given in Example 9.23, using the rule (9.42):

$$\begin{aligned}E(Y) &= (\tfrac{1}{27} \cdot 0) + (\tfrac{2}{27} \cdot 1) + (\tfrac{2}{27} \cdot 1) + (\tfrac{4}{27} \cdot 2) + (\tfrac{2}{27} \cdot 0)\\ &\quad + (\tfrac{4}{27} \cdot 1) + (\tfrac{4}{27} \cdot 1) + (\tfrac{8}{27} \cdot 2)\\ &= \tfrac{36}{27} = \tfrac{4}{3}.\end{aligned}$$

As a general rule, it is easier to compute the expectation of a random variable using the density function than the sample space.

Sometimes we will need to calculate expectations of *functions* of random variables. For example, suppose X is a random variable with range $\{-2, -1, 0, 1, 2\}$, whose density function is as follows:

x	$f(x)$
-2	$\frac{1}{3}$
-1	$\frac{1}{4}$
0	$\frac{1}{5}$
1	$\frac{1}{6}$
2	$\frac{1}{20}$

What is $E(X^2)$? The random variable X^2 assumes the values $\{0, 1, 4\}$:

x	x^2	$f(x)$
-2	4	$\frac{1}{3}$
-1	1	$\frac{1}{4}$
0	0	$\frac{1}{5}$
1	1	$\frac{1}{6}$
2	4	$\frac{1}{20}$

And $E(X^2)$ can be computed from the table using a sum-of-products rule:

$$E(X^2) = \left(4 \cdot \frac{1}{3}\right) + \left(1 \cdot \frac{1}{4}\right) + \left(0 \cdot \frac{1}{5}\right) + \left(1 \cdot \frac{1}{6}\right) + \left(4 \cdot \frac{1}{20}\right) = \frac{117}{60} = 1.95.$$

[Note that $E(X) = -\frac{2}{3} - \frac{1}{4} + \frac{1}{6} + \frac{2}{20} = -\frac{39}{60}$, so $E(X^2) \neq (E(X))^2$.] In general, if X is a random variable with density function $f(x)$, and if $\phi(X)$ is a function of the random variable X, then

(9.44)
$$E(\phi(X)) = \sum_x \phi(x) f(x).$$

EXAMPLE 9.28 Let X denote the number of successes in three Bernoulli trials, with probability of success $\frac{3}{4}$ on each trial. Find $E(X)$ and $E(X^3)$.

SOLUTION The random variable X has the binomial density (see Example 9.21) with $n = 3$, $p = \frac{3}{4}$, and $q = \frac{1}{4}$, so

$$P\{X = 0\} = \left(\frac{1}{4}\right)^3 = \frac{1}{64},$$

$$P\{X = 1\} = \binom{3}{1}\frac{3}{4}\left(\frac{1}{4}\right)^2 = \frac{9}{64},$$

$$P\{X = 2\} = \binom{3}{2}\left(\frac{3}{4}\right)^2\frac{1}{4} = \frac{27}{64},$$

$$P\{X = 3\} = \left(\frac{3}{4}\right)^3 = \frac{27}{64}.$$

By (9.43),

$$E(X) = \sum_{x=0}^{3} xf(x) = 0\left(\frac{1}{64}\right) + 1\left(\frac{9}{64}\right) + 2\left(\frac{27}{64}\right) + 3\left(\frac{27}{64}\right)$$

$$= \frac{144}{64} = \frac{9}{4} = 2.25.$$

By (9.44),

$$E(X^3) = \sum_{x=0}^{3} x^3 f(x) = 0^3\left(\frac{1}{64}\right) + 1^3\left(\frac{9}{64}\right) + 2^3\left(\frac{27}{64}\right) + 3^3\left(\frac{27}{64}\right)$$

$$= \frac{954}{64} = 14.91.$$

[Notice that in Example 9.28, $E(X)$ turns out to be equal to the *number of trials* (3) times the *probability of success* ($\frac{3}{4}$). In general, if X denotes the number of successes in n Bernoulli trials where the probability of success is p, then $E(X) = np$. See Problem 33 for a proof of this result.]

We'll conclude this section with a brief discussion of the significance of the expectation of a random variable. The expectation has both a superficial and a deep significance. Superficially, $E(X)$ is a single number that gives us a feeling for the typical values of X. For example, if we know the average temperature of a certain locality, we get at least a general feeling for the climate at that locality.

But $E(X)$ has a much deeper mathematical significance, which is, however, not as easy to appreciate at first. It goes something like this.

Suppose an experiment involving an observation of X is repeated n times under identical circumstances, and let $X_1, X_2, \ldots, X_n$ be the n values of X that are observed. The experimental average of the n observations is then $(X_1 + X_2 + \cdots + X_n)/n$. Experience shows that this experimental average is usually quite near to the expectation $E(X)$. For example, if the experiment is flipping a fair coin, with $X = 1$ representing heads and $X = 0$ for tails, then $(X_1 + X_2 + \cdots + X_n)/n$ represents the fraction of the time that heads appears, and is very likely to be near .5, which is the value of $E(X)$.

It turns out that there are theorems in probability theory that perfectly confirm such experimental evidence. The theorems are called *laws of large numbers*, and they say in a precise way that after a large number of trials, the experimental average is unlikely to differ significantly from the expectation $E(X)$. The laws of large numbers are the most important results in probability theory, and form the cornerstone of all advanced courses in the subject.

Problems for Section 9.5

1. In Example 9.20, let Y be the number of tails. Find the density function of Y.

2. In Example 9.20, let Z be the number of times a head is followed immediately by a tail; for example, $Z(\text{HHT}) = 1$, $Z(\text{THT}) = 1$, and $Z(\text{THH}) = 0$. Find the density function of Z.

3. As in Example 9.21, consider a sequence of Bernoulli trials, and let X be the number of successes. If $n = 6$, $p = .4$, and $q = .6$, find $\Pr\{3 \le X \le 5\}$.

4. In Problem 3, find $\Pr\{X \ge 1\}$.

5. If two fair dice are rolled and X_i is the face appearing on the ith die, $i = 1, 2$, find $\Pr\{X_1^2 + X_2^2 < 10\}$.

6. In Problem 5, calculate $\Pr\{X_1 X_2 > 24\}$.

7. Explain why, in Example 9.23, the row sums give the density function of Y, and the column sums give the density function of X.

8. If two fair dice are rolled and X is the larger of the two numbers, and Y is the smaller ($X = Y$ if the two results are equal), find the joint density of X and Y.

9. Verify that the construction given in Example 9.24 is correct; i.e., that the joint density of the random variables X and Y constructed in the solution is as given in the example statement.

10. Let X and Y be two random variables whose joint density function is given below:

		Y		
		0	1	2
X	−1	$\frac{1}{9}$	$\frac{2}{9}$	0
	0	0	0	$\frac{1}{27}$
	1	$\frac{1}{9}$	$\frac{1}{9}$	$\frac{4}{27}$
	2	$\frac{1}{9}$	$\frac{1}{9}$	$\frac{1}{27}$

Construct a probability space which yields the given joint density function, and calculate the individual density functions of X and Y.

11. Find a formula for the average of the n numbers in the set $S_n = \{1, 2, \ldots, n\}$.

12. In Example 9.24, find $E(X)$ and $E(Y)$.

13. In Example 9.24, find $E(Y^2)$.

14. In Example 9.23, compute $E(X + Y)$ and compare with $E(X) + E(Y)$. Is the result a coincidence?

15. Suppose an urn contains eight red and four green balls, and a sample of size 4 is selected. Let X denote the number of red balls selected. Find the density function of X under the following conditions:
(a) The sampling is done *with* replacement.
(b) The sampling is done *without* replacement.

16. (The **Hypergeometric Density**.) Generalizing Problem 15, suppose the urn contains a red and b green balls, and a sample of size r is selected (no replacement.) Let X denote the number of red balls selected. Calculate the density function of X.

17. In Example 9.23, let $Z = X + Y$. Calculate the density function of Z.

18. Consider the experiment of flipping a fair coin ($\Pr\{H\} = \Pr\{T\} = \frac{1}{2}$) four times. Define

$$X = \begin{cases} 0, & \text{if 0 or 1 heads appear;} \\ 1, & \text{if 2 heads appear;} \\ 2, & \text{if 3 or 4 heads appear.} \end{cases}$$

Calculate the density function of X.

19. Find a solution to Example 9.24, in which the corresponding probability space has only four atoms.

20. If A is an event in a probability space, its *indicator function* is a random variable X_A defined as follows:

$$X_A(\omega) = \begin{cases} 1, & \text{if } \omega \in A; \\ 0, & \text{if } \omega \notin A. \end{cases}$$

Compute the density function for X_A.

21. Consider the following nine-point sample space:

ω	$\Pr\{\omega\}$	X	Y	Z
(0, 0)	$\frac{1}{6}$	0	0	0
(0, 1)	$\frac{1}{12}$	1	0	1
(0, 2)	$\frac{1}{12}$	2	0	2
(1, 0)	$\frac{1}{6}$	1	0	1
(1, 1)	$\frac{1}{12}$	2	1	3
(1, 2)	$\frac{1}{12}$	3	2	5
(2, 0)	$\frac{1}{6}$	2	0	2
(2, 1)	$\frac{1}{12}$	3	2	5
(2, 2)	$\frac{1}{12}$	4	4	8

(a) Compute the density functions of X, Y, and Z.
(b) Compute the *joint* density functions of (X, Y) and (X, Y, Z).

22. Consider a random variable X whose density function is given below:

x	$f(x)$
−2	$\frac{1}{3}$
−1	$\frac{1}{4}$
0	$\frac{1}{5}$
1	$\frac{1}{6}$
2	$\frac{1}{20}$

Calculate the following:
(a) $E(X)$ (b) $E(X^3)$ (c) $E(X + 1)$.

23. If X is the random variable defined in Problem 22, find the density function for X^2. Use this density function to compute $E(X^2)$.

24. Consider the experiment of randomly placing three balls into three urns. Let X_i denote the number of balls in the ith urn, and let N denote the number of *empty* urns.
(a) Construct an appropriate probability space.
(b) Calculate the density functions for X_1, X_2, X_3, and N.

25. In Problem 24, calculate the joint density function for X_1 and N.

26. In Problem 24, calculate the joint density function for X_1, X_2, and X_3.

27. Let X, Y, and Z be random variables with the following density functions:

t	$f_X(t)$	$f_Y(t)$	$f_Z(t)$
-2	$\frac{1}{5}$	0	$\frac{2}{5}$
-1	$\frac{1}{5}$	0	0
0	$\frac{1}{5}$	$\frac{1}{5}$	$\frac{1}{5}$
1	$\frac{1}{5}$	$\frac{2}{5}$	0
2	$\frac{1}{5}$	$\frac{2}{5}$	$\frac{2}{5}$

Construct a probability space on which X, Y, and Z are all defined, and which yields the given three density functions.

28. An urn contains 10 balls labelled $1, 2, \ldots, 10$. One ball is removed, its number (X) noted, and replaced. A second ball is then removed, and its number (Y) is noted. Let Z denote the *larger* of the two numbers. Calculate the density function of Z.

29. Consider n Bernoulli trials, with $\Pr\{S\} = p$ and $\Pr\{F\} = q$. Let X denote the total number of S's and Y the number of S's among the first $n - 1$ trials. Calculate the joint density function of X and Y.

30. In a given probability space, the function $\Pr\{\cdot\}$ is *itself* a random variable, since it is a function defined on the sample space. Find the density function of $\Pr\{\cdot\}$ for the sample space that describes one roll of a pair of dice. [*Hint*: For one die, the answer is $f(\frac{1}{6}) = 1$, and $f(x) = 0$ for all $x \neq \frac{1}{6}$.]

31. (The game of *Yahtzee*.)
(a) You are given a single die, and allowed to roll it three times. Let $X_1 = 1$ if at least one of the rolls is a 6; and $X_1 = 0$ otherwise. Calculate the density function of X_1.
(b) Suppose now you are given *three* dice, and allowed to roll each one up to three times in an attempt to get a 6. (After a 6 is obtained, no more rolls of the particular die are allowed.) Let X denote the number of 6s you eventually get. (Thus X is one of the numbers 0, 1, 2, or 3.) Calculate the density function for X.

32. Consider a spinner which is divided into five equal sectors. If the spinner is spun five times (independently), let Z denote the number of *different* sectors which result.
(a) Calculate the density function for Z.
(b) Generalize to the case of m sectors and n spins.

33. Let X be the number of successes in n Bernoulli trials, with probability of success p on a given trial.
(a) Show that $E(X) = \sum_{k=0}^{n} k\binom{n}{k} p^k q^{n-k}$.
(b) Show that $k\binom{n}{k} = n\binom{n-1}{k-1}$, and use this, along with (a), to obtain

$$E(X) = np \sum_{k=1}^{n} \binom{n-1}{k-1} p^{k-1} q^{n-k}.$$

(c) Use the binomial theorem, along with (b), to show that $E(X) = np$.

Summary

In this chapter we have learned how discrete probability spaces can be used to model experiments whose outcomes are subject to random variations. They key concepts are as follows:

- A **discrete probability space** is a **sample space** with a **probability assignment**.
- An **event** is a subset of a probability space.
- Computing the probability of an event in a **uniform probability space** amounts to a counting problem.
- **Conditional probabilities**: $\Pr\{A|B\} = \Pr\{A \cap B\}/\Pr\{B\}$.
- **Independent events**: $\Pr\{A \cap B\} = \Pr\{A\}\Pr\{B\}$.
- **Mutually independent events.**

- **Bernoulli trials**: The probability of exactly k successes in n independent trials is $\binom{n}{k}p^k q^{n-k}$, where p is the probability of success on each trial, and $q = 1 - p$.
- Analysis of **dependent** trials via **tree diagrams**.
- **Theorem of total probability**; **Bayes' rule**.
- A **random variable** is a function on a probability space.
- **Density function** and **expectation** (average) of a random variable.

CHAPTER TEN

FINITE DIFFERENCE EQUATIONS

In this final chapter we will study the theory of *finite difference equations*, which are sometimes also called *linear recurrence relations*. Finite difference equations occur constantly in practical applications, most often in the analysis of certain discrete-time systems (e.g., digital filters), analysis of algorithms, error-correcting codes, etc. (But see Problems 10.2.38 and 10.2.39 for the most practical application of all!) In Section 10.1, we will study the simplest kind of difference equation—*homogeneous difference equations*. In Section 10.2, we will study a more difficult class—*inhomogeneous difference equations*. Finally, in Section 10.3, we will introduce you to one of the most powerful tools in all of discrete mathematics—the *generating function*, which can be used to solve not only difference equations, but many other problems in discrete mathematics as well.

10.1 Homogeneous Difference Equations

The famous **Fibonacci numbers**, which we encountered briefly in Section 1.5, are the numbers in the sequence

$$0, 1, 1, 2, 3, 5, 8, 13, 21, 34, 55, \ldots,$$

where each term is the sum of the previous two. If we denote the nth Fibonacci number by F_n, then the defining relationship can be written as

$$F_n = F_{n-1} + F_{n-2}.$$

In this chapter we will describe a method for producing an algebraic formula for the nth Fibonacci number, or indeed for the nth term in *any* sequence of numbers

$$x_0, x_1, x_2, x_3, \ldots,$$

in which the terms are related to each other by an equation of the form

$$x_n = a_1 x_{n-1} + a_2 x_{n-2} + \cdots + a_k x_{n-k}, \tag{10.1}$$

or

$$x_n = a_1 x_{n-1} + a_2 x_{n-2} + \cdots + a_k x_{n-k} + f_n, \tag{10.2}$$

where $\{a_1, a_2, \ldots, a_k\}$ is a fixed set of constants, and (f_n) is a given sequence of numbers (sometimes called the **forcing sequence**) satisfying certain conditions.

A sequence satisfying a relationship like (10.1) or (10.2) is called a kth-order **linear recurring sequence**; the relationship itself is called a **recurrence relation**. (The phrase "kth order" means that each term in the sequence depends only on the k previous terms; and "linear" is to contrast it with "nonlinear" recurrences like $x_n = x_{n-1}^2 + x_{n-2}^2$ or $x_n = -2x_{n-1}x_{n-2}x_{n-3} + 3x_{n-2}^{-1}$ in which quadratic, cubic, or other higher-power terms appear.) Sometimes Equations (10.1) and (10.2) are written instead as

$$x_n - a_1 x_{n-1} - a_2 x_{n-2} - \cdots - a_k x_{n-k} = 0, \tag{10.1'}$$

$$x_n - a_1 x_{n-1} - a_2 x_{n-2} - \cdots - a_k x_{n-k} = f_n. \tag{10.2'}$$

An equation of the form (10.1′) is called a (kth-order, linear) **homogeneous difference equation** (HDE), and (10.2′) is called a (kth-order, linear) **inhomogeneous difference equation** (IDE), with **forcing sequence** f_n.*

For example, the Fibonacci numbers, which as we have seen satisfy the second-order linear recurrence relation $F_n = F_{n-1} + F_{n-2}$ [this is Equation (10.1) with $k = 2$ and $a_1 = a_2 = 1$], also satisfy the following second-order HDE:

$$F_n - F_{n-1} - F_{n-2} = 0, \qquad \text{for } n \geq 2.$$

The restriction "for $n \geq 2$" is necessary, since with $n = 1$, the HDE says that $F_1 - F_0 - F_{-1} = 0$; but F_{-1} isn't defined. This restriction can be removed if we write the HDE in the alternative form

$$F_{n+2} - F_{n+1} - F_n = 0,$$

which now holds for all $n \geq 0$.

In this section, we will learn how to solve HDEs, and in Section 10.2 we will learn how to solve IDEs. We will begin with an example which shows how an HDE might arise in practice (it is also algebraically a little easier than the Fibonacci numbers, which we postpone to Example 10.2).

EXAMPLE 10.1 Recall language A from Section 7.1, which consists of all strings that can be built by concatenating together copies of the strings *a*, *ab*, and *bc*, in any order. For example, *aabc*, *bcaabbc*, and *ababab aab*, are strings in this language, of lengths 4, 7, and 9, respectively. *Question*: How many strings are there of length 20?

* To emphasize that the constants a_i do not depend on n, the words *constant coefficient* are sometimes added to these long-winded names. A second-order linear homogeneous difference equation like $x_n - x_{n-1} - nx_{n-2} = 0$, for example, which has *nonconstant* coefficients, cannot be solved by the methods of this chapter.

SOLUTION The answer, as we shall see, is 699051. But let's begin less ambitiously, and just list the strings of lengths 1, 2, 3, directly:

n	Strings of length n
1	*a*
2	*aa, ab, bc*
3	*aaa, aab, abc, aba, bca*

If we denote the number of strings of length n by x_n, we now have $x_1 = 1$, $x_2 = 3$, $x_3 = 5$. What about x_4? We could continue as above, but instead, let's argue indirectly. Any string of length 4 must begin with *a*, *ab*, or *bc*. If it begins with *a*, the rest of the string must be one of the five strings of length 3. On the other hand, if it begins with *ab* or *bc*, the rest of the string must be one of the three strings of length 2. That is, every string of length 4 must have one of the following forms:

a[string of length 3]: 5 possibilities
ab[string of length 2]: 3 possibilities
bc[string of length 2]: 3 possibilities.

(Notice that these possibilities are disjoint, and, for example, a string of the form *a*[string of length 3] cannot coincide with a string of the form *ab*[string of length 2].) Therefore, without having to write them all down, we can see that there are $5 + 3 + 3 = 11$ strings of length 4.

This argument can be generalized. Suppose we already know $x_1, x_2, \ldots, x_{n-1}$, and wish to compute x_n. Each string of length n must begin with *a*, *ab*, or *bc*, and so must have one of the following forms:

a[string of length $n-1$]: x_{n-1} possibilities,
ab[string of length $n-2$]: x_{n-2} possibilities,
bc[string of length $n-2$]: x_{n-2} possibilities.

Therefore $x_n = x_{n-1} + 2x_{n-2}$, or

(10.3) $$x_n - x_{n-1} - 2x_{n-2} = 0, \qquad \text{for} \qquad n \geq 3,$$

which is a second-order HDE satisfied by the sequence x_n.

We will soon see how to convert the HDE (10.3) into an explicit formula for x_n, but before that, let's notice that in some ways, (10.3) is just as good as, or even better than, an explicit formula. Why? Because starting with the simple facts $x_1 = 1$, $x_2 = 3$, and armed with a pocket calculator, we can use (10.3) to see that the sequence of x_n's is

$$1, 3, 5, 11, 21, 43, 85, 171, 341, 683, \ldots.$$

Admittedly it's not much fun to compute x_{20} this way, and x_{200} or x_{2000} would be even worse. But if you used a *computer* instead of a calculator, it would be easy to write a program to calculate as many x_n's as you might want. The point is that while a general formula is nice, and often very useful, just as often the difference equation itself is the best tool to use to compute specific values. It will be good to keep this fact in mind as we get involved in the details of solving difference equations.

We're almost ready to solve the HDE in Equation (10.3). Almost but not quite, since there are unfortunately *infinitely many* solutions to (10.3)! To be sure, the sequence 1, 3, 5, 11, 21, etc., that we began to compute above is one solution, but so are these:

$n = 1$	2	3	4	5	6	7	...
$x_n = 1$	2	4	8	16	32	64	...
5	0	10	10	30	50	110	...
1	-1	1	-1	1	-1	1	...
0	0	0	0	0	0	0	...
$\sqrt{2}$	π	$\pi + 2\sqrt{2}$	$3\pi + 2\sqrt{2}$	$5\pi + 6\sqrt{2}$	$11\pi + 10\sqrt{2}$	$21\pi + 22\sqrt{2}$	...
$\vdots$							

In fact, for *any* choice of x_1 and x_2, if we define $x_3 = x_2 + 2x_1$, $x_4 = x_3 + 2x_2$, etc., using (10.3), the resulting sequence will satisfy (10.3). We must know the "start-up" values x_1 and x_2 before we can hope to solve a given problem. In mathematical parlance, x_1 and x_2 are called the **initial conditions** for the HDE. In Example 10.1, the initial conditions are $x_1 = 1$ and $x_2 = 3$, and so we are required to solve not (10.3), but (10.3) subject to the *initial conditions*

$$x_1 = 1, \qquad x_2 = 3. \tag{10.4}$$

To derive the promised explicit formula for x_n, we assume, without any real justification yet, that $x_n = \lambda^n$ for some nonzero constant λ. Given this assumption, Equation (10.3) becomes

$$\lambda^n - \lambda^{n-1} - 2\lambda^{n-2} = 0,$$

$$\lambda^{n-2}(\lambda^2 - \lambda - 2) = 0,$$

$$\lambda^2 - \lambda - 2 = 0. \tag{10.5}$$

The final equation, Equation (10.5), is called the **characteristic equation** of the difference Equation (10.3), and the polynomial $\lambda^2 - \lambda - 2$ is called the **characteristic polynomial**. Notice that the coefficients of the characteristic equation are the same as those of the difference equation:

$$\begin{array}{ll} \text{difference equation:} & x_n - x_{n-1} - 2x_{n-2} = 0 \\ \text{characteristic equation:} & \lambda^2 - \lambda \quad\; - 2 \quad\;\; = 0. \end{array}$$

In general, for the HDE described in Equation (10.1′), we have

$$\begin{array}{ll} \text{difference equation:} & x_n - a_1x_{n-1} - a_2x_{n-2} - \cdots - a_kx_{n-k} = 0 \\ \text{characteristic equation:} & \lambda^k - a_1\lambda^{k-1} - a_2\lambda^{k-2} - \cdots - a_k \quad\;\; = 0. \end{array}$$

In the case of Equation (10.5), the characteristic equation has two solutions, $\lambda = 2$ and $\lambda = -1$, so that both $x_n = 2^n$ and $x_n = (-1)^n$ are solutions to the HDE in Equation (10.3). This illustrates an important rule about HDEs in general, which we call HDE Rule 1.

HDE RULE 1

If λ is any root of the characteristic equation of a given HDE, then

$$x_n = \lambda^n$$

is a solution to the HDE.

Unfortunately, however, in the case of HDE (10.3), neither of the solutions given by HDE Rule 1 is the answer to the sentence-counting problem, since neither agrees with the initial conditions (10.4). To proceed further, we must use the following rule, sometimes called the **linearity rule** for HDEs.

HDE RULE 2 (THE LINEARITY RULE)

If (x_n) and (y_n) are both solutions to a given HDE, then for any constants A and B,

$$z_n = Ax_n + By_n$$

is also a solution.

To see why Rule 2 is correct for our example, consider the situation for the HDE in Equation (10.3), by supposing that x_n and y_n are both solutions to (10.3). Then $z_n = Ax_n + By_n$ also satisfies (10.3), by the following reasoning:

$$\begin{aligned} z_n - z_{n-1} - 2z_{n-2} &= (Ax_n + By_n) - (Ax_{n-1} + By_{n-1}) - 2(Ax_{n-2} + By_{n-2}) \\ &= A(x_n - x_{n-1} - 2x_{n-2}) + B(y_n - y_{n-1} - 2y_{n-2}) \\ &= A \cdot 0 + B \cdot 0 = 0. \end{aligned}$$

A general proof of HDE Rule 2 can be given along the same lines; see Problem 32.

Applying Rule 2 to the problem at hand, we see that for any constants A and B,

(10.6) $$x_n = A \cdot 2^n + B \cdot (-1)^n$$

is a solution to (10.3). In fact, *every* solution to (10.3) is of the form (10.6)—see Problem 30—and so by using Rules 1 and 2, we haven't missed any solutions. For this reason, (10.6) is called the **general solution** to the HDE (10.3).

To get the solution to the sentence-counting problem, we must choose the constants A and B in (10.6) so that the initial conditions in (10.4) are satisfied:

$$x_1 = \begin{cases} 1, & \text{from Equation (10.4)} \\ 2A - B, & \text{from Equation (10.6)} \end{cases}$$

$$x_2 = \begin{cases} 3, & \text{from Equation (10.4)} \\ 4A + B, & \text{from Equation (10.6).} \end{cases}$$

Solving the resulting equations $2A - B = 1$ and $4A + B = 3$, we get $A = \frac{2}{3}$ and $B = \frac{1}{3}$, and plugging these values back into Equation (10.6), we arrive at the exact formula

$$x_n = \tfrac{2}{3} \cdot 2^n + \tfrac{1}{3} \cdot (-1)^n, \tag{10.7}$$

for the number of strings of length n. For $n = 20$, Equation (10.7) gives $x_{20} = \frac{2}{3} \cdot 2^{20} + \frac{1}{3}(-1)^{20} = 699051$, as promised. ■

In the next example, we will see how the techniques developed in Example 10.1 can be used to produce an exact formula for the nth Fibonacci number F_n.

EXAMPLE 10.2 Solve the Fibonacci HDE, i.e.,

$$F_0 = 0, \qquad F_1 = 1, \tag{10.8}$$

and for $n \geq 2$,

$$F_n - F_{n-1} - F_{n-2} = 0. \tag{10.9}$$

SOLUTION The first step is to form the characteristic equation corresponding to the given HDE. In this case it is

$$\lambda^2 - \lambda - 1 = 0. \tag{10.10}$$

Using the quadratic formula, we find that the two roots to this equation are

$$\lambda_0 = \frac{1 + \sqrt{5}}{2} \qquad \text{and} \qquad \lambda_1 = \frac{1 - \sqrt{5}}{2}.$$

It follows then that the *general* solution to (10.9) is

$$F_n = A\lambda_0^n + B\lambda_1^n, \tag{10.11}$$

where A and B are constants that must be determined by the initial conditions. In this case, the initial conditions are $F_0 = 0$ and $F_1 = 1$, and so from (10.11) we have

$$0 = A + B \qquad \text{and} \qquad 1 = A\lambda_0 + B\lambda_1.$$

It is a simple matter to solve these two equations for A and B. The result is

$$A = \frac{1}{\sqrt{5}}, \qquad B = \frac{-1}{\sqrt{5}}.$$

Therefore the nth Fibonnaci number is given explicitly by the remarkable formula

$$F_n = \frac{1}{\sqrt{5}}\left[\left(\frac{1 + \sqrt{5}}{2}\right)^n - \left(\frac{1 - \sqrt{5}}{2}\right)^n\right]. \tag{10.12}$$

■

The procedure we used to solve Examples 10.1 and 10.2 can be used to solve most HDEs. Starting with the general HDE (Equation 10.1′), we first form the corresponding characteristic equation

(10.13) $$\lambda^k - a_1\lambda^{k-1} - \cdots - a_k = 0,$$

and find its roots.* If $\lambda_1, \lambda_2, \ldots, \lambda_m$ are the distinct roots of (10.13), then for any set of constants $A_1, A_2, \ldots, A_m$,

(10.14) $$x_n = A_1\lambda_1^n + A_2\lambda_2^n + \cdots + A_m\lambda_m^n$$

will be a solution to the given HDE. The roots λ_i can be positive, negative, or complex. In the examples considered in the main body of this chapter, however, the roots will never be complex. The complex case can involve slightly harder algebra, and is treated briefly in Section 10.4.

Taken together, Rules 1 and 2 give many solutions to a given HDE, and usually they give *all* solutions. An exception occurs when the characteristic equation has *repeated* roots. For example, if the characteristic equation is $(\lambda - 2)(\lambda - 4)^3 = 0$, the root $\lambda = 4$ is a repeated root of multiplicity 3. When repeated roots occur, Rule 1 needs to be supplemented.

HDE RULE 1 (SUPPLEMENT)

If λ is a root of the characteristic equation of a given HDE with multiplicity m, then

$$x_n = \lambda^n, n\lambda^n, \ldots, n^{m-1}\lambda^n$$

are all solutions to the HDE.

Thus if the characteristic equation is as above $(\lambda - 2)(\lambda - 4)^3$, then

$$x_n = 2^n, \qquad x_n = 4^n, \qquad x_n = n \cdot 4^n, \qquad x_n = n^2 \cdot 4^n,$$

are all solutions to the corresponding HDE.

Luckily, this is as complicated as it gets: *All* solutions to *any* HDE can now be found by combining Rules 1 (as supplemented) and 2. (A proof that the supplement to Rule 1 is correct requires calculus and will be omitted.)

EXAMPLE 10.3 Solve the HDE

(10.15) $$x_n = 2x_{n-1} - x_{n-2}$$

subject to the initial conditions

(10.16) $$x_0 = 1, \qquad x_1 = 3.$$

* For most HDEs of practical interest, this is the step which will prove to be the most difficult. However, techniques for finding the roots of polynomials do not fall under the heading of discrete mathematics, so throughout this chapter, we will be working with examples for which the roots are easy to find. In the real world, however, you might need to use techniques from the theory of equations, or more likely, a computer, to find the roots.

SOLUTION Equation (10.15) is a second-order HDE whose characteristic polynomial is $\lambda^2 - 2\lambda + 1 = (\lambda - 1)^2$. Thus the characteristic equation has only the root 1, and Rule 1 only gives us solutions $x_n = A$, for any constant A. These *are* solutions, to be sure, but none of them satisfy the given initial conditions (10.16). This is a job for the supplement to Rule 1! Since $\lambda = 1$ is a root of the characteristic equation with multiplicity 2, $x_n = 1$ and $x_n = n$ are both solutions to the HDE (10.15), and so by Rule 2,

(10.17) $$x_n = A + Bn$$

is a solution, for all A and B. [In fact *every* solution to (10.15) has the form (10.17); see Problem 31.] To find A and B, we use the initial conditions (10.16):

$$n = 0: \qquad A = x_0 = 1$$

$$n = 1: \qquad A + B = x_1 = 3.$$

Hence, $A = 1$ and $B = 2$, and *the* solution to Equation (10.15), given the initial conditions in (10.16), is

(10.18) $$x_n = 2n + 1.$$

ANOTHER SOLUTION Example 10.3 can also be solved by another powerful technique—guessing! Indeed, by using the initial conditions (10.16) and the recursion (10.15), we can easily generate the first few terms in the sequence (x_n):

$$1, 3, 5, 7, 9, 11, \ldots.$$

It's not hard to guess the next term in this sequence! And having *guessed* that $x_n = 2n + 1$ is the solution, the guess can be rigorously verified by mathematical induction; see Problem 8. Guessing is worth a try on any problem; unfortunately, however, it's often hard to see a pattern emerging, and we must fall back on Rules 1 and 2, which aren't as much fun as guessing, but which always work.

Our next example shows how to solve HDEs when, instead of initial conditions, we are given separated values of the unknown sequence (sometimes called **boundary values**).

EXAMPLE 10.4 Solve the third-order HDE

(10.19) $$y_n = 3y_{n-1} - 3y_{n-2} + y_{n-3}$$

subject to the boundary values

(10.20) $$y_0 = 0, \qquad y_3 = 3, \qquad y_5 = 10.$$

SOLUTION The characteristic equation associated with (10.19) is

$$\lambda^3 - 3\lambda^2 + 3\lambda - 1 = 0,$$

i.e.,

$$(\lambda - 1)^3 = 0,$$

which has $\lambda = 1$ as a repeated root with multiplicity 3. It follows then, from Rule 1 (supplemented) and Rule 2, that for any three constants A, B, and C, the sequence

(10.21) $$y_n = A + Bn + Cn^2$$

is a solution to (10.19). To determine the constants A, B, and C, combine (10.21) with the three boundary values in (10.20):

$$n = 0: \qquad 0 = A$$

$$n = 3: \qquad 3 = A + 3B + 9C$$

$$n = 5: \qquad 10 = A + 5B + 25C.$$

Solving these three equations for A, B, and C, we find

$$A = 0, \qquad B = -\tfrac{1}{2}, \qquad C = +\tfrac{1}{2},$$

and so the solution to (10.19) subject to (10.20) is

(10.22) $$y_n = \frac{n(n-1)}{2}.$$

We conclude this section with a probability problem (one version of the famous "random walk" problem), whose solution involves a simple HDE.

EXAMPLE 10.5 (One-dimensional Random Walk.) There are two walls, located N feet apart. A malfunctioning robot is located exactly n feet from the left wall. (See Figure 10.1.) Every second the robot takes a step of length 1 foot, either to the right or to the left. It is equally likely to move in either direction. What is the probability that it will crash into the *right* wall before the *left* wall?

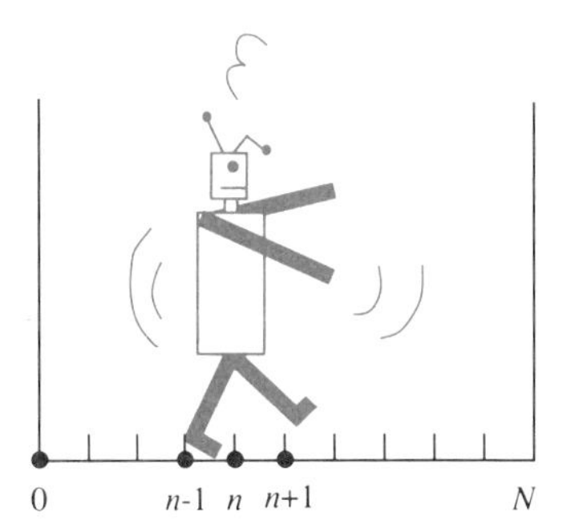

Figure 10.1 The runaway robot.

SOLUTION We denote this unknown probability by p_n. We wish to find a formula for the $N + 1$ probabilities $p_0, p_1, \ldots, p_N$. We assume that if $n = 0$ or $n = N$ a crash has already occurred, so that $p_0 = 0$ and $p_N = 1$.

It turns out that by using the ideas in Section 9.4 we can derive a homogeneous difference equation for p_n! We do this by viewing the robot's walk as a two-stage process. The first stage is the *first step*; it is either to the left, in which case the robot arrives at location $n - 1$, or to the right, in which case it arrives at location $n + 1$. The second stage is the *rest of the journey*, which must either terminate at the right wall (R) or left (L). Figure 10.2 shows the corresponding tree diagram.

The label on the branch from n to $n - 1$ is $\frac{1}{2}$, as is that on the branch from n to $n + 1$, since the robot is equally likely to move left or right. The label from $n - 1$ to R is *by definition* p_{n-1} (the probability of hitting the right wall first, starting at $n - 1$); similarly the label from $n + 1$ to R is p_{n+1}. Thus by the theorem of total probability (Equation 9.33), we can calculate the probability of R, starting

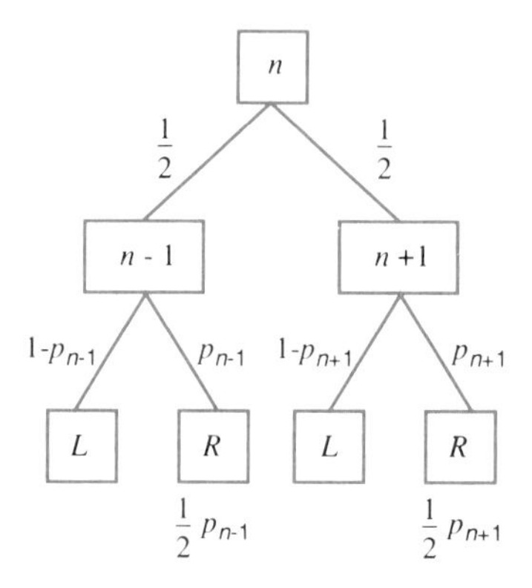

Figure 10.2 A tree diagram for Example 10.5.

at n, as follows:

$$\begin{aligned} p_n &= \Pr\{\mathrm{R}\} \\ &= \Pr\{\text{first step left}\} \Pr\{\mathrm{R} | \text{first step left}\} \\ &\quad + \Pr\{\text{first step right}\} \Pr\{\mathrm{R} | \text{first step right}\} \\ &= \tfrac{1}{2} \cdot p_{n-1} + \tfrac{1}{2} \cdot p_{n+1}. \end{aligned}$$

This yields the HDE

(10.23) $$p_{n+1} - 2p_n + p_{n-1} = 0,$$

which is valid for $n = 1, 2, \ldots, N - 1$. We can now apply the methods of this section. The characteristic polynomial of the HDE (10.23) is $\lambda^2 - 2\lambda + 1$, which factors as $(\lambda - 1)^2$. Thus by HDE Rules 1 and 2, we know that the solution to (10.23) is of the form

(10.24) $$p_n = an + b$$

for certain constants a and b. To determine a and b, we use the boundary values $p_0 = 0$ and $p_N = 1$:

$$p_0 = \begin{cases} 0 & \text{(boundary condition)} \\ b & \text{(from Equation 10.24)} \end{cases}$$

$$p_N = \begin{cases} 1 & \text{(boundary condition)} \\ aN + b & \text{(from Equation 10.24).} \end{cases}$$

This implies $a = 1/N$ and $b = 0$, and so the final answer is

(10.25) $$p_n = \frac{n}{N}.$$

Thus, for example, if the two walls are 60 feet apart and the robot starts exactly 4 feet from the left wall, the probability that it will reach the right wall first is only $\frac{1}{15}$. (You might have guessed this answer; but in Problems 38–43, we give some variations of this example for which the answer can not be guessed.) ■

This concludes our discussion of homogeneous difference equations. The important thing to remember is that *all* solutions to *any* HDE can be found by combining Rule 1 (using the supplement, if the characteristic equation has multiple roots) with Rule 2.

Problems for Section 10.1

1. In the text we noted that the HDE satisfied by the Fibonacci numbers $F_0, F_1, F_2, \ldots$ could be written as either

$$F_n - F_{n-1} - F_{n-2} = 0, \qquad \text{for} \qquad n \geq 2,$$

or as

$$F_{n+2} - F_{n+1} - F_n = 0, \qquad \text{for} \qquad n \geq 0.$$

Which of the following are also valid descriptions of the Fibonacci recursion?
(a) $F_{n+1} - F_n - F_{n-1} = 0$, for $n \geq 1$.
(b) $F_{n-1} - F_{n-2} - F_{n-3} = 0$, for $n \geq 3$.
(c) $F_{n+3} - F_{n+2} - F_{n+1} = 0$, for $n \geq -1$.
(d) $F_{n-2} - F_{n-3} - F_{n-4} = 0$, for $n \geq 1$.

2. In Example 10.1, write down all 11 strings of length 4.

3. Derive Equation (10.3) by observing that a string of length n must *end* with *a*, *ab*, or *bc*.

4. In Example 10.1, we asserted that the solution to the simultaneous equations $2A - B = 1$ and $4A + B = 3$ is $A = \frac{2}{3}$ and $B = \frac{1}{3}$. Verify this.

5. In Example 10.2, we asserted that the solution to the simultaneous equations $A + B = 0$ and $\lambda_0 A + \lambda_1 B = 1$ is $A = 1/\sqrt{5}$ and $B = -1/\sqrt{5}$. Verify this.

6. Define the "modified" Fibonacci numbers $f_0, f_1, f_2, \ldots$ by $f_0 = 0$, $f_1 = 2$, and $f_n = f_{n-1} + f_{n-2}$ for $n \geq 2$.
(a) Find an explicit expression for f_n.
(b) What is the relationship between f_n and the ordinary Fibonacci number F_n?

7. In Example 10.3, Equation (10.15), assume that $x_{n-1} = 2(n-1) + 1 = 2n - 1$ and $x_{n-2} = 2(n-2) + 1 = 2n - 3$, and prove that $x_n = 2n + 1$.

8. Use Problem 7 to give a proof by mathematical induction that the solution to Example 10.3 is $x_n = 2n + 1$.

In Problems 9–15, find the general solution.

9. $x_n - 3x_{n-1} - 10x_{n-2} = 0$.

10. $x_{n+2} + 3x_{n+1} - 4x_n = 0$.

11. $x_{n+2} + 6x_{n+1} + 9x_n = 0$.

12. $x_n - 4x_{n-i} + x_{n-2} = 0$.

13. $2x_n + 2x_{n-1} - x_{n-2} = 0$.

14. $x_n - 3x_{n-1} = 0$.

15. $x_{n+2} - 3x_{n-1} - 4x_{n-2} = 0$.

In Problems 16 and 17, solve by inspection and then solve again (overkill!) with the method of this section.

16. $x_{n+1} - x_n = 0$, with initial condition $x_1 = 4$.

17. $ax_{n+2} + bx_{n+1} + cx_n = 0$, with initial conditions $x_0 = 0$ and $x_1 = 0$.

In Problems 18–24, find the *general solution*. Also find a *particular solution* which satisfies the given initial conditions.

18. $y_{n+2} + 2y_{n+1} - 15y_n = 0$; $y_0 = 0$, $y_1 = 1$.

19. $y_{n+2} - 8y_{n+1} + 16y_n = 0$; $y_0 = 0$, $y_1 = 8$.

20. $y_{n+1} - y_{n-1} = 0$; $y_0 = 0$, $y_1 = 1$.

21. $y_n - 3y_{n-1} + 3y_{n-2} - y_{n-3} = 0$; $y_1 = 0$, $y_2 = 1$, $y_3 = 0$.

22. $y_{n+1} - y_n = 0$; $y_3 = 4$.

23. $y_{n+2} - 2y_{n+1} + y_n = 0$; $y_0 = 1$, $y_1 = 2$.

24. $y_{n+2} - 2y_{n+1} - 3y_n = 0$; $y_0 = 1$, $y_1 = 0$.

25. Use Equation (10.12) (and a pocket calculator?) to estimate F_{1000}. Compare this estimate with the estimate $[(1 + \sqrt{5})/2]^{n-1}$ given in Example 1.28.

26. Solve the HDE $x_n = 2x_{n-1} - x_{n-2}$ subject to the initial conditions $x_0 = 1$ and $x_1 = 1$.

27. Solve the HDE $y_n = 3y_{n-1} - 3y_{n-2} + y_{n-3}$ subject to the boundary values $y_1 = 0$, $y_3 = 3$, and $y_5 = 9$.

28. Referring to Example 10.1 in the text: How many digits does x_{100} have?

29. Consider the sequence $0, 1, \frac{1}{2}, \frac{3}{4}, \frac{5}{8}, \frac{11}{16}, \ldots$, in which each term is the *average* of the previous two terms; e.g., the next term will be $\frac{1}{2}(\frac{5}{8} + \frac{11}{16}) = \frac{21}{32}$. Find a formula for the general term.

30. Show that *every* solution to the HDE (10.3) has the form (10.6). [*Hint*: Use (10.6) to solve for A and B in terms of the initial conditions x_1 and x_2.] Use your technique to solve (10.3) with the initial conditions $x_1 = -1$ and $x_2 = 0$.

31. Show that every solution to the HDE (10.15) has the form (10.17).

32. Verify the *linearity rule* (Rule 2) for a general HDE.

33. Here is a difference equation of a different kind:

$$x_n = x_{n-1} + x_{n-2} + \cdots + x_1 + x_0,$$

i.e., each term in the sequence $(x_0, x_1, \ldots, x_n, \ldots)$ equals the sum of *all* the previous terms. Let $x_0 = 1$. Guess the value of x_n, and prove that your guess is correct by induction.

34. In discussing the HDE in Equation (10.3), we saw that there was a solution with $x_1 = \sqrt{2}$, $x_2 = \pi$, $x_3 = \pi + 2\sqrt{2}, \ldots$. Find a formula for the coefficient of π in the nth term of this sequence.

35. Consider the HDE with *nonconstant* coefficients $x_n = nx_{n-1}$, $n = 1, 2, \ldots$, with initial condition $x_0 = 1$. Calculate enough terms to see a pattern, and confirm your guess using mathematical induction.

36. Solve the HDE $M_n = 3M_{n-1} - 2M_{n-2}$ subject to the initial condition $M_0 = 0$ and $M_1 = 1$.
(a) By guessing; (b) Using Rules 1 and 2.

37. Consider the *nonlinear* recurrence relation $\phi_n = \phi_{n-1} \cdot \phi_{n-2}$, with initial conditions $\phi_0 = 1$ and $\phi_1 = 2$.
(a) Calculate ϕ_2, ϕ_3, ϕ_4, and ϕ_5.
(b) Find a general formula for ϕ_n.

Problems 38–44 are related to the random walk of Example 10.5.

38. Consider again the robot of Example 10.5, only now the probability is $\frac{1}{3}$ that it moves to the left and $\frac{2}{3}$ that it moves to the right. Set up the difference equation and boundary values for p_n.

39. Find the general solution to the difference equation obtained in Problem 5.

40. Using the general solution found in Problem 39, and the boundary values found in Problem 38, find the probabilities p_n.

41. Assume now that the robot goes to the left with probability p and to the right with probability $q = 1 - p$, where $0 < p < 1$. If p_n is the probability of crashing into the right wall before the left wall, show that p_n satisfies the difference equation

$$qp_{n+1} - p_n + pp_{n-1} = 0, \qquad n = 1, 2, \ldots, N-1,$$

with boundary conditions $p_0 = 0$ and $p_N = 1$.

42. Continuing Problem 41:
(a) Show that $(p - q)^2 = 1 - 4pq$. [Note that $1 = (p + q)^2 = p^2 + 2pq + q^2$, and $(p - q)^2 = p^2 - 2pq + q^2$; subtract one equation from the other.]
(b) Use part (a) to show that the roots of the characteristic equation of the difference equation for p_n are

$$\frac{1 \pm |p - q|}{2q}.$$

(c) In (b), show that the roots are 1 and p/q.

43. Continuing Problem 42, show that if $p \neq q$, then

$$p_n = \frac{1 - \left(\dfrac{p}{q}\right)^n}{1 - \left(\dfrac{p}{q}\right)^N}.$$

44. (Gambler's Ruin.) A certain gambler has a stake of \$50 and is playing roulette. At each play, she bets \$1 on "black." With probability $\frac{18}{37}$ she wins \$1, but with probability $\frac{19}{37}$ she loses the dollar. What is the probability that she goes broke before her stake totals \$100?

10.2 Inhomogeneous Difference Equations

Recall that a sequence of numbers $x_0, x_1, x_2, \ldots$ is said to satisfy an **inhomogeneous difference equation** (IDE) if there are constants $a_1, a_2, \ldots, a_k$ such that

$$x_n - a_1 x_{n-1} - \cdots - a_k x_{n-k} = f_n,^*$$

where the sequence (f_n) is called the **forcing sequence**. In this section we will see that there is a simple procedure for solving any IDE for which the forcing sequence itself satisfies an HDE. We begin with a simple example, which is closely related to Example 10.1.

* We also recall that technically this is called a "kth-order linear inhomogeneous difference equation with constant coefficients."

EXAMPLE 10.6 Solve the second-order IDE

(10.26) $$x_n - x_{n-1} - 2x_{n-2} = 1, \qquad n \geq 3$$

subject to the initial conditions

(10.27) $$x_1 = 1, \qquad x_2 = 3.$$

SOLUTION The only difference between this problem and the problem in Example 10.1 is the forcing term, "1," which appears on the right side of (10.26). We know from Example 10.1 that the homogeneous form of (10.26), viz.

(10.28) $$y_n - y_{n-1} - 2y_{n-2} = 0, \qquad n \geq 3$$

has the general solution (cf. Equation 10.6)

(10.29) $$y_n = A \cdot 2^n + B \cdot (-1)^n.$$

This is apparently not much help in solving (10.26), however. We can use a hint.

Hint: One solution to (10.26) is

(10.30) $$p_n = -\tfrac{1}{2}, \qquad n = 1, 2, \ldots.$$

We'll see later where the hint came from. For now, let's just be grateful and verify that (10.30) is in fact a solution to (10.26):

$$p_n - p_{n-1} - 2p_{n-2} = -\tfrac{1}{2} - (-\tfrac{1}{2}) - 2(-\tfrac{1}{2}) = -\tfrac{1}{2} + \tfrac{1}{2} + 1 = 1,$$

so the hint is correct! Unfortunately, however, (10.30) still doesn't solve the given problem, since the initial conditions (10.27) aren't satisfied: $p_1 = -\frac{1}{2}$ and $p_2 = -\frac{1}{2}$; but we want $x_1 = 1$ and $x_2 = 3$. Do we need another hint? No! Now we can find *all* solutions to (10.26), and in particular *the* solution that satisfies the given initial conditions, by reasoning as follows: Let (x_n) be a solution to (10.26), and consider the new sequence (y_n) defined by

$$y_n = x_n - p_n = x_n + \tfrac{1}{2}.$$

Then (y_n) satisfies the homogeneous form of Equation (10.26), because we have

$$\begin{aligned} y_n - y_{n-1} - 2y_{n-2} &= (x_n + \tfrac{1}{2}) - (x_{n-1} + \tfrac{1}{2}) - 2(x_{n-2} + \tfrac{1}{2}) \\ &= (x_n - x_{n-1} - 2x_{n-2}) - (-\tfrac{1}{2} + \tfrac{1}{2} + 1) \\ &= 1 - 1 = 0. \end{aligned}$$

So the solution to the homogeneous form of (10.26) comes in handy after all, since now we can say that the sequence (y_n) must be of the form given in Equation (10.29), i.e.

$$\begin{aligned} y_n &= A \cdot 2^n + B \cdot (-1)^n \\ x_n + \tfrac{1}{2} &= A \cdot 2^n + B \cdot (-1)^n. \end{aligned}$$

And so the *general solution* to the IDE (10.26) is

(10.31) $$x_n = A \cdot 2^n + B \cdot (-1)^n - \tfrac{1}{2}.$$

To satisfy the initial conditions (10.27), we just need to choose A and B so that

$$1 = x_1 = 2A - B - \tfrac{1}{2} \qquad \text{and} \qquad 3 = x_2 = 4A + B - \tfrac{1}{2}.$$

This forces $A = \frac{5}{6}$ and $B = \frac{1}{6}$, and thus the solution to Example 10.6 is

(10.32) $$x_n = \tfrac{5}{6} \cdot 2^n + \tfrac{1}{6}(-1)^n - \tfrac{1}{2}.$$ ■

In Example 10.6, we saw how to take one **particular solution** to a given IDE and produce *all* solutions. The technique we used was a special case of the following general rule.

IDE RULE 1

The *general solution* to a given IDE is obtained by adding the general solution to the corresponding HDE to any *particular* solution of the IDE.

It is not difficult to give a proof of Rule 1 along the lines suggested by the argument used in Example 10.6; see Problem 41.

Rule 1 for IDEs will allow us to find the *general* solution to any IDE, provided we can find, by hook or crook, just *one* solution. But we can't always expect to be *given* a particular solution, as in Example 10.6. We come therefore to the second important rule for IDEs, a rule that is usually a reliable way to find a particular solution to a given IDE, *if the forcing sequence itself satisfies* an HDE.

IDE RULE 2

A *particular solution* to a given IDE can usually be found by appropriately selecting the constants in the *general solution* to the HDE satisfied by the forcing sequence.

The method for finding a particular solution to an IDE suggested by IDE Rule 2 is sometimes called the **trial sequence method**, or the **annihilator method**.

To illustrate this method, we return to Example 10.6, where the forcing sequence is the constant sequence $f_n = 1$. By HDE Rule 1, this sequence satisfies an HDE whose characteristic polynomial is $(\lambda - 1)$, i.e.,

(10.33) $$p_n - p_{n-1} = 0.$$

The *general solution* to (10.33) is, by the results of Section 10.1,

(10.34) $$p_n = A,$$

for an arbitrary constant A. IDE Rule 2 tells us to use this general solution as a *trial sequence*, and to try to select the constant A so that the original IDE (10.26) is satisfied. If we substitute the sequence $p_n = A$ into (10.26), we obtain

$$\begin{aligned} p_n - p_{n-1} - 2p_{n-2} &= 1 \\ A - A - 2A &= 1 \\ -2A &= 1 \\ A &= -\tfrac{1}{2}. \end{aligned}$$

Thus IDE Rule 2 has led us to the same particular solution, viz. $p_n = -\frac{1}{2}$, as we were given in the hint. The next example shows IDE Rule 2 in action again.

EXAMPLE 10.7 Solve the IDE

$$x_n - 2x_{n-1} = n, \qquad n \geq 1, \tag{10.35}$$

subject to the initial condition

$$x_0 = 1. \tag{10.36}$$

SOLUTION The general solution to the homogeneous form of (10.35) is

$$y_n = C \cdot 2^n. \tag{10.37}$$

Hence by IDE Rule 1, the general solution to (10.35) is

$$x_n = C \cdot 2^n + p_n, \tag{10.38}$$

where (p_n) is a particular solution to (10.35). In order to find such a solution, we use IDE Rule 2. In this case the forcing sequence is $f_n = n$, and by the results of Section 10.1 (the supplement to HDE Rule 1), this sequence satisfies an HDE whose characteristic polynomial has $\lambda = 1$ as a root with multiplicity 2. Thus the characteristic polynomial is $(\lambda - 1)^2 = \lambda^2 - 2\lambda + 1$, and the corresponding HDE is $p_n - 2p_{n-1} + p_{n-2} = 0$. The *general* solution to this HDE is

$$p_n = An + B, \tag{10.39}$$

where A and B are arbitrary constants. IDE Rule 2 tells us to use the sequence (10.39) as a *trial sequence*, and to try to choose the constants A and B so that (10.35) is satisfied:

$$\begin{aligned} p_n - 2p_{n-1} &= n && \text{(this is Equation 10.35)} \\ (An + B) - 2(A(n-1) + B) &= n && \text{(from 10.39).} \end{aligned}$$

This becomes, after a little algebra,

$$n(-A) + (2A - B) = n.$$

For this equation to be true for all $n \geq 1$, we must have $-A = 1$, and $2A - B = 0$, i.e., $A = -1$ and $B = -2$. Therefore

$$p_n = -n - 2 \tag{10.40}$$

is a particular solution to (10.35), and so by IDE Rule 1, all solutions to (10.35) are of the form

$$x_n = A \cdot 2^n - n - 2,$$

where A is a constant to be determined by the initial conditions. To satisfy the initial condition (10.36), we must have

$$1 = x_0 = A - 2,$$

i.e.,

$$A = 3.$$

Hence finally we have the solution

(10.41) $$x_n = 3 \cdot 2^n - n - 2.$$

After a little practice with IDE Rule 2, you will find that it's not necessary to write down the HDE satisfied by the forcing sequence. Instead, it's easy to go directly from the forcing sequence to the trial sequence by *assuming a solution of the same general form as the forcing sequence.* In fact, in Table 10.1 we have summarized just about everything you will ever need to know about the trial sequence method.

In Table 10.1, the lowercase letters *a*, *b*, etc., represent *specific* constants, which will be given as part of the forcing sequence, and the uppercase letters, *A*, *B*, etc., represent *arbitrary* constants in the trial sequence, which will be determined when the trial sequence is substituted into the IDE. The next example illustrates the use of Table 10.1.

EXAMPLE 10.8 For each of the following forcing sequences, determine the corresponding trial sequence.

(a) $f_n = 1$.
(b) $f_n = n$.
(c) $f_n = 2^n$.
(d) $f_n = 3n^2 + 5$.
(e) $f_n = 2^n + 4(-5)^n$.
(f) $f_n = n \cdot 2^n + 1$.
(g) $f_n = 2 \cdot 7^n + 3n + 4$.
(h) $f_n = 6n^2 3^n$.

TABLE 10.1 Some trial sequences corresponding to certain forcing sequences

Forcing Sequence (f_n)	Trial Sequence (p_n)
a	A
$an + b$	$An + B$
$\vdots$	$\vdots$
$a_r n^r + a_{r-1} n^{r-1} + \cdots + a_0$	$A_r n^r + A_{r-1} n^{r-1} + \cdots + A_0$
$a\lambda^n$	$A\lambda^n$
$(an + b)\lambda^n$	$(An + B)\lambda^n$
$\vdots$	$\vdots$
$(a_r n^r + \cdots + a_0)\lambda^n$	$(A_r n^r + \cdots + A_0)\lambda^n$

SOLUTION

(a) and (b) These forcing sequences are just the ones we considered in Examples 10.6 and 10.7, and as we have seen, the corresponding trial sequences are $p_n = A$ and $p_n = An + B$, respectively. These two examples correspond to the first two entries in Table 10.1.

(c) This forcing sequence is of *exponential* form, and according to Table 10.1, the corresponding trial sequence is $p_n = A \cdot 2^n$.

(d) This forcing sequence is a *second-degree polynomial in n*, and according to Table 10.1, the corresponding trial sequence is $p_n = An^2 + Bn + C$. Notice that the trial sequence must include the term Bn, even though the forcing sequence does not have a first-degree term in n.

(e) This forcing sequence is the *sum* of two exponential forcing sequences, viz. $f'_n = 2^n$ and $f''_n = 4(-5)^n$. The trial sequences corresponding to these forcing sequences are by Table 10.1 $p'_n = A \cdot 2^n$ and $p''_n = B \cdot (-5)^n$. It is a general rule that whenever the forcing sequence for an IDE is a sum of several sequences, a trial sequence can be obtained by adding together the corresponding trial sequences. In this case, then, the desired trial sequence is $p_n = A \cdot 2^n + B \cdot (-5)^n$.

(f) In this case the forcing sequence is the sum of the two sequences $f'_n = n \cdot 2^n$ and $f''_n = 1$, which, by Table 10.1, correspond to the trial sequences $p'_n = (An + B) \cdot 2^n$ and $p''_n = C$, respectively. It follows that the needed trial sequence is $p_n = (An + B) \cdot 2^n + C$.

(g) This time, $f_n = (2 \cdot 7^n) + (3n + 4)$ is the sum of two simpler forcing sequences, with trial sequences $A \cdot 7^n$ and $Bn + C$. Therefore, $p_n = A \cdot 7^n + Bn + C$ is the desired trial sequence.

(h) Here the forcing sequence is a second-degree polynomial times an exponential, and so by Table 10.1 the corresponding trial sequence is $p_n = (An^2 + Bn + C) \cdot 3^n$.

We have summarized the solution to this example in Table 10.2. ■

As mentioned, IDE Rule 2 *usually* gives a particular solution to the IDE under investigation. There is, however, one situation in which Rule 2 may fail:

TABLE 10.2 Solution to Example 10.8

	Forcing Sequence (f_n)	Trial Sequence (p_n)
(a)	1	A
(b)	n	$An + B$
(c)	2^n	$A \cdot 2^n$
(d)	$3n^2 + 5$	$An^2 + Bn + C$
(e)	$2^n + 4(-5)^n$	$A \cdot 2^n + B(-5)^n$
(f)	$n \cdot 2^n + 1$	$(An + B)2^n + C$
(g)	$2 \cdot 7^n + 3n + 4$	$A \cdot 7^n + Bn + C$
(h)	$6n^2 3^n$	$(An^2 + Bn + C)3^n$

When the forcing sequence f accidentally satisfies the homogeneous form of the IDE. When this happens, a modification to IDE Rule 2 is needed.

IDE RULE 2 (SUPPLEMENT)

If IDE Rule 2 fails, take the trial function which didn't work, and *multiply by n.* Try again. Repeat this procedure as often as necessary to find a particular solution.

With this modification, IDE Rule 2 will always succeed in finding a particular solution to a given IDE—provided, of course, that the forcing sequence satisfies an HDE!

EXAMPLE 10.9 Find the general solution to the IDE

(10.42) $$x_{n+2} - 4x_{n+1} + 4x_n = 2^n, \qquad n \geq 0.$$

Also, find *the* solution which satisfies the initial conditions $x_0 = x_1 = 0$.

SOLUTION The HDE corresponding to (10.42) is $x_{n+2} - 4x_{n+1} + 4x_n = 0$; its characteristic polynomial is $\lambda^2 - 4\lambda + 4 = (\lambda - 2)^2$, and so by the rules for HDEs in Section 10.1, the general solution to this HDE is

(10.43) $$x_n = A2^n + Bn2^n.$$

To complete the solution to (10.42), we need to find a particular solution and we look to Table 10.1 for guidance. Table 10.1 suggests we look for a particular solution of (10.42) of the form $p_n = k2^n$ for some constant k. Unfortunately, this attempt is doomed to failure, since we know from (10.43) that this sequence satisfies the homogeneous form of (10.42). Just to be sure, though, let's try it:

$$\begin{aligned} p_{n+2} - 4p_{n+1} + 4p_n &= 2^n \\ k \cdot 2^{n+2} - 4k2^{n+1} + 4k2^n &= 2^n \\ k \cdot 2^n(4 - 8 + 4) &= 2^n \\ 0 &= 2^n. \ ??? \end{aligned}$$

Too bad. IDE Rule 2 has failed us. But undiscouraged, we apply the supplement to IDE Rule 2, which says to try not $k \cdot 2^n$ but $kn2^n$. Unfortunately this attempt too will fail, because Equation (10.43) tells us that this new trial sequence will also satisfy the HDE, and we will once again get $0 = 2^n$ if we try it. (Try it!) Still undaunted, we apply IDE Rule 2's supplement again, and propose the trial sequence $p_n = kn^2 2^n$. This time we are sure to succeed, since this choice for p_n definitely won't satisfy the HDE. [It isn't of the form prescribed by Equation

(10.43).] To determine the constant k, however, we need to do some work:

$$\begin{aligned} p_{n+2} - 4p_{n+1} + 4p_n &= 2^n \\ k(n+2)^2 \cdot 2^{n+2} - 4k(n+1)^2 2^{n+1} + 4kn^2 2^n &= 2^n \\ k2^n(4(n+2)^2 - 8(n+1)^2 + 4n^2) &= 2^n \\ k2^n(4n^2 + 16n + 16 - 8n^2 - 16n - 8 + 4n^2) &= 2^n \\ 8k \cdot 2^n &= 2^n. \end{aligned}$$

Success! We see that by choosing $k = \frac{1}{8}$, this equation will be satisfied. Hence, $p_n = \frac{1}{8}n^2 2^n$ is a particular solution to (10.42), and so by IDE Rule 1 the general solution to (10.42) is

(10.44) $$x_n = A \cdot 2^n + Bn \cdot 2^n + \tfrac{1}{8}n^2 \cdot 2^n.$$

To find *the* solution which satisfies $x_0 = x_1 = 0$, we substitute $n = 0, 1$ into (10.44):

$$0 = x_0 = A$$

$$0 = x_1 = 2A + 2B + \tfrac{1}{4}.$$

Solving these two equations, we find $A = 0$ and $B = -\frac{1}{8}$, and so the required solution is

(10.45) $$x_n = n(n-1)2^{n-3}.$$ ■

Problems for Section 10.2

1. In Example 10.6, find the terms x_7, x_8, and x_9.

2. Re-solve Example 10.6, assuming now that the forcing sequence is $f_n = n$.

3. In Example 10.7, find x_2, x_3, and x_4 directly from the difference equation, and check that your answers agree with (10.41).

4. Re-solve Example 10.7, assuming now that the forcing sequence is $f_n = n^2$.

5. In Example 10.9, modify the initial conditions so that the solution is simply $x_n = \frac{1}{8}n^2 2^n$.

6. Re-solve Example 10.9, assuming now that the forcing sequence is $f_n = 3^n$.

In Problems 7–12, find an HDE satisfied by the given sequence.

7. $f_n = n$, $n = 0, 1, 2, \ldots$.

8. $f_n = (-1)^n$.

9. $f_n = 5$.

10. $f_n = 2^{-n}$. [Note: 2^{-n} is the same as $(\frac{1}{2})^n$.]

11. $f_n = 2^{-n} + n^4$.

12. $f_n = 3 \cdot 6^{-n} + 7 \cdot (-4)^n$.

In Problems 13–22, find the trial sequence that corresponds to the given forcing sequence.

13. $f_n = -1$.

14. $f_n = -n + 1$.

15. $f_n = \begin{cases} -1 & \text{if } n \text{ is odd} \\ +1 & \text{if } n \text{ is even.} \end{cases}$

16. $f_n = \binom{n}{2}$.

17. $f_n = \sum_{k=0}^{n-1} (2k+1)$.

18. $f_n = 2^n + (-2)^n$.

19. $f_n = -9n^4 \cdot 2^n$.

20. $f_n = \binom{n}{3} \cdot 2^n$.

21. $f_n = 2^{-n} + n^4$.

22. $f_n = 0$.

In Problems 23–34, find the *general solution*, and a *particular solution* that satisfies the given initial conditions.

23. $x_n - 2x_{n-1} = 6n$; $x_1 = 2$.

24. $x_{n+2} - x_{n+1} - 2x_n = 1$; $x_1 = 1$, $x_2 = 3$.

25. $x_{n+2} + 2x_{n+1} - 15x_n = 6n + 10$; $x_0 = 1$, $x_1 = -\frac{1}{2}$.

26. $x_{n+2} - x_{n+1} - 6x_n = 18n^2 + 2$; $x_0 = -2$, $x_1 = 0$.

27. $x_{n+1} + 2x_n = 3 + 4^n$; $x_0 = 2$.

28. $2x_{n+1} - x_n = (\frac{1}{2})^n$; $x_1 = 2$.

29. $x_{n+2} - 2x_{n+1} + x_n = 1$; $x_0 = 1$, $x_1 = \frac{1}{2}$.

30. $x_{n+2} - 8x_{n+1} + 16x_n = (-1)^n$; $x_0 = 0$, $x_1 = 0$.

31. $x_{n+1} + x_n = 5$; $x_0 = 1$.

32. $x_n - 3x_{n-1} + 3x_{n-2} - x_{n-3} = -1$; $x_0 = 1$, $x_1 = 1$, $x_2 = 0$.

33. $x_n - 3x_{n-1} + 2x_{n-2} = n^2$; $x_0 = 0$, $x_1 = 0$.

34. $x_{n+2} - 3x_{n+1} + 2x_n = 1$; $x_0 = 1$, $x_1 = 0$.

35. Let $x_0 = 0$, and for $n \geq 1$, define

$$x_n = \begin{cases} 2x_{n-1} & \text{if } n \text{ is even} \\ 2x_{n-1} + 1 & \text{if } n \text{ is odd.} \end{cases}$$

Find a general formula for x_n.

36. Consider strings of length n built from the three symbols A, B, and C. Let y_n be the number of strings with an *even* number of A's. For example, $y_1 = 2$ (the strings B and C have an even number of A's, namely 0); $y_2 = 5$ (BB, BC, CB, CC, AA).

(a) Show that $y_n = (3^{n-1} - y_{n-1}) + 2y_{n-1}$, or $y_n - y_{n-1} = 3^{n-1}$.

(b) Find an explicit formula for y_n ($n = 1, 2, \ldots$).

37. (The Legendary Towers of Hanoi.) Suppose that n rings are placed on a peg, the rings increasing in size from top to bottom. Two initially empty pegs are also available (see the accompanying diagram). We wish to move the rings one at a time from peg 1 to peg 2 (using peg 3 for storage), under the constraint that no ring can ever be placed on a smaller ring. If y_n is the minimum number of moves required, set up and solve a difference equation to find y_n. (*Hint*: When the largest ring is ready to be moved to peg 2, where must the other $n - 1$ rings be?)

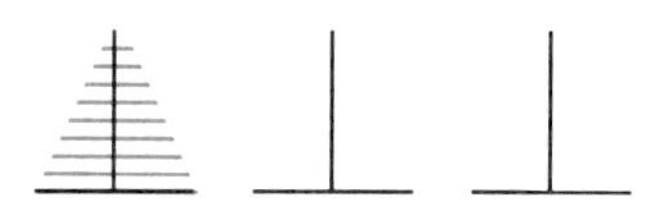

38. Suppose you borrow \$10000 from the bank to buy a new computer, at a 12% interest rate, and repay the loan in 36 equal monthly installments of D dollars each. What is the value of D? [*Hint*: Let y_n represent the *unpaid balance* after n payments have been made. Then just before the $(n + 1)$st payment, the new balance will be $1.01 \cdot y_n$, since 12% annual interest is the same as 1% per month, and just after the $(n + 1)$st payment the unpaid balance will be $1.01y_n - D$. Thus the sequence (y_n) satisfies the IDE $y_{n+1} = 1.01y_n - D$.]

39. Generalize Problem 38, and find the amount of the monthly payment if P dollars are borrowed at interest rate i and must be repaid in N monthly installments.

Problems 40–41 show how to prove that IDE Rule 1 is correct.

40. Let (x_n) and (x'_n) be any two solutions to the IDE $x_n - a_1x_{n-1} - \cdots - a_kx_{n-k} = f_n$. Show that the sequence $(x_n - x'_n)$ satisfies the HDE $x_n - a_1x_{n-1} - \cdots - a_kx_{n-k} = 0$.

41. Use the result of Problem 40 to prove the correctness of IDE Rule 1. [*Hint*: Let p_n be a particular solution to the IDE, and let x_n be any other solution. What can you say about the sequence $(x_n - p_n)$?]

42. Use difference equations to find a formula for $S(n) =$ the sum of the first n squares. [Let $S(1) = 1^2$, $S(2) = 1^2 + 2^2$, $S(3) = 1^2 + 2^2 + 3^2, \ldots$. Write a difference equation for $S(n)$ and solve it.]

10.3 The Generating-Function (z Transform) Approach

In this section we will see how difference equations can be solved using the powerful *generating-function method.* Generating functions are an important tool in discrete mathematics, and their use is by no means confined to the solution of difference equations.

The basic idea behind generating functions is quite simple. If $a_0, a_1, a_2, \ldots, a_n$ is a finite sequence of numbers, the **generating function** for the a_n's is the polynomial

$$G(z) = a_0 + a_1z + a_2z^2 + \cdots + a_nz^n = \sum_{k=0}^{n} a_kz^k, \tag{10.46}$$

where z is an indeterminate (i.e., an abstract symbol). For example, the generating function for the sequence $(3, -4, 7)$ is the polynomial $3 - 4z + 7z^2$, and the generating function for the sequence $(7, -4, 3, 0, 1)$ is $7 - 4z + 3z^2 + z^4$.

Similarly, if we are given an *infinite* sequence of numbers $a_0, a_1, a_2, \ldots$, its generating function is defined to be

$$G(z) = a_0 + a_1 z + a_2 z^2 + \cdots = \sum_{n \geq 0} a_n z^n. \tag{10.47}$$

The sum in (10.47) is an *infinite* sum, and we must admit that infinite sums are not as simple to deal with as finite sums. However, in this section we will adopt an informal approach to infinite sums, deliberately ignoring certain mathematical subtleties.*

The symbol z is just the name given to a variable, and has no special significance. However, z is more often used for generating functions than any other letter, and for this reason $G(z)$ is also sometimes called the **z transform** of the sequence (a_n). The generating function is just a way of tying together all the terms in the given sequence. The hope is always that, when viewed as a function of the variable z, $G(z)$ will be simple in some sense, so that useful conclusions can be drawn from that in (10.46) or (10.47)

$$a_n = \text{coefficient of } z^n \text{ in } G(z). \tag{10.48}$$

Normally the process of extracting the coefficients (a_n) from a closed-form expression for $G(z)$ will require techniques from calculus. Since this book does not presuppose any knowledge of calculus, we will not be able to say very much about this important reverse process. Still, we will be able to get a feeling for why generating functions are useful in solving difference equations and other important problems in discrete mathematics.

The grandmother and grandfather of all generating functions are given in the following two examples.

EXAMPLE 10.10 Let $G_n(z)$ denote the generating function for the *binomial coefficients* we encountered in Chapter 2:

$$G_n(z) = \binom{n}{0} + \binom{n}{1}z + \binom{n}{2}z^2 + \cdots + \binom{n}{n}z^n. \tag{10.49}$$

Find a simple formula for $G_n(z)$.

SOLUTION We recall from Section 2.3 that the *binomial coefficients* appear in the *binomial theorem*:

$$(x + y)^n = \sum_{k=0}^{n} \binom{n}{k} x^k y^{n-k}. \tag{10.50}$$

If in (10.50) we replace y with 1, and x with the indeterminate z, we obtain

$$(z + 1)^n = \sum_{k=0}^{n} \binom{n}{k} z^k = \binom{n}{0} + \binom{n}{1}z + \cdots + \binom{n}{n}z^n = G_n(z).$$

* These subtleties, which have caused unnecessary confusion among discrete mathematicians for years, are explained perfectly in the first few pages of Richard Stanley's book *Enumerative Combinatorics* (Monterey, Calif.: Brooks/Cole, 1986).

Therefore the required simple formula for the given generating function is

$$G_n(z) = (1 + z)^n.$$

EXAMPLE 10.11 Find the generating function for the infinite sequence $1, \lambda, \lambda^2, \lambda^3, \ldots$, where λ is a fixed constant.

SOLUTION Let us denote the required generating function by $G(z)$. Then

$$G(z) = 1 + \lambda z + \lambda^2 z^2 + \lambda^3 z^3 + \cdots. \tag{10.51}$$

We could derive a closed-form expression for $G(z)$ by invoking the formula for the sum of a geometric series (see Problem 8 of this section, and also Example 1.30), but we prefer the following indirect approach, which is typical of the informal way discrete mathematicians manipulate generating functions.

We begin by subtracting 1 from both sides of (10.51):

$$G(z) - 1 = \lambda z + \lambda^2 z^2 + \lambda^3 z^3 + \cdots. \tag{10.52}$$

Next, we notice that every term on the right side of (10.52) is a multiple of λz. Therefore if we divide both sides of (10.52) by λz, we obtain

$$\frac{G(z) - 1}{\lambda z} = 1 + \lambda z + \lambda^2 z^2 + \cdots, \tag{10.53}$$

which is just the original generating function $G(z)$ back again! Therefore (10.53) becomes

$$\frac{G(z) - 1}{\lambda z} = G(z),$$

an equation which can be solved for $G(z)$ as follows:

$$G(z) - 1 = \lambda z G(z)$$

$$G(z)(1 - \lambda z) = 1$$

$$G(z) = \frac{1}{1 - \lambda z}.$$

The required generating function is therefore $1/(1 - \lambda z)$.

Of course, the result of Example 10.10 tells us nothing new about the binomial coefficients; and the result of Example 10.11 tells us nothing new about the simple sequence $(1, \lambda, \lambda^2, \ldots)$. Still, it is interesting to see how neatly the terms of these sequences are tied together by their generating functions. Notice also that the second sequence $(1, \lambda, \lambda^2, \ldots)$ satisfies an HDE, viz. $x_n - \lambda \cdot x_{n-1} = 0$.

It turns out that *any* sequence satisfying an HDE has a generating function which is a **rational function** of z, i.e., the quotient of two polynomials. For example, the generating function for $(1, \frac{1}{2}, \frac{1}{4}, \ldots)$ is $1/(1 - \frac{1}{2}z)$. The next example illustrates this phenomenon in a more interesting case.

EXAMPLE 10.12 Find the generating function for the *Fibonacci numbers*, i.e., the sequence $(F_n)_{n\geq 0}$ defined by

$$F_0 = 0, \qquad F_1 = 1 \tag{10.54}$$

$$F_n = F_{n-1} + F_{n-2}, \qquad n \geq 2. \tag{10.55}$$

SOLUTION The generating function is

$$F(z) = F_0 + F_1 z + F_2 z^2 + \cdots = \sum_{n\geq 0} F_n z^n. \tag{10.56}$$

To calculate the generating function $F(z)$ explicitly, we use a trick which is usually described by the phrase "take the z transform of both sides of the HDE." In this case we multiply both sides of (10.55) by z^n, and sum over all $n \geq 2$, obtaining

$$\sum_{n\geq 2} F_n z^n = \sum_{n\geq 2} F_{n-1} z^n + \sum_{n\geq 2} F_{n-2} z^n. \tag{10.57}$$

Each of the three sums in (10.57) is closely related to the generating function $F(z)$ defined in (10.56). For example, the first sum is

$$\sum_{n\geq 2} F_n z^n = F_2 z^2 + F_3 z^3 + F_4 z^4 + \cdots = F(z) - F_0 - F_1 z.$$

Similarly, the second sum is

$$\begin{aligned}\sum_{n\geq 2} F_{n-1} z^n &= F_1 z^2 + F_2 z^3 + F_3 z^4 + \cdots = z(F_1 z + F_2 z^2 + F_3 z^3 + \cdots)\\ &= z(F(z) - F_0).\end{aligned}$$

Finally, the third sum is

$$\begin{aligned}\sum_{n\geq 2} F_{n-2} z^n &= F_0 z^2 + F_1 z^3 + F_2 z^4 + \cdots = z^2(F_0 + F_1 z + F_2 z^2 + \cdots)\\ &= z^2 F(z).\end{aligned}$$

Substituting these three expressions into (10.57) we get

$$F(z) - F_0 - F_1 z = z(F(z) - F_0) + z^2 F(z).$$

After a little algebra, this becomes

$$F(z)(1 - z - z^2) = F_0 + z(F_1 - F_0). \tag{10.58}$$

Now $F_0 = 0$ and $F_1 = 1$ according to the initial conditions (10.54). Hence, (10.58) becomes $F(z)(1 - z - z^2) = z$, i.e.,

$$F(z) = \frac{z}{1 - z - z^2}. \tag{10.59}$$

Equation (10.59) is the promised representation of the generating function for the Fibonacci numbers as a rational function. ∎

The technique used in Example 10.12 (take the z transform of both sides of the HDE) can be used to find the generating function for any sequence satisfying an HDE. The next example illustrates this.

EXAMPLE 10.13 Find the generating function of the sequence $y_0, y_1, y_2, \ldots$ defined as follows:

(10.60) $$y_0 = 0, \qquad y_1 = 1,$$

(10.61) $$y_n + 2y_{n-1} - 15y_{n-2} = 0, \qquad \text{for } n \geq 2.^*$$

SOLUTION Exactly as in Example 10.12, take the z transform of both sides of (10.61) by multiplying by z^n and summing over all $n \geq 2$. We obtain

$$G(z) - y_0 - y_1 z + 2z[G(z) - y_0] - 15z^2 G(z) = 0.$$

After a little rearranging, this becomes

$$G(z)(1 + 2z - 15z^2) = y_0 + (2y_0 + y_1)z.$$

By (10.60) the right side is simply z, so the generating function is given by

(10.62) $$G(z) = \frac{z}{1 + 2z - 15z^2}.$$

The generating functions we found in Examples 10.12 and 10.13 (Equations 10.59 and 10.62) are simple and perhaps even elegant, but they do not appear to shed much light on the original sequences (F_n) and (y_n). However, it is possible to derive *exact expressions* for these sequences from their generating functions, using a technique called the *partial-fraction expansion* or *decomposition.* This technique is studied in detail in calculus courses, but it's really not difficult to learn, and we will illustrate it here in the case of Example 10.13.

In Example 10.13, the denominator of $G(z)$ is $1 + 2z - 15z^2$, which factors nicely as $(1 - 3z)(1 + 5z)$, and so from (10.62) we have

(10.63) $$G(z) = \frac{z}{(1 - 3z)(1 + 5z)}.$$

An expression like (10.63) is called a **proper fraction**, meaning simply that the degree of the numerator is less than the degree of the denominator. It turns out that any proper fraction can be decomposed into a sum of what are called **partial fractions**. In this case, the partial-fraction decomposition of $G(z) = z/(1 - 3z)(1 + 5z)$ is

(10.64) $$\frac{z}{(1 - 3z)(1 + 5z)} = \frac{A}{1 - 3z} + \frac{B}{1 + 5z},$$

where A and B are constants that need to be determined. There are several possible ways of determining A and B (see Problem 23), but our favorite way is the following: If we multiply both sides of (10.64) by $(1 - 3z)$, we get

(10.65) $$\frac{z}{1 + 5z} = A + \frac{B(1 - 3z)}{1 + 5z}.$$

* This HDE can also be written as $y_{n+2} + 2y_{n+1} - 15y_n = 0$ for $n \geq 0$, and in a variety of other equivalent ways (cf. Problem 10.1.1), but the form (10.61) will allow us to repeat the analysis of Example 10.12 most easily.

We can isolate A in (10.65) by substituting $z = \frac{1}{3}$, which makes the term $(1 - 3z)$ zero:

$$\frac{\frac{1}{3}}{1 + \frac{5}{3}} = A.$$

Therefore $A = \frac{1}{8}$. Similarly, if we multiply (10.64) by $(1 + 5z)$ we get

(10.66) $$\frac{z}{1 - 3z} = \frac{A(1 + 5z)}{1 - 3z} + B.$$

If we set $z = -\frac{1}{5}$ in (10.66) the term $(1 + 5z)$ will equal zero, and we get

$$\frac{-\frac{1}{5}}{1 + \frac{3}{5}} = B.$$

Therefore, $B = -\frac{1}{8}$, and the partial-fraction expansion of $G(z)$ is

(10.67) $$G(z) = \frac{\frac{1}{8}}{1 - 3z} - \frac{\frac{1}{8}}{1 + 5z}.$$

Using the result of Example 10.11, we have

(10.68) $$G(z) = \tfrac{1}{8}(1 + 3z + 3^2z^2 + \cdots) - \tfrac{1}{8}(1 - 5z + 5^2z^2 - 5^3z^3 + \cdots).$$

It follows from (10.68) that the coefficient of z^n in $G(z)$ is $\frac{1}{8} \cdot 3^n - \frac{1}{8}(-5)^n$, i.e.,

$$y_n = \tfrac{1}{8} \cdot 3^n - \tfrac{1}{8}(-5)^n,$$

a result we could also have obtained via the methods of Section 10.1. The next example will give us more practice with the method of partial fractions.

EXAMPLE 10.14 Convert the generating function (10.59) into an explicit expression for F_n, the nth Fibonacci number.

SOLUTION The denominator $1 - z - z^2$ in (10.59) can be factored as $(1 - \lambda z)(1 - \bar{\lambda} z)$, where

$$\lambda = \frac{1 + \sqrt{5}}{2}, \qquad \bar{\lambda} = \frac{1 - \sqrt{5}}{2}.$$

The plan is to decompose the generating function $F(z)$ as follows:

(10.69) $$F(z) = \frac{z}{(1 - \lambda z)(1 - \bar{\lambda} z)} = \frac{A}{1 - \lambda z} + \frac{B}{1 - \bar{\lambda} z},$$

where A and B are unknown constants that need to be determined. To find A, we multiply both expressions for $F(z)$ in (10.69) by $(1 - \lambda z)$,

$$\frac{z}{1 - \bar{\lambda} z} = A + B \cdot \frac{1 - \lambda z}{1 - \bar{\lambda} z},$$

and substitute $z = 1/\lambda$:

$$\frac{1/\lambda}{1 - \bar{\lambda}/\lambda} = A.$$

From this we conclude that $A = 1/(\lambda - \bar{\lambda}) = 1/\sqrt{5}$. Similarly, to find B we multiply both expressions in (10.69) by $(1 - \bar{\lambda}z)$, and substitute $z = 1/\bar{\lambda}$. This gives

$$\frac{1/\bar{\lambda}}{1 - \lambda/\bar{\lambda}} = B,$$

i.e., $B = 1/(\bar{\lambda} - \lambda) = -1/\sqrt{5}$. Thus the complete partial-fraction decomposition for $F(z)$ is

(10.70)
$$F(z) = \frac{1}{\sqrt{5}} \cdot \frac{1}{1 - \lambda z} - \frac{1}{\sqrt{5}} \cdot \frac{1}{1 - \bar{\lambda} z}.$$

The result of Example 10.11 applied to (10.70) now gives the following formula for the nth Fibonacci number:

(10.71)
$$F_n = \frac{1}{\sqrt{5}}\lambda^n - \frac{1}{\sqrt{5}}(\bar{\lambda})^n,$$

which is the same formula we obtained in Example 10.2. ■

The generating-function method (combined with the partial-fraction expansion) can be used to solve HDEs. It works equally well on IDEs, but we will not pursue this very far since we already learned one method of solving IDEs in Section 10.2. However, the next example illustrates the important point that the generating-function method can often be used to solve IDEs for which the forcing sequence does not satisfy an HDE.

EXAMPLE 10.15 Let $(b_n)_{n \geq 0}$ be a sequence of integers satisfying the third-order IDE

(10.72)
$$b_0 = b_1 = 0, \qquad b_2 = 2$$

(10.73)
$$b_n - 2b_{n-1} - b_{n-2} + 2b_{n-3} = \begin{cases} 1 & \text{for } n = 3 \\ 0 & \text{for } n \geq 4. \end{cases}$$

Find the generating function for the sequence (b_n).

SOLUTION The point of this example is that the forcing sequence $(1, 0, 0, 0, \ldots)$, though very simple, does not satisfy an HDE, and so the methods given in Section 10.2 do not apply directly. However, the generating-function approach works like a charm. Thus let

$$\begin{aligned} B(z) &= b_0 + b_1 z + b_2 z^2 + b_3 z^3 + \cdots \\ &= \qquad\qquad 2z^2 + 5z^3 + \cdots \end{aligned}$$

be the generating function for the sequence (b_n). Taking the z transform of both sides of (10.73), we obtain

$$\sum_{n \geq 3} z^n(b_n - 2b_{n-1} - b_{n-2} + 2b_{n-3}) = 1 \cdot z^3 + 0 \cdot z^4 + 0 \cdot z^5 + \cdots = z^3.$$

Converting this to information about the generating function $B(z)$, we get

$$[B(z) - b_0 - b_1 z - b_2 z^2] - 2z[B(z) - b_0 - b_1 z] - z^2[B(z) - b_0] + 2z^3 B(z) = z^3.$$

Using the initial conditions (10.72), this equation becomes

$$B(z)(1 - 2z - z^2 + 2z^3) = 2z^2 + z^3.$$

Thus the generating function for the sequence (b_n) is given by

(10.74)
$$B(z) = \frac{2z^2 + z^3}{1 - 2z - z^2 + 2z^3}.$$

(In Problem 22 we will ask you to convert this generating function into an explicit formula for b_n.) ■

The method of generating functions can be applied to many areas of discrete mathematics besides difference equations, and we cannot resist the temptation to tell you about just one such application—an application to a simple problem in combinatorics.

Thus let A and B be two sets of nonnegative integers. If n is a nonnegative integer, we ask the following question: How many solutions to the equation

(10.75)
$$a + b = n$$

are there, if a must be an element of A, and b must be an element of B? For example, if $A = \{0, 1, 2, 3, 4\}$, $B = \{2, 3, 4\}$, and $n = 4$, the answer is *three*, as the following table shows:

a	+	b	=	n
0	+	4	=	4
1	+	3	=	4
2	+	2	=	4

Let us denote the number of solutions to (10.75) with $a \in A$ and $b \in B$ by g_n. Then with $A = \{0, 1, 2, 3, 4\}$ and $B = \{2, 3, 4\}$, we have just seen that $g_4 = 3$; and here are the other values of g_n:

(10.76)

n	0	1	2	3	4	5	6	7	8	9	10	11	12	$\cdots$
g_n	0	0	1	2	3	3	3	2	1	0	0	0	0	$\cdots$

Computing the g_n's for a given A and B isn't really hard, but it can be a bit tedious. However, there is a really easy way to obtain the *generating function* for the g_n's, which is of course defined as

(10.77)
$$G(z) = g_0 + g_1 z + g_2 z^2 + \cdots.$$

Here's how it works. We define the *generating function for A* as

$$G_A(z) = \sum_{a \in A} z^a.$$

For example, if $A = \{0, 1, 2, 3, 4\}$, then $G_A(z) = 1 + z + z^2 + z^3 + z^4$. Similarly, the generating function for B is defined as

$$G_B(z) = \sum_{b \in B} z^b.$$

Thus if $B = \{2, 3, 4\}$, we have $G_B(z) = z^2 + z^3 + z^4$. Now the remarkable thing is that *the generating function for the g_n's is the product of $G_A(z)$ and $G_B(z)$*:

(10.78) $$G(z) = G_A(z) \cdot G_B(z).$$

We will see why this is so shortly. For now, let's just verify that (10.78) is true in the special case where $A = \{0, 1, 2, 3, 4\}$ and $B = \{2, 3, 4\}$:

$$\begin{aligned} G_A(z) \cdot G_B(z) &= (1 + z + z^2 + z^3 + z^4) \cdot (z^2 + z^3 + z^4) \\ &= z^2 + 2z^3 + 3z^4 + 3z^5 + 3z^6 + 2z^7 + z^8, \end{aligned}$$

and indeed these coefficients agree with the g_n's tabulated in (10.76). The result in (10.78) can be generalized to any number of sets of nonnegative integers, as follows: Let $A_1, A_2, \ldots, A_M$ be M sets of nonnegative integers, and let g_n denote the number of solutions to the equation

(10.79) $$a_1 + a_2 + \cdots + a_M = n,$$

where each integer a_i is required to lie in the corresponding set A_i. Then the generating function for the g_n's is given by

(10.80) $$G(z) = G_{A_1}(z) G_{A_2}(z) \cdots G_{A_M}(z),$$

where $G_{A_1}(z), G_{A_2}(z), \ldots, G_{A_M}(z)$ are the generating functions for the sets $A_1, A_2, \ldots, A_M$, respectively. The following example illustrates the use of (10.80).

EXAMPLE 10.16 Let g_n denote the number of solutions in nonnegative integers to the equation

(10.81) $$x_1 + x_2 + x_3 + x_4 = n,$$

subject to the restrictions

(10.82) $$4 \le x_2 \le 7$$

(10.83) $$2 \le x_3 \le 6$$

(10.84) $$x_4 \ge 13.$$

Find the generating function for the g_n's.

SOLUTION This problem is a special case of (10.79) with $M = 4$ sets of nonnegative integers. There are no restrictions on x_1, and so the set A_1 is just the set of all nonnegative integers, viz.

(10.85) $$A_1 = \{0, 1, 2, 3, \ldots\}.$$

The restriction (10.82) says that

(10.86) $$A_2 = \{4, 5, 6, 7\};$$

the restriction (10.83) says that

(10.87) $$A_3 = \{2, 3, 4, 5, 6\};$$

and finally (10.84) says that

(10.88) $$A_4 = \{13, 14, 15, 16, \ldots\}.$$

Now we have to compute the generating functions for the four sets A_1, A_2, A_3, and A_4. For A_1, we have

$$\begin{aligned} G_{A_1}(z) &= 1 + z + z^2 + z^3 + \cdots && \text{(from 10.85)} \\ &= \frac{1}{1-z}. && \text{(from Example 10.11)} \end{aligned}$$

For A_2, we have from (10.86)

$$G_{A_2}(z) = z^4 + z^5 + z^6 + z^7.$$

For A_3, we have from (10.87)

$$G_{A_3}(z) = z^2 + z^3 + z^4 + z^5 + z^6.$$

And finally, for A_4 we have

$$\begin{aligned} G_{A_4}(z) &= z^{13} + z^{14} + z^{15} + \cdots && \text{(from 10.88)} \\ &= z^{13}(1 + z + z^2 + \cdots) && \text{(algebra)} \\ &= z^{13} \cdot \frac{1}{1-z} && \text{(from Example 10.11).} \end{aligned}$$

It follows then from the general result (10.80) that the required generating function $G(z)$ is given by

$$\begin{aligned} G(z) &= G_{A_1}(z)G_{A_2}(z)G_{A_3}(z)G_{A_4}(z) \\ &= \frac{1}{1-z} \cdot (z^4 + z^5 + z^6 + z^7) \cdot (z^2 + z^3 + z^4 + z^5 + z^6) \cdot \frac{z^{13}}{1-z} \\ &= \frac{z^{19}(1 + z + z^2 + z^3)(1 + z + z^2 + z^3 + z^4)}{(1-z)^2}. \end{aligned}$$

■

We will conclude this section with a short proof of the important relationship described in (10.78). We begin with a specific example, again taking $A = \{0, 1, 2, 3, 4\}$ and $B = \{2, 3, 4\}$. Equation (10.78) asserts that the coefficient of z^5 in the product

$$G_A(z)G_B(z) = (1 + z + z^2 + z^3 + z^4) \cdot (z^2 + z^3 + z^4)$$

is equal to the number of solutions to the equation $a + b = 5$, with $a \in A$ and $b \in B$. Why should this be so? A typical term in the product $G_A(z) \cdot G_B(z)$ is obtained by multiplying one of the powers of z in $G_A(z)$ (viz. $1, z, z^2, z^3, z^4$) by one of the powers of z in $G_B(z)$ (viz. z^2, z^3, z^4). The important thing to notice is that when powers of z are *multiplied*, the exponents are *added*, so that to get a z^5 term in the product $G_A(z) \cdot G_B(z)$ we must multiply a term from $G_A(z)$ and a term from $G_B(z)$ where the exponents add to 5. There are three ways this can happen:

Term in G_A	Term in G_B	Product
z^1	z^4	z^5
z^2	z^3	z^5
z^3	z^2	z^5

These three possibilities are in one-to-one correspondence with the three solutions to the equation $a + b = 5$, with $a \in A$ and $b \in B$:

a	$+$	b	$=$	5
1	+	4	=	5
2	+	3	=	5
3	+	2	=	5

And that is why the coefficient of z^5 in $G_A(z) \cdot G_B(z)$ is equal to the number of solutions to $a + b = 5$.

Exactly the same argument shows why (10.78) is true in general. The product $G_A(z) \cdot G_B(z)$ can be written symbolically as

$$\left(\sum_{a \in A} z^a\right) \cdot \left(\sum_{b \in B} z^b\right).$$

A typical term in this product is of the form $z^a \cdot z^b = z^{a+b}$. Therefore the number of terms in this product equal to z^n, i.e., the coefficient of z^n in $G_A(z)G_B(z)$, is equal to the number of ways of choosing $a \in A$ and $b \in B$ so that $a + b = n$. This is exactly what Equation (10.78) says.

Problems for Section 10.3

In Problems 1–7 find the generating function for the given sequences.

1. $a_n = 3^n$.

2. $a_n = (-2)^n$.

3. $0, 0, 1, 1, 1, 1, \ldots$.

4. $0, 0, 0, 1, 1, 1, 1, \ldots$.

5. $0, 1, 2, 3, 4, \ldots$. (*Hint*: What HDE does the sequence satisfy?)

6. $a_n = 1$ if n is even; $a_n = 0$ if n is odd.

7. $1, 0, -1, 0, 1, 0, -1, 0, 1, 0, -1, 0, \ldots$.

8. Find a closed-form expression for the generating function for the finite sequence $1, \lambda, \lambda^2, \ldots, \lambda^{N-1}$.

9. If (a_n) has generating function $A(z)$ and (b_n) has generating function $B(z)$, and we take the sum of the sequences, i.e., define $c_n = a_n + b_n$, show that the corresponding generating function is $C(z) = A(z) + B(z)$.

10. Find the generating function of the sequence $2^n + 3^n$, $n \geq 0$.

11. Find the generating function of $n + 3$, $n \geq 0$ (see Problem 5).

12. Define

$$a_n = \begin{cases} 1 & \text{if } n \text{ is a multiple of 3} \\ 0 & \text{if not.} \end{cases}$$

Find the generating function of (a_n).

Each of the next six problems is identical to a problem from Section 10.1, only now you are asked to find the *generating function* for the sequence satisfying the given initial conditions.

13. Problem 10.1.18.

14. Problem 10.1.19.

15. Problem 10.1.20.

16. Problem 10.1.23.

17. Problem 10.1.24.

18. Problem 10.1.26.

In Problems 19–22, use the partial-fraction technique to derive an exact formula for the nth term in the given generating function.

19. $G(z) = \dfrac{1}{1 - z^2}$.

20. $G(z) = \dfrac{1}{1 - 3z + 2z^2}$.

21. $G(z) = \dfrac{1}{1 - 6z + 11z^2 - 6z^3}$.

22. The generating function $B(z) = (2z^2 + z^3)/(1 - 2z - z^2 + 2z^3)$ from Example 10.15.

23. In the text we discussed one possible way of determining the constants in the partial-fraction decomposition [see, e.g., the steps leading from (10.64) to (10.67)]. In this problem, we will see another way.

(a) Multiply both sides of Equation (10.64) by $(1 - 3z)(1 + 5z)$. What values must A and B have in order for the resulting equation to be valid?

(b) Use the method suggested in part (a) to determine the constants A and B in (10.69).

24. In the solution to Example 10.15 we claimed that "the forcing sequence $(1, 0, 0, 0, \dots)$ does not satisfy an HDE." Explain why, using the HDE rules from Section 10.1.

In Problems 25–27, express your answer using generating functions.

25. Find the number of nonnegative integer solutions of $x_1 + x_2 + x_3 = 9$.

26. Repeat Problem 25 with the constraints $x_1 \le 5$, $3 \le x_2 \le 7$, and $x_3 \le 2$.

27. Refer to Example 2.20. A bakery sells chocolate chip cookies, peanut butter cookies, sugar cookies, and oatmeal cookies. In how many ways can we buy 12 cookies if the number of chocolate chip cookies must be between 2 and 4, and the number of oatmeal cookies must be between 3 and 6 (inclusive)?

28. Let $(F_n)_{n \ge 0}$ be the sequence of Fibonacci numbers. Define a new sequence (D_n) (the *double* Fibonacci numbers) by

$$D_0 = 0, \qquad D_1 = 1,$$

and for $n \ge 2$

$$D_n - D_{n-1} - D_{n-2} = F_n.$$

Find the generating function for (D_n).

Problems 29–34 develop some further theory of z transforms.

29. If $a_0, a_1, a_2, \dots$ and $b_0, b_1, b_2, \dots$ are sequences of numbers, we define the **convolution** of the two sequences as the sequence $c_0, c_1, c_2, \dots$ with

$$c_n = a_0 b_n + a_1 b_{n-1} + a_2 b_{n-2} + \cdots + a_n b_0 = \sum_{k=0}^{n} a_k b_{n-k}.$$

If (a_n), (b_n), and (c_n) have z transforms $A(z)$, $B(z)$, and $C(z)$, respectively, show that

$$C(z) = A(z)B(z);$$

thus the z transform of a convolution is the product of the z transforms. *Hint*: Write

$$\sum_{n \ge 0} \left(\sum_{k=0}^{n} a_k b_{n-k} \right) z^n = \sum_{k \ge 0} a_k z^k \left(\sum_{n \ge k} b_{n-k} z^{n-k} \right).$$

30. Apply Problem 29 with $A(z) = B(z) = 1/(1 - \lambda z)$ to show that the generating function $1/(1 - \lambda z)^2$ corresponds to the sequence $c_n = (n + 1)\lambda^n$.

31. Now apply Problem 29 with $A(z) = 1/(1 - \lambda z)^2$ and $B(z) = 1/(1 - \lambda z)$ to show that the generating function $1/(1 - \lambda z)^3$ corresponds to the sequence

$$d_n = \sum_{k=0}^{n} (k + 1)\lambda^k \lambda^{n-k}.$$

32. Evaluate the sum in Problem 31 explicitly to show that

$$d_n = \frac{(n + 1)(n + 2)}{2} \lambda^n.$$

33. Consider Example 10.9:

$$x_0 = x_1 = 0,$$

and

$$x_{n+2} - 4x_{n+1} + 4x_n = 2^n, \qquad n \ge 0.$$

Take the z transform of both sides of the IDE by multiplying by z^{n+2} (*not* z^n, which we would use if the equation were written as $x_n - 4x_{n-1} + 4x_{n-2} = 2^{n-2}$, $n \ge 2$) and summing over all $n \ge 0$. If $G(z)$ is the generating function of (x_n), show that

$$G(z) - x_0 - x_1 z - 4z[G(z) - x_0] + 4z^2 G(z) = \sum_{n \ge 0} z^{n+2} 2^n = \frac{z^2}{1 - 2z}.$$

Use the initial conditions $x_0 = x_1 = 0$ to obtain

$$G(z) = \frac{z^2}{(1 - 2z)^3}.$$

34. Continuing Problem 33, use Problem 32 to show that

$$\frac{1}{(1 - 2z)^3} = \sum_{n \ge 0} \frac{(n + 1)(n + 2)}{2} 2^n z^n;$$

hence

$$G(z) = \sum_{n \ge 2} (n - 1)n 2^{n-3} z^n.$$

Conclude that the solution to the difference equation is $x_n = n(n - 1)2^{n-3}$ as found in Example 10.9.

Each of the next four problems is identical to a problem from Section 10.2, only now you are asked to find the generating function of the sequence satisfying the given initial conditions. (If you are familiar with partial-fraction expansion, continue on to find the sequence itself.)

35. Problem 10.2.25.

36. Problem 10.2.27.

37. Problem 10.2.29.

38. Problem 10.2.31.

10.4 Difference Equations Whose Characteristic Polynomials Have Complex Roots

In Sections 10.1 and 10.2 we saw that the key to solving any HDE or IDE is finding the roots of its characteristic equation. These roots can be real or complex. Since we wanted to make the contents of this chapter accessible to students without prior knowledge of complex numbers, in all of the examples we have considered so far, these roots have been real. However, in practice it often happens that the roots are complex, and in this section we will briefly consider the complex case. We begin with a quick review of complex numbers.

Rectangular Form of a Complex Number

Expressions of the form $a + bi$ where a and b are real numbers and $i^2 = -1$ are called **complex numbers**. The **real part** is a and the **imaginary part** is b. If the imaginary part is 0, then the number is real, so the complex numbers include the real numbers as a special case.

Addition, Multiplication, Division

Familiar rules of arithmetic apply. For example, if $z = 2 + 3i$ and $w = 7 - 8i$ then

$$z + w = 9 - 5i$$

$$zw = (2 + 3i)(7 - 8i) = 14 - 24i^2 + 21i - 16i = 38 + 5i$$

$$\frac{z}{w} = \frac{2 + 3i}{7 - 8i} = \frac{2 + 3i}{7 - 8i} \cdot \frac{7 + 8i}{7 + 8i} = \frac{-10 + 37i}{113} = \frac{-10}{113} + \frac{37i}{113}.$$

Magnitude and Angle

The number $x + iy$ is pictured as the point (x, y) in the plane and is said to have **magnitude** r and **angle** θ (see Figure 10.3), where r and θ are defined by

$$r = \sqrt{x^2 + y^2}, \qquad \tan\theta = \frac{y}{x}.$$

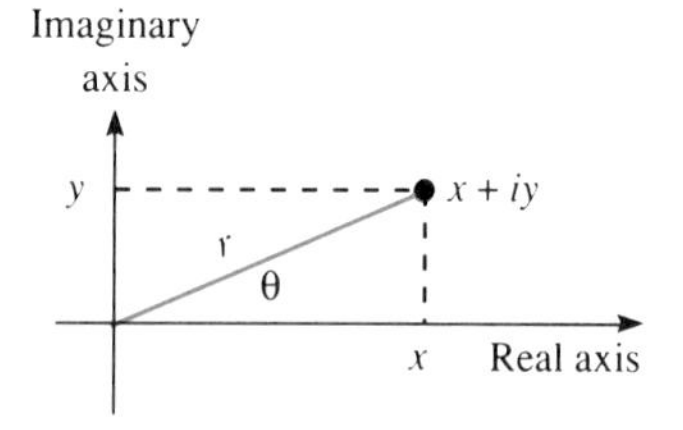

Figure 10.3 Magnitude and angle of a complex number.

Note that r and θ are the polar coordinates of the point (x, y).

For example:

If $z = -2 + 2i$ then $r = \sqrt{8}$ and $\theta = 3\pi/4$.
If $z = 6i$ then $r = 6$ and $\theta = \pi/2$.
If $z = -3 - 2i$ then $r = \sqrt{13}$, $\tan\theta = \frac{2}{3}$, and $\theta \approx 214°$.

Polar Form of a Complex Number

If $x + iy$ has magnitude r and angle θ, then the polar coordinates r, θ and the rectangular coordinates x, y are related by $x = r\cos\theta$, $y = r\sin\theta$, so

$$x + iy = r(\cos\theta + i\sin\theta)$$

which is known as the **polar form** of $x + iy$. For example,

$$-2 + 2i = \sqrt{8}\left(\cos\frac{3\pi}{4} + i\sin\frac{3\pi}{4}\right)$$

Magnitude and Angle of Products, Quotients, and Powers (DeMoivre's* Theorem)

If z_1 has magnitude r_1 and angle θ_1, and z_2 has magnitude r_2 and angle θ_2, then the following hold:

$z_1 z_2$ has magnitude $r_1 r_2$ and angle $\theta_1 + \theta_2$ (multiply magnitudes, add angles).

z_1/z_2 has magnitude r_1/r_2 and angle $\theta_1 - \theta_2$ (divide magnitudes, subtract angles).

z_1^n has magnitude r_1^n and angle $n\theta$ (raise magnitude to nth power, multiply angle by n).

For example, if z_1 has magnitude $\sqrt{2}$ and angle 225°, z_2 has magnitude $2\sqrt{5}$ and angle −60°, and z_3 has magnitude 1 and angle −90°, then the product $z_1 z_2 z_3$ has magnitude $2\sqrt{10}$ and angle 75°, i.e., $z_1 z_2 z_3 = 2\sqrt{10}(\cos 75° + i\sin 75°)$. Here is a proof of the product formula:

$$\begin{aligned} z_1 z_2 &= r_1(\cos\theta_1 + i\sin\theta_1)\cdot r_2(\cos\theta_2 + i\sin\theta_2) \\ &= r_1 r_2[\cos\theta_1\cos\theta_2 - \sin\theta_1\sin\theta_2 + i(\sin\theta_1\cos\theta_2 + \cos\theta_1\sin\theta_2)] \\ &= r_1 r_2[\cos(\theta_1 + \theta_2) + i\sin(\theta_1 + \theta_2)]. \end{aligned}$$

To establish the z_1^n formula, apply the product formula repeatedly with $z_1 = z_2$. For the z_1/z_2 formula, write $z_1 = (z_1/z_2)\cdot z_2$ and use the product formula on z_1/z_2 and z_2.

After that brief review of complex numbers, we will consider two typical HDEs, whose characteristic equations have complex roots.

EXAMPLE 10.17 Solve the HDE

(10.89) $$x_{n+1} = x_n - x_{n-1},$$

subject to the initial conditions

(10.90) $$x_0 = 0, \qquad x_1 = 1.$$

* Abraham DeMoivre (1667–1754). A gifted Protestant mathematician who emigrated from France to England following the revocation of the Edict of Nantes in 1685. His major contributions are all in probability theory; he was one of the earliest to understand the law of large numbers, which is the cornerstone of modern probability theory (see Chapter 9). He published "DeMoivre's theorem" in 1722.

SOLUTION The characteristic equation associated with the HDE (10.89) is

(10.91) $$\lambda^2 - \lambda + 1 = 0.$$

The roots of this equation are $\lambda_0 = (1 + \sqrt{-3})/2$ and $\lambda_1 = (1 - \sqrt{-3})/2$, both complex numbers. It follows then from HDE Rules 1 and 2 that for any complex constants A and B, the sequence

(10.92) $$x_n = A\lambda_0^n + B\lambda_1^n$$

will be a solution to (10.89). We now choose the constants A and B so the initial conditions (10.90) will be satisfied:

$$n = 0: \qquad A + B = x_0 = 0;$$

$$n = 1: \qquad A\lambda_0 + B\lambda_1 = x_1 = 1.$$

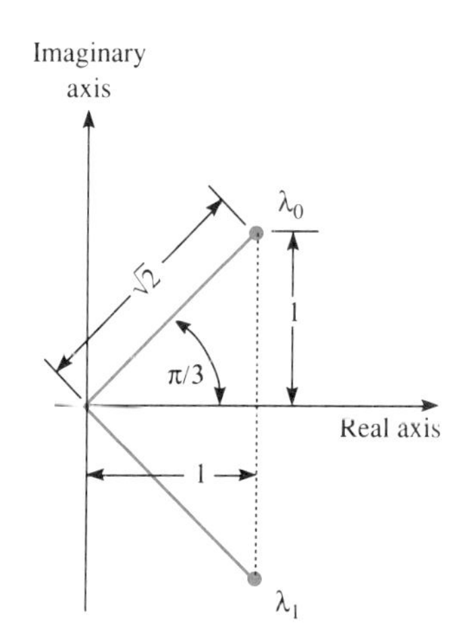

Figure 10.4 The complex roots of the equation $\lambda^2 - \lambda + 1 = 0$.

Solving these two equations we find that $A = 1/\sqrt{-3}$ and $B = -1/\sqrt{-3}$, and so the answer to Example 10.17 is

(10.93) $$x_n = \frac{1}{\sqrt{-3}}\left(\frac{1+\sqrt{-3}}{2}\right)^n - \frac{1}{\sqrt{-3}}\left(\frac{1-\sqrt{-3}}{2}\right)^n.$$

This expression can be considerably simplified, if we use some of the geometric properties of complex numbers. Notice (see Figure 10.4) that λ_0 and λ_1 each have magnitude 1, with angles $\pi/3$ and $-\pi/3$, respectively. Therefore

$$\lambda_0 = \cos\frac{\pi}{3} + i\sin\frac{\pi}{3}, \qquad \lambda_1 = \cos\frac{\pi}{3} - i\sin\frac{\pi}{3}.$$

Since by DeMoivre's theorem $(\cos\theta + i\sin\theta)^n = \cos n\theta + i\sin n\theta$, we have

(10.94) $$\lambda_0^n = \cos\frac{n\pi}{3} + i\sin\frac{n\pi}{3}, \qquad \lambda_1^n = \cos\frac{n\pi}{3} - i\sin\frac{n\pi}{3},$$

and (10.93) becomes, after a little algebra (Problem 11)

(10.95) $$x_n = \frac{2}{\sqrt{3}}\sin\frac{n\pi}{3}.$$

Substituting the first few values of n into Formula (10.95), we obtain the following sequence:

n	0	1	2	3	4	5	6	7	...
x_n	0	1	1	0	−1	−1	0	1	...

Incidentally, this listing shows something very interesting; the sequence (x_n) *repeats periodically.* This is because, as we see, $x_6 = 0$ and $x_7 = 1$, the same two values as the initial conditions x_0 and x_1. Thus the difference equation "starts over" once it reaches the sixth term; we say that (x_n) is *periodic* of period 6.

The next example illustrates how to handle a case when the magnitude of the complex root of the characteristic equation is not 1.

EXAMPLE 10.18 Solve the HDE

$$x_n = 2x_{n-1} - 2x_{n-2} \qquad (n \geq 2),$$

subject to the initial conditions

$$x_0 = 1, \qquad x_1 = 2.$$

SOLUTION In this case the characteristic equation is $\lambda^2 - 2\lambda + 2 = 0$, which has the complex roots $1 \pm \sqrt{-1}$, which can be written as (see Figure 10.5)

$$\lambda_0 = \sqrt{2}\left(\cos\frac{\pi}{4} + i\sin\frac{\pi}{4}\right), \qquad \lambda_1 = \sqrt{2}\left(\cos\frac{\pi}{4} - i\sin\frac{\pi}{4}\right).$$

By DeMoivre's theorem we have

$$\lambda_0^n = 2^{n/2}\left(\cos\frac{n\pi}{4} + i\sin\frac{n\pi}{4}\right), \qquad \lambda_1^n = 2^{n/2}\left(\cos\frac{n\pi}{4} - i\sin\frac{n\pi}{4}\right).$$

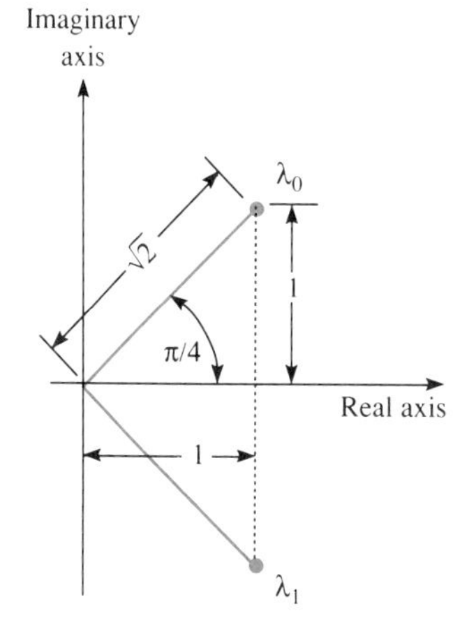

Figure 10.5 The complex roots of $\lambda^2 - 2\lambda + 2 = 0$.

It follows that, for certain (possibly complex) constants A and B, we have

(10.96)
$$\begin{aligned} x_n &= A\lambda_0^n + B\lambda_1^n \\ &= 2^{n/2}\left\{(A+B)\cos\frac{\pi n}{4} + i(A-B)\sin\frac{\pi n}{4}\right\}. \end{aligned}$$

We could now solve for the complex constants A and B using the initial conditions $x_0 = 1$ and $x_1 = 2$, as in Example 10.17. However, we can simplify the arithmetic considerably by taking advantage of the fact that x_n is real for all n, and predicting that x_n is given by a formula of the following type, which is called the **general real solution** to the HDE:

(10.97)
$$x_n = 2^{n/2}\left\{C\cos\frac{\pi n}{4} + D\sin\frac{\pi n}{4}\right\},$$

for certain real constants C and D. Now if we substitute $n = 0$ and 1 in Equation (10.97), we have

$$\begin{aligned} x_0 &= C && [\text{since } \cos(0) = 1, \sin(0) = 0], \\ x_1 &= C + D && [\text{since } \cos(\pi/4) = \sin(\pi/4) = \sqrt{2}/2]. \end{aligned}$$

Since $x_0 = 1$ and $x_1 = 2$, we find that $C = D = 1$. Plugging these values back into (10.97), we obtain

(10.98)
$$x_n = 2^{n/2}\left\{\cos\frac{\pi n}{4} + \sin\frac{\pi n}{4}\right\}.$$

The first few values of this sequence are 1, 2, 2, 0, −4, −8, −8, 0, ...

Problems for Section 10.4

1. Find $(2 + 6i)(8 - 3i)$.

2. Find $1/(8 + 3i)$.

3. Find $(2 + 9i)/(4 - i)$.

In Problems 4–8, find r and θ for the given complex numbers.

4. $-7i$. **6.** $-4 + 3i$. **8.** $10 - 10i$.

5. $4 - 4\sqrt{3}i$. **7.** -7.

9. Find $(-1 + i)^5$ using DeMoivre's theorem.

10. Find $(\sqrt{3} + i)^3$ using DeMoivre's theorem.

11. Derive (10.95) from (10.93).

12. In Example 10.17, show that *any* solution to the HDE is periodic with period 6 [use Formulas (10.92) and (10.94)].

13. Refer to Example 10.18. Given the expression (10.96) for x_n and the initial conditions $x_0 = 1$ and $x_1 = 2$, find the complex constants A and B.

14. In Example 10.18, find a general relation between the complex constants A and B and the real constants C and D.

15. Find the general solution in real form to $x_{n+2} + 2x_{n+1} + 2x_n = 0$.

16. Find the general solution in real form to $x_{n+2} + x_{n+1} + x_n = 0$.

17. Solve $x_{n+2} + 4x_{n+1} + 8x_n = 0$ with initial conditions $x_0 = 0$ and $x_1 = 2$.

18. Solve $x_{n+2} + 4x_n = 0$ with initial conditions $x_0 = 0$ and $x_1 = 1$.

In Problems 19–22, the roots of the characteristic equation of an HDE are given. Write the general solution in (a) complex form and (b) real form, using sines and cosines, as in Equation (10.97).

19. $\lambda = 6i, -6i$ (i.e., there are two roots, $6i$ and $-6i$).

20. $\lambda = 3 + 3i, 3 - 3i$.

21. $\lambda = -3, 4, 4$ (i.e., 4 is a root of multiplicity 2), $-3 + i$, $-3 - i$.

22. $\lambda = 1, 2, -2, 3, 2i, 2i, -2i, -2i$.

23. Show that the general solution (real form) of the HDE of Example 10.18 is given by

$$x_n = 2^{n/2}\left[x_0 \cos\frac{n\pi}{4} + (x_1 - x_0)\sin\frac{n\pi}{4}\right],$$

where x_0 and x_1 are the initial conditions.

24. A certain language contains the five strings, *a*, *bcd*, *cde*, *dcd*, *edc*, and all strings that can be built by concatenating these words together. If x_n is the number of strings in the language of length n, set up a difference equation for x_n.

25. In Problem 24, compute x_n explicitly for $n = 1, 2, 3, 4, 5$, and 6.

26. In Problem 24, an estimate of the form $x_n = A2^n$ (for some constant A) will be approximately correct for large n. Explain why this is so, and indicate how to find A.

27. Find a general real solution to $x_{n+4} - 16x_n = n + 3^n$.

28. Solve $x_{n+2} - x_{n+1} + x_n = 3 \cdot 2^n$ with initial conditions $x_0 = 1$ and $x_1 = 3$.

29. If an IDE has forcing function $f_n = 5\cos n\pi/2$ and the roots of the characteristic equation (of the homogeneous part) are $\lambda = \pm i$, what particular solution p_n would you try?

30. Repeat Problem 29 with $f_n = 5\cos n\pi/2$ and $\lambda = \pm 2i$.

Summary

In this Chapter we have learned how to solve homogeneous and inhomogeneous difference equations (HDEs and IDEs). The basic rules are summarized below:

- HDE Rule 1: If λ is a root of the characteristic equation of a given HDE, then $x_n = \lambda^n$ is a solution to the HDE.
- HDE Rule 1 (supplement): If λ is a root of the characteristic equation of a given HDE with multiplicity m, then $x_n = \lambda^n, n\lambda^n, \ldots, n^{m-1}\lambda^n$ are all solutions to the HDE.
- HDE Rule 2: If x_n and y_n are both solutions to a given HDE, then for any constants A and B, $Ax_n + By_n$ is also a solution.

- IDE Rule 1: The general solution to a given IDE is obtained by adding the general solution of the corresponding HDE to any particular solution of the IDE.
- IDE Rule 2: To find a particular solution to an IDE, try something that looks like the forcing sequence.
- IDE Rule 2 (supplement): If IDE Rule 2 fails, multiply by n and try again.

These rules will enable you to solve any HDE, and any IDE whose forcing function satisfies an HDE.

We also discussed the generating function method for solving difference equations. The generating function for a sequence $a_0, a_1, a_2, \ldots$ is the expression

$$G(z) = a_0 + a_1 z + a_2 z^2 + \cdots.$$

We found that the generating function for a sequence satisfying a HDE always turns out to be a rational function, i.e., a quotient of polynomials.

APPENDIX

EXPRESSING ALGORITHMS IN PSEUDOCODE

Since one of the key features of modern discrete mathematics is its emphasis on algorithms, it is important to have a consistent and clear way of describing algorithms. Unfortunately, however, there as yet is no agreement about the best way to do this. In this book, we have chosen to describe most of our algorithms in two ways: **flowcharts** and **pseudocode**. The flowchart method has been used by programmers since the earliest days of the computer. Flowcharts have the advantage of being self-explanatory and easy to learn, and many experienced programmers still use them. However, they have several disadvantages, the most important being that they can lead to bad programming habits, and are hard to read, if the algorithm is large and complex. For this reason, many modern workers use pseudocode instead. In this appendix, we will give a brief description of our own favorite dialect of pseudocode, which is what we have used throughout the book.

In the pseudocode method, algorithms are written in a primitive structured programming language. This language is much too simple to work on any known computer, however, and when compared with real programming languages like BASIC or FORTRAN or C or Pascal, it is so embarrassingly rudimentary and incomplete that it is called a *pseudolanguage.* A program written in such a pseudolanguage is also said to be written in *pseudocode.* There are many different pseudolanguages, but the one we shall use is based on the popular and powerful language "C," and so we call it "pseudo-C."

It is possible to give precise rules for pseudo-C, and indeed pseudo-C can be thought of as a formal language of the kind we study in Chapter 7. However, we won't spend much time worrying about the precise rules of the language, since the object of using pseudo-C (or any other pseudolanguage) is only to give a clear description of the operations involved in a particular algorithm. Just remember that no matter how careful you try to be, a program written in pseudo-C won't run!

A program in pseudo-C is by definition simply a series of valid **statements** enclosed in braces. (For now we won't say exactly what valid statements are.) Thus, if *statement 1*, *statement 2*, and *statement 3*, are valid statements, the following are two valid pseudo-C programs. (Anything enclosed between "/*" and "*/" is just a *comment* which is not an official part of the program.)

```
/* A Program with Only One Statement */
        {
            statement 1
        }
/* A Program Containing Three Statements */
        {
            statement 1
            statement 2
            statement 3
        }
```

It might be helpful to think of the opening brace "{" as the word "begin," and the closing brace "}" as the word "end." The exact location of the braces and the line breaks is not important, but it is traditional to place them as we have shown. In Figure A.1, we show the flowchart form for programs 1 and 2.

Now we must say exactly what we mean by a pseudo-C "statement." There are three kinds of statements: action statements, control statements, and compound statements.

An **action statement** tells the program to do a specific thing. We won't make any kind of restriction on what kind of action can be performed by an action statement—this is the great advantage of pseudocode over a real programming language! There is one technicality with action statements in pseudo-C—they

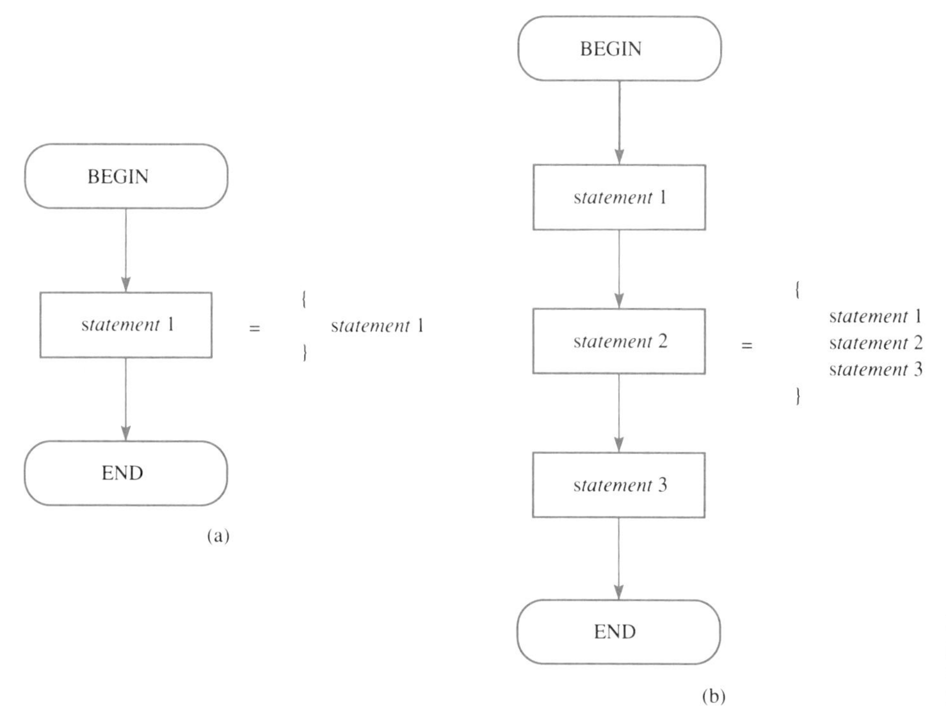

Figure A.1 Two simple programs in flowchart and in pseudo-C form: (a) A one-statement program; (b) a three-statement program.

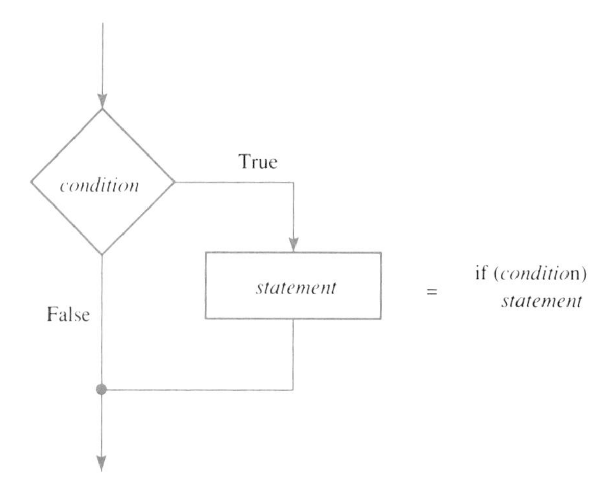

Figure A.2 The if statement.

must end with a semicolon. Thus

```
x ← x + 1;
delete the smallest element from the set X;
add the cheapest edge to the growing tree;
print n;
```

are all valid (action) statements in pseudo-C.

A **control statement** is one that allows the program to make decisions, or to repeat some procedure. In pseudo-C, there are five kinds of control statements: `if`, `if–then`, `while`, `do–while`, and `for` statements. These five statements can all be described with flowcharts, as shown in Figures A.2–A.6. Note that in each of these statements, a "condition" appears. This is supposed to be an expression which the program can evaluate as either *true* or *false.* If the condition is true, the program takes the "true" branch on the corresponding flowchart; if the condition is false, the program takes the "false" branch. (In the `for` statement, the condition is of the special form `x = EOL`, where `EOL` stands for *end of list.*)

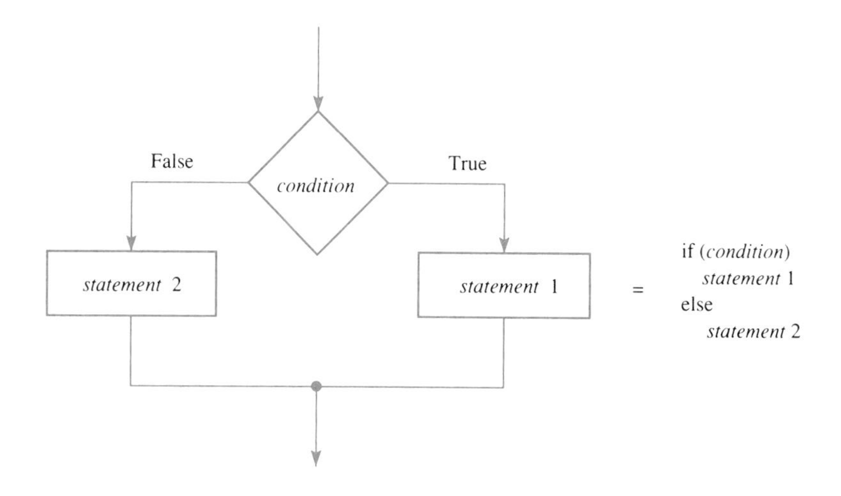

Figure A.3 The if–then statement.

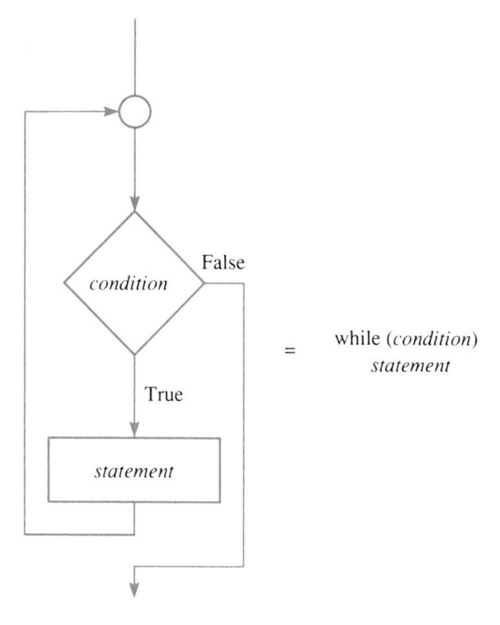

Figure A.4 The `while` statement.

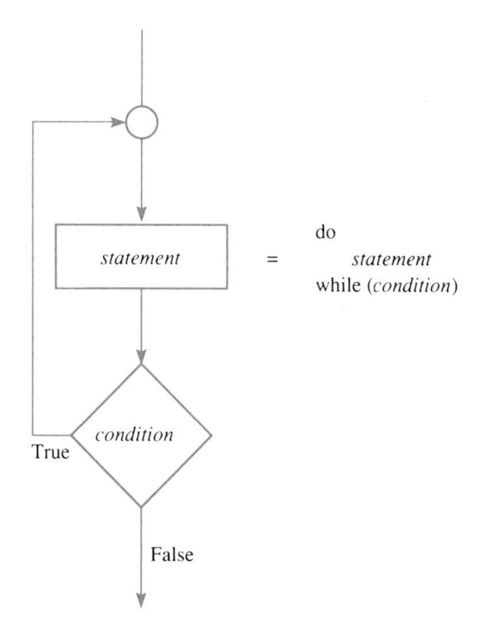

Figure A.5 The `do-while` statement.

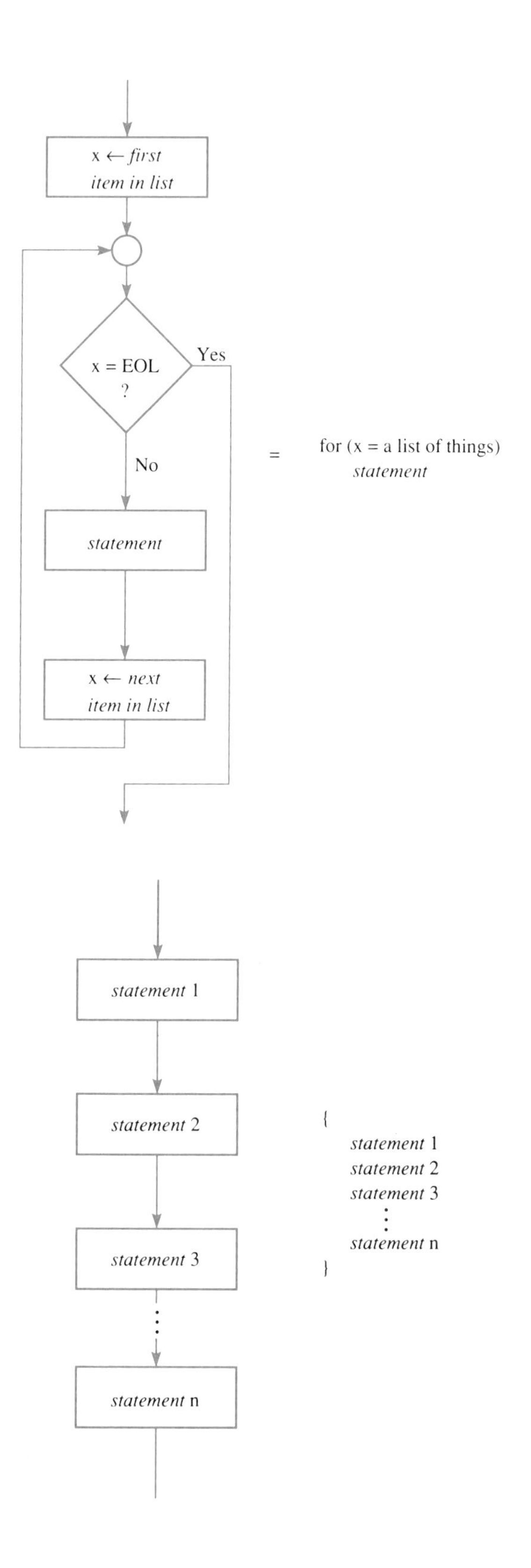

Figure A.6 The for statement. (*Note*: EOL means "end of list.")

Figure A.7 A compound statement.

A **compound statement** is simply a statement which is composed of two or more simpler statements enclosed in braces. For example, if *statement 1, statement 2,* and *statement 3* are all valid statements, then the following pseudo-C construction counts as a single valid statement:

```
/* This is a Compound Statement, Built from Three Simpler Statements */
                        {
                            statement 1
                            statement 2
                            statement 3
                        }
```

This general idea is illustrated in Figure A.7. For example,

```
{
    x ← x + 1;
    delete the smallest element from the set X;
    add the cheapest edge to the growing tree;
    print n;
}
```

is a valid statement in pseudo-C. (Note that by this definition, any program in pseudo-C is itself just a compound statement.) Again, we note that the positioning of the braces and the line breaks isn't critical, and so another valid way to write the above statement is

```
{x ← x + 1; delete the smallest element from the set X;
 add the cheapest edge to the growing tree;}
```

although most good programmers would prefer the first method.

Note that the `for` statement described in Figure A.6 isn't really necessary, since it is exactly the same as the compound statement

```
{
    x ← first item in list;
    while (x ≠ EOL) {
        statement
        x ← next item in list;
    }
}
```

However, this particular kind of compound statement occurs so often in practice that every programming language has a `for` statement, or something very similar. For example, the following program prints the numbers from 1 to 10:

```
/* Prints 1 Through 10 */
{
    i = 1;
    while (i ≤ 10) {
      print i;
      i ← i + 1;
    }
}
```

and so does this:

```
/* Also Prints 1 Through 10 */
{
    for (i = 1 to 10)
        print i;
}
```

It turns out that armed with only action, control, and compound statements, we can write programs in pseudo-C that will describe any possible algorithm. To illustrate this fact, we will now present several simple examples (the text contains many more!).

Here is a program written in pseudo-C that prints the first 13 powers of 2 ($1, 2, 4, 8, \ldots, 4096$). (In this program, the statement `print newline;` is supposed to cause the printer to move to the next line.)

```
/* Powers of Two Program */
{
    y ← 1;
    while (y ≤ 4096) {
        print y;
        print newline;
        y ← 2 * y;
    }
}
```

If we ran this program on a pseudo-computer, the output would look like this:

```
   1
   2
   4
   8
  16
  32
  64
 128
 256
 512
1024
2048
4096
```

Using the `for` statement, the powers-of-2 program could be written in the following alternative form:

```
/* Powers of Two Program, Version 2 */
{
    for (y = all powers of 2 less than 5000) {
        print y;
        print newline;
    }
}
```

If we wanted to get fancier, we could also write a program to print the powers of 2 in ordinary form and in Roman numerals:

```
/* Powers of Two in Roman Numerals */
{
  y ← 1;
  do {
        x ← y;
        print x;
        print blank space;
        while (x ≥ 1000) {
              print 'm';
              x ← x − 1000;
        }
      if (x ≥ 500) {
            print 'd';
            x ← x − 500;
      }
      while (x ≥ 100) {
            print 'c';
            x ← x − 100;
      }
      if (x ≥ 50) {
            print 'l';
            x ← x − 50;
      }
      while (x ≥ 10) {
            print 'x';
            x ← x − 10;
      }
      if (x ≥ 5) {
            print 'v';
            x ← x − 5;
      }
      while (x ≥ 1) {
            print 'i';
            x ← x − 1;
      }
      print newline;
      y ← 2*y;
   }
   while (y ≤ 5000);
}
```

This program consists of just two statements, the action statement `y ← 1;` and a `do–while` statement. The statement executed by the `do–while` statement, however, is a compound statement, composed of two assign statements, three `print` statements, four `while` statements, and three `if` statements. Each of the `while` and `if` statements, in turn, executes a compound statement, composed of one `print` statement, and an assign statement like `x ← x − 1;`. When run on a pseudo-computer, this program should print the following pseudo-output:

```
 1 i
 2 ii
 4 iiii
 8 viii
16 xvi
32 xxxii
64 lxiiii
```

```
 128 cxxviii
 256 cclvi
 512 dxii
1024 mxxiiii
2048 mmxxxxviii
4096 mmmmlxxxxvi
```

(Note that these are actually pseudo-Roman numerals, in that "4," for example, is written as "iiii" rather than "iv." To get real Roman numerals, we'd have to work harder.)

ANSWERS TO ODD-NUMBERED PROBLEMS

Chapter 1

Section 1.1

1. $A \cup C = \{1, 3, 5\} = A$ $\quad A \cap C = \{1, 5\} = C$
   $A \cup D = \{1, 2, 3, 4, 5\}$ $\quad A \cap D = \{3\}$
3. $\{1, 3, 5, 7, 8\}$
5. (a) $\{0, 2, 4, 6, 7, 8, 9\}$ (c) $\varnothing$
   (b) $\{0, 1, 3, 5, 6, 9\}$
7. (a) $\{1, 3, 5\}$ (b) $\{4, 8, 10\}$ (c) $\{8, 10\}$
9. (a) $\varnothing, \{2\}, \{4\}, \{2, 4\}$
   (b) Any subset which contains either a 2 or a 4 and is missing at least one of 1, 3, 5 will do. There are 21 such subsets.
11. $A = \{1, 2\}$; $A = \{1\} \cup \{2\}$
13. $|A \cup B \cup C| = 7$ $\quad |A \cap (B \cup C)| = 3$
    $|A \cap B' \cap C| = 1$
15.

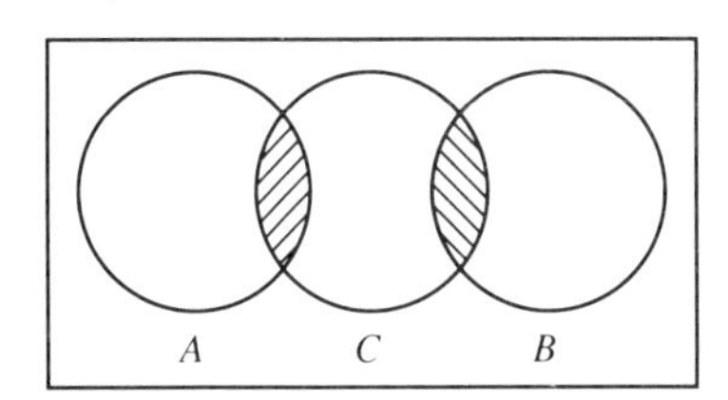

17. $N = 4$. An example is $\{1, 2\}, \{1, 3\}, \{1, 4\}, \{1, 5\}$.
19. (a) True (c) False (e) False
    (b) True (d) True (f) False
21. $A \cup B = \{1, 4, 8, \{1, 4, 8\}, \{3\}\}$
    $A \cap B = \{4\}$
23. $|A \cap B\} = |\{1\}| = 1$ $\quad$ 25. $\{\varnothing\}$
27. (a) $P(\{0, 1, 2\}) = \{\varnothing, \{0\}, \{1\}, \{2\}, \{0, 1\}, \{0, 2\}, \{1, 2\}, \{0, 1, 2\}\}$
    (b) $P(\varnothing) = \{\varnothing\}$ (note that this is not $\varnothing$)
29. $[0, 1)$ $\quad$ 31. $(0, 1]$
33. $A \cap B$ is a (possible empty) closed interval. $A \cup B$ is not necessarily a closed interval.

Section 1.2

1. See Figure 1.2(a).
3. See Figure 1.3.
5.

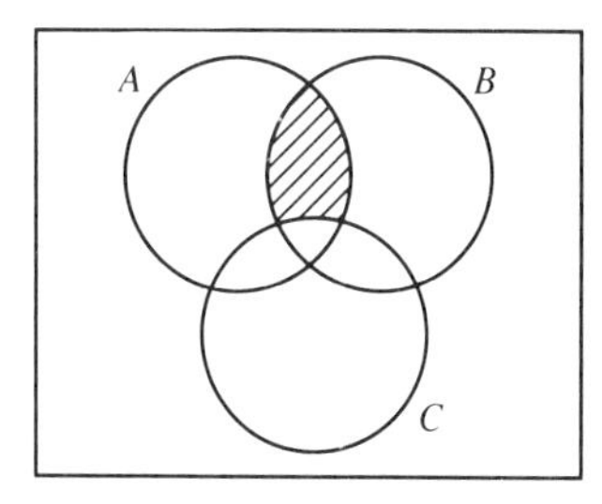

7.

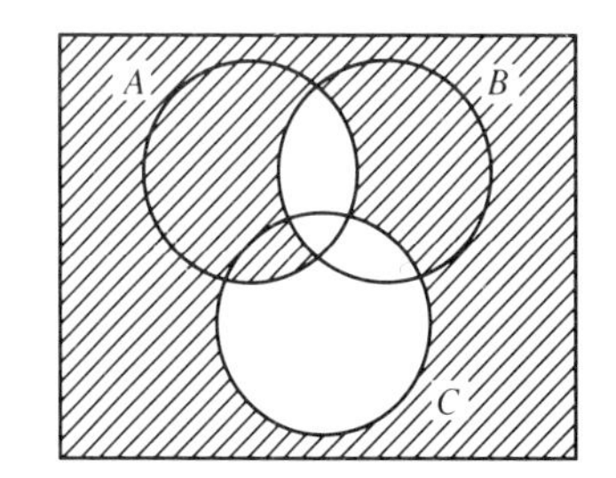

9.

A	B	$A \cap B$	$(A \cap B)'$	A'	B'	$A' \cup B'$
T	T	T	F	F	F	F
T	F	F	T	F	T	T
F	T	F	T	T	F	T
F	F	F	T	T	T	T

11.

A	B	C	$A \cup B$	$A \cup C$	$(A \cup B) \cap (A \cup C)$	$B \cap C$	$A \cup (B \cap C)$
T	T	T	T	T	T	T	T
T	T	F	T	T	T	F	T
T	F	T	T	T	T	F	T
T	F	F	T	T	T	F	T
F	T	T	T	T	T	T	T
F	T	F	T	F	F	F	F
F	F	T	F	T	F	F	F
F	F	F	F	F	F	F	F

13.

A	B	C	$B \cup C$	$A - (B \cup C)$	$A - B$	$(A - B) - C$
T	T	T	T	F	F	F
T	T	F	T	F	F	F
T	F	T	T	F	T	F
T	F	F	F	T	T	T
F	T	T	T	F	F	F
F	T	F	T	F	F	F
F	F	T	T	F	F	F
F	F	F	F	F	F	F

15. The statements "$x \in A$ or ($x \in B$ or $x \in C$)" and "($x \in A$ or $x \in B$) or $x \in C$" are equivalent.

17. We have $x \in (A')'$ if and only if $x \notin A'$ if and only if $x \in A$.

19. Here $x \in (A \cap B \cap C)'$ if and only if $x \notin A \cap B \cap C$ if and only if x does not belong to all three sets A, B, C if and only if $x \notin A$ or $x \notin B$ or $x \notin C$ if and only if $x \in A'$ or $x \in B'$ or $x \in C'$ if and only if $x \in A' \cup B' \cup C'$.

21. If $x \in A \cap (B - C)$ then $x \in A$ and $x \in B - C$, so $x \in A$ and $x \in B$ and $x \notin C$. Thus ($x \in A$ and $x \in B$) and ($x \in A$ and $x \notin C$), so $x \in A \cap B$ and $x \notin A \cap C$—therefore $x \in (A \cap B) - (A \cap C)$. Conversely, if $x \in (A \cap B) - (A \cap C)$, then $x \in A \cap B$ and $x \notin A \cap C$. Now $x \in A \cap B$ implies that $x \in A$ and $x \in B$. Since $x \in A$, the only way $x \notin A \cap C$ can happen is to have $x \notin C$. Thus $x \in A$ and ($x \in B$ and $x \notin C$), which says $x \in A \cap (B - C)$.

23. (a) Insufficient information
   (b) $x \notin A$
   (c) $x \in A$
   (d) Insufficient information

25. Clearly $A \cap B \subseteq A$ is always true. For the inclusion in the other direction, assume $x \in A$. Since $A \subseteq B$, $x \in B$ also, so $x \in A \cap B$.

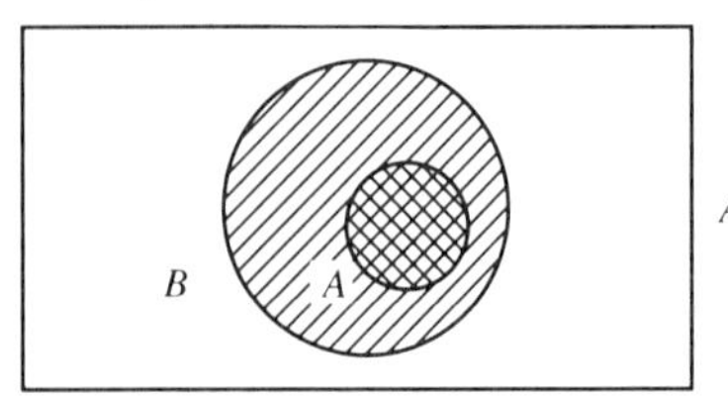

$A \cap B$ is cross hatched

27. No. An example is $A = B = C = \{1\}$.

29. It is obvious that $A_1 \cap \cdots \cap A_n \subseteq A_n$ and $A_1 \subseteq A_1 \cup \cdots \cup A_n$ for all sets A_i. For the reverse inclusions, first suppose $x \in A_n$. Since A_n is contained in all the A_i, $x \in A_1 \cap \cdots \cap A_n$, proving the first equality. For the second, note that if $x \in A_1 \cup \cdots \cup A_n$, then x belongs to some A_i. But since A_1 contains all the other A_i, $x \in A_1$.

31.

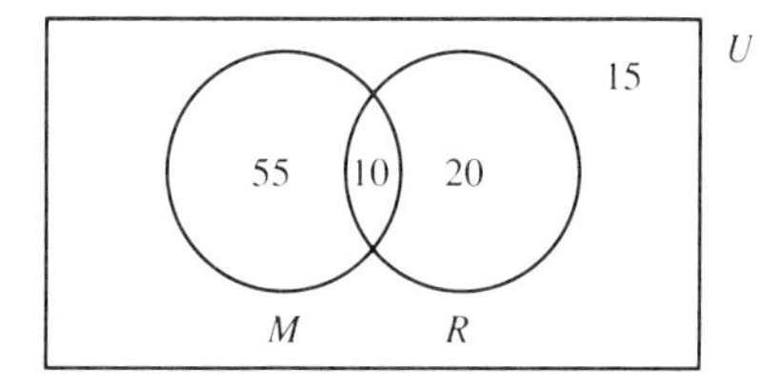

The number of nonregistered women is

$$|R' \cap M'| = |(R \cup M)'| = 15.$$

33.

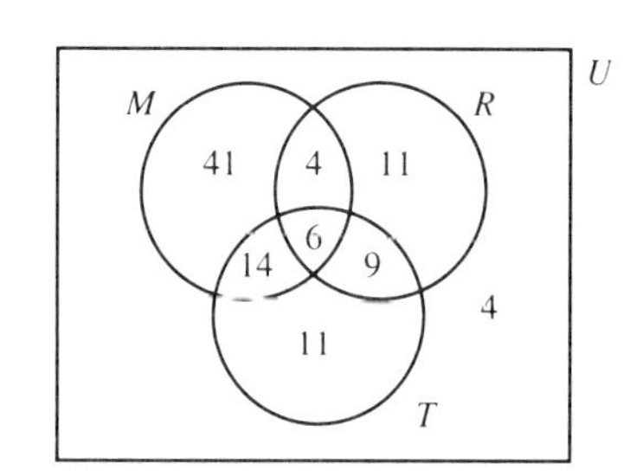

The number of short, nonregistered women is

$$|T' \cap R' \cap M'| = |(T \cup R \cup M)'| = 4.$$

Section 1.3

1.

x	f_1	f_2	f_3	f_4
1	a	a	b	b
2	a	b	a	b
One-to-one?	No	Yes	Yes	No
Onto?	No	Yes	Yes	No

3.

x	f_1	f_2	f_3	f_4	f_5	f_6	f_7	f_8
a	1	1	1	1	2	2	2	2
b	1	1	2	2	1	1	2	2
c	1	2	1	2	1	2	1	2
One-to-one?	No	No	No	No	No	No	No	No
Onto?	No	Yes	Yes	Yes	Yes	Yes	Yes	No

5. $h(x) = g(f(x)) = \cos(x^3)$
   $h'(x) = f(g(x)) = (\cos x)^3$

7. (a) If $f(x) \neq f(y)$ then $x \neq y$.
   (b) If $f(x) \neq f(y)$ then $x = y$.

(c) If $f(x) = f(y)$ then $x \neq y$.
(d) If you will not do your homework, you are dumb.
(e) If elephants do not have wings, then $1 + 1 \neq 3$.

9.

x	f_1^{-1}	f_2^{-1}	f_3^{-1}	f_4^{-1}	f_5^{-1}	f_6^{-1}
1	1	1	2	3	2	3
2	2	3	1	1	3	2
3	3	2	3	2	1	1

11. $A \times B = \{(1, a), (1, b), (2, a), (2, b), (3, a), (3, b)\}$

13.

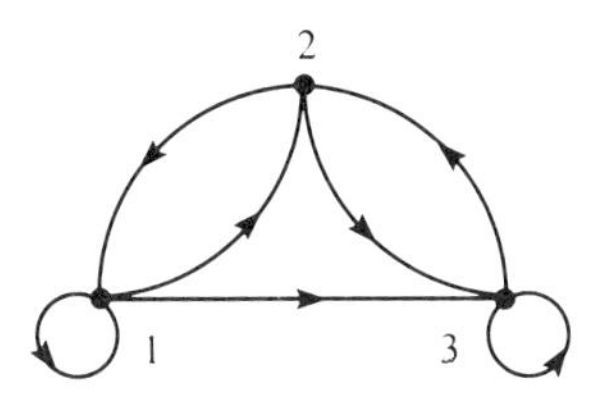

15. $R = \{(1, 1), (2, 2), (3, 3), (4, 4)\}$

17. For Figure 1.16,
$R^{-1} = \{(b_1, a_1), (b_1, a_4), (b_2, a_1), (b_2, a_2), (b_2, a_3), (b_3, a_4)\}$.
For Figure 1.17,
$R^{-1} = \{(2, 1), (3, 1), (1, 2), (4, 2), (3, 3), (4, 3), (1, 4)\}$.

19. 81 21. 0 23. 36

25. $2^{12} = 4096$

27. (a)

x	1	2	3	4
$(f \circ f)(x)$	4	3	2	1

(b)

x	1	2	3	4
$(f \circ f \circ f)$	3	1	4	2

(c)

x	1	2	3	4
$(f \circ f \circ f \circ f)(x)$	1	2	3	4

(d) They will simply repeat the earlier "powers."

29. If $(g \circ f)(x) = (g \circ f)(y)$, then $g(f(x)) = g(f(y))$, so $f(x) = f(y)$ (since g is one-to-one) and therefore $x = y$ (since f is one-to-one).

31. Yes. If $f(x) = f(y)$, then $g(f(x)) = g(f(y))$ so $(g \circ f)(x) = (g \circ f)(y)$ and therefore $x = y$ (since $g \circ f$ is one-to-one).

33. No. Define $f: \{1\} \to \{1, 2\}$ by $f(1) = 1$, and $g: \{1, 2\} \to \{1\}$ by $g(1) = g(2) = 1$. Then $g \circ f$ is onto, but f is not.

35. If $|A| > |B|$, there are no onto functions from B to A.

37. Since $f_n = f_{n-1} \circ f = i$, which is one-to-one, f is one-to-one by #31. Since $(f_{n-1} \circ f)(x) = f_n(x) = x$ and $(f \circ f_{n-1})(x) = f_n(x) = x$, $f^{-1} = f_{n-1}$.

39. (d)

41. Take the pigeonholes $\{1, 6\}$, $\{2, 5\}$, $\{3, 4\}$. Since we are selecting at least 4 values, some two must land in the same pigeonhole, and these two add to 7.

43. With n odd we have $(n + 1)/2$ pigeonholes:
$$\{1, n\}, \{2, n - 1\}, \ldots, \left\{\frac{n-1}{2}, \frac{n+3}{2}\right\}, \left\{\frac{n+1}{2}\right\}.$$
Thus we need $k \geq [(n + 1)/2] + 1$.

Section 1.4

1. (a) R is reflexive and antisymmetric since
$$R \cap \{(y, x) : (x, y) \in R\} = \{(x, x) : x \in A\}.$$
The only nontrivial instances of transitivity which need to be checked arise from $(1, 3), (3, 2) \in R$ and $(4, 3), (3, 2) \in R$. But $(1, 2) \in R$ and $(4, 2) \in R$ so R is a partial ordering. The Hasse diagram is

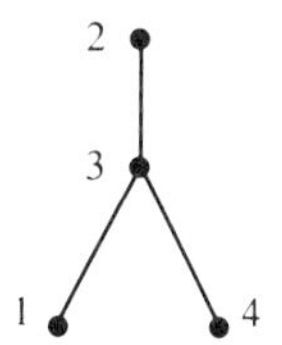

(b) Similar to (a). The Hasse diagram is

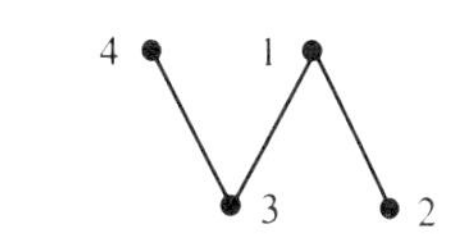

3.

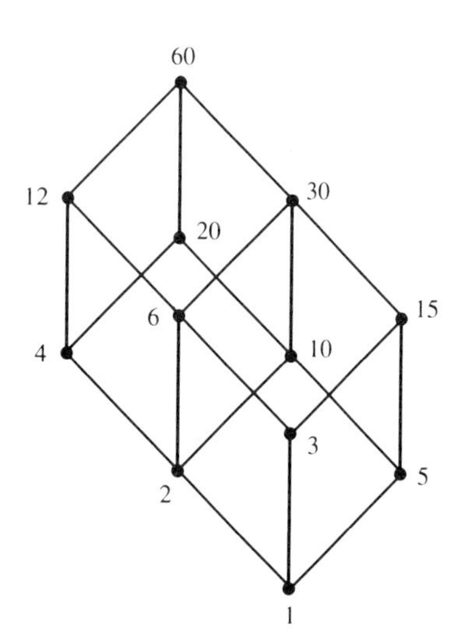

5. $\{(K, \Lambda'), (K, \bar{\Lambda}), (K, N), (K, \Lambda), (\Lambda', \bar{\Lambda}), (\Lambda', N), (\Lambda', \Lambda), (\bar{\Lambda}, N), (\Lambda, N), (K, K), (\Lambda', \Lambda'), (\bar{\Lambda}, \bar{\Lambda}), (\Lambda, \Lambda), (N, N)\}$
7. Because x is connected to u but there is an element between x and u (namely y).
9. No. A given partial ordering may have many different Hasse diagrams. For example, the diagrams

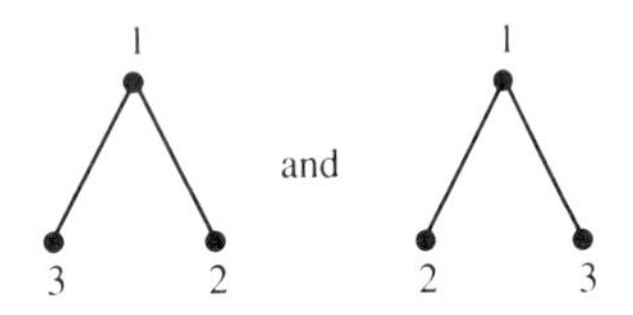

both represent the same partial order $\{(1, 1), (2, 2), (3, 3), (3, 1), (2, 1)\}$ and the second of these is in the list of Figure 1.24. The same analysis holds for the student's other example.
11. 19.
13. $\sim$ is reflexive since any word has the same number of letters as itself. It is symmetric since if WORD_1 has the same number of letters as WORD_2 then WORD_2 has the same number of letters as WORD_1. Finally, if WORD_1 has the same number of letters as WORD_2 and WORD_2 has the same number of letters as WORD_3, then WORD_1 has the same number of letters as WORD_3, so $\sim$ is transitive. The equivalence class of "student" consists of all 7-letter words in the English language.
15. See the discussion following Example 1.19. There are seven equivalence classes, one for each $r = 0, \ldots, 6$. The equivalence class corresponding to r consists of all integers which leave a remainder of r when divided by 7.
17. $x \sim y$ if and only if $x \equiv y \pmod 4$
19. (a) 2 (c) 4 (e) 2
    (b) 1 (d) 0
21. No. A_i must be the equivalence class of some x, and this x must belong to A_i.
23. Yes. $>$ is not a partial ordering (not reflexive).
25. (a) Transitive, antisymmetric
    (b) Reflexive, symmetric, transitive
    (c) Reflexive, symmetric
    (d) Reflexive
    (e) Reflexive, symmetric, transitive
27. (a) False (b) False
29. A maximal element has no connections leading up (and a minimal element has no connections leading down). A greatest element is connected to every other element by a path leading down (and a least element is connected to every other element by a path leading up).
31.

	$\varnothing$	$A \times A$
Reflexive	No	Yes
Symmetric	Yes	Yes
Antisymmetric	Yes	No*
Transitive	Yes	Yes

*Provided $|A| > 1$.

33. If R is such a relation on A, then $\{(x, x): x \in A\} \subseteq R$ by the fact that R is reflexive. If $(x, y) \in R$, then $x\,R\,y$, so $y\,R\,x$ by symmetry and therefore $x = y$ by antisymmetry. Thus $R = \{(x, x): x \in A\}$, i.e., the equality relation. On the set $\{1, 2, 3\}$, the only such relation is $\{(1, 2), (2, 2), (3, 3)\}$.
35. *Reflexive*: Clearly every element of $\mathscr{A}$ is a subset of some element of $\mathscr{A}$ (namely itself).
Antisymmetric: If $\mathscr{A} \leq \mathscr{B}$ and $\mathscr{B} \leq \mathscr{A}$, let $X \in \mathscr{A}$. Then there exists some $Y \in \mathscr{B}$ such that $X \subseteq Y$, and hence there exists some $Z \in \mathscr{A}$ such that $Y \subseteq Z$. But then X and Z are two blocks of the partition $\mathscr{A}$ which overlap, so we must have $X = Z$. Since Y is trapped in between, $X = Y \in \mathscr{B}$. Thus $\mathscr{A} \subseteq \mathscr{B}$. Likewise $\mathscr{B} \subseteq \mathscr{A}$ so $\mathscr{A} = \mathscr{B}$.
Transitive: Suppose $\mathscr{A} \leq \mathscr{B}$, $\mathscr{B} \leq \mathscr{C}$, and let $X \in \mathscr{A}$. Then there is a $Y \in \mathscr{B}$ such that $X \subseteq Y$ and a $Z \in \mathscr{C}$ such that $Y \subseteq Z$. Hence $X \subseteq Z \in \mathscr{C}$, so $\mathscr{A} \leq \mathscr{C}$. The required Hasse diagram is shown below.

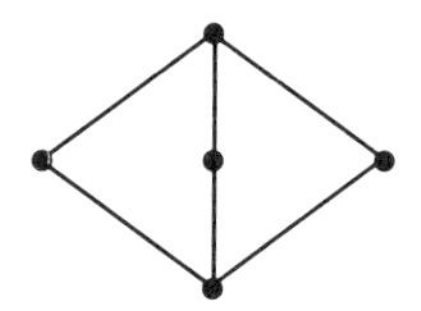

Section 1.5

1. In Problem 1, the algebra is slightly simpler if we execute step I2 by going from $(n - 1)$ to n rather than from n to $n + 1$. For the basis, simply observe that $1 = \frac{1}{3} + \frac{1}{2} + \frac{1}{6}$. Assume then that the assertion holds for $n - 1$. Then
$$\begin{aligned} &1^2 + 2^2 + \cdots + (n-1)^2 + n^2 \\ &\quad = \tfrac{1}{6}[(n-1)((n-1)+1)][2(n-1)+1] + n^2 \\ &\quad = \tfrac{1}{6}[(n-1)(n)(2n-1) + 6n^2] \\ &\quad = \tfrac{1}{6}(2n^3 + 3n^2 + n) = \tfrac{1}{6}[n(n+1)(2n+1)]. \end{aligned}$$
3. Certainly the assertion holds for $n = 0$ and $n = 1$ since a set with fewer than two elements can have no 2-element subsets. Assume that the formula holds for n

and let $S = \{a_1, \ldots, a_n, a_{n+1}\}$ be a set with $n + 1$ elements. The 2-element subsets of S are of two types, those which are subsets of $\{a_1, \ldots, a_n\}$ [and there are $n(n-1)/2$ of these, by the induction hypothesis] and those which contain the element a_{n+1} (and there are n of these, one for each element a_i, $i = 1, \ldots, n$). Thus the total number of 2-element subsets is

$$\frac{n(n-1)}{2} + n = \frac{n^2 - n + 2n}{2} = \frac{n^2 + n}{2} = \frac{(n+1)(n)}{2},$$

as required.

5. The assertion is certainly true for $n = 1$. Assuming it is true for n, $n^3 - 4n + 6 = 3k$ for some k. Then

$$\begin{aligned}(n+1)^3 - 4(n+1) + 6 &= (n^3 + 3n^2 + 3n + 1) - 4n - 4 + 6 \\ &= n^3 + 3n^2 - n + 3 \\ &= (n^3 - 4n + 6) + 3n^2 + 3n - 3 \\ &= 3k + 3n^2 + 3n - 3 \\ &= 3(k + n^2 + n - 1),\end{aligned}$$

which is divisible by 3.

7. $2^1 > 1$ is true. Assuming $2^n > n$, we have

$$2^{n+1} = 2 \cdot 2^n > 2n = n + n \geq n + 1.$$

9. We need to show that $F_{3n} = 2k$ for some k. As the problem statement shows, this is true for $n = 1, 2$, and 3. Assuming $F_{3n} = 2k$, we have

$$\begin{aligned}F_{3(n+1)} = F_{3n+3} &= F_{3n+2} + F_{3n+1} \\ &= F_{3n+1} + F_{3n} + F_{3n+1} = 2(F_{3n+1} + k),\end{aligned}$$

which is even.

11. The general pattern is

$$1 - 2^2 + 3^2 - \cdots + (-1)^{n+1}n^2 = (-1)^{n+1}(1 + \cdots + n).$$

Since $1 = 1$ forms the basis for the induction, assume that the assertion holds for n. Then

$$\begin{aligned}&1 - 2^2 + 3^2 - \cdots + (-1)^{n+1}n^2 + (-1)^{n+2}(n+1)^2 \\ &= (-1)^{n+1}(1 + \cdots + n) + (-1)^{n+2}(n+1)^2 \\ &= (-1)^{n+2}[(n+1)^2 - (1 + \cdots + n)] \\ &= (-1)^{n+2}\left[(n+1)^2 - \frac{n(n+1)}{2}\right] \\ &= (-1)^{n+2}\left\{\frac{(n+1)[2(n+1) - n]}{2}\right\} \\ &= (-1)^{n+2}\left[\frac{(n+1)(n+2)}{2}\right] \\ &= (-1)^{n+2}[1 + 2 + \cdots + (n+1)].\end{aligned}$$

13. Since k is a constant, it may be computed by setting $n = 1$. This yields $1 = \frac{1}{5} + \frac{1}{2} + \frac{1}{3} + k$, so $k = -\frac{1}{30}$. If the result holds for n, we must verify that

$$\begin{aligned}&\frac{n^5}{5} + \frac{n^4}{2} + \frac{n^3}{3} - \frac{n}{30} + (n+1)^4 \\ &= \frac{(n+1)^5}{5} + \frac{(n+1)^4}{2} + \frac{(n+1)^3}{3} - \frac{(n+1)}{30}\end{aligned}$$

which may be accomplished by expanding both sides and comparing.

15. $$\begin{aligned}&(A_1 \cup A_2 \cup \cdots \cup A_n \cup A_{n+1})' \\ &= ((A_1 \cup \cdots \cup A_n) \cup A_{n+1})' \\ &= (A_1 \cup \cdots \cup A_n)' \cap A'_{n+1} \\ &= A'_1 \cap \cdots \cap A'_n \cap A'_{n+1}\end{aligned}$$

17. Since $2^2 - 1 = 3$ is divisible by 3, we have a basis for the induction. Assuming the result holds for n, $2^{2n} - 1 = 3k$ for some k. Then

$$\begin{aligned}2^{2(n+1)} - 1 = 2^{2n+2} - 1 &= 4 \cdot 2^{2n} - 1 \\ &= 4(2^{2n} - 1 + 1) - 1 \\ &= 4(3k + 1) - 1 \\ &= 12k + 3 = 3(4k + 1),\end{aligned}$$

which is divisible by 3.

19. (a) $1^2 + 5 \cdot 1 + 1 = 7$, which is not even.
(b) Assuming $n^2 + 5n + 1 = 2k$ is even, we have

$$\begin{aligned}(n+1)^2 + 5(n+1) + 1 &= n^2 + 2n + 1 + 5n + 5 + 1 \\ &= (n^2 + 5n + 1) + (2n + 6) \\ &= 2(k + n + 3),\end{aligned}$$

which is also even.

21. Suppose that n¢ worth of postage can be paid for with a 9¢ stamps and b 5¢ stamps. If $a > 0$, then $(n + 1)$¢ worth of postage can be paid for with $(a - 1)$9¢ stamps and $(b + 2)$5¢ stamps. If $a = 0$, then $(n + 1)$¢ worth of postage can be paid for with $(b - 7)$5¢ stamps and 4 9¢ stamps.

23. Assuming that $2^n \geq n^3$ and $n \geq 10$, we have

$$\begin{aligned}2^{n+1} &= 2^n + 2^n \geq n^3 + n^3 \geq n^3 + 10n^2 \\ &= n^3 + 3n^2 + 7n^2 \geq n^3 + 3n^2 + 70n \\ &= n^3 + 3n^2 + 3n + 67n > n^3 + 3n^2 + 3n + 1 \\ &= (n+1)^3.\end{aligned}$$

25. (a) $S(0) = 41$, $S(1) = 41$, $S(2) = 43$, and $S(3) = 47$ are all prime.
(b) It must fail because $S(41)$ is false: $S(41) = 41^2$ is not prime.

27. Suppose A does not have a smallest element. Then $S(1)$ is true, for if it were false we would have $1 \in A$ and this would clearly be the smallest element of A. Assume then that $n > 1$ and $S(1), \ldots, S(n-1)$ all hold. If $n \in A$, then n would be the smallest element of A since none of $1, \ldots, n-1$ can belong to A. Thus $n \notin A$ and $S(n)$ holds.

29. 144

31. The ratio of the bound to F_n approaches 1.382. An exact formula is

$$F_n = \frac{1}{\sqrt{5}}\left[\left(\frac{1+\sqrt{5}}{2}\right)^n - \left(\frac{1-\sqrt{5}}{2}\right)^n\right].$$

(See Example 10.2 (p. 422), where this formula is derived.)

Section 1.6

1. 8 5. $\frac{3^n - 1}{2}$ 9. 14 13. $\left[\frac{n(n+1)}{2}\right]^2$

3. $\frac{341}{16}$ 7. $\frac{9(x^{10} - 1)}{x - 1}$ 11. 42 15. 14400

17. 24 19. $\frac{1}{n+1}$

21. The statement holds if and only if $\sum_{x \in A \cap B} a_x = 0$. (In particular, it holds if A and B are disjoint.)

23. $\frac{n+1}{2n}$

25. The analogous result is

$$\prod_{i=1}^{n} (x_i y_i) = \left(\prod_{i=1}^{n} x_i\right)\left(\prod_{i=1}^{n} y_i\right).$$

The basis is easy. For the induction step we have

$$\begin{aligned}\prod_{i=1}^{n} (x_i y_i) &= \left(\prod_{i=1}^{n-1} (x_i y_i)\right)(x_n y_n) \\ &= \left(\prod_{i=1}^{n-1} x_i \prod_{i=1}^{n-1} y_i\right)(x_n y_n) \\ &= \left(\prod_{i=1}^{n-1} x_i\right)x_n\left(\prod_{i=1}^{n-1} y_i\right)y_n \\ &= \left(\prod_{i=1}^{n} x_i\right)\left(\prod_{i=1}^{n} y_i\right).\end{aligned}$$

27. $2 + 3 + 4 + 5 + 3 + 4 + 5 + 6 + 4 + 5 + 6 + 7 = 54$.

29. There are 64 terms, each of the form

$$f(1, i)f(2, j)f(3, k),$$

where i, j, and k are between 1 and 4.

31. Both are equal to A_1.

33. (a) False (b) True

Section 1.7

1. 0000000 0101110

3. 0100101 1110100

5. The received word is interpreted by the decoder as 1001100. This differs from the transmitted codeword in three locations: 2, 3, and 4. The decoder believes the message word was 1001.

7. The decoder will flip bit 2, introducing another error.

9. It will always make another error.

11. When the word 1100010 is received, all three of the sets A, B, and C have even parity, so the decoder accepts this as a valid codeword and decodes it as 1100. At least no new errors are introduced.

13. The decoding algorithm from the text will simply ignore the 8th bit, yielding the same result as before.

15. If the overall parity is 1, an odd number of errors have occurred. Supposing that only a single error is involved, if the parity checks on the first 7 bits all work out, the error was in bit 8 and the message can be accepted. Otherwise the parity checks on the first 7 bits will find and correct the error. If the overall parity check is 0, an even number of errors has occurred. Supposing that no more than two errors have happened, look at the decoding of the first 7 bits. If this shows an error, two errors have occurred. Otherwise no error has occurred.

17. The decoding of 10011110 is 100001110. Two errors have occurred in 11101110.

19. The only one of the four possibilities which is actually a codeword is 1101001, which is therefore accepted.

21. The decoding algorithm is indicated by Problem 19. Look at the possibilities for the received word and pick the only codeword.

23. Replace the erased bit with the value needed to make the overall parity even. If all parity checks work out, we are done. Otherwise change the erased bit and decode as usual.

25. Yes

27. There are 7 distinct columns and 7 distinct nonzero binary sequences of length 3. In the given matrix, a 1 corresponds to being in the set at the left and a 0 corresponds to not being the set at the left. Hence, 100 corresponds to being in A but not in B or C.

29. Check A:

$$0 + 0 + 0 + 0 + 0 + 0 + 0 + 1 = 1. \quad \text{(Odd)}$$

Check B:

$$0 + 0 + 0 + 1 + 0 + 1 + 0 + 0 = 2. \quad \text{(Even)}$$

Check C:

$$0 + 0 + 0 + 1 + 0 + 1 + 1 + 0 = 3. \quad \text{(Odd)}$$

Check D:

$$0 + 0 + 0 + 1 + 0 + 0 + 1 + 0 = 2. \quad \text{(Even)}$$

Chapter 2

Section 2.1

1. $(2)(3)(4) = 24$ 3. $26^4 = 456976$

5. The elements of $S \times T$ are in one-to-one correspondence with the cartoons, but technically are not the same as the cartoons.

7. $(10)(9)(8)(7)(6)(5) = 151200$

9. $2^8 = 256$

11. (a) $(1)(26)(26)(26) = 17576$
 (b) $(1)(26)(26)(1) = 676$
 (c) $(1)(25)(25)(25) = 15625$

13. $10^6 = 1000000$

15. (a) $(5)(4) = 20$
 (b) $(5)(4)(4)(5) = 400$
 (c) $(5)(4)(3)(4) = 240$

17. $(9)(8)(7)(6)(5)(1) + (8)(8)(7)(6)(5)(1) = 28560$

19. $2^5 = 32$

21. $(3)(2)(1)(4)(3)(2)(1) = 144$

23. Yes. There are $(26^3)(10^3) = 17576000$ plates.

25. $n(n-1)$ 27. $(2)(2) = 4$

29. If n is even, there are $(9)(10^{(n/2-1)})$ n-digit palindromes. If n is odd, there are

$$(9)(10^{[(n-1)/2-1]})(10) = (9)(10^{[(n-1)/2]})$$

n-digit palindromes.

31. The 3 children in each unordered arrangement can be lined up in $(3)(2)(1) = 6$ ways to form ordered arrangements. Thus there are 6 times as many ordered arrangements as unordered ones, and the number of unordered arrangements is $210/6 = 35$.

Section 2.2

1. Ordered, with repetition:

 AA AB AC BA BB BC CA CB CC.

 Ordered, no repetition:

 AB AC BA BC CA CB.

 Unordered, with repetition:

 AA AB AC BB BC CC.

 Unordered, no repetition:

 AB AC BC.

3. $5^6 = 15625$

5. (a) $52^3 = 140608$
 (b) $(52)(51)(50) = 132600$

7. None 9. 4

11. (a) $(4!)(7!)(3!) = 725760$
 (b) $(3!)(4!)(7!)(3!) = 4354560$
 (c) $(4 + 7 + 3)! = 14! = 87178291200$

13. 6^n

15. (a) $(15)(14)(13)(12) = 32760$
 (b) $(15)(14)(13)(12) = 32760$
 (c) $15^4 = 50625$

17. $21! = 51090942171709440000$

19. $(6)(5)(21)(20)(19)(18)(17) = 73256400$

21. The ith ball, $i = 1, \ldots, k$ can be placed in any of n different ways. Thus there are n^k ways in which the sequence of choices can be made.

23. $(7!)(8)(7)(6) = 1693440$

25. $(2)(4)(5)(6) = 240$

27. Since there are $|Y|^{|X|}$ functions from X to Y, $|Y|^{|X|} = 103$. This can only happen if $|Y| = 103$ and $|X| = 1$.

29. If $k = 0$, there is one way of selecting a sample of size k from a set with n elements. If $n = 0$ and $k \neq 0$, there are no selections.

31. (a) $3! = 6$
 (b) Yes; "Permutations," perhaps?

Section 2.3

1. $\{1, 2\}, \{1, 3\}, \{1, 4\}, \{1, 5\}, \{2, 3\}, \{2, 4\}, \{2, 5\}, \{3, 4\}, \{3, 5\}, \{4, 5\}$. The subset rule gives $5(4)/2! = 10$.

3. $\binom{55}{11} = 119653565850$

5. $\binom{n}{n} = \frac{n!}{n!0!} = 1$ $\binom{n}{1} = \frac{n!}{1!(n-1)!} = n$

7. $\binom{52}{5} = 2598960$ $\binom{51}{4} = 249900$

9. (a) $\binom{50}{3} = 19600$ (b) $\binom{48}{3} = 17296$

11. $\binom{6}{1}\binom{7}{2}\binom{8}{2} = 3528$ 13. $\binom{10}{4} = 210$

15. (a) $\binom{10}{6} = 210$ (b) $\binom{10}{6} - \binom{8}{4} = 140$

17. (a) $4 \cdot \binom{13}{5} = 5148$ (including royal flushes)

 (b) $(13)\binom{4}{3}(12)\binom{4}{2} = 3744$ (c) $(9)\binom{4}{1}^5 = 9216$

19. $\binom{n}{2} = \frac{n(n-1)}{2}$

21. (a) $\binom{15}{5} = 3003$ (d) $\binom{15}{2} = 105$

(b) $\binom{15}{4} = 1365$ (e) $\binom{3}{1}\binom{15}{4} = 4095$

(c) $\binom{7}{2}\binom{10}{2} = 945$

23. $\binom{50}{25} \cdot 2^{25} = 4241636097794311716864$

25. (a) $\binom{20}{5} = 15504$ (c) $\binom{18}{5} + \binom{18}{3} = 9384$

(b) $\binom{20}{5} - \binom{18}{3} = 14688$

27. $\binom{n_1}{k_1}\binom{n_2}{k_2}\binom{n_3}{k_3}$ 29. $\binom{70}{10} = 396704524216$

31.
$$\binom{n-1}{k} + \binom{n-1}{k-1} = \frac{(n-1)!}{k!(n-k-1)!} + \frac{(n-1)!}{(k-1)!(n-k)!}$$
$$= \frac{(n-1)!(n-k) + (n-1)!(k)}{k!(n-k)!}$$
$$= \frac{(n-1)!(n-k+k)}{k!(n-k)!}$$
$$= \frac{n!}{k!(n-k)!}$$
$$= \binom{n}{k}$$

33. By throwing k balls into n boxes so that they land in k different boxes, you are choosing k of the n boxes. This can be done in $\binom{n}{k}$ ways.

Section 2.4

1. (a) $5^4 = 625$; $7^8 = 5764801$ (b) $5 \cdot 4 \cdot 3 \cdot 2 = 120$; 0

(c) $\binom{5}{4} = 5$; 0 (d) $\binom{8}{4} = 70$; $\binom{14}{8} = 3003$

3. (a) $\binom{11}{6} = 462$ (b) $\binom{8}{3} = 56$

(c) $\binom{6}{3}\binom{5}{3} = 200$

5.

x_1	x_2	x_3	Unordered Sample
5	0	0	$x_1x_1x_1x_1x_1$
4	1	0	$x_1x_1x_1x_1x_2$
4	0	1	$x_1x_1x_1x_1x_3$
3	2	0	$x_1x_1x_1x_2x_2$
3	1	1	$x_1x_1x_1x_2x_3$
3	0	2	$x_1x_1x_1x_3x_3$
2	3	0	$x_1x_1x_2x_2x_2$
2	2	1	$x_1x_1x_2x_2x_3$
2	1	2	$x_1x_1x_2x_3x_3$
2	0	3	$x_1x_1x_3x_3x_3$
1	4	0	$x_1x_2x_2x_2x_2$
1	3	1	$x_1x_2x_2x_2x_3$
1	2	2	$x_1x_2x_2x_3x_3$
1	1	3	$x_1x_2x_3x_3x_3$
1	0	4	$x_1x_3x_3x_3x_3$
0	5	0	$x_2x_2x_2x_2x_2$
0	4	1	$x_2x_2x_2x_2x_3$
0	3	2	$x_2x_2x_2x_3x_3$
0	2	3	$x_2x_2x_3x_3x_3$
0	1	4	$x_2x_3x_3x_3x_3$
0	0	5	$x_3x_3x_3x_3x_3$

7. There are as many possibilities as there are nonnegative integer solutions to $x_1 + x_2 + x_3 + x_4 = 1000$. This is $\binom{1003}{1000} = 167668501$.

9. (a) $\binom{14}{10} = 1001$ (b) $\binom{9}{5} = 126$ (c) $\binom{7}{3} = 35$

11. $\binom{13}{10} = 286$

13. (a) $\binom{10}{6} = 210$ (b) $\binom{5}{2}\binom{5}{4} = 50$ (c) $\binom{5}{1} = 5$

15. $\binom{17}{14} = 680$ 17. $\binom{23}{19} = 8855$ 19. $\binom{n+k-1}{k}$

21. $\binom{22}{15} = 170544$

(b) $8^{15} = 35184372088832$

23. (a) $\binom{4}{1}\binom{5}{2}\binom{6}{3}\binom{8}{5} = 44800$

(b) $\binom{3}{1}\binom{3}{1}\binom{5}{3} + \binom{3}{1}\binom{3}{1}\binom{4}{2}\binom{7}{5}$

$+ \binom{3}{1}\binom{4}{2}\binom{7}{5} + \binom{3}{1}\binom{4}{2}\binom{3}{1} = 1656$

25. (a) $\binom{6}{4}\binom{5}{2} = 150$

(b) $\binom{6}{1} + \binom{6}{2}\binom{5}{4} + \binom{6}{3}\binom{5}{3}$

$+ \binom{6}{4}\binom{5}{2} = 431$

27. (a) $\binom{10}{6} = 210$ (b) $\binom{10}{2} = 45$ (c) $\binom{11}{2} = 55$

29. Each solution of $x_1 + x_2 + x_3 < 14$ in nonnegative integers yields a solution of $x_1 + x_2 + x_3 + x_4 = 14$ in nonnegative integers with $x_4 > 0$ by setting $x_4 = 14 - x_1 - x_2 - x_3$. The correspondence is one-to-one.

31. $\binom{17}{14} = 680$

Section 2.5

1. (a) $\binom{9}{3, 2, 1, 1, 1, 1} = 30240$

(b) $\binom{6}{3, 2, 1} = 60$

3.

$\varnothing$	0 0 0 0
$\{A\}$	1 0 0 0
$\{B\}$	0 1 0 0
$\{C\}$	0 0 1 0
$\{D\}$	0 0 0 1
$\{A, B\}$	1 1 0 0
$\{A, C\}$	1 0 1 0
$\{A, D\}$	1 0 0 1
$\{B, C\}$	0 1 1 0
$\{B, D\}$	0 1 0 1
$\{C, D\}$	0 0 1 1
$\{A, B, C\}$	1 1 1 0
$\{A, B, D\}$	1 1 0 1
$\{A, C, D\}$	1 0 1 1
$\{B, C, D\}$	0 1 1 1
$\{A, B, C, D\}$	1 1 1 1

5. $\binom{20}{3, 2, 6, 7, 1, 1} = 55870214400$

7. (a) $2^{50} = 1125899906842624$

(b) $\binom{50}{20} = 47129212243960$

9. $\binom{20}{3, 3, 4, 4, 2, 4} = 2444321880000$

11. $\binom{20}{3, 3, 4, 4, 2, 4} \cdot \left(\frac{1}{3!2!1!}\right) = 203693490000$

13. (a) $(3!)(2!) = 12$

(b) $(3!)(2!) = 12$

15. (a) $\binom{12}{3, 2, 2, 2, 1, 1, 1} = 9979200$

(b) $2\binom{11}{3, 2, 2, 2, 1, 1} = 1663200$

(c) $\binom{6}{3, 2, 1}\binom{7}{2, 2, 1, 1, 1} = 75600$

17. (a) Each word is counted twice. Choosing first Z's and then D's is the same as choosing first D's and then Z's.

(b) $\binom{26}{2}\binom{4}{2} = 1950$

19. $\binom{30}{7, 4, 4, 8, 2, 5} = 9442245542069340000$

21. $\binom{21}{2, 2, 2, 2, 2, 2, 2, 2, 2, 2, 1} = 49893498214560000$

23. (a) $\binom{26}{8, 7, 6, 5} = 22969641895200$

(b) Same as (a)

25. $\binom{15}{5}\binom{10}{5}\frac{1}{2!} = 378378$

27. Each factor of $(x_1 + x_2 + x_3)$ contributes a single x_i to each term in the expansion. The coefficient of $x_1^2x_2x_3 = x_1x_1x_2x_3$ equals the number of permutations of $x_1x_1x_2x_3$, i.e.,

$$\binom{4}{2, 1, 1} = 12.$$

29. A term like $12x_1x_2^2x_3$ corresponds to the unordered sample $x_1x_2x_2x_3$. The number of unordered samples of size 4 from the alphabet x_1, x_2, x_3 is

$$\binom{6}{4} = 15.$$

31. $(x_1 + \cdots + x_n)^k = \sum \binom{k}{k_1, \ldots, k_n} x_1^{k_1} \cdots x_n^{k_n}$,

where the sum is taken over all nonnegative integers $k_1, \ldots, k_n$ whose sum is k. To prove it, note that in the expansion of the left-hand side, a typical term has the form $x_1^{k_1} \cdots x_n^{k_n}$ and the number of such terms is given by the rule of generalized permutations.

33. Each partition corresponds to a division of n people into m teams where team i contains k_i members. If $k_1, \ldots, k_m$ are distinct, the number of partitions is

$$\binom{n}{k_1, \ldots, k_m}.$$

If, however, there are t_i teams of size s_i, $i = 1, 2, \ldots, q$, the number of partitions is

$$\binom{n}{k_1, \ldots, k_m} \frac{1}{t_1!\, t_2! \cdots t_q!}.$$

Section 2.6

1. $70 + 40 - 20 = 90$
3. $100 - 25 - 35 - 45 + 15 + 5 = 15$
5. $\binom{m}{1}\binom{n-m}{k-1} + \binom{m}{2}\binom{n-m}{k-2} + \cdots + \binom{m}{m}\binom{n-m}{k-m}$
7. $5^6 - 5(4^6) + \binom{5}{2}(3^6) - \binom{5}{3}(2^6) + \binom{5}{4}(1^6)$

   $= 1800$
9. (a) $\binom{4}{3}\binom{48}{2} + \binom{4}{2}\binom{48}{3} - \binom{4}{3}\binom{4}{2}$

   $= 108264$

   (b) $\binom{4}{3}\binom{48}{2} + \binom{4}{3}\binom{48}{2} - 0 = 9024$
11. $\binom{13}{5} + \binom{51}{4} + \binom{51}{4} - \binom{12}{4} - 0$

    $- \binom{50}{3} + 0 = 480992$
13. $26! - (24! + 24! - 22!)$

    $= 402051688323866934312960000.$
15. $\binom{6}{3} + \binom{7}{3} + \binom{8}{3} = 111$
17. $|A - B| = |A| - |A \cap B|$. If $B \subseteq A$, $|A - B| = |A| - |B|$. If $A \subseteq B$, $|A - B| = 0$. If A and B are disjoint, $|A - B| = |A|$.
19.

Term	Region 1	2	3	4	5	6	7	Unmarked
$\lvert S\rvert$	1	1	1	1	1	1	1	1
$-\lvert A\rvert$	−1	0	−1	0	−1	0	−1	0
$-\lvert B\rvert$	0	−1	−1	0	0	−1	−1	0
$-\lvert C\rvert$	0	0	0	−1	−1	−1	−1	0
$\lvert A \cap B\rvert$	0	0	1	0	0	0	1	0
$\lvert A \cap C\rvert$	0	0	0	0	1	0	1	0
$\lvert B \cap C\rvert$	0	0	0	0	0	1	1	0
$-\lvert A \cap B \cap C\rvert$	0	0	0	0	0	0	−1	0
Totals	0	0	0	0	0	0	0	1

21. $|A \cup B \cup C| = |(A \cup B) \cup (A \cup C)|$
    $= |A \cup B| + |A \cup C| - |(A \cup B) \cap (A \cup C)|$
    $= |A \cup B| + |A \cup C| - |A \cup (B \cap C)|.$

    Since

    $$A \cap B \cap C \subseteq A \cup (B \cap C),$$
    $$|A \cap B \cap C| \le |A \cup (B \cap C)|,$$

    so

    $$-|A \cup (B \cap C)| \le -|A \cap B \cap C|.$$

    Thus

    $$|A \cup B \cup C| \le |A \cup B| + |A \cup C| - |A \cap B \cap C| \le |A \cup B| + |A \cup C|$$

    and both statements are always true.
23. The number who lost all four is at least $100 - 20 - 15 - 30 - 25 = 10$.
25. $|X| = 4$, $|Y| = 3$.
27. nofix(n) (see Problem 26)
29. By Problem 28, onto$(n, m) = m! \cdot S(n, m)$.
31. The n distinguishable balls correspond to the integers $1, \ldots, n$ and the m indistinguishable boxes to m pairwise disjoint subsets in a partition of $\{1, \ldots, n\}$.

Chapter 3

Section 3.1

1. The essentially different cycles are 1231, 1241, 1341, 2342, 12341, 12431, and 13241.
3. v_1•——•v_2, v_3•——•v_4 •v_5
5. Take $n + 1$ points and join them in all possible ways.
7. In a tree, there is a unique path joining any two vertices (see the daisy chain theorem in Section 3.2). If a new edge e from v to w is added to a tree T, any cycle in the resulting graph must contain e (since T has no cycles). A cycle must thus have the form eP where P is a path (in T) from w to v. There is a unique such P, so there is just one cycle.
9.

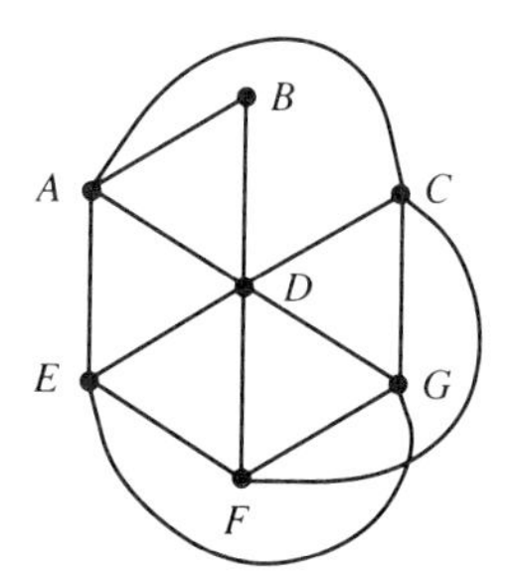

An Euler cycle is AEGFCABDCGDFEDA.

11.

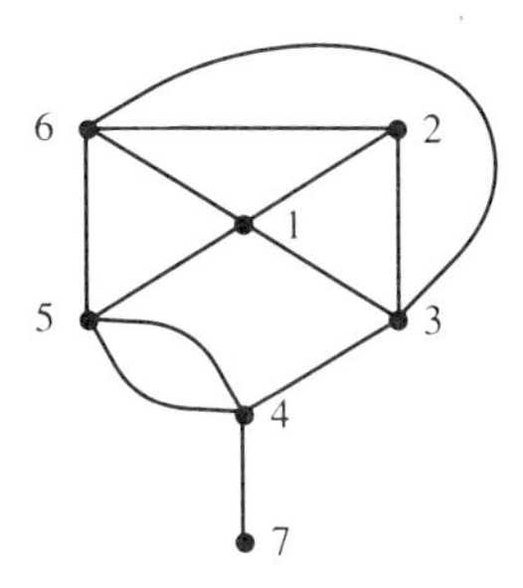

The graph has no Euler cycle, but 7454315612632 is an Euler path.

13. No.

15. Longest cycle: ACADBA. Longest path: CABDCAB.

17. Modify condition (a) to say that G contains at most one connected component with more than one vertex.

19. If every vertex were of even degree, then since trees are connected, the tree would have an Euler cycle. Since trees have no cycles at all, this cannot happen.

21. Figure 3.2:

	A	B	C	D
A	0	2	2	1
B	2	0	0	1
C	2	0	0	1
D	1	1	1	0

Figure 3.3(a):

	1	2	3	4	5	6	7	8
1	0	1	0	1	1	0	0	0
2	1	0	1	0	0	1	0	0
3	0	1	0	1	0	0	1	0
4	1	0	1	0	0	0	0	1
5	1	0	0	0	0	1	0	1
6	0	1	0	0	1	0	1	0
7	0	0	1	0	0	1	0	1
8	0	0	0	1	1	0	1	0

Figure 3.3(b):

	a	b	c	d	e	f
a	0	1	1	1	1	0
b	1	0	0	0	0	1
c	1	0	0	0	0	1
d	1	0	0	0	0	1
e	1	0	0	0	0	1
f	0	1	1	1	1	0

Figure 3.3(c):

	a	b	c	d	e	f
a	0	1	1	0	0	0
b	1	0	1	0	0	0
c	1	1	0	0	0	0
d	0	0	0	0	1	1
e	0	0	0	1	0	1
f	0	0	0	1	1	0

Figure 3.3(d):

	0	1	2	3	4	5	6	7	8
0	0	1	1	0	0	0	0	0	0
1	1	0	0	0	0	0	0	0	0
2	1	0	0	1	0	0	1	0	0
3	0	0	1	0	1	1	0	0	0
4	0	0	0	1	0	0	0	0	0
5	0	0	0	1	0	0	0	0	0
6	0	0	1	0	0	0	0	1	1
7	0	0	0	0	0	0	1	0	0
8	0	0	0	0	0	0	1	0	0

Figure 3.3(e):

	x	y	z
x	2	1	0
y	1	0	0
z	0	0	0

23. An edge from v to w contributes 1 to $\deg(v)$ and 1 to $\deg(w)$, and so is counted twice in the computation of S. Hence S is twice the number of edges (and therefore even). If T denotes the sum of the degrees of the vertices of even degree, then S and T are both even, and $S - T$ is an even number which is the sum of the degrees of the vertices of odd degree. But for the sum of a list of odd numbers to be even, the list must contain an even number of items, so N is even.

25. $(2n)!$

27. An Euler cycle is 10042146773525631.

29. $2^{\binom{n}{2}}$

31. $2^{n+\binom{n}{2}}$

Section 3.2

1.

3.

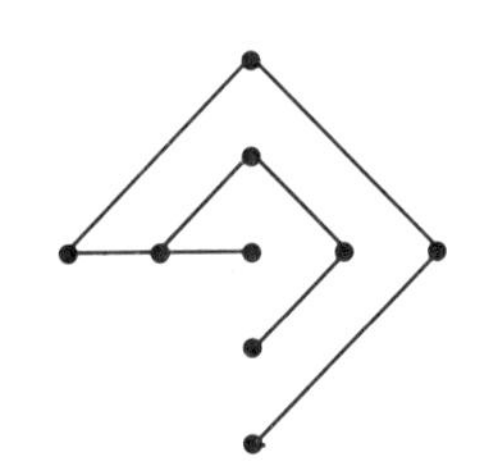

5.

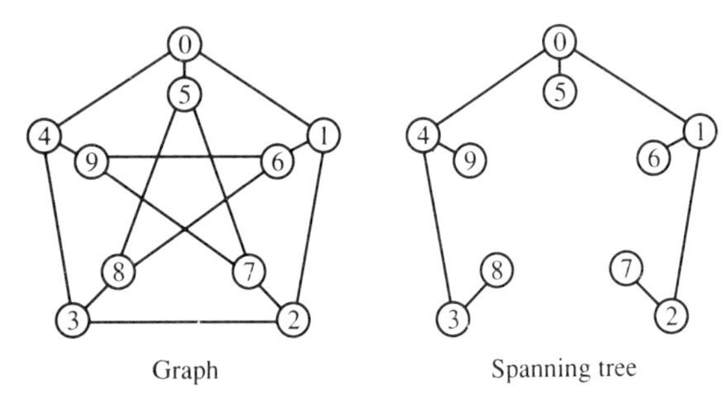

Graph Spanning tree

7. There are 125 such spanning trees, 60 of which look like

60 of which look like

and 5 which look like

9. The graph has 23 essentially different cycles—3 of length 3, 5 of length 4, 6 of length 5, 5 of length 6, and 4 of length 7. If all rotations and reversals of a cycle are counted as distinct, there are 234 cycles.

11. You get the spanning tree with edges 12, 23, 24, 25, and 46.

13. Yes

15. Yes

17. (a) This follows from either b or c of the DCT.
 (b) By definition the spanning tree must include all n vertices. By either b or c, it will have $n - 1$ edges.

19. False

21. True. By part b of the DCT, G is a tree and so is connected.

23. True. If it contained no cycles it would be a tree and so could not have the same number of vertices as edges.

25. If T is a tree, it is connected, so there is a path between any two vertices u and v. If there is a second path from u to v, the two paths must diverge at some point, and meet at some later point, producing a cycle in T, which is a contradiction.

27. A spanning tree for G has $n - 1$ edges. Since it contains no cycles, it is not all of G. Thus G must contain at least one more edge.

29. This works since removing an edge from a cycle can never disconnect a graph. At the end we will be left with a graph containing all the vertices of the original graph, still connected, and having no cycles—that is, a spanning tree.

31. This works. When the algorithm terminates, all vertices will be connected to each other (otherwise another edge can be added without creating a cycle), and there will be no cycles. So the terminal graph is a spanning tree. (This algorithm is called *Kruskal's algorithm* for MSTs.)

33. 191

Section 3.3

1. The MST has weight 129.
3. The MST has weight 8.
5. The MST has weight 10.
7.

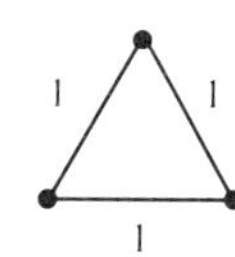

9. The MST consists of the edges $(0, 1), (1, 2), \ldots, (n-2, n-1)$ for a total weight of $n - 1$.

11. If MAX is the largest weight of any edge in the graph, we can assign the edges which are supposed to have infinite weight the value MAX + 1. They will never be selected by the algorithm.

13. The tree has weight 255.

15. At least one endpoint must be in G_k to maintain connectedness. If both were inside, we would form a cycle and not have a tree.

Section 3.4

1.

Vertex	LCHILD	RCHILD
→A	B	C
B	D	E
C	—	F
D	—	G
E	—	—
F	H	—
G	—	—
H	I	J
I	—	—
J	—	—

3.

Vertex	LCHILD	RCHILD
→A	B	C
B	D	E
C	F	G
D	H	I
E	J	K
F	L	M
G	N	O
H	—	—
I	—	—
J	—	—
K	—	—
L	—	—
M	—	—
N	—	—
O	—	—

5.

Vertex	LCHILD	RCHILD
→A	B	H
B	C	D
C	—	—
D	E	F
E	—	—
F	—	G
G	—	—
H	I	J
I	—	—
J	K	L
K	—	—
L	—	—

7.

Vertex	LCHILD	RCHILD
→A	B	C
B	D	E
C	—	—
D	F	—
E	—	G
F	—	H
G	—	—
H	—	—

9.

11.

13.

15.

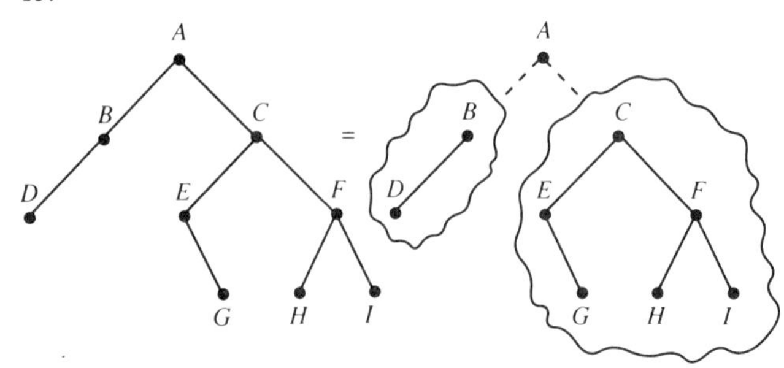

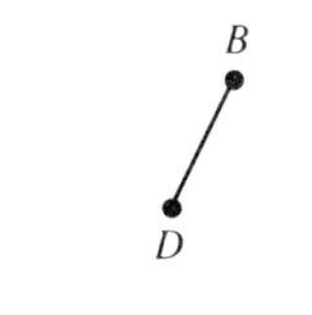

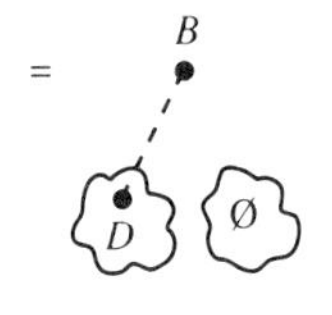

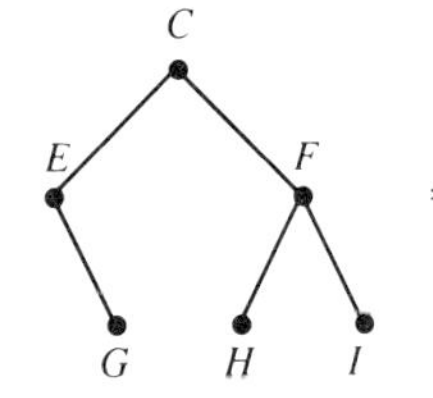

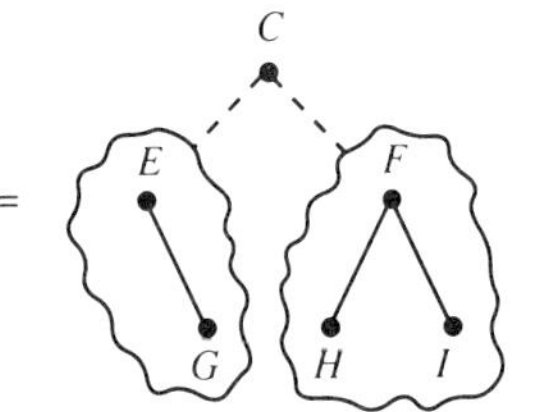

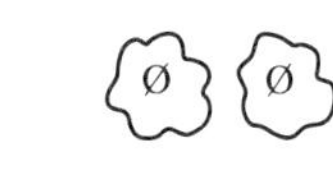

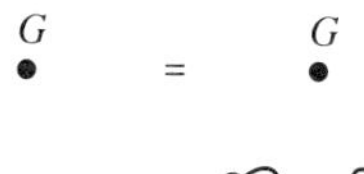

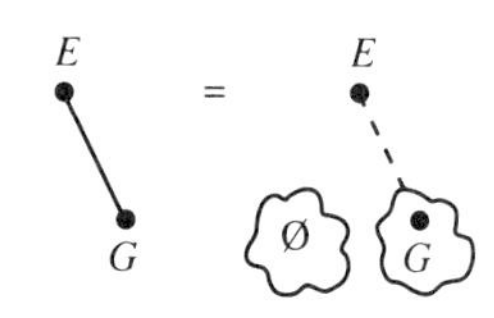

H = H

Ø Ø

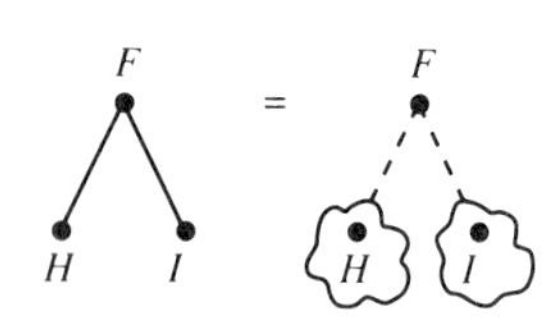

I = I

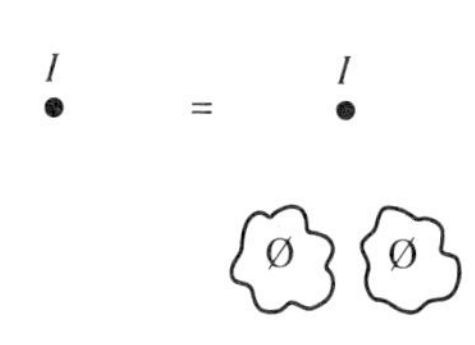

17. *x, y, s, u, v, w, z, t, j, k*
19. *a, b, d, f, c, e, g*
21. *A, B, C, D, E, F, G*
23. *A, B, C, D*
25. *A, B, C, D, E*
27. *E, A, C, D, B*
29. *v, w, u, s, y, k, j, t, z, x*
31. *f, d, b, g, e, c, a*
33. *B, D, F, G, E, C, A*
35. *D, C, B, A*
37. *E, D, C, B, A*
39. *C, A, B, D, E*
41. *y, v, u, w, s, x, z, k, j, t*
43. *f, d, b, a, c, e, g*
45. *B, A, D, C, F, E, G*
47. *A, B, C, D*
49. *A, B, C, D, E*
51. *A, C, E, B, D*
53.

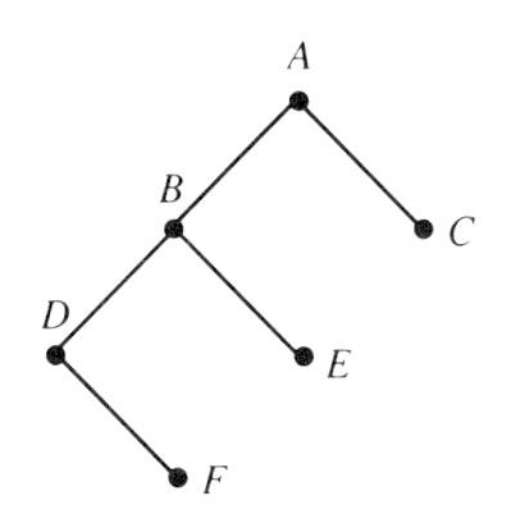

55.

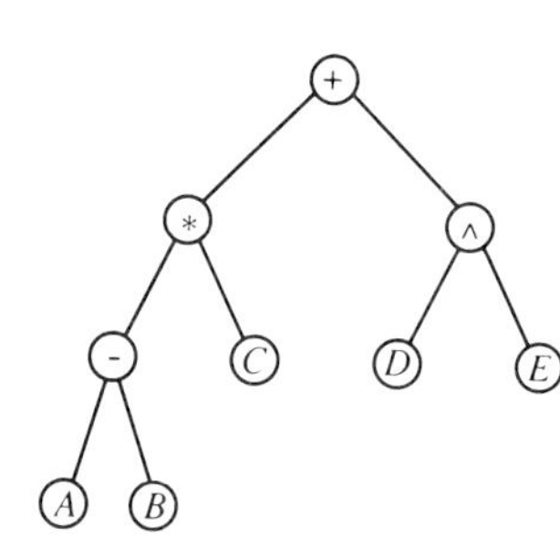

$AB - C * DE \wedge +$

57.

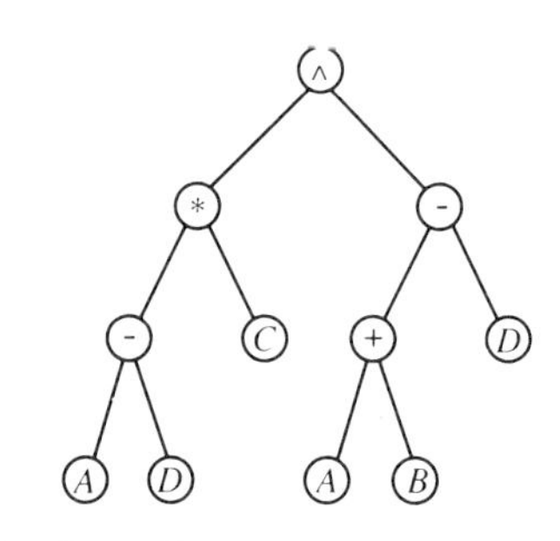

$AD - C * AB + D - \wedge$

59.

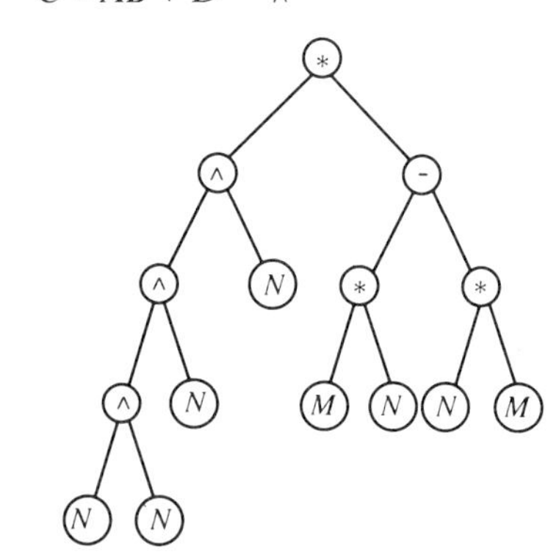

$NN \wedge N \wedge N \wedge MN * NM * - *$

61. 36
63. 38

Section 3.5

1. (a)

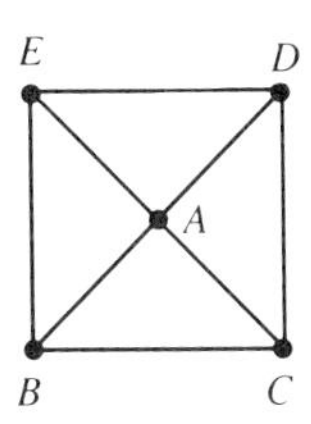

$V = 5,\ E = 8,\ F = 5$

(b)

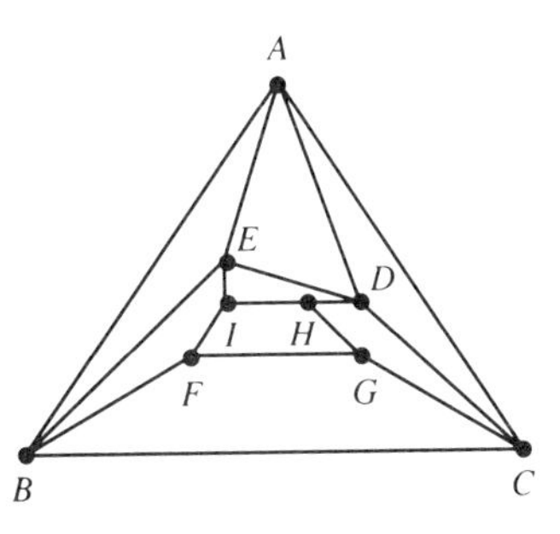

$V = 9,\ E = 16,\ F = 9$

(c)

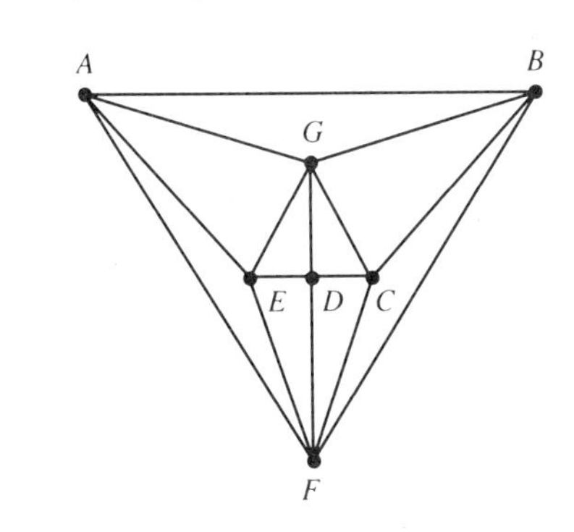

$V = 7,\ E = 15,\ F = 10$

(d)

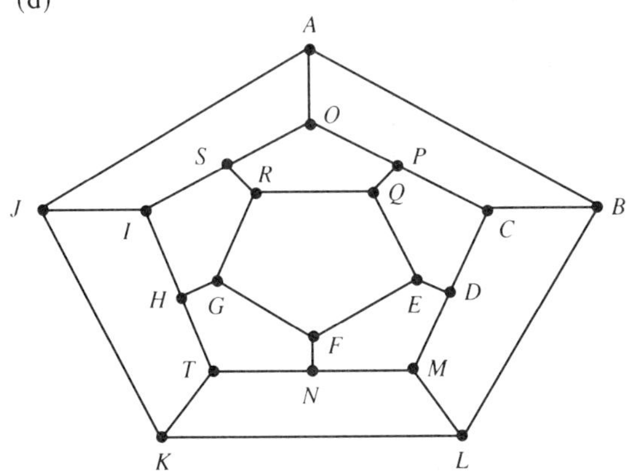

$V = 20,\ E = 30,\ F = 12$

3. $E = 14,\ F = 7$

5. (a) and (b) are two formulations of the same result. Suppose we have a planar graph G each of whose vertices has degree 6 or more. Then the sum of the degrees of the vertices is at least $6V$, so $E \geq 6V/2 = 3V$. On the other hand, since the graph is planar with no loops or multiple edges, the edge-vertex inequality holds. Moreover, the quantity $g/(g-2)$ is no bigger than 3, so $E \leq 3(v-2)$. But then we have $3V \leq E \leq 3(V-2)$ which implies $3V \leq 3(V-2)$, a contradiction.

7.

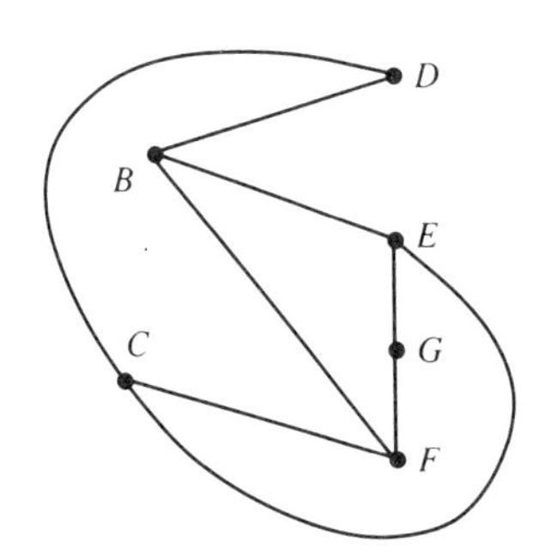

9. girth (3.39a) = 3, girth (3.39b) = 3, girth (3.39c) = 5, girth (3.39d) = 5, girth (3.39e) = 4, girth (3.39f) = 6, girth (3.40a) = 3

11. No, because the girth is 2.

13.

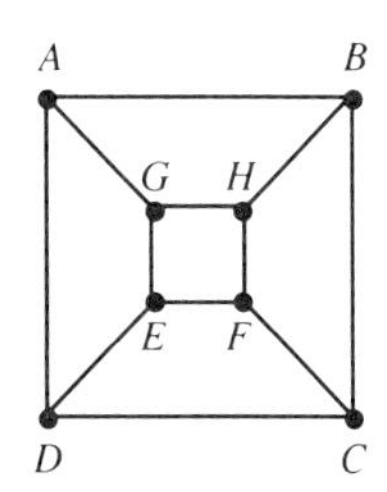

15. The theorem is clearly true for trees. For nontrees, the girth is at least 3 (since there are no loops or multiple edges), so $3F \leq gF \leq 2E$ by the result established in the proof of the edge-vertex inequality.

17. $V = n,\ E = \binom{n}{2},\ g = 3$

19.

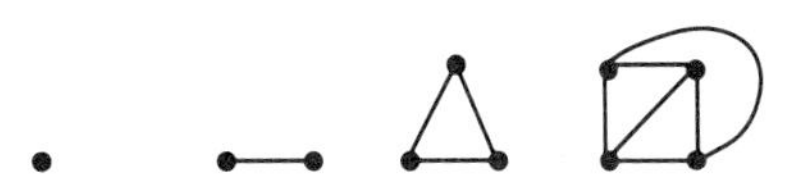

For larger values of n, K_n contains a copy of K_5 and is therefore nonplanar by Kuratowski's theorem.

21.

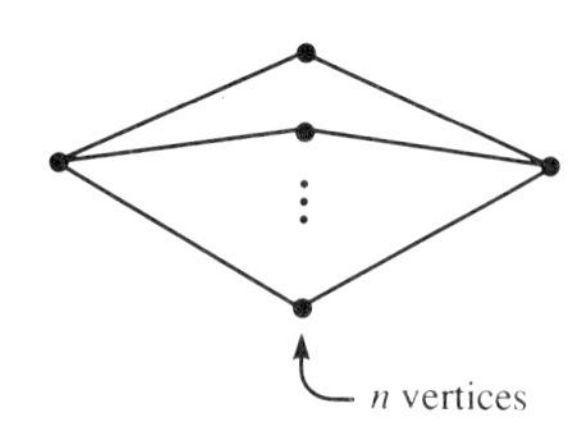

23. (a) $\ell n + \ell m + nm$
 (b) 3
 (c) $K_{\ell,m,n}$ is planar for $\ell = m = 1$ and arbitrary n, and for $\ell = 1$, $m = n = 2$ (where subscripts are rearranged so that $\ell \le m \le n$).

25. Technically no, since all the vertices of the Perterson graph have degree 3 and the vertices of K_5 are of degree 4. However, if edges $\overline{af}$, $\overline{bg}$, $\overline{ch}$, $\overline{di}$, $\overline{ej}$, are "shrunk to a point," the resulting graph is K_5.

27. No

29. $K_{m,n}$ has $\binom{m}{2}\binom{n}{2}$ 4-cycles.

31. $V - E + F = 9 - 18 + 9 = 0$. Euler's theorem does not apply to solids with holes.

Chapter 4

Section 4.1

1.

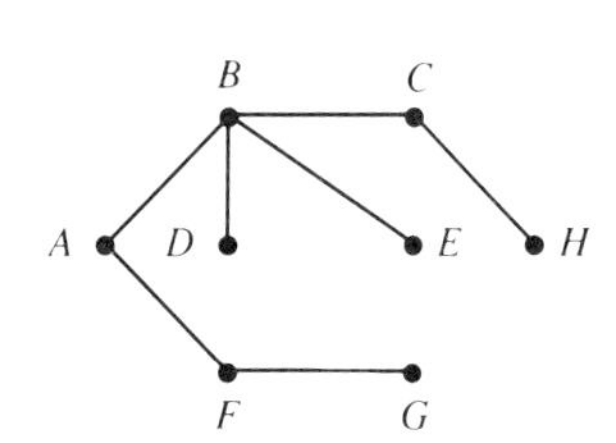

3.

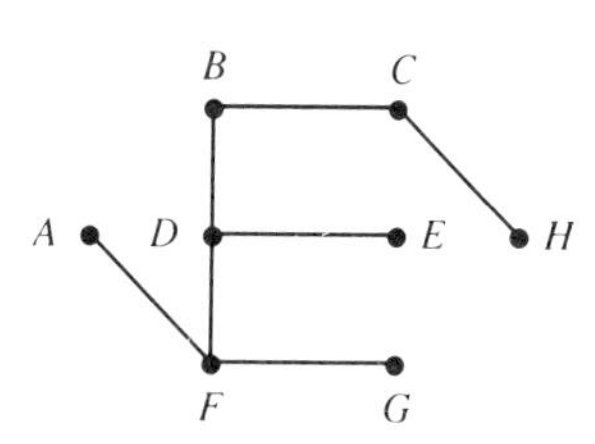

5.

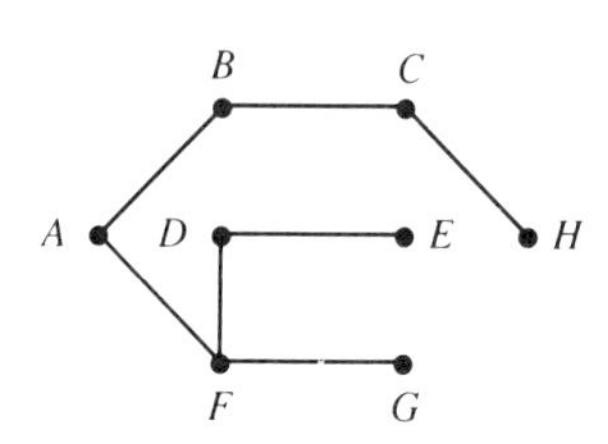

7.

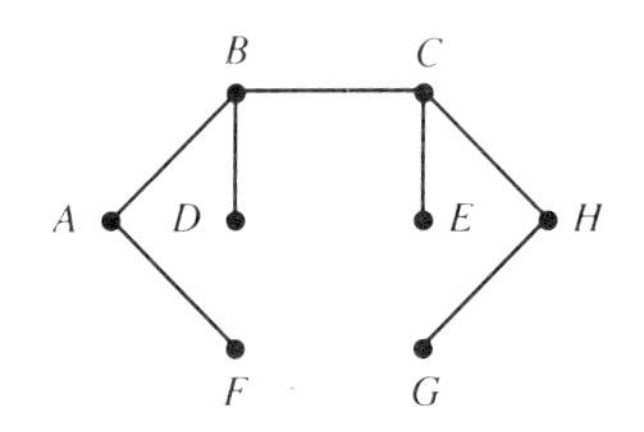

9. The shortest path has length 14.

11. (a) BCDE length 5
 (b)

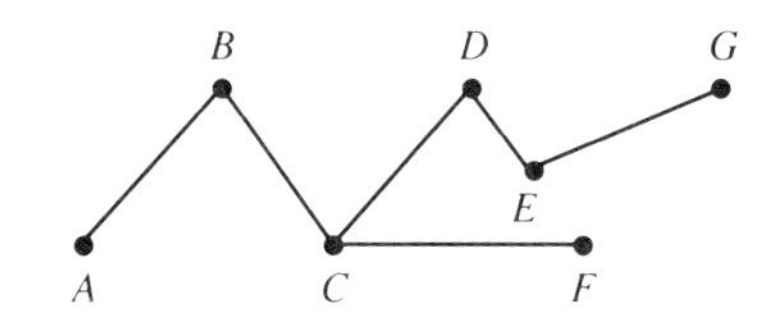

13.

Vertex	A	B	C	D	E	F	G
dist	5	2	0	1	3	2	4
next	B	C	*	C	D	C	E

15.

Vertex	A	B	C	D	E	F	G
dist	8	5	3	2	0	2	1
next	B	C	D	E	*	E	E

17.

Vertex	A	B	C	D	E	F	G
dist	8	5	3	3	1	1	0
next	B	C	F	E	G	G	*

19. 14 length 2.0

21. ABD length 4

23. PEFD length 15

25. Step 6 will fail (assuming, of course, that it is implemented in such a way as to never allow a point at infinite distance to be chosen as v^*).

27. Once each

29. (a) and (b) certainly hold when step 5 is reached for the first time. The question is whether executing the body of the `while` loop can affect their validity. Now `dist`$[v]$ and `next`$[v]$ will only be changed if `dist`$[v] >$ `dist`$[v^*] + \ell(v^*, v)$—i.e., if the shortest path from v to v_0 leads through v^*. The change `next`$[v] \leftarrow v^*$ ensures that (b) will hold the next time through step 5. Also, the next time through step 5, the question "`dist`[`next`$[v]$] < `dist`$[v]$?" becomes "`dist`$[v^*]$ < `dist`$[v]$?". This is clearly true since `dist`$[v]$ is `dist`$[v^*]$ plus the positive quantity $\ell(v^*, v)$.

31. If steps 7 and 8 are interchanged, the algorithm still works, but is less efficient.

33. (a) M certainly holds after the `for` loop has executed once since then `min` $= x_1$ and $\{x_1, \ldots, x_{i-1}\} = \{x_1\}$. On later passes through the loop, if $x_i \ge$ `min`, then `min` does not change and $\text{Min}(x_1, \ldots, x_i) =$

$\text{Min}(x_1, \ldots, x_{i-1}) = \texttt{min}$. If $x_i < \texttt{min}$, then $\text{Min}(x_1, \ldots, x_i) = x_i$, but assertion M continues to hold since min is changed to x_i.

(b)
```
{
   min ← ∞;
   index ← 0;
   for(i=1 to n)
      if(x_i < min){
        min ← x_i;
        index ← i;
      }
}
```

35.

Vertex	1	2	3	4	5	6	7
Status	?	!	?	?	?	?	!
next	2	7	7	7	7	2	*
Cost	3	9	10	∞	∞	8	0

Vertex	1	2	3	4	5	6	7
Status	!	!	?	?	?	?	!
next	2	7	1	1	7	1	*
Cost	3	9	4	5	∞	7	0

Vertex	1	2	3	4	5	6	7
Status	!	!	!	?	?	?	!
next	2	7	1	1	3	1	*
Cost	3	9	4	5	8	7	0

Vertex	1	2	3	4	5	6	7
Status	!	!	!	!	?	?	!
next	2	7	1	1	3	1	*
Cost	3	9	4	5	8	7	0

Vertex	1	2	3	4	5	6	7
Status	!	!	!	!	?	!	!
next	2	7	1	1	3	1	*
Cost	3	9	4	5	8	7	0

Vertex	1	2	3	4	5	6	7
Status	!	!	!	!	!	!	!
next	2	7	1	1	3	1	*
Cost	3	9	4	5	8	7	0

Section 4.2

1.
$$A_0 = \begin{pmatrix} 0 & 7 & \infty & 2 \\ 2 & 0 & 4 & \infty \\ \infty & \infty & 0 & 1 \\ \infty & 3 & \infty & 0 \end{pmatrix}$$

$$A_4 = \begin{pmatrix} 0 & 5 & 9 & 2 \\ 2 & 0 & 4 & 4 \\ 6 & 4 & 0 & 1 \\ 5 & 3 & 7 & 0 \end{pmatrix}$$

3.
$$A_0 = \begin{pmatrix} 0 & 12 & 10 & \infty \\ \infty & 0 & \infty & \infty \\ \infty & 1 & 0 & 3 \\ \infty & \infty & \infty & 0 \end{pmatrix} \qquad A_4 = \begin{pmatrix} 0 & 11 & 10 & 13 \\ \infty & 0 & \infty & \infty \\ \infty & 1 & 0 & 3 \\ \infty & \infty & \infty & 0 \end{pmatrix}$$

5.

	A	P	C	D	E	F	G	H
A	0	∞	3	∞	∞	∞	∞	∞
P	4	0	∞	31	∞	∞	∞	∞
C	∞	1	0	∞	8	∞	∞	∞
D	∞	∞	2	0	∞	∞	∞	∞
E	∞	∞	∞	∞	0	8	3	∞
F	∞	∞	∞	∞	∞	0	∞	6
G	∞	∞	∞	∞	∞	4	0	∞
H	∞	∞	∞	2	∞	∞	1	0

$$A_8 = \begin{pmatrix} 0 & 4 & 3 & 26 & 11 & 18 & 14 & 24 \\ 4 & 0 & 7 & 30 & 15 & 22 & 18 & 28 \\ 5 & 1 & 0 & 23 & 8 & 15 & 11 & 21 \\ 7 & 3 & 2 & 0 & 10 & 17 & 13 & 23 \\ 22 & 18 & 17 & 15 & 0 & 7 & 3 & 13 \\ 15 & 11 & 10 & 8 & 18 & 0 & 7 & 6 \\ 19 & 15 & 14 & 12 & 22 & 4 & 0 & 10 \\ 9 & 5 & 4 & 2 & 12 & 5 & 1 & 0 \end{pmatrix}$$

7.
$$\text{Adj} = \begin{pmatrix} 0 & 1 & 0 & 1 \\ 1 & 0 & 1 & 0 \\ 0 & 0 & 0 & 1 \\ 0 & 1 & 0 & 0 \end{pmatrix} \qquad \text{Reach} = \begin{pmatrix} 1 & 1 & 1 & 1 \\ 1 & 1 & 1 & 1 \\ 1 & 1 & 1 & 1 \\ 1 & 1 & 1 & 1 \end{pmatrix}$$

9.
$$\text{Adj} = \begin{pmatrix} 0 & 1 & 1 & 0 \\ 0 & 0 & 0 & 0 \\ 0 & 1 & 0 & 1 \\ 0 & 0 & 0 & 0 \end{pmatrix} \qquad \text{Reach} = \begin{pmatrix} 0 & 1 & 1 & 1 \\ 0 & 0 & 0 & 0 \\ 0 & 1 & 0 & 1 \\ 0 & 0 & 0 & 0 \end{pmatrix}$$

11.
$$\text{Adj} = \begin{pmatrix} 0 & 0 & 1 & 0 & 0 & 0 & 0 & 0 \\ 1 & 0 & 0 & 1 & 0 & 0 & 0 & 0 \\ 0 & 1 & 0 & 0 & 1 & 0 & 0 & 0 \\ 0 & 0 & 1 & 0 & 0 & 0 & 0 & 0 \\ 0 & 0 & 0 & 0 & 0 & 1 & 1 & 0 \\ 0 & 0 & 0 & 0 & 0 & 0 & 0 & 1 \\ 0 & 0 & 0 & 0 & 0 & 1 & 0 & 0 \\ 0 & 0 & 0 & 1 & 0 & 0 & 1 & 0 \end{pmatrix}$$

$$\text{Reach} = \begin{pmatrix} 1 & 1 & 1 & 1 & 1 & 1 & 1 & 1 \\ 1 & 1 & 1 & 1 & 1 & 1 & 1 & 1 \\ 1 & 1 & 1 & 1 & 1 & 1 & 1 & 1 \\ 1 & 1 & 1 & 1 & 1 & 1 & 1 & 1 \\ 1 & 1 & 1 & 1 & 1 & 1 & 1 & 1 \\ 1 & 1 & 1 & 1 & 1 & 1 & 1 & 1 \\ 1 & 1 & 1 & 1 & 1 & 1 & 1 & 1 \\ 1 & 1 & 1 & 1 & 1 & 1 & 1 & 1 \end{pmatrix}$$

13. The shortest i-path from x to y does not pass through vertex i.

15. Both row k and column k will consist of ∞'s.

17. If $A_{i-1}[x, i] = \infty$, then $A_i[x, y] = A_{i-1}[x, y]$.

19. If $a\,R\,b$, there is an edge from a to b, so there is a path of length 1 from a to b, so $a\,R^*\,b$.

21. If $a\,R^*\,b$, there is a path from a to b and by Problem 20, $a\,T\,b$, so $R^* \subseteq T$.

23. In view of Problem 22, all that needs to be shown is that $R \subseteq T^*$. But this is clear since R is a subset of each T_i in the intersection making up T^*.

25. By Problems 21 and 24, $T^* = R^*$.

Section 4.3

1. •————• (five disjoint edges)
   •————•
   •————•
   •————•
   •————•

3. $\{(1, 2), (2, 3), (3, 1), (4, 4)\}$, $A' = \{1, 2, 3, 4, 5, 6\}$

5. $\{(1, B), (2, C), (3, D), (4, E), (5, A)\}$

7. No. ($\{B, D, E, F, H\}$ still like only $\{b, c, d, h\}$).

9. (a) Has 7 matchings and 2 maximal matchings.
   (b) Has 8 matchings and 1 maximal matching.
   (c) Has 6 matchings and 5 maximal matchings. (Don't forget to count the empty matching!)

11. The Hungarian algorithm yields a complete matching: $\{(G_1, B_2), (G_2, B_4), (G_3, B_5), (G_4, B_3), (G_5, B_1)\}$.

13. The given matching is maximal. A trouble spot is $\{A, B, C, D, E, F\}$.

15. (a) $(1, 5, 2, 4, 6)$ is an SDR.
    (b) No SDR exists.
    (c) No SDR exists. (Sets S_2, S_4, S_5, S_6, S_7, S_8 together contain only 5 elements: a, b, c, d, h.)

17. $2^{|A| \cdot |B|}$

19. Let S be a subset of the girls and let T be the set of boys liked by at least one girl in S. The total number of edges leaving S is $4|S|$, and the total number of edges leaving T is $4|T|$. But every edge leaving S is also an edge leaving T, so $4|S| \leq 4|T|$, i.e., $|S| \leq |T|$. Therefore by Hall's theorem, a complete matching exists.

21. The first situation violates the fact that the a-matching is a matching. The second violates the assumption that the b-matching is a matching.

Section 4.4

1.

n	(4.4)	(4.5)
1000	1749671469.95	1750000000
10000	1749967014699.95	1750000000000
100000	1749996700146999.95	1750000000000000

3. $K = 4$

5. First observe that $(1.5)^{24} \approx 16834$ and $24^3 = 13824$, so $(1.5)^n > n^3$ holds for $n = 24$. To see that it holds for all $n \geq 24$, we use induction:

$$(1.5)^{n+1} = 1.5(1.5)^n > 3n^3/2$$

by the inductive assumption, so it suffices to show that $3n^3/2 > (n + 1)^3$ for $n \geq 24$. But $3n^3/2 > (n + 1)^3$ is equivalent to

$$1 > \frac{6}{n} + \frac{6}{n^2} + \frac{2}{n^3}$$

which is certainly true for $n \geq 24$.

7. This is straight computation.

9. 10^6 seconds (about $11\frac{1}{2}$ days).

11. The order n task sizes are multiplied by 10. The order n^2 task sizes are multiplied by $\sqrt{10} \approx 3.16$. The order n^3 task sizes are multiplied by $\sqrt[3]{10} \approx 2.15$. The order 2^n task sizes have $(\log 10)/(\log 2) \approx 3.32$ added to them.

13. $1 + 2 + \cdots + (n - 1) = (n - 1)(n)/2 \leq \frac{1}{2}n^2$ (see Section 1.5, Example 1.23)

15. There are $(n - 1) + (n - 2) + \cdots + 1$ comparisons, so the algorithm is $O(n^2)$.

17. Assume that finding the least element of T takes a constant amount of time. This operation must be performed $n - 1$ times, so the algorithm is $O(n)$.

19. Change the $>$ in Algorithm 2 to a $<$ and it will find the smallest element. Apply this to the k vertices which are not already in the growing tree and it will run in time $O(k)$.

21. $O(n^3)$

23. (a) $\lim_{n\to\infty} \frac{n}{n^2} = \lim_{n\to\infty} \frac{1}{n} = 0$

    (b) $\lim_{n\to\infty} \frac{n^2}{n^3} = \lim_{n\to\infty} \frac{1}{n} = 0$

    (c) $\lim_{n\to\infty} \frac{2^n}{3^n} = \lim_{n\to\infty} \left(\frac{2}{3}\right)^n = 0$ since $\frac{2}{3} < 1$.

Chapter 5

Section 5.1

1. (a) No (b) Yes (c) No (d) Yes

3. $A \to B$ corresponds to $A' \cup B$:

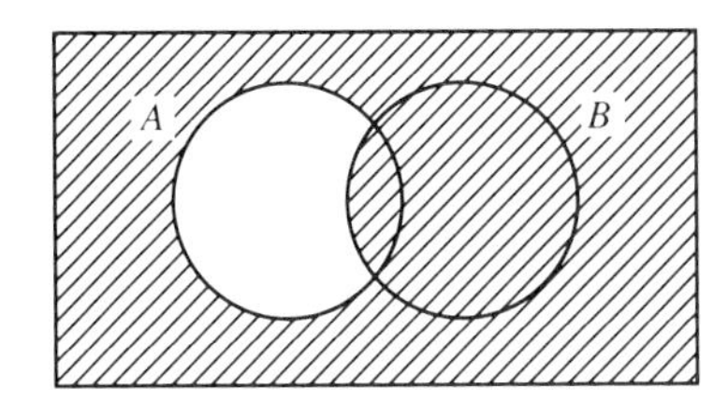

$A \leftrightarrow B$ corresponds to $(A \cap B) \cup (A' \cap B')$:

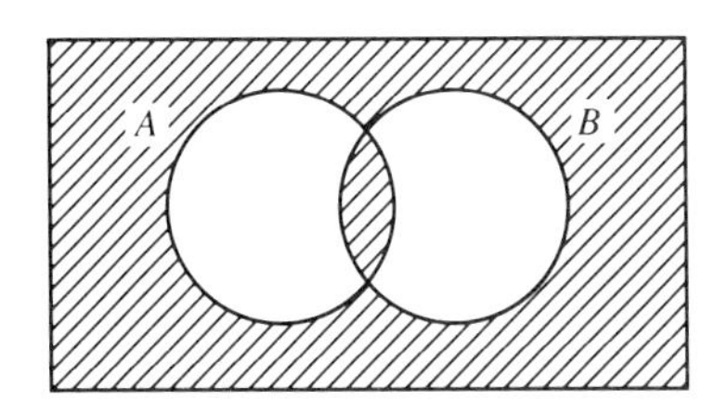

5. (a)

P	Q	$((P \leftrightarrow \sim Q) \vee Q)$
T	T	T
T	F	T
F	T	T
F	F	F

(b)

P	Q	R	$((\sim P \wedge Q) \to (\sim Q \wedge R))$
T	T	T	T
T	T	F	T
T	F	T	T
T	F	F	T
F	T	T	F
F	T	F	F
F	F	T	T
F	F	F	T

7. (a) They are equivalent.
   (b) They are equivalent.

9. (a) They are equivalent.
   (b) They are not equivalent.

11.

x_1	x_2	x_3	$((x_1 \wedge x_2) \wedge x_3)$	$(x_1 \wedge (x_2 \wedge x_3))$
T	T	T	T	T
T	T	F	F	F
T	F	T	F	F
T	F	F	F	F
F	T	T	F	F
F	T	F	F	F
F	F	T	F	F
F	F	F	F	F

13. Any WFP has the same truth table as itself. If P and Q have the same truth table then Q and P have the same truth table. If P has the same truth table as Q and Q has the same truth table as R, then P has the same truth table as R.

15.

x	y	$(x \to \sim y)$
T	T	F
T	F	T
F	T	T
F	F	T

17.

x	$\sim x$	$\sim\sim x$	$(x \leftrightarrow \sim\sim x)$
T	F	T	T
F	T	F	T

19.

x	y	$(x \vee y)$	$(x \to (x \vee y))$
T	T	T	T
T	F	T	T
F	T	T	T
F	F	F	T

21.

x	y	$(x \to y)$	$(\sim y \to \sim x)$	$((x \to y) \leftrightarrow (\sim y \leftarrow \sim x))$
T	T	T	T	T
T	F	F	F	T
F	T	T	T	T
F	F	T	T	T

23.

x	y	$(x \vee y)$	$((x \vee y) \wedge \sim y)$	$(((x \vee y) \wedge \sim y) \to x)$
T	T	T	F	T
T	F	T	T	T
F	T	T	F	T
F	F	F	F	T

25.

x	y	z	$(x \to y)$	$(y \to z)$	$(x \to z)$	$(((x \to y) \wedge (y \to z)) \to (x \to z))$
T	T	T	T	T	T	T
T	T	F	T	F	F	T
T	F	T	F	T	T	T
T	F	F	F	T	F	T
F	T	T	T	T	T	T
F	T	F	T	F	T	T
F	F	T	T	T	T	T
F	F	F	T	T	T	T

27.

x	y	$(\sim(x \vee y))$	$(\sim x \wedge \sim y)$	$(\sim(x \vee y) \leftrightarrow (\sim x \wedge \sim y))$
T	T	F	F	T
T	F	F	F	T
F	T	F	F	T
F	F	T	T	T

29. (a) It is a theorem. (c) It is a theorem.
    (b) It is not a theorem.

31. (a) It is not a theorem. (b) It is not a theorem.

33. B_n is a theorem if and only if n is odd. The proof boils down to the observations that $A \to A$ is a theorem and $T \to A$ is equivalent to A.

35. Valid 39. Not valid 43. Not valid

37. Valid 41. Valid 45. Valid

47. (a)

x	y	$(x \wedge y)$	$\sim(x \wedge y)$
T	T	T	F
T	F	F	T
F	T	F	T
F	F	F	T

(b) (i)

x	$\sim x$	$(x\vert x)$
T	F	F
T	T	T

(ii)

x	y	$(x \wedge y)$	$(x\vert y)$	$((x\vert y)\vert(x\vert y))$
T	T	T	F	T
T	F	F	T	F
F	T	F	T	F
F	F	F	T	F

(iii)

x	y	$(x \vee y)$	$(x\vert x)$	$(y\vert y)$	$((x\vert x)\vert(y\vert y))$
T	T	T	F	F	T
T	F	T	F	T	T
F	T	T	T	F	T
F	F	F	T	T	F

(iv)

x	y	$(x \rightarrow y)$	$(y\vert y)$	$(x\vert(y\vert y))$
T	T	T	F	T
T	F	F	T	F
F	T	T	F	T
F	F	T	T	T

(c) $(((x|(y|y))|(y|(x|x)))|((x|(y|y))|(y|(x|x))))$

Section 5.2

1. In the second column from the right, 1 + 1 yields 0 plus a carry, which must be added to the 0 and 1 in the next column.

3.

x	y	z	c
0	0	0	1
0	0	1	1
0	1	0	0
0	1	1	0
1	0	0	0
1	0	1	0
1	1	0	0
1	1	1	0

5.

w	x	y	z	c
0	0	0	0	0
0	0	0	1	0
0	0	1	0	0
0	0	1	1	0
0	1	0	0	0
0	1	0	1	0
0	1	1	0	0
0	1	1	1	0
1	0	0	0	0
1	0	0	1	0
1	0	1	0	0
1	0	1	1	0
1	1	0	0	0
1	1	0	1	0
1	1	1	0	0
1	1	1	1	1

7. Always false 11. XOR

9. NAND 13. OR

15.

17.

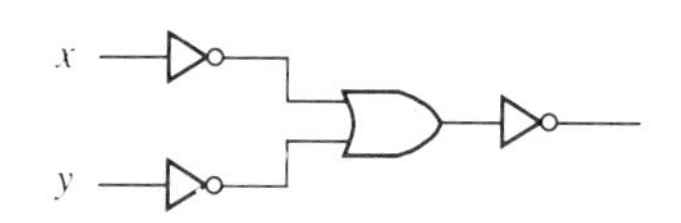

19.

21. The "?" gate is an OR gate.

23.

x — yields all zeros

x — yields all ones

25. $2^4 = 16$.

27. Certainly $x(yz) = (xy)z$ always holds, but $1 - (1 - 1) \neq (1 - 1) - 1$ and $(2/2)/2 \neq 2/(2/2)$.

29. If we let $h(x, y)$ represent OR, $h(x, y) = 0$ if and only if $x = y = 0$. A brief argument shows that $h(h(x, y), z)$ and $h(x, h(y, z))$ both yield 0 if and only if $x = y = z = 0$.

31. Giving one's name to something; named after.

Section 5.3

1. $\overline{(x + y)}\ \overline{(yz)}$ 3. $((xy)z)w$ 5. F

7.

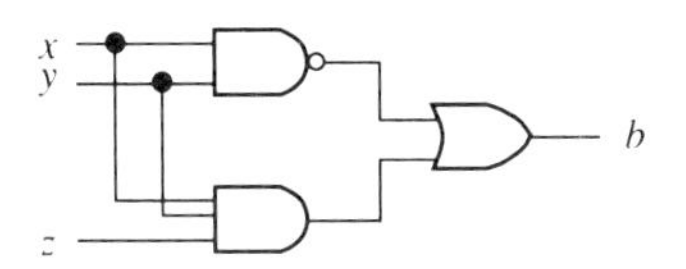

9.

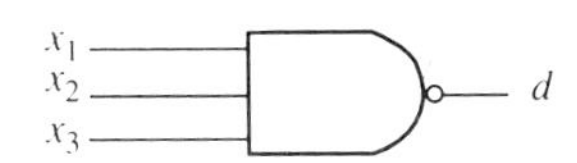

11.

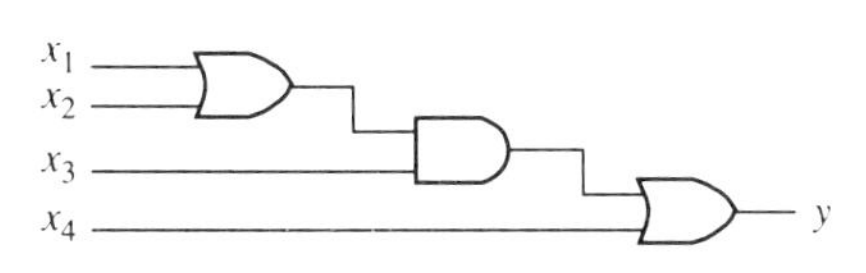

13.

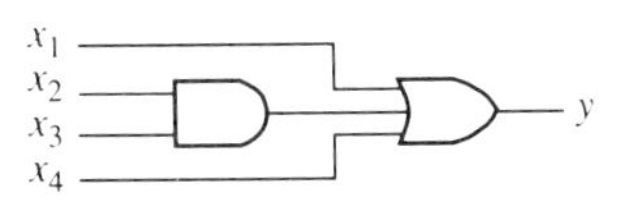

15. The truth table for $z = \overline{(xy)}$ is

x	y	xy	$\overline{(xy)}$
0	0	0	1
0	1	0	1
1	0	0	1
1	1	1	0

which is the same as that for NAND.

17. The truth table for $z = (x\bar{y}) + (\bar{x}y)$ is

x	y	$\bar{x}$	$\bar{y}$	$x\bar{y}$	$\bar{x}y$	$(x\bar{y}) + (\bar{x}y)$
0	0	1	1	0	0	0
0	1	1	0	0	1	1
1	0	0	1	1	0	1
1	1	0	0	0	0	0

which is the same as XOR.

19. $(((x_1x_2)x_3)x_4)$ and $((x_1x_2)x_3x_4))$

21.

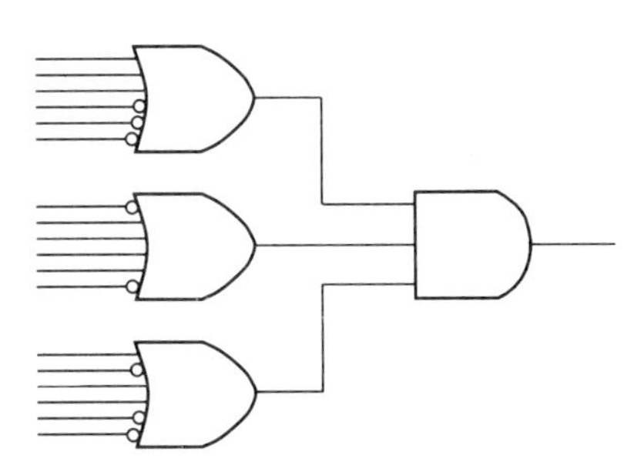

23. The circuit

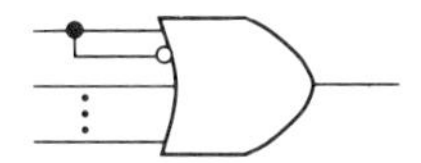

accepts all inputs. Change the OR gate to an AND gate to get one which rejects all inputs.

25. Two

27. $a = (x_1 + x_2 + x_3)(x_1 + \bar{x}_2 + \bar{x}_3) \times (\bar{x}_1 + x_2 + \bar{x}_3)(\bar{x}_1 + \bar{x}_2 + x_3)$

$b = \overline{(x_1 +}\, x_2 + x_3)(x_1 + x_2 + \bar{x}_3) \times (x_1 + \bar{x}_2 + x_3)(\bar{x}_1 + x_2 + x_3)$

29. The circuit accepts everything.

31. The truth table for $y = (x_1 + x_3)\bar{x}_2$ is

x_1	x_2	x_3	$x_1 + x_3$	$\bar{x}_2$	$(x_1 + x_3)\bar{x}_2$
0	0	0	0	1	0
0	0	1	1	1	1
0	1	0	0	0	0
0	1	1	1	0	0
1	0	0	1	1	1
1	0	1	1	1	1
1	1	0	1	0	0
1	1	1	1	0	0

which is the same as Table 5.7.

33. A minterm accepts the smallest possible number of input combinations (just one). A maxterm accept many (all but one).

Section 5.4

1. There are 5 expressions. With outer parentheses omitted, they are $((x_1x_2)x_3)x_4$, $(x_1x_2)(x_3x_4)$, $(x_1(x_2x_3))x_4$, $x_1((x_2x_3)x_4)$, and $x_1(x_2(x_3x_4))$.

3. (a) {000}
   (b) {0000, 0001, 0010, 0011, 0100, 0101, 0110, 1000, 1001, 1010, 1011, 1100, 1101, 1110, 1111}
   (c) {000, 001, 010, 011, 110, 111}

5. (a) $0(x + y)$ (c) $\overline{(x_1 + \cdots + x_n)}$
   (b) $1(\bar{x}_1(x_2 + x_3))$

7.

x	y	$x + y$	$y + x$
0	0	0	0
0	1	1	1
1	0	1	1
1	1	1	1

9. (a)

x	y	xy	$y + xy$
0	0	0	0
0	1	0	1
1	0	0	0
1	1	1	1

(b) $y + xy = 1y + xy = (1 + x)y = 1y = y$

11.

x	$x + x$
0	0
1	1

13.

x	0	$x + 0$
0	0	0
1	0	1

15. $A' \cap B \cap C' \cap D' \cap E$; $A \cup B' \cup C \cup D' \cup E'$;

$\sim x_1 \wedge x_2 \wedge \sim x_3 \wedge \sim x_4 \wedge x_5$; $x_1 \vee \sim x_2 \vee x_3 \vee \sim x_4 \vee \sim x_5$

17. (a) Involution refers to raising a quantity or expression to a given power.
(b) The Latin stem *idem* means "the same as previously given or mentioned."
(c) Redundant refers to the extra or unneeded term.
(d) Consensus refers to a "general agreement."

19. Involution is self-dual.

21. $(f + \bar{g})(f + \bar{h})(g + h) = (ff + f\bar{h} + \bar{g}f + \bar{g}\bar{h})(g + h)$
$= (f(1 + \bar{h} + \bar{g}) + \bar{g}\bar{h})(g + h) = (f1 + \bar{g}\bar{h})(g + h)$
$= (f + \bar{g}\bar{h})(g + h) = f(g + h) + \bar{g}\bar{h}(g + h)$
$= f(g + h) + \bar{g}\bar{h}g + \bar{g}\bar{h}h = f(g + h) + 0 + 0$
$= f(g + h)$

23. $h(\overline{f + g}) + \bar{f}gh = h\bar{f}\bar{g} + \bar{f}gh = \bar{f}h(\bar{g} + g) = \bar{f}h$

25. $(f + g + h)\overline{fgh} = (f + g + h)(\bar{f} + \bar{g} + \bar{h})$
$= f\bar{f} + f\bar{g} + f\bar{h} + g\bar{f} + g\bar{g} + g\bar{h} + h\bar{f} + h\bar{g} + h\bar{h}$
$= f\bar{g} + f\bar{h} + \bar{f}g + g\bar{h} + \bar{f}h + \bar{g}h$
$= f\bar{g} + fg\bar{h} + f\bar{g}\bar{h} + \bar{f}gh + \bar{f}g\bar{h} + g\bar{h} + \bar{f}h + f\bar{g}h + \bar{f}\bar{g}h$
$= f\bar{g}(1 + \bar{h} + h) + g\bar{h}(f + \bar{f} + 1) + h\bar{f}(g + 1 + \bar{g})$
$= f\bar{g} + g\bar{h} + h\bar{f}$

27. (a) No more than one f_i is ever true.
(b) Exactly one f_i is always true.
(c) Some f_i can be true.
(d) Some f_i can be false.
(e) All the f_i's can't be true.
(f) f_i implies f_j, i.e., if $f_i(x_1, \ldots, x_n) = 1$, then $f_j(x_1, \ldots, x_n) = 1$.

29. Suppose first that $f \leq g$. If $g(x_1, \ldots, x_n) = 0$ then $f(x_1, \ldots, x_n) = 0$ so $f + g = g$. If $g(x_1, \ldots, x_n) = 1$, then $f + g = g$ since $x + 1 = 1$ for all x. Now assume $f + g = g$. Clearly $f \leq g$ is true whenever $f(x_1, \ldots, x_n) = 0$. If $f(x_1, \ldots, x_n) = 1$ then

$$g(x_1, \ldots, x_n) = f(x_1, \ldots, x_n) + g(x_1, \ldots, x_n) = 1 + g(x_1, \ldots, x_n) = 1$$

so $f \leq g$ again.

31. *Set Theory*: If, in a valid set identity, you replace all intersections by unions, all unions by intersections, all empty sets by the universal set, and all universal sets by empty sets, you obtain a valid set identity. *Propositional Calculus*: If, in a theorem of the propositional calculus, you replace all $\wedge$'s by $\vee$'s, all $\vee$'s by $\wedge$'s, all T's by F's and all F's by T's, you obtain another theorem of the propositional calculus. [Here T is the universally true WFP, and F is the universally false WFP, e.g., we may regard T as an abbreviation for $(x \vee \sim x)$, and F as an abbreviation for $(x \wedge \sim x)$.]

33. (a) Positive logic:

(b) Negative logic:

35. No. If $x = 1$, the identity always holds.

37. (a) True (redundancy)
(b) False ($x = 0, y = z = 1$)
(c) False ($x = 1, y = z = 0$)

39. The strong absorption law. The given WFP is logically equivalent to $((x \wedge y) \rightarrow x)$, and the analogous Boolean expression is $xy \leq x$, i.e. (see Problem 29), $xy + x = x$.

Section 5.5

1. (a) Distributive and commutative
(b) Distributive
(c) Complementarity and identity
(d) Redundancy

3.

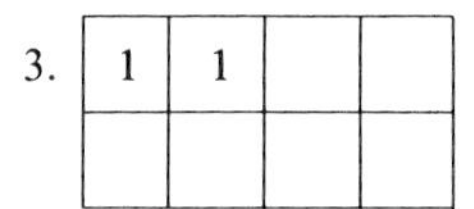

5.

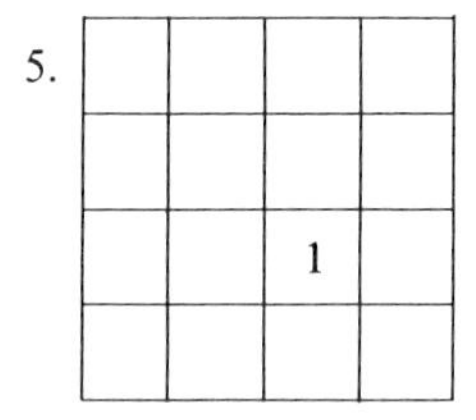

7.

9. (a) $\bar{x}_2$ (d) x_2x_3 (g) $x_1x_2x_3$
   (b) x_1 (e) $x_2\bar{x}_3$ (h) x_1x_3
   (c) $\bar{x}_2x_3$ (f) $\bar{x}_1\bar{x}_2$ (i) $x_1\bar{x}_3$

11. (a) x_2x_4 (d) $\bar{x}_2x_3$ (g) $x_1x_2x_3x_4$
    (b) $\bar{x}_1x_2\bar{x}_4$ (e) $\bar{x}_3x_4$ (h) $x_2\bar{x}_4$
    (c) $\bar{x}_4$ (f) $\bar{x}_1x_3$

13. (a) $\bar{x}_1\bar{x}_2\bar{x}_3 + \bar{x}_1\bar{x}_2x_3 + x_1\bar{x}_2\bar{x}_3 + x_1\bar{x}_2x_3$
    (b) $x_1\bar{x}_2\bar{x}_3 + x_1\bar{x}_2x_3 + x_1x_2\bar{x}_3 + x_1x_2x_3$
    (c) $\bar{x}_1\bar{x}_2x_3 + x_1\bar{x}_2x_3$ (g) $x_1x_2x_3$
    (d) $\bar{x}_1x_2x_3 + x_1x_2x_3$ (h) $x_1\bar{x}_2x_3 + x_1x_2x_3$
    (e) $\bar{x}_1x_2\bar{x}_3 + x_1x_2\bar{x}_3$ (i) $x_1\bar{x}_2\bar{x}_3 + x_1x_2\bar{x}_3$
    (f) $\bar{x}_1\bar{x}_2\bar{x}_3 + \bar{x}_1\bar{x}_2x_3$

15. (a) 256
    (b) 26
    (c) About 10%

17. (a) $x_1x_4 + x_2\bar{x}_3\bar{x}_4$
    (b) $x_1x_2 + x_1\bar{x}_3\bar{x}_4 + x_1x_3x_4 + x_2x_3\bar{x}_4 + x_2\bar{x}_3x_4$
    (c) $x_1\bar{x}_2 + \bar{x}_2x_3 + \bar{x}_1x_2\bar{x}_3$
    (d) $x_1x_4 + x_2\bar{x}_3x_4 + x_1x_2x_3$
    (e) $x_1x_2\bar{x}_3 + \bar{x}_1x_2x_4 + \bar{x}_1x_2x_3$
    (f) $\bar{x}_1x_4 + \bar{x}_2x_3x_4 + \bar{x}_1\bar{x}_2\bar{x}_3$
    (g) $\bar{x}_1\bar{x}_3 + x_1x_4 + \bar{x}_1x_2x_3\bar{x}_4$
    (h) $\bar{x}_1\bar{x}_2 + x_1x_2 + x_2\bar{x}_3$
    (i) $\bar{x}_1 + x_2$
    (j) $x_4 + x_1\bar{x}_3 + x_2x_3 + \bar{x}_1\bar{x}_2\bar{x}_3$

19. $x + y + z$

21.

Part	Minterm Total	Karnaugh Total
(a)	8	2
(b)	17	5
(c)	14	6
(d)	30	3
(e)	17	17
(f)	54	16

23. $x_1\bar{x}_3 + \bar{x}_3x_4 + \bar{x}_1\bar{x}_2\bar{x}_4 + x_2x_3\bar{x}_4$

25. $\bar{x}_1\bar{x}_3 + x_2\bar{x}_3x_4 + x_1\bar{x}_2\bar{x}_4 + x_2x_3\bar{x}_4$

27. $\bar{x}_2\bar{x}_4 + x_1x_3 + \bar{x}_1\bar{x}_2\bar{x}_3$

29. $\overline{(\bar{x}_1x_3 + x_2x_3)}$

31. $\overline{(\bar{x}_1x_2x_3 + x_1\bar{x}_2\bar{x}_3)}$

33. $\overline{(\bar{x}_1\bar{x}_2\bar{x}_3\bar{x}_4 + \bar{x}_1x_2x_3\bar{x}_4 + x_1x_2\bar{x}_3x_4 + x_1\bar{x}_2x_3 + \bar{x}_2x_3x_4)}$

35. $\overline{(\bar{x}_2\bar{x}_3\bar{x}_4\bar{x}_5 + x_3x_4\bar{x}_5 + x_2\bar{x}_4\bar{x}_5 + x_2x_4\bar{x}_5 + \bar{x}_1\bar{x}_3 + \bar{x}_1\bar{x}_2)}$

37. 8

39. (a) Since the "covered by" relation is equivalent to a subset relation (P is covered by Q if the set of minterms of P is a subset of the set of minterms of Q), this follows from the fact that the subset relation is a partial ordering.
    (b)

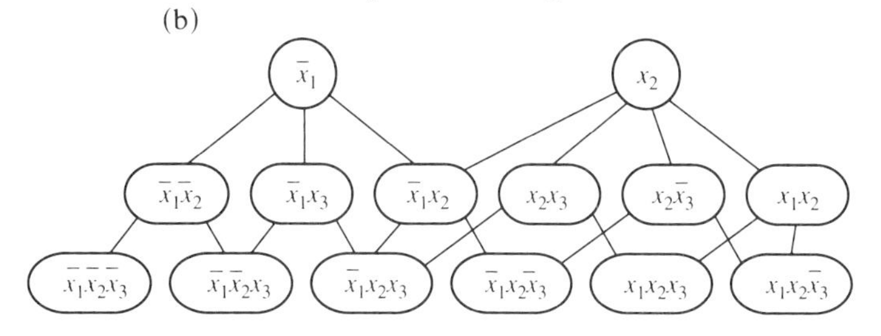

    (c) $\bar{x}_1$ and x_2

Chapter 6

Section 6.1

In deriving equations for sequential circuits, the following thought process may be helpful. If a circuit has input x, output y, and a flip-flop with state s, then after the arrival of the $(n + 1)$st clock pulse, the input is x_{n+1}, the output is y_{n+1}, the output of the flip-flop is s_n, and its input (as well as its state) is s_{n+1}.

1. ?01101101 10 . . .

3. (a) $y_{n+1} = s_n$; $s_{n+1} = x_{n+1} + s_n$
   (b) 00000111111 . . .

5. (a) 000000 . . .
   (b) $y_{n+1} = s_0x_1 \ldots x_n = 0$
   (c) $s_0 = 1$. Constant functions are not very interesting.

7. (a) $y_{n+1} = \overline{(s_nx_{n+1})}$; $s_{n+1} = x_{n+1}$
   (b) $y_4 = \overline{(x_3x_4)}$

9. (a)

Present State	Next State $x = 0$	Next State $x = 1$	Output $x = 0$	Output $x = 1$
00	10	10	0	0
01	10	00	1	1
10	11	11	0	0
11	11	01	1	1

(b)

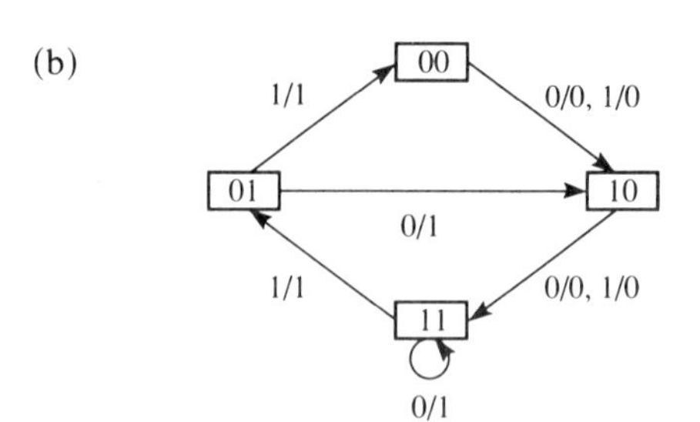

(c) 001110 . . .

11. (a)

Present State	Next State $x = 0$	Next State $x = 1$	Output $x = 0$	Output $x = 1$
00	00	10	0	0
01	01	11	0	0
10	10	01	0	0
11	11	00	0	1

(b)

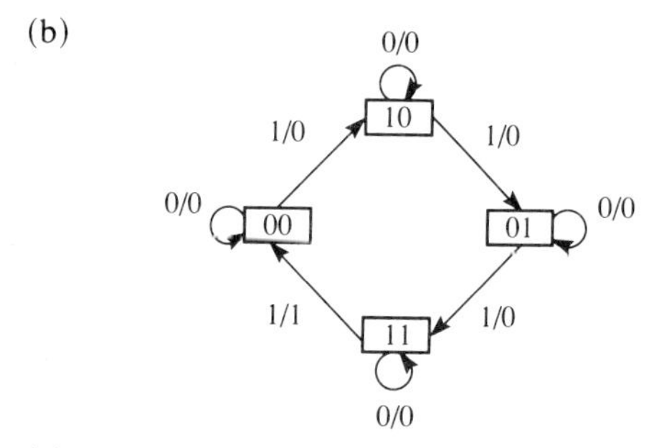

(c) 000100

13.

00/0, 01/0 (0) —10/0→ (1) 00/1, 10/1
01/1

15. $y_{n+1} = s_n$; $s_{n+1} = x_{n+1} \oplus s_n \bar{x}'_{n+1}$

17. $y_{n+1} = s_n$; $s_{n+1} = \bar{s}_n x_{n+1} + s_n \bar{x}'_{n+1} + x_{n+1} \bar{x}'_{n+1}$

19. (a) 1011100001
(b) T

21. (a) $y_{n+1} = a_n \bar{x}_{n+1}$ (c) $y_{n+1} = x_n \bar{x}_{n+1}$
(b) $a_{n+1} = x_{n+1}$ (d) 010010010; no

23. (a) $y_{n+1} = x_{n-2}$
(b)

Present State	Next State $x = 0$	Next State $x = 1$	Output
000	000	100	0
001	000	100	1
010	001	101	0
011	001	101	1
100	010	110	0
101	010	110	1
110	011	111	0
111	011	111	1

(c)

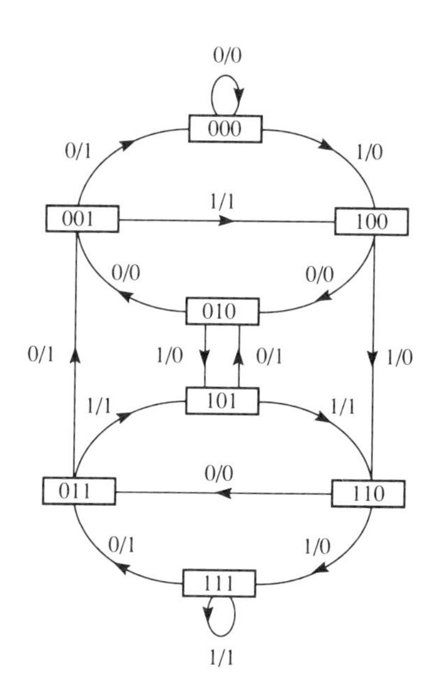

25. (a)

Present State	Next State 00	01	10	11	Output 00	01	10	11
00	00	01	10	11	000	000	101	101
01	00	01	10	11	001	001	100	100
10	00	01	10	11	010	010	111	111
11	00	01	10	11	011	011	110	110

(b)

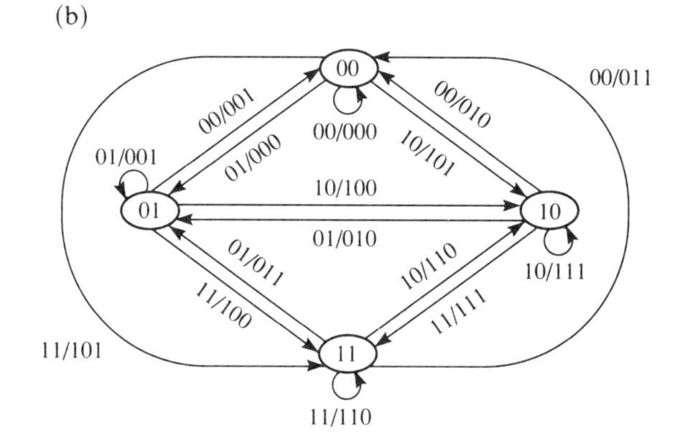

(c) (000), (100), (010), (101)(011), (100), (010), (101), (011), (100), (010), (101), . . . (repeats), . . .

Section 6.2

1.

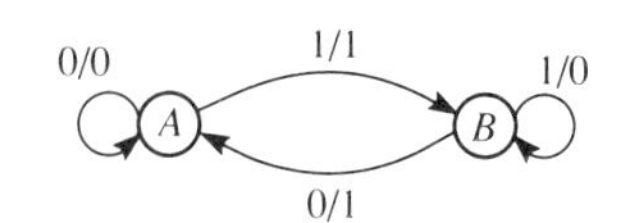

Present State	Next State		Output	
	$x = 0$	$x = 1$	$x = 0$	$x = 1$
A	0	1	0	1
B	0	1	1	0

3.

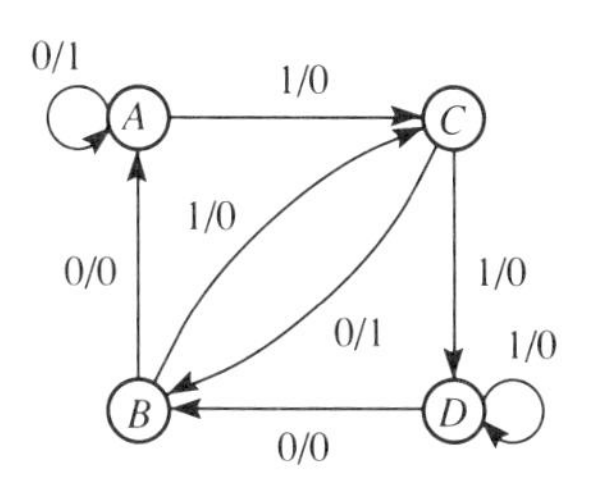

Present State	Next State		Output	
	$x = 0$	$x = 1$	$x = 0$	$x = 1$
A	A	C	1	0
B	A	C	0	0
C	B	D	1	0
D	B	D	0	0

5.

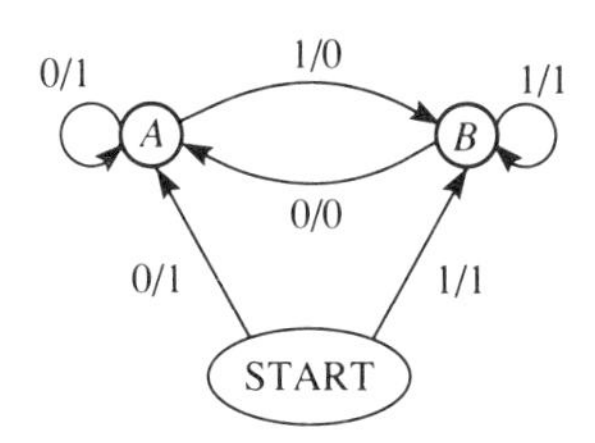

7.

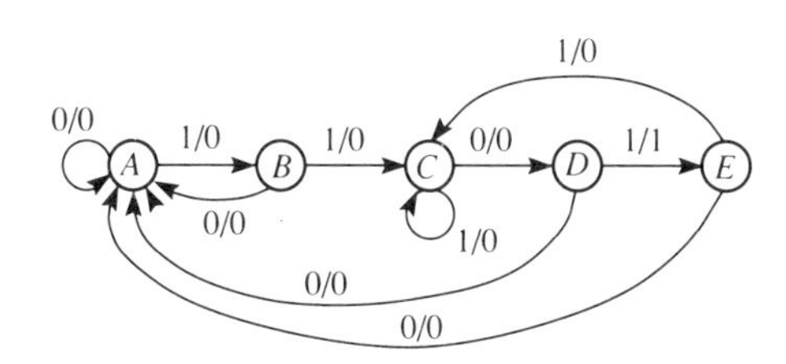

9.

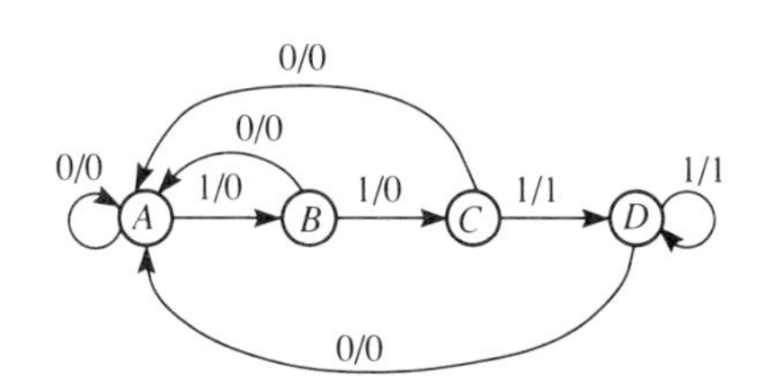

11.

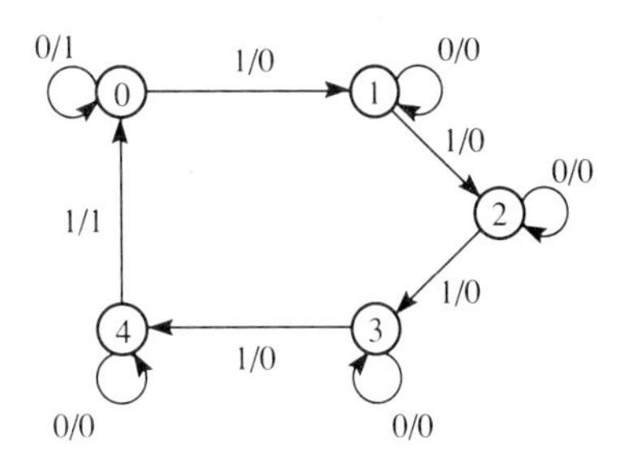

13. (a)

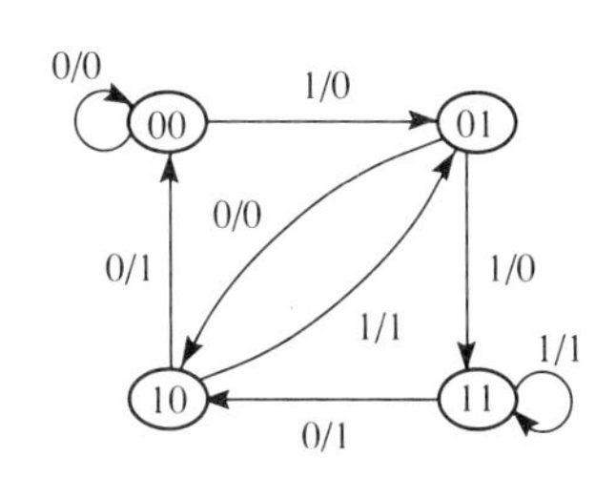

(b)

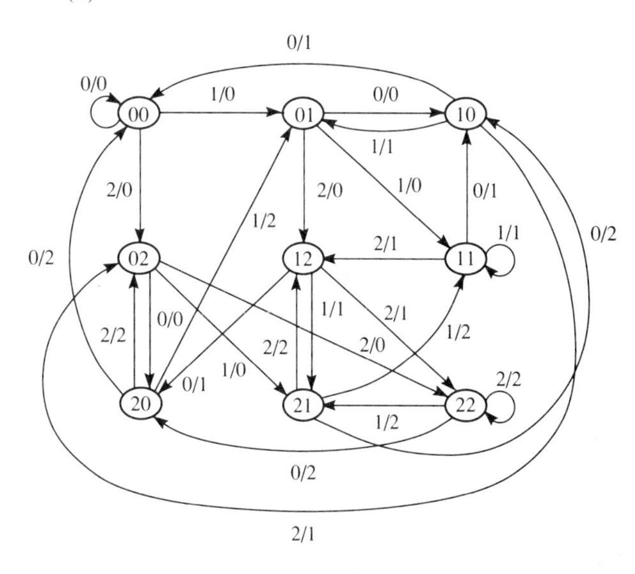

(c) N^2

15.

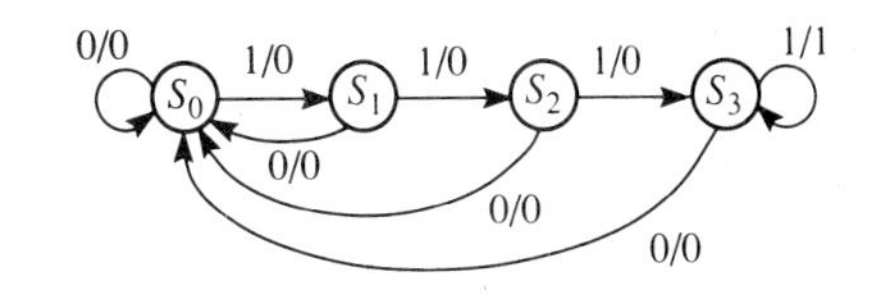

17. Example 6.5

Section 6.3

1. There are 15 altogether: $\{A\}, \{B\}, \{C\}, \{D\}$; $\{A, B, C, D\}$; six of type $\{A, B\}, \{C\}, \{D\}$; three of type $\{A, B\}\{C, D\}$; and four of type $\{A, B, C\}, \{D\}$.

3. $|A_1| \cdot |A_2| \cdots |A_r|$

5.

Present State	Next State		Output	
	$x = 0$	$x = 1$	$x = 0$	$x = 1$
A	B	H	0	0
B	G	D	0	0
D	A	G	0	1
G	D	B	1	0
H	D	A	0	0

7.

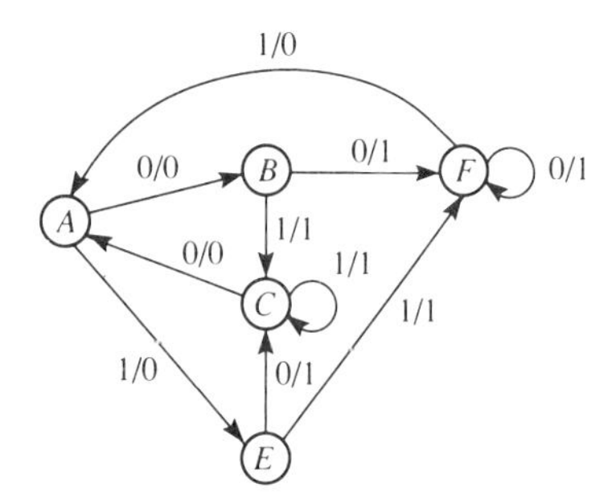

9. The equivalence classes are $\{B, E\}, \{H\}, \{A\}, \{F\}$, $\{D, G\}$, and $\{C\}$.

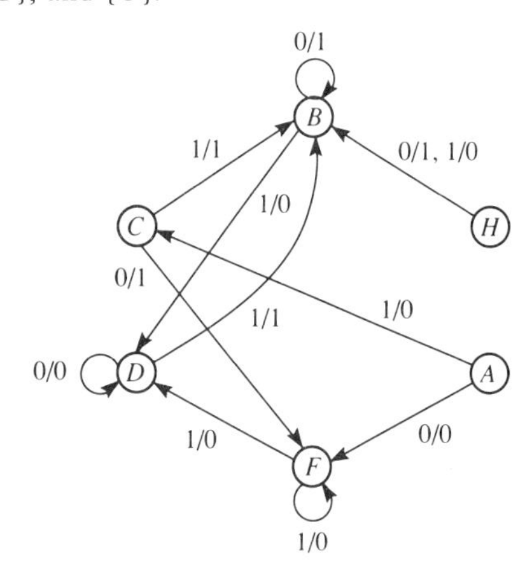

11. It is already reduced.

13. It is already reduced.

15. It is already reduced.

17. It is already reduced.

19. (a)

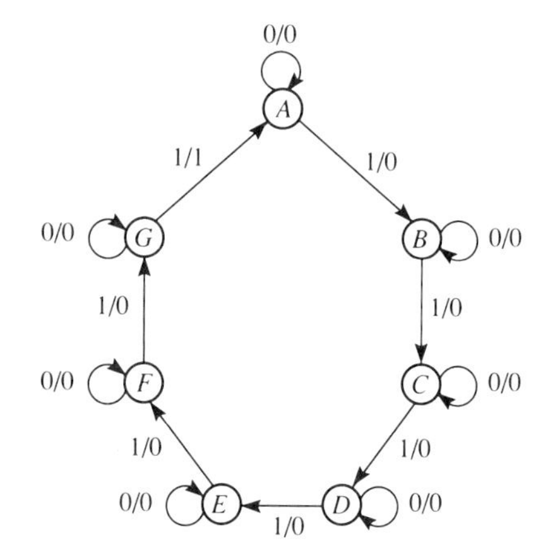

Present State	Next State		Output	
	$x = 0$	$x = 1$	$x = 0$	$x = 1$
A	A	B	0	0
B	B	C	0	0
C	C	D	0	0
D	D	E	0	0
E	E	F	0	0
F	F	G	0	0
G	G	A	0	1

(b) The machine is already reduced—all the equivalence classes are singletons.
(c) 5

21. It will reduce to a machine with a single state.

23. The number of sets in a partition increases after the `REFINE PARTITION` step.

25. After at most $n - 1$ trips through the `REFINE PARTITION` block, the partition must be consistent (otherwise it could be refined further). The algorithm will then terminate when the `YES` branch from the `CONSISTENT`? test is taken.

27. It clearly suffices to show that S_1 and T_1 (i) produce the same output for any input x and (ii) go to equivalent (by the second definition) states under any input x. (An easy induction then establishes the result.) But if S_1 and T_1 are in the same block of the final SE partition, they had to be in the same block of the initial SE partition, so (i) holds. If S_1 and T_1 went to inequivalent states under some input x, the partition would not be consistent, so (ii) holds.

Section 6.4

1.

Present State	Next State $x = 0$	$x = 1$
S_0	S_0	S_1
S_1	S_0	S_2
S_2	S_0	S_3
S_3	S_0	S_4
S_4	S_0	S_4

3. Just interchange the accepting and the nonaccepting states in the machine of Figure 6.22(a).

5. 2^{n-2}

7. Those strings having an even number of 0s and an even number of 1s.

9.

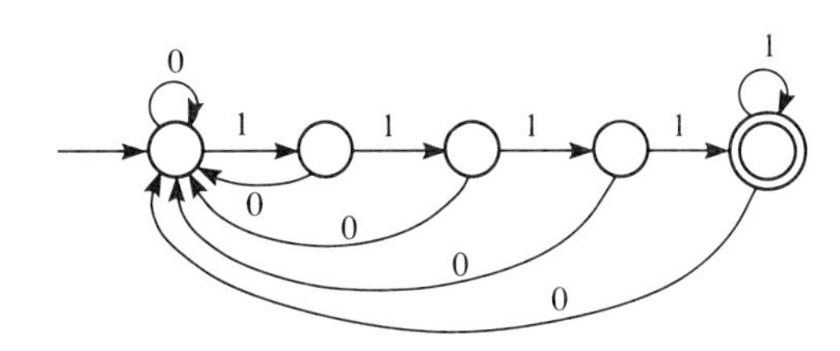

11.

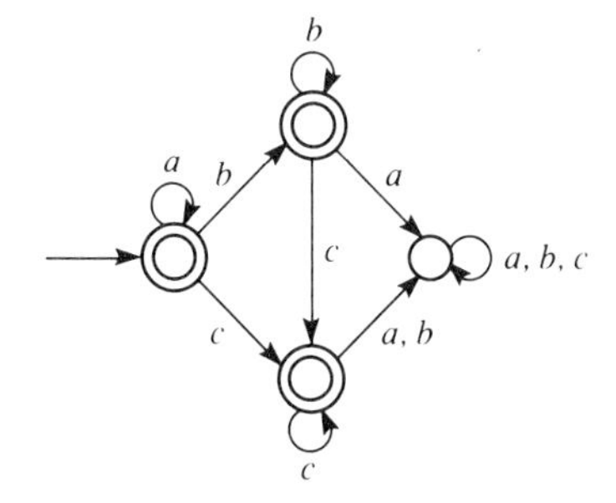

13. (a) No.
(b) Only the string 101 is accepted.

15. Only the string *po* is accepted. There needs to be an arrow leading from state 3 to itself for each letter of the alphabet.

17. 28

19.

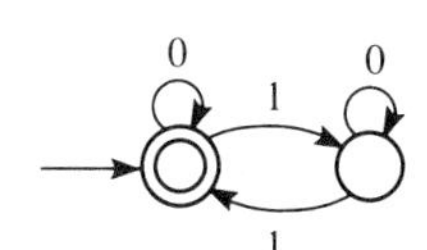

21.

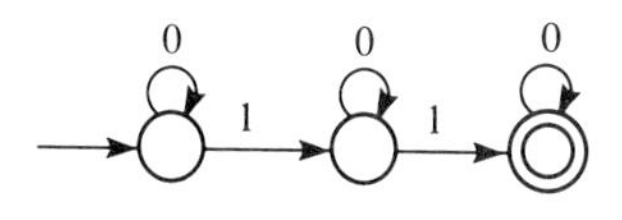

23.

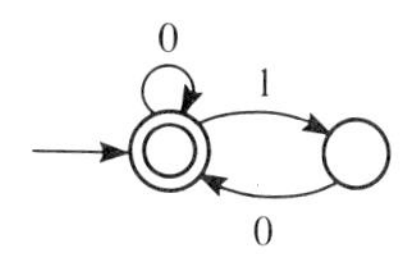

25.

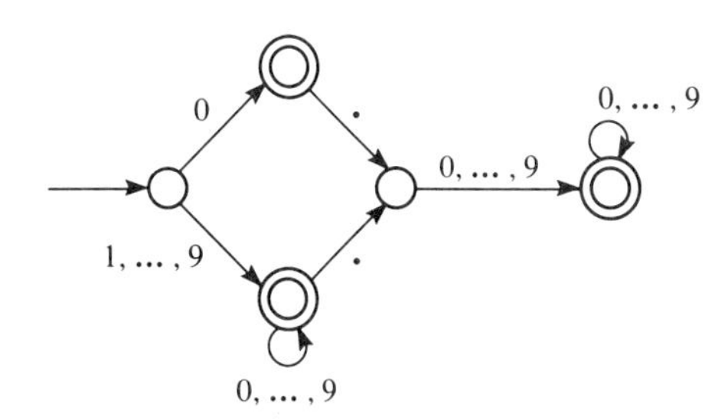

27. If we regard two states as "related" if the second state can be reached from the first by a single ε transition, then the set of states which are ε-reachable from S is just the transitive closure of this relation, which is precisely what Warshall's algorithm computes.

29.

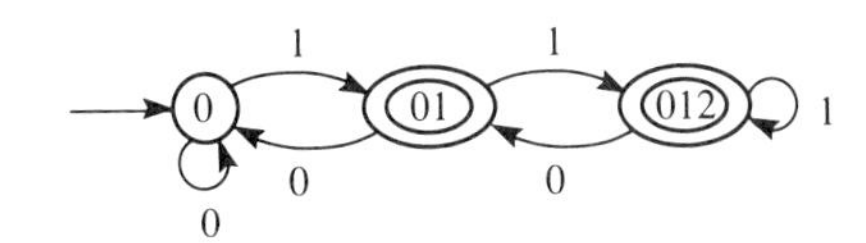

31.

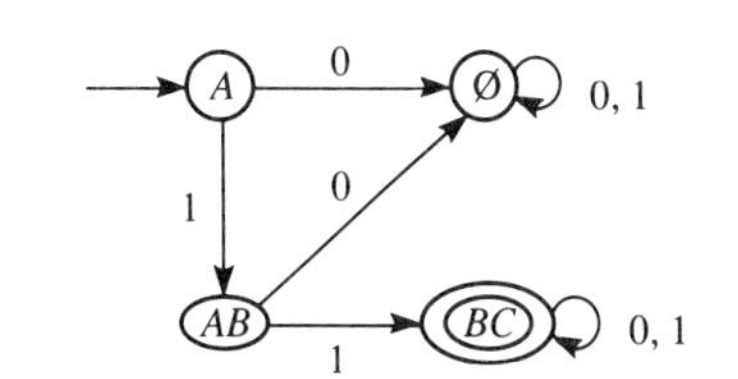

33. See Problem 11.

35. (a)

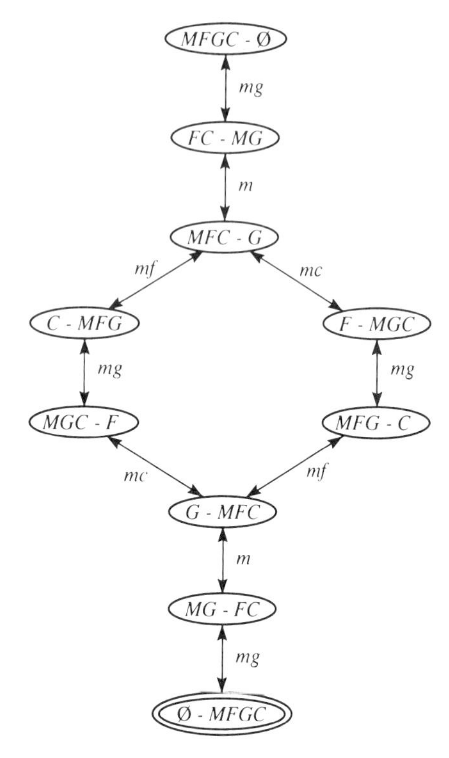

(b) From the above diagram, there are two essentially different solutions, one of which is

$$(mg)(m)(mf)(mg)(mc)(m)(mg).$$

Section 6.5

1. $11\cdots1000\cdots$ where there are $2n$ 1's at the beginning of the string.

3. Just take states 0 through $M-1$ of the machine in Problem 1.

5. It would erase the rightmost 1, move one square right, and halt.

7.

	Next State		Output		Operation	
	$x=1$	$x=b$	$x=1$	$x=b$	$x=1$	$x=b$
Q_0	Q_1	Q_0	1	Even	R	H
Q_1	Q_0	Q_1	1	Odd	R	H

9. The basis is certainly true since the machine starts in state Q_0 and at that time has read no 1's. Suppose the result holds for n. If n is even, the machine is in state Q_0 after reading n 1's. When the $(n+1)$st 1 is read, the rule $Q_0 1 : Q_1 1 R$ causes the machine to move to state Q_1, which is appropriate since it has now read an odd number of 1's. Similarly for the case when n is odd.

11. $Q_0 0 : Q_0 Y R$ (Q_0 is the start state)
 $Q_0 1 : Q_0 X R$
 $Q_0 b : Q_0 b H$

13. It blanks out every other 1, beginning with the leftmost 1.

15.

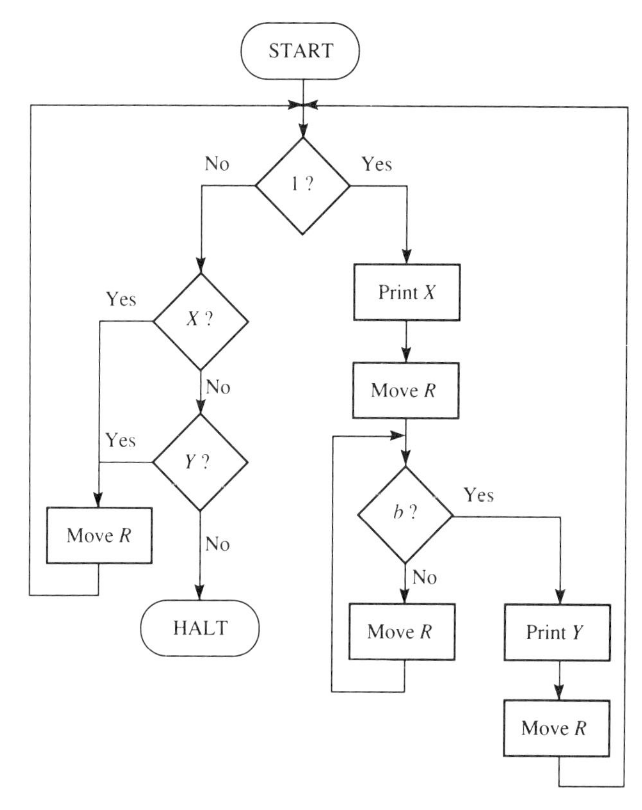

17. Not enough memory to remember all the rooms which have been visited.

19. $A1 : A1H$ $Ab : B1R$ $B1 : B1H$ $Bb : CbL$
 $C1 : D1L$ $D1 : D1H$ $Db : EbL$ $E1 : E1H$
 $Eb : FbR$ $Fb : G1R$ $G1 : G1R$ $Gb : HbR$
 $H1 : H1H$ $Hb : KbL$ $Kb : L1L$ $L1 : L1L$
 $Lb : EbL$

21. The machine cannot be reading a 1 while in state Q_1.

Section 6.6

1. (a) 000000111010001010 . . .
 (b) 11010001000100 . . .
 (c) 1110010101 . . .

3. Figure 6.42(a):

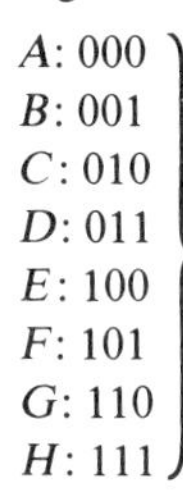

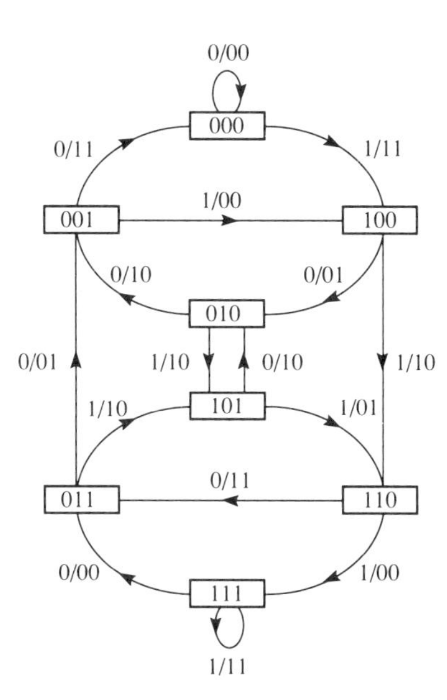

Figure 6.42(c):

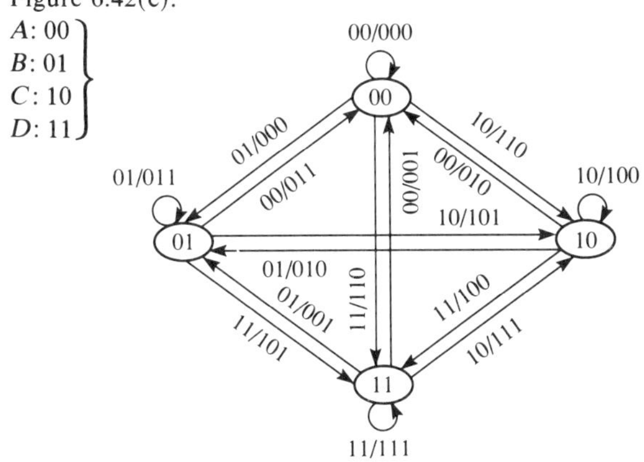

5. Figure 6.42(a): Assuming you have already computed x_{n-2}, x_{n-1} may be computed via $x_{n-1} = y_n \oplus z_n \oplus x_{n-2}$. Figure 6.42(c): We have $a_n = c_n$ and $b_n = e_{n+1}$.

7. (a) 00 00 00 11 10 10 00 10 10 ⋯

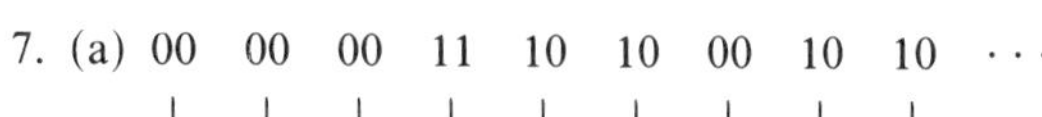

↓ ↓ ↓ ↓ ↓ ↓ ↓ ↓ ↓

0 0 0 0 1 1 0 1 1 ⋯

(b) 11 01 00 01 00 01 00 ⋯

↓ ↓ ↓ ↓ ↓ ↓ ↓

0 1 0 1 0 1 0 ⋯

(c) 11 10 01 01 01 01 ⋯

↓ ↓ ↓ ↓ ↓ ↓

0 1 1 1 1 1 ⋯

9. (a) and (e)

11. (a) (Figure 6.42a):

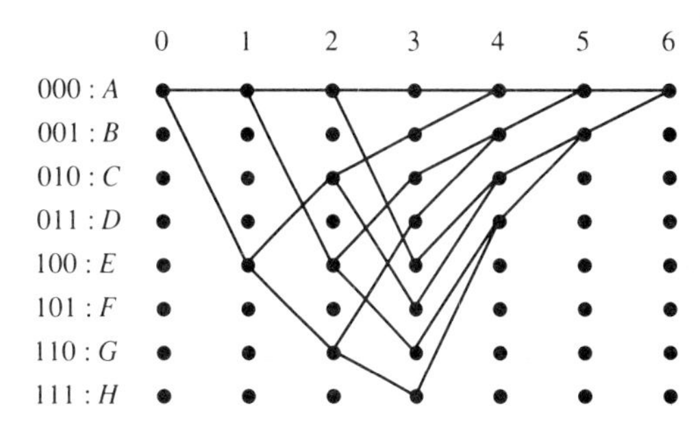

(b) (Figure 6.42c):

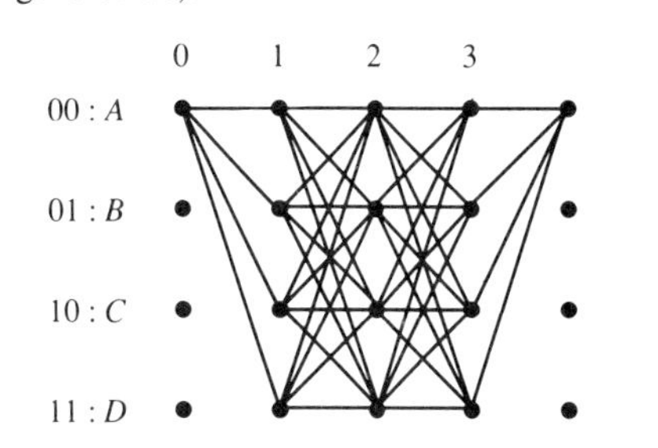

13.

15.

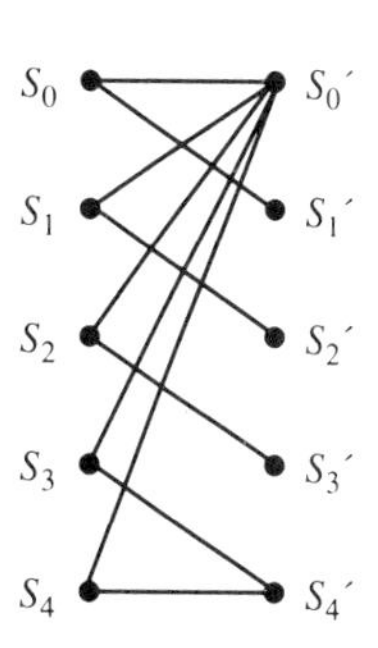

17.

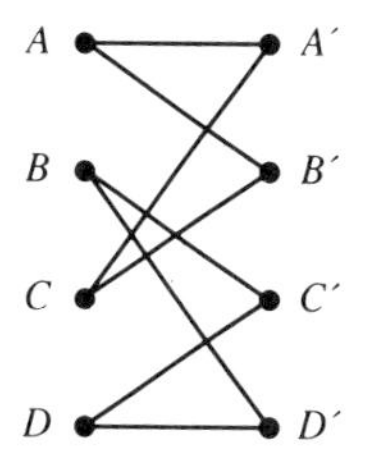

19.

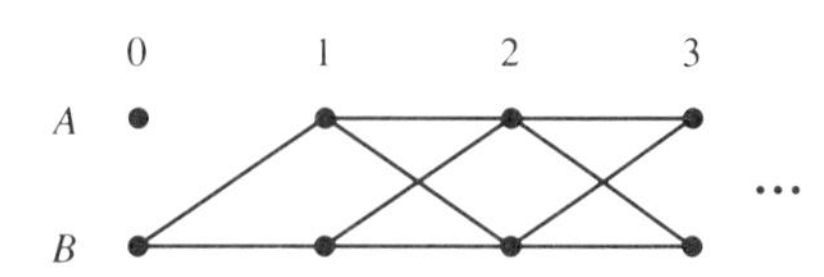

21.

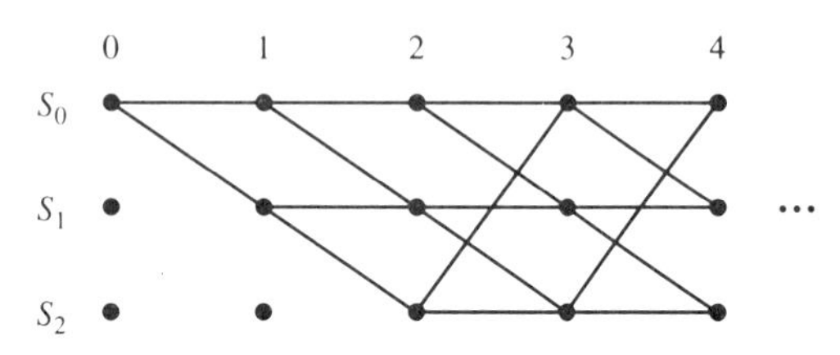

23. If both y_n and z_n are in error, the results will still be valid.

25. $f(1) = 2$; $f(2) = 6$; $f(n) = 6n - 6$, $n \geq 3$. The basis is established by observation. To verify the induction step, note that $f(n+1) = f(n) + 6$ since there are 6 edges in the trellis module.

27. (a) 28
    (b) $12 + 8(L - 2)$

29. (a) 1000(00) (distance 2)
    (b) 1010(00) (distance 2)
    (c) 1111(00) (distance 0)

31. 11 10 01 01 10 11 [input bits: 1111(00)] (least like R = most like $\bar{R}$)

Chapter 7

Section 7.1

1. (a)

3. 3; 5

5. Use the basic algorithm of Figure 7.1, with these changes: Change "suffix" to "prefix," and look for prefixes in the list (*abc*, *ab*, *a*, *bc*) in that order. Thus for example if the string begins with *abc*, you remove *abc* rather than just the *a*.

7. Use the algorithm of Figure 7.1, with this change: If *ba* is a suffix, remove it, even though *a* is also a suffix.

9. (a) ( )( ) · · · ( ) (n pairs)
   (b) (( )( )( ))
   (c) (((· · ·))) ($n - 1$ of each)
   (d) (((( )( ))( ))( ))

11. n; $n/2$

13. Note that COUNT is always $L - R$ where L is the number of left parentheses scanned and R is the number of right parentheses scanned. When $n = L + R$ is odd we must have $L \neq R$ (one must be odd, the other even) and therefore COUNT $\neq 0$ at the end of the string.

15. Since the empty string satisfies (i) and (ii) and these properties are clearly preserved by rules ($B1$) and ($B2$), each string in language B satisfies (i) and (ii). To go in the other direction, assume w is a string satisfying (i) and (ii). If w is empty it is in language B. Thus assume $w \neq \varepsilon$. The count referred to in (ii) begins and ends at 0. If the count stays positive until all of w is exhausted, w has the form (x) where x also satisfies (i) and (ii) and by induction belongs to language B. By ($B1$), w also belongs to B. If the count reaches 0 at some intermediate point (say after x has been examined), $w = xy$ where x and y satisfy (i) and (ii). Using the induction hypothesis and ($B2$), we see that $w \in B$.

17. (a) None of them
    (b) The same as algorithm B, except that we insist that COUNT stay positive until the entire string has been read.

19. No

21. We know that $MI^1 = MI^{(2^0)}$ is born on day 0 by ($C0$). Assuming that MI^{2^k} is born on day k, on day $k+1$ we see the birth of $MI^{(2^k)}I^{(2^k)} = MI^{(2^k+2^k)} = MI^{(2^{k+1})}$ by ($C2$).

23. No

25. Clearly Theorem 1 holds for MI, the only string born on day 0. Assuming it holds for strings born on day n, to see its truth for those born on day $n+1$ we need only observe that none of the rules ($C1$)-($C4$) change the first character of the string or introduce any new M's.

27. 1

29. $2^3 - 2 = 6$

31. See Problem 21.

33. (a) $MI^{32} \in C$ by Problem 31. Then by Problem 32 so are MI^{29}, MI^{26}, MI^{23}, MI^{20}, MI^{17}, MI^{14}, and MI^{11}.
    (b) By following the example of part (a), all we need to do is find a power of 2 which is bigger than N and leaves the same remainder as N when divided by 3. The remainder must be 1 or 2 since N is not a multiple of 3. Then simply note (or prove by induction) that 2^k leaves a remainder of 2 if k is odd and 1 if k is even.

Section 7.2

1. $\frac{1}{6}\binom{10}{5} = 42$; $\frac{1}{7}\binom{12}{6} = 132$

3. (()()()), ((())()), (()(())), ((()())), (((()))), ()()()(), ()(())(), (())()(), ()()(()), ()(())(), ()(()()), (()())(), ()((())), ((()))()

5. ( ) ( ))
   1 0 1 0 −1
   ( )) (( )
   1 0 −1 0 1 0
   (( )) (( )) ( )
   1 2 1 0 1 2 1 0 1 0
   (( ) (( )) ( ))
   1 2 1 2 3 2 1 2 1 0
   ((( ) (( )) ( )
   1 2 3 2 3 4 3 2 3 2
   ( )) ( )) (( ) (
   1 0 −1 0 −1 −2 −1 0 −1 0

7. In order for there to be a lattice path from (x, y) to (x', y') we must have $x \le x'$; $x' - x$ and $y' - y$ must be of the same parity; and $y' - y \le x' - x$. If this is the case, let u be the number of upward edges, and v the number of downward edges. Then $u + v = x' - x$, $u - v = y' - y$, so $u = ((x' - x) + (y' - y))/2$. The number of paths is $\binom{x'-x}{u}$.

9. $\frac{1}{8}\binom{14}{7} \Big/ \left(\frac{14!}{7!\,7!}\right) = .125$

11. (a)

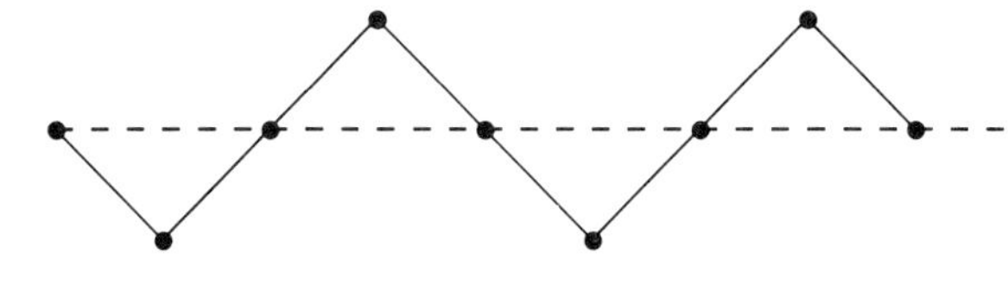

(b)

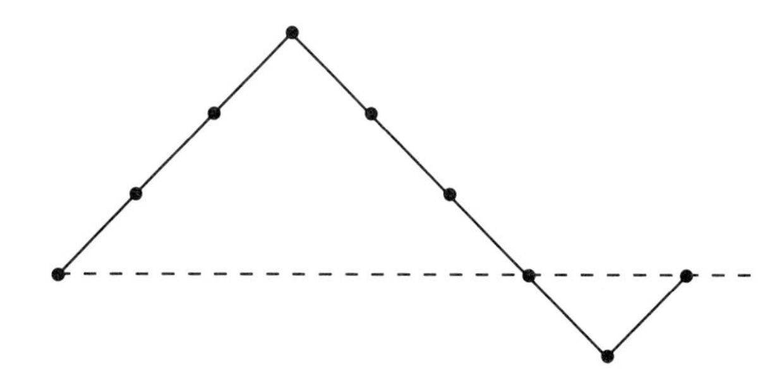

(c)

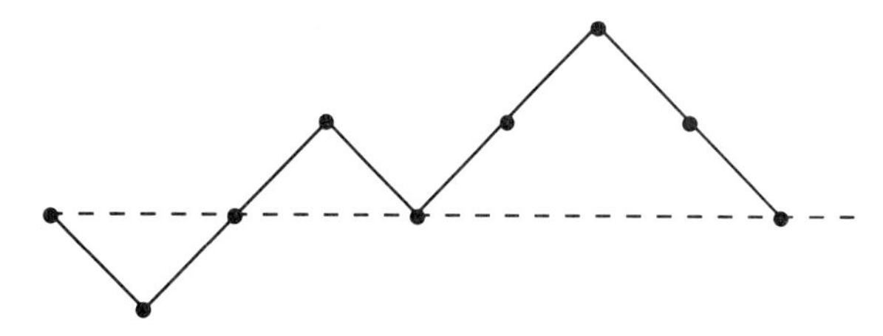

13. The answer is 0 unless the coordinates satisfy the restrictions mentioned in Problem 7. If those conditions are met, the number of paths that touch or cross the x-axis equals the total number of paths from $(x, -y)$ to (x', y'), namely

$$N_1 = \binom{x' - x}{\dfrac{(x' - x) + (y' + y)}{2}}.$$

The total number of paths from A to B is

$$N = \binom{x' - x}{\dfrac{(x' - x) + (y' - y)}{2}},$$

and the answer is $N - N_1$.

Section 7.3

1. For (7.15), 0 and 1 are regular expressions (atoms). Then by rule 2, so are (01), (00), and (11). By rule 1, ((00) + (11)) is a regular expression. Finally, by rule 2, ((01)((00) + (11))) is a regular expression. For (7.17), 0 and 1 are again atoms and (0 + 1) and (00) are regular expressions by rules 1 and 2, respectively. Rule 3 applied to (0 + 1) yields the regular expression ((0 + 1))*. Finally, rule 2 yields that (((0 + 1))*(00)) is a regular expression.

3. No

5. No. A counterexample is $x = a$, $y = aa$.

7. (01101)*. All strings which consist of 0 or more copies of 01101.

9. (a) Not in general (b) Yes (c) Yes

11. (a) and (c)

13. (a) Let $t = s + \varepsilon$. Then the identity becomes $(rt)^*r = r(tr)^*$. A typical term on the left has the form $(r_1t_1)(r_2t_2)\cdots(r_nt_n)r = r_1(t_1r_2)(t_2r_3)\cdots(t_nr)$ which is the form of a typical term on the right, so the identity holds.
    (b) It is false. For example, 01 belongs to the language represented by $(0 + 1)^*$ but it doesn't belong to the language represented by $0^* + 1^*$.

(c) It is false. Terms on the left begin with an s while those on the right begin with an r.

15. Because it is always possible to get from a vertex to itself without consuming any input, even if there is no ε label.

17. Since 0 is the only atom in the given expression, its language must be a subset of 0^*. On the other hand by taking the ε choices in the first and last term of the product and the 0 choice in the middle term, we see that its language includes 0^*.

19. (a) $R_1[1, 3] = R_0[1, 3] + R_0[1, 1]R_0[1, 1]^*R_0[1, 3]$
$= \varnothing + (\varepsilon + 0)(\varepsilon + 0)^*\varnothing = \varnothing$

(b) $R_1[2, 1] = R_0[2, 1] + R_0[2, 1]R_0[1, 1]^*R_0[1, 1]$
$= 0 + 0(\varepsilon + 0)^*(\varepsilon + 0)^* = 0 + 0(0^*)(0^*) = 00^*$

(c) $R_1[2, 3] = R_0[2, 3] + R_0[2, 1]R_0[1, 1]^*R_0[1, 3]$
$= 1 + 0(\varepsilon + 0)^*\varnothing = 1 + \varnothing = 1$

(d) $R_1[3, 1] = R_0[3, 1] + R_0[3, 1]R_0[1, 1]^*R_0[1, 1]$
$= \varnothing + \varnothing(\varepsilon + 0)^*(\varepsilon + 0) = \varnothing + \varnothing = \varnothing$

(e) $R_1[3, 2] = R_0[3, 2] + R_0[3, 1]R_0[1, 1]^*R_0[1, 2]$
$= \varnothing + \varnothing(\varepsilon + 0)^*1 = \varnothing + \varnothing = \varnothing$

(f) $R_1[3, 3] = R_0[3, 3] + R_0[3, 1]R_0[1, 1]^*R_0[1, 3]$
$\varepsilon + 0 + 1 + \varnothing(\varepsilon + 0)^*\varnothing = \varepsilon + 0 + 1 + \varnothing$
$= \varepsilon + 0 + 1$

21. $0^* + 0^*1(\varepsilon + 00^*1)^*00^*$

23. $1 + (\varepsilon + 00^*1)(\varepsilon + 00^*1)^*1$

25. $R_3[3, 2]$ and $R_3[3, 1]$

27. $(0 + 1)(0 + 1)^*$

29. $0^* + (0 + 1)^*00$

31. We need $R_3[1, 1]$ and $R_3[1, 3]$, which require $R_2[1, 1]$, $R_2[1, 3]$, $R_2[3, 3]$, and $R_2[3, 1]$. All entries of the matrices R_1 and R_0 are needed.

Section 7.4

1. (7.20) Is the string empty? (7.21) Is the string equal to 0? (7.22) Is the string equal to 01? (7.23) Is the string equal to 00 or 11? (7.24) Is the string equal to 0100 or 0111? (7.25) Answer YES. (7.26) Does the string end in 00?

3. (a)

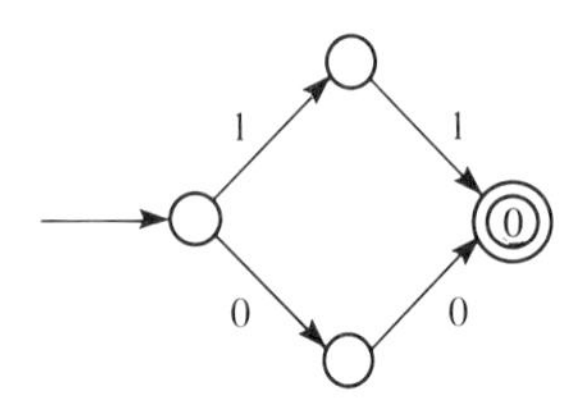

(b)

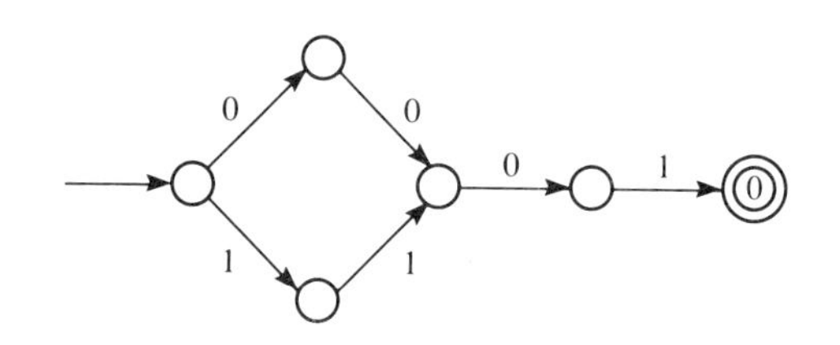

(c)

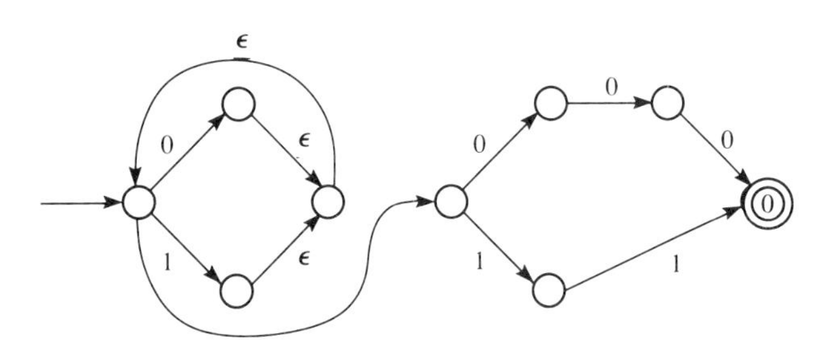

(d)

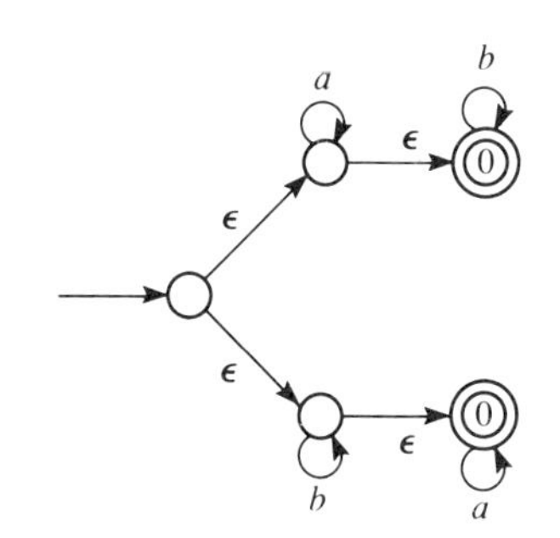

(e)

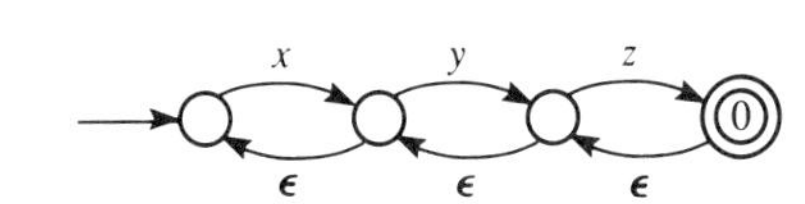

(f)

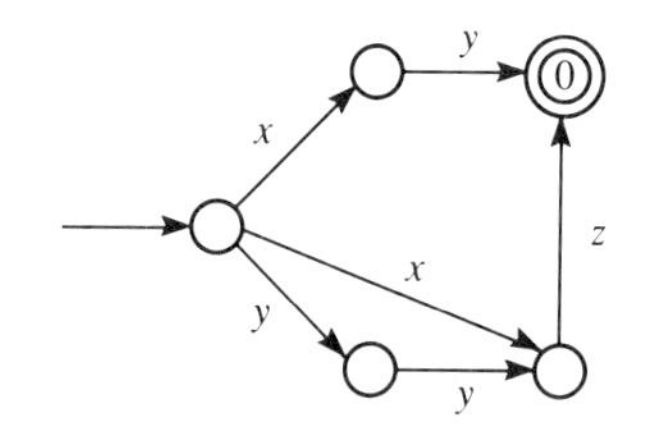

(g)

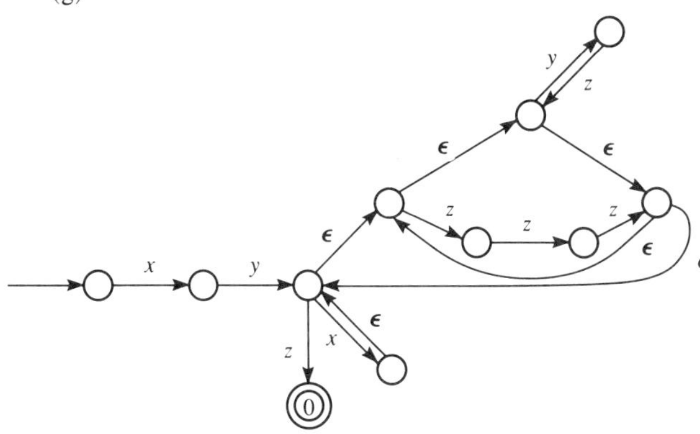

(h)

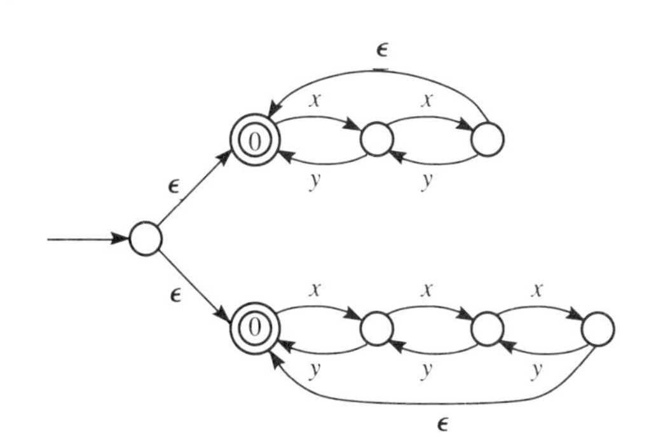

5.

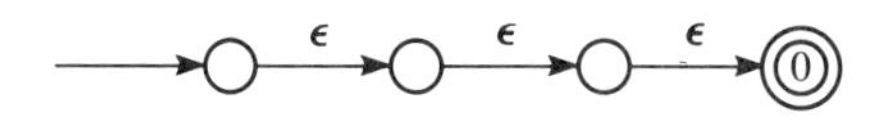

7.

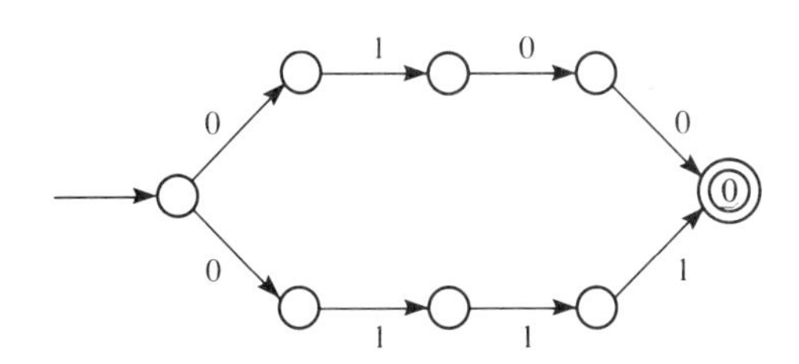

9. Because you can either accept the empty string by a single ε transition or accept a nonempty string of letters by following a series of letter ε paths.

11. The student. A string of the form $w_1w_2w_3w_4$, where w_1 leads from I_1 to A_1 in SD_1, w_2 leads from I_2 back to I_2 in SD_2, w_3 leads from A_1 back to A_1 in SD_1, and w_4 leads from I_2 to A_2 in SD_2, will be erroneously accepted.

13. (a)

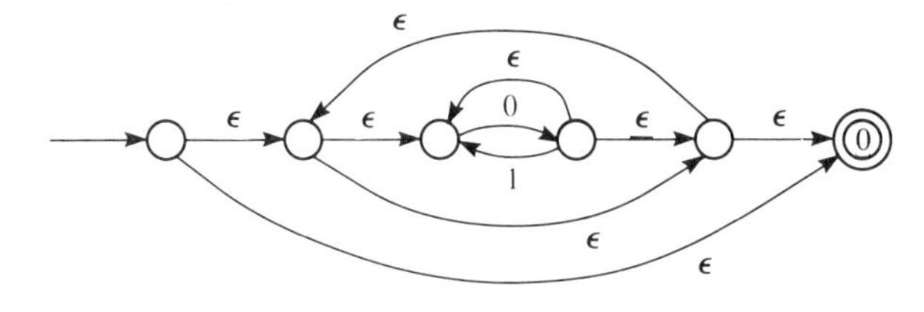

(b)

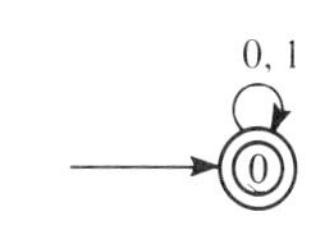

15.

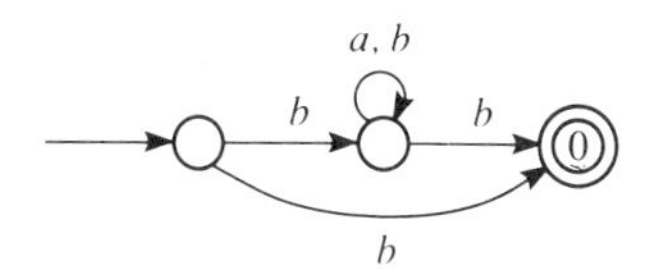

17.

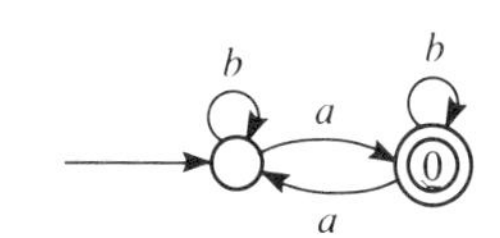

19.

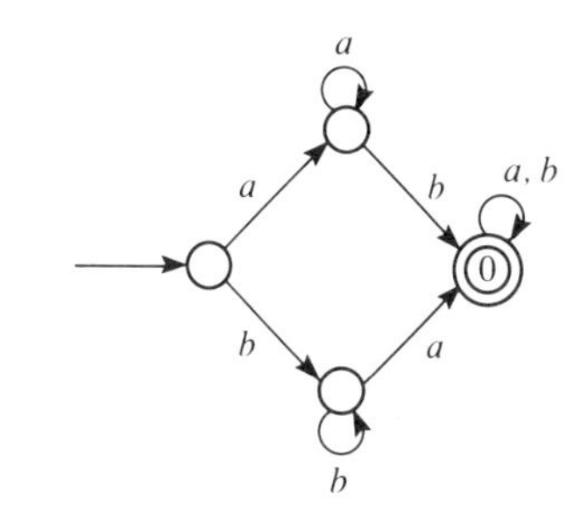

21.

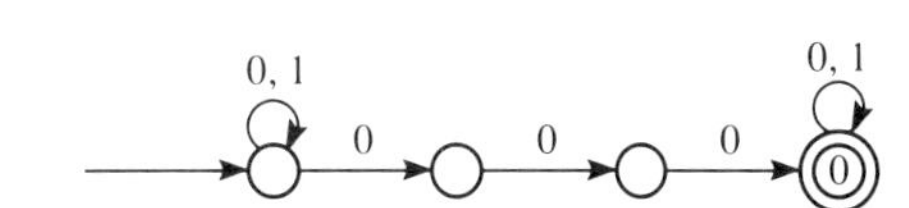

Chapter 8

Section 8.1

1. (a) 29 (b) 0 (c) 106928
3. See Problem 1.

5. The Nim-sum is 6. The possible Nim-sums which can result from a legal move are 0, 1, 2, 3, 4, 5, and 7.

7. Decrease the 17-heap to size 3.

9. (a) Reduce the 17-heap to 3.
   (b) Reduce the 11-heap to 2 or the 9-heap to 0, or the 10-heap to 3, or the 8-heap to 1.
   (c) Remove 9937 from the 10000-heap.

11. (a) Take 5 from the 17-heap.
    (b) No good moves
    (c) Reduce the 100000-heap to 10000.

13. Cora's first two moves are good, and Dave's are helpless. But Cora's third move is bad and Dave's is good. Cora resigns because Dave can now copy her moves and win.

15. 64

17. It is fuzzy because all the powers of 2 involved are distinct, so the Nim-sum is the ordinary sum, namely, $2^{n+1} - 1$. The unique winning move is to take 1 from the 2^n pile.

19. Because the second method is simply a shorthand form of the first

21. Just Nim-add b to both sides of the equation.

23. $a \stackrel{*}{+} b = c \Rightarrow a \stackrel{*}{+} b \stackrel{*}{+} b = c \stackrel{*}{+} b \Rightarrow a \stackrel{*}{+} 0$
    $= b \stackrel{*}{+} c \Rightarrow a = b \stackrel{*}{+} c.$
    In the other direction,
    $a = b \stackrel{*}{+} c \Rightarrow b \stackrel{*}{+} a = b \stackrel{*}{+} b \stackrel{*}{+} c \Rightarrow a \stackrel{*}{+} b = 0 \stackrel{*}{+} c = c.$

25. The difference between binary Nim-addition and ordinary addition is that carries can occur in ordinary addition but not in Nim-addition. Since carries increase the size of the answer the ordinary sum is always bigger than or equal to the Nim-sum. The two sums $a + b$ and $a \stackrel{*}{+} b$ are equal if an only if there is no position in which the binary decompositions of a and b both contain a 1.

27. Betty

29. XOR

31. The leftmost binary position in which a and $a \stackrel{*}{+} b$ can differ is the position which holds the leftmost 1 in the binary representation of b. In this position, $a \stackrel{*}{+} b$ contains a 1 while a contains a 0, so $a \stackrel{*}{+} b \geq a$. Equality ($a \stackrel{*}{+} b = a$) holds if and only if $b = 0$ (and in this case the highest power of 2 involved in b is undefined).

33. $1/2^m$

35. A rook in row h, column k corresponds to two Nim-piles of sizes h and k. One good move is to move the rook at (4, 2) to (1, 2).

37. (a) By leaving an odd number of counters with your opponent to move, you can be sure he will have to take the last counter since each player will remove just one counter per turn.
    (b) (0, 3, 0, 1) to (0, 0, 0, 1)
    (1, 2, 1, 1) to (1, 0, 1, 1)
    (1, 1, 1)—no good move
    (2, 2, 1, 1, 1) to (2, 2, 1, 1, 0)
    (c) On move 6 she should move to (0, 0, 0, 1).

Section 8.2

1. (a) 0 (b) 0 (c) 1

3. $g(A) = 0$; $g(B) = 1$; $g(C) = 1$; $g(D) = 0$; $g(E) = 2$; $g(F) = 1$; $g(G) = 1$; $g(H) = 1$; $g(I) = 2$; $g(J) = 0$

5. They come out the same.

7. 10

9. $x + y + \min(x, y)$

11. $g(A) = 2$; $g(B) = 1$; $g(C) = 0$

13. Yes; no

15. Row 8: 8, 6, 7, 10, 1, 2, 5, 3, 4, 15, 16
    Row 9: 9, 10, 11, 12, 8, 7, 13, 14, 15, 16, 17
    Row 10: 10, 11, 9, 8, 13, 12, 0, 15, 16, 17, 14

17. Eliminate vertex J and make vertices G and H winning positions. The new Nim-values are $g(A) = 1$, $g(B) = 0$, $g(C) = 0$, $g(D) = 2$, $g(E) = 1$, $g(F) = 0$, $g(G) = 0$, $g(H) = 0$, and $g(I) = 1$. (Another method is to add a new vertex K and an edge directed from J to K.)

19.

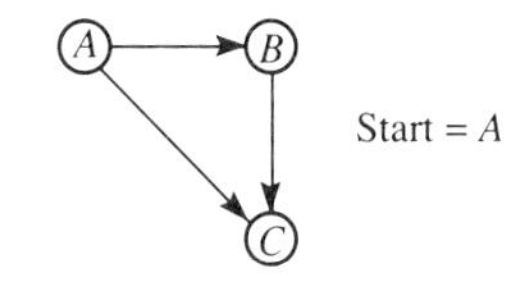

21.

	0	1	2	3	4	5	6	7
0	0	0	1	1	0	0	1	1
1	0	0	2	1	0	0	1	1
2	1	2	2	2	3	2	2	2
3	1	1	2	1	4	3	2	3
4	0	0	3	4	0	0	1	1
5	0	0	2	3	0	0	2	1
6	1	1	2	2	1	2	2	2
7	1	1	2	3	1	1	2	0

23. (a)

	0	1	2	3	4	5	6	7
0	0	1	2	3	4	5	6	7
1	1	0	3	2	5	4	7	6
2	2	3	0	1	6	7	4	5
3	3	2	1	0	7	6	5	4
4	4	5	6	7	0	1	2	3
5	5	4	7	6	1	0	3	2
6	6	7	4	5	2	3	0	1
7	7	6	5	4	3	2	1	0

(b) Move to [4, 4].

25. The algorithm fails after only E and F have been numbered. At this stage, any unnumbered vertex leads to the A-B-C cycle. The algorithm fails because the ending condition does not hold.

27. Since each position has only one option, the only computations involved are $\text{mex}\{0\} = 1$ and $\text{mex}\{1\} = 0$.

29. $g(0, 0, 0) = 0$; $g(0, 0, 1) = \text{mex}\{g(0, 0, 0)\} = \text{mex}\{0\} = 1$; $g(0, 0, 2) = \text{mex}\{g(0, 0, 0), g(0, 0, 1)\} = \text{mex}\{0, 1\} = 2$; $g(0, 1, 1) = \text{mex}\{g(0, 0, 1)\} = \text{mex}\{1\} = 0$; and so on. The answers come out to be $g(n_1, n_2, n_3) = n_1 \ddagger n_2 \ddagger n_3$, of course.

Section 8.3

1. Move from (5, 4) to (2, 4); (5, 4) to (5, 1); (7, 7) to (4, 7); (7, 7) to (7, 4); (7, 7) to (0, 0); (3, 1) to (0, 1); (4, 7) to (4, 1); (4, 7) to (1, 4). Notice that the two good moves from (4, 7) *increase* the size of one of the Bogus Nim-piles.

3. Move from (5, 5) to (4, 5).

5. Move from (3, 3) to (0, 0).

7. Cover cells 2 and 3.

9. No good move

11. Move to [4, 4, 1, 3, 3].

13. Move to [17, 13, 5, 19, 20].

15. (a) 8 (b) 0 (c) 1 (d) 7

17. Move from (3, 4) to (1, 3).

19. For example, $g_7 = \text{mex}\{g_5, g_1 \ddagger g_4, g_2 \ddagger g_3\} = \text{mex}\{0, 0 \ddagger 2, 1 \ddagger 1\} = \text{mex}\{0, 2, 0\} = 1$. The others are similar.

21. $G_{21} = 4$; $G_{22} = 3$; $G_{23} = 0$; $G_{24} = 4$; $G_{25} = 3$

23. (a) $D_0 = 0$; for $n > 0$, $D_n = 0$ if n is odd and $D_n = 1$ if n is even.
(b) All moves from a fuzzy position are good, so no brains are necessary to play.

25. G_n = the remainder when n is divided by 4

27. (a) Move to cell 4 in row 3.
(b) Same as (a)
(c) Move so as to restore the original gap.

29. (a)

n	1	2	3	4	5	6	7	8	9	10	11	12
K_n	1	2	3	1	4	3	2	1	4	2	6	4

(b) No good move is possible.

33. The relationship to Nim becomes obvious when you realize that the closed curves divide the plane into regions. The dots in a region behave like Nim-heaps. A legal move in the game is to remove one or more counters from a heap and then, if you want, divide the remaining heap into two heaps (any size). A player in a fuzzy position can ignore the second option and produce a zero position. In a zero position, the first option will lead to a fuzzy position, and the second option can never create a zero position. For example, if there are four piles of 7, 4, 2, 1 we may remove 2 from the 7-pile to reach (5, 4, 2, 1) with Nim-sum 2. No matter how the 5-pile is split the Nim-sum can never be changed to 0. In the given position, a good move is to draw a closed curve through 4 of the 5 dots at the left.

Chapter 9

Section 9.1

1. $\{H, T, E\}$; $P(H) = P(T) = 99/200$, $P(E) = 1/100$

3. $\{a\}, \{a, b\}, \{a, d\}, \{a, b, d\}, \{c, e\}, \{b, c, e\}, \{c, d, e\}, \{b, c, d, e\}$

5. 1/649740 7. $\frac{6}{27}$

9. $P\{D\} = \frac{1}{3}$; $P\{E\} = \frac{5}{36}$

11. $\leq \frac{1}{2}$

13. The same as the probability of having no fixed points in a permutation of 10 integers, or about .3679

15. (a) $\frac{1}{24}$ (b) 0 (c) $\frac{1}{4}$ (d) $\frac{1}{3}$

17. (a) 60 (b) $\frac{2}{5}$ each time

19. 31/105

21. $\left(11\binom{4}{1}^5 - 1\right) \Big/ \binom{52}{5} \approx .00433$

23. $\binom{13}{2}\binom{4}{2}\binom{4}{2}\binom{44}{1} \Big/ \binom{52}{5} \approx .0475$

25. About .0738

27. $\frac{1}{8}$

29. It holds in general—there are $(n-1)!$ permutations which leave a given one of the n integers $1, \ldots, n$ fixed, making $n(n-1!) = n!$ fixed points in all.

31. (a) $a^2/(a+b)(a+c)$
(b) $a(b+c)/(a+b)(a+c)$
(c) $a/(a+b)(a+c)$

33. Observing that $A_1 \cup \cdots \cup A_n$ and A_{n+1} are disjoint, for the inductive step we have

$$\begin{aligned}\Pr\{A_1 \cup \cdots \cup A_n \cup A_{n+1}\} &= \Pr\{(A_1 \cup \cdots \cup A_n) \cup A_{n+1}\} \\ &= \Pr\{A_1 \cup \cdots \cup A_n\} + \Pr\{A_{n+1}\} \\ &= \Pr\{A_1\} + \cdots + \Pr\{A_n\} + \Pr\{A_{n+1}\}.\end{aligned}$$

35. $\Pr\{A_1 \cup \cdots \cup A_n\} = \Pr\{A_1\} + \cdots + \Pr\{A_n\} - \Pr\{A_1 \cap A_2\} - \cdots - \Pr\{A_{n-1} \cap A_n\} \cdots \pm \Pr\{A_1 \cap \cdots \cap A_n\}$

37. It is less than or equal to $\binom{n}{k}\Big/6^k$.

39. $\binom{n}{r}\sum_{k=0}^{n-r}\frac{(-1)^k}{k!}$

Section 9.2

1. $\frac{1}{6}$ 5. $\frac{1}{2}$ 9. 1 13. $\frac{1}{2}$

3. 1 7. $\frac{1}{6}$ 11. $\frac{10}{17}$

15.

i	1	2	3
$\Pr\{U_i, W\}$	0	$\frac{1}{6}$	$\frac{2}{6}$
$\Pr\{U_i, B\}$	$\frac{2}{6}$	$\frac{1}{6}$	0

17. $\frac{1}{11}$ 19. $\frac{1}{18}$ 21. $\frac{3}{17}$

23. In a uniform space

$$\begin{aligned}\Pr(A \cap B)/\Pr(B) &= (|A \cap B|)/|\Omega|)/(|B|/|\Omega|) \\ &= |A \cap B|/|B|.\end{aligned}$$

25. $\frac{1}{4}$

27. B can be any nonempty set having exactly half its elements in A.

29. $\frac{1}{2}$ 31. $\frac{5}{8}$

Section 9.3

1. $\Pr\{A\} = \Pr\{(H, H), (H, T)\} = \frac{1}{2}$
$\Pr\{B\} = \Pr\{(H, H), (T, H)\} = \frac{1}{2}$
$\Pr\{D\} = \Pr\{(H, T), (T, H)\} = \frac{1}{2}$
$\Pr\{A \cap B\} = \Pr\{(H, H)\} = \frac{1}{4} = \frac{1}{2} \cdot \frac{1}{2} = \Pr(A)\Pr(B)$
$\Pr\{A \cap D\} = \{(H, T)\} = \frac{1}{4} = \frac{1}{2} \cdot \frac{1}{2} = \Pr(A)\Pr(D)$
$\Pr\{B \cap D\} = \{(T, H)\} = \frac{1}{4} = \frac{1}{2} \cdot \frac{1}{2} = \Pr\{B\}\Pr\{D\}$

3. $\Pr\{A \cap B \cap D\} = \Pr\{\varnothing\} = 0 \neq \Pr\{A\}\Pr\{B\}\Pr\{D\} = \frac{1}{8}$

5. $\Pr\{A \cap B\} = \Pr\{A\}\Pr\{B\}$
$\Pr\{A \cap C\} = \Pr\{A\}\Pr\{C\}$
$\Pr\{A \cap D\} = \Pr\{A\}\Pr\{D\}$
$\Pr\{B \cap C\} = \Pr\{B\}\Pr\{C\}$
$\Pr\{B \cap D\} = \Pr\{B\}\Pr\{D\}$
$\Pr\{C \cap D\} = \Pr\{C\}\Pr\{D\}$
$\Pr\{A \cap B \cap C\} = \Pr\{A\}\Pr\{B\}\Pr\{C\}$
$\Pr\{A \cap B \cap D\} = \Pr\{A\}\Pr\{B\}\Pr\{D\}$
$\Pr\{A \cap C \cap D\} = \Pr\{A\}\Pr\{C\}\Pr\{D\}$
$\Pr\{B \cap C \cap D\} = \Pr\{B\}\Pr\{C\}\Pr\{D\}$
$\Pr\{A \cap B \cap C \cap D\} = \Pr\{A\}\Pr\{B\}\Pr\{C\}\Pr\{D\}$

7. The space consists of all 3-tuples (C, D_1, D_2) where C is a card and the D_i are outcomes from rolling a die. It is uniform.

9. $\frac{1}{2} \cdot \frac{1}{2} \cdot \frac{1}{2} = \frac{1}{8}$ 11. $10p^2q^3$

13. $q^6 + 6pq^5 + 15p^2q^4 + 20p^3q^3$

15. A and C

17. (a) The space is $\{(1, H), (2, T), (3, H), (4, H), (5, T), (6, T)\}$. It is uniform.
(b) Yes; no

19. $P\{0 \text{ successes}\} = (1-p)^8$
$P\{1 \text{ success}\} = 8p(1-p)^7$
$P\{2 \text{ successes}\} = 28p^2(1-p)^6$
$P\{3 \text{ successes}\} = 56p^3(1-p)^5$
$P\{4 \text{ successes}\} = 70p^4(1-p)^4$
$P\{5 \text{ successes}\} = 56p^5(1-p)^3$
$P\{6 \text{ successes}\} = 28p^6(1-p)^2$
$P\{7 \text{ successes}\} = 8p^7(1-p)$
$P\{8 \text{ successes}\} = p^8$

21. (a) $(.9)^{10}$
(b) $(.1)^{10}$
(c) $1 - (.9)^{10}$

23. $$\begin{aligned}\Pr\{A \times B\} &= \sum_{(a,b)\in A\times B} \Pr\{a, b\} \\ &= \sum_{a\in A}\sum_{b\in B} \Pr_1\{a\}\Pr_2\{b\} \\ &= \left(\sum_{a\in A} \Pr_1\{a\}\right)\left(\sum_{b\in B} \Pr_2\{b\}\right) \\ &= \Pr_1\{A\}\Pr_2\{B\}\end{aligned}$$

25. If A depends only on the first experiment and B depends only on the second, then

$$\begin{aligned}\Pr\{(A \times \Omega_2) \cap (\Omega_1 \times B)\} &= \Pr\{A \times B\} \\ &= \Pr_1\{A\}\Pr_2\{B\} \\ &= \Pr\{A \times \Omega_2\}\Pr\{\Omega_1 \times B\}.\end{aligned}$$

27. $\sum_{(\omega_1,\omega_2)\in\Omega_1\times\Omega_2} \Pr\{(\omega_1,\omega_2)\} = \sum_{\omega_1\in\Omega_1}\sum_{\omega_2\in\Omega_2} \Pr_1\{\omega_1\}\Pr\{\omega_2\}$

$= \left(\sum_{\omega_1\in\Omega_1} \Pr_1\{\omega_1\}\right)\left(\sum_{\omega_2\in\Omega_2} \Pr_2\{\omega_2\}\right) = 1 \cdot 1 = 1$

The generalization to n spaces is immediate.

29. Yes. Consider the experiment of drawing a ball from an urn containing 4 black balls, 1 red, white, and blue striped ball, 2 red and white striped balls, and 5 blue and white striped balls. Let A be the event "the ball has red stripes," B be "the ball has blue stripes," and C be "the ball has white stripes."

31. $\dfrac{4 \cdot 3^n - 6 \cdot 2^n + 4}{4^n}$

33. (a) $\Pr\{A \text{ beats } B\} = .56$
$\Pr\{A \text{ beats } C\} = .51$
$\Pr\{B \text{ beats } C\} = .6178$
(b) $\Pr\{A \text{ beats } B \text{ and } C\} = .2856$
$\Pr\{B \text{ beats } A \text{ and } C\} = .3322$
$\Pr\{C \text{ beats } A \text{ and } B\} = .3822$
(c) C

35. (a) $1 \cdot \left(1 - \frac{1}{n}\right)\left(1 - \frac{2}{n}\right) \cdots \left(1 - \frac{k-1}{n}\right)$
(b) $k = 22$ (c) About .71

37. (a) $\Pr\{b_1b_3b_3b_2b_1\}$
$= \Pr\{b_1\}\Pr\{b_3\}\Pr\{b_3\}\Pr\{b_2\}\Pr\{b_1\}$
$= p_1p_3p_3p_2p_1 = p_1^2p_2p_3^2$
(b) $\binom{5}{2}\binom{3}{1}\binom{2}{2} = \frac{5!}{2!3!} \cdot \frac{3!}{1!2!} \cdot \frac{2!}{2!0!} = \frac{5!}{2!1!2!}$,
or by direct appeal to multinomial coefficients.
(c) By the analysis of part (a) each such sequence has probability $p_1^2p_2p_3^2$ and by part (b) there are $5!/2!1!2!$ of them.

3. $\Pr\{U_1|WW\} = 0$; $\Pr\{U_2|WW\} = \frac{1}{5}$

5.

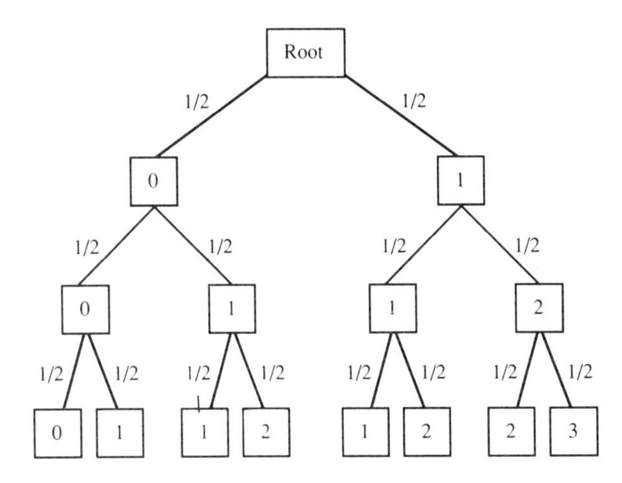

7. $\frac{1}{18}$

9. $\Pr\{M_1|D\} = \frac{6}{17}$;
$\Pr\{M_2|D\} = \frac{8}{17}$;
$\Pr\{M_3|D\} = \frac{3}{17}$

11. $\frac{2}{3}$ 13. $\frac{1}{3}$ 15. $\frac{7}{30}$ 17. .6

19. $\Pr\{U_1|BBW\} = 0$; $\Pr\{U_2|BBW\} = .45$;
$\Pr\{U_3|BBW\} = .4$; $\Pr\{U_4|BBW\} = .15$;
$\Pr\{U_5|BBW\} = 0$

21. 4, 6, 8, and 10 are all tied as most likely.

23. $\frac{1}{5}$

25. (a) $\frac{1}{2}$ (b) $\frac{1}{3}$ (c) $\frac{1}{6}$

Section 9.4

1.

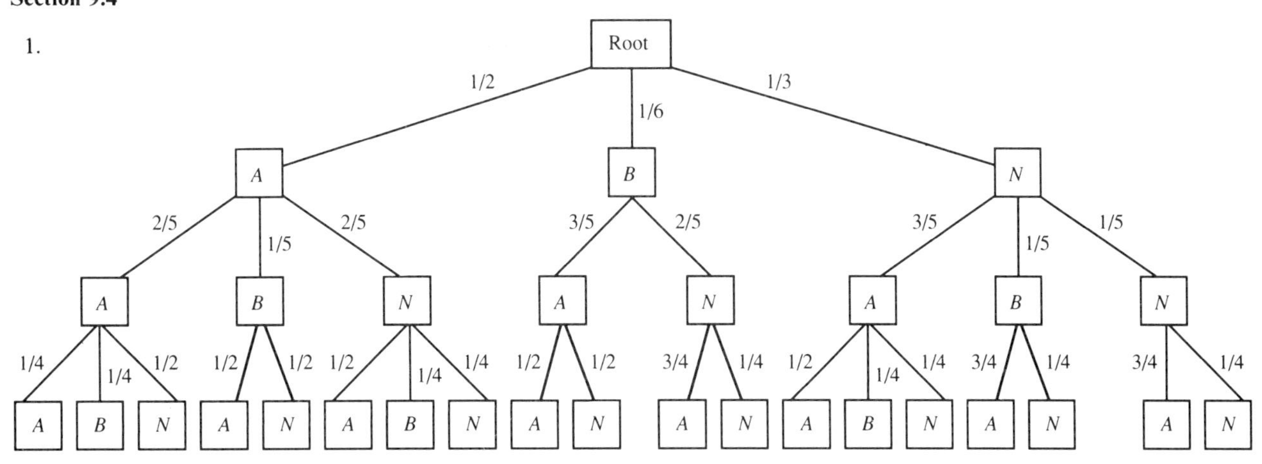

Section 9.5

1.

y	0	1	2	3
$f(y)$	$\frac{1}{27}$	$\frac{6}{27}$	$\frac{12}{27}$	$\frac{8}{27}$

3. .451584

5. $\frac{1}{9}$

7. The row sums yield the density of Y because each event in the range of Y must occur in conjunction with one and only one event in the range of X. Similarly for column sums.

9. This is a virtual tautology:

$$\Pr\{X = -1, Y = 1\} = \Pr\{(-1, 1)\} = 0;$$
$$\Pr\{X = -1, Y = 2\} = \Pr\{(-1, 2)\} = \tfrac{1}{8};$$
$$\Pr\{X = -1, Y = 3\} = \Pr\{(-1, 3)\} = \tfrac{1}{2};$$
$$\Pr\{X = 1, Y = 1\} = \Pr\{(1, 1)\} = \tfrac{1}{4};$$
$$\Pr\{X = 1, Y = 2\} = \Pr\{(1, 2)\} = \tfrac{1}{8};$$
$$\Pr\{X = 1, Y = 3\} = \Pr\{(1, 3)\} = 0.$$

11. $[n(n+1)/2]/n = (n+1)/2$

13. $\frac{23}{4}$

15. (a)

x	0	1	2	3	4
$f(x)$	$\frac{1}{81}$	$\frac{8}{81}$	$\frac{8}{27}$	$\frac{32}{81}$	$\frac{16}{81}$

(b)

x	0	1	2	3	4
$f(x)$	$\frac{1}{495}$	$\frac{32}{495}$	$\frac{56}{165}$	$\frac{224}{495}$	$\frac{70}{495}$

17.

x	0	1	2	3	4	5
$f(x)$	0	0	$\frac{2}{3}$	$\frac{1}{3}$	0	0

19. Simply eliminate the two atoms of probability 0.

21. (a) Density of X:

x	0	1	2	3	4
$f(x)$	$\frac{1}{6}$	$\frac{1}{4}$	$\frac{1}{3}$	$\frac{1}{6}$	$\frac{1}{12}$

Density of Y:

x	0	1	2	4
$f(x)$	$\frac{2}{3}$	$\frac{1}{12}$	$\frac{1}{6}$	$\frac{1}{12}$

Density of Z:

x	0	1	2	3	5	8
$f(x)$	$\frac{1}{6}$	$\frac{1}{4}$	$\frac{1}{4}$	$\frac{1}{12}$	$\frac{1}{6}$	$\frac{1}{12}$

(b)

Joint Density of (x, y)		**Joint Density of (x, y, z)**	
(x, y)	$f(x, y)$	(x, y, z)	$f(x, y, z)$
$(0, 0)$	$\frac{1}{6}$	$(0, 0, 0)$	$\frac{1}{6}$
$(1, 0)$	$\frac{1}{4}$	$(1, 0, 1)$	$\frac{1}{4}$
$(2, 0)$	$\frac{1}{4}$	$(2, 0, 2)$	$\frac{1}{4}$
$(2, 1)$	$\frac{1}{12}$	$(2, 1, 3)$	$\frac{1}{12}$
$(3, 2)$	$\frac{1}{6}$	$(3, 2, 5)$	$\frac{1}{6}$
$(4, 4)$	$\frac{1}{12}$	$(4, 4, 8)$	$\frac{1}{12}$

23.

x	0	1	2
$f(x)$	$\frac{1}{5}$	$\frac{5}{12}$	$\frac{23}{60}$

$E(X^2) = \frac{39}{20}$

25.

	N		
x_1	0	1	2
0	0	$\frac{2}{9}$	$\frac{2}{27}$
1	$\frac{2}{9}$	$\frac{2}{9}$	0
2	0	$\frac{2}{9}$	0
3	0	0	$\frac{1}{27}$

27. Let the experiment consist of randomly choosing an integer T between -2 and 2 and define $X = T$, $Y = |T|$, and

$$Z = \begin{cases} -2 & \text{if } T < 0, \\ 0 & \text{If } T = 0, \\ 2 & \text{if } T > 0. \end{cases}$$

29. $\Pr\{X = k+1, Y = k\} = \binom{n-1}{k} p^k q^{n-1-k} p$

$\Pr\{X = k, Y = k\} = \binom{n-1}{k} p^k q^{n-1-k} q$

for $k = 0, 1, \ldots, n-1$. The joint density is 0 in all other cases.

31. (a) $f(1) = \dfrac{91}{216}$; $f(0) = \dfrac{125}{216}$

(b) $f(0) = \dfrac{1953125}{10077696}$; $f(1) = \dfrac{4265625}{10077696}$;
$f(2) = \dfrac{3105375}{10077696}$; $f(3) = \dfrac{753571}{10077696}$

33. (a) $E(X) = \sum_{k=0}^{n} k \Pr\{X = k\}$. The result follows from Equation (9.31).

(b) $k\binom{n}{k} = \dfrac{kn!}{(n-k)!k!}$

$$= \frac{n(n-1)!}{((n-1)-(k-1))!(k-1)!}$$

$$= n\binom{n-1}{k-1}$$

$$E(X) = \sum_{k=0}^{n} k\binom{n}{k} p^k q^{n-k}$$

$$= \sum_{k=1}^{n} k\binom{n}{k} p^k q^{n-k}$$

$$= \sum_{k=1}^{n} n\binom{n-1}{k-1} \cdot p \cdot p^{k-1} q^{n-k}$$

$$= np \sum_{k=1}^{n} \binom{n-1}{k-1} p^{k-1} q^{n-k}$$

(c) Setting $j = k - 1$ in the right-hand side of (b), we get

$$E(X) = np \sum_{j=0}^{n-1} \binom{n-1}{j} p^j q^{(n-1)-j}$$

$$= np(p+q)^{n-1} = np.$$

Chapter 10

Section 10.1

1. (a), (b), and (c)
3. There are x_{n-1} strings of length n which end in a, x_{n-2} which end in ab, and x_{n-2} which end in bc. Thus $x_n = x_{n-1} + x_{n-2} + x_{n-2}$, as required.
5. $\dfrac{1}{\sqrt{5}} + \left(-\dfrac{1}{\sqrt{5}}\right) = 0$

$$\left(\frac{1+\sqrt{5}}{2}\right)\left(\frac{1}{\sqrt{5}}\right) + \left(\frac{1-\sqrt{5}}{2}\right)\left(-\frac{1}{\sqrt{5}}\right)$$

$$= \frac{1}{2\sqrt{5}} + \frac{1}{2} - \frac{1}{2\sqrt{5}} + \frac{1}{2} = 1$$

7. $x_n = 2(2n-1) - (2n-3)$
   $= 4n - 2 - 2n + 3 = 2n + 1$
9. $x_n = A \cdot 5^n + B \cdot (-2)^n$
11. $x_n = A(-3)^n + B(n)(-3)^n$
13. $x_n = A[(-1+\sqrt{3})/2]^n + B[(-1-\sqrt{3})/2]^n$
15. $x_n = A(4^n) + B(-1)^n$
17. $x_n = 0$
19. $y_n = A(4^n) + B(n)(4^n)$; $y_n = 2n \cdot 4^n$
21. $y_n = A + B(n) + C(n^2)$; $y_n = -3 + 4n - n^2$
23. $y_n = A + Bn$; $y_n = 1 + n$
25. Approximating F_n by $G_n = [(1+\sqrt{5})/2]^n$, we find $F_{1000} \approx 4.35 \times 10^{208}$. If $H_n[(1+\sqrt{5})/2]^{n-1}$ is the estimate of Example 1.28, then

$$\frac{G_n}{H_n} = \frac{1}{\sqrt{5}}\left(\frac{1+\sqrt{5}}{2}\right),$$

which is approximately .724.

27. $y_n = -\frac{3}{8} + (\frac{3}{8})n^2$
29. $x_n = (-\frac{2}{3})(-\frac{1}{2})^n + \frac{2}{3}$
31. The solution to (10.15) is unique, for given values of x_1 and x_2. Since there *is* a solution to (10.15) of the form (10.17) for any given values of x_1 and x_2 (choose $A = x_0$ and $B = x_1 - x_0$), every solution must be of this form.
33. $x_n = 2^{n-1}$, $n \geq 1$. In a formal verification, the inductive step is the critical one:

$$x_{n+1} = x_n + (x_{n-1} + \cdots + x_0) = x_n + x_n = 2x_n.$$

35. $x_1 = 1$, $x_2 = 2 \cdot 1$, $x_3 = 3 \cdot 2 \cdot 1, \ldots, x_n = n!$
37. (a) $\phi_2 = 2$, $\phi_3 = 4$, $\phi_4 = 8$, $\phi_5 = 32$
    (b) $\phi_n = 2^{e_n}$, where $e_0 = 0$, $e_1 = 1$, and

$$e_n = \frac{1}{\sqrt{5}}\left[\left(\frac{1+\sqrt{5}}{2}\right)^n - \left(\frac{1-\sqrt{5}}{2}\right)^n\right].$$

39. The HDE of Problem 38 has characteristic polynomial

$$2\lambda^2 - 3\lambda + 1 = (2\lambda - 1)(\lambda - 1) = 0$$

with roots $\frac{1}{2}$ and 1. The general form of the solution is $A(\frac{1}{2})^n + B(1)^n$.

41. The boundary conditions $p_0 = 0$ and $p_N = 1$ hold for the same reasons as in Example 10.5. In order to reach the right wall from location n, the robot must either move left and get there from location $n - 1$ or move right and get there from location $n + 1$. This leads to the equation $p_n = pp_{n-1} + qp_{n+1}$.
43. By Problem 42(c), the general form for p_n is $p_n = A + B(p/q)^n$. Then $p_0 = 0$ implies $B = -A$ and $p_N = 1$ implies $A = 1/[1-(p/q)^N]$. Thus

$$p_n = \frac{1}{1-(p/q)^N} - \frac{(p/q)^n}{1-(p/q)^N} = \frac{1-(p/q)^n}{1-(p/q)^N}$$

Section 10.2

1. $x_7 = 106$, $x_8 = 213$, $x_9 = 426$
3. $x_2 = 8 = 3 \cdot 2^2 - 2 - 2$
   $x_3 = 19 = 3 \cdot 2^3 - 3 - 2$
   $x_4 = 42 = 3 \cdot 2^4 - 4 - 2$
5. $x_0 = 0$; $x_1 = \frac{1}{4}$

7. $f_n - 2f_{n-1} + f_{n-2} = 0$

9. $f_{n+1} - f_n = 0$

11. $f_{n+6} - (\frac{11}{2})f_{n+5} + (\frac{25}{2})f_{n+4} - 15f_{n+3} + 10f_{n+2} - (\frac{7}{2})f_{n+1} + (\frac{1}{2})f_n = 0$

13. $p_n = A$

15. $p_n = A(-1)^n$

17. $p_n = An^2 + Bn + C$

19. $p_n = (An^4 + Bn^3 + Cn^2 + Dn + E)2^n$

21. $p_n = A \cdot 2^{-n} + Bn^4 + Cn^3 + Dn^2 + En + F$

23. $x_n = A \cdot 2^n - 6n - 12$
$x_n = 10 \cdot 2^n - 6n - 12$

25. $x_n = A(-5)^n + B \cdot 3^n - (\frac{1}{2})n - 1$
$x_n = (\frac{5}{8})(-5)^n + (\frac{11}{8})3^n - (\frac{1}{2})n - 1$

27. $x_n = A(-2)^n + 1 + (\frac{1}{6})4^n$
$x_n = (\frac{5}{6})(-2)^n + 1 + (\frac{1}{6})4^n$

29. $x_n = A + Bn + (\frac{1}{2})n^2$; $x_n = 1 - n + (\frac{1}{2})n^2$

31. $x_n = A(-1)^n + \frac{5}{2}$; $x_n = (-\frac{3}{2})(-1)^n + \frac{5}{2}$

33. $x_n = A \cdot 2^n + B - (\frac{1}{3})n^3 - (\frac{5}{2})n^2 - (\frac{49}{6})n$
$x_n = 11 \cdot 2^n - 11 - (\frac{1}{3})n^3 - (\frac{5}{2})n^2 - (\frac{49}{6})n$

35. $x_n = (\frac{2}{3})2^n - (\frac{1}{6})(-1)^n - \frac{1}{2}$

37. $y_n = y_{n-1} + 1 + y_{n-1} = 2y_{n-1} + 1$; $y_n = 2^n - 1$

39. $y_{n+1} = (1 + i/12)y_n - D$ is the general form of the difference equation. Use the constraints $y_0 = P$ and $y_N = 0$.

41. If x_n and p_n are solutions to the same IDE then $x_n - p_n$ must be a solution to the corresponding HDE by Problem 40.

Section 10.3

1. $G(z) = \dfrac{1}{1 - 3z}$

3. $G(z) = \dfrac{z^2}{1 - z}$

5. $G(z) = \dfrac{z}{(1 - z)^2}$

7. $G(z) = \dfrac{1}{1 + z^2}$

9. Take the z transform of $c_n = a_n + b_n$ to find that $\sum c_n z^n = \sum a_n z^n + \sum b_n z^n$, rso $C(z) = A(z) + B(z)$.

11. $G(z) = \dfrac{z}{(1 - z)^2} + \dfrac{3}{1 - z}$

13. Write the equation as $y_n + 2y_{n-1} - 15y_{n-2} = 0$, which is identical to Example 10.13.

15. $G(z) - y_0 - y_1 z - z^2 G(z) = 0$, $G(z) = \dfrac{z}{1 - z^2}$

17. $G(z) = \dfrac{1 - 2z}{(1 - 3z)(1 + z)}$

19. $G(z) = \dfrac{\frac{1}{2}}{1 - z} + \dfrac{\frac{1}{2}}{1 + z}$
$a_n = \frac{1}{2} + \frac{1}{2}(-1)^n$

21. $G(z) = \dfrac{\frac{1}{2}}{1 - z} - \dfrac{4}{1 - 2z} + \dfrac{\frac{9}{2}}{1 - 3z}$
$a_n = \frac{1}{2} - 4 \cdot 2^n + (\frac{9}{2})3^n$

23. (a) $A = \frac{1}{8}$, $B = -\frac{1}{8}$
(b) $A = 1/\sqrt{5}$, $B = -1/\sqrt{5}$

25. 55

27. 66

29. Write

$$\sum_{n \geq 0} \left(\sum_{k=0}^{n} a_k b_{n-k} \right) z^n = \sum_{n \geq 0} \sum_{k=0}^{n} (a_k b_{n-k}) z^k z^{n-k}$$

and reverse the order of summation. Since $k \leq n$ is the same as $n \geq k$, we obtain

$$\sum_{k \geq 0} a_k z^k \left(\sum_{n \geq k} b_{n-k} z^{n-k} \right)$$

as desired.

31. $d_n = \sum_{k=0}^{n} a_k b_{n-k}$, where $a_k = (k + 1)\lambda^k$ and $b_{n-k} = \lambda^{n-k}$.

33. $\sum (x_{n+2} z^{n+2}) - x_0 - x_1 z$
$- 4z(\sum x_{n+1} z^{n+1} - x_0) + 4z^2 \sum x_n z^n$
$= \sum 2^n z^{n+2}$
$G(z) - x_0 - x_1 z - 4z[G(z) - x_0]$
$+ 4z^2 G(z) = z^2 \sum 2^n z^n$
$= \dfrac{z^2}{1 - 2z}$
Set $x_0 = x_1 = 0$.

$$G(z) = \frac{z^2}{(1 - 2z)^2}$$

35. $G(z)[1 + 2z - 15z^2] = 1 + \dfrac{3}{2} z + \dfrac{6z^3}{(1 - z)^2}$

37. $G(z) = \dfrac{1 - \frac{5}{2}z + \frac{5}{2}z^2}{(1 - z)^3}$

Section 10.4

1. $34 - 42i$

3. $-\dfrac{1}{17} + \dfrac{38i}{17}$

5. $r = 8$, $\theta = -\pi/3$

7. $r = 7$, $\theta = \pi$

9. $(-1 + i)^5 = 4\sqrt{2}\left[\cos\left(-\frac{\pi}{4}\right) + i \sin\left(-\frac{\pi}{4}\right)\right] = 4 - 4i$

11. $x_n = \left(\frac{1}{i\sqrt{3}}\right)\left[\cos\left(\frac{n\pi}{3}\right) + i\sin\left(\frac{n\pi}{3}\right)\right]$

$= -\left(\frac{1}{i\sqrt{3}}\right)\left[\cos\left(\frac{n\pi}{3}\right) - i\sin\left(\frac{n\pi}{3}\right)\right]$

$= \left(\frac{2}{\sqrt{3}}\right)\sin\left(\frac{n\pi}{3}\right)$

13. $A = 1/2 - i/2,\ B = 1/2 + i/2$

15. $x_n = 2^{n/2}\left[C\cos\left(\frac{3n\pi}{4}\right) + D\sin\left(\frac{3n\pi}{4}\right)\right]$

17. $x_n = (2\sqrt{2})^n \sin\left(\frac{3n\pi}{4}\right)$

19. (a) $x_n = A(6i)^n + B(-6i)^n$

(b) $x_n = 6^n\left[C \cdot \cos\left(\frac{n\pi}{2}\right) + D \cdot \sin\left(\frac{n\pi}{2}\right)\right]$

21. (a) $x_n = A(-3)^n + B \cdot 4^n + Cn4^n + D(-3+3i)^n + E(-3-3i)^n$

(b) $x_n = A(-3)^n + B \cdot 4^n + Cn4^n + (3\sqrt{2})^n\left[D\cos\left(\frac{3n\pi}{4}\right) + E\sin\left(\frac{3n\pi}{4}\right)\right]$

23. Setting $n = 0$ and $n = 1$ in the general equation yields $x_0 = C$ and $x_1 = C + D$. Thus $D = x_1 - x_0$ and

$$x_n = 2^{n/2}\left[x_0\cos\left(\frac{n\pi}{4}\right) + (x_1 - x_0)\sin\left(\frac{n\pi}{4}\right)\right].$$

25. $x_1 = 1;\ x_2 = 1;\ x_3 = 5;\ x_4 = 9;\ x_5 = 13;\ x_6 = 33$

27. $x_n = A \cdot 2^n + B(-2)^n + 2^n\left[C\cos\left(\frac{n\pi}{2}\right) + D\sin\left(\frac{n\pi}{2}\right)\right] - \frac{4}{225} - \left(\frac{1}{15}\right)n + \left(\frac{1}{65}\right)3^n$

29. $p_n = An\cos\left(\frac{n\pi}{2}\right) + Bn\sin\left(\frac{n\pi}{2}\right)$

INDEX